The Biology of Human Starvation

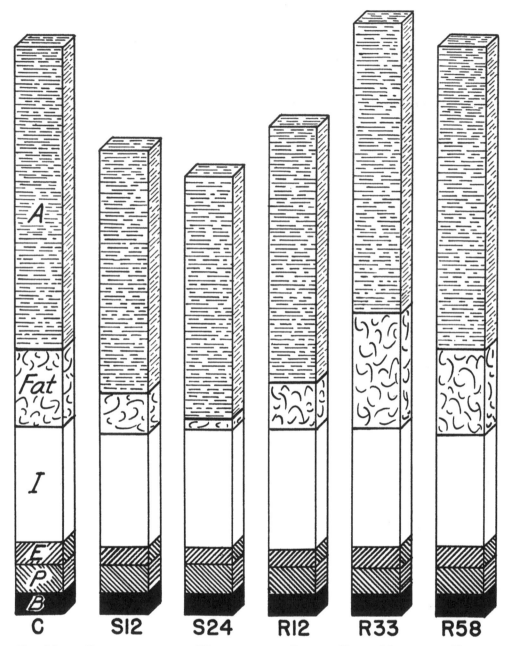

THE MAJOR COMPARTMENTS, AS WEIGHT, OF THE BODY OF YOUNG MEN IN THE NORMAL STATE OF NUTRITION, IN SEMI-STARVATION, AND IN SUBSEQUENT REHABILITATION. *Columns*: C = control (pre-starvation); S12 and S24 = 12 and 24 weeks of semi-starvation; R12, R33, and R58 = 12, 33, and 58 weeks of rehabilitation. *Compartments*: B = bone mineral; P = blood plasma; E = erythrocytes; I = interstitial fluid (thiocyanate space less plasma volume); A = active tissue (total body weight less the other indicated compartments). See Chapter 15.

The Biology of

HUMAN STARVATION

by

ANCEL KEYS

JOSEF BROŽEK   AUSTIN HENSCHEL

OLAF MICKELSEN

HENRY LONGSTREET TAYLOR

WITH THE ASSISTANCE OF

Ernst Simonson, Angie Sturgeon Skinner, and Samuel M. Wells

OF THE LABORATORY OF PHYSIOLOGICAL HYGIENE
SCHOOL OF PUBLIC HEALTH, UNIVERSITY OF MINNESOTA

With Forewords by

J. C. Drummond, Russell M. Wilder, and Charles Glen King
and Robert R. Williams

VOLUME I

THE UNIVERSITY OF MINNESOTA PRESS · MINNEAPOLIS

LONDON · GEOFFREY CUMBERLEGE · OXFORD UNIVERSITY PRESS

PRINTED AT THE NORTH CENTRAL PUBLISHING COMPANY, ST. PAUL

Contents for Volume I

CONTENTS FOR VOLUME I

Biochemistry

Physiology

Contents for Volume II

Forewords

"*I am sensible, that, notwithstanding all my care and attention in making both the Observations and Experiments, not only many inaccuracies, but mistakes must have escaped me; which, those will most readily excuse, who having themselves made researches of this kind, are not unacquainted with the difficulties attending them. Yet, however imperfect these sheets may be, I may hope they will serve as a foundation for others to go upon; who, by making improvements on this subject, will concur with me, in attempting to draw from the calamities of war some benefit to mankind.*"

JOHN PRINGLE, in *Observations on the Diseases of the Army* (2d edition, 1753), p. xv.

Foreword

by SIR JACK DRUMMOND, D.Sc., F.R.I.C., F.R.S.
Director of Research, Boots The Chemists, England

I HAVE vivid recollection of a bitterly cold day in January 1945. In a bare, unheated room in Eindhoven, not long liberated from the German grip, a group of Dutch, United States, and British public health officers and nutrition advisers were listening to the grim story of the sufferings of the starving people in the towns to the west of the German lines. We were brought together to plan relief measures, measures which it was feared might be impossible to apply for months to come, by which time inevitably the plight of the starving Dutch would certainly be desperate.

It was frightening to realise how little any of us knew about severe starvation. In our lifetime millions of our fellow men had died in terrible famines, in China, in India, in the U.S.S.R., without these tragedies having yielded more than a few grains of knowledge how best to deal with such situations on a scientific basis.

Neither academic treatises, such as the classic study of Benedict and his colleagues, nor the records of famine relief missions gave much help. There was no alternative but to make plans and make them quickly, for time was short, on foundations that most of us then recognised as being very insecure.

How misleading were not a few of our assumptions was quickly demonstrated when the plans were put into operation on the collapse of German resistance in the western Netherlands four months later.

Looking back on those exciting days, a crowd of memories comes to one's mind. Two impressions dominate the picture; the immense importance of the psychological aspect of inanition and the comparative simplicity of the nutritional and biochemical problem. Formed in the hard school of experience, those impressions are both clear and deep. But ever the thought recurs that so much time would have been saved, so much better planning would have been possible, so much more comprehensive an approach to the problem could have been made had we had access to the wealth of knowledge of inanition in its every aspect that Professor Ancel Keys with his group of distinguished colleagues and self-sacrificing volunteers acquired from the remarkable study that is recorded in these pages.

He and they have made an outstanding contribution to the science of human nutrition. It will become a classic from the day it is published. Everywhere it will arouse deep interest, stimulate even wider exploration, and, I trust, provoke vigorous controversy.

I venture to hope that it will be read side by side with the official report of the Royal Netherlands Government on starvation as it afflicted the western part of their country in the grim months from September 1944 until relief came in May 1945.* In many respects the two records will be found complementary, because the tragedy of western Holland made it possible to study in detail the later and the terminal stages of starvation, which, for obvious reasons, were outside the scope of the Minnesota Experiment.

Between them, these two works cover almost every aspect of inanition and starvation.

I have said that an outstanding impression gained in western Holland in 1945 was of the importance and significance of the psychological consequences of food shortage. In almost every respect our observations coincide with those reported by Professor Keys and his collaborators. From the grumbling and grousing that are inevitably provoked when the energy intake is deficient to the extent of 15–20 per cent, to the apathy and dissolution of higher human qualities that come with severe starvation, there is a wide variety of psychological reactions to hunger, many of which are almost, of themselves, diagnostic of the level of calorie intake.

This fact was never more clearly apparent than when there was sharp conflict of opinion concerning the calorie value of the food being eaten by the people of the liberated towns of France and Belgium and, more recently, by the Germans. Time and time again, it was claimed that people were unable to obtain significant quantities of food over and above the official rations, providing about 1000–1400 calories a day. Apart altogether from considerations of body weight and facial appearance, the psychological picture presented by these people itself often revealed the truth that they were getting not a little extra food by hook or by crook, mainly by crook.

One of the curious, and rather disconcerting, psychological manifestations of starvation, seen repeatedly in Western Europe, was the unresponsive and uncooperative attitude of those to whom relief was brought. It disappeared without trace when calorie intakes rose above 1500–1800 a day.

But I am not a psychologist, and so I must leave those who work in that field to give thought to the significance of the wealth of psychological observations recorded during the Minnesota Experiment. Their conclusions will be of far-reaching value to students of sociology, and of industrial welfare.

It is as a biochemist that I am qualified to pay a tribute, and it is a rich tribute, to the admirable and comprehensive study begun with sturdy determination in 1944, in the face of no little discouragement, and carried forward with perseverance and enthusiasm for three years in the cavernous underworld of the vast Sports Stadium of the University of Minnesota. On how many fundamental biochemical processes of the human body does this investigation throw fresh

* Burger, G. C. E., Drummond, J. C., and Sandstead, H. R., eds., *Malnutrition and Starvation in Western Netherlands, September 1944–July 1945.* Parts I and II. General State Printing Office, The Hague, Netherlands, 1948.

light! We are given a new and unusually comprehensive analysis of the basal metabolic rate. If the final word has not yet been said on the relation of the total metabolism to the mass of the "active tissue," it is because there is still much to be discovered about the biochemical activity of tissue cells exposed to conditions produced in the body as a whole by inanition and starvation. To what extent deficiencies, qualitative or quantitative, in the dietary affect the amount or the activity of the enzymes in cells is, at the present time, more a matter of conjecture than of factual knowledge. It seems plausible to assume that severe protein deprivation would eventually deplete the tissues of enzymes, especially those of the external secretions. Yet among many surprising facts revealed in Holland, none was more astonishing than the ability of those in the last stages of starvation to swallow, digest, and assimilate considerable quantities of protein, given in the form of a thick cream prepared from separated (skimmed) milk powder and water. The Dutch scientists recorded that these patients had, in most cases, pepsin in their gastric juice although probably in reduced amount, and that the enzyme content of the duodenal juice was within the normal range.

In this connexion, it is not unimportant to admit that we were quite wrong in making our original plans for relief, in assuming that predigested food (hydrolysed proteins) would be required for resuscitating severe cases of starvation. Whether given orally or by intravenous injection these preparations showed no superiority over separated milk powder in the treatment of the most desperate cases.

Experience in Western Europe threw another light on the significance of vitamins in inanition than that reflected by the records set out in these pages. The character of the food which people were forced to eat under conditions of severe restriction in France, Belgium, Holland, and, later, Germany ensured that vitamin deficiency would not be a serious complication of any state of inanition that might arise.

Wholemeal breads, every vegetable that could be acquired, and, most important, potatoes, provided, in general, nutritionally protective intakes of most of the alphabet of vitamins, even if they were not always adequate by normal standards. Signs of scurvy or of characteristic deficiencies of thiamine, riboflavin, and niacin were notable for their rarity in the famine zones of Holland in the winter of 1944–45. Intakes of thiamine were, of course, low, but related to the calories, enough was there to allow for normal utilisation according to the generally accepted theory of their relationship.

This brings me to the vital question of adaptation. To me, having seen so much undernutrition, it is at once gratifying that Professor Keys and his colleagues have clearly demonstrated that adaptation to low calorie intakes occurs. The calculation that an economy of the order of 600 calories daily might follow adaptation related to a reduction of weight such as was shown by Professor Keys' volunteers is certainly compatible with the general picture presented by the populations of the larger towns of Europe, where food restrictions reduced the average energy intake to something between 1500 and 2000 calories.

The Minnesota Experiment and the field trials carried out by Dr. John Beattie and his colleagues in Holland and Germany are very important contributions to our knowledge of protein metabolism and nitrogen requirements, but, unfortunately, they do not provide the long-sought answer to the vexing question of what are human needs for protein. Of course, neither study was directed primarily at this problem, but it was a reasonable hope that close study of metabolism during and on recovery from inanition would clarify to some extent ideas on the body's essential needs for protein. In this field, much remains to be discovered. It is impossible impartially to review the scope of our present knowledge without having to fall back on the discouraging conclusion that the estimates so uncritically used today have no more precise derivation than had Voit's 118 g. of protein a day, a dogmatic pronouncement that dominated nutritional theories for over half a century.

I feel greatly honoured that I was invited to contribute a foreword to this admirable volume. I had the privilege of visiting Professor Keys in 1946 and of seeing for myself how the experiment was being conducted. The occasion made a deep impression on my mind, for rarely had I come in touch with an investigation so meticulously planned and so enthusiastically carried through. My admiration was tinted with only one regret, that the investigation had not been begun three years earlier. But that is small matter now that the full results are appearing. This volume will soon take its place among the classic treatises on nutrition, a fitting reward for the years of thought and labour that Professor Ancel Keys and his "band of brothers" gave to their formidable task.

February 15, 1950
Nottingham, England

Foreword

by PROFESSOR RUSSELL M. WILDER, M.D., Ph.D.
Chief of the Department of Medicine, Mayo Clinic

"FROM lightning and tempest; from earthquake, fire and flood; from plague, pestilence and *famine*; from battle and murder and from sudden death, Good Lord deliver us." Prayers like this prayer from the Litany have been on the tongues of man from earliest recorded times. Even today, despite the "prodigies" of agriculture whereby in the more progressive areas of the earth one man can produce more food than ten could produce before, the world is confronted constantly with famine or the threat of famine, with the diseases for which famine paves the way, with tempest, fire, and flood, but above all, with war, which from the dawn of history has led to the neglect or the destruction of the means available for production of food. Starvation thus remains a subject of extreme significance, and this comprehensive work of Keys and his associates, reviewing critically as they have done the extensive literature on the subject and supplementing that with the data of their beautifully controlled investigation, the Minnesota Experiment, represents a contribution of the greatest possible importance to human welfare. It is reasonable to suppose that this monograph will always be considered a landmark, and will serve for many years as a *vade mecum* for physicians and others who will be called on to alleviate the damaging effects of famine.

To physicians, furthermore, study of these volumes will prove rewarding through revealing, as they do, the importance of combating the starvation which accompanies much disease. It is not always possible to feed sick people effectively. The appetite fails in many diseases, and the patient simply cannot cope with food. Nausea or even vomiting interferes. However, with more attention to the preparation of food in as palatable a form as possible, to judicious selection of food, and, when necessary, to suitable intravenous feeding, much more could be accomplished than frequently is accomplished. The physician could with great advantage demand an accurate accounting of the food actually consumed by patients, but this he seldom does. Physicians are paying increasingly greater attention in specific cases to the drawing of a balance of the intake and output of water, chloride, and sodium. Of equal importance in instances of calorie or protein deficit would be the drawing of an accurate balance for calories and nitrogen.

Reference is made in this monograph to investigations of Strang, McCluggage, and Brownlee, who studied the basal metabolic rate, the nitrogen balance, and the body weight of 18 undernourished patients before and during treatment with high calorie diets. The daily intake of nitrogen averaged 10.96 gm., and the total excretion of nitrogen 8.70 gm. Possibly more nitrogen would have been stored had the intake of nitrogen been greater, but the study is the type of study which if pursued with diligence undoubtedly will prove to be rewarding. There is much interest at present in the therapeutic value of diets high in protein, but little advantage is to be obtained from use of such diets unless the calories also are sufficient, for otherwise the protein given is simply used as fuel and storage of nitrogen is minimized.

Keys, Brožek, Mickelsen, Henschel, and Taylor also emphasize that much must still be learned about the effects of prolonged deprivation of calories and about effective means for the rehabilitation of the famished. Their experiments relate to young men who were vigorous and in excellent health beforehand and the period of semi-starvation was limited to twenty-four weeks. As is pointed out by the authors, extrapolation from these results requires caution. Men of older ages could react much differently; so also could women, children, and persons afflicted by disease of one kind or another. The authors plead, therefore, for much more study of clinical cachexia of various origins. Anorexia nervosa has had some attention (Chapter 44); the cachexia of occasional cases of stricture of the esophagus also has been investigated to advantage, but by and large the potentialities of readily available clinical material in hospitals have not been realized. For example, every hospital for the mentally diseased contains many starving patients whose nutritional conditions frequently are almost wholly subordinated to study of the psychiatric status. Cancer presents a special problem. A large proportion of patients with cancer also suffer from semi-starvation, yet little attention customarily is paid to its effects. To what degree does the cancer itself contribute to the cachexia, and how much of the disability is owing to provision or absorption of insufficient calories? Patients with ulcerative colitis, sprue, and other chronic diarrheal diseases usually are emaciated. Some of them have been grossly underweight for years, certainly for a much longer period of time than the subjects of the Minnesota Experiment. Do they accommodate themselves to semi-starvation in other ways than those now recognized? Many supposedly normal people are very thin and their basal metabolic rates usually have been normal, in contrast to the marked depression of the basal metabolic rate observed in persons, other than the obese, who are losing weight because of restriction of calories. It is interesting that the obese man or woman can be placed on a diet as low as 600, or even 400, calories without loss of body nitrogen or other ill effects and without depression of the basal metabolic rate. In these respects obese persons and persons of normal weight differ strikingly in their response to a low intake of calories.

The severely diabetic patient before the days of insulin presented a picture of extreme cachexia with practically all the features observed in famine plus a degree of acidosis of more or less severity. The basal metabolic rate, however, was normal or even elevated; this state has been attributed to the accompanying acidosis. The feeding of protein to one such patient, on whom I made extensive studies years ago, with Boothby and Beeler, was much more stimulating to the basal metabolic rate than it is in normal subjects. The patient who had typhoid fever, in the days when the typhoid patient was starved for fear of intestinal perforation, also presented a picture of severe starvation. The influence of infection has to be considered, but I think it not improbable that the deaths of many of these patients in those former days ought to have been attributed to starvation.

In the interval between World War I and World War II a great deal of attention was devoted by students of nutrition to deficiencies of vitamins. The relationship of ascorbic acid to scurvy, of vitamin D to rickets and osteomalacia, of vitamin A to keratomalacia, of thiamine to beriberi and of niacin to pellagra was established. It, furthermore, was recognized that these diseases could develop even when the patients were receiving what apparently represented an adequate supply of calories. It also came to be suspected strongly that degrees of deficiency of several of the vitamins which were not severe enough to provoke frank disease, such as scurvy and pellagra, would give rise to symptoms of poor health accompanied frequently by superficial signs of abnormality, such as dry, rebellious, lusterless hair, dry skin, follicular keratosis, erosions at the angles of the mouth, atrophy of the papillae of the tongue, and many others. Thus we came to think of "hidden hungers" and to recognize that malnutrition could result from poorly proportioned diets even when the total food supply was adequate.

In the light of observations of vitamin deficiencies in persons who are otherwise well fed, it came as a surprise to many students of nutrition that evidence of deficiency of vitamins was encountered so infrequently among the famished populations of the areas of active warfare in World War II. Explanations commonly considered are as follows: It has long been known that fat babies are much more susceptible to deprivation of vitamin D than thin babies and that the healing process can be started in a rickety infant simply by withholding calories. It also has been known that experimental animals and men deprived of thiamine will go in and out of conspicuous degrees of thiamine deficiency, depending on the amount of carbohydrate in the diet. The basal metabolic rate in semi-starvation may fall as much as 50 per cent below normal standards based on energy exchange per square meter of surface area. In large part this is due to wastage of the active tissues, but Keys and his associates have shown that even the oxygen utilization of the diminished mass of active tissue is depressed, not greatly but significantly. Furthermore, the total energy exchange, the basal plus the energy of work, is profoundly lowered because of the fatigue and listlessness which accompanies starvation. It is probable, furthermore, although not established, that enzymes containing certain vitamins are released from wasting tissue to become available elsewhere in the body.

For these and other reasons the requirements for several of the vitamins may be diminished in starvation, while at the same time the foods that people eat in times of famine — garden truck, potatoes, unmilled or undermilled grains, and the like — are relatively better sources of certain of the vitamins than are the processed foods, such as sugar and white flour, which constitute a large proportion of Western European and American diets in times of peace.

There are, however, some exceptions to the general rule that vitamin deficiencies are not conspicuous in cachexia. There seems to be less documented evidence of vitamin deficiencies in Europe than in Eastern Asia, where not only beriberi but various neurologic abnormalities were definitely associated with a combined deficiency of calories, protein, and vitamins. Supplies of sources of preformed vitamin A, such as cream, butter, eggs, and organ meats, are usually very short in times of famine; likewise fats are scarce, and with a diet low in fat the provitamin A, carotene, is likely not to be absorbed effectively. It will be remembered that in World War I Danish children developed xerophthalmia and that this was attributed at the time to the export of most of the Danish butterfat to Germany and England. So, even though the vitamin A content of the blood of the subjects of the Minnesota Experiment was satisfactory at the end of the period of twenty-four weeks of semi-starvation, it is not to be assumed that longer periods of starvation will not result in vitamin A deficiencies. It also is conceivable that lack of calories and the accompanying retardation of the circulation may predispose to deficiency of certain vitamins, so that a relative deficiency of vitamin A could have played a part in the occurrence of the xerosis and follicular keratosis which developed in the Minnesota subjects. Another abnormality observed in many of the Minnesota subjects whom twice I was privileged to examine was brownish pigmentation of the skin of the face, especially below the eyes and around the mouth. Such pigmentation also was reported as occurring frequently in the population of the famine areas in the war. It appeared to me to be identical with what had been observed in Newfoundland in 1944, where the diet, although adequate in calories, was very poor in several vitamins. In Newfoundland four years later the incidence of this abnormality in a comparable sample of the population was much lower. The population in the interval had received additional riboflavin, niacin, and thiamine through fortification of the flour and additional vitamin A through fortification of the margarine.

In any case, the question of the relationship of a low intake of calories or protein versus vitamin deficiency to the development of several of the superficial signs of malnutrition requires further study, as does also the relationship of protein, of vitamins, and of calories to wound healing and convalescence. Here are fields in which clinical investigation can be very helpful. Above all, however, physicians who read this study of starvation will be compelled to realize that starvation if allowed to go too far is a cause of death, whether it occurs in a concentration camp or in a modern hospital. When this is fully recognized the diagnosis "starvation" will be recorded much more frequently than it has been on death certificates.

February 9, 1950
Rochester, Minnesota

Foreword

by PROFESSOR C. G. KING, Ph.D.
Scientific Director, The Nutrition Foundation

and

DR. ROBERT R. WILLIAMS, Sc.D.
Chairman of the Williams-Waterman Fund

THE science of nutrition is generally regarded as having first reached a quantitative and sound philosophic basis in the studies of the French chemist, Lavoisier. These discoveries, toward the end of the eighteenth century, were remarkably significant in the sense of revealing a direct relationship between the quantity of food burned inside the body and the amount of heat produced, while oxygen from the air was consumed and carbon dioxide was exhaled. Hence food came to be regarded as fuel almost as soon as man learned the nature of what was happening when a candle or splint was burned in the air.

Despite the intervening century and a half of progress in clarifying the early concept of energy sources in the human body, and the past half-century of exciting discoveries concerning the respective roles of sugars, fats, vitamins, amino acids, and trace mineral elements, the record now being published by Professor Ancel Keys and his associates will be recognized as a major contribution.

Among the reasons why the present report is highly significant, scientists will quickly recognize the following:

(1) The studies were conducted under controlled conditions, with a sufficient number of carefully selected test subjects and over a long enough period of time to permit in nearly all instances a satisfactory statistical treatment of the data — and therefore the data permit a corresponding degree of confidence in the conclusions.

(2) Appraisals of performance of the test subjects included a battery of techniques based upon different professional disciplines, thereby bringing into the study a critical approach to variations that might be observed (a) in the diets consumed, (b) in physical performance capacity, (c) in psychologic factors at play, by design or accident, and (d) in metabolic disturbances that might be identified by chemical analysis — using objective rather than subjective evidence whenever possible.

(3) Both in the planning of the work and in its interpretation there was a timely sense of relationship to actual human need and experience. For the rec-

ord, this item was of major importance in making research subjects and all other personnel available.

Large sections of the world's population are recurrently or chronically subjected to varying degrees of starvation and possible convalescence. There is no reason to believe that escape from such experiences will be possible in the immediate future — too many scientific, economic, and social problems are in sight to permit a soft optimism in appraising the future.

Whether one regards starvation primarily as a characteristic experience induced by war, or as one of the basic causes of war, there is need for broad and accurate appraisal of its effects. Similarly, there is need of information regarding the steps by which convalescence can be accomplished most satisfactorily. Lack of such guidance during and after World War II led to tragic failures which cannot be reviewed here.

Less dramatic circumstances of starvation are common to hospital experience in normal times, especially among patients who for one reason or another cannot ingest or assimilate food. The studies reported by Professor Keys and his associates will find application in this area of day-to-day living, too, independently of mass tragedies.

From a public health point of view, there is some risk that readers will overinterpret the findings recorded in the present report relative to the effects of different levels of intake of specific nutrients. The tests were designed to measure and interpret the effects of deficiencies over limited periods of time — perhaps long periods from an experimental point of view, but short in terms of nutrition research with small animals having short life spans. Within such periods, the effects of partial caloric restriction are dominant. It is a well-known principle also that caloric restriction tends to lessen the body's demand for specific nutrients such as the vitamins.

A different type of undernutrition that can be superimposed unwittingly on a generous caloric intake is thought to be common in many sections of the world, in which individuals are subjected to partial deficiencies and imbalances of specific nutrients. These conditions may exist at levels of intake that are high enough to prevent specific signs or symptoms of deficiency disease, but that nevertheless affect health adversely over a period of many years. Conclusive data based on human experience over long periods of time are difficult to obtain, but there is a considerable body of pertinent information at hand. The Minnesota studies were not designed primarily to throw light on that problem. They have only a minor bearing, for example, upon the Recommended Dietary Allowances published by the National Research Council, except as the observations tend to show that through short periods, impairment of body functions was not as demonstrable as had been expected on the basis of vitamin and protein studies in other laboratories.

On one point there will be complete agreement — the new information provided in the present publication will have widespread use. The report will have fruition in human betterment and is likely to become the leading scientific reference in its field.

February 16, 1950
New York, New York

Authors' Foreword

THE need for an inclusive critical treatise on human starvation and undernutrition became abundantly clear to us when, after several years of work on the immediate problems of military subsistence, we began to ponder the larger questions of the feeding of peoples. Elementary calculations showed that not only were the world's stocks of foods dwindling, but also that the war was devouring the resources for food production and distribution. We were astonished to discover the paucity of knowledge on the effects of simple undernutrition, though here surely is the oldest and one of the most persistent of the disabilities which plague mankind. We were dismayed to contemplate the empty hands which science would bring to meet the manifold problems of a disorganized world, where government decree would have to substitute for the good sense of the body in apportioning the daily bread, where for many millions the first step of recovery must be nutritional rehabilitation.

Early in 1944 we urged that, even in the midst of active war, the Allies must make serious preparations to meet the inevitable consequences of dedication to physical conflict. In such preparation food should loom large and a major requisite would be to get far more precise information on the effects of semi-starvation and the problems of refeeding. But other priorities held the stage and it was clear that research in this area would have to be started by private initiative. It was on this basis that we set out, with small encouragement from official quarters, on what has come to be known as the Minnesota Experiment. A short sketch of the history and support of the experiment is given at the end of this Foreword.

The problems of famine and subsequent relief as we expected them to develop in Europe determined the form of the research plan but the generality of the phenomena to be studied obviously has a broader significance. The world has its share of hungry people at all times. Moreover, general or caloric undernutrition plays an important role in a large proportion of all persons who are ill, even in times of plenty. And, whether the origin is primary or secondary, the presence of undernutrition makes a special kind of person, different morphologically, chemically, physiologically, and psychologically from his well-fed counterpart. Some of this we could understand from the widely scattered and inadequate literature available in 1944.

As the work progressed we attempted to provide hastily assembled progress reports to meet practical needs of the moment. These were in great demand — our privately printed copies disappeared rapidly before a mounting flood of requests — but they neither critically presented our own findings nor placed them in proper perspective with the literature. Besides, only a small portion of our results could be worked up rapidly enough to be presented at the time.

The scope of the present book was decided on long before the last observations were finished in the Minnesota Experiment. With more ambition than wisdom, perhaps, it was agreed that, with the exception of purely sociological, economic, and agricultural concerns, the whole scope of the biology of human starvation would be presented in a critical analysis of all the available knowledge and that this would include the full material from the Minnesota Experiment.

The accomplishment of this formidable task was not rendered easier by the fact that the Laboratory, and thereby the authors, had to earn a living in the meantime. The present system of project sponsorship as the major means of research support is not designed to encourage exhaustive scholarship or sustained critical thought. It is now (March 1, 1950) almost 4 years since the "completion," and this means sponsorship as well as laboratory studies, of the Minnesota Experiment. Much of the analysis of the data, the search of the world literature, and the synthesis of the whole into some kind of form, as well as the actual writing, has had, perforce, to be an avocation, a rather fatiguing substitute for recreation to be done "after hours."

Within the past 2 years there has been a large accession to the literature, including the valuable books of Hottinger *et al.* (1948), Simonart (1948), Lamy, Lamotte, and Lamotte-Barrillon (1948), and de Castro (1948). During this time we also first received a copy of the remarkable study made at Warsaw (Apfelbaum, ed., 1946) and a number of lesser publications which had been missed. With many parts of the present book completed, it was difficult to incorporate, without evidences of abrupt transition, the materials contained in these 2000 pages.

Finally, on January 23, 1950, we received the long-awaited Report of the Dutch Government on the 1944–45 famine in the western Netherlands (Burger, Drummond, and Sandstead, eds., 1948). At that time we were engaged in reading proofs and it was out of the question to insert a review and discussion of this work of 486 pages. The Dutch Report is unique; no other famine has been studied in such detail. Fortunately, some accounts of the Dutch famine had been published earlier (cf. e.g. Dols and van Arcken, 1946; Boerema, ed., 1947; Burger, Sandstead, and Drummond, 1945; Cardozo and Eggink, 1946; Bigwood, ed., 1947). Moreover, some of the material had been made available to us previously in the form of field reports and surveys (referred to in the present book as "Surveys"). But the final report contains much new information. We can only urge that the Dutch Report be studied in parallel with the present book.

Originally we had set our goal as a complete and rigorously critical rendition of world knowledge on human starvation. The preceding two paragraphs will indicate our recognition of failure to achieve completeness. But by now the authors are resigned to less exalted ideals than had been entertained originally. The pursuit of perfection must have its rewards in the effort rather than the accomplishment.

The bibliography is large but it represents the residue of a much larger list of publications consulted. We have been at some pains to consult the originals, having suffered a substantial amount of exasperation with misquotations and erroneous citations. That errors persist here we have no doubt, but at least they have not arisen from the pernicious habit of compiling an impressive bibliography at second hand. Where original data were published in the older works we have frequently recalculated them; the discovery of "original sin" occurred often enough to provide some recompense for this labor — which also is recommended as a sure way to discover the virtues as well as the faults of scientific papers.

One of the thorniest problems raised by the modern teamwork system of scientific research is the apportionment of individual credit. The general philosophy of the Laboratory of Physiological Hygiene has been discussed in two papers on interdisciplinary research (Brožek and Keys, 1945, 1946). The work of the Minnesota Experiment, including the writing of this book, was, like all the activities of the Laboratory, a cooperative venture in the fullest sense. Though each Staff member has an area of specialization, there are no hard and fast lines, and mutual consultation proceeds throughout the days (and nights) of close association.

In the Minnesota Experiment all of the five major authors shared in the general and detailed planning as well as in the numerous problems of housekeeping, diets, regimen, methods, and recording. Dr. Taylor took a major responsibility in recruiting the subjects and played an important role in maintaining the morale and active interest of the subjects. Together with Dr. Henschel, he concentrated on physical performance, respiration, and postural tests. He made the electrophoresis studies. Dr. Henschel, besides joining with Dr. Taylor as indicated above, had charge of blood morphology and carried the large burden of scheduling all tests and measurements. Dr. Mickelsen had general charge of the chemical work and took the leading role in all dietary matters, including day-to-day supervision of the diet kitchen and its staff. Dr. Brožek had charge of all psychological work, including the special senses and psychomotor tests, as well as the anthropometric measurements. He also exercised general supervision over the extensive statistical analyses and directed the work of the computers. Dr. Keys exercised general supervision over the entire project, had charge of the X-ray work, and carried the administrative burden.

Among the three assisting authors, Mrs. Skinner supervised the general housekeeping, assembled the records, and directed the staff of technical assistants. She actively participated in the blood studies and in the work on basal metabolism. Dr. Wells was the physician in residence and participated in several phases of the laboratory program. Dr. Simonson was not directly involved in the planning or the experimental work, but, together with Dr. Wells, he aided in the search and analysis of the older literature on metabolism and clinical matters. He analyzed the electrocardiograms.

In the writing there is hardly a chapter that can be credited solely to any single individual; there was constant collaboration at all stages and reference material was pooled at the outset. The senior author acted as general editor throughout.

We regret, of course, that this work was not published when the hunger of the world was more acute than now. But we are not so sanguine as to suppose that these problems have been permanently banished, though we rejoice at their present amelioration. The fact is that a large portion of the earth's population is always underfed. At the present moment there are most ominous reports from China where a major famine is feared. Nor can we forget that the whole of Southeast Asia is now, as for many centuries past, frighteningly close to starvation disaster. The current existence of overeating as a major problem in some areas, such as the United States, should not distract us from the world picture and the longer view. We trust that this work will aid in the recognition and evaluation of these problems as well as in the general understanding of man.

April 3, 1950
Minneapolis, Minnesota

Acknowledgments

FINANCIAL sponsorship and aid to the Minnesota Experiment came from the most diverse sources. The project was conceived at the start of 1944 as a result of casual discussions between the persons then working together in the Laboratory of Physiological Hygiene on controlled experiments in human vitamin nutrition. Drs. *Brožek* and *Taylor* of the regular Staff and Dr. *Harold Guetzkow* of the Civilian Public Service Unit in the Laboratory took a leading part in these discussions and, together with the Director of the Laboratory, undertook explorations as to the feasibility of a realistic but thoroughly scientific study on human semi-starvation. From the outset it was planned to use volunteers from Civilian Public Service as subjects. Mr. *Harold Row* of the Brethren Service Committee was an early convert and gave much aid in enlisting the cooperation of the "peace churches" and the National Committee for Religious Objectors.

Initially it was hoped to start the research, designed to last a year, early in 1944, but failure to obtain the necessary financial aid from the responsible authorities in Washington caused a delay. Financial aid was guaranteed, however, by the *Brethren Service Committee of the Church of the Brethren,* by the *Service Committee of the American Society of Friends,* and by the *Mennonites Central Committee.* Permission was granted by the donors to divert to this program the research funds granted to the Laboratory by the *Sugar Research Foundation,* New York, and the *National Dairy Council,* operating on behalf of the *American Dairy Association,* Chicago. Requests for funds were granted by the *Home Missions' Board of the Unitarian Society,* Boston, and the *John and Mary R. Markle Foundation,* New York. The *University of Minnesota* gave permission to use for this project the facilities and the regular budget of the Laboratory, then mainly derived from *Athletic Funds.* Permission to set up a Civilian Public Service Unit and to recruit volunteer subjects from CPS was obtained through Mr. *R. C. Imrie,* director of CPS camps for the *Selective Service System,* after endorsement by the *Committee on Medical Research* of the *Office of Scientific Research and Development.*

Actual operations began with the arrival of the subjects at Minneapolis in November 1944. Subsequently, important financial support was provided under the terms of a contract, recommended by the *Committee on Medical Research,* between the *Regents of the University of Minnesota* and the *Office of Scientific Research and Development.* This contract was eventually (December 1945) transferred to the *Office of the Surgeon General, U.S. Army.* In the meantime, the project was aided by a grant from the *Graduate School Research Fund of the University of Minnesota.* Even with all these sources of aid the project would have had serious financial difficulty if it had not been for the willingness of the Laboratory Staff to undertake a burden of labor which should have called for more workers, and that of the CPS men to help actively as well as to accept a passive subject role. The men were housed in the University Stadium and a host of helping hands made the rather primitive quarters both effective and pleasant. The Unit took care of its own housekeeping, including laundry, and aided in many aspects of the laboratory operations.

A last financial difficulty arose when the manuscript of this book was ready for the press. It was obvious that without substantial subsidy the book would have to carry a selling price which would effectively prevent it from having the distribution, especially foreign, which was hoped for it. This problem was solved by generous grants from the *Nutrition Foundation, Inc.,* New York, and the *Williams-Waterman Fund,* New York.

The subjects were selected, from many volunteers in CPS, by Drs. Taylor, Brožek, Henschel, and Guetzkow. All subjects were fully informed as to what to expect; the men knew they were to face a long period of discomfort, severe restriction of personal freedom, and some real hazard. Besides the opportunity to lend themselves for work which they believed would have significance for starving people elsewhere, they were promised an educational program designed to prepare them for relief work abroad. It is a pleasure to record here the whole-hearted cooperation of these men in accepting the most important but the least enviable role of the project. Their names are listed in the roster of the CPS Unit given at the end of these Acknowledgments.

The CPS Unit included 18 men who were not subjects but who provided full-time assistance, in some cases of a highly technical nature, and who, with a few exceptions, also lived in the Laboratory with the starvation subjects. Their names are listed here in the roster of the CPS Unit. The CPS educational program was directed by Mr. *Paul Hoover Bowman,* whose services were provided for some six months by the Brethren Service Committee.

The junior members of the regular Laboratory Staff carried a large load with notable devotion and skill. For two years it was taken for granted that everyone should work harder and more effectively than would have been thought possible. Following that period a large task of clerical and typing work developed. The persons who shouldered these burdens are listed at the end of these acknowledgments.

Technical assistance, cooperation, and advice in the Minnesota Experiment were generously given by a large number of persons of whom the following should be noted: Dr. *H. H. Mitchell* of the University of Illinois provided bomb calorimeter analyses and other help in checking the caloric values of the foods used; Dr. *Russell M. Wilder* of the Mayo Clinic made several special trips to participate in clinical examinations; Dr. *Dorothy Sundberg* of the University of Minnesota made the bone marrow examinations and assisted in the preparation of the section on bone marrow (Chapter 14); Dr. *J. B. Wilkie* of the Food and Drug Administration and Dr. *Otto Bessey,* now of the University of Illinois, carried out analyses for vitamin A and carotene; Drs. *Harold Hume* and *George Rogers* of the University of Minnesota made the semen examinations, and the former aided in the preparation of the manuscript on this material; Drs. *Hamilton*

Montgomery and *O. E. Okuly* of the Mayo Clinic took skin biopsies and worked up the material; Drs. *Thomas Lowry, Cecil J. Watson,* and *Frederick Hoffbauer* of the University of Minnesota aided in problems of medical care of the Minnesota subjects; Dr. *Wallace D. Armstrong* of the University of Minnesota provided analyses for sodium and potassium in the blood; Dr. *Burtrum C. Schiele* of the University of Minnesota aided in the psychological and psychiatric evaluations; Dr. *Leo Rigler* of the University of Minnesota cooperated in radiological work and diagnosis; Dr. *Pauline Beery Mack* and her colleagues of Pennsylvania State College gave much labor to radiological estimations of bone density.

In March 1946 Drs. Keys and Taylor again examined 21 of the subjects, who were admitted to several hospitals for this purpose. For these hospital facilities and many courtesies we are indebted to Dr. *Robert W. Keeton,* Mr. *Nathaniel Glickman,* and others of the staff at the University of Illinois Medical School, to Dr. *Winfred Overholser* and Drs. *Eldridge, Farley,* and *Katzenelbogen* at St. Elizabeth's Hospital, Washington, D.C., and to Dr. *Isaac Starr* and Drs. *Austin, Black, Buerki,* and *Mayock* at the University of Pennsylvania Hospitals.

Several of the men of the Civilian Public Service Unit continued to provide assistance in certain special phases of the work after their discharge from CPS: Drs. *Howard Alexander* and *Harold Guetzkow,* Messrs. *Joseph C. Franklin, Richard Seymour, Glenn Fisher, Walter Carlson, Gerald Wilsnack,* and *Max M. Kampelman.* Mr. Kampelman provided valuable references and notes used in Chapter 1.

Several commercial firms generously provided special materials used in the rehabilitation phase of the Minnesota Experiment: Purified Soy Protein, Archer Daniels Midland Co., Minneapolis, Minnesota (Dr. *James Hayward*); "Casec," Mead Johnson and Co., Evansville, Indiana (Dr. *W. M. Cox*); "Hexavitamin" Tablets and Placebos, Strong, Cobb and Co., Cleveland, Ohio (Dr. *William Hosler*); "Essenamine," Frederick Stearns and Co., Detroit, Michigan (Dr. *Earl Burbridge*).

The persons who aided in the analysis of the Minnesota Experiment, the collection of literature, and the interpretation of many problems are very numerous. Among those who scrutinized data or provided personal information and advice, or both, should be mentioned: Drs. *Marvin Corlett, John B. Youmans, V. P. Sydenstricker, Harold R. Sandstead, R. K. Salmon,* and *Richard Scammon,* all of the United States; Drs. *H. E. Magee, John Beattie, H. M. Sinclair,* and Sir *Jack Drummond,* all of England; Drs. *A. G. van Veen, M. van Eckelin, M. J. L. Dols, B. C. P. Jansen,* and *J. Groen,* all of the Netherlands; and Dr. *Hugues Gounelle* of France. The exchange of information with Dr. Beattie was particularly valuable. Dr. *O. Gsell* of St. Gallo, Switzerland, provided an advance copy of the valuable monograph *Hungerkrankheit, Hungerödem und Hungertuberkulose* by Hottinger, Gsell, Uehlinger, Salzmann, and Labhart. Dr. *Maurice Lamotte* of Paris, France, sent an advance copy of the large treatise *La Dénutrition* by Lamy, Lamotte, and Lamotte-Barrillon. Dr. *A. Simonart* of Louvain, Belgium, supplied his book *La Dénutrition de Guerre* as well as separate photographs illustrating famine edema. We are grateful to Dr. *John M. Berkman* of the Mayo Clinic, Rochester, Minnesota, for the photographs illustrating anorexia nervosa.

For permission to reproduce, in various tables and figures, materials published elsewhere, we are obliged to:

Academic Press, New York

Acta Medica Belgica, Brussels, Belgium

Acta Medica Scandinavica Forlag, Stockholm, Sweden

Agricultural Experiment Station, University of Missouri, Columbia, Mo.

American Jewish Joint Distribution Committee, Inc., New York

American Medical Association, Chicago, Ill.

American Physiological Society, Washington, D.C.

American Public Health Association, New York

American Society of Biological Chemists, New Haven, Conn.

American Society for Clinical Investigation, Cincinnati, Ohio

J.-B. Baillière et Fils, Ed., Paris, France

J. F. Bergmann, Munich, Germany

British Medical Association, London, England

Canadian Medical Association, Montreal, Canada

Carnegie Institution of Washington, D.C.

Child Development Publications, Evanston, Ill.

E. P. Dutton and Co., New York

Journal of the Mt. Sinai Hospital, New York

The Lancet, London, England

Lea and Febiger, Philadelphia, Pa.

J. F. Lehmanns Verlag, Munich, Germany

Masson et Cie, Ed., Paris, France

Milbank Memorial Fund, New York

C. V. Mosby Co., St. Louis, Mo.

Einer Munksgaard, Copenhagen, Denmark

National Tuberculosis Association, Baltimore, Md.

New England Journal of Medicine, Boston, Mass.

New York Academy of Medicine, New York

Rockefeller Institute for Medical Research, New York

Benno Schwalbe and Co., Basel, Switzerland

Society for Experimental Biology and Medicine, New York

Springer Verlag, Berlin, Germany

Georg Thieme Verlag, Stuttgart, Germany

Williams and Wilkins Co., Baltimore, Md.

Wistar Institute of Anatomy and Biology, Philadelphia, Pa.

Yale University Press, New Haven, Conn.

Staff of the Laboratory of Physiological Hygiene

During the experimental portion of the Minnesota Experiment the following persons of the regular Staff provided invaluable aid:

MARIETTA ANDERSON, Dietitian

LUCILE CARLSON, Senior secretary

HOWARD CONDIFF, Junior chemist

DORIS DOEDEN, Technologist

LESTER ERICKSON, Research assistant

NEDRA FOSTER, Administrative technologist

DORIS FREDSON, Technologist

LORRAINE HERTZ, Technician

LUELLA HONG, Secretary

ERSAL KINDEL, General mechanic

BETTY KRUEGER, Laboratory helper

ERMA V. O. MILLER, Assistant scientist

MARY NEILSON, Stenographer

NELLIE OLESON, Laboratory helper

MILDRED OLSON, Assistant technician

SIGNE HAUGEN OLSON, Dietitian

LORRAINE RENSTROM, Senior clerk typist

MARY HOCKMAN SMITH, Stenographer

DELPHINE SWANSON, Senior secretary

LAURA WERNER, Technologist

University of Minnesota Civilian Public Service Unit

Starvation Subjects. The core of the Minnesota Experiment consisted of the 36 volunteer subjects who are identified by code numbers elsewhere in this work:

WILLIAM F. ANDERSON, II, Nashville, Tennessee

HAROLD BLICKENSTAFF, Chicago, Illinois

WENDELL BURROUS, Peru, Indiana

EDWARD COWLES, Port Ludlow, Washington

GEORGE EBELING, Moylan, Pennsylvania

CARLYLE FREDERICK, Nappanee, Indiana

JASPER GARNER, Okeechobee, Florida

LESTER J. GLICK, Sugar Creek, Ohio

JAMES E. GRAHAM, Madison, Wisconsin

EARL HECKMAN, Rocky Ford, Colorado

ROSCOE HINKLE, Hershey, Pennsylvania

MAX M. KAMPELMAN, New York, New York

SAMUEL B. LEGG, Brandon, Vermont

PHILIP LILJENGREN, Chicago, Illinois

HOWARD T. LUTZ, Lansdowne, Pennsylvania

ROBERT P. McCULLAGH, Monrovia, California

WILLIAM T. McREYNOLDS, Salem, Oregon

DAN J. MILLER, LaVerne, California

L. WESLEY MILLER, Enid, Oklahoma

RICHARD MUNDY, Bloomington, Indiana

DANIEL J. PEACOCK, Richmond, Indiana

JAMES PLAUGHER, Fresno, California

WOODROW RAINWATER, Fort Worth, Texas

DONALD G. SANDERS, Sherrill, New York

CEDRIC SCHOLBERG, Bingham Lake, Minnesota

CHARLES D. SMITH, Merchantville, New Jersey

WILLIAM STANTON, Woodbury, New Jersey

RAYMOND SUMMERS, E. Dearborn, Michigan

MARSHALL SUTTON, Clintondale, New York

KENNETH D. TUTTLE, Cleveland, Ohio

ROBERT VILLWOCK, Toledo, Ohio

WILLIAM WALLACE, San Anselmo, California

FRANKLIN WATKINS, Pittsburgh, Pennsylvania

W. EARL WEYGANDT, Clarksville, Michigan

ROBERT WILLOUGHBY, Harrisburg, Pennsylvania

GERALD WILSNACK, Malverne, New York

CPS Assistants. These men devoted full-time effort and thought to their several responsibilities in the Minnesota Experiment:

HOWARD ALEXANDER, Statistician

WALTER CARLSON, Laboratory assistant

W. O. CASTER, Chemist

GLENN FISHER, Laboratory assistant

JOSEPH FRANKLIN, Psychologist

ANDREW GIBAS, Chemist

HAROLD GUETZKOW, Psychologist

W. JARROTT HARKEY, Unit leader

WILBUR HELD, Cook

DON MARTINSON, Assistant cook

WILLIAM R. MICHENER, Clerk

JOHN N. PHILLIPS, Clerk

NORRIS SCHULTZ, Clerk

ARTHUR SNOWDON, Physicist

ROBERT STEVENS, Educational assistant

EUGENE SUNNEN, Chemist

MARVIN VAN WORMER, Bacteriologist

GRANT WASHBURN, Assistant cook

For the final preparation of the typed manuscript the major burden fell on Mrs. Eileen Ladd, Mrs. Lorraine Renstrom, Mrs. Helen Rasmusson, Mrs. Margaret Barnum, Mrs. Mae Carter, and Mrs. Ernestine Hagard. Some of the later computational work was done by Heman Gursahaney.

We are indebted to the staff of the University of Minnesota Press for the care with which the work was shepherded through the printing process and their patience with the vagaries of the authors.

Finally, the authors are mindful of the forbearance of their wives throughout the years of dedication of time and effort to the present work.

Background

Chapter 1. THE HISTORY OF STARVATION

Chapter 2. THE SOURCES OF EVIDENCE AND INFORMATION

Chapter 3. THE CARNEGIE NUTRITION LABORATORY EXPERIMENT

Chapter 4. GENERAL FEATURES OF THE
MINNESOTA EXPERIMENT

"*Many have unskillfully written upon the preservation of health, particularly by attributing too much to the choice, and too little to the quantity of meats.*"

FRANCIS BACON, in *The Advancement of Learning,*
The Fourth Book, Chapter 2.

CHAPTER 1

The History of Starvation

A FULL account of human experience with starvation would cover all of history and would penetrate every phase of human affairs. As some scholars have pointed out, the history of man is in large part the chronicle of his quest for food (cf. Prentice, 1939). Hunger, or the fear of it, has always played a major role in determining the actions and the attitudes of man. In every age and every land people have starved, and the 20th century is no exception. Two thirds of the population of the world is now engaged in agriculture but still half the world is chronically underfed. What can we expect of the future?

Technological improvement in food production continues apace, and even now we are not primarily limited by physical nature. The full application of present knowledge alone would suffice to provide food in abundance for the entire world population. Basically, neither production nor transportation capacity is theoretically limiting. But this fact is no guarantee that famines will not continue or that chronic undernutrition will soon vanish.

Distribution is and always has been a major factor, but this does not mean that a more equal distribution of the world's food would solve the problem. At the present level of world food production, equal distribution might prevent anyone from starving to death but it would mean that everyone would be undernourished. Obviously a general increase in food production would be necessary.

At the Hot Springs Conference of June 1943, it was concluded that "The first cause of hunger and malnutrition is poverty." We might equally well define poverty as that condition in which all food needs are not met. The Hot Springs statement implies that people are underfed because they have insufficient money to buy food. This is true only in a very limited sense. The "poverty" explanation does not suffice for the major recent cases where large numbers of people have starved — the Ukraine in 1933, Greece in 1942, the Netherlands in 1945. The major cause of chronic undernutrition may be purely economic but the primary cause of modern starvation is political strife, including war.

For the past 300 years the population of the world has been increasing at a tremendous rate and this growth has not ceased. We cannot but wonder about the limits of food productivity, yet the fact seems to be that man, on the average, is steadily achieving a better diet. If we cannot be unconcerned about the limits of food production, it is equally true that we need not despair. The first goal must be the final elimination of famine and chronic undernutrition. This would be only the beginning on the long road to complete conquest of food wants, but even this is a task to challenge us all.

A natural failure of food supply is not always the cause for men being underfed. Deliberate enforced starvation has too often been a political tool, though seldom in history has it been applied with the brutality used by the Nazis. Far more common is the semi-starvation associated with disease. Most serious illnesses result in a deficient food intake, and some degree of cachexia is extremely common in hospital patients. Until recently, little attention has been given to the problems of diagnosis and therapy raised thereby. This is so large a subject in itself that we can do little more than call attention to it here.

Prehistoric Man

From the earliest archeological evidence it is clear that paleolithic man competed with animals on a personal and equal level; they killed and ate one another (Lechler, 1943). Man also ate such fruits and other wild produce as he found — and the analogy of primitive peoples today suggests that his tastes were indeed catholic — but the precariousness and sparseness of the food supply prevented any organization of man into units much larger than the family. Whether prehistoric man killed and ate his own kind with any frequency is not certain; Furnas and Furnas (1937) suggest that Cro-Magnon man quite probably ate the Neanderthal man whom he displaced in Europe.

Perhaps about 10,000 B.C. Neolithic man began to domesticate herbivorous animals and thus ushered in a pastoral economy with its much greater assurance of food. Man was firmly tied to his animals and was forever on the move with them in their need for forage. The tribe, with its larger number of people, could subsist on the flocks and some specialization of tasks became possible. At the same time the vital necessity of grazing land drove man to long wanderings and, on occasion, to conflict between tribes as they sought to use the same lands.

The Development of Agriculture

The real beginning of civilization came with the development of agriculture. Agriculture certainly could only have arisen in those areas, like the Nile Valley and parts of Asia Minor, where pastoral man could stay in one place long enough to see the cycle from seed to fruition. Somewhere between 5000 and 6000 B.C. the practice of planting and reaping began. It seems unlikely that this was the result of experimentation, as is sometimes suggested, nor can we credit the likelihood that any one person really "discovered" agriculture, even though an individual was immortalized by the Egyptians (Isis), the Greeks (Demeter), and the Romans (Ceres) as the goddess of agriculture and the produce of the earth.

In any case, agriculture first permitted the growth of large communities, the storage of large amounts of foods (grains) against times of want, and the real specialization of human occupations. The dependence for food of one segment of the population on another began, the basis of hoarding and complex economies was laid, and the possibilities of wars, mass starvation, and other dangers of organized society arose.

One early result of a more stable food supply and a fixed home was a rapid increase in the population. There began the race between population demand and increased food supply which was to be emphasized by Malthus. Productive land became property to be defended by the possessor and to be fought for by others. Agricultural man lost his mobility when he committed himself to the produce of a particular spot which he cultivated. On the other hand, those peoples who were still in the pastoral state could be envious of the fertile cultivated fields and, being mobile, could raid the larder of the agriculturist. The latter, in turn, had to stand his ground and was forced to build fixed defenses, much like our American ancestors built blockhouses. In another area the great agricultural civilization of China built a wall 1700 miles long to keep out the raiding nomads.

The Beginning of History

The "stele of famine" is among the earliest authentic records of history. It was discovered carved on a granite tomb on the Island of Sahal in the first cataract of the Nile. Its exact antiquity is unknown but there is evidence to show it was chiseled in the time of Tcheser (or Tosorthus) about 2000 years before the story of Abraham (Graves, 1917). Its tale is one of woe: "I am mourning on my high throne for the vast misfortune, because the Nile flood in my time has not come for seven years! Light is the grain; there is lack of crops and of all kinds of food. Each man has become a thief to his neighbor. They desire to hasten and cannot walk. The child cries, the youth creeps along, and the old man; their souls are bowed down, their legs are bent together and drag along the ground, and their hands rest in their bosoms. The counsel of the great ones in the court is but emptiness. Torn open are the chests of provisions, but instead of contents there is air. Everything is exhausted."

This record from the remote past is of interest because it clearly conveys the picture of the physical and moral deterioration of a starved people and because the famine was apparently not caused by man himself but by an act of nature. This causation of famine by unkind nature was by far the most common until fairly modern times, when man's dominance of nature allowed him to assume the role of creator of his own misery.

The Old Testament has many references to starvation and famine, beginning in Genesis 12:10 in the time of Abraham, but there are few details of importance. At the time of Joseph the development of the granary had progressed so far that its control was of immense political consequence and was used to reduce a people to slavery. Through control of the grain the Pharaohs of Egypt became great landowners and rented the land to the peasants, who thereby became completely subservient to the monarchs. The process was described by Flavius Josephus, one of the great historians of early Rome, and has been corroborated by modern Egyptologists; the date at which this central control of the grain in Egypt became effective is between 1500 and 2082 B.C., probably in the reign of Apholis at the end of the Seventeenth Dynasty (Lacy, 1923).

The Record of Famines

In the Appendix of this book we have tabulated some of the major famines for which there is reasonably good historical evidence (cf. Short, 1749; Walford, 1879). The list, though only partial, is far too long to discuss here. It may be noted, however, that the vast majority of these famines had their immediate origins in crop failures from drought, excessive rain, or unseasonable frosts.

The later annals of the Bible provide the first record of a general famine which is well confirmed from non-biblical sources. This occurred during the reign of Claudius and is notable also in that it elicited a concerted effort at relief as related by Luke (Acts 11:28–30): "Then the disciples, every man according to his ability, determined to send relief unto the brethren which dwelt in Judea. Which also they did, and sent it to the elders by the hands of Barnabas and Saul."

Throughout later recorded history famines have been frequent in almost every land. A single bad crop season, a few weeks of drought, a hailstorm at a critical time, or a local flood was enough to produce famine. Wars produced starvation also, but generally this was on a local scale in a besieged city, in the actual zone of battle, or immediately along the line of march of an army. The great famines occurred when the size of the population grew to the point where it could just barely be supported in ordinary times; a slightly bad year inevitably brought some starvation, and a succession of bad years depopulated large areas and dehumanized the survivors. Such conditions still arise in China and India today.

In appraising the severity and frequency of famines in the past it must be remembered that until very recently every community always had some persons who were on the verge of starvation. The term *famine* was applied only when there were riots about food, the disposal of the dead became a problem, and the price of grain suddenly rose to three or more times the ordinary rate. The old chronicles seldom estimated the deaths; only occasionally do we find a remark that "thousands starved" or some equally laconic statement without further explanation. One criterion of the severity of past famines is the stark one of whether or not people resorted to cannibalism.

Cannibalism

One of the first accounts of cannibalism appears rather incidentally in the Bible (II Kings 6:26–29). Ben-hadad, king of Syria, was besieging Samaria in the 9th century B.C. When the king of Israel was inspecting the walls a woman complained to him: "This woman said unto me, Give thy son that we may eat him today and we will eat my son tomorrow. So we boiled my son, and did eat him; and I said unto her the next day, Give thy son that we may eat him; and she hath hid her son."

Cannibalism is mentioned again and again in records of famines but always as a sporadic and exceptional phenomenon (Sorokin, 1942). People ate grass and leaves, they ate their domestic pets, they ate all manner of vermin and filth, but only rarely did they eat human flesh. When they did resort to cannibalism, it was usually to eat part of the dead bodies of those who had already succumbed, or, in besieged towns, of captured enemies. In Ireland in 1316 at the siege of Carrickfergus 8 captured Scots were eaten (Walford, 1879). Whenever many people are starving there is apt to be some cannibalism. Evidences of it were seen in 1945 when the Allied armies captured Belsen and other monuments to German culture. Markowski (1945) stated that cannibalism was not infrequent among the Russian and Polish prisoners of war at the New Brandenburg Camp. It is noteworthy, however, that in spite of the tremendous scope of modern famines in India and China there are very few authentic records of cannibalism in those countries; perhaps they escape notice because of the secrecy usually practiced, but it is also possible that basic religious beliefs are responsible for a real rarity of cannibalism in those areas.

In the Russian famine of 1921–22 there were many rumors of cannibalism and some of them were tracked down (see Fisher, 1927, p. 436). Eating of dead bodies was apparently rather common in all districts. Professor Frank of Kharkov University was able to establish the authenticity of 26 cases in which human beings were killed and eaten by their murderers. In 7 cases murder was committed and the flesh, disguised in sausages, was sold on the open market. In Orenburg a notorious case resulted in an order by the city authorities forbidding the sale of meat balls and all forms of ground and chopped meats (Fisher, 1927, p. 109). In many areas the cemeteries had to be guarded to prevent the exhumation of freshly buried corpses (Sorokin, 1942). In the Ukrainian famine of 1933 cannibalism again appeared.

Famine and Food Prices

In many records, such as that of Penkethman (1748), the severity of famine has been indicated by quoting the changes in the price of grain. Thus Walford (1879, p. 10) noted a famine in England in 1437 which he described only in the following succinct statement: "Wheat rose from its ordinary price of 4s to 4s6d per quarter to 26s8d." The relatively free operation of the law of supply and demand meant that even a temporary and local shortage of food resulted in tremendous rises in the price of staples; the hungry man is by no means a free bargaining unit when he has to pay the asking price or starve to death.

Famine has often occurred in places where there were abundant stores of food in the hands of the rich, who not infrequently profited largely. One of the bitterest complaints of the Irish during the great famines in Ireland in the 19th century was the fact that the English and the great landowners benefited while thousands were dying in the streets (Emmett, 1903).

Gapp (1935, p. 261) stated: "In the ancient world, as in the modern, famine was always essentially a class famine. Since the poor and improvident never had large reserves either of money or food, they suffered immediately upon any considerable rise in the cost of living. The rich, on the other hand, had large reserves both of money and of hoarded grain, and rarely, if ever, experienced hunger during famine. Thus, while all classes of society suffered serious economic discomfort during a shortage of grain, the actual hunger and starvation were restricted to the lower classes. As a famine became more severe, the distress mounted higher and higher in the social structure."

In the Mediterranean world, where the major production of grain was concentrated in a few places, a failure in the harvest in one of these areas would affect the price of grain in the whole territory. Until relatively modern times, this was particularly the case with Egypt, which ordinarily exported enormous quantities of grain and depended upon the good behavior of the Nile (Graves, 1917). Although the total supply of food in the area was large even in famine times, only a small part of it was constantly available on the market; at the first sign of a bad harvest the rich increased their stores for personal security and the speculators naturally began to manipulate the supply to their own best advantage.

Where so large a proportion of the population lived from hand to mouth, a minor diminution in the supply could work tremendous hardships.

In modern times improvements in transportation have resulted in so-called "world prices" for basic commodities which move in international trade. The effects of local crop failure on general food prices have been thereby reduced and the possibilities of famine have been diminished accordingly. More and more it has been the case that famines tend to be restricted to those regions where transportation is least developed. Walford (1879) argued effectively that transportation deficiencies were a fundamental factor in the continued frequency of famines in India, and Mallory (1926) came to the same conclusion about China.

Starvation in a system of free exchange is closely related to local food prices, and local food prices cannot deviate greatly from world prices if there is effective transportation. This presumes, of course, that there are no other hindrances to trade. Unfortunately, it seems that as transportation limitations per se are eliminated other obstacles to trade are multiplied. Import and export tariffs, quota systems, cartel agreements, and the like can be just as effective impediments as the physical lack of railways and shipping. Price and food availability are related only so long as free exchange exists.

Attempts at regulating the prices of food have been, until recently, highly unsuccessful. Both price and wage regulation efforts extend back to very early history; even in Babylonian times governments were concerned about these questions. In 1314 the English Parliament passed a measure limiting the price of provisions (Walford, 1879) but it seems to have been totally ineffective. Queen Elizabeth published a proclamation and a "Booke of Orders" for relief of the poor during the famine of 1586 (this was the start of the poor law), "notwithstanding all which the excessive prices of graine still encreased" (Penkethman, 1748).

Even in World War I price regulation by government was not greatly developed or very effective. In Russia between the world wars there was rigid price fixing of rationed foods, but the acquiescence of the government in the price disparity and open operation of the "free" markets merely accentuated the inadequacy of an artificial price structure. In World War II price regulation was enforced with great vigor in many areas. In general this meant that there were both official and black market prices, and though the regulation of prices was extremely useful, it was not perfect as a means of ensuring all people the basic necessities of life. Probably more effective in this regard were the various rationing systems almost universally adopted in World War II.

Rationing

Rationing unquestionably prevented intolerable injustices in the supply of basic necessities in both world wars. In no country was it perfect and in every country a considerable proportion of the total food consumption was obtained from non-rationed items, from private hoards, and from the black market. Absolute, rigid, total rationing could, of course, allow no adequate adjustment for individual peculiarities of either need or taste and could be obtained only in a highly efficient police state.

Equal distribution of some food items, notably luxuries or nonessentials, is generally accepted as necessary and reasonable, if not really equitable, in times of community stress and temporary emergency. The real problem arises when a major share of the total food supply must be rationed. Obviously different classes of people have different needs, and the community must set up priorities in food allocation. In World War II practically all countries recognized the special needs and vulnerability of children and pregnant and nursing women, and effectively protected them in one way or another. The result was that these "vulnerable" persons were maintained with less nutritional damage than the rest of the community (cf. Dols and van Arcken, 1946; Magee, 1946). In most countries patients with diabetes mellitus were allowed special rations and persons with other types of diseases that required some kind of special diet were granted supplementary rations on submitting an application supported by a certificate from the attending physician.

The most troublesome problem in food rationing is the provision of equitable rations to persons of different metabolic needs. Individual equity in these respects is simply too complicated to be achieved by official action, and only crude attempts in this direction were made in World War II. In many areas, such as the United States, where ample calories from non-rationed foods were readily available, there was no attempt to adjust the rations to the variable needs of individuals other than some few sick people. In most of the European countries there were several sizes of ration allowances depending on age and occupation. In the Netherlands, for example, all adults received a "normal consumer" ration and supplementary rations were provided to workers on long hours and night work, to workers on heavy work, to workers on very heavy work, and to miners. Whereas the "normal consumer" in the Netherlands in 1943 averaged 1700 to 1800 Cal. per day, the workers on heavy work received about 2400 Cal. in their rations, and the miners averaged about 3500 Cal. daily (Dols and van Arcken, 1946). In no case was allowance made for differences in body size, so that in general big individuals were penalized for their size.

Under the severest famine conditions all rations would, presumably, be used and every person would draw his rations in full unless prevented by lack of money. There may be considerable differences between the metabolic need and the ability to pay even under rationing. In India, for example, the ration system was applied primarily to the basic grains, which provide the bulk of the calories in the usual diet there, and even when the ration allotment was small many persons did not take all they were entitled to. To some extent this was simply an indication that the ration card holders had too little money. In Cochin the rice ration averaged perhaps 8 oz. (about 230 gm.) per adult daily and about 80 per cent of this was actually "demanded," that is, taken up (and paid for); in 1945 the rice ration went up 50 per cent and the amount purchased was only 57 to 60 per cent (Sivaswamy, 1946). Since the grain ration formed by far the largest item of the diet of these people, it can scarcely be believed that they really did not want more than about 6 oz. of rice per day; this amounts to barely 600 Cal. Obviously even rationing plus price regulation cannot ensure adequate food to the destitute.

Malthus

In the analysis of the effects of food shortage in Cochin, India, in 1942–43, Sivaswamy (1946) noted that normally (as in 1940–41) there was an excess of births over deaths amounting to 4.1 per 1000 of population, but that in 1942–43 the deaths exceeded the births by 1.2 per 1000 of population. While Sivaswamy's argument that the food shortage was responsible for a net loss in population of 5.3 persons per 1000 is sustained, the problem is obviously posed of a permanent increase of population in an area already at the limit of subsistence production. This is essentially the problem which Thomas Robert Malthus recognized and analyzed in England at the start of the 19th century. Before Malthus there had been occasional consideration of the general question of population pressure, notably by Süssmilch in Germany.

Malthus arrived at the conclusion that the population tends to increase geometrically, or at least indefinitely, unless checked by war, disease, or food shortage. England, with a population of about 11 million in 1830, could be expected to double its population every 25 years or so to reach 176 million in 100 years (i.e., by 1930), unless checked. Malthus also estimated that England could actually furnish food for not more than 55 million, even at the low standards of the 1820s, so that 100 million would have to starve. Malthus further argued that the industrial revolution in England had effectively increased the area from which England could draw subsistence, but that as the population rose and as other areas utilized the new techniques of production this temporary advantage would disappear. The majority of men would therefore be condemned to a hopeless round of toil, starvation, war, and disease, for either populations will become so great that food supply is outstripped and starvation ensues, or war and disease will operate to bring about the required decrease in population.

Great support for the Malthusian doctrine was drawn from his analysis of the events in ancient Greece and Rome. In both cases relative success in meeting the basic needs of the local population quickly resulted in an increase in that population and an expanding search for more subsistence. In that search and its ultimate failure were the bases for disintegration and collapse. Malthus had only to point to the rate of population increase in England at the time to convince his readers that such an unbridled natural increase could not be maintained indefinitely. Furthermore, he could triumphantly demand to be shown nations where the population maintained stability in the face of ample food supplies.

Perhaps the greatest controversy raised by Malthus was the solution he proposed. He could see no alternative but to keep wages down to the subsistence level where they would just ensure the rearing and training of labor replacements. Besides this, he advocated, without much hope, the exercise of moral factors in advancing the marriage age and in the practice of continence. Birth control in the modern sense was unknown then and undoubtedly would have outraged the Reverend Malthus.

Toward the end of the 19th century Malthus' theory was much discounted because of its failure in quantitative prediction of population changes in that period and because of the increasingly obvious role of noneconomic, or at least nonfood, factors which affect the population level. The gradual disappearance of

European famines naturally led Europeans to question his premises, though they might have been less hasty had they carefully observed the developments in Asia and Russia. With World War I Malthus was reread, and current thinking is again concerned about the relation between growth in populations and growth in food resources and the visible means for their expansion. Much of the history of the world is certainly in keeping with some features of the Malthusian doctrine. Can the future be predicted on this basis?

Present Population Tendencies

A saturation point of population can be reached under quite different economic and food supply conditions in different countries. A rapid growth of the population and its demand for food may be checked by devastating famine and mass migration, as in Ireland in the 19th century, by reaching the lowest denominator of subsistence, with endemic famine, as in China, or by mass insistence on a high standard of living, as in France, Sweden, and the United States.

Notestein (1945) believed that any analysis of the future of the population and its food needs must count on a world population of at least 3000 million by the year 2000; this represents a minimum estimate of a 20 per cent increase in about 50 years. But by that time, possibly as early as 1970, the rate of population increase should be declining. Notestein emphatically argued that high birth rates are characteristic of areas where there is a high death rate or where the land is comparatively new and empty. As the death rate falls and the land becomes at least relatively fully populated, the birth rate declines, but this decline is long delayed after the time when maintenance of the birth rate was useful in guaranteeing perpetuation. The rise in world population, which began to be prominent in the 17th century, has been, according to Notestein, solely a reflection of a declining death rate, the latter in turn resulting from an era of peace and domestic order, a series of agricultural innovations, and sanitary and medical advances.

Governmental policies deliberately calculated to affect the rate of reproduction have been used in any real degree only to *increase* the population, and it is questionable whether these policies have been independently effective or have merely coincided with the current natural tendency of the population. Germany, Italy, Russia, and Japan are countries in which bounties and inducements to reproduce have been intensively used in modern times. The policy seemed to be most effective in Japan and Russia but in both these countries the already existent trend was not appreciably altered by the official policy. The population of European Russia has been rising sharply for at least 150 years and that of Japan for about 100 years. In France, on the other hand, where the trend was downward, every one of the innumerable governments in power between the two world wars attempted to increase the birth rate without avail.

Many of the countries which are now increasing in population owe this as much to a fall in the death rate as to any maintained or increased birth rate. Sir William Beveridge's analysis in his *Economica* showed that the true "fertility rate" in many European countries began to fall between 1860 and 1880 and steadily declined thereafter. The populations, however, continued to increase, or at least did not decrease, because of the constant reduction in the death rate. Even in India the death rate is diminishing and it may be that the fertility rate is also falling; the continued increase of population in India is in considerable part a reflection of a falling death rate alone.

The present world population is estimated at something like 2500 million. This may be contrasted with Michelot's estimate of a trifle over 1000 million in 1845; it is not likely that there will be another 150 per cent rise in the next century, but the world may hold another 1000 million persons by then. It is comforting to realize that the world's food productivity has more than kept pace with the phenomenal population rise of the past 100 years; certainly the people of today are, on the average, somewhat better fed than they were 100 years ago. We may hope that agriculture will continue its prodigies.

Tendencies in Food Production

It would be foolhardy not to recognize some of the limitations now looming up in food production which were scarcely operative in the time of Malthus. The largest supply of the world's grain today comes from areas which had barely felt the plow in 1830 — Canada, the United States, Australia, and the Argentine. The world does not possess much more untilled land of this sort, though there may be considerable potentialities in Siberia, Rhodesia, and some few other areas. More intensive cultivation and better methods have, of course, been responsible for much of the increased food production. But we must also note that food production per acre has not risen the world over as fast as the total production. In many parts of the Far East and Europe, as a matter of fact, the production per acre long ago reached fairly high levels for many crops.

Perhaps the major problem confronting food production is the task of improving the diet of people beyond preventing simple starvation. There have been many claims, mostly based on entirely inadequate evidence and the unjustifiable use of arbitrary "standards" of food intake, that a large proportion of the population of the United States is badly fed. The correction of these nutritional "defects" in the United States would, if some enthusiasts are followed, demand large increases in the production of milk, eggs, fresh vegetables, fruits, and probably meats. Even though this might be attended by a corresponding reduction (equivalent in calories) in the production of other items like grains, potatoes, and sugar, the net effect would be a very considerable increase in the demanded agricultural production. It takes much more acreage and agricultural manpower to produce the "protective" foods than the cheap staples.

If it is true that such a change in agriculture in the United States is desirable, what may be the case elsewhere? Only a small fraction of the world's population is as well fed as the people of the United States. At least two thirds of the population of the world subsists at a far lower level and to bring these people to a dietary standard comparable to that now existing in the United States would demand almost fantastic increases in agricultural productivity.

Bennet (1941) listed countries outside Europe where the diets are made up of 80 to 100 per cent cereals and potatoes; these include India, the U.S.S.R., China, Manchuria, Ceylon, Indo-China, the Philippines, Madagascar, and Nigeria — in which live something like 60 per cent of the people in the world. Lindberg (1945) estimated that in these countries the ratio of calories from animal sources to those from vegetable sources is about 5:95; that is, about 95 per cent of all calories consumed by the people are strictly vegetable. In contrast, this ratio in countries that have "adequate national diets" he estimated to be approximately 40:60. Tolley's (1945) estimates are similar. To bring the low ratio of the "potato-cereal" countries to something like the ratio in the United States, or even in Western Europe, would require a great increase in primary calorie production. For example, Lindberg (1945) calculated that, at a constant average human consumption of 3000 Cal. per day, an animal-vegetable ratio of 10:90 requires 4800 Cal. of agricultural production, and that at a ratio of 50:50 the primary production would have to be about 12,000 Cal. per person daily. As a rough average 1 Cal. from animal sources requires perhaps 7 Cal. of primary food production.

It is possible to make conservative estimates as to the total primary caloric production needed to make a substantial but by no means luxurious improvement in the world dietary. If we take, as a world average, 3000 Cal. daily as the desired human intake, we may estimate that this level is reached at present by half the population and that the other half actually gets perhaps 2000 Cal. There is needed, then, about 20 per cent more total calories to bring the average world consumption level to 3000 Cal. But if also the ratio of animal to vegetable calories is to be improved, for half the population of the world, from a value of only 5:95 to one of 30:70, the primary caloric production for these people will necessarily be increased from an average of 2600 Cal. (for a 2000 Cal. intake and 5:95 ratio) to one of 8400 Cal. (for a 3000 Cal. intake and 30:70 ratio). Primary food productivity for this half of the world would have to be increased more than 300 per cent, and that for the world as a whole more than 150 per cent. These are more or less minimal estimates to achieve moderately good, but not excellent, universal nutrition.

There is no possible improvement to be achieved by increasing the sheer manpower devoted to agriculture; already some two thirds of the people of the world are engaged in agriculture. Only technological improvements of great magnitude could be efficacious. The Food and Agriculture Organization of the United Nations is beginning to grapple with these problems, but as yet it has only started the machinery from which will come the analysis of the present facts. It is hoped that the effective presentation of this analysis to the member governments will bring real action.

The Recognition of Food Shortages and Famine

In modern industrial states the statistics on agricultural production, imports, and exports are reasonably complete and current. From these it is possible to detect and even to predict important changes in the available food supply, though these data do not provide information on the actual level of individual consumption. In countries where the centralization of distribution is less advanced, knowledge of the available food supply is much less complete and accurate, though some help in evaluating the seriousness of a food shortage can be gained by surveys at the production level. In all countries, however, it should be realized that when real food shortages develop, the associated conditions make the collection of useful data increasingly difficult and the data themselves progressively less reliable.

Various surveys of the population itself can be made to estimate the actual food consumption and the frequency and severity of the signs of undernutrition. Many criteria for the latter might be suggested but few of them are adequately objective, informative, and easy of application in the field. Simple clinical impressions have their use but are obviously difficult to standardize and evaluate. The degree of undernutrition is not recognizable at a glance and may even elude accurate estimation by fairly elaborate inspection methods. Emaciation may be somewhat obscured by edema and the lack of standards for the normal condition of the population may prevent anything but the roughest of approximations.

Body weight measurements have been extensively used in the past year or two in Central Europe and this seems to be a highly useful step. There are difficulties, however. In the first place it may often happen, especially in the more backward countries, that the average normal body weights for the population are unknown. In the second place it is difficult to ensure that the population is properly sampled. Weighing people at random in the streets gives proper values for the street population but not necessarily for the population as a whole. The more undernourished and feeble members of the population are least likely to be walking about in the streets.

The Stigmata of Starvation

A major task of the present book is to describe the morphological and functional changes wrought by starvation and undernutrition. These changes are numerous and, in many instances, of large order, but in general they refer to the changes of the individual as he passes from a well-nourished state through successive stages to extreme starvation. To describe, in general terms, a severely starved population is relatively easy, but to estimate the degree of undernutrition of a population is another matter, particularly when the pre-starvation state of that population has not been adequately characterized. The more important changes are loss of weight, weakness, depression, polyuria, bradycardia, and increased relative hydration of the body. These are stigmata of starvation but they are not readily classifiable into a quantitative measure of the degree of starvation. We have noted the use of measurement of the body weight in samples of the population. This is a most valuable simple measure but it is only a recent innovation and has notable limitations in the recovery phase. Fat, water, and muscle may weigh alike but they have quite different functional significances.

Edema as a Sign of Starvation

Severe underfeeding generally, and probably universally, results in a relative increase in the water content of the body. The actual total water content may not rise but it does not decrease in proportion to the loss of tissue, so there is a relative increase in hydration. At first this is not visible in ordinary clinical examination, though it may be detected by complicated tests, but eventually it is recognizable as a puffiness of the ankles and face and may progress to severe generalized edema. In an underfed population, then, the frequency with which cases of edema are seen may be taken as an indication of the severity of the food deficiency. It is necessary, of course, to exclude renal and cardiac causes which produce edema in well-fed persons.

The edema of starvation, frequently called famine, hunger, or war edema, is discussed in detail in Chapter 43. Here it is enough to note that even in severe famine it is not a universal phenomenon and its appearance is dependent on ample supplies of water and salt as well as inadequate food. In the individual the severity of clinical edema is not by itself a reliable guide to his nutritional state. Where the complications of malaria, dysentery, anemia, and vitamin deficiencies exist, the edema appears more readily and in more extreme form. In the last stages of starvation the edema may disappear and the starved person may die in a dehydrated state. Finally, edema may be produced by protein deficiency alone, even though the caloric intake is reasonably high. But in spite of these limitations, records of edema frequency are valuable in following the course of famine and the progress of relief measures. In an area of food shortage the mere finding of an appreciable number of edema cases is objective evidence that the food deficiency is really serious.

Famine Edema in World War I

By the beginning of 1915 a dropsical condition, which we now recognize as famine edema, was seen among the distressed Polish population of the eastern battle zone (Budzynski and Chelchowski, 1915, 1916). Of the 224 cases reported from one small region, about half were children from 2 to 10 years of age; in the entire group there were 118 males and 106 females. The dietary conditions were deplorable and most of the patients had subsisted for months on little else than meager quantities of potatoes. Edema was soon common in many parts of Poland, Galicia, and Silesia, especially among the captured Russians and displaced Poles (Rumpel, 1915; Strauss, 1915). At the time the full significance of these findings was not appreciated and for several years there was argument about the possible role of infection as well as undernutrition.

The prevalence of famine edema among prisoners of war was so great that several German authors in 1916 discussed the condition under the title of the "edema disease of the prison camps" (e.g. Jürgens, 1916). But by the winter of 1916–17 there were many edema cases in the German and Austrian military hospitals in and near the eastern battle zone. Franke and Gottesmann (1917) observed more than 300 cases in a single military hospital at Lemberg.

Maase and Zondek (1920) saw their first cases among civilians in Berlin in January 1917, and in the succeeding months about a third of all the polyclinic patients were cases of famine edema. In the spring of 1917 cases began to be frequent in Vienna, Schiff's (1917a) first patients there being seen in February 1917. In the period from May 15 to October 15, 1917, there were 824 cases of disability on this account in Vienna, collected by Schiff (1917b). Knack and Neumann (1917) began to see edema cases in Hamburg in the winter of 1916–17; between April and July of 1917 they personally studied almost 50 cases. In the Rhineland, also, famine edema was seen early in 1917 (Döllner, 1917).

By 1918 famine edema was common in all parts of Central Europe, though the conditions varied from region to region depending on the supply of food from local production. In Bohemia, von Jaksch (1918) counted 22,842 cases, of whom 1028 died. Nor was the condition limited to Central and Eastern Europe. Famine edema in Italy was chiefly seen in returned prisoners of war, but there were also cases among older people of the civilian population in the northern part of the country (Pighini, 1918). In the occupied area of eastern France famine edema was common; Prince (1921) reported 45 autopsies in a small town in Alsace. Vandervelde and Cantineau (1919) saw 200 edematous patients in the St. Pierre Hospital in Brussels. In Holland famine edema began to be seen in 1917 but the condition was not common at any time (Beyerman, 1920). Famine edema in Turkey affected both the military and civilian populations (His, 1918; Bigland, 1920; Enright, 1920). Conditions were particularly bad in the besieged garrison of Kut and edema was very common (Hehir, 1922).

It thus appears that from the records of edema alone a useful estimate could be made of the seriousness of food shortages in Europe during World War I. The worst conditions were found in Eastern Europe — Poland, Bohemia, and Silesia. By 1917 conditions became bad everywhere in Germany and Austria. Less severe but still serious underfeeding occurred in the Allied areas close to the battle zones. The absence of edema in England and in most of France, Belgium, and Italy shows that the food shortage in these areas was not extreme.

Starvation in World War II

Starvation conditions appeared in many parts of the world as a direct result of World War II. The Spanish Civil War, which was a prelude to the more general conflict, produced much starvation and a normal nutritional state had not been achieved in Spain in 10 years. The Japanese invasion of China produced immediate food shortages there because of the disruption of transportation, the interference with agriculture, mass migration, and the requisitioning of food by the Japanese Army. Similar effects, but on a smaller and less catastrophic scale, occurred in Southeast Asia when the Japanese campaigns extended into that territory. The same general picture emerged in Russia and Eastern Europe when the German Army turned east and south. We have as yet no detailed information about the extent of starvation in western Russia and Poland, but even the fragmentary data indicate that conditions were deplorable over the whole area and particularly at Leningrad and Stalingrad, where there were prolonged sieges.

More definite information is available for Greece and the Netherlands. In Greece the acute manifestations of famine covered the period from May 1941 to April 1943. Out of a population of less than a million, the total losses from increased mortality and decreased births in Athens and Piraeus were estimated at over 60,000 and the total loss for the country at about 450,000 (Valaoris, 1946); it is of interest that no severe epidemics were involved.

In the Netherlands the food shortage was serious but not at the famine level until late in 1944. Famine conditions prevailed in western Holland for about 6 months; in this period there were some 16,000 excess deaths in 12 of the larger municipalities (Dols and van Arcken, 1946).

In France, Belgium, and Italy there were food shortages but no real mass starvation during or immediately after World War II. Apparently conditions were much worse in Yugoslavia but there are no authentic details. There was seemingly no famine, except possibly in isolated spots, in Rumania or Bulgaria. In England the food supply was most precarious but admirable distribution aided in preventing anything worse than very slight underfeeding at any time; the excellent health record of England during the war is impressive (cf. Magee, 1946). Norway suffered far more serious food shortages but again accurate details are lacking. Denmark, Switzerland, Sweden, and Portugal were the only European countries which had relatively good food supplies both during and immediately after the war.

In Germany, Austria, and Hungary food was strictly rationed during the war but the nationals of those countries had enough to eat at all times, largely owing to the policy of plundering the occupied countries. With the end of the war, however, this advantage disappeared and a period of severe food shortage set in. In Italy also serious food shortages did not appear until the war's end.

In the Far East the Japanese occupation of Burma, Indo-China, and the Netherlands East Indies prevented the export of food from these areas of high productivity. The effects were rapidly seen in India, where many areas of food shortage developed and where famine conditions arose in Travancore, Cochin, and the Calcutta area.

World War II was remarkable for the enormous numbers of people who were kept in prisons or concentration camps or were placed under conditions of slavery by the Germans and Japanese. In general, all these people — and their total must be counted in millions — were ill-fed at best and deliberately starved to death at worst. In this respect the Germans operated on a larger scale and with more deliberate intent than did the Japanese, in spite of the shocking conditions at Santo Tomas, Cabanatuan, and other camps in the Philippines. The Japanese food provisions to prisoners and internees produced starvation in many instances simply because the Japanese refused to recognize that the amount of food which might suffice for an Oriental laborer could not possibly maintain an Occidental with his larger body size and higher metabolism.

The Germans, of course, used starvation as one method of exterminating Jews, Poles, and politically troublesome persons, although this method apparently did not have the official sanction accorded to the pure murder camps and the gas chambers. Chronic starvation in the concentration camps seems, at least in part, to have been produced by the graft and sadism of the local officials.

In German prisoner-of-war camps the food conditions varied from tolerable to extremely bad, but in general the Russians were given much worse treatment than the British and Americans; the food packets of the International Red Cross were a tremendous help to those who received them but the Russians received no such relief (Leyton, 1946). In the last months of the war the developing chaos in Germany produced starvation among the British and American prisoners who had been recently captured and not yet removed into the interior; this was not the result of deliberate policy but of sheer failure of food and transportation.

CHAPTER 2

The Sources of Evidence and Information

As IN any scientific analysis, the conclusions in the present work are only as good as the available evidence. In succeeding chapters details of the sources of information are presented together with indications of their specific limitations, but it is desirable at the outset to describe the general character and validity of the evidence. The total relevant material is extensive but the degree to which it is applicable to the several problems is variable and, in many cases, somewhat uncertain. The amount of direct quantitative data obtained under reasonably well-controlled conditions is not great.

The center of interest in the present work is the condition of human starvation as it occurs most commonly under natural circumstances. It rarely happens that the food intake is abruptly decreased to zero; almost always there is simply a prolonged period of caloric deficit and the result is chronic undernutrition or semistarvation. Famines usually are measured in months or at most a year or two; the progressive death toll, together with various adaptive changes, reduces the food demand after not many months. Food shortages of only a few weeks, on the other hand, are generally insufficient to exhaust various hoards and stores, including the fat stores in the body itself. Most human beings can tolerate a weight loss of 5 to 10 per cent with relatively little functional disorganization. At the other extreme, save for exceptional individuals, human beings do not survive weight losses greater than 35 to 40 per cent. Severe famines are commonly attended by weight losses of about 15 to 35 per cent, and this degree of undernutrition is of the most obvious interest and importance for this reason, as well as because the functional changes are greatest in this range. Finally, we should note that our present concern is primarily in the quantitative aspects of function, since these have direct relevance to human life and behavior.

The foregoing considerations should make it clear that, for the present analysis, studies on acute total starvation, on brief or slight degrees of undernutrition, and on animals have small utility. Clinical impressions and simple observations on people in famine conditions are useful but only to a limited extent.

Studies on Actual Famine Victims

There are numerous small bits of information in the old literature but the first substantial study of famine victims is that of Porter (1889). Besides some discussion of the general medical and social problems of famine, Porter's main contribution was in providing details on 459 autopsies performed in the relief camps which were attempting to care for many thousands of starving Indians in the

Madras area. To this day this is the most extensive series of post-mortem reports, but Porter collected almost no data on function.

Later studies of famines in India have been devoted almost wholly to questions of economics, relief administration, elementary medical care, and communicable diseases. For our present interests there are almost no data of value from the famines in China and very little from the Russian famines of 1920–21 and 1933.

A large amount of useful data from the field was collected by the Germans during World War I, although this was limited to the more ordinary items of medical examinations. Interest in the "edema disease," which was so common in much of Central Europe, produced many reports. The deficiencies of this material are glaring, however. Many modern methods of examination and measurement were not then available, of course, and the World War I studies were notably deficient on such things as work capacity and psychological features. It is curious that the Germans paid little attention to anthropometric matters or to histological details; it is difficult even to estimate weight losses from the German material. Allowance must be made for the postwar dislocation in Central Europe, but the larger German summaries and syntheses do not add much to the briefer reports published during the actual war period (cf. Schittenhelm and Schlecht, 1919; Maase and Zondek, 1920; Jansen, 1920).

All the reports from World War II are not yet published but a great many have appeared and they are often disappointing. Systematic, quantitative studies are conspicuously rare. There are several valuable reports from the public health standpoint (Dols and van Arcken, 1946; Valaoris, 1946; Bigwood, 1947). An excellent report on data collected as an inmate in a German prisoner-of-war camp was published by Leyton (1946). The decline of German medical science since World War I is indicated by the practical absence of useful observations made by the Germans in World War II. The medical and nutrition officers and teams of the Allied armies collected a variety of spot information as the Americans and British broke into "Festung Europa" in 1944 and 1945, but their responsibilities for hasty appraisal and immediate relief precluded detailed and careful work. In the Far East there were apparently no plans for scientific study of starvation conditions at the time the Japanese capitulated. A number of studies were made on repatriated Allied military personnel (RAMPS) and civilians but these were almost all begun some weeks after rescue when uncontrolled rehabilitation was well started (cf. e.g. Letterman, 1945; Morgan et al., 1946).

In the following pages we have provided notes on some of the major studies and publications relating to undernutrition in World War II. The details of the findings in these works are given in the several chapters where the particular aspects have relevance, but it appeared desirable to give in this chapter a synopsis of the material on which these works are based so as to avoid the necessity of repeatedly recapitulating the background.

Western Europe in World War II

Under the title *Enseigements de la Guerre 1939–1945 dans le Domaine de la Nutrition,* a volume edited by Bigwood (1947) summarizes the material pre-

sented at a symposium held at Liége in October 1946. Although of an international character, the symposium was heavily weighted on Belgium and France,
both by attendance and by the character of the published papers. The published
work contains a full-length presentation of Govaert's work on the role of colloid
osmotic pressure in the pathogenesis of famine edema, statistical studies on food
consumption, growth of children, and mortality in France (by J. Trémolières), a
valuable summary of observations by the Oxford Nutrition Survey in Holland
and Germany (by H. M. Sinclair), a short but useful review of the state of
health in western Holland from May 1941 to August 1945 (by Steijling), 7 contributions on the problem of wartime bread, 4 valuable papers on the nutrition
of domestic animals in wartime, and a large number of shorter papers on famine
edema, on rationing, and on clinical experience. In most cases the individual papers represent only rather extended abstracts of the findings of the authors on
particular aspects of the subjects, but the volume is a mine of facts and opinions
not available elsewhere.

Belgium, 1940–1944

During the German occupation of Belgium, Dr. Lucien Brull and his colleagues carried out systematic studies of the effects of undernutrition on the
population of Liége (Brull et al., 1945). For almost 4 years there were constant
moderate food shortages but the worst period was the year 1941 and early 1942.
Famine conditions affected isolated individuals and the inmates of some institutions (see Simonart, 1948), but the general population carried on with a moderate loss of weight, constant hunger, and discomfort. The work by the Liége
group covered characteristics exhibited by the most severely undernourished
persons who entered the hospital or appeared at the polyclinic, statistical studies
on body weight and on the medical findings in children, the nutritional properties of the wartime bread, the question of vitamin deficiencies, the incidence
and characteristics of common diseases, and some special experiments designed
to throw light on the cause of famine edema.

The Belgian workers had the advantages of fairly good technical facilities and
libraries and of no breakdown in sanitation and civil order. For contributing to
the knowledge on semi-starvation, however, they suffered from the constant surveillance and interference by the Germans and the limited extent of the malnutrition they observed. They made no general nutrition surveys. They quickly discovered that the shift in the character of the Belgian diet automatically
averted the major vitamin deficiencies they had expected might appear. But the
total caloric supply of the country was substantially reduced, and proportionally
there was perhaps a more important reduction in proteins. Rationing, as elsewhere, averted wholesale inequalities and especially protected the children. The
more resourceful elements of the population, by personal effort or by greater
wealth, subsisted with no more than discomfort, but the poor, the aged, and the
infirm suffered severe privation, much as in the Netherlands and France at the
same time. The 16 separate papers assembled by Brull et al. (1945) are not a
fully integrated whole, but they, together with the other papers from Belgium
in Bigwood (1947), provide a valuable picture of nutrition in Belgium in World
War II.

Central Civilian Prison at Louvain, Belgium

Simonart (1948) in his capacity as a prison physician made an extensive series of observations on the effects of semi-starvation. The study is of interest because it was carried out under conditions almost approaching an experimentally controlled situation so far as food intake is concerned. In the summer of 1940 the calculated daily food intake of the prisoners was suddenly reduced to 1700–1800 Cal., and this level was maintained for about 18 months before it was somewhat increased (see Appendix Tables). This is to be compared with the prewar level of about 3400 Cal. These figures seem to be based on the gross value of food supplies as issued to the kitchen. The actual intake during the period of semi-starvation was estimated as 1400–1500 Cal.

The prison population numbered about 400. No information as to the rate of turnover is given but there is no doubt that a considerable portion stayed long enough to develop marked symptoms of semi-starvation; in June 1941 clinical edema was present in about 40 per cent of the inmates. There are no data on the over-all changes in body weight; it seems that the author missed a unique opportunity to get information on the rate of decrement of body weight as related to the duration and extent of semi-starvation on a constant and known diet. In a group of 5 patients the average weight decrement from the pre-starvation weight, after the elimination of clinical edema, was 24.9 per cent with a range of 10.8 to 36.1 per cent (see Table 35). But this is not necessarily a proper description of Simonart's material and is probably a considerable overestimate of the average weight loss. In the largest series of men for whom the requisite data are given (Simonart, 1948, p. 221), we have calculated the deviation from normal standards for the 21 men concerned. This proves to indicate a mean of 13.2 ($SD=\pm10.8$) per cent under normal weight with an extreme of -34.9 per cent; 6 of the 21 men were 20 per cent or more underweight. For this calculation we have used the standards of the Medico-Actuarial Investigations of 1912. Application of these same standards to the data for 11 men called normal by Simonart indicates that these averaged 100.1 ($SD=\pm4.4$) per cent of the standard.

Edema, its genesis and treatment, was the focus of the author's interest, but he considered also the effects of semi-starvation on the major physiological systems and functions (circulation, digestion, excretion, and basal metabolism). An attempt was made to measure endurance in muscular work but the technique fails to satisfy elementary requirements of standardization. From reported observations made on one patient, Simonart concluded that the lack of strength of semi-starved patients is relieved more effectively by thiamine therapy than by the caloric improvement of the dietary regimen. This is contradicted by the experience of all other investigators who have been active in this field.

The biochemical data given by Simonart are of dubious value. A good deal of stress was placed on pyruvic acid but the data cited merely prove that the method was totally unreliable, both for famine patients and for normally nourished people. Similar strictures must be made against the data on respiratory metabolism. The preoccupation of the author with his theory of thiamine deficiency colors the entire work, and supporting citations are often misread or so selected as to be seriously misleading. In spite of these and other glaring faults, some of the findings are of much interest, notably on the magnitude of sudden weight variations reflecting the state of edema.

The Warsaw Ghetto, 1942

The majority of detailed scientific reports on the effects of famine on man have been based on observations made during the phase of refeeding or attempted rehabilitation. Though a few days or weeks of medical care and extra alimentation may not greatly alter the picture, the situation in the active phase of starvation is of peculiar interest. For this reason, among others, the studies made in the Warsaw Ghetto in 1941 and 1942 are of unusual value. The volume *Maladie de Famine*, edited by Apfelbaum (1946), contains a remarkable amount of objective clinical, anatomical, and physiological data on the Jews whose protracted severe undernutrition ended, in general, with death either from starvation or in the extermination camps. The technical superiority of the latter as a means of mass murder is attested by the fact that, whereas starvation directly or indirectly took 43,000 lives in a year in the Warsaw Ghetto, the gas chambers accounted for 250,000 in 2 months. The appalling circumstances under which the authors of *Maladie de Famine* worked are difficult to describe. As one of them wrote, "Trop pâle et trop pauvre est notre langue pour pouvoir exprimer l'infini de notre malheur" (Milejowski, 1942).

In spite of these conditions, which were eventually fatal to most of the investigators as well as to the patients, the technical work at Warsaw was surprisingly sound and has the great advantage that it concerns a population group of all ages and both sexes as seen in the two hospitals established within the isolated ghetto. For the first year of famine there were 768 autopsies; the next year, 1941, provided 1934 autopsies; in the final period of 6 months ending July 22, 1942, there were 956 autopsies. In the total of 3658 autopsies there were 251 males and 241 females representing "pure," uncomplicated starvation. The more important part of the Warsaw studies, however, were the tests and measurements made on the living, who were subsisting on an estimated 600–800 Cal. daily for adults (Apfelbaum-Kowalski *et al.*, 1946). A considerable number of special tests and experiments were made on groups of 10 to 20 or more persons who represented simple, semi-starvation without signs of other disease or disorder. It is regrettable that no psychological or psychiatric data are provided. The dietary details are not available, but the general pattern appears to have been that of the usual north and central European famine — potatoes, turnips, cabbage, whole-grain wheat and rye, with extremely little animal protein or fat.

Siege of Leningrad, 1941–1942

The German blockade of Leningrad was imminent in the late summer of 1941 and became effective in September 1941. Serious food shortages began in November, became severe in December 1941, and continued to April 1942, when the food supplies improved somewhat. The winter was unusually harsh and the lack of fuel caused much hardship. Due to the shortage of fuel the electrical power supply was reduced and streetcar service was discontinued. The water supply and sewage disposal systems were damaged as a result of air raids. Semi-starvation was widespread and reached an alarming degree. Complicating diseases like dysentery, bronchopneumonia, and tuberculosis were rampant.

Nearly all the data on semi-starvation in Leningrad were obtained on patients seen in the clinics and the hospitals. There is little detailed, factual information on the prevalence, degree, and effects of semi-starvation in the population as a whole. We have been unable to obtain even crude estimates of the caloric intake, but patients exhibiting the classical symptoms of semi-starvation began to be admitted to the hospitals by the end of November 1941 after some 4 weeks of food restriction. By December the hospital records indicate deaths due to semi-starvation; frequently there was a coma and sudden exitus. Many patients died during the first 24 hours after being brought to the hospital. No over-all mortality figures are available but it is of interest that the mortality among hospitalized men was highest during December 1941 and January 1942; the peak in the mortality of women occurred later, in March and April. In a sample of 48 cases of relatively uncomplicated semi-starvation examined in the spring of 1942, the most frequent weight loss fell in the range of 20 to 25 per cent of the pre-starvation value, the total range being from 10 to 33 per cent of the pre-starvation weight.

In the first three months of semi-starvation, November 1941 to January 1942, the dry, cachectic form of the "famine disease" was dominant; edema was seen rarely. Edema and ascites began to appear in February 1942 and continued to increase throughout the spring of 1942. For men the peak in the incidence of severe edema occurred during April, for women during May, of 1942. It appears that the development of ascites required some months of chronic undernutrition upon which further nutritional insult, most frequently in the form of diarrhea, had been superimposed.

By the spring of 1942 scurvy was seen in an increasing number of patients, with the peak during April and May. This condition was not closely correlated with the degree of semi-starvation. The first cases of pellagra were reported in February 1942 and the peak was reached during the early part of the summer; the symptoms of dermatitis, glossitis, and diarrhea were present, but mental changes, characteristic of endemic pellagra, were largely absent. Night-blindness occurred infrequently during March and April 1942.

Numerous biochemical and metabolic investigations were made, mostly during the spring and summer of 1942, the early period of recovery. Because of fuel and power failure and consequent disruption, few such data were recorded during the most severe phase of semi-starvation.

Different medical aspects of the Leningrad semi-starvation episode were presented in a series of independent, frequently somewhat repetitive articles published in Russian in Vol. 3 (1943) and Vol. 5 (1944) of the periodical *Studies of Leningrad Physicians during the Years of Patriotic War*, issued by the Leningrad branch of the State Medical Publishing House (Medgiz), and in a separate publication, *Alimentary Distrophy and Avitaminoses* (Leningrad, Medgiz, 1944). Some of this material was reviewed by Brožek, Wells, and Keys (1946). The substantial improvement of the diet in 1943 was paralleled by an epidemic incidence of hypertension. This phase of developments was described in Vol. 7 (1945) and Vol. 8 (1946) of the *Studies of Leningrad Physicians* and in Vol. 3

of the series *Problems of Pathology of the Blood and Blood Circulation* (Leningrad, Medgiz, 1946). The post-starvation hypertension in Leningrad was discussed, along with data from other sources, by Brožek, Chapman, and Keys (1948).

A French Mental Hospital, 1941–1942

Beginning in 1941, a valuable series of studies were made on the patients in a mental hospital of the Department of the Seine. H. Gounelle had general direction of the work, which was reported in a series of articles by him and his colleagues (Gounelle *et al.*, 1941; Gounelle, Marche, and Bachet, 1942; Gounelle and Marche, 1946; Bachet, 1943). A large part of the findings, together with many details not available elsewhere, is given in a thesis by Bachet (1943).

The conditions, desperate as they were, seem to have been almost ideal for scientific study. The patients were isolated from the rest of the world; they had a good diet until July 1940, at which time it was sharply reduced to reach a level of around 1500–1800 Cal. a day in August, and this general level was maintained for about 20 months. Prior to food restriction the hospital had had about 1 death a month; on the semi-starvation regimen the death rate rose steadily, reaching a peak of 11 deaths a month from October 1941 to March 1942, inclusive. Dietary improvement began at the start of 1942 and there was an average of 3 deaths a month from June to October of that year.

There were several epidemics of diarrhea, and tuberculosis made a steady increase in the hospital, but otherwise the phenomena observed in these men were directly attributable to undernutrition with a minimum of complications. The daily routine of their lives was little changed. In the first months of dietary restriction those who died seemed to succumb to ordinary disease but at an accelerated rate; later there were many cases in which death in hypoglycemic coma would appear to have been the direct consequence of starvation. The incidence and severity of edema steadily increased but the individual patients often had remissions, sometimes associated with severe diarrhea.

Among 120 patients who never received any supplementary food there were 47 who had very severe edema, 36 who had moderate to slight edema, and 37 who were edema-free. The group of 47 men who showed the severest edema included 5 patients with general and extreme paralysis, 25 deteriorated cases of schizophrenia (hebephrenia) who were immobile and completely withdrawn, and 2 patients with senile dementia; 32 of these men died. In contrast, the men who never had edema included 2 cases of extreme paralysis and 3 deteriorated schizophrenics, and the remainder was made up of earlier and less advanced cases of psychosis. In general, there was a relation between the severity and duration of the psychosis and the gravity of the response to semi-starvation.

An important feature of the studies at this mental hospital was the systematic trial of the effect on edema of drugs, endocrine preparations, vitamins, fats, proteins, and carbohydrates. In general, the edema responded to foods such as butter, milk, soybeans, and casein, but not to plain sugar or any of the vitamins, drugs, or endocrine preparations. In many cases the edema could be reduced or even caused to disappear with a chloride-free diet but it reappeared promptly when salt was allowed unless a substantial dietary improvement had also been made.

Western Holland, September 1944–May 1945

The story of events leading to the Dutch famine of the winter 1944–45 is clear (cf. Dols and van Arcken, 1946). With the outbreak of the war in 1939, the import-export movement of foodstuffs and the import of feed for livestock was reduced. From May 15, 1940, the date of the occupation of the country by the Germans, no supplies of fats, grains, or other raw materials could be obtained from abroad. This led to a reduction in the number of hogs and poultry that could be raised, and the production of meat, eggs, and milk was also reduced. The diet became richer in carbohydrates but poorer in proteins and fats. Up to 1943 the official basic ration for adults supplied about 1800 Cal. per day. The workers received about 2100 Cal., while those engaged in heavy and very heavy work received about 2400 and 3300 Cal. in their official rations. In 1944 the official caloric level for adults on general rations was reduced to about 1600 Cal. This was due to a number of factors such as lack of fertilizers and shortage of agricultural machinery and labor, as well as to the increasing black market. It is essential to realize that the official rations in the Netherlands, as in many other areas during World War II, indicate only grossly the actual food intake.

Conditions became radically worse in the fall of 1944. The Allied offensive near Arnheim and Nijmegen, the general railroad strike declared on September 17, 1944, a German embargo on the shipment of food from northern and eastern Holland to the densely populated west, all resulted in food shortages and a severe disruption of food distribution. The embargo was lifted on November 8, 1944, but the frost set in before adequate food stocks could be amassed in the western provinces of Noord-Holland, Zuid-Holland, and Utrecht. Famine was an inevitable result. The official adult rations for the last quarter of 1944 declined to 1035 Cal. per day. The values for the first and second quarters of 1945 were 619 and 1376 Cal. The desperate situation was relieved somewhat by the food provided by the Swedish Red Cross in January, and later by the International and Swiss Red Cross. Late in April 1945 the Allies started to drop large amounts of food from the air and additional relief material was provided by land and sea. After the German capitulation in the early part of May, the daily ration was raised to 2400 Cal.

The severity of famine is indicated by the rise in mortality. For 12 cities in western Holland, including Amsterdam, Rotterdam, The Hague, Delft, and Leiden, the number of deaths during the first half of the year was 13,155 in 1944 and 29,122 in 1945; the latter figure represents 221 per cent of the 1944 value. Dols and van Arcken (1946, p. 353) noted a marked sex difference in the mortality increase, with the percentage values of 269 for men and 179 for women.

The effects on health of wartime nutrition in Holland, including the 1944–45 famine, were discussed from the point of view of medical specialties in a monograph edited by Boerema (1947). It was unfortunate that frequently the authors covered the material only up to 1943 or the first half of 1944. This is true, for example, of the sections on neurology and psychiatry, surgery, pediatrics, and tuberculosis and for much of the statistical chapter on mortality and morbidity.

In other words, the majority of the discussions refer only to consequences of the mild undernutrition characteristic of the period preceding the famine of 1944–45. The effects of real semi-starvation on the human organism, as seen in the famine period, are treated systematically in a short chapter by de Jongh, who provided a synopsis of the general clinical picture (Boerema, 1947, pp. 233–43).

Patients from German Concentration Camps — Switzerland, 1945

In the latter part of May 1945 arrangements were completed for the transfer to the military hospital at Herisau, Switzerland, of a group of about 300 Allied deportees who had been held in Germany. The idea was to rehabilitate them sufficiently that they could be transported to their respective homelands. The project "Hospitalization Allies" was completed by August 1945, at which time the patients who needed further treatment were transferred to military Alpine stations in Davos, Arosa, and Leysin. Out of the total number of 296 patients at Herisau, 10 died soon after arrival, lung tuberculosis being the most frequent cause of death. In the course of the year 1945–46, tuberculosis claimed 11 other victims. Except for the presence of pyodermia in 269 patients out of the 296, tuberculosis of the lungs was the most frequent disease observed in these semi-starved people; the diagnosis of active tuberculosis was made in 124 cases, of inactive tuberculosis in 27. In the total group there were 271 men and 25 women. In terms of nationality, 116 were French, 50 Dutch, 37 Belgians, and the rest from other European nations.

By the end of 1944 a few former inmates of German concentration camps had managed to get across the Swiss frontier, and they crossed in greater numbers in the spring of 1945. An assembly camp was set up at St. Gallen; 57 individuals who were seriously ill were hospitalized in the medical clinic of the Cantonal Hospital, also at St. Gallen. In spite of all the care, 22 of them died.

The clinical, pathophysiological, and pathological anatomical studies and observations made on these patients were assembled in a monograph that is an important link in the chain of recent publications on semi-starvation (Hottinger, Gsell, Uehlinger, Salzman, and Labhart, 1948). In our text, references are generally made to the specific chapters, which are largely independent of one another. The book contains excellent pictorial material which provides a rich documentation of many aspects of semi-starvation (general appearance, skin, histological sections, and X-ray pictures of the chest). There is a short historical chapter on "hunger disease"; curiously enough, both the experimental work of Benedict et al. (1919) and Jackson's (1925) exhaustive summary of the whole field escaped the authors' attention.

Victims from German Concentration Camps — French Experience

When the German concentration camps were opened by the Allies in 1945, the French necessarily assumed great and difficult responsibilities for the rescue of the human wreckage that emerged. Among other steps immediately taken was the establishment, on the initiative of General Melnotte of the First French Army, of a research and treatment center at the Hospital of Mainau in Germany. Lamy, Lamotte, and Lamotte-Barrillon (1948) have summarized the research at that center in an extensive treatise in which their own findings were placed against the background of the general literature and the experience of the authors during 5 years of observation of variable degrees of undernutrition in France. Special attention was given to 40 cases of the most severe undernutrition in which no concomitant signs of infection or other disease were recognized. In several cases, however, tuberculosis was diagnosed not long after the patients were accepted into the special study group.

The studies at Mainau covered the period from the first week in June 1945 until death or discharge from the hospital — a period of from a few days to a maximum of about 12 weeks. In general, then, these studies refer to the early period of rehabilitation following the most extreme undernutrition. At Mainau 13 autopsies were made immediately after death and some 300 tissue blocks were prepared for microscopic study.

A special virtue of the studies of Lamy, Lamotte, and Lamotte-Barrillon was the effort to carry out special tests and measurements, notably on renal function and on the physical chemistry of the blood serum. A serious limitation, however, was the fact that their "uncomplicated" cases turned out to be quite unrepresentative of simple chronic starvation. Instead of the usual bradycardia and hypothermia, their men had marked tachycardia and most of them were febrile. The evidence for renal damage also puts these men in a special class.

A much less ambitious but valuable general picture of concentration camp victims as seen by French physicians was provided by Debray et al. (1946), who examined 771 patients.

Netherlands East Indies

As soon as the Netherlands were liberated, preparations were made to provide help to the people of the Netherlands East Indies, and the first members of a Netherlands Red Cross Nutrition Research Team arrived at Batavia early in October 1945. Studies began on both natives and Dutch from the Japanese internment camps before very extensive and protracted refeeding had occurred. However, the data definitely pertain to the period when, in most cases, some substantial rehabilitation was under way. The results of some 8 months' work by this team in the field comprise valuable data on the aftereffects of both general and specific dietary inadequacies in the Netherlands East Indies (Netherlands Red Cross Feeding Team, 1948). The general situation in the Japanese camps in the Netherlands East Indies was a chronic but not extreme shortage of calories and protein, much more frequent deficiency of the B vitamins than in Europe, and all the complications of very bad sanitation in a tropical climate. The team made extensive studies of vitamins in the blood and urine and carried on a number of feeding experiments. As was found elsewhere in World War II, it was far easier to attribute to vitamin deficiencies the manifold abnormalities found than to prove the relationship or to cure the conditions with selected vitamin therapies.

The Studies of Beattie, Herbert, and Bell

Beattie, Herbert, and Bell (1948) made studies on severely undernourished persons in the Netherlands and in Germany. In the Netherlands the main work

was carried out in May and June of 1945 — that is, immediately after the relief
of the famine in western Holland — on 17 men and women who had sustained
weight losses of 12 to 31.5 kg. (20.5 to 69.5 lbs.). In Germany the subjects were
male civilians who had been in prison for at least 12 months with an average intake of about 1600 Cal. daily. With these men it was possible to continue the
semi-starvation regimen for some weeks and then to increase the diet by controlled steps for about 2 months of further study.

These investigations are notable for the choice of subjects who were really
free from complicating diseases, the maintenance of controlled conditions, and
the care used in the measurements of metabolic rate, nitrogen balance, and the
fluid compartments of the body.

The Nutritional Situation in Japan

In modern times Japan has been increasingly dependent upon large imports of
food to maintain a semblance of nutritional adequacy; these were maintained
with little change through 1942, but in 1943 and 1944 there was a sharp decline
and in 1945 the imports fell precipitously. Rationing began in April 1941. Obviously the nutritional results should be of much interest. Practically the only recent information available about both food and nutrition in Japan is contained
in the U.S. Strategic Bombing Survey (1947) which devoted 62 large pages to
the subject. Unfortunately, though the sections concerned with food and fertilizers and with vital statistics are apparently fairly reliable, the nutritional discussion is replete with conclusions that are obviously at variance with the "facts"
cited in their support; and the "facts" themselves are often enough in conflict.

In 1942, and for some years previous, imported foods supplied about 20 per
cent of the total calories and 15 per cent of the protein consumed in the main
islands. By 1945 imports of calories had fallen to a fourth and of protein to a
half of the 1941 figure. With no change in domestic production, this would mean
a reduction of the total supply for domestic consumption by some 15 per cent in
calories and about 8 per cent in protein. The domestic food production was
fairly well maintained in the all-important cereals — rice, wheat, and barley. For
the 2 years 1940–41 (1941 was a bad rice year) the yearly average domestic
production for these grains was 12,906,000 metric tons; for 1944, the last year of
full record, the total was 12,856,000. The fish catch decreased sharply. On the
other hand, it is probable that vegetables, including potatoes, which were grown
everywhere, were substantially increased. The U.S. Strategic Bombing Survey
estimated the total calories available for civilian food consumption as 2007, 2041,
and 2013 in 1941, 1942, and 1943, respectively, with a fall to 1895 Cal. in 1944
and finally to 1680 in 1945.

The estimated actual caloric intakes by factory workers indicated declines of
600–1000 Cal. daily from 1938 to 1943. However, body weight changes over this
period were of the order of 0.5 kg. or less in spite of the fact that the workers
were maintaining an 11-hour day 13 days out of every 14. Low-salaried government employees were stated to have fared badly but even they lost averages of
only 1.5 to 3.9 kg. in the various groups weighed. Clearly, the calorie estimates
are unreliable.

The true situation in Japan cannot be evaluated by equating absenteeism in the bombed factories of Tokyo with malnutrition; this is reminiscent of similar excesses in absenteeism familiar in the United States during the war. Actually, the best appraisal of the condition in Japan would be that the war provided some exacerbation of a rather unsatisfactory chronic situation with regard to nutrition. Serious malnutrition was rare from all evidence. In the worst year in the worst district — that is, in Tokyo from January to September 1945 — the Keio University Clinic counted 54 cases of malnutrition with edema out of 6630 ambulatory outpatients. The survey experts' anxiety to demonstrate severe undernutrition may be judged from their conclusion that the failure of half the district physicians' associations to report increased noninfectious diarrhea "probably should be regarded as an oversight on their part" (U.S. Strategic Bombing Survey, 1947, p. 87). Similarly it was explained that malnutrition increased the susceptibility to infection, but the failure of this to be reflected in the statistics was in turn explained as the result of the destruction of the pathogens and their vectors by fire bombs.

Experiments on Acute Starvation in Man

In textbooks of physiology, biochemistry, and medicine the discussions of starvation are almost exclusively based on experiments with a few individuals in total fasting, and such studies, incidentally, are largely devoted to certain aspects of metabolism. The limited objectives of these experiments, combined with the question of whether the rather peculiar subjects involved can be accepted as valid "samples" of human populations, considerably restrict their value.

In any case there are very important differences between the results of total abstinence from food and those from a prolonged period of caloric deficiency. Here it may be enough to point out three: (1) in total fasting the hunger sensation almost disappears after a few days, but it is progressively accentuated in prolonged undernutrition; (2) ketosis is a typical result of fasting but it does not develop in semi-starvation; (3) famine edema has never been reported in total starvation. There are important similarities — bradycardia and lowered metabolism are two of these — but it is obvious that the two situations are neither quantitatively nor qualitatively identical.

Animal Experiments

The majority of animal experiments on starvation have been devoted to total fasting and therefore suffer the limitations to be expected. But the most serious objection to placing much reliance on animal experiments in the present discussion is simply the fact that they cannot possibly provide quantitative information which has direct application to man. How much underfeeding in a dog corresponds to the situation in which a man loses 20 per cent of his weight in a year? What quantitative changes of strength and endurance in a semi-starved man can be predicted from observations on an underfed animal? Can any useful generalizations be drawn regarding changes in the skin, hair, and blood morphology? Obviously, behavioral and emotional aspects of the problem escape analysis in the animal, or at least have little in common with the problems in man.

Some important generalizations result from a comparison of the effects of starvation in men and animals but these achieve validity only when they are proved to apply to man. Such is the case, for example, with the general tendency to an increase in the relative hydration of the body. We shall have occasion many times to make reference to findings on animals, but in general these are confined to the purpose of establishing the universality of particular phenomena associated with starvation.

Clinical Cachexias

A great deal of information about starvation and undernutrition could be obtained from clinical cachexias of various origins, and we have attempted to use what there is available from this source. Anorexia nervosa (Chapter 44) has had some attention and has many similarities to the conditions of famine, though the emotional and psychological aspects are not comparable. Occasional cases of stricture of the esophagus have also been studied to much advantage, but by and large the potentialities of readily available clinical material in the hospitals have not been realized. A large percentage of cases of cancer of the gastrointestinal tract are also cases of semi-starvation but they have had small attention in this respect. Every insane asylum has its quota of emaciated patients whose psychiatric status is usually the only item to be examined by the medical and scientific staff. Tuberculous cachexia is no longer so commonly seen as formerly and the modifying effects of the infection must be considered, but such patients could obviously be used in the study of starvation effects as well as in studying some of the problems of rehabilitation.

Human Experiments on Semi-Starvation

Experimental studies on controlled semi-starvation in man are very few indeed; practically speaking, there are only two real experiments to be considered. At the end of World War I F. G. Benedict and his colleagues at the Carnegie Nutrition Laboratory carried out an ambitious and valuable study on young men volunteers. The most serious limitation in this work was the small degree of undernutrition that was involved. The weight losses were only some 10 per cent, that is, less than to the point where many important changes become clear. This "Carnegie Experiment" is well known in the sense that everyone knows it took place, but the method of presentation was such that very few persons have actually read the report. We have summarized the Carnegie Experiment in some detail in a separate chapter here (Chapter 3).

A major purpose of the present work is to present the results of the Minnesota Experiment, which attempted to simulate the nutritional picture of severe famine. The organization and general arrangement of the Minnesota Experiment is presented in Chapter 4.

Both the Carnegie and Minnesota experiments involved only normal, healthy young men, and the results can be applied to heterogeneous populations only with reservations. We are in possession of no information from which we can estimate accurately the probable or possible differences which would obtain with children, women, older people, and persons with various abnormalities.

Rehabilitation Following Semi-Starvation

Information regarding rehabilitation of man after semi-starvation and prolonged undernutrition is very scanty. At the end of World War I there were some studies on metabolism in the early stages of refeeding (e.g. von Hoesslin, 1919) but practically nothing on the recovery of normality in terms of function. The Carnegie Experiment did not touch these questions. By default, then, it is necessary to rely on the Minnesota Experiment for almost all quantitative information on rehabilitation processes and for the majority of conclusions about the rate and character of recovery. Differences between persons of different ages and between men and women may be large in regard to the recovery processes, so only very tentative general conclusions can be drawn from the evidence now at hand.

Caloric Intake and the Level of Undernutrition

With some notable exceptions, there is surprisingly little evidence as to the levels of caloric intake that have produced given degrees of undernutrition — undernutrition being identified by weight loss, the incidence of famine edema, and other objective measures. In the Appendix to this work there are summaries of some of the more significant dietary data from World War II but even the best of these are open to some question as to actual calories consumed. Moreover, the effect of a given reduction in the diet is largely dependent upon the energy demands on the individuals, and these demands are rarely specified with any degree of precision. Where personal volition controls the situation, a spontaneous reduction in physical activity tends to offset mild to moderate dietary limitations. But too often the very conditions that bring about a food shortage also demand a high rate of physical work. In many of the German and Japanese prison camps the caloric supply would not have been so deficient if the prisoners had not been kept at forced labor.

During the war, and for some time afterward, too much credence was given to official ration lists and to the accounts of nonscientific reporters, with the result that many of the contemporary analyses were grossly erroneous. In a few cases the actual food intakes were overestimated because no allowance was made for failure to honor rations and for spoilage and unavoidable food wastage, but in most cases analyses based on the official ration lists greatly underestimated the caloric consumption. The black market was very large in some areas, notably France, Italy, and Greece. In Italy it was estimated that only 5 per cent of the expenditure for food was on rationed items (Metcoff and McQueeney, 1946). In Athens in 1942–43 the rations provided some 700–1000 Cal. per day and about 1000 Cal. more came from the black market (Logaras, 1946). In the regions occupied by the Germans self-protection demanded a maximum of concealment of stores and current production of food; the Dutch were very successful at this. In all parts of Europe the local production of vegetables was an important source of non-rationed food.

In the German and Japanese prison camps food intakes were generally much less than indicated by the official rations. Wholesale graft by the local camp officials was so prevalent that the official records are for the most part worthless.

Not only were the allowances unfulfilled; much of the food actually supplied
was so poor in quality that it could not be eaten even by starving persons. On
the other hand, the recollections and crude computations of the prisoners themselves naturally tended to minimize the amounts of food they did eat.

For some areas where the food shortages were not too severe and where there
was little serious undernutrition in World War II, the average actual intakes for
various towns and districts have been estimated with considerable accuracy. The
analysis for Switzerland is particularly thorough (Fleisch, 1946, 1947). In Switzerland, as in England, an average reduction of caloric intake of the order of 10
per cent was associated with a small but nonprogressive loss of weight, particularly in older adults, and no evidence for real health impairment. In England,
Switzerland, and the Netherlands the reduction in total food supply was accompanied by some improvement in quality and the elimination of many of the prewar inequalities of distribution (Marrack, 1947a; Fleisch, 1946, 1947; Boerema
et al., 1947). The situation in Oslo, Norway, seems to have been similar (Strøm,
1948).

The food distribution programs in practically all parts of Europe in World
War II provided special care for what were considered the more vulnerable
groups of the population, and country-wide averages for food intake must be interpreted with this in mind. Moreover, it was universally true that the more
energetic and ingenious persons and those with convertible wealth managed to
supplement their food supplies far more effectively than the rest of the population. Finally, within the family the division of the combined food resources
probably rarely corresponded to a simple per capita equality. For all these reasons, then, the best average figures for population groups are not apt to allow
any precise calculation of the relation between caloric intake and the progression
of undernutrition.

After careful consideration of all these factors, the authors have decided to
place little emphasis on reported or estimated food intakes. The best definition
of food deficiency is to be found in the consequences of it. However, there are
some instances where the food intake was really known, both before and during
the period of restriction, and where the general activity level was not greatly
changed. Besides the Minnesota Experiment, we may cite the Carnegie Experiment (Benedict et al., 1919) and studies on a few isolated groups, including
patients in a mental hospital in the Department of the Seine (Bachet, 1943; Gounelle, 1947) and civilian prisoners in Louvain, Belgium (Simonart, 1948).

In many places in the present work we have given references to estimates and
summaries of dietary intakes in various regions of the world during and shortly
after World War II. In addition there are a great many other dietary reports, of
varying degrees of excellence and utility, which have not been discussed elsewhere. A few of these are listed here: Abramson (1947, Sweden); Adcock et al.
(1948, Great Britain); Aykroyd (1948, India, Burma, and Ceylon); Branion
et al. (1947, Canadian Air Force); C.N.R.R.A.-U.N.R.R.A. (1946, China);
Cruickshank and Stewart (1947, maternity in Great Britain); Cuthbertson (1947,
Newfoundland); Drummond (1947, Great Britain); Bransby et al. (1948, Great
Britain); Fehily (1947, Hong Kong); Fortuin (1947, workers in Holland);

Hynes, Ishaq, and Verma (1946, Indian Army); Kark and Doupe (1946, Indians in Burma); Ke (1947, students in China); Ministry of Education (1948, children in Great Britain); Bihar (1947, India); Pett (1945, dietary policy); Peretti (1943, Sardinia); Platt (1946, British West Indies); Platt (1947, British colonies); Pyke *et al.* (1947, old people in London); Reid and Wilson (1947, British in the Orient); Smart *et al.* (1948, workers in Belgium); Stare (1947, ideal intakes); U.S. Strategic Bombing Survey (1947, Japan); van Veen (1946a, Dutch East Indies); Verma, Dilwali, and Thomson (1947, Indian Army); Wan (1947, Chinese Army); Wan and Chen (1946, Chinese medical students); Whitfield (1947, Japanese prison camps); de Wijn (1947, Dutch East Indies); Youmans (1946, general).

CHAPTER 3

The Carnegie Nutrition Laboratory Experiment

In 1919 Benedict, Miles, Roth, and Smith published a 700-page volume, in large octavo, under the title *Human Vitality and Efficiency under Prolonged Restricted Diet*. This was a detailed account of an experiment at the Carnegie Nutrition Laboratory at Boston. The general nature of this study and some of the findings have been widely discussed and quoted, but the experiment has never been summarized nor has it been critically evaluated. The great bulk and prolixity of the original publication have undoubtedly deterred most investigators from a detailed examination of it.

The Carnegie Experiment resembles the Minnesota Experiment (Chapter 4) in the use of young men as volunteer subjects and in the application of multiple methods in the attempt to evaluate a major sample of the whole complex of human function and behavior in prolonged undernutrition. A practical goal was the effort to throw light on problems of famine, relief, and food shortage.

General Approach

In spite of these similarities between the Carnegie and Minnesota experiments there are important differences, not only in technical details but also in general purpose, approach, and organization. Comparison of the two experiments is instructive but the Carnegie results are valuable in themselves and must be examined separately.

In the Carnegie program the goal was tentatively set at a total weight reduction of 10 per cent to be achieved over a period of some months. It was believed that such a loss would be enough to demonstrate the changes in metabolism and various functions which characterize famine and semi-starvation conditions, and that a greater degree of undernutrition would entail undue time, discomfort, and labor as well as introduce the danger of some real disability. Benedict and his colleagues seem to have been so anxious to avoid discomfort and inconvenience that even in its planning the program failed to provide conditions properly comparable to those seen in severe food shortages such as existed in Central Europe in 1916–19. It was repeatedly stressed that no hazard to health was involved and that, for example, the 10 per cent reduction in body weight would not be a reason for refusal of life insurance. In some respects the entire work has the appearance of a defense of the thesis that a considerable degree of undernutrition, or less than ordinary food intake, is compatible with the maintenance of good health and vigor.

Benedict and his colleagues were impressed with A. E. Taylor's report (quoted by Benedict *et al.*, 1919) that the curtailment of the average diet to 1800 Cal. in Germany did not necessarily produce "cataclysmic changes"— and they saw in that a challenge for explanation to the scientific world. The philosophy underlying their approach can be best illustrated by the statement at the end of their book:

"To instill into the world at large a belief that a pronounced lowering of rations is not necessarily accompanied by a complete disintegration of the organism and collapse of mental and physical powers may, after all, be of real service . . . Experimental evidence has accumulated in sufficient amounts to justify a serious consideration of a material reduction in the intake of protein, which is one of the most expensive factors in human food. . . . For all practical purposes it is clear that the so-called low protein diet is perfectly justifiable as a war measure and in all probability is a logical procedure that cannot be accompanied with any untoward effects, even by long-continued practice. . . . We may say, in summarizing, that protein curtailment is an assured and physiologically sound procedure, and a reduction in calories is possible for long periods, but definite and significant disturbances of blood composition, normal sex expression, neuromuscular efficiency, and the appearance of mental and physical unrest are deterrent factors in too sweeping generalizations as to the minimum calories being synonymous with an optimum level."

It can be seen that, as a whole, the Carnegie Experiment was directed toward proving adaptation of the organism to a restricted diet rather than toward documenting quantitatively the disability, distress, or discomfort resulting from undernutrition. We must consider carefully how the Carnegie Experiment's results bear out the far-reaching conclusions.

At the end of and after World War I, more or less at the same time as the appearance of the Carnegie report, a number of papers appeared with clinical observations on malnourished persons in Central Europe. (These papers are reviewed in other chapters of this work.) As a rule, the patients came to observation in a state of malnutrition, so that the main problem was rehabilitation. This is, indeed, the problem of malnutrition as it would appear to the average physician or to relief organizations. In contrast, Benedict and his colleagues were interested only in the development of undernutrition, and rehabilitation was neither controlled nor systematically observed. In the Minnesota Experiment roughly equal importance was given to events in rehabilitation and in semi-starvation.

Although a variety of functions were studied by Benedict and his colleagues, their main interest was centered on the metabolic rate. When they started their series, they regarded as still unsettled the question of the response of metabolism to a loss of body weight due to undernutrition. Therefore, their experiment was expected to decide the following questions: (1) Is it possible to alter the basal metabolism by a reduced ration? (2) Can the body be held in nitrogen and carbon equilibrium at the lower level? (3) If such a lowering in basal metabolism is obtained by a reduction in diet, will it be in proportion or out of proportion to the loss of body weight? (4) Since the body material lost would presumably be in greater part fat, and thus supposedly inactive in metabolism, will the basal metabolism of the remaining tissues increase with the loss of weight, as might be expected, or will it decrease? (5) Will superimposed muscular work be done at a higher or lower cost of energy when the level of basal metabolism is altered? (6) Will the stimulating effect of foodstuffs, primarily that of protein, be the same with reduced body weight as with the normal body weight?

Since the total weight reduction was arbitrarily set at 10 per cent, the foregoing questions were really limited to what the response of the metabolic rate would be at that level of reduction. The authors thought it entirely feasible (Benedict *et al.*, 1919, p. 40) to answer these questions by producing a body weight loss of 10 per cent by complete fasting for about 14 days, but they agreed it was impractical to ask a group of men to sacrifice their entire time to a test of this kind and to undergo such a rigorous experience. This statement is surprising, since Benedict was, according to his literature review, well aware of the difference between complete starvation and semi-starvation.

Because of this "practical" consideration, it was decided to produce the loss in weight by a diet reduced to between 50 and 70 per cent of the estimated food requirements. After the reduction of 10 per cent in body weight had been reached, the basal ration was then to be supplemented with sufficient food materials to obtain caloric and nitrogen equilibrium at the lower level of body weight. The concentration on this point, and the basic idea involved, is the greatest single difference from the arrangement and procedure in the Minnesota Experiment. Obviously such a supplementation period is neither strict semi-starvation nor rehabilitation, since features of both are incorporated.

Specific Organization, Arrangement, and Procedure

The experiments were carried out on two groups of men, Squad A and Squad B, both consisting initially of 12 subjects. Another 3 subjects served as members of Squad B only a part of the time, and not during the weeks of diet reduction. All subjects were students of the International Young Men's Christian Association College in Springfield, Massachusetts. Their ready cooperation and high moral character, combined with intellectual and physical fitness, appeared to be advantages enough to compensate for the inconvenience of carrying out most of the investigations at Springfield rather than at the laboratory. All of the subjects continued their usual college work during the period of the experiments. Because of the limitations of methods which could be used in Springfield, the subjects were taken to Boston on alternating weekends (the train trip required 2 hours and 15 minutes), mainly for psychological tests, work tests, and metabolic investigations. The authors realized that the success of the experiment depended on the veracity of each subject and stated that there was but one suspected violation early in the series. However, as will be seen, the arrangment was such that only very gross violations could be easily recognized.

The main experiment (120 days on a reduced diet) was carried out on the 12 men in Squad A, who ranged in age from 20 to 44 years. After a brief standardization period from September 22 to October 3, 1917, the diet was reduced, with several interruptions, from October 4, 1917, to February 2, 1918. Although some observations on a few functions were made after this date, the diet was no longer controlled, so that for all practical purposes the experiment ended with the last day of semi-starvation. Squad B, with an age range from 18 to 29, was used from October to January 6 as a normal control group. It was placed on a restricted diet of 1400 Cal. for 20 days (January 7–27, 1918).

The standardization period for Squad A was so short that for several important functions only one control (and trial) value was available. Several tests were introduced only during the period of reduced diet; for these items no control values on the same subjects were available. Therefore, the results were compared not only with those obtained with Squad B but also with those obtained on other small normal groups available in the Carnegie Nutrition Laboratory. For evaluation of the results on Squad B a long standardization period of 3 months was available. Also, the dietary conditions were better controlled during the experiment. Consequently, Squad A, on whom the major experiment was performed, was far less well controlled than Squad B, who submitted to only 20 days of semi-starvation.

In order to enlist the cooperation of the subjects, several concessions were made which were regrettable for the consistency, accuracy, and meaning of the observations. The experiments were interrupted for a Thanksgiving recess of 4 days and a Christmas recess of more than 2 weeks (from December 20, 1917, to January 6, 1918). During those periods the diet was not controlled and the authors had to rely on the subjects' reports as to what they had eaten. The same was true for every second Sunday, which most of the subjects spent away from the college. During these free periods the men were asked to control their protein and fat intake, with the result that there was an excessive consumption of cakes, candies, and pies, especially during the Christmas recess. Under the pressure of relatives and friends at home the Sunday meals were often excessive in both nitrogen and caloric content. During the Christmas vacation the loss of body weight was largely recovered, as will be discussed later. The authors recognized later that the periods of uncontrolled food intake amounted to a serious break in the experiment: "Subsequent inspection of the data returned by these men as the record for the uncontrolled meals on Sundays made us regret extremely that we did not urge more strongly the desirability of complete control throughout every meal, even at the sacrifice of shortening the entire experiment." Squad B, who were on a restricted diet for only 20 days, had no uncontrolled meals.

The men were fully advised that loss of body weight could be achieved by exercise as well as by reduction of diet, but their physical activity was not controlled. However, crude records of daily physical activity were made, based on personal statements and on pedometer readings. Each man was given a pedometer which he carried with him during the day. Since the men were aware of the effect of exercise on body weight, they invariably took considerable exercise on the days following the uncontrolled Sundays and after the Thanksgiving and Christmas recesses, in order to offset the excess intake. The authors regard this compensation of excess food intake by increased physical activity as desirable for the continuity and consistency of the experiment, but objections can be made that physiologically the two procedures are not equivalent. In any case, with the frequent interruptions in the schedule of restricted diet and the considerable variations in physical activity, it appears that the experimental conditions had only a superficial similarity to those prevailing in Central Europe at that time.

Of the original 12 subjects in Squad A, only 9 men went through the entire period without a break. In many of the tests performed, the number of subjects varied from time to time; not every subject was present every time the test was given. Some of the tables look quite fragmentary, so that the process of averaging is doubtful. No standard deviations were calculated for the physiological data although they are given for the psychomotor and sensory tests. The formula used for the standard deviations does not appear adequate today; neither the t-test nor similar statistical tests of significance were available at that time. We have calculated the statistical significance for a few items but this could not be done for all tables.

Diet

Throughout the entire experiment the subjects received the regular college mess hall food, which was essentially the same as that of the other students, but in reduced amount. The only qualitative difference was the substitution of grape or apple jelly for butter during certain periods and the addition of considerable amounts of spinach and other bulky food material. The food was by no means comparable in kind to that ordinarily eaten under famine conditions. It was well prepared and served at a special table, but unfortunately in the general mess hall, so that the subjects could compare their reduced diet with the liberal amounts of food served at the other tables. The subjects complained frequently about this serious psychological handicap but nothing could be done about it.

The caloric intake was curtailed by serving one half to one third of the regular portions. Samples were analyzed in the Carnegie Nutrition Laboratory as to nitrogen and energy content but no attempt was made to secure either high or low protein in the diet, the adjustment being made wholly upon the caloric content. The actual protein intake could be estimated for most days from the nitrogen content in the food. Fats and carbohydrates were not measured but careful inspection indicated that the proportions of fats and carbohydrates were not abnormal. Total fats and carbohydrates were analyzed on 3 days, and the proportion of fat to carbohydrate was found to vary between 1:4 and 1:5.

Because of the procedure of diet reduction, the caloric allotment was not regular from day to day; it was in large part decided by the character of food served in the regular mess hall on that particular day. Wide fluctuations from day to day of the caloric as well as the nitrogen content could not be avoided; in fact, there were probably not two days during the whole period of diet restriction with exactly equal caloric and nitrogen intake. But in spite of the wide fluctuations from day to day, the average for a week or ten days remained fairly uniform. In general, the average caloric intake during the reduction period started with 2200 Cal. from October 4 to 15, was decreased to 1800 Cal. on October 15, and was reduced further to 1600–1700 Cal. from November 1 to Thanksgiving. Since the length of periods for which the average caloric intake is given varies between 3 and 18 days and does not always coincide for the different subjects, no exact averages can be given.

After the Thanksgiving recess the caloric and nitrogen intake was increased in order to obtain body weight and nitrogen equilibrium at the reduced level of body weight. This was done empirically by a gradual increase of the diet during this period until the body weight appeared to become constant. The assistant in charge of the apportionment at no time determined the exact caloric content of the food allotted. In general, the average diet between Thanksgiving and the Christmas recess amounted to about 2200 Cal. On the return of the men to college in January, all subjects received a low diet of approximately 1500 Cal. in order to compensate for the body weight gain during the Christmas vacation; the diet was increased later to approximately 2000 Cal. following the same procedure as that used in December. Thus the level in the maintenance periods, both between Thanksgiving and Christmas and in January, was adjusted without a previous knowledge of the caloric requirement.

Squad B was given a controlled diet stated to provide approximately 1400 Cal. daily for 20 days (Benedict et al., 1919, p. 225). According to Table 38 (ibid., p. 297), however, the average caloric intake from January 15 to January 23 was 1534 Cal. The nitrogen content of the diets will be discussed in a separate section.

Body Weight

The initial, the minimum, and the final weights for Squad A and for Squad B are shown in Table 1. Individual data for Squad A are given in Table 2. The average greatest loss for Squad A was 12.2 per cent of the initial body weight, and the final loss was 10.7 per cent. Squad B lost 6.6 per cent.

Typical curves of body weight loss in Squad A are given in Figures 1 and 2. Both curves show the more or less regular, almost linear, drop during the first weeks of diet reduction, the maintenance of the body weight during the weeks preceding the Christmas recess, the large gain of body weight during the Christmas recess, the renewed drop after January 6, the final level during the last period, and the steep slope of body weight gain during the uncontrolled rehabilitation period. The rather large fluctuations, especially during the maintenance periods, are quite typical and might be due to the irregular food intake on different days or to changing levels of physical activity. Figure 3 shows the drop in body weight of 3 subjects of Squad B.

The body weight of Squad A could be maintained at the lowest weight level (early in December) for approximately 2 to 3 weeks with an average intake of 1967 Cal. (see Table 3); the individual differences, however, are rather large (between 1600 and 2500 Cal.). Since the original energy consumption of Squad A was estimated to be approximately 3800 Cal., it appeared that the body weight could be maintained at a level 12 per cent below the initial weight with approximately half the original caloric intake. The authors were aware that the maintenance periods were extremely short, so that it is doubtful whether actual equilibrium was obtained, particularly since the fluctuations of body weight during the maintenance period were considerable. However, in the second maintenance period, in January, separated from the first by 4 to 6 weeks, it was believed that the newly reduced weight level was maintained with the same caloric

TABLE 1

AVERAGE BODY WEIGHTS, in kg., just before (Control), at the end of semi-starvation (End), and at the point of least weight (Minimum). Compiled from Benedict, Miles, Roth, and Smith (1919).

	Control	Minimum	End	Maximum Loss		Final Loss	
				Kg.	%	Kg.	%
Squad A	67.0	58.8	59.8	8.2	12.2	7.2	10.7
Squad B	67.9	63.4	63.4	4.5	6.6	4.5	6.6

TABLE 2

BODY WEIGHT, in kg., of Squad A before diet restriction (Control), at the lowest point (Minimum), at the end of diet restriction (End), and 6 weeks after the end (Recovery). In the final column are given the differences between the Recovery and the Control.

Subject	Control	Minimum	End	Recovery	Difference
Bro	61.8	54.0	54.4	62.5	0.7
Can	79.8	68.8	69.3	81.0	1.2
Kon	69.0	60.3	61.5		
Gar	71.3	62.3	63.0	72.5	1.2
Gul	66.8	59.0	61.0	69.5	2.7
Mon	68.8	59.5	60.6	70.0	1.2
Moy	63.5	56.0	57.8	70.0	6.5
Pea	69.3	60.0	61.3	74.0	4.7
Pec	64.3	57.8	59.1	71.5	7.2
Tom	59.5	54.3	55.1	62.0	2.5
Vea	65.8	58.3	58.5	69.3	3.5
M	67.3	59.1	60.1	70.2	2.9

TABLE 3

ESTIMATE OF THE NET CALORIC INTAKE REQUIRED FOR MAINTENANCE AT LOW WEIGHT LEVEL for men in Squad A.

Subject	Calories	Subject	Calories
Bro	2000	Pea	2400
Can	2500	Pec	1600
Kon	1600	Spe	2200
Gar	2000	Tom	1600
Gul	1800	Vea	1900
Mon	2000		
Moy	2000	M	1967

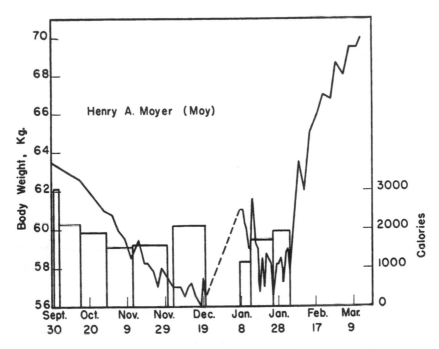

FIGURES 1 and 2. Body Weight Curves of Subjects Moy and Pea (Squad A). Solid lines = body weight during the restricted diet and rehabilitation; broken lines = body weight gain during Christmas recess. The average caloric intakes for the consecutive experimental periods are blocked in.

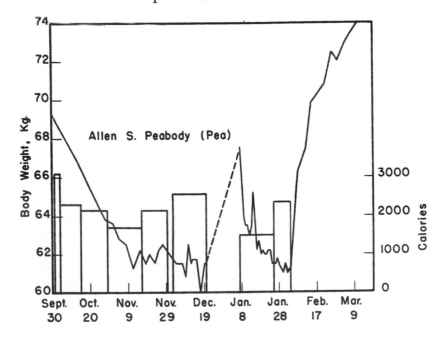

41

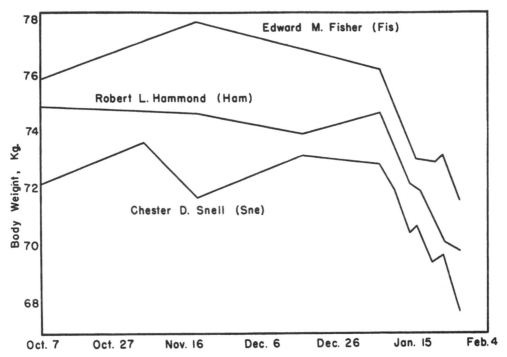

FIGURE 3. BODY WEIGHT CURVES OF 3 SUBJECTS OF SQUAD B during the preliminary period of uncontrolled diet (October 7 to January 6) and during the restricted diet of 1400 Cal. (from January 6).

intake. Therefore, the authors considered that the uniform results obtained in 2 repeat experiments gave reasonable assurance that the maintenance values were correct.

Anthropometric Data

A considerable number of anthropometric measurements were made on Squad A at the start, mid-point, and end of the dietary restriction. Measurements on Squad B were made at the start and end only. These measurements were made for the purpose of computing the surface area of the body and comparing this result with that obtained with the linear formula of DuBois. Some of these measurements, however, are of interest in showing the changes in various parts of the body produced by dietary restriction. The items of most value in this respect are summarized in Table 4. All the circumferences diminished, but the changes in forearm and ankle measurements were small, both in absolute and in percentage terms. It will be noted that there was no indication of edema in the ankle measurements.

The body surface was calculated also from another DuBois formula based only on height/weight charts, and, finally, it was computed from silhouette photograph areas. The results obtained with the three different methods agreed reasonably well. Table 5 shows the average decrease of the body surface during the diet reduction.

The photographs of Squad A at the end of starvation do not show any consid-

TABLE 4

CIRCUMFERENCES OF THE LIMBS AND THE TRUNK, in cm., in the
Carnegie Experiment.

Period	Upper Arm	Forearm	Nipples	Umbilicus	Hips	Thigh	Ankle
Squad A							
Start	30.91	26.93	90.09	75.47	92.12	55.42	21.76
Mid-point .	27.87	25.28	84.72	69.06	87.15	50.85	20.97
End	28.31	25.63	85.10	70.69	88.09	50.97	20.95
Squad B							
Start	30.60	27.40	91.48	74.57	94.13	56.17	21.30
End	28.71	26.18	87.58	70.64	90.37	52.46	20.56

TABLE 5

BODY SURFACE measured by different methods. Mean values, in square
meters, for 11 men of Squad B. The average body weight on January 5
was 68.9 kg. and on January 27, 64.0 kg. Compiled from Benedict,
Miles, Roth, and Smith (1919).

Method	January 5	January 27	Difference Sq. M.	Difference %
DuBois' linear formula...........	1.845	1.768	0.077	4.17
DuBois' height/weight chart......	1.813	1.767	0.046	2.54
Photographic area..............	1.786	1.723	0.063	3.53

erable degree of emaciation; without knowledge of the experiment they would
appear to be quite normal. The authors commented that at the end of the experiment it would have been very difficult to distinguish the subjects from other college men on the campus.

Thermoregulation — Rectal Temperature

Out of 11 subjects in Squad A only 3 showed a decrease in body temperature
of 0.9° and 0.8° F. at the end of the experiment, while another subject showed
an increase of 1.1° F. The average temperatures were 97.6, 97.5, 97.8, 97.5, 97.3,
97.2, and 97.4 on November 11, 25, December 9, 20, January 13, 27, and February 3, respectively. The differences are only slight and obviously not significant.

In contrast, the averages of Squad B show a distinct tendency to decrease
during the period of diet reduction, while there is no such trend during the November 18 to January 6 period when this group was used as control without diet
reduction. The average values of rectal temperature were 97.9, 97.7, and 97.7 on
November 18, December 16, and January 6 and decreased to 97.4, 97.2, and 96.8
on January 14, 20, and 28. The authors conclude that with so low a diet (1400
Cal.) a slight real reduction in body temperature may have resulted. We have
calculated the statistical significance of the temperature changes in Squad B
from January 6 to January 28 by means of the t-test; although slight, the changes
were highly significant (less than 1 per cent probability that the differences
could be due to chance distribution).

It is interesting that Squad A showed no significant changes at a greater loss of body weight, while Squad B showed a highly significant change at a smaller loss of body weight. Two explanations for the discrepancy are possible: (1) the rate of weight reduction, that is, the degree of diet reduction, is an important factor for body temperature regulation; or (2) the fluctuations of caloric intake and activity in Squad A interfered with the decrease of rectal temperature which might have occurred under more rigidly controlled conditions.

Thermoregulation — Skin Temperature

Measurements of skin temperature on the back of the hands and on the forehead were started at the end of the experiment (January 19 and 27) with Squad A. Therefore, no control values are available for the same subjects. Squads A and B were compared with groups of 2 to 6 normal subjects. Table 6 shows the

TABLE 6

Mean Skin Temperatures, in degrees centigrade, of Squads A and B during periods of reduced diet, compared with mean temperatures of normal subjects.

Group	Number	Room Temperature	Right Hand	Left Hand	Fore-head
		January 26			
Squad A.......	10	22.8	29.09	28.83	32.58
Normals	4	22.8	31.17	31.03	34.03
		February 2			
Squad A.......	10	23.6	32.04	31.26	34.20
Normals	6	23.6	33.10	33.26	34.09
		January 19			
Squad B.......	12	25.2	31.34	30.95	32.70
Normals	2	25.2	30.28	30.02	33.93
		January 27			
Squad B.......	12	27.3	32.50	32.42	34.43
Normals	4	27.3	33.41	33.30	34.55

average results. The change in room temperature obviously obscured any change in skin temperature which might have occurred in the interval between the two measurements; with both Squad A and Squad B the room temperature as well as the skin temperature was higher on the days of the second measurements. Also, the higher skin temperature of Squad B on January 27 as compared to Squad A is probably due, at least in part, to the higher room temperature. It is to be noted, however, that the trend with regard to room temperature is only slight and not very consistent in the normal groups. An exact comparison of the normal groups is impossible because different combinations of subjects were used on each of the 4 measurement days.

The values for Squad B on January 19 were higher than those for Squad A on

February 2, the difference in room temperature being only 1.6° C. Compared with the normal groups, the values for Squad A were consistently lower except for the forehead temperature on February 2. The hand temperature of Squad B was higher than that of the control group on January 19 and lower on January 27. The control group had a slightly higher forehead temperature on both dates. The authors believed that the forehead is by far the most reliable location for the measurement of skin temperature because it shows less fluctuation than any other part. This, however, does not necessarily mean that the forehead is the most sensitive area in which to detect alterations of skin temperature due to changes of internal or external conditions. In fact, the differences in hand temperature between Squad A and the controls are more consistent and greater than the differences in forehead temperature.

The authors concluded that the data did not indicate uniformly lower surface temperature for most of the men, although for certain members of the 2 squads the skin temperature measured on the back of the hands was definitely lower than the normal values, especially in those who apparently suffered most from cold hands. In an attempt to discover the facts in these confusing data we have subjected them to detailed analysis. We found all differences between the experimental and control groups to be statistically not significant except for the forehead temperature of Squad A on January 26. The forehead temperature difference of Squad A on February 2 compared to that of Squad B on January 19 was statistically highly significant, while the normal groups showed no significant difference.

Sensitivity to Cold

In spite of the absence — as the authors believed — of a distinct decrease in body or skin temperature, the extreme sensitivity of the subjects to cold was a noticeable feature of the experiment. In Squad A this became evident about the middle of November and grew more pronounced toward the end of the experiment. Also in Squad B an increased sensitivity to cold was reported. The increased sensitivity to cold was evidenced in several ways in addition to the subjective reports: nearly all the men wore heavier underclothing and overclothing, their bedclothing was frequently very noticeably increased, they were inclined to gather about the steam radiators whenever possible, and they avoided swimming in the indoor pool although the water felt comfortably warm to their fellow students. The authors saw a contradiction between the reduction of heat loss and the increased cold sensation. They concluded that the process of weight reduction necessarily demands increased clothing for insulation to retard as far as possible the loss of heat.

We do not think there is necessarily a contradiction between increased cold sensation and reduced heat loss. Rather it seems logical that, other things being equal, a reduced heat production and heat loss would lead to reduced body and skin temperature and an increased sensation of, and sensitivity to, cold. Without change in environmental temperature, the increased heat production in exercise produces a feeling of warmth, partly because the skin capillary bed is increased. Conversely, one might expect that at a reduced metabolic rate the peripheral capillary bed would be reduced. Obviously a study of the peripheral circulation is necessary for proper interpretation of the relationship between metabolic rate, skin temperature, and sensitivity to cold.

Basal Metabolism

In the control period before the diet reduction, the basal metabolic rate of Squad A was between —6.2 and +12.5 per cent of the normal standards of Harris and Benedict. These values were obtained in individual experiments with the usual procedure. In addition, the whole group was investigated during sleep in the big respiration chamber in the Carnegie Nutrition Laboratory. The heat computed per hour was 1.1 Cal. per kg. and 40.8 Cal. per square meter of body surface for Squad A. The initial value for Squad B (October 6 and 7) was practically identical (1.1 Cal. per kg. and 40.5 Cal. per sq. m.), but there was a significant trend downward during the subsequent control period, which ended on January 6 for this group. The values were: November 3–4, 1.10 Cal. per kg., 40.7 Cal. per sq. m.; November 17–18, 1.08 Cal. per kg., 40.3 Cal. per sq. m.; December 15–16, 1.06 Cal. per kg., 39.6 Cal. per sq. m.; January 5–6, 0.98 Cal. per kg., 36.8 Cal. per sq. m.

It is interesting that the downward trend of the B.M.R. in Squad B was not interrupted by the Christmas vacation, in contrast to the trend in Squad A. In any case, there was undoubtedly a downward "seasonal" trend of the B.M.R. in the control group. Using the values on October 7 and November 4 as basal values, on January 9 the decrease of heat production amounted to 11 per cent per kg. and about 9 per cent per square meter of body surface.

During the period of diet reduction all subjects of Squad A showed a continuous decline of the B.M.R., interrupted only by the Christmas recess. Table 7 shows the group averages of heat production for both individual experiments and group experiments in the respiration chamber. The minimal weight was attained before the Christmas vacation; it is interesting that the B.M.R. was somewhat lower at the end of the experiment when the body weight was somewhat higher (i.e., body weight and B.M.R. showed different trends after Christmas). The reduction of the B.M.R. exceeded the reduction of the body weight and, even more so, that of the body surface. The maximum reduction of the B.M.R. as determined in the group respiration chamber was greater than that recorded in the individual experiments; it amounted to 11.5 per cent (individual) and 18.2 per cent (respiration chamber) per kg. and 16.2 per cent (individual) and 22.1 per cent (respiration chamber) per square meter of surface. The absolute heat production, determined in individual experiments, decreased from an average of 1686 Cal. per 24 hours in the control period to 1367 Cal. per 24 hours at the end of the experiment — a reduction of 19 per cent.

In their final conclusions (Benedict et al., 1919, p. 694), the authors estimated the average reduction of heat production per kg. of body weight and per square meter of body surface as approximately between 15 and 20 per cent. The seasonal downward trend of the B.M.R. revealed in Squad B was not taken into account, although it was quite considerable compared to the changes observed in Squad A. If allowance were made for the seasonal change, the drop of the B.M.R. due to diet reduction alone would be about 8 to 10 per cent. While it is debatable whether or not the changes of the B.M.R. in the control group should be considered, the main reason for the parallel maintenance of experimental and control groups is negated by such disregard of the control data.

In any case, the decrease of the basal metabolism in Squad A did not exceed 15 to 20 per cent and was possibly much less. This is hard to reconcile with the conclusions about the diet necessary for maintenance at the minimum level of body weight. At 88 per cent of the original body weight the maintenance level of the diet was estimated to be approximately one half of the diet in the control period. This discrepancy can hardly be explained by the energy consumed in muscular work, since there was a similar reduction in the energy expended for rest and work and it was stated that the level of physical activity did not change. The authors concluded that the body weight data used for comparison covered altogether too short a period, but if this is the case, then the statement emphasized in numerous places in the book that the body weight can be maintained at 2000 Cal. is entirely unjustifiable.

TABLE 7

HEAT PRODUCTION, Squad A. Mean values, in calories per 24 hours, obtained with portable Benedict apparatus on the single subjects (Benedict) and in the group respiration chamber (Group). Body surface calculated for height/weight.

	Per Kilogram		Per Square Meter	
	Benedict	Group	Benedict	Group
During normal diet................	25.2	26.4	940	979
At period of minimum weight.......	23.1	22.6	817	792
At end of semi-starvation..........	22.3	21.6	788	763
Decrease in semi-starvation (%).....	11.5	18.2	16.2	22.1

The downward trend of the B.M.R. of Squad B during the control period makes it difficult to evaluate the decrease during the 3 weeks of diet reduction (1500 Cal. per day). As stated before, the group average on January 6 was 0.98 Cal. per kg. and 36.8 Cal. per square meter. The basal energy expenditure in the semi-starvation period was: January 14, 0.97 Cal. per kg., 35.5 Cal. per sq. m.; January 20, 0.90 Cal. per kg., 32.6 Cal. per sq. m.; January 28, 0.85 Cal. per kg., 30.5 Cal. per sq. m.

The B.M.R. showed a rather sharp drop from December 16 to January 6, the last day of the control period, and it is hard to say whether this drop would have continued. The authors regretted that measurements were not taken after January 6, before the diet restriction was begun, but assumed that a further fall without diet reduction was not probable since the fall during the first week of diet reduction (January 14) was slight. The total reduction of the B.M.R. from October 7 to January 28 amounted to 22.7 per cent per unit of weight and 24.7 per cent per unit of body surface. The decrease from January 7 to January 26 was 11 and 9 per cent, respectively.

The energy expenditure (in calories per kilogram of body weight) of Squad B in the standing position was 14.1 per cent lower at the end of 3 weeks of diet restriction; this decrease is approximately of the same magnitude as that in the supine position. No comparison of the energy expenditure in the standing position before and at the end of diet restriction was made in Squad A.

Energy Expenditure during Exercise

The increment of the energy expenditure during walking over that in the standing position was calculated for the horizontal transport of 1 kg. of body weight per 1 meter distance. The speed of the treadmill was 69.5 meters per minute, or 2.6 miles per hour. The main experiments were carried out on Squad B, who were compared on January 6 and 28. In Squad A the values were obtained only on the last day of semi-starvation (February 3). Table 8 shows the

TABLE 8

METABOLIC COST OF WORK AFTER DIETARY RESTRICTION. Per cent change is calculated by reference to Squad B, Normal. Weight, in kg., includes clothes, electrodes, etc. Speed of walking (Speed) is in meters per minute, and the amount of work (Work) is expressed as horizontal kg. meters per minute. Energy expenditure is in calories per minute. Total heat production is calculated for walking a distance of 1 kilometer. All values are averages.

	Weight	Speed	Work	Energy Rate		Total Heat	
				Cal.	% Change	Cal.	% Change
Squad B, Normal....	70.5	69.4	4894	0.597		626	
Squad B, End......	66.6	69.7	4641	0.562	−6.0	533	−14.8
Squad A, End......	63.4	69.6	4410	0.522	−12.6	484	−22.7

average data for both squads. The initial (January 6) energy expenditure of Squad B was found to coincide with values obtained in another group of normal subjects; for this reason it was believed that they probably would represent also the initial values for Squad A. No controls were made to show whether the rate of energy expenditure during walking showed a downward trend similar to that during rest. Under the assumption that the initial values were probably the same in both groups, the percentage of decrease was calculated. It can be seen that the absolute energy cost for walking decreased by 14.8 per cent for Squad B and 22.7 per cent for Squad A, or approximately in the same proportion as the B.M.R. Also, the average heat output per horizontal kilogram meter was found to be decreased at the end of semi-starvation by 6 per cent in Squad B and 12.6 per cent in Squad A, indicating an increased efficiency. The authors regard this result to be of such great physiological importance that "it is only with considerable reserve that one should draw deductions from it."

We have calculated the statistical significance of individual differences in walking efficiency for Squad B (comparing the values of January 6 and 28) and of the group differences between Squad A at the end of the experiment (February 3) and the initial values of Squad B, using the t-test. None of the differences were found to be statistically significant, not even at the 10 per cent level.

In the walking experiments, the total number of steps were counted and the average length of the steps was calculated. There was a slight tendency for the rate of stepping to increase during diet reduction, that is, the stride tended to shorten. These differences are hardly significant and are too small to be considered in the calculation of efficiency.

Nitrogen Balance

The nitrogen intake in food and the nitrogen excretion in urine and feces were determined for each experimental day. The average nitrogen intake for Squad A during the control period varied between 14.90 and 16.01 gm. per day, compared with an average nitrogen excretion in the urine of 12.19 gm. per day, varying in 5 days between 10.53 and 13.15 gm. Considerable fluctuations were observed in the individual nitrogen intake as well as in nitrogen excretion during the period of diet restriction but the group averages for the single days were remarkably constant. In the first week of diet reduction the nitrogen intake varied between 9.33 and 13.82 gm., the nitrogen excretion in urine between 11.73 and 12.79 gm. During the maintenance period of December 5–18 the nitrogen intake varied between 8.00 and 13.52 gm., the nitrogen excretion in urine between 9.8 and 12.21 gm. The average nitrogen excretion in urine for October was 11.47 gm., for November 10.60 gm., for 3 weeks in December 10.87 gm., and for January 10.34 gm. It appears that in spite of the great alterations in the diet the average nitrogen excretion was little affected. At the end of the experiment, in the middle of January, low figures appear more frequently, but the over-all differences in urinary nitrogen excretion were slight, and the nitrogen excretion in the last week of October was essentially the same as that in the last week in January. During the last week of January, however, the diet was increased in order to maintain the body weight.

TABLE 9

Total and Average Daily Losses of Nitrogen, in gm., for Squad A, October 4–January 27, inclusive. (The data for September 29 and February 2 indicate a weight decrement from 67.0 kg. to 61.0 kg.)

Subject	Total Loss of Nitrogen	Number of Days	Average Loss of Nitrogen per Day
Bro	153.48	83	1.85
Can	155.77	84	1.85
Fre	32.29	20	1.61
Kon	233.08	57	4.09
Gar	168.95	86	1.96
Gul	162.45	86	1.89
Mon	134.07	86	1.56
Moy	230.31	83	2.77
Pea	206.14	86	2.40
Pec	252.85	87	2.91
Spe	130.15	61	2.13
Tom	48.66	78	0.62
Vea	159.70	86	1.86
M	159.07	75.6	2.12

Nitrogen balances were calculated from the sum of nitrogen output in feces and urine and the nitrogen content of the food. The nitrogen excretion from the skin was neglected; probably 0.4 gm. of nitrogen loss should be added to obtain a closer estimate of the nitrogen balance. On the other hand, the excess nitrogen intake on the uncontrolled days was not considered either, on the assumption that the two factors would compensate.

The nitrogen balance for the single days varied roughly between ±3 gm., but in general there is a great predominance of minus figures. Throughout the period of weight loss there was a pronounced tendency for every subject to lose nitrogen. During the weight maintenance periods in December and the last week in January, positive figures for the nitrogen balance appeared more frequently. Table 9 shows the accumulated loss of nitrogen during the experiment and the average daily nitrogen loss for the subjects of Squad A. The total loss of nitrogen is not quite comparable for the different subjects because of the variation in the periods of diet restriction, as shown in Table 9. The average daily loss (except for subject Fre, who quit in October) was 2.16 gm.; however, the average loss per day showed wide individual variations which are obviously not correlated with the duration of the experiment. Figure 4 shows the remarkable parallelism between the accumulated nitrogen loss and the body weight loss for 9 subjects.

Because of the larger and better controlled reduction of the diet of Squad B, there was a distinct nitrogen loss from the beginning to the end of the experiment. The average daily nitrogen loss was 3.1 gm., with individual limits between 1.63 and 4.50 gm.

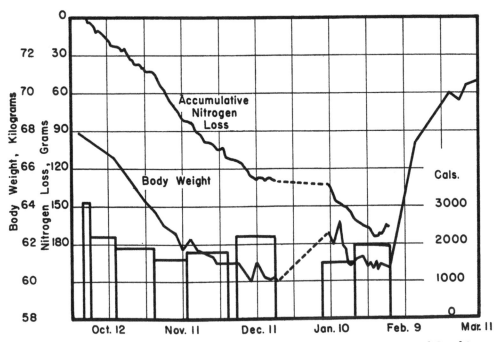

FIGURE 4. ACCUMULATIVE NITROGEN LOSS AND BODY WEIGHT, average of 9 subjects. Broken lines = Christmas recess. The average caloric intakes for the consecutive periods are indicated by the bars.

In spite of the parallelism between nitrogen loss, body weight loss, and the degree of diet reduction, the caloric value of the protein corresponding to the nitrogen loss cannot account for any appreciable percentage of the total energy lost from the body.

Utilization of Food

The nitrogen and energy utilization of the food was calculated for Squad A for 5 periods of 4 to 16 days, the first period being in the control period before diet reduction with 4 more periods during the diet restriction (October 8–12, October 17–21, October 31–November 4, November 12–18, January 14–30). The nitrogen utilization was obtained from the comparison of food intake and fecal elimination. The energy content of the feces was determined by means of direct calorimetry. The energy content of the urine was obtained by multiplication of the nitrogen content with the factor 8.0. The energy utilization was then calculated as the difference of energy intake minus energy of feces plus urine, expressed as a percentage of the energy content of food.

TABLE 10

NITROGEN AND ENERGY UTILIZATION OF FOOD in the Carnegie
Experiment. Average for 10 subjects.

		Nitrogen			Energy	
	Food	Absorbed (gm.)	Used (%)	Food	Absorbed (cal.)	Used (%)
Control	15.47	13.94	90.5	3333.9	3095.8	92.8
Second week..	12.26	11.00	89.7	2308.1	2058.5	89.3
Final period...	9.85	8.60	87.1	1894.7	1654.3	87.3

In Table 10, Benedict's Table 37 has been condensed to show the percentage of nitrogen and energy utilization in the control period, the second week of semi-starvation, and the last experimental period (January 14–30, 1918). For the calculation of averages, 3 subjects were omitted because their data were incomplete. The difference of food nitrogen minus feces nitrogen is designated as nitrogen absorbed. Similarly, the difference of total food calories minus calories in feces and urine energy is designated as energy used.

It can be seen that in semi-starvation there was a decline in the utilization of both nitrogen and energy, especially the latter. This was true for all subjects. The decreased utilization was apparent in the second week of semi-starvation, although it was more pronounced in the final period. During the greater part of the experiment, however, no definite trend was discernible, so it is not possible to correlate the decreased utilization with a well-defined period of the experiment or physiological state. The authors discuss the possibility that a greater proportion of bran in the reduced diet might be, in part at least, responsible for the reduction in the food utilization, but comparison of individual results does not support this assumption. However, the extensive use of bulky leafy vegetables might well account for at least some of the changes. Although the degree of deterioration was not great and the absolute utilization was still within normal limits, the change is statistically highly significant. The result is not easy to explain, but obviously it is not compatible with the authors' concept of the perfection of metabolic adaptation to semi-starvation.

Respiratory Quotient

The average basal R.Q. of Squad A was 0.80 in the control period. During the period of semi-starvation the R.Q. was frequently found to be high after the days of uncontrolled diet but, with these exceptions, it was fairly constant at about 0.79. In a few instances the R.Q. was below 0.73. The authors concluded that there was a tendency toward decreased use and probably decreased storage of carbohydrates. Inspection of the tables and of the very slight difference in the average level of the R.Q. fails to support this statement. In Squad B the average R.Q. during walking was 0.80 before the diet reduction and dropped to 0.78 at the end of the experiment. The difference was statistically not significant.

Respiration Rate, Tidal Volume, and Alveolar Air

The average respiration rate at rest did not change significantly in Squad A during the period of semi-starvation. The same is true for the respiration rate in the standing position as well as during walking for Squad B. No data from Squad A were available for the standing position or during walking. The data of Squad A were so irregular that no definite trend in tidal volume during the semi-starvation period can be recognized. Alveolar air samples showed no significant change in carbon dioxide tension in Squad A. No data were obtained on Squad B for these last two variables.

Ventilation Rate and Respiratory Efficiency

From the Carnegie data we have calculated the average rate of pulmonary ventilation and the respiratory efficiency for Squad A for the control period and for the last day of semi-starvation. The mean pulmonary ventilation decreased from 5.09 to 4.49 liters per minute (-11.8 per cent), while the oxygen consumption per minute fell 18.4 per cent (from 239 to 195 cc. of oxygen per minute). If we consider respiratory efficiency to be measured by the amount of oxygen removed per 100 cc. of ventilation, it appears that the control value was 4.70 cc. of oxygen and this declined to 4.34 cc. at the end. If we consider the relation between ventilation and oxygen demand to have been normal and adequate in the control period, then we must conclude that the ventilation was relatively excessive in semi-starvation.

Benedict and his colleagues observed that the absolute respiratory minute volume was decreased in semi-starvation but that there was no change in the alveolar tension of carbon dioxide. They interpreted this to mean that the sensitivity of the respiratory center was decreased. Clearly the data available are not adequate to decide the point without evidence on the alkaline reserve of the blood and the strength of the respiratory muscles. Since the alveolar tension was so regulated as to remain essentially constant, we must conclude that there was probably little if any actual change in the sensitivity of the respiratory center. Also, there was no trend toward a change in the rate of respiration.

Circulation — Blood Pressure

The first blood pressure measurement in Squad A was made at the end of the second semi-starvation week. The group averages for the systolic, the diastolic, and the pulse pressure were 115, 81, and 34 mm. Hg., respectively. At the end of semi-starvation the average values were 95, 64, and 31 mm. Hg., representing a 20 mm. drop of the systolic blood pressure and a 17 mm. Hg. drop of the diastolic blood pressure. In Squad B the average systolic blood pressure decreased during the period of restricted diet from 120 to 94 mm. Hg., the diastolic pressure from 83 to 64 mm. Hg., and the pulse pressure from 37 to 30 mm. Hg. The pressures in Squad A were at the lowest points in the middle of December (mean systolic pressure 89 and diastolic pressure 71 mm. Hg.). It is possible that the Christmas recess spoiled the consistency of blood pressure changes during semi-starvation. The authors inferred from the decrease of blood pressure that the subjects would be distinctly unable to withstand surgical shock; it is now known that predisposition to surgical shock cannot be judged from blood pressure alone. The blood pressure was also measured immediately after walking in both groups on the last day of restricted diet; since no control values were obtained, the results will not be discussed.

Circulation — Pulse Rate in Rest

There was a considerable variation in the number of subjects on each day the pulse rate was taken so that the group averages for any particular day are not exactly comparable. However, the tendency toward a pronounced decrease during the period of restricted diet is quite obvious. The group average on the 7 control days prior to diet restriction varied between 58 and 53 beats per minute, compared to values between 38 and 45 in the last week of semi-starvation. This low level was reached by the middle of November, after 6 weeks of restricted diet; it increased after the Christmas recess and declined again in the last 2 weeks. Because of the questionable nature of the averages for each day, the average for the whole normal period (7 days) was compared with the period from November 17 to 25, when in general the lowest level was observed. The individual averages decreased from 5 to 21 beats, while the group average showed a decrease of 14 beats. In a few subjects values as low as 29 or 30 beats per minute were occasionally observed. The pulse rate of Squad B decreased from an average of 56 to 48 at the end of the first week of restricted diet, to 43 at the end of the second week, and to 40 at the end of the third week. The pulse rates in the sitting position and during meals also declined during semi-starvation. In the standing position the pulse rate was close to 20 beats per minute faster than in the supine position and this differential was unchanged in semi-starvation.

Circulation — Pulse Rate in Exercise

The pulse rate during a short static type of exercise (maintaining the chinning position on a bar) was determined with an electrocardiograph. The results are given in Table 11. The cycle length was measured before (at rest) and during the exercise. At the end of semi-starvation for Squad B, the cycle length was prolonged both before and during the exercise, but the percentage of shortening during the exercise is the same. The same proportion of shortening was observed in Squad A at the end of semi-starvation and in a normal group. It was concluded that the state of semi-starvation did not interfere with the ability for adaptive increase of the heart rate in muscular exercise.

In these studies it was observed that variations in the cycle length in Squad A during the rest period before the exercise were greater than those of the normal group used for comparison. From this it was inferred that semi-starvation accentuates the natural arrhythmia. Such a difference might well reflect individual differences not related to the state of semi-starvation. In Squad B the curve on January 20, at the end of 2 weeks of semi-starvation, is at least as smooth as those taken in the control period (November 18 and December 16). We do not think, therefore, that the claim of increased tendency to arrhythmia in semi-starvation is supported by the actual evidence. The slope of recovery of the cycle length after the exercise was not changed in semi-starvation (Squad B).

TABLE 11

SUMMARY FOR PULSE CHANGES OCCASIONED BY SHORT PERIODS
OF EXERTION.

Groups of Subjects and Conditions Compared	Pulse Rate		Percentage Change
	Rest	Exercise	
Squad A, low diet................	54	78	44.4
Squad B, normal.................	65	90	38.5
Squad B, low diet................	58	79	36.2
Normal group...................	72	100	38.9

The rate of change in the pulse rate during the first phases of exercise was determined electrocardiographically for both the short static exercise and for walking. For the short static exercise there was no change during semi-starvation.

During the initial phases of walking, in many individual curves a peak (i.e., minimum cycle length) was reached within 10 to 20 seconds, after which the pulse rate dropped again, sometimes sharply, followed by a slower secondary rise. This phenomenon was observed at the end of semi-starvation in 8 out of 11 subjects of Squad B and in 8 out of 10 subjects of Squad A. This phenomenon was not observed in a group of 5 normal subjects. Also, fluctuations during the further course of exercise were smaller in the normal than in both experimental groups. The maximum percentage variation of cycle length in the normal group was 17.7 per cent; those of Squads B and A at the end of semi-starvation were 23.6 and 28.2 per cent. While we could not regard the evidence for the claim of increased arrhythmia at rest as conclusive, it may be different for exercise. The overshooting pulse rate increase in the beginning of exercise might be due to excessive sympathetic impulses or vagal disinhibition. Assuming that in the state of semi-starvation the slow basal pulse rate is produced by vagotonia, it would appear possible that inhibition of the vagus tone is associated with some imbalance of the vegetative regulation of the pulse rate. The comparison of the normal group with both experimental groups at the end of diet restriction shows that there was probably no decrease in the promptness with which the heart responds to exercise.

The pulse rate was taken at 1 to 3 minute intervals during 25 minutes of walking and was followed through 9 minutes of recovery. Table 12 shows the average results. In Squad B the pulse rate at the end of semi-starvation was lower at rest, in standing, during exercise, and in recovery. The recovery of the pulse rate after walking was not changed in semi-starvation.

TABLE 12

MEAN PULSE RATES, in beats per minute, seated, standing, and after 1, 6, 12, and 24 minutes of walking, taken in the Carnegie Experiment at the end of dietary restriction, compared with the means for Squad B before dietary restriction and with the means for another group of 5 men in a normal nutritional state.

Group	Seated	Standing	Minutes			
			1	6	12	24
Squad A, end.......	52	60	72	69	72	73
Squad B, end.......	46	52	69	64	69	69
Squad B, before.....	72	79	88	85	85	89
Normal group.......	68	76	85	85	87	85

The pulse rate was also measured during recovery after 1 minute of work on the bicycle ergometer at a rate equal to the performance of 67 kilogram meters of work per minute. In Squad A the absolute pulse rates were lower in the first recovery minute during the semi-starvation period, but the percentage increase above the resting pulse rate was greater. In Squad B, the absolute pulse rate was also lower during the first recovery minute in the semi-starvation period, while the percentage increase was the same or lower in most subjects. The return of the pulse rate to the pre-exercise level was not significantly changed during the period of restricted diet.

Circulation — The Electrocardiogram

The electrocardiograms of all subjects (3 standard leads) were stated to be within normal limits during semi-starvation, except for the pronounced sinus-bradycardia. No evidence of heart block was found.

Blood Morphology

At the end of semi-starvation the average number of red blood cells for Squad A was 4.5 million per cu. mm. of blood, with a Sahli hemoglobin content of 76 per cent. In Squad B the red blood cell count dropped from a pre-starvation level of 5.52 to 4.45 million per cu. mm. during semi-starvation, and the hemoglobin value from 87 to 81 per cent. The mild anemia was associated with slight histological abnormalities of the red cells such as hypochromasia, anisocytosis, and occasionally polychromatophilia.

The average white count of Squad A during semi-starvation was 9500 per cu. mm., which is slightly higher than the normal average. There was no change in

Squad B. Although the number of leucocytes was not greatly altered, the absolute number (3400) and the percentage (36 per cent) of lymphocytes were definitely high in Squad A, and the same tendency was noted in Squad B. The polymorphonuclear neutrophile cells were slightly decreased (56 per cent), while the other white cells showed no significant alteration.

Strength and Speed

The grip strength was measured in the morning and in the evening. No normal control values are available for Squad A, but on May 21, 11 weeks after the end of the semi-starvation period, the measurements were repeated on 6 of the men. Figure 5 shows the morning values of handgrip strength for right and left arm in Squad B (broken lines) and the 6 subjects of Squad A. There was a significant group difference throughout the duration of the experiment; Squad B had consistently higher values. No definite change can be seen in Squad B as a result of semi-starvation. There was a rather consistent drop in Squad A from November 10 to December 19, more so in the left hand. The left hand recovered somewhat during the Christmas recess, while the right hand stayed on the same level. In the rehabilitation period there was a pronounced increase of both right and left handgrip strength, distinctly above the initial values (which are not the normal values). The steepness of the rehabilitation gain is exaggerated because the time scale is not uniform.

In the evening tests both squads showed a drop of grip strength, which was, curiously enough, somewhat reversed at the end of the restriction period, January 12–27 (see Figure 6). It is interesting that in Squad B the evening tests re-

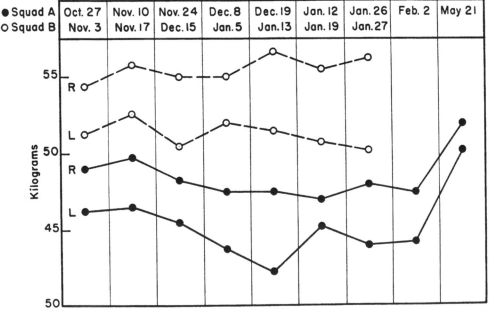

FIGURE 5. AVERAGE HANDGRIP STRENGTH of 6 subjects of Squad A (solid lines) and of Squad B (broken lines), determined at the evening sessions.
R = right hand; L = left hand.

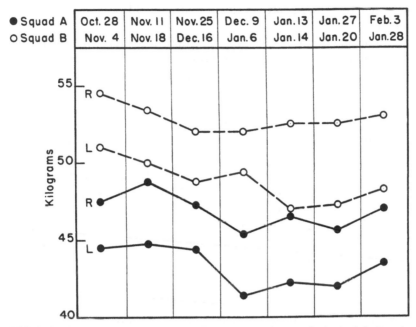

FIGURE 6. Average Handgrip Strength of Squad A (solid lines) and Squad B (broken lines), determined in the morning.
R = right hand; L = left hand.

vealed a loss of strength, while the morning tests did not. The average gain of the 6 subjects of Squad A in the rehabilitation period was 4.5 kg., at an average level of 50 kg. It is estimated that the reduction of grip strength during semi-starvation was about 8 or 9 per cent.

The latent period, refractory period, and amplitude of the patellar reflex did not show any distinct trend during the semi-starvation period. The authors, however, had the impression that there was a tendency toward decreased reflex irritability during the restricted diet.

The maximum rate of finger movements performed in 10 seconds was measured in the morning and in the evening. For Squad A a control value is available only for the evening tests. The average rate for 10 subjects of Squad A before diet restriction was 68.7, a value which was not attained at any time during the period of semi-starvation. This, however, was true only for the average; in 5 subjects the control value was occasionally exceeded during semi-starvation. The over-all low diet average was 65.1 movements per 10 seconds for Squad A.

In Squad B the group averages for 5 evening determinations varied within the narrow range from 64.7 to 66.0 movements in 10 seconds (over-all average 65.6). The average rate dropped to 64.2, 62.4, and 61.4 (over-all low diet average 62.7) at the end of the first, second, and third weeks of semi-starvation. In this group the results were more consistent in that values exceeding the pre-starvation control levels were found in only 2 subjects during the period of diet restriction. The results obtained in the morning sessions with Squad B revealed the same and rather consistent downward trend during semi-starvation; the average rate decreased from the control value of 66.4 to 61.9 at the end of the second semi-starvation week.

The rate of finger movements was recorded in 5 successive 2-second periods. The drop during the 10 seconds of performance was taken to indicate the fatigue trend. Figure 7 shows the results for both Squads A and B; it will be seen that the drop of the rate is a rather linear function of time. For both groups the control values are higher than the low diet values obtained in semi-starvation but the "fatigue curves" are parallel. This means that the fatigability was not changed during semi-starvation, in spite of the general depression of the rate.

The time required for making a horizontal sweep of the eye through a visual angle of 40 degrees increased progressively for movement of the eye from right to left in Squad A during semi-starvation; the total change at the end of semi-starvation amounted to an increase of 9.3 per cent (see Figure 8). Interestingly enough, there was no significant trend in the speed of the eye movements from left to right. It was concluded that eye movements from right to left are more easily disturbed than those from left to right. No consistent trend was apparent in Squad B.

Motor Coordination

The accuracy of movement in tracing a line with a pencil was measured in errors (the number of deviations) per line. The test blank consisted of 5 lines. Squad A showed, with some fluctuations, an improvement in accuracy during the period of semi-starvation, but so did the control group, Squad B, without diet restriction. The improvement in Squad B continued during the first 2 semi-starvation weeks, but there was a slight drop at the end of the third semi-starvation week. The improvement due to training was obviously a much more important factor for the outcome of the test than the diet restriction. The authors stated: "It seems justifiable to conclude, although the results for Squad B . . . will not entirely support the statement, that with Squad A, particularly during October and November, the motor functions involved in steadiness in tracing were interfered with by the reduced diet, since the Squad as a whole did not make such rapid improvement in the test as would have been expected of them under normal conditions." We do not think there is sufficient evidence for this conclusion. The training curves were too irregular to allow accurate extrapolation or prediction. It is regrettable that in this test as well as in the other motor and sensory tests the experiments were not carried out at the maximum training level (plateau).

A similar test was performed with a rectangular maze consisting of parallel lines with an interspace of 2 mm. The speed of tracing, performed with a pencil, was controlled by a metronome. In this test, too, the improvement due to training overshadows any possible effect of the diet restriction.

Sensory Threshold — Visual Acuity

The average minimum visual angle required for recognition declined progressively in Squad A from 86.2 seconds of arc in the first week of semi-starvation to 66.3 seconds before the Christmas recess. It remained at that level until the end of the experiment. The progressive improvement of visual acuity was found

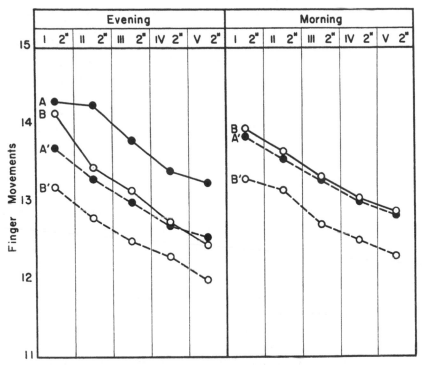

FIGURE 7. Number of Finger Movements Performed in 5 Successive 2-Second Intervals. A and B = Squads A and B on normal diet;
A′ and B′ = on semi-starvation diet.

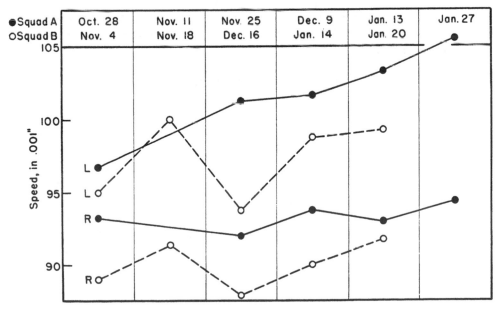

FIGURE 8. Speed of Eye Movements. Time (in milliseconds) necessary for a 40-
degree sweep of the eyes to the left (L) and to the right (R).

in 10 out of 12 subjects. We found this change to be satistically significant. In Squad B no consistent trend can be seen; there was a deterioration from November 3 to December 15 with subsequent improvement throughout the rest of the control period and the 3 weeks of restricted diet. At the end of the restricted diet the average acuity was slightly better than the initial value. The results with Squad B do not necessarily contradict the conclusion that there was a gradual improvement of visual acuity in Squad A. This improvement was overlooked by the authors, probably because they expected the opposite result. They stated that the results did not definitely prove that the reduced diet raised the visual threshold with Squad A.

Acoustical Threshold for Pitch Discrimination

An evaluation was made of the percentage of correct judgments of pitch differences of 30, 23, 17, 12, 8, 5, 3, 2, and 1 vibrations lower than the standard pitch of 435 oscillations per second. All subjects were able to discriminate without error pitch differences over 8 oscillations. The percentage of correct judgments by Squad A increased sharply for the 1, 2, 3, and 5 oscillation differences during October and the first half of November, probably because of training. From the middle of November there was a deterioration which continued to December 19. After the Christmas recess the results were not consistent. In Squad B there was a drop for the 1 oscillation pitch difference at the beginning of semi-starvation. With the pitch difference of 2 oscillations a decline was apparent by November 17, 9 weeks before the beginning of diet reduction, and this continued until the end of the second starvation week, with a slight improvement during the last week. The 3 and 5 oscillation pitch differences show minor fluctuations without any clear relationship to the diet. The authors interpreted these results as evidence of a general decline in the accuracy of pitch difference discrimination associated with the diet reduction. We do not think the actual results support this conclusion.

Sensory Threshold to Electrical Shock

No trend in the sensory threshold to electrical stimulation was apparent in Squad A during semi-starvation. The absence of a practice effect was believed to be a result of the reduced diet, since "it would be surprising if the physiological threshold were reached without practice in this case of electrical stimulation." Squad B showed some improvement from October 6 to December 15, followed by a rise of threshold until January 19, the end of the second semi-starvation week. The increase of the threshold can hardly be due to the restricted diet since it started 3 weeks earlier; furthermore, there was a marked improvement in the last week of restricted diet. We do not think the actual data warrant the conclusion of the authors that the data with both squads indicate an increase of the electrical threshold due to diet restriction.

Reaction Time

The reaction time for turning the eye to a new point of attention appearing suddenly somewhere in the field of vision was measured by a photographic method such as is used for the recording of eye movements. There was a decline of the average reaction time in Squad A from 0.231 seconds in the third semi-starvation week to 0.199 seconds in the ninth week, probably due to training. After the Christmas recess the average reaction time increased to 0.210 seconds. In Squad B the training improvement observed in the control period continued during semi-starvation. It appears that the eye reaction time was not significantly changed by the reduced diet. Also, the reaction time for speaking 4-letter words visually presented in chance order was not significantly changed during semi-starvation.

Working Capacity, Endurance, and Activity

The number of chin-ups were counted in all 12 subjects of Squad A at the end of semi-starvation, but only in 7 was the best previous record known. The data for these subjects are given in the accompanying tabulation. In all but one subject a marked drop of endurance is evident. Endurance in a static type of work, that is, holding the arms at the level of the shoulders pointing forward at an angle of 45°, was not significantly altered in semi-starvation.

Subject	Previous Record	At End of Semi-Starvation
Bro	12	12
Gul	24	14
Moy	12	8
Pea	18	15
Pec	5–10	5
Tom	12	7
Vea	15	5

Each man was given a pedometer which he carried at all times. From the pedometer records the average distance walked during a day was estimated. The average values for Squad A, excluding the Thanksgiving and Christmas recesses, were: October, 7.0 miles; November, 6.4 miles; December, 6.8 miles; January, 6.2 miles. Since the subjects continued their normal college activity during the experiment, other factors such as free time available or weather conditions influenced the results. The same is true for other forms of voluntary physical activity, which were determined from subjective estimates as recorded in a detailed log.

In the evaluation of these records it should be remembered that the subjects were encouraged to compensate for the weight gains during the days or periods of uncontrolled diet by increased physical activity. Therefore physical activity was not a true independent variable. The authors are probably correct in concluding that only minor differences in the level of physical activity occurred among their subjects during semi-starvation, but it is questionable how far these results can be generalized.

The authors made elaborate estimates of the energy expenditure per day, assigning what they believed to be reasonable figures for the caloric equivalents of sitting, walking, and so on. They were able to arrive at an average of 2245 Cal. for the period of lowest weight, and this seems to be in good agreement with their estimate of caloric requirements from nitrogen and weight balance. It must be emphasized, however, that the assumptions involved were large and numerous, and such agreement must be considered fortuitous.

In contrast to the pedometer and written activity records, the subjective reports appear to indicate a loss of the working capacity or at least of the will power to work. Complaints of weakness, lack of "pep," and fatigue are fairly common. A very frequent complaint was weakness of the legs, particularly in stair climbing. Professors Affleck, Johnson, and Berry, who were continually observing the subjects in the gymnasium and in other athletic activities, unanimously expressed the opinion that their performance, endurance, alertness, and will power were less than in previous years and also less than that of their fellows.

Intellective Performance and Personality

Several tests involving intellective performance (addition, number cancellation, memory, arrangement of four- or five-digit numbers, description of the location of certain items in work sheets, etc.) showed considerable practice effects without any clear indication of effects of the diet. Scholastic performance was evaluated by comparing the grades of the subjects before and during the semi-starvation period with the grades of their non-starving classmates. No trends or differences were seen, the relative positions of the men in their classes remaining unchanged.

The men were commonly found to be somewhat irritable during the semi-starvation period. All the subjects reported a distinctly decreased sex interest and a reduced frequency of erections while on the restricted diet. The normal state was quickly regained on the uncontrolled diet after the experiment.

CHAPTER 4

General Features of the Minnesota Experiment

In planning the Minnesota Experiment it was recognized that to obtain the most generally applicable data on the effects of semi-starvation and subsequent rehabilitation the use of a well-defined sample of the general population would be highly desirable. It would certainly be expected that the degree of deterioration and the adjustments to the processes of semi-starvation and rehabilitation would differ with age and sex and with the physiological, psychological, and biochemical normality of the subjects.

Among the more apparent limitations in securing a sample of experimental subjects who would provide the most comprehensive information are the following: (1) the difficulty inherent in determining a true cross-sectional sample of the population; (2) the difficulty of securing the long-time, exclusive services as subjects of those persons who might constitute such a sample; (3) the danger of possible temporary and/or permanent disability which might occur as a result of the stress.

Under normal circumstances it might have been impossible to secure a sufficient number of acceptable subjects for this study. The risks, necessity for surveillance 24 hours a day, undeviating adherence to the regimen, withdrawal from ordinary pursuits with drastic changes in most aspects of living, and physical discomfort to be expected as an unescapable consequence of participation would no doubt have deterred the enlistment of most of the candidates.

Several circumstances early in 1944 encouraged and facilitated the institution of an experimental study of human semi-starvation and rehabilitation. Reports arriving from the occupied areas of Europe and from prison camps indicated that starvation was present in many places and that millions of people were in grave danger of mass famine. It was recognized that with the liberation of the occupied areas and the cessation of hostilities large-scale relief feeding would be necessary. It was equally well recognized that there was little reliable quantitative information upon which to base an efficacious relief program. The need for a controlled experiment to determine the changes induced by semi-starvation in man and the best type of rehabilitation diet was apparent and urgent.

A potential source of subjects for such an experiment were the conscientious objectors who, as drafted civilians, were under the direction and control of the Selective Service System for the duration of the war (Eisan, 1948). Many of these men were genuinely interested in the problems of relief and rehabilitation and said they would welcome the opportunity of serving as subjects for the semi-starvation experiment. A brochure describing the proposed experiment and setting forth the conditions under which human "guinea pigs" would serve as subjects was transmitted with a call for volunteers to the various Civilian Public Service camps and units, after Selective Service approval was obtained and the cooperation of the historic peace churches (Brethren, Friends, and Mennonites) had been enlisted. Besides the subjects a number of other conscientious objectors who had special technical skills were to be brought to the Laboratory to assist in conducting the experiment by serving as assistants and technicians.

Criteria for Selecting the Subjects

It was fully realized from the beginning that the success or failure of the experiment would depend to a large degree upon the caliber of the subjects. The selections were made on the basis of 4 major criteria:

(1) All subjects must be in good health. They should neither have a history of any serious disabling disease. nor show clinical evidence of any but minor disorders. No physical disabilities or handicaps should be present since many of the standardized tests to be used demanded both the ability and the willingness to perform strenuous and sustained physical work.

(2) The mental health of the subjects should be capable of supporting full cooperation in the experimental work and adherence to the regimen with a minimum of direct supervision.

(3) It was mandatory that the subjects have the ability to get along reasonably well with others under the difficult and trying condition of living together without privacy and with greatly reduced personal freedom. It was necessary that they be willing to subordinate personal interests and activities to the group as a whole and to the experimental program.

(4) It was decided that subjects should be selected who had a real interest in relief and rehabilitation, and an educational program was developed to furnish special training in that field. It was believed that full commitment to the exigencies of the experiment with the maintenance of optimal motivation and cooperation was more likely to be achieved by subjects having a personal sense of responsibility in bettering the nutritional status of famine victims.

Methods of Selecting the Subjects

Individuals in the various CPS camps and units who were definitely interested in being subjects for the experiment sent written applications to the Laboratory. When it became apparent that a sufficient number of men were interested, special arrangements were made to interview those applicants who had been particularly recommended by their churches and the camp directors. Most of the interviews were held at the various camps, but a few of the men came directly to the Laboratory. More than 100 men made applications to serve as subjects.

During the preliminary screening interviews the applicants were given a superficial physical examination and were questioned in regard to food habits, attitude toward physical work, and past and present health status. They were specifically cautioned about the possible dangers and the physical and mental discomfort inherent in the experiment. The Selective Service medical record of each man was carefully studied. They were further questioned about general mental health, emotional balance, personal relationships with others in their group, and motives for wanting to be a subject in the experiment. Finally all the applicants were given the Minnesota Multiphasic Personality Inventory.

The applicants who were obviously not acceptable were dismissed from further consideration. Camp superintendents and project supervisors were asked to give their appraisal of each potential subject, and all individuals who had made poor camp adjustments were eliminated. All the data obtained from the interviews were brought back to the Laboratory of Physiological Hygiene where each application was carefully considered at a Senior Staff meeting. About 40 of the applicants who seemed to meet all requirements were then requested to come to the Laboratory for complete clinical, radiological, biochemical, physiological, and psychological examinations, and on the basis of the results from the tests, 36 of the volunteers were finally selected to make up the subject group for the experiment.

Motives for Volunteering

A brief discussion of the reasons given by the applicants for volunteering to serve in the Minnesota Experiment may offer some insight into the psychological constitution of the subjects. Neither the motives nor the strength of motivation were the same for all subjects. For some, personal reasons predominated, while for others a genuine concern about the state of scientific knowledge of the effects of semi-starvation and of the best methods for rehabilitation of the starved was of primary importance. The following motives were most frequently cited by the subjects as reasons for volunteering:

(1) Some men felt that their previous assignments under Selective Service were "made work" that offered no adequate opportunities for service and filled no real need. To them the Minnesota Experiment appeared to be a means of contributing substantially to the general welfare of the nation and the world.

(2) For some the experiment was an opportunity to participate in and be associated with activities which more closely paralleled the discomforts, risks, and sufferings of the men in the armed forces and the civilians of war-ravaged areas.

(3) A powerful motive was the expected contribution to the body of scientific knowledge on the effects of semi-starvation and the best methods of rehabilitation, especially in so far as the knowledge would expedite famine relief. Many of the subjects had previously volunteered to serve abroad in relief operations but had been unable to do so while under Selective Service jurisdiction as conscientious objectors.

(4) There were many personal reasons, such as "Can I take it?" or "I can take it and this will prove it to others." Some considered the experiment a means of allaying guilt feelings in regard to their pacifist position; being a subject would bring them nearer to recognition and/or approval by exhibiting such positive and acceptable behavioral traits as self-sacrifice and fortitude.

(5) The educational opportunities provided were also important. A relief and rehabilitation training course together with a program of guest speakers and lecturers was offered. The subjects were also permitted to enroll in courses at the University of Minnesota.

(6) The cultural and social advantages of a large urban community appealed

to men who had been relatively isolated in camps in rural areas for a considerable period of time.

(7) Curiosity concerning what it feels like to starve and the self-disciplinary
value of the experimental regimen were important motives for some.

Anthropometric Characterization of the Subjects

The characterization of the subject group during the control period is based
on data derived from the 32 men who completed the 6 months of semi-starvation and the first 12 weeks of rehabilitation. The 2 men who were dropped from
the experiment during the semi-starvation period and the 2 men dropped at the
end of of semi-starvation are not included. The complete anthropometric data for
the 32 men are given in the Appendix Tables.

The mean age as of January 1, 1945, for this group of adult males was 25.5 ±
3.47 (standard deviation), with a range of 20 to 33 years. The distribution according to 5-year age groups was 13 men from 20 through 24, 14 from 25 through
29, and 5 from 30 through 34.

The mean height of the subject group was 178.8 ± 5.77 cm. (70.4 ± 2.27 in.)
with a range of 167.5 to 191.9 cm. (65.9 to 75.6 in.). The body weight as of February 5–11, 1945, averaged 69.39 ± 5.85 kg. (152.7 ± 12.87 lbs.) with a range
of 62.0 to 83.6 kg. (136.4 to 183.9 lbs.). The mean weight/height index was 38.8
± 2.8 with a range of 35.0 to 54.3.

TABLE 13

DEVIATIONS FROM STANDARD WEIGHT FOR A GIVEN HEIGHT in the subjects of the Minnesota Experiment at the end of the control period.

| | Deviation as Percentage of the Standard Weight | | | | | |
	−17.50 to −12.51	−12.50 to −7.51	−7.50 to −2.51	−2.50 to +2.50	+2.51 to +7.50	+7.50 to +12.50
Number of subjects...	2	8	8	8	4	2
Per cent of subjects...	6.25	25.00	25.00	25.00	12.50	6.25

The deviations in the weight of the subjects from normal weight for a given
height based on the Selective Service registrants data are presented in Table 13.
The data indicate that during the control period the Minnesota Experiment subjects were, on the average, a little below normal weight for their height.

Table 14 provides data for comparing the Minnesota Experiment subjects with
the general population of young adult males represented by the Selective Service
registrants. Both the mean ages and the age range coincide closely in the two
groups, as would be expected because the Minnesota Experiment subjects were
a sample of the larger Selective Service registration group. The Minnesota Experiment subjects were 4.7 cm. taller but only slightly heavier than the average
registrant. The Pignet and weight/height indexes show that the subjects were,
on the average, slightly taller and relatively thinner than would be expected in
a random sample of adult males of similar age in the United States.

TABLE 14

COMPARISON OF THE MINNESOTA EXPERIMENT SUBJECTS AND 99,172
WHITE SELECTIVE SERVICE REGISTRANTS (U.S. Selective
Service System, 1943).

Characteristic	Minnesota Experiment Subjects		White Selective Service Registrants	
	M	SD	M	SD
Age (years)...............	25.5	3.5	26.0	
Weight (kg.)..............	69.4	5.8	69.0	10.4
Height (cm.)..............178.6	5.9	173.9	7.1	
Chest girth (cm.)*........	89.3	3.3	90.8	6.3
Umbilical girth (cm.).......	78.1	4.4	78.1	7.7
Weight/height index†.......	38.8	2.8	39.7	
Pignet's index†............	20.0	8.4	14.1	

*In the Minnesota Experiment the chest circumference was measured at the end of normal expiration. For the Selective Service registrants the values are the average of the chest circumference at the end of normal inspiration and expiration.

†Pignet's index = height in cm. — (chest girth in cm. + weight in kg.); the lower value indicates a more stout body build. The higher the weight/height index the more stout is the body build.

The subject group encompassed a rather wide range of physical work performance capacity and athletic habitus. In general the subjects were in good physical condition as exemplified by their performance on the Harvard Fitness Test (Johnson et al., 1942). The average time of the run was 241.6 ± 62.9 seconds with a range of 134 to 300 seconds. Their mean "fitness" score was 64.1 ± 17.2 and the range was 26 to 90.

Intellective Characterization of the Subjects

The general intellective capacity of the subjects may be indicated on the basis of the years of formal education completed and by the scores on the Thorndike CAVD Test (Thorndike et al., 1935) and the Army General Classification Test (form 1A) (Staff, Personnel Research Section, 1945). The mean years of formal education were 15.4 ± 1.42 with a range of 13 to 19 years. All the subjects had had at least one year of college education and 18 of the 32 men held college degrees.

The results of the Thorndike CAVD Test showed the subjects to be of high intellective capacity. The group mean score was 426 with the range from 402 to 454. The range of scores was almost identical with the scores of a graduate class in advanced statistics at Columbia University (mean 427, range 410 to 447) and surpassed the Master of Arts candidates at the Teachers' College, Columbia University (mean 410, range 388 to 445) (Institute of Educational Research, Columbia University, no date).

The mean standard score on the Army General Classification Test (form 1A) was 138.7 ± 8.5 and the range was 122 to 154. In comparison with approximately 10 million Selective Service inductees, 28 of the 32 subjects fell within

Army Group I and 4 within Army Group II. The percentile range of the subject group was from 84 to 100. The mean score of the subject group (138.7) was about 2 standard deviations above the mean for Selective Service inductees. The significance of the Army General Classification Test has been discussed by Bingham (1946).

Personality Characterization of the Subjects

The emotional and personality characteristics of the subjects were obtained by use of the Minnesota Multiphasic Personality Inventory (Hathaway and McKinley, 1943) and are summarized in Table 15.

TABLE 15

Minnesota Multiphasic Personality Inventory Scores for 32 subjects of the Minnesota Experiment during the control period. All values are the average of two administrations of the Inventory. The "normal" (mean) values are 50 ± 10. Hs = hypochondriasis, D = depression, Hy = hysteria, Pd. = psychopathic deviation, Pa = paranoia, Pt = psychasthenia, Sc = schizophrenia, and Ma = hypomania.

	Hs	D	Hy	Pd	Pa	Pt	Sc	Ma
M	45.7	54.3	59.0	52.2	53.5	45.7	47.5	51.0
SD	3.42	6.49	6.13	7.55	5.01	7.39	7.55	7.08
Range ...	40–52	43–67	42–70	31–68	42–64	37–76	39–79	40–64

The group means and standard deviations are well within the normal limit for all the items. With the exception of a single subject in the schizophrenia and psychoasthenia categories, the individual scores were not outside the normal range. Even this single case would hardly be considered abnormal on the basis of the two moderately high scores. In general the subjects were emotionally well balanced and mature and had made as good an adjustment to life as people usually do.

Subject Duties and Activities

According to his special training, abilities, or desires, each subject was assigned to a specific project job that required about 15 hours a week. Included in the project work were such tasks as general maintenance of the Laboratory and living quarters, laundry, laboratory assistance, shop duties, and clerical and statistical work. In addition to the project work, each subject was required to walk 22 miles out-of-doors per week and for a half-hour each week on a motor-driven treadmill at 3.5 miles per hour on a 10 per cent grade. Walking to and from the dining hall added an extra 2 to 3 miles per day. The physical activity program was rigidly maintained until the later part of the semi-starvation period when increasing weakness and apathy made strict adherence difficult. The outdoor and treadmill walks, however, were completed each week by every subject throughout the experiment; it was the project work that suffered through neglect in most cases during the last month of semi-starvation.

The extensive educational program that was set up within the unit occupied about 25 hours per week of each subject's time. Instruction was offered in foreign languages, sociology, political science, and relief and rehabilitation methods. Special lectures and discussions were given by visiting experts, guest lecturers, and Staff members. Many of the subjects enrolled in regular academic courses at the University. The educational program was well maintained until the last month of semi-starvation when it too largely succumbed to the general apathy. A few of the subjects sustained their educational pursuits throughout the experiment. One subject completed the course requirements for the master's degree.

The project work, educational program, and testing schedule were designed to require 48 hours per week of the subjects' time; being a subject was a full-time job. During the evenings and on Sundays there were no scheduled activities. Many of the subjects participated actively in church and young people's groups and in settlement house activities. The cultural and recreational facilities of the University and the surrounding urban and rural regions were used extensively. Provisions were made for recreation and the pursuit of personal hobbies and interests within the housing quarters.

During the control period and the first few weeks of semi-starvation no special restraints were placed upon the subjects other than adherence to the activity schedule and taking no food except that provided in the dining hall. But soon after the food intake was cut to the low semi-starvation levels it became apparent that special precautions would be desirable. The physical and mental stress of continuous hunger when food was available at any restaurant or store was sufficient to strain the will power of the most resolute. In order to lessen to some extent the personal responsibility of not taking extra food, a "buddy system" was put into effect on April 16, 1945 (the beginning of the tenth week of semi-starvation). From that date through September 13, 1945 (6½ weeks of rehabilitation), no subject was allowed outside the Laboratory or living quarters unless accompanied by one or more of the subjects or another responsible person. The buddy system seriously curtailed personal freedom and free-time activities and was often a source of considerable resentment and frustration. Everyone, however, felt it was a valuable aid in resisting the overwhelming urge to get food and accepted it as a necessary inconvenience.

Chronology of the Minnesota Experiment

The sequence of the major events in the Minnesota Experiment is summarized in Table 16.

The 32 subjects were in continuous residence at the Laboratory of Physiological Hygiene from November 19, 1944, through October 20, 1945, including the 12-week control period, 24 weeks of semi-starvation, and 12 weeks of restricted rehabilitation. Twelve of the subjects remained for an additional 8 weeks of unrestricted rehabilitation, and follow-up examinations were made on more than half the subjects after about 8 and 12 months of post-starvation recovery.

As shown in Table 16, the major testing periods were during the last two weeks of control, at 12 and 24 weeks of semi-starvation, and at 6, 12, 20, 33, and 55 to 58 weeks of rehabilitation. Some tests, however, such as basal metabolism, work pulse rates, Harvard Fitness, and gross body reaction time, were repeated at more frequent intervals.

TABLE 16

The Chronology of the Major Events in the
Minnesota Experiment.

Control Period—12 Weeks

November 19, 1944, through February 11, 1945
Control testing period (C), January 29, 1945, through February 10, 1945

Semi-Starvation Period—24 Weeks

February 12, 1945, through July 28, 1945
Mid-starvation testing period (S12), April 23, 1945, through May 5, 1945
End-starvation testing period (S24), July 16, 1945, through July 28, 1945

Restricted Rehabilitation Period—12 Weeks

July 29, 1945, through October 20, 1945
Mid-rehabilitation testing period (R6), September 3, 1945, through September 8, 1945
End-rehabilitation testing period (R12), October 8, 1945, through October 20, 1945

Unrestricted Rehabilitation Period—8 Weeks

October 21, 1945, through December 20, 1945
Testing period (R20), December 10, 1945, through December 20, 1945

Follow-Up Testing Periods

March 18, 1946, through March 23, 1946 (R33)
August 26, 1946, through September 14, 1946 (R55–R58)

Special Events

April 16, 1945, buddy system started
May 20, 1945, group picnic
May 26, 1945, relief meal
September 9, 1945, relief meal
September 13, 1945, buddy system ended

Clinical Security and Examinations

Thorough medical care and supervision of the subjects was maintained throughout the experiment. All illnesses and complaints, regardless of how minor or trivial they appeared to be, were reported to the Laboratory Staff internist who, after examining the individual, decided whether he should be sent to the Students' Health Service of the University of Minnesota for further observation, treatment, or hospitalization. All the members of the Laboratory Staff had daily contact with the subjects and were constantly on the alert to observe any signs of developing illnesses. On the few occasions when hospitalization was required, the subject in each case continued to receive the diet being served to the group at the time. Fortunately no disorders developed which would have seriously interfered with the experimental program.

Complete daily records were kept on the health status of each subject, including type, severity, and duration of all illnesses. After the appearance of the first signs of developing edema, all the subjects were frequently examined to determine the presence and degree of the edema. Special clinical examinations, with particular reference to the skin, hair, mouth, and eyes, were made at the end of the semi-starvation period by Dr. V. P. Sydenstricker, Dr. Russell M.

Wilder, Colonel John B. Youmans, Major Marvin Corlette, and the Staff of the Laboratory. These examinations were repeated by Dr. Wilder and the Laboratory Staff at 6 weeks of rehabilitation (R6). Skin biopsies were taken for histological study on a few of the subjects with the more pronounced skin changes at S24 and R10. The details of the special skin examinations are discussed in Chapter 13.

Teleroentgenograms (72-inch) of the chest were taken routinely on all the subjects at 3-month intervals (C1, C12, S12, S24, R12, and R20). The chest plates were carefully examined by Dr. Leo Rigler, Department of Radiology, University of Minnesota Hospitals, for any signs of developing or active pulmonary tuberculosis. Suggestive signs of early pulmonary tuberculosis appeared in one subject at R12 after the testing period had been completed. After expert consultation this subject was sent to his home state for treatment, where he achieved a rapid recovery.

Complete dental care, including full mouth X-rays and prophylactic and corrective procedures, was furnished to all the subjects at the Students' Health Service, University of Minnesota. Extensive dental care was not required by any of the subjects. The cost of the dental work was financed by a special dental fund set aside for that purpose by the Laboratory.

It was apparent during the second month of semi-starvation that emotional and personality changes were developing in the subjects which would yield important information and might, if they progressed far enough, require expert psychiatric treatment. Consequently, a series of thorough psychiatric interviews were given most of the subjects during the last 4 months of semi-starvation and throughout the first 20 weeks of rehabilitation. The psychiatric aspects of the Minnesota Experiment are discussed in full detail in Chapter 38 (see also Schiele and Brožek, 1948, and Franklin et al., 1948).

Tests and Measurements

All the physiological tests and most of the psychological tests were made in a suite of air-conditioned rooms maintained at 78° ± 2° F. and 50 per cent relative humidity. All tests were made in a standard relation to the last meal. Physical activity preceding the tests was controlled and every effort was made to ensure a constant emotional and mental set. The subject always maintained the same order in the testing schedules so that a specific test was made at the same hour of the day during the several testing periods.

The program of tests and measurements during the control period provided for sufficient repetitions to ensure familiarity with the test procedures, for training to plateaus on those functions in which training was essential, and for complete characterization of the subjects. Most of the tests and measurements were repeated on two occasions during the last 2 weeks of the control period to furnish the base-line data for all subsequent observations, at 12 and 24 weeks of semi-starvation, and at 12 weeks of rehabilitation. On most of the other testing periods single determinations were made.

The tests and measurements used in the Minnesota Experiment are listed in the Appendix, together with a detailed description of the experimental procedures and the analytical techniques. The entire group of tests was not used throughout the experiment. Some tests were dropped at the end of semi-starvation when it became apparent that they were not sufficiently sensitive to the stress. Other tests were added when the condition of the subjects promised to yield valuable data.

Rehabilitation Grouping of the Subjects

The 32 subjects were divided at the beginning of the 12-week rehabilitation period into 3 sets of sub-groups. The caloric, vitamin, and protein content of the rehabilitation diet was different for each sub-group. The design of the grouping and the number of subjects in each group are given in Table 17.

An attempt was made to match the subjects in each sub-group in such a way that the central tendency (means) for the caloric, vitamin, and protein groups would be identical. In view of the large number of characteristics and functions measured, it was impossible to obtain identical mean values for the groups on all the measurements made. The group matching was done mainly on the basis of a few of the functions that showed the greatest change during semi-starvation.

The statistical tests carried out on a representative set of functions indicated that at the beginning of the rehabilitation regimen the differences between the means of the 4 caloric groups in no case reached the level of statistical significance. The differences between the protein sub-groups were significant on the 5 per cent level for three of the tests and those between the vitamin sub-groups were similarly significant on one test. The results of the F-test for bias between the 4 caloric groups, the protein sub-groups, and the vitamin sub-groups are given in Table 18. The F-test values marked with an asterisk [*] denote a between-group difference at the 5 per cent level of significance.

Minnesota Experiment Diets — Control

Complete control over the diet kitchen was maintained from the beginning of the control period through the end of restricted rehabilitation (R12). Part of the facilities of one of the University of Minnesota food service centers was exclusively used for the Minnesota Experiment. The food was prepared and served by a full-time cook and two assistants under the direct and constant supervision of a trained dietitian. Carefully weighed portions of the food items were served to each subject, the amount being adjusted to the individual. At frequent intervals a complete day's serving of all food items, including fluids, was collected, weighed, ground, and thoroughly mixed, and samples were taken for chemical analysis. The samples were stored at −20° C. until analyzed. Further details on the collection and analysis of the food samples are included in Chapter 20.

The basic diet for the control period consisted of menus devised to be calorically adequate and normal in regard to variety of food items, palatability, and specific nutrients as eaten under good economic circumstances in the United States. The adequacy and character of the control diets may be gauged from the menus and the nutritional analyses given in the Appendix. The basic control diet provided 3492 Cal. per day with 112 gm. of protein, 124 gm. of fat, and 482 gm. of carbohydrates. The minerals and vitamins were considered entirely adequate.

Quantitative adjustments of the basic control diet were made on an individual

TABLE 17

Design for the Grouping of the 32 Subjects for the First 12 Weeks of Rehabilitation. Z = basal diet ("zero" supplement), L = "basal diet + 400 Cal." ("low" supplement), G = "basal diet + 800 Cal." ("good" supplement), T = "basal diet + 1200 Cal." ("top" supplement). U = protein-unsupplemented ("unsupplemented"), Y = protein-supplemented ("yes"), P = vitamin-unsupplemented ("placebo"), H = vitamin-supplemented ("Hexavitamin tablet"). The number of subjects in each sub-group is set in parentheses. (Minnesota Experiment.)

Calories:	Basal = Z(8)				+400 = L(8)			
Protein:	U(4)		Y(4)		U(4)		Y(4)	
Vitamin:	P(2)	H(2)	P(2)	H(2)	P(2)	H(2)	P(2)	H(2)

Calories:	+800 = G(8)				+1200 = T(8)			
Protein:	U(4)		Y(4)		U(4)		Y(4)	
Vitamin:	P(2)	H(2)	P(2)	H(2)	P(2)	H(2)	P(2)	H(2)

TABLE 18

F-Test for Significance of the Differences between the 4 Caloric Groups, the 2 Protein Sub-Groups, and the 2 Vitamin Sub-Groups at the beginning of the rehabilitation period. The F values were obtained as ratios of the variance between groups to the variance within groups. For the caloric groups the degrees of freedom were $(4 - 1)$ and $4 \times (8 - 1)$; $F_{0.05} = 2.95$, $F_{0.01} = 4.57$. For the protein and vitamin sub-groups the degrees of freedom were $(2 - 1)$ and $2 \times (16 - 1)$; $F_{0.05} = 4.17$, $F_{0.01} = 7.56$. (Minnesota Experiment.)

Function Tested	Caloric Groups	Protein Groups	Vitamin Groups
Harvard Fitness Test			
Time	1.41	0.33	0.00
Score	1.23	0.94	0.01
Kg. m. work	0.81	0.61	0.05
Hemoglobin (gm.)	0.06	0.40	0.03
Basal pulse (beats/min.)	0.56	2.29	1.31
Basal O₂ consumption (cc./min.)	0.17	0.00	0.21
Basal blood pressure (mm. Hg.)	0.55	0.10	5.75[*]
Basal pulse pressure (mm. Hg.)	0.40	5.34[*]	1.44
Aerobic work O₂ (cc./min.)	0.97	0.39	0.99
Aerobic work pulse (beats/min.)	1.68	2.54	0.52
Body fat (kg.)	0.65	2.33	4.01
Hand dynamometer (kg.)	0.46	0.90	0.11
Reaction time (.01 sec.)	0.88	6.05[*]	0.47
Pattern tracing			
Number of contacts	0.79	1.73	1.12
Time of contacts	0.60	6.17[*]	0.46
Ball-pipe test	0.35	3.68	1.49
M.M.P.I.			
Hypochondriasis	1.80	2.22	0.00
Hysteria	1.32	0.50	0.43
Depression	0.41	1.92	0.11

[*] In this and the subsequent tables, one asterisk in brackets [*] attached to an F value indicates statistical significance at the 5 per cent level.

basis according to body size and relative obesity. The objective was to adjust the diet for each individual to maintain caloric balance in those who did not depart widely from "ideal" weight and to bring the thin and obese members of the group closer to their "ideal" weights. The average weight change from the beginning to the end of the control period was −0.80 kg. (1.76 lbs.), with a range of −5.68 kg. (12.5 lbs.) to +3.8 kg. (8.4 lbs.). Substantial adjustments toward the "ideal" weight occurred in some of the extreme cases but on the average the group ended the control period slightly below their "ideal" weight.

Minnesota Experiment Diets — Semi-Starvation

During the semi-starvation period only two meals a day were served, at 8:30 A.M. and 5:00 P.M. The semi-starvation diet consisted of 3 basic menus repeated in rotation. These menus and their nutritional content (both calculated and analyzed) are given in the Appendix. The major food items served were whole-wheat bread, potatoes, cereals, and considerable amounts of turnips and cabbage. Only token amounts of meats and dairy products were provided. The diet was designed to represent as nearly as possible the type of foods used in European famine areas. Some famine diets, both past and recent, are presented in the Appendix Tables for comparison with the conditions of the Minnesota Experiment, in which the average daily intake was 1570 Cal. and included about 50 gm. of protein and 30 gm. of fat.

It was planned at the outset to attempt to produce a roughly equivalent starvation stress in each of the subjects. In order to do so it was necessary to adjust the rate and extent of the weight loss according to the metabolic demands and the initial nutritional status of the individual. The metabolic need of each subject was determined empirically from the caloric intake required to maintain a relatively constant body weight during the last 2 weeks of the control period. A body weight loss of 20 per cent in a thin person, for example, may produce as much general deterioration as would a 30 per cent weight loss in an obese individual. Though none of the subjects was grossly obese or strikingly thin at the beginning of semi-starvation, there were considerable individual differences, and in order to approach a relative constancy of starvation stress, the individual body weight loss goals were set at from 19 to 28 per cent, with a group average of 24 per cent. The individual rate and extent of weight loss was accomplished by adjusting the caloric value of the semi-starvation diet furnished to each individual.

A prediction weight loss curve, based on the concept that the rate of weight loss would progressively decrease and would reach a relative plateau at S24, was calculated for each subject, taking into account age, weight, body type, and relative obesity. Mathematically, the general curve required for weight versus time is represented by a parabola with vertical axis and zero slope at S24. If Wx = the weight at t time, Wo = the initial weight, Wf = the final predicted body weight, P = the desired total per cent of body weight loss, and t = time in weeks, then the equation can be written:

$$Wx = Wf + K(24 - t)^2$$

For each case the constant K is obtained: $K = Wo/100 \times P/24^2$. The prediction weight loss curve for each subject was calculated according to the above formula; the prediction weight loss curve for 25 per cent weight loss and the actual mean weight loss for the 32 subjects during the 24 weeks of semi-starvation are shown in Figure 9.

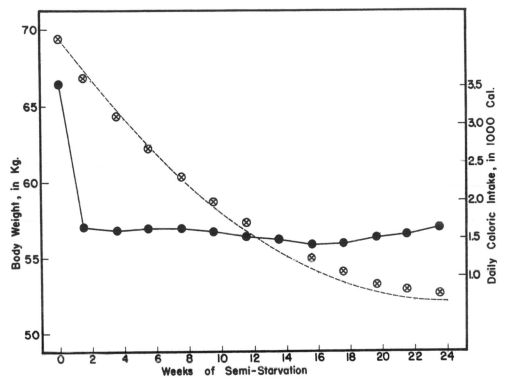

FIGURE 9. THE PREDICTED 25 PER CENT WEIGHT LOSS CURVE (BROKEN LINE), THE ACTUAL MEAN BODY WEIGHTS (CROSSED CIRCLES), AND THE MEAN DAILY CALORIC INTAKES (SOLID CIRCLES) for the 32 subjects at 2-week intervals during the 24 weeks of semi-starvation (Minnesota Experiment).

All the subjects received the same amounts of all the food items except bread and potatoes; the amounts served of these two foods were varied to secure the individual dietary adjustments. Besides the individual differences in the amount of the basic semi-starvation diet served, weekly adjustments were made according to the weight records of the previous week as compared with the individual weight prediction curve. If the weight loss for an individual in any week fell short of his predicted weight loss, the servings of bread and/or potatoes allowed him during the following week were decreased in an amount calculated to bring the weight back to the prediction curve; conversely, if the weight loss was too great, extra bread and potatoes were added. In general this type of empirical adjustment proved entirely satisfactory; weights for most of the subjects closely followed the prediction weight loss curves with only minor adjustments in food intake. The weight loss problem was complicated, however, by the occurrence of clinical edema and the relative increase in interstitial fluid volume.

In order to bolster morale and break the monotony of the semi-starvation and restricted rehabilitation diets, "relief" meals were served on May 26 and September 9, 1945. The food items used in the relief meals were mainly those chosen by the subjects as being most desirable. The size of the servings was adjusted so as not to furnish a great excess of calories for those days. The composition of the relief meals and the menus are given in the Appendix.

Minnesota Experiment Diets — Rehabilitation

As indicated in Table 17, during the first 12 weeks of rehabilitation the 32 subjects were divided into 4 groups of 8 men each, with the caloric value of the basic rehabilitation diets supplied to the groups differing by approximately equivalent steps. It should be noted that the plan was to give to each subject in rehabilitation an amount of food adjusted to his estimated minimum balance requirement at the end of starvation, so that the men in group Z would have a daily supplement for rehabilitation amounting in the first 6 weeks to about 400 Cal., with corresponding supplements for the men in groups L, G, and T of about 800, 1200, and 1600 Cal. Actually, subsequent analyses of the diets served indicated some discrepancies, generally resulting in smaller supplements than originally planned.

Each caloric group was subdivided into 2 groups of 4 men each; one group received extra protein and the other group received just the protein present in the basic diet. The protein sub-groups were divided into 2 groups of 2 men each; one of the sub-groups received daily one Hexavitamin tablet and the other sub-group received placebos. By means of this factorial design it was possible to determine what effect, if any, the caloric level of refeeding had on the rate of rehabilitation, whether extra proteins were beneficial, and/or whether the addition of extra vitamins to a refeeding diet was desirable.

The composition of the basic rehabilitation menus is given in detail in the Appendix Tables. The rehabilitation diets were in general the same ones used in the semi-starvation period except that more of each food item was given. The extra protein given to the protein-supplemented subjects was mainly casein and a purified soybean protein; these proteins were incorporated in the bread and were effectively disguised. For the vitamin-supplemented group the Hexavitamin tablets supplied each day 2500 I.U. of vitamin A, 1.0 mg. of thiamine, 1.5 mg. of riboflavin, 10.0 mg. of niacin amide, 37.5 mg. of ascorbic acid, and 200 I.U. of vitamin D.

The average actual caloric intake of the various groups during rehabilitation is presented in Table 19. During the first 6 weeks of the rehabilitation period the average caloric intake was 2449 Cal., with a range of 1931 Cal. in the lowest caloric group (Z) to 2944 Cal. in the highest caloric group (T). The caloric groups differed by an average of 338 Cal., and there was only a small caloric difference (82 Cal.) between the protein-supplemented and protein-unsupplemented sub-groups. The protein, fat, and carbohydrate content of the diets varied with the caloric value of the diet. Within each caloric group the protein-supplemented subjects received, on the average, 12.1 gm. of protein more than did the protein-unsupplemented subjects (see Table 20).

At the end of the sixth week of rehabilitation it became apparent that only a very small degree of rehabilitation was being achieved in the subjects comprising the 2 lower caloric groups. The subjects were becoming discouraged and problems of morale were arising. Consequently, with the start of the seventh week of rehabilitation, the intake of all 4 caloric groups was increased by an average of about 800 Cal. At the same time the protein differential between the protein-supplemented and protein-unsupplemented sub-groups was increased to 24.5 gm. The caloric intake of the protein-supplemented groups then averaged 193 Cal. more than the protein-unsupplemented groups.

The caloric intake was again increased an average of 259 Cal. for the last 2 weeks of restricted rehabilitation (R11 and R12) and the protein differential was increased to an average of 38.4 gm.

Strict dietary control ended with the completion of the testing program at R12. A small group of the subjects (12) consented to remain at the Laboratory for an

TABLE 19

AVERAGE ACTUAL DAILY CALORIC INTAKE DURING REHABILITATION for the 4 caloric groups, Z, L, G, and T, and the protein and vitamin supplemented (Y and H) and unsupplemented (U and P) sub-groups (Minnesota Experiment).

Special Supplements		Caloric Groups				
Prot.	Vit.	Z	L	G	T	Average
R1–R6						
Basal (U)	Basal (P)	1877	2279	2596	2828	2395
Basal (U)	Extra (H)	1927	2179	2646	2928	2420
Extra (Y)	Basal (P)	1984	2286	2804	3085	2540
Extra (Y)	Extra (H)	1934	2236	2654	2935	2440
Average		1931	2245	2675	2944	2449
R7–R10						
Basal (U)	Basal (P)	2654	3056	3374	3606	3173
Basal (U)	Extra (H)	2704	2956	3424	3706	3198
Extra (Y)	Basal (P)	2823	3125	3643	3925	3379
Extra (Y)	Extra (H)	2773	3075	3493	3775	3279
Average		2739	3053	3484	3753	3257
R11–R12						
Basal (U)	Basal (P)	2942	3344	3662	3894	3461
Basal (U)	Extra (H)	2992	3244	3712	3994	3486
Extra (Y)	Basal (P)	3056	3358	3876	4158	3612
Extra (Y)	Extra (H)	3006	3308	3726	4008	3512
Average		2999	3314	3744	4014	3518
Weighted m Values, R1–R12						
Basal (U)	Basal (P)	2314	2716	3033	3265	2832
Basal (U)	Extra (H)	2364	2616	3083	3365	2857
Extra (Y)	Basal (P)	2442	2744	3262	3544	2998
Extra (Y)	Extra (H)	2392	2694	3112	3394	2898
Average		2378	2692	3123	3392	2896

TABLE 20

Protein Content of the Basic Diet for the protein-supplemented (Extra) and unsupplemented (Basal) sub-groups within the 4 caloric groups. The values are given as gm. of protein per day. (Minnesota Experiment.)

Protein Group	Caloric Groups				Average
	Z	L	G	T	
R1–R6					
Extra (Y).......	73.1	81.2	89.5	98.8	85.65
Basal (U)	61.0	69.1	77.4	86.7	73.55
R7–R10					
Extra (Y).......	104.3	112.4	120.7	130.0	116.85
Basal (U)	79.8	87.9	96.2	105.5	92.35
R11–R12					
Extra (Y).......	126.3	134.4	142.7	152.0	138.85
Basal (U)	87.9	96.0	104.3	113.6	100.45
Weighted m Values, R1–R12					
Extra (Y).......	92.4	100.5	108.8	118.1	104.9
Basal (U)	71.7	79.8	88.1	97.4	84.3

additional 8 weeks (from R12 through R20). During these 8 weeks the diet was unrestricted but the amount and type of all food eaten was carefully recorded. The data on the food eaten during this period are given in the Appendix Tables. Extremely high food intakes — as high as 7000–10,000 Cal. per day — were achieved by most of the subjects during the first 2 weeks of ad lib eating, but the intakes tended to level off at more reasonable values of about 3200 to 4500 Cal. per day.

Morphology

"Morphology is not only a study of material things and the form of material things, but has its dynamic aspect, under which we deal with the interpretation, in terms of force, of the operations of Energy."

D'ARCY WENTWORTH THOMPSON, in On Growth and Form (1943), Chapter 1.

CHAPTER 5

Significance and Limitations of Morphological Data

DETAILED information on the morphological changes induced in man by semi-starvation is necessary, along with the physiological, biochemical, and psychological data, for a complete understanding of the alterations resulting from dietary restrictions. The functions and responses of the intact organism are the composite results of the activities of its individual cells, tissues, and organ systems; deviations in function from the normal are a reflection of altered activity and integration of the component parts of the organism. Morphological data, then, should furnish important, if incomplete, clues for the description and explanation of the changes in behavior and activity that occur in man during periods of semi-starvation and famine.

Morphological data can be obtained mainly from anthropometric measurements (body weight, body density, and body dimensions), the analysis of body fluids, and the histological examination of biopsy or post-mortem tissues and organs. As amply illustrated by much of the morphological material presented in the literature, the problems of obtaining reliable data on changes that occur in man during semi-starvation are often complicated and at times nearly insurmountable. The gathering and interpretation of such seductively simple measurements as body weights demand technical facilities that are generally not available in field surveys in famine areas. Consequently, the data obtained under such conditions are often of limited scientific value. Except in the grossly emaciated person, the body weight is meaningful only in terms of relative changes that have occurred over a known time interval; it is necessary either to know the body weight before the period of starvation or to possess reliable "normal standard" data for the population in question. In few instances do reports in the literature meet either of these requirements.

To interpret the anthropometric changes properly, more information than gross changes in body weight and body dimensions is necessary. The first important question is, In what tissues or organs has the change in weight occurred? The loss of 20 pounds of adipose tissue in an obese individual would have little significance but the loss of an equal amount of weight of active tissue, such as muscular or glandular tissue, might well result in drastic alterations in essential body functions. Fully as important as the relative changes in body fat and active tissue, which in the living animal can be determined only by body density measurements, are the changes in body hydration that frequently occur in semi-starvation. Loss of fat and active tissue may be masked or exaggerated by shifts in plasma and extracellular fluid volumes. True body weight determinations must, therefore, take cognizance of the body water.

The ease with which blood samples can be secured has made the cellular elements of the blood a favorite object of study in semi-starvation. Although there is rarely any evidence for questioning the accuracy of the techniques, in many cases the interpretation of the data reported in the literature is not acceptable as a true representation of the morphological alterations of the blood produced by a low caloric intake. The hematological changes observed in famine victims are frequently complicated by shifts in body hydration and by disease processes which in themselves may vastly change the composition of the blood irrespective of the changes resulting directly from semi-starvation. In most reports no information is given on the state of body hydration, nor are the duration, severity, and type of concomitant disease processes adequately described.

The semi-starvation literature is practically devoid of any histological and cytological data based on biopsy material. The tissues and organs of the body that can be investigated by this method are, of course, limited, but the data obtained would be extremely valuable in explaining the changes that occur within those organs during semi-starvation. The decrease in muscular strength and endurance and the suspected hypofunction of some of the endocrine glands, for example, might possibly become evident and understandable from histological alterations in the tissues and cells.

Little of the morphological data from autopsy material can be accepted without reservation. In other words, the tissues represent the combined results of starvation and pathological processes, and these are further confused by postmortem changes. In the first place, death, though indirectly caused by undernutrition, commonly results from a decreased capacity to withstand disease or injury; the immediate cause of death is pneumonia, malaria, typhus, traumatic shock, or something of the kind. Every histologist recognizes that in order to obtain acceptable material the specimen must be fixed with the least possible delay after circulation to the organ is stopped and that the specimen must be handled with the utmost care. In most autopsy material neither of these requirements is met; hours and even days elapse between the death of the patient and the time the autopsy is performed, little care is taken in handling the specimens, and even the size of the specimen taken is seldom adjusted to give the best histological fixation or to be representative of the organ from which it is taken. For gross pathological changes in the tissues the usual autopsy methods are no doubt satisfactory, but for detailed cytological studies more exacting methods are demanded. It is indeed unfortunate that most of even the small amount of data available on the morphological changes in tissues occurring during semi-starvation are at best only indicative and certainly cannot be taken as absolute evidence.

Even more frustrating than the lack of good morphological data is the problem of interpretation of morphological changes in terms of altered physiological function. Grossly degenerated cells would, no doubt, be incapable of performing their normal functions, but only slight alterations and variations in cytological detail would probably have no demonstrable effect. It is unlikely, for example, that the size or shape of the red blood cell influences the oxygen-carrying capacity of the hemoglobin in the cell, but the size and shape might easily change the diffusion rate of oxygen and the physiological life and efficiency of the cell.

On the other hand, it is clear that many important functional alterations cannot be related to histological characteristics currently recognized and measured. The whole field of functional morphology, both normal and pathologic, is vague and imperfectly understood, but it does offer a wide opportunity for basic fundamental research which should, in the future, make it possible to correlate cellular morphology with cellular physiology.

By and large, most functional alterations, including those in starvation, are quantitative and not qualitative; the machinery works in the same fashion but at a different rate or efficiency. The necessity of correlation between quantitative morphology and quantitative physiology is obvious. The quantitative vagueness of histology is, however, notorious. In gross morphology quantitative measures are indeed regularly employed, but whether the most fundamental features are measured is highly questionable.

Strictures on current and past practices of morphologists are not useful for the present purposes except to indicate the limitations of the material presented in succeeding chapters and, possibly, to stimulate more and better investigations in the future. It must be understood that this entire book is devoted to the presentation and analysis of the known and reported facts of both morphology and function. There will appear a host of associations; in starvation there are characteristic changes in almost all items explored, but the separation of cause and effect from mere parallelism is often impossible.

CHAPTER 6

Body Weight

BODY weight is one of the basic anthropometric characteristics and might well be discussed together with the dimensions of the body. However, the amount of information obtained in the Minnesota Experiment and provided by other sources is relatively large, and it appeared desirable to devote a separate chapter to this topic.

It deserves emphasis that before the body weight can become a physiologically meaningful variable we have to know, at least approximately, the value of its components: active tissue, fat, bone, and extracellular fluid. The changes in these components during starvation are discussed in Chapters 12, 14, 15, and 43. The data available in the literature on starvation and famine present in no case the information which would be necessary for a truly physiological analysis. Accordingly, the present chapter will be limited largely to a detailed consideration of the gross body weight, and its general tenor will be descriptive rather than interpretative.

Evaluation of the Normality of Body Weight and of the Degree of Weight Loss

The problem of evaluating quantitatively the degree of weight loss is related to the much broader question of the "normality" of weight. This latter question, which is both of practical significance and of theoretical interest, is in an unsatisfactory state as to actual practice, clarification of the basic concepts, and availability of the empirical data necessary for establishing the norms, especially for adults. One not infrequently encounters the use of the same height/weight/age tables for both men and women. The height/weight tables resulting from the *Medico-Actuarial Mortality Investigations,* published in 1912 by the Association of Life Insurance Medical Directors and the Actuarial Society of America, are reprinted from textbook to textbook, frequently without reference to the original source and without consistent specification of the correction, if any, for the weight of clothing. The distributions of weight in samples of the American population tend to be skewed, the higher values predominating over the lower values; under such conditions the mode rather than the mean would appear to be the more correct "standard."

Sporadic efforts have been made (see e.g. Metropolitan Life Insurance Co., 1942, 1943) to consider the body frame in determining the weight, "as ordinarily dressed," for a given height, "with shoes"; unfortunately, no quantitative criteria for classifying the body frames as small, medium, or large were given.

Even when such criteria are made available, the estimation of the nutritional status of an individual on the basis of external dimensions and weight will remain a very crude procedure. The fundamental requirement for real progress in the question of body weight is to go beyond the gross weight and the gross anthropometric data and undertake analysis of weight in terms of its components. Differentiating between the contribution of muscles and that of fat tissues to the total body weight is of particular importance. It is of profound physiological consequence whether "overweight" is due to an unusual muscular development or to excessive deposits of fat.

Standard weights are usually defined as the average weights obtained for a representative sample of individuals of a given height, age, and sex. It is an elementary but not always appreciated fact that in the general population there is a large amount of variation around the average. This fact is brought out in Table 21. In non-starved individuals 11 per cent of this population sample deviated below the group mean by 15 per cent or more of the "standard" weight.

TABLE 21

DISTRIBUTION OF WEIGHT IN 14,882 WHITE DRAFT REGISTRANTS who received physical examinations at local draft boards from November 1940 through September 1941. Height = 68 in. Average weight for this group = 151 lbs. From U.S. Selective Service System (1943, p. 38).

Weight Class (lbs.)	Upper Limit of Weight Class as Percentage of Average Weight	Frequency		Cumulative Frequency (%)
		N	%	
Under 100	66.2	2	0.01	0.01
100–9	72.8	46	0.31	0.32
110–19	79.5	300	2.02	2.34
120–29	86.1	1287	8.65	10.99
130–39	92.7	2939	19.75	30.74
140–49	99.3	3562	23.93	54.67
150–59	106.0	2832	19.03	73.70
160–69	112.6	1800	12.09	85.79
170–79	119.2	914	6.14	91.93
180–89	125.8	528	3.55	95.48
190–99	132.4	285	1.92	97.40
200–9	139.1	104	1.10	98.50
210–19	145.7	90	0.60	99.10
220–29	152.3	56	0.38	99.48
230–39	158.9	33	0.22	99.70
240–49	165.6	13	0.09	99.79
250 and over		31	0.21	100.00

Some of these men were rejected for selective service, and legitimate doubts may be raised concerning the clinical "normality" of a part of the sample, although it is probable that rejections of overweight individuals were as frequent as of underweight men. There is a possibility — but not a probability — that some of the men were suffering from caloric underfeeding and that the age heterogeneity of the sample may have accentuated the magnitude of individual differences. Consequently, it appeared worth while to include the distribution of relative weights for a sample of 646 adult men, 45 to 54 years old, who were available for study at the Laboratory of Physiological Hygiene of the University of Minnesota. The data are presented in Table 22.

Even in this sample 8.1 per cent were 15 per cent or more below their "standard" weight. It may be noted that occupationally the subjects belonged to the clerical, business, and executive groups. They were carrying on successfully their respective jobs. In no case was their food intake limited by the availability of food or by financial considerations.

TABLE 22

DISTRIBUTION OF DEVIATIONS OF THE ACTUAL BODY WEIGHT FROM THE "STANDARD" BODY WEIGHT, expressed as percentage of the "standard," in a group of 646 men, aged 45 to 54 years, studied at the Laboratory of Physiological Hygiene, University of Minnesota.

Weight Class	Percentage Deviation from Standard	Frequency		Cumulative Frequency (%)
		N	%	
A........	−15.00 and below	52	8.05	8.05
B........	−14.99 to −5.00	130	20.12	28.17
C........	−4.99 to +4.99	211	32.66	60.83
D........	+5.00 to +14.99	189	29.26	90.09
E........	+15.00 and above	64	9.91	100.00

The concept of standard weight is derived from statistical considerations. In its interpretation the mythology of the "golden middle" creeps subtly in. It is obvious that "ideal" weights cannot be derived by means of statistical manipulation of the weight data alone but must be referred to external criteria of "goodness," such as minimal morbidity and maximal longevity.

These are just some of the disturbing aspects of the problem of normal weight and deviations from normality. However, the simpler problem of how to characterize the degree of weight changes in starvation has its thorny facets as well. In evaluating the severity of *loss* of weight in an individual, the *pre-starvation weight* serves as a legitimate and more sensitive reference point; in evaluating the *degree of starvation* as indicated by the starvation weight, the *standard weight* should be used as the criterion. Unfortunately, this procedure is not as satisfactory as it might appear. In field studies the information on the patient's usual or previous weight is, as a rule, unreliable or lacking; this fact means that we should abstain from speculation about weight loss and should simply describe the relative weight as a percentage of the standard. Only when we deal with large groups of starved individuals for whom the pre-starvation (actual) weight and the standard weight would have been close to each other is it legitimate to express the mean weight decrement as a percentage of the standard. The procedure may be grossly misleading when applied to individual cases.

Loss of Body Weight and Caloric Intake

The question of caloric requirements is discussed in detail in Chapter 18. However, a few points may be stressed in the present context. The existing lack of reliable information on the relationship between reduction of caloric intake and weight loss is a result of several factors. First, there is the fact that under conditions of natural starvation only the official rations, but not the actual caloric intakes, are known. Second, for the same diet the picture of caloric insufficiency will vary widely with caloric expenditure resulting from physical activity. As this variable in an ambulant population eludes all attempts to measure it, except under highly artificial experimental conditions, the problem of the relationship of decreased caloric intake and the rate of weight loss in semi-starvation defies solution. The situation is more favorable in patients undergoing treatment for obesity. Under such conditions it is possible to predict the amount of weight loss with satisfactory accuracy.

The first point may deserve some elaboration. As a result of inaccurate information on the total dietary intake, scientific analysis of the relation between body weight and caloric intake is often clearly impossible even in those instances where data on both variables are reported and where the caloric expenditure may be presumed to remain relatively constant. The situation in Japan in World War II is a good case in point; the data are provided by the U.S. Strategic Bombing Survey (1947, Tables 55–59, inclusive). According to this report the average actual caloric intake in 44 factories in the Tokyo area steadily declined from a level of 3012 Cal. daily in 1938 to an average of 2301 Cal. daily in 1943. But over these same years the average body weights of the men in these factories declined only 0.80 kg. and the women workers *gained* an average of 0.45 kg. Measurements on the same 3661 workers over 30 years of age showed an average body weight of 53.68 kg. in 1938 and 53.70 in 1942; for 1095 workers aged 17 to 30 who were weighed both in 1938 and in 1942, the averages were 53.00 and 52.87 kg., respectively One can only conclude that the caloric intake values reported are meaningless, and one clue to the discrepancy is given in the statement that in 1944 "about 50 per cent of all food purchased was obtained by 'free purchase,' that is, through the black market" (*ibid.,* p. 72). The most lamentable feature of such reports as those on Japan is the frequent zeal of the nutritionist to prove serious and widespread malnutrition in spite of the evidence or the lack of it.

Extent of Loss of Body Weight in Animals

The total loss of body weight in hibernating animals was estimated by Morgulis (1923, p. 26) not to exceed as a rule 20 to 25 per cent of the initial weight. It is of interest that after hibernation the weight of depot fat in a marmot had decreased by 99.3 per cent of the original value.

The losses of body weight obtained under natural and experimental conditions will vary within wide limits from species to species. The maximal loss of gross body weight observed in a mammal is that of a dog who died after he lost 65 per cent of his initial body weight in the course of a 93-day fast (Preobrajensky and Baranova, 1932).

Extent of Body Weight Decrement in Natural Starvation and Famine

Porter (1889) published the results of his extensive autopsy observations made during the Madras famine of 1877. As no pre-starvation weights are known, an estimate of the standard weight must be used. Martin (1928, Vol. I, esp. p. 314) presents data which provide a basis for arriving at such an estimate, gross as it may be. The mean weight of the upper caste Hindus is given as 53.2 kg. and of the lower caste Hindus as 48.7 kg.; the average of these two figures, 51.0 kg., appears to be a reasonable estimate for the pre-starvation weight for Porter's sample of 30 men without edema. For another Asiatic people, the Javanese, the mean height was indicated as 163.3 cm. and the mean weight as 50.2 kg. (Martin, *op. cit.*, p. 321).

The weight decrement of the 30 Madras men who exhibited atrophy without noticeable edema was 28.4 per cent of the "standard" body weight. In the case of 13 men with pronounced starvation edema the percentage loss of gross weight was negligible and by no means indicated the serious disturbance of the nutritional status of these patients (see Table 23).

TABLE 23

WEIGHT OF MEN WHO DIED OF STARVATION DURING THE MADRAS FAMINE OF 1877. The case material, summarized in the form of mean values in this table, was presented by Porter (1889).

	No Edema (30 cases)	Marked Edema (13 cases)
Age (yrs.)	47.8	48.3
Height (cm.)	164.6	161.8
Body weight (kg.)	36.5	47.0
"Standard" weight (kg.)	51.0	49.5
Difference (kg.)	−14.5	−1.5
Difference as percentage of "standard" weight	−28.4	−3.0

The average weight of men registered for emergency relief during the 1877 famine in southern India was given by Digby (1878, Vol. II, p. 197) as 94.3 lbs. (42.8 kg.), that of women as 77.6 lbs. (35.2 kg.). Using the above estimates of the standard weight, the ratio of starvation to standard weight for men was 0.84. No data on standard weights are available on women. The number of individuals weighed was not recorded.

During the siege of Kut, from December 1915 to April 1916, the British troops lost on the average 12.5 lbs. per man and the Indian troops 17 lbs., which was less than 10 per cent and about 14 per cent, respectively, of their pre-siege weight (Hehir, 1922). In a platoon of Rajpoots the average weight decreased to 110 lbs. and in a platoon of Jats to 108.5 lbs., as compared with normal weights of 129 and 127 lbs., respectively; the decrements thus were 14.7 and 14.6 per cent of "normal." It was noted that the loss of weight was maximal in those officers and men who began the siege with a superabundance of fat.

During World War I the caloric value of the official daily rations allowed Germans was about 3200 Cal. in 1915; it declined precipitously during the year 1916 and remained below 1500 Cal. for the whole of 1917 (Rubner, in Bumm, 1928, Vol. I, p. 72). No over-all data on weight changes during this period are available; actually, the average figure would be of little value. The undernourishment was most severe in various public institutions, such as homes for the aged and prisons, which had to rely fully on the rations as issued. The decrease of body weight in a German prison during the years 1914–17 is indicated in Figure 10. The men were weighed every three months. The population was not stationary, and the number of prisoners or the rate of turnover is not known. The average body weight decreased from 67.5 kg. to 52.0 kg.; the decrement was 23 per cent of the 1914 value. It is unfortunate that the caloric intake and the amount of activity were not included. Rubner stated in a general way that the curve of the weight loss was parallel to the decrease in food rations for the region of Frankfurt am Main in which the prison was apparently located.

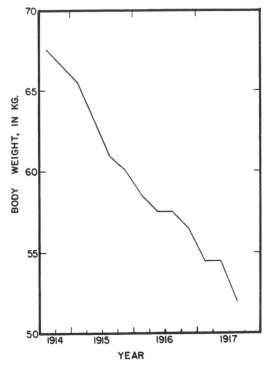

FIGURE 10. CHANGES IN THE BODY WEIGHT OF A GERMAN PRISON POPULATION during the years 1914–17. The number of prisoners was not given. (Rubner, in Bumm, 1928, Vol. I, p. 73.)

Data obtained in a German institution for epileptics at this same time and based on the same population weighed at intervals show a similar trend and magnitude of changes. In 1914 the average weight for men was 66.3 kg.; in 1917 it was only 53.3 kg., indicating a decrement of about 20 per cent of the initial weight. The distribution of the weight in 1914, 1917, and 1920 is reproduced in Table 24. Brugsch (in Bumm, 1928, Vol. I, p. 275) gave very similar figures for men in a Berlin institution. The average weight was 68 kg. in 1912–13, 65 kg. in 1915–16, and 53 kg. at the start of the year 1919.

Rubner (in Bumm, 1928, Vol. II, p. 20) presented data on weight decrements obtained in an unspecified institution and broken down according to weight classes. The material is presented in Table 25. As would be expected on the basis of approximately equal rations, the absolute weight decrement in the heavy persons is larger than in the light individuals. Assuming that the material was large enough, the mid-point of the class intervals may serve as an estimate of the average initial weight of the 4 sub-groups. Using these figures as the reference point, the percentage decrement, calculated in Table 25, was larger for the heavier women. The error involved in using the mid-points of the class intervals rather than the actual, but to us unavailable, mean values is on the conservative side. If the total distribution approached statistical "normality," the mean of the

TABLE 24

PERCENTAGE DISTRIBUTION OF BODY WEIGHT FOR INMATES OF A GERMAN
INSTITUTION FOR EPILEPTICS during World War I. (From Rubner, in
Bumm, 1928, Vol. I, p. 74.)

	Weight Class (kg.)						
	Below 40	41–50	51–60	61–70	71–80	81–90	Above 90
Men							
August 1914......		8.9	35.4	35.4	17.1	2.5	0.7
August 1917......	6.8	47.3	39.6	5.6	0.7		
January 1920.....	5.2	14.6	35.3	27.3	2.1		
Women							
August 1914......	0.4	6.0	24.0	44.0	17.0	7.0	
August 1917......	2.5	47.2	40.2	9.4	0.7		
January 1920	3.9	10.7	36.9	36.0	6.3	0.5	

TABLE 25

WEIGHT DECREMENTS IN INMATES OF A GERMAN INSTITUTION during
World War I (1914–17). The individuals were grouped according to
their initial weight. (Tabulated from Rubner, in Bumm,
1928, Vol. II, p. 20.)

	Initial Weight Class (kg.)			
	50–59	60–69	70–80	80–95
Average weight decrement (kg.)	10.7	14.5	19.5	24.0
Maximal weight decrement (kg.)	24.5	34.0	32.5	31.5
Average weight decrement as percentage of initial weight	19.5	22.3	26.0	27.4

50–59 class would be somewhat higher and the mean of the 80–95 class somewhat lower than the mid-point values. Consequently, the percentage decrement
would have been slightly smaller than 19.5 per cent and slightly larger than 27.4
per cent.

Next to institutionalized persons it was the populations of large cities and
industrial centers that felt the inadequacy of food in World War I most acutely.
The weight losses were more marked in men than in women, and more in the
older age groups than in the younger individuals. There was no pronounced emaciation and Rubner comments that the undernourishment was noticeable only
when the patients were examined without clothes. Rubner (in Bumm, 1928, Vol.
II, p. 21) recorded that his body weight in the summer of 1917 was 20.9 per
cent below its prewar level. The weight of Loewy, another German nutritionist,
decreased from 64.0 kg. in 1941 to 51.2 kg. in February 1918 (Zuntz and
Loewy, 1918). These changes appear to be unusually large. Jansen's (1917) 6
male medical students weighed on the average at the start of the war 63.5 kg.,
in 1917 58.7 kg.; the decrement was 7.6 per cent of the initial value. It may be

noted that at the same time the two female students *increased* in weight from 52.5 to 61.3 kg. and from 58.0 to 60.4 kg. No edema was involved.

Small towns in agricultural areas were affected only slightly. Brugsch (in Bumm, 1928, Vol. I, p. 276) cited Blum's data, obtained in a Bavarian locality, for healthy adult individuals. From 1914 to 1917 the weight decreased in 250 men from 74.2 to 65.2 kg. (a decrement of 12.2 per cent) and in 170 women from 57.2 to 53.3 kg. (a decrement of 6.4 per cent). The rural population largely had sufficient food through 1917.

Petényi (1918), working with undernourished members of a labor battalion, measured the height and weight in two groups of men. He obtained a calculated average weight loss of 14.7 per cent of the standard weight in one group of 13 men and a loss of 10.7 per cent in another group of 15. The true weight losses were masked, to an unknown degree, by the presence of edema in some of the men.

In examining living children who had suffered from starvation Nicolaeff (1923) found decreases of gross body weight falling most frequently in the range from 25 to 40 per cent. In the cadavers of children the weight loss was still greater, reaching close to 50 per cent of the normal body weight. In general the relative decrements of body weight increased with age.

According to Ivanovsky's observations (1923), made during the Russian famine following World War I, the weight diminished rapidly during the first months of the famine, reaching the lowest point during the second year. In the third year it is said to have shown only small variations. There was little change in weight even in those cases where somewhat increased amounts of food became available; this parallels our observations on the lowest caloric group during the first 6 weeks of rehabilitation in the Minnesota Experiment. In Ivanovsky's material the weight of obese individuals decreased more than the weight of thin persons. Younger people lost weight more rapidly than older persons (above 40 years of age); this was true of both sexes. At the same age, the diminution appeared to be more rapid in women than in men. The latter two observations are in contrast to the majority of the field reports. The extent of the loss of body weight in different racial and national groups and in different provinces during the famine in Russia following World War I is given in Table 26.

Data for infants weighed in Tsaritsin by the medical personnel of the American Relief Administration are presented in Table 27. Within this range the weight decrements tended to become somewhat more severe in the older age groups.

For the semi-starvation which afflicted the population of Leningrad during the German siege, 1941–42, only data on hospitalized patients are available. According to Tushinskij, Aleshina, and Zeits (1943), in a sample of 48 semi-starved patients examined in April and May 1942, some 6 months after the drastic reduction in food rations, the weight losses ranged from about 10 to 35 per cent of the pre-starvation weights; the majority of the cases fell in the range of a 20–25 per cent weight loss. No correction was made for the amount of edema, which is one of the factors that widens the range of the weight losses. It may be noted that the symptoms of semi-starvation in Leningrad became manifest in about 4

TABLE 26

Diminution of Weight during the Russian Famine of 1920–22. The tabular values indicate percentages of the population exhibiting a particular weight decrement. From Ivanovsky (1923).

Group and Region	Decrement as Percentage of Pre-Famine Weight							
	Males				Females			
	1–10	11–20	21–30	31–40	1–10	11–20	21–30	31–40
Great Russians (provinces)								
Tver	26	48	22	4	14	66	20	
Riazan	30	44	24	2	26	44	30	
Koursk	32	43	27		28	41	31	
Erivan	24	36	33	7	18	45	34	3
Average for Great Russians.	28	43	26	3	22	49	28	1
Ukrainians (provinces)								
Kiev	19	58	23		22	64	14	
Ekaterinoslave	16	54	30		30	46	24	
Tauride	22	66	12		34	52	14	
Average for Ukrainians....	19	59	22		29	54	17	
White Russians, province of Minsk	24	54	20	2	28	56	16	
Armenians, province of Erivan..	18	66	14	2	34	46	20	
Georgians, province of Tiflis....	32	42	24	2	30	52	18	
Tartars, Crimea...............	36	46	18					
Zyrians, province of Oust-Dvinsk	34	48	18		40	52	8	
Permiaks, province of Perm.....	37	51	12		46	45	9	
Bashkirs, province of Orenburg.	39	44	17		43	51	6	
Kalmucks, province of Astrakhan	31	41	28		38	49	13	
Kirghiz	33	47	20		41	45	14	

TABLE 27

Actual, "Standard," and Relative (Actual as Percentage of Standard) Weights of Infants in Tsaritsin, in the Volga region, during 1922 (American Relief Administration Bulletin, Ser. 2, No. 26, July 1922, p. 21).

Age (mo.)	Actual (gm.)	Standard (gm.)	Relative (%)
3	4550	5330	85.4
6	6032	6900	86.3
9	7080	8590	82.4
12	7730	9970	77.5
24	9950	12365	80.4

weeks, that is, by the end of November 1941. A large number of starvation deaths occurred as early as December 1941, but the level of weight losses at that time is not known to us.

In Holland many people are said to have lost as much as 5 kg. of body weight in the first months of World War II; since the food supply was not greatly deficient, the weight loss was ascribed to psychological factors (Banning in Boerema, 1947, p. 32). Later when food became scarcer, weight losses of 10 to 15 per cent were reported for the country as a whole.

It is unfortunate that there are no good figures on body weight in the western provinces, which suffered a severe famine from September 1944 to May 1945 resulting in a large number of deaths from starvation. Some indications about the degree of starvation may be obtained from records concerning the 38,000 individuals, mostly children, who were cared for at special feeding centers in The Hague during the first 3 months of 1945. According to Banning (*ibid.*, p. 27) 29.2 per cent were considered undernourished by more than 20 per cent; 31.5 per cent fell in the range from 15 to 20 per cent underweight; 27.3 per cent were in the range from 7 to 15 per cent underweight; and 12 per cent were underweight by less than 7 per cent. The group was certainly far from being a representative sample for the population as a whole. In addition, it should be kept in mind that the degree of underweight was obtained probably by reference to standard norms and that it does not represent an actual decrement from the individuals' pre-starvation level. The records are not fully clear on this point.

The question of weight decrement in Holland was also discussed by De Jongh (in Boerema, 1947, p. 236). Out of a group of 800 men and 400 women in an outpatient clinic, about one half had weights which were more than 30 per cent below their "normal." It was mentioned that there were patients whose weight was as much as 50 per cent below their "normal" but no frequency of these cases was indicated. The data are inadequate in themselves and are practically useless for characterizing the degree of starvation in the general population.

TABLE 28

MEAN BODY WEIGHTS OF 6000 WORKERS in the textile union at Paris. The values pertain to the first half of each year indicated, except for 1939 when weights were obtained at various times of the year. Data from Trémolières (1947).

	Men		Women	
	Kg.	% Change from 1939	Kg.	% Change from 1939
1939	70.2		62.6	
1943	61.8	−12.0	54.1	−13.6
1944	62.5	−11.0	55.1	−12.0
1945	63.1	−10.1	56.8	−9.3
1946	62.4	−11.1	55.4	−11.5

The changes in body weight of textile workers in Paris during the early part of World War II were recorded by Trémolières (see Table 28). The 1939 values appear to be larger than one would expect for this section of the French population, even when one takes into account the added weight of clothes. From the author's description it is not clear whether the same group of individuals was weighed on the dates indicated.

The changes in the body weight of the population of Liége, Belgium, are reflected in the data collected by Neuprez (1945), as presented in Table 29. The data are based on patients reporting to the Medical Clinic and are not represen-

tative of the population as a whole. As a rule, these patients were not seriously
ill. They were recruited from the poorer socioeconomic strata. The material was
broken down according to age, but not according to sex. The data, presented in
Table 30, indicate that the older the patients the higher the weight loss. The

TABLE 29

MEAN BODY WEIGHT OF 9265 PATIENTS in the Medical Clinic at Liége,
Belgium (Neuprez, 1945).

	Body Weight (kg.)	Decrement (kg.)	Decrement as Percentage of the 1940 Value
1940	64.0		
1941	56.3	7.7	12.0
1942	52.7	11.3	17.7
1943	57.0	7.0	10.9
1944	59.0	5.0	7.8

TABLE 30

AVERAGE WEIGHT LOSSES IN PATIENTS examined at the Medical Clinic at
Liége, Belgium, according to age groups (Neuprez, 1945).

	Age Group					
	20–30	30–40	40–50	50–60	60–70	70+
Maximal loss during 1940 to 1944 (%)...	4.6	11.0	15.7	15.4	17.1	18.3
Lasting loss at the start of 1944 (%)...	0.0	6.9	7.9	7.9	12.4	13.6

table takes into account the maximal decrement occurring during the period
from 1941 through 1943 as well as the status at the start of 1944. By this time the
youngest group, 20–30 years, had recovered all its weight loss while the oldest
group remained markedly underweight. Probably more than one factor is responsible for this phenomenon. For one thing, more of the older patients had to
rely on the official rations, which provided 1300–1500 Cal. per day in May 1941,
while the younger patients were able to procure additional food from the country and by growing vegetables in their own gardens. Neuprez mentions also the
possibility that the digestion and assimilation of the wartime diet, with its high
content of carrots and turnips, may have been less efficient in the older patients.

It is instructive to compare the age/weight curves obtained by Neuprez in
May 1940 and May 1941. In 1940 the weight was increasing up to the age of
40–50 and declined slowly in the higher age groups. In 1941 the normally observed increase in weight beyond the age of 20–30 years was absent, the weight
being approximately constant between 20–30 and 50–60 years. The slope of the
decline of weight in the groups beyond 60 years of age was parallel to that observed at times of abundant food supply.

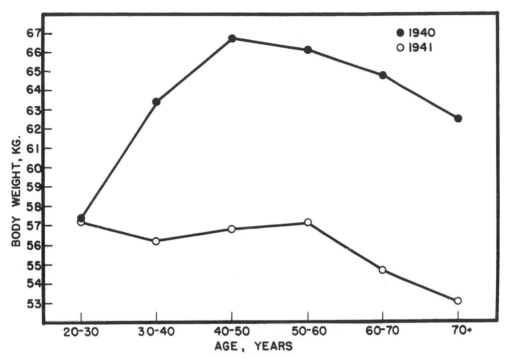

FIGURE 11. MEAN WEIGHT IN RELATION TO AGE under conditions of normal food supply (1940) and during rationing (1941) (Neuprez, 1945, p. 26).

Extent of Weight Losses in Concentration Camps and among Prisoners of War

Zimmer, Weill, and Dubois (1944) presented data on the body weight in a semi-starved population of the internment camps in southern France. They expressed the values of the actual body weight as obtained in June 1942 — that is, after some 9 months of inadequate food intake — in reference to the "physiological body weight" for a given height. Because of the frequent occurrence of scoliosis, the pre-internment heights were used as the basis for computing the deviation of the weight from the norm. The physiological (normal) value of the weight/height index postulated by the authors for both sexes was 41.0. This value is somewhat high for the men and definitely too high for the women if compared with the probable "normal average" values. Thus in a group of Central European factory workers measured in the 1930s, ages 26–45, the mean value of the weight/height index was 38.0 for men ($N=10,396$) and 37.0 for women ($N=955$) (Brožek, unpublished data). For American white draft registrants with an average weight of 69 kg. and a height of 173.9 cm., the index is 39.7 (U.S. Selective Service System, 1943). For the sample of men in the Minnesota Experiment the mean value of the weight/height index was 38.8 in the pre-starvation period. No control values for the internment camp population were given.

In Camp Gurs, where conditions were particularly bad, the average decrease below the postulated physiological norm, calculated from Figures 1 and 2 in Zimmer, Weill, and Dubois (1944), was 18.8 per cent for men and 22.9 per cent for women. A more detailed breakdown of the population of this camp according to the nutritional state is given in Table 31. In all the camps taken together conditions were more favorable, and of the 9000 inmates examined, only 331 persons were classified as cachectic, with 839 pre-cachectic, 4000 threatened, and 3830 normal.

TABLE 31

NUTRITIONAL STATE OF SEMI-STARVED INMATES OF AN INTERNMENT CAMP in southern France (Camp Gurs), June 1942. From Zimmer, Weill, and Dubois (1944).

Inmates of Camp Gurs	Nutritional State According to Weight/Height Index				
	Above Normal (Wt./Ht. 41.6 to 50.5)	Normal (Wt./Ht. 40.6 to 41.5)	Threatened (Wt./Ht. 31.6 to 40.5)	Pre-Cachectic (Wt./Ht. 28.6 to 31.5)	Cachectic (Wt./Ht. 20.6 to 28.5)
Men (%).........	2.6	1.2	63.2	24.7	8.3
Women (%).....	2.7	0.9	44.8	27.4	24.2

By definition, the weight/height index gives the average weight of a slice of the body 1 cm. in thickness. This value differentiated clearly between the groups of different nutritional status, with the minimum of 206 gm. observed in the cachectic group and the maximum of 505 gm. in the overweight group. Deviations from the "physiological" weight up to 50 per cent were observed. Schwarz (1945), who worked in the same area, reported cases in which the patients lost up to 53 per cent of their original body weight.

Data on the changes of body weight among Americans interned by the Japanese in the Philippines (at Santo Tomas, Hay Holmes, Bilibid, and Los Banos), with some indications of the caloric intake, were presented by Butler et al. (1945). The intake figures given represent the basic rations and do not include the food purchased at canteens, provided by Red Cross food packages, or raised in gardens. This introduces a serious error for which no valid correction can be attempted. Information for one of the camps, Los Banos, indicates that this supplementary food supply was by no means insignificant. The daily caloric intake from canteen food at this particular camp was estimated at 1500 Cal. per person from June to December 1943, at 900 Cal. from January to August 1944, and at 150 Cal. from September 1944 to December 1945. This food, added to the basic ration of 1800 Cal., would have provided abundant calories up to August 1944. No comparable data for the other camps were made available.

It was estimated that during 1942 and 1943 not more than 10 per cent of the internees at Los Banos and Santo Tomas suffered seriously from malnutrition. The reason for the bad state of nutrition in some individuals was not that food was lacking but that they failed to adjust emotionally to the available food and to general conditions in the internment camps or were unable to digest and assimilate the food because of illness. By the beginning of 1944 the mean weight losses, determined for an unspecified number of cases, were 8 lbs. for women and 17.5 lbs. for men. In February 1944 the supply of all extra food from the outside was cut off and the rate of weight loss was drastically accelerated. By August 1944 the average weight loss for women increased to 16 lbs. and to 26 lbs. for men. At the end of 1944 the respective figures for weight loss were 32 and 51 lbs.

In 22 patients, a small sample of the men who were admitted for malnutrition to the 47th British General Hospital, Singapore, in September and October 1945 (Mitchell and Black, 1946), the average weight before capture had been 155 lbs. but on admission it was 120 lbs., yielding an average loss of 35 lbs. (22.6 per cent of the pre-internment value). Six of the patients had slight edema and eight had moderate edema. In another series, for which neither the size of the sample nor the weight before capture was given, the weight loss during internment was

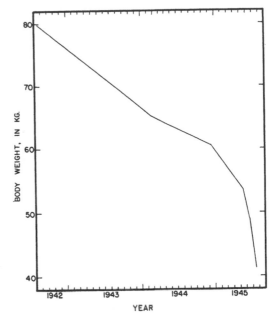

FIGURE 12. WEIGHT DECREMENT IN A 31-YEAR-OLD PRISONER OF WAR (Price, 1946).

estimated as 41 lbs. The weight losses observed in a 31-year-old man who was captured in Singapore in February 1942 and put to work on the Burma-Siam railway are given in Figure 12.

Individual records of changes in body weight for 3 men for whom exact weight data were obtained by Beattie (1947) are given in Figure 13; the dietary intake for the period included in the graph was calculated from food supplies issued to the kitchen of the institution, corrected for wastage.

In Indian prisoners of war evacuated from Japanese prison camps the gross body weight during internment was reduced on the average to 75 per cent of the original weight (Walters, Rossiter, and Lehmann, 1947b). The authors emphasize that the reduction of body components other than water must have been considerably larger than 25 per cent since many patients were grossly edematous.

Leyton (1946) observed in the German camps for Allied prisoners of war a decrease of body weight up to 20 per cent of the original weight over the first 6 months of underfeeding. An approximate equilibrium was established by the end of the 6 months; the rate of the loss in weight decreased markedly, and the continued changes were evident only when the weights were compared at long intervals (from 1 to 6 months). The weight losses are said to have been more severe in large or fat men. No estimate of terminal weight losses is given. Leyton comments that only a few patients died of uncomplicated starvation. The majority of deaths were due to intercurrent infectious diseases, especially rapidly progressing tuberculosis. Coigly and Stephenson (1946) stated that at the Belsen concentration camp losses of body weight up to 50 per cent or more were observed, but no actual data were presented.

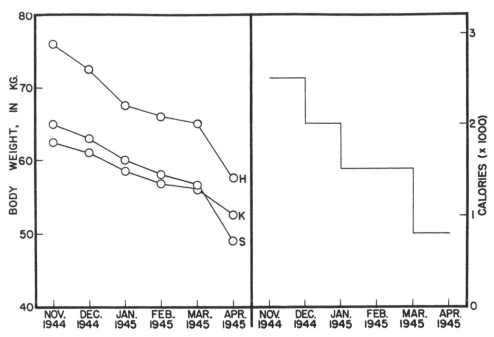

FIGURE 13. Weight and Estimated Caloric Intakes of 3 Adult Males
(Beattie, 1947, Warmsvald series).

In Frenchmen transferred from German concentration camps losses of 30 kg. of body weight, or about 40 per cent, were "not infrequent" (Debray *et al.*, 1946). Occasionally losses of body weight amounting to more than half the original level were noted. One man, 19 years old and 172 cm. tall, with a predicted standard weight of about 65 kg., weighed 28 kg. or 43 per cent of the standard. It is unfortunate that the actual pre-starvation value of the body weight was not given. Since it may well be that this patient was at the start below the standard weight as predicted on the basis of the age/height/weight tables, the weight loss of 57 per cent calculated on the basis of the "standard" weight would exaggerate the apparent tolerance of the human organism to the reduction of food intake and to the loss of body tissues. This limitation, and it is a severe one, is inherent in most of the clinical records.

Weight Loss in Man during Experimental Starvation and Undernutrition

The data on weight losses in man during experimental fasting were summarized by Morgulis (1923, p. 90). The record is held by Succi who lost 25.3 per cent of his initial weight in a 40-day fast; Levanzin in 31 days of fasting lost 21.9 per cent of his original body weight.

In the undernutrition experiment by Benedict *et al.* (1919) the weight losses were relatively small. For Squad A the average maximum loss was 12.1 per cent. The final weight obtained on February 3, 1918, was 59.8 kg. as compared with the initial weight of 67.0 kg. recorded on September 30, 1917; the difference of 7.1 kg. represents 10.6 per cent of the control weight. Squad B was maintained on a caloric intake reduced to about 1400 Cal. per day for 3 weeks, during which time the average body weight of the 12 men decreased from 67.9 kg. to 63.4 kg., or by 6.5 per cent of the pre-experimental weight. The caloric intake in Squad A was irregular and any attempt to relate the weight loss to caloric deficit would be hopeless.

Weight Loss in Anorexia Nervosa

Perhaps the most impressive degree of emaciation may be seen in some cases of anorexia nervosa. The absence of physical and climatic stresses and of complicating diseases, together with supporting medical and nursing care, allows the body to tolerate greater degrees of tissue wastage than could be borne under conditions of famine or catastrophe in the field. In any case, astonishingly low levels of weight in patients with anorexia nervosa have been recorded by competent observers.

Stephens (1895) measured the cadaver of a 16-year-old girl 56 hours after death. The body was 64 in. (162.6 cm.) long and weighed 49 lbs. (22.2 kg.); these figures indicate a weight of only 40.8 per cent of standard, or 59.2 per cent underweight. Since some dehydration must have occurred after death, the actual emaciation is overestimated by these figures. A woman of 61 years of age weighed 59.5 lbs. (27 kg.) with a body length of 59.75 in. (151.8 cm.) at necropsy (Oppenheimer, 1944); this weight is 44.7 per cent of the normal standard.

Similar degrees of emaciation have been reported for living patients. Bruckner, Wies, and Lavietes (1938) had 2 female patients with body weights amounting to 48 per cent of the normal standards. In the first case, age 15, the weight was 48.5 lbs. (22 kg.) and the height was 55.5 in. (141 cm.); in the second case, age 18, the weight was 54 lbs. (24.5 kg.) and the height was 60 in. (152.4 cm.). Farquharson (1941) reported his most extreme case in a series of 13 patients with anorexia nervosa to be a woman of 26 years who weighed 56 lbs. (25.4 kg.) with a height of 61 in. (154.9 cm.); this is 46.6 per cent of standard weight.

Berkman, Weir, and Kepler (1947) reported 4 out of 31 recent cases (1941–47) with body weights of 48.0 per cent or less of the normal standards. These are summarized in Table 32. Pictures of patient No. 7, taken upon admission and during recovery, are given in Figure 14. Some of the other patients in this series may actually have suffered equally severe tissue losses, but the presence of edema may have masked the effect as judged by weight. Three of these women (Cases 9, 10, and 11) are included in Table 32.

Berkman et al. (1947) said that a 50 per cent reduction below normal standard weight "seems to be the critical level" with regard to the possibility of a successful clinical recovery. In this connection the history of some of the patients listed in Table 32 is of interest. Cases 1, 9, and 18 did not remain at the clinic for treatment. Case 2 died unexpectedly in an early stage of treatment after gaining about 13 pounds without exhibiting edema. Case 7 was not edematous on admission but on treatment she developed generalized grade 3 edema which lasted about 40 days and then gradually subsided; in 82 days she was greatly improved and had gained 20.5 lbs. (9.3 kg.). Case 10 developed further edema on

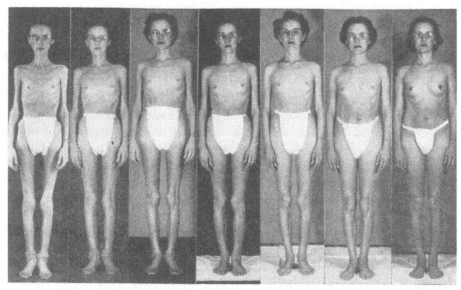

Feb. 2	Mar. 17	Apr. 14	May 12	June 12	July 12	Sept. 19
59½ lbs.	75 lbs.	78 lbs.	83 lbs.	88 lbs.	98 lbs.	100 lbs.

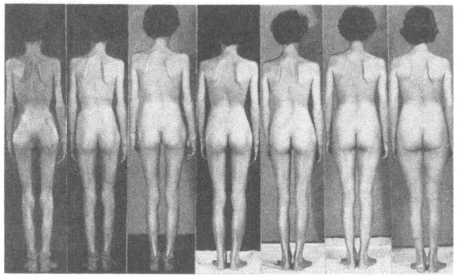

Feb. 19	Mar. 17	Apr. 14	May 12	June 12	July 12	Sept. 19
59½ lbs.	75 lbs.	78 lbs.	83 lbs.	88 lbs.	98 lbs.	100 lbs.

(All data for 1947)

FIGURE 14. Recovery from Anorexia Nervosa. The pictures were kindly supplied by Dr. John M. Berkman, Division of Medicine, Mayo Clinic, Rochester, Minnesota (see Berkman *et al.*, 1947).

TABLE 32

EXTREME DEGREES OF EMACIATION, FEMALE PATIENTS WITH ANOREXIA
NERVOSA. Data from Berkman, Weir, and Kepler (1947). "Standard"
weights from Association of Life Insurance Medical Directors (1912).
Edema refers to the presence or absence of pitting edema at the time
of measurement.

Case	Age	Height (in.)	Weight (lbs.)	Standard (lbs.)	Weight (%)	Edema	Duration of Illness (yrs.)
1 ...	24	62.0	50.0	121.0	41.3	0	1.5
2 ...	18	63.5	51.0	121.5	41.9	0	4.5
7 ...	32	62.5	59.5	126.5	47.0	0	14
18 ...	31	62.0	60.0	125.0	48.0	0	5
9 ...	51	64.0	70.0	144.0	48.6	+	4
10 ...	30	62.0	72.0	124.0	58.0	+	2
11 ...	23	66.0	79.0	142.0	55.6	+	3

treatment; on dismissal in 6 weeks she had gained 17 lbs. (7.8 kg.). Case 11 at
first developed increased edema and then gradually became edema-free; in 4
months she was greatly improved and had gained 25 lbs. (11.3 kg.).

The patient with anorexia nervosa, in great contrast to the ordinary famine
victim, often has a very low intake of water and liquids as well as of regular
food. In these cases there is a tendency to dehydration; the water deficiency limits or prevents the edema that would be expected otherwise. The result is apt to
be a degree of underweight which may be disproportionate to the actual tissue
wastage. In a case of anorexia nervosa a reduction of body weight to 60 per cent
of the normal standard is serious but may have a far less grave connotation than
an equal loss in a famine victim.

Extent of Cachexia in Simmonds' Disease

Very drastic reductions in weight are obtained in cases of Simmonds' disease and of a similar syndrome described by Kylin (1937). Weights for 3 patients are given in Table 33, the maximal weight loss being 58 per cent.

TABLE 33

CHANGES IN BODY WEIGHT IN THREE FEMALE PATIENTS WITH SYMPTOMS OF
"LATE-PUBERTY EMACIATION" (Kylin, 1937).

Age (yrs.)	Height (cm.)	Pre-Illness Date	Weight (kg.)	Minimal Date	Weight (kg.)	Difference	Difference (%)
18	170	1934	60	June 1936	25.0	−35.0	58.3
24	170	1929	58	June 1936	28.8	−29.2	50.3
23	166	1930	70	August 1935	30.5	−39.5	56.4

Weight Losses in Obesity Reduction

Obesity may be defined physiologically as a surplus of deposited fat. In sedentary individuals positive deviations from the standard weight are likely to be
due to the accumulation of fat rather than muscular tissue, and the identification
of obesity with marked "overweight" is largely, though only approximately, cor-

rect. In obesity reduction the fatty tissue is the primary and frequently the only component of the body being used up as "fuel." This differentiates obesity reduction sharply from semi-starvation, in which muscular and glandular tissues as well are drawn upon as a source of calories.

The magnitude of weight losses in obesity treatment may be illustrated by Mason's (1927) data presented in Table 34. The losses ranged from 18 to 29 per cent of the initial body weight.

TABLE 34

LOSS OF WEIGHT DURING REDUCTION OF OBESITY IN WOMEN (Mason, 1927, p. 96).

Age (yrs.)	Weight on Admission		Weight on Discharge		Total Loss		Percentage Loss	Number of Days	Loss per Day (kg.)
	Kg.	Lbs.	Kg.	Lbs.	Kg.	Lbs.			
36	112.5	248.0	92.4	203.7	20.1	44.3	17.9	63	0.318
51	172.9	381.2	138.2	304.7	34.7	76.5	20.2	90	0.386
50	149.1	328.7	124.8	275.1	24.3	53.6	16.3	64	0.380
31	123.6	272.5	87.5	192.9	36.1	79.6	29.2	108	0.334
29	109.9	242.3	78.6	173.3	31.3	69.0	28.5	117	0.264

Evans and Strang (1931) reported data obtained by the use of low calorie diets in the routine treatment of 187 obese patients. On the average they lost 30 lbs. (13.6 kg.); no initial average weight was given. The average duration of the dieting period was about 9 weeks. In the large majority of cases the weight loss was more rapid in the first 4 weeks than in the later weeks. As to sex, 26 of the patients were men, 161 women. The men tended to respond to the diet reduction more rapidly, a fact related to their higher energy expenditure.

Among the large weight losses Evans and Strang cite a decrement of 44 lbs. (from 208 to 164) within 20 weeks in a 12-year-old girl; of 80 lbs. (from 223 to 143) within the same period in a 16-year-old girl; and of 96 lbs. (from 278 to 182) within 24 weeks in a 67-year-old woman.

It is natural that the largest absolute values of weight decrement will be found in grossly overweight individuals. As far as we are aware the record is held by Newburgh's patient whose weight decreased from 225 kg. (560 lbs.) in July 1931 to 124 kg. (274 lbs.) a year later, and to 88 kg. (194 lbs.) in July 1933. The total loss of weight was 137 kg. (366 lbs.), and a follow-up in March 1934 disclosed that the patient was maintaining his reduced weight. The pictures presented by Newburgh (1942, p. 1094) tell the story more dramatically than the data on changes in body weight.

Edema and Body Weight in Semi-Starvation

Edema as a characteristic of prolonged semi-starvation is discussed in detail in Chapter 43. In this context we shall consider edema simply as a factor influencing the gross body weight and masking, in part, the loss of body tissues. Jansen (1920), working with undernourished institutionalized patients for whom he had reliable weight records previous to the reduction of food intake, observed that in *all* cases there was an initial decrease in body weight; in the heavier individuals the loss was greater on both the absolute and the percentage basis. With the onset of edema and without any improvement of the nutritional situation, the weight, especially in lighter individuals, tended to remain constant or even to increase somewhat.

Nicolaeff (1923) was concerned with the problem of edema in the examination of Russian children who died of starvation. He realized clearly that edema and ascites partially mask the loss of body tissues. No systematic analysis was made but the possible magnitude of the error is indicated by the fact that he extracted from the abdominal cavities of 13- and 14-year-old children as much as 7 to 8 liters of edema fluid.

TABLE 35

DECREASE IN THE BODY WEIGHT THROUGH ELIMINATION OF THE EDEMA FLUID in semi-starved male prisoners (Simonart, 1948, pp. 35–54).

Patient	Age	Pre-Starvation Weight (kg.)	Duration of Treatment (days)	Weight (kg.)		Difference (kg.)
				At Start of Treatment	At End of Treatment	
No. 1..	32		17	66.0	60.0	6.0
No. 2..	55	78	11	76.0	59.0	17.0
No. 3..	25	65	5	60.5	58.0	2.5
No. 4..	58	103	6	80.0	75.0	5.0
No. 5..	57		6	64.0	60.5	3.5
No. 6..	49	75	6	65.0	58.0	7.0
No. 7..	47		12	78.0	61.5	16.5
De C..		(72)	23	52.0	46.0	6.0
S	49		11	64.5	51.5	13.0

The concealment of the destruction of tissue by retention of water has been observed repeatedly in the course of obesity treatment. Newburgh and Johnston (1930, esp. p. 819) encountered periods during which the subjects failed to lose weight even though they were definitely in a negative caloric balance. These periods lasted for a number of days and were followed in all cases by a rapid fall in body weight. In one case, presented by Newburgh in detail, the patient was estimated to burn daily 150 gm. of his body tissue, yet for 10 days the gross body weight did not reflect any tissue destruction. In the subsequent 3 days the weight loss was rapid and by the thirteenth day the body weight had reached the value predicted on the basis of a continuous utilization of the body tissues at the estimated daily amount.

Very large values were reported for the edema fluid by Simonart (1948) in his study on semi-starved civilian prisoners at Louvain, Belgium. Results obtained on 9 patients by injection of thiamine (patients No. 1–No. 7) and ingestion of brewer's yeast (patient De C) and of cod liver oil (patient S) are summarized in Table 35.

Massive edema of the legs and a generalized edema, observed in Simonart's patients, together with the appearance after elimination of the edema, are pictured in Figures 15 and 16. The curve of the body weight of one of these patients is presented in Figure 17.

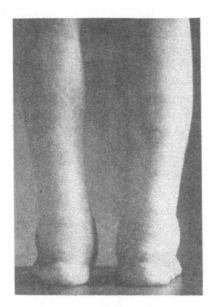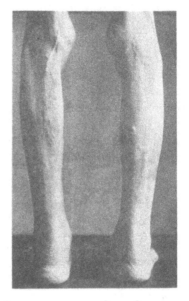

FIGURE 15. LEGS OF A PRISON INMATE SUFFERING FROM SEMI-STARVA-
TION. Pre-starvation weight, 80 kg.; weight at the time of death, 37 kg.
Left, before elimination of edema; *right*, after elimination of edema.
(Simonart, 1948, pp. 76, 77.)

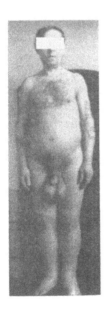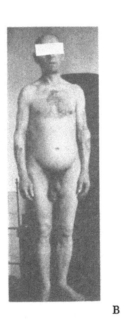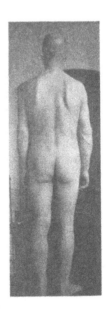

A   B

FIGURE 16. ANASARCA IN A SEMI-STARVED PRISON INMATE. Age, 47 years; height,
167 cm.; approximate standard weight, 68 kg. A. Body weight plus edema, 80 kg.;
B. body weight without clinical edema, 61.5 kg. (Simonart, 1948, p. 105.)

104

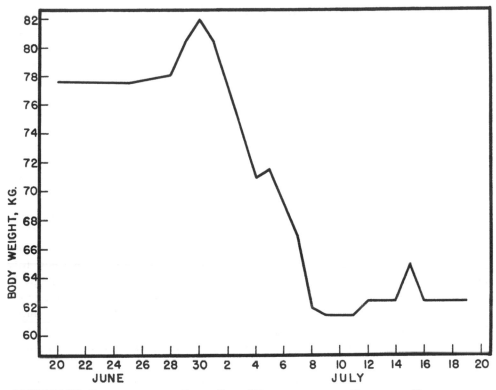

FIGURE 17. CHANGES IN THE GROSS BODY WEIGHT THROUGH DRAMATIC FLUCTUATIONS
IN THE AMOUNT OF EDEMA (Simonart, 1948, p. 39).

The Concept of a Lethal Level of the Loss of Body Weight

Not infrequently one encounters the opinion that there is a rather definite
level of the loss of body weight which is lethal. The idea was held by Krieger
among others; he stated (1921, p. 92): "The loss of weight suffered by the body
cannot go beyond a certain limit [*eine bestimmte Grenze*] without resulting in
death." The author set the lethal level of the loss of original body weight at 20
per cent for the young, at 40 per cent for adults exposed to acute starvation, and
up to 50 per cent in cases of semi-starvation. But Krieger's own material demonstrated that the concept is not valid except for general trends. The deviations below a computed standard weight, averaged for adult cases of cachexia of the same
etiology, varied from 36 to 48 per cent, with the estimated weight decrements
reaching as much as 55 per cent in individual cases of inanition not complicated
by chronic organic disease.

Zimmer, Weill, and Dubois (1944) disagree sharply with the idea of a definite lethal level. They stated: "In contrast to the general belief, there is no state,
except the final coma, when the disease (i.e., starvation) is irreversible." These
authors reported recovery in some patients who had suffered a striking loss of
body weight and who exhibited profound asthenia and a general poor condition
of health.

Kerpel-Fronius (1947) reported that with the help of adequate diet and

blood and plasma transfusions, continued daily for 10 to 20 days, infants reduced
to 50 per cent and in some cases even to 40 per cent of the standard weight
would recover.

On the basis of all the evidence available at present the concept of a definite,
generally valid lethal level of weight loss must be relegated to the rich store of
scientific mythology.

Over-All Trends of Weight Changes in the
Minnesota Experiment

In the Minnesota Experiment it was desired that a decrement in gross body
weight be obtained which would be comparable to values found in severe famine. Observations made in famine areas indicate that a weight loss of 25 to 30
per cent of the original body weight represents a severe degree of semi-starvation. The loss of about one quarter of the body weight appears to be fairly characteristic of persons who have been starving for many months but who have not
become irretrievably cachectic. Reports of greater weight losses are frequently
complicated by dehydration or pre-starvation obesity. When the decrement in
food intake is sudden (as was the case during the 1941–42 siege of Leningrad)
and when there is no striking reduction in physical activity, the rate of weight
loss is greatest at the start, decreases steadily, and frequently reaches a temporary plateau. Information from famine areas, together with suggestions provided

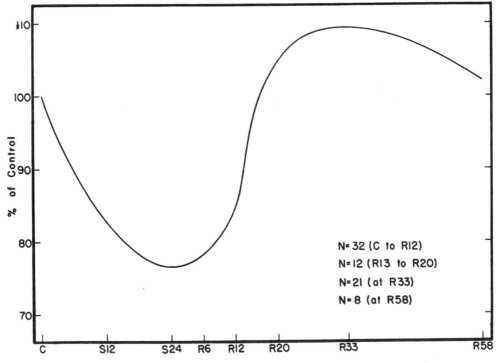

FIGURE 18. OVER-ALL CHANGES IN BODY WEIGHT IN THE MINNESOTA EXPERIMENT,
expressed as percentage of the control value for the particular group of subjects.

by experimental work on animals, indicates that a weight loss of some 25 per cent of the pre-starvation body weight is about as much as is needed to reproduce conditions of severe semi-starvation and as much as can be attempted with reasonable safety in 6 months in subjects not obese initially. Information on the speed of recovery, scant as it is, indicates a rather rapid restoration of body weight which often rises above the pre-starvation value, provided the food supplies are not limited.

Before entering upon a more detailed discussion of weight changes in the different phases of the Minnesota Experiment it may be useful to trace the general course of the average body weight throughout the 24 weeks of starvation and the 58 weeks of nutritional rehabilitation. The data based on observations made at C, S12, S24, R6, R12, R20, R33, and R58 are summarized in Figure 18. There was a decelerating loss of weight during starvation, the curve having a zero slope at S24. The recovery was relatively slow up to the twelfth week of rehabilitation. The most dramatic rise in weight took place from R12 to R20 and it was not until about R30 that the weight began to decrease. By R58, the last date at which systematic observations were made, the body weight was still somewhat above the respective control value.

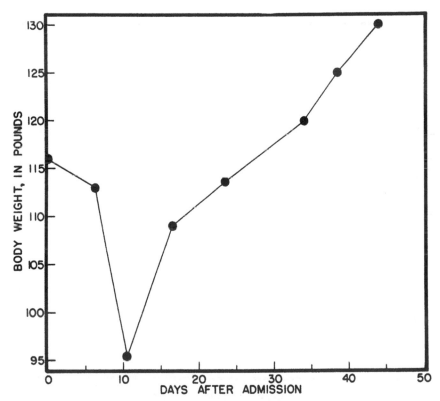

FIGURE 19. INITIAL LOSS OF GROSS BODY WEIGHT IN AN EDEMATOUS PATIENT DURING THE EARLY STAGES OF REFEEDING. The prewar weight was 162 lbs. (Mitchell and Black, 1946.)

Paradoxical Loss of Weight in the Early Phase of Refeeding

In individuals with disturbed water balance the shifts in body water can affect markedly the apparent nutritional status as judged on the basis of body weight. The hydration, taking at times the form of a clinical edema, masks the degree of starvation. On the other hand, in the early phase of refeeding a loss of body weight may take place, simulating a further "deterioration" of the nutritional condition of the patient, although actually the return toward normal water balance is a sign of recovery. Sometimes just the opposite thing happens and on refeeding the edema develops or, when present, temporarily increases. This, of course, will give a very erroneous picture of an improved "nutritional status." In reference to the starvation victims in southern India who were admitted to relief camps, it was stated: ". . . many of the people who came into camps appear to be filling out and fattening, when in reality they are getting dropsical and in a fair way to die" (Digby, 1878, Vol. II, p. 196).

It is obvious that changes in body weight in the early phases of rehabilitation are complex phenomena depending on such factors as the degree of edema and the caloric level of refeeding. In this section we are concerned with the weight loss on refeeding. In the Minnesota Experiment weight losses rather than weight gains were observed in 6 men during the first week of rehabilitation (losses rang-

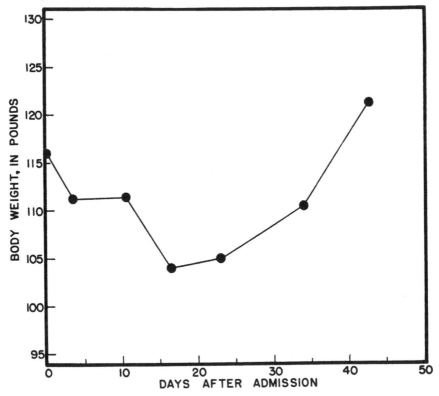

FIGURE 20. INITIAL LOSS OF GROSS BODY WEIGHT IN AN EDEMATOUS PATIENT DURING THE EARLY STAGES OF REFEEDING. The prewar weight was 152 lbs. (Mitchell and Black, 1946.)

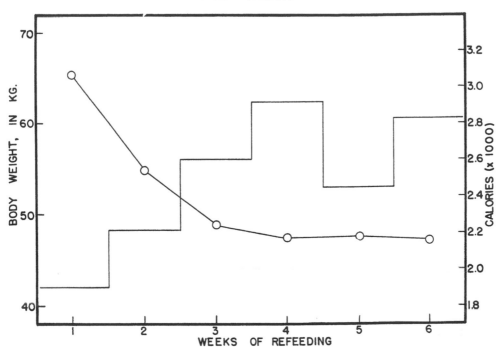

FIGURE 21. Loss of Body Weight on Refeeding. The patient was a Dutch woman, 77 years old, weighing normally 66 kg. (in July 1944); she suffered from emaciation combined with a massive edema. (Beattie, 1947.)

ing from —0.1 kg. to —0.5 kg.); in 7 men during the second week; and in 8 men during the third week. Of the 8 men, 4 were in the lowest (Z) caloric group, 2 in the L group, 1 in the G group, and 1 in the T group. By R6 there were only 1 man in the L group and 3 men in the Z group whose weights were 0.3 to 1.1 kg. below the semi-starvation (S24) value. By R10 all subjects registered weight gains.

An initial loss of body weight during nutritional rehabilitation has been reported frequently in the literature, and this will be commented upon later in connection with clinical and field data on refeeding after starvation. Here simply the magnitude of the possible weight losses will be illustrated. Weight records for 2 of the patients seen by Mitchell and Black (1946) are given in Figures 19 and 20. A striking case of a loss of body weight on refeeding was observed in June and July 1945 in Holland by Beattie (1947) (see Figure 21).

Weight Changes on a Controlled Caloric Intake — Minnesota Experiment

In determining the amount of total weight loss desired by S24, the relative weight during the control period was taken into account in the attempt to equalize, if possible, the biological stress of semi-starvation for the individual subjects. For subjects whose actual weight was equal to their standard (or "ideal") weight, the weight decrement was set at 25 per cent of the control value; the overweight individuals were to lose more, the underweight subjects to lose less.

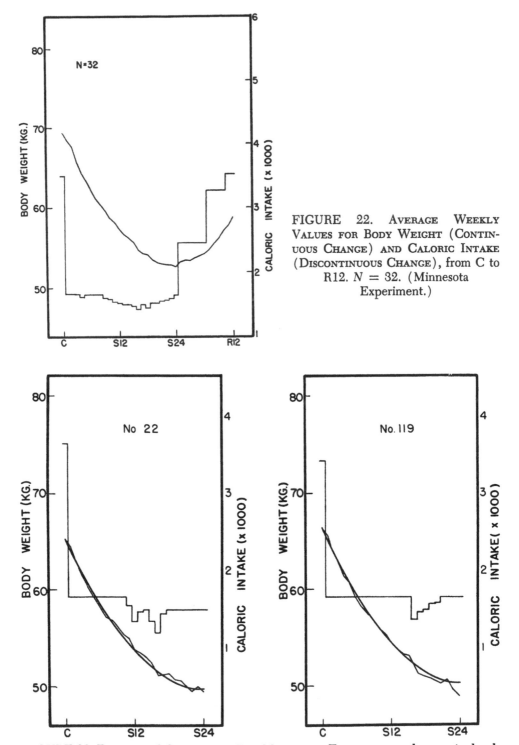

FIGURE 22. AVERAGE WEEKLY VALUES FOR BODY WEIGHT (CONTINUOUS CHANGE) AND CALORIC INTAKE (DISCONTINUOUS CHANGE), from C to R12. $N = 32$. (Minnesota Experiment.)

FIGURE 23. DATA FOR 2 SUBJECTS IN THE MINNESOTA EXPERIMENT who required only a minimal adjustment of the dietary intake and who followed very closely the predetermined curve of the weight loss.

As the subjects were, on the average, somewhat under the standard weight corresponding to their height, the average desired weight loss was slightly less than 25 per cent of the control weight. The actual average values of body weight were 69.39 kg. for the last week of the control period and 52.57 kg. for the last week of semi-starvation (S24); the decrement of 16.82 kg. was 24.24 per cent of the control value, indicating that the total desired weight loss was achieved.

The desired rate of weight loss was to be rapid at first, decreasing as time went on and reaching zero value at S24. The mathematical form of the desired weight loss curve was described in Chapter 4. An attempt was made to counteract individual weight deviations from the desired curve of weight loss by weekly adjustments of the potato and bread rations. As indicated in Figure 9 in Chapter 4, the weight loss curve obtained by averaging the individual values followed very closely the desired rate and amount of weight loss, and the average values of the weekly food adjustments for caloric intake were small.

These facts are interesting, but the close correspondence of the actual and "predicted" curves of weight loss, attained with a minimum of dietary manipulations, should not be overrated. Information available at the time the experiment was planned did not allow for more than an intelligent guess. Even at present the ability to predict changes in weight in an ambulatory population is severely limited as long as an important component of the predictive equation, the amount of physical activity, remains uncontrolled.

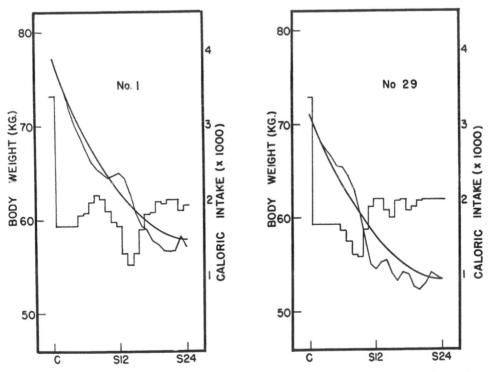

FIGURE 24. DATA FOR 2 SUBJECTS IN THE MINNESOTA EXPERIMENT who showed a large amount of fluctuation in their weight loss curve and who required a considerable adjustment in their caloric intake.

TABLE 36

Body Weight, in kg. and as percentage of control, for the 8-man caloric groups (Z, L, G, T) at the various stages of the Minnesota Experiment; also the decrement in starvation (dS24) and the increment in rehabilitation (ΔR12).

	Z		L		G		T		All Groups	
	Kg.	%C	Kg.	%C	Kg.	%C	Kg.	%C	Kg.	%C
C	68.32	100.00	69.41	100.00	72.30	100.00	67.51	100.00	69.39	100.00
S12	56.91	83.30	56.50	81.40	59.68	82.54	56.05	83.02	57.28	82.55
S24	51.99	76.10	51.74	74.54	54.48	75.35	52.09	77.16	52.57	75.76
R6	52.04	76.17	53.35	76.86	56.45	78.08	55.05	81.54	54.22	78.14
R12	55.42	81.12	56.98	82.09	61.72	85.37	60.91	90.22	58.76	84.68
dS24	−16.34		−17.68		−17.82		−15.42		−16.82	
$\dfrac{\text{dS24}}{\text{C}}$	23.9		25.5		24.6		22.8		24.2	
$\triangle$R6	+0.05		+1.61		+1.97		+2.96		+1.65	
$\triangle$R12	+3.44		+5.24		+7.25		+8.82		+6.19	
$\dfrac{\triangle \text{R6}}{\text{dS24}} \times 100$	0.3		9.1		11.1		19.2		9.8	
$\dfrac{\triangle \text{R12}}{\text{dS24}} \times 100$	21.0		29.6		40.7		57.2		36.8	

112

The average weekly weights and the weekly caloric intakes for the group of 32 men during starvation and controlled rehabilitation are given in Figure 22; during the period from R1 to R12 no weekly individual readjustments were made.

It has been pointed out that the desired and the actual curves of weight loss for the group as a whole were practically identical and that the changes in the caloric intake were not very marked. In a large number of individual cases the two weight curves, actual and desired, were essentially superimposed on each other and the amount of dietary readjustment was minimal (see Figure 23). In a few instances the weight fluctuations, due to accumulation of edema and rapid diuresis, were marked and led to drastic but not always successful attempts to bring the actual weight closer to the desired weight level (see Figure 24). The weight records for the 4 men who did not follow strictly the semi-starvation eating regimen are discussed in Chapter 41.

Recovery of Body Weight in Early Rehabilitation — Minnesota Experiment

The values of body weight, given both in kg. and as a percentage of the control value, for the 8-man caloric groups at important stages of the experiment are indicated in Table 36. In order to facilitate the process of relating the observed weight changes to caloric intake, Table 37 is included. The weekly incre-

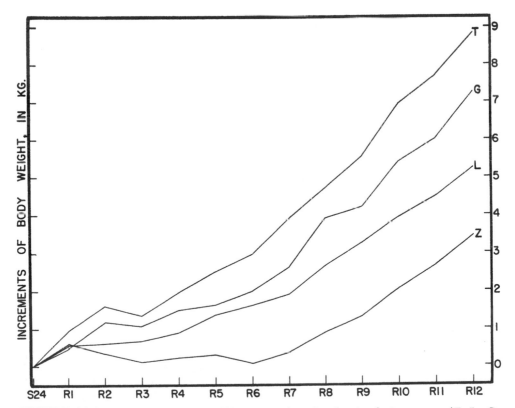

FIGURE 25. INCREMENTS OF BODY WEIGHT, in kg., for the 4 caloric groups (Z, L, G, and T) during the controlled phase of rehabilitation in the Minnesota Experiment. The separate curves were plotted for each caloric group.

TABLE 37

ACTUAL DAILY CALORIC INTAKE IN DIFFERENT PHASES OF
THE MINNESOTA EXPERIMENT.

	Z	L	G	T	All Groups
C	3521	3449	3536	3464	3492
S1	1658	1658	1658	1658	1658
S6	1629	1637	1658	1633	1639
S12	1558	1561	1493	1489	1525
S18	1442	1503	1331	1462	1435
S24	1578	1648	1760	1582	1642
S1–S24 ..	1544	1571	1619	1544	1570
R1–R6	1930	2245	2675	2944	2448
R7–R10	2738	3053	3484	3753	3257
R11–R12 ...	2999	3314	3744	4014	3518

TABLE 38

STATISTICAL SIGNIFICANCE OF THE DIFFERENCES IN WEIGHT GAINS, in
kg., for the caloric groups at R6. The F values were computed as the
ratio of the between-group variance, V_{bGr}, to the replicate variance,
$V_{rep} = 1.15$. (Minnesota Experiment.)

	Groups Compared					
	Z vs. L	Z vs. G	Z vs. T	L vs. G	L vs. T	G vs. T
V_{bGr}	9.76	14.82	33.93	0.53	7.29	3.90
F	8.49[**]	12.89[**]	29.50[**]	0.46	6.33[*]	3.39

[**] In this and the subsequent tables, two asterisks in brackets [**] attached to an F value indicate statistical significance at the 1 per cent level.

TABLE 39

STATISTICAL SIGNIFICANCE OF THE DIFFERENCES IN WEIGHT GAINS, in
kg., for the caloric groups at R12. $V_{rep} = 3.02$ (Minnesota Experiment).

	Groups Compared					
	Z vs. L	Z vs. G	Z vs. T	L vs. G	L vs. T	G vs. T
V_{bGr}	12.96	58.14	116.10	16.20	51.48	9.92
F	4.29	19.25[**]	38.44[**]	5.36[*]	17.05[**]	3.28

ments in body weight, plotted both for the individual caloric groups and for
single weeks, are represented in Figures 25 and 26.

The differential trend of the mean gains in body weight became apparent in
the second week of rehabilitation. By R6 the gains for groups Z, L, G, and T represent 0.3, 9.1, 11.1, and 19.2 per cent of the weight lost in semi-starvation. By
R12 the differentiation is still more clear-cut, with respective relative weight
gains of 21.0, 29.6, 40.7, and 57.2 per cent of the semi-starvation loss. The statistical significance of the differences is indicated in Tables 38 and 39. At R12
the absolute differences between the caloric groups are larger than at R6 but the

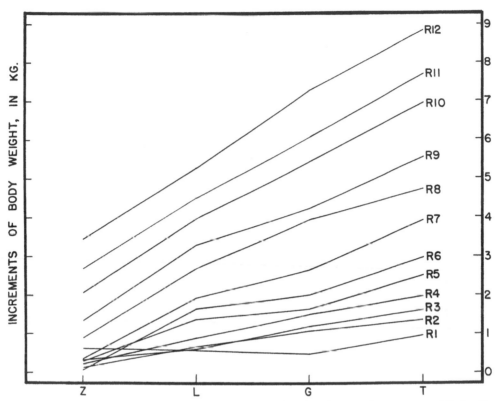

FIGURE 26. INCREMENTS OF BODY WEIGHT, in kg., for the 4 caloric groups (Z, L, G, and T) during the controlled phase of rehabilitation in the Minnesota Experiment. Separate curves were plotted for each week.

replicate variance, used statistically as the measure of the "experimental error," also increased. This explains the shift in the significance of some of the differences.

It should be noted that the above F-tests were computed for a particular single week of the rehabilitation period. The values for all weeks, starting with R2, are in the same direction, as is evident from Figures 25 and 26. This consid-erably strengthens our confidence that the differences in the recovery of weight obtained for the 4 levels of caloric refeeding are not due to chance fluctuations.

Relative Effectiveness of Caloric Supplement in Rehabilitation — Minnesota Experiment

What was the *relative effectiveness* of the calories added to the starvation diet in bringing about the gain in weight? The daily caloric supplements above the intake for the period S21–S24, serving as a base line, are indicated in Table 40. The average daily weight gains during the periods R1–R6, R7–R10, and R11–R12 for the 4 caloric groups are given in Table 41. The average daily body weight gains in gm., per 1000 Cal. of daily supplements, are presented in Table 42.

It is evident that both the caloric level and the duration of refeeding play a

TABLE 40

ACTUAL DAILY CALORIC SUPPLEMENTS received by the 4 groups (Z, L, G, T) above their basal intake for S21–S24 (Minnesota Experiment).

	Z	L	G	T	All Groups
Daily caloric intake					
S21–S24	1514	1570	1717	1515	1579
Daily caloric supplement					
R1–R6	416	675	979	1429	869
R7–R10	1224	1483	1757	2238	1678
R11–R12	1485	1744	2045	2499	1939
R1–R12	863	1122	1416	1869	1317

TABLE 41

AVERAGE DAILY WEIGHT GAINS, in gm., at different periods of the controlled rehabilitation in the Minnesota Experiment.

	Z	L	G	T	All Groups
R1–R6	1.2	38.3	46.9	70.5	39.3
R7–R10	71.1	83.2	122.1	141.1	104.3
R11–R12	100.0	92.9	132.9	136.4	115.7
R1–R12	41.0	62.4	86.3	105.0	73.7

TABLE 42

RATIO OF DAILY WEIGHT GAINS, IN GM., TO DAILY CALORIC SUPPLEMENTS, IN 1000 CAL., in the controlled phase of rehabilitation in the Minnesota Experiment.

	Z	L	G	T	All Groups
R1–R6	2.9	56.7	47.9	49.3	45.2
R7–R10	58.1	56.1	69.5	63.0	62.2
R11–R12	67.3	53.3	65.0	60.6	59.7
R1–R12	47.5	55.6	60.9	56.2	56.0

part in determining the optimal amount of caloric supplementation. From R1 to R6 the optimum lies in group L; during R7 to R10 it shifts to group G; and in the last 2 weeks, R11 and R12, it is in the Z group. For the total period of 12 weeks of rehabilitation, the optimum was attained by group G; there was a decline in the relative effectiveness of refeeding in both the higher (group T) and the two lower (groups L and Z) levels of caloric supplementation.

The recovery of body weight is neither a simple nor an all-important criterion of the effective process of nutritional rehabilitation. There is the complicating factor of changes in body hydration. Furthermore, from the point of view of functional recovery it is important not only how much tissue is being laid down but also what kind of tissue. The differentiation between fatty and non-fatty (primarily muscular) tissue is of fundamental importance.

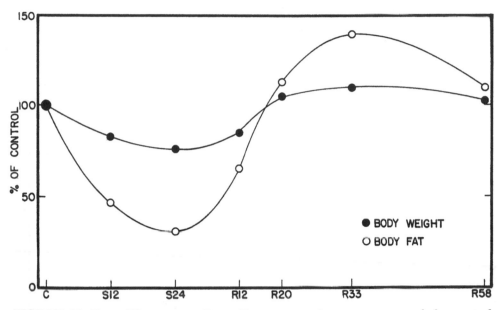

FIGURE 27. BODY WEIGHT AND BODY FAT, expressed as percentages of the control values (Minnesota Experiment).

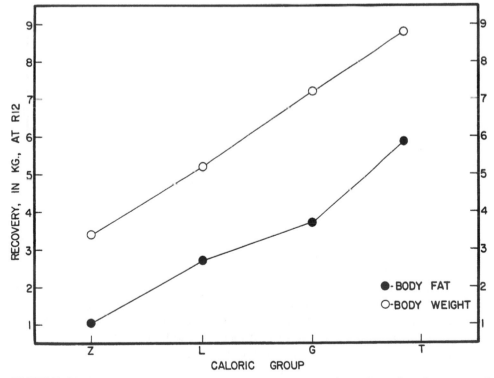

FIGURE 28. RECOVERY OF BODY TO WEIGHT AND FAT after 12 weeks of nutritional rehabilitation (Minnesota Experiment).

Body Fat and Changes in Body Weight — Minnesota Experiment

The question of body fat is discussed in detail in Chapter 8. Here the changes in body fat will be related to changes in the gross body weight. Figure 27 clearly indicates that the changes in body fat were more dramatic than the changes in body weight, both characteristics being expressed as percentages of the respective control values; there was both a larger decrement of fat in semi-starvation and a greater increment during rehabilitation.

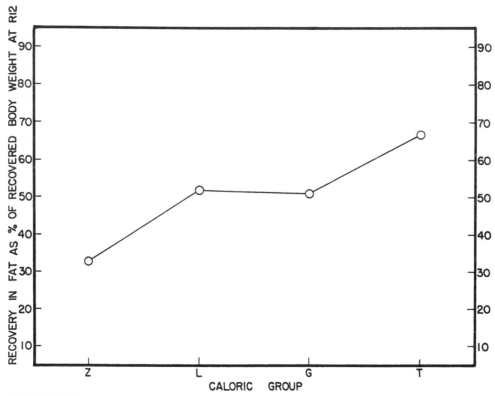

FIGURE 29. RECOVERY OF BODY FAT, expressed as percentage of body weight recovered after 12 weeks of rehabilitation (Minnesota Experiment).

The amount of body weight and body fat in kg. regained in the course of the first 12 weeks of rehabilitation is indicated in Figure 28. In both characteristics the relationship to the level of caloric refeeding can be well approximated by a straight line, and the 2 lines are nearly parallel. The recovery of fat is expressed as a percentage of recovered body weight in Figure 29. The higher the caloric level the greater the amount of recovery and, at the same time, the greater the percentage of recovered weight which can be accounted for as body fat. Whereas this percentage was 32.7 in group Z, it increased to 66.7 in group T.

Relating the amount of recovery to the losses suffered during semi-starvation results in a similar picture, except that the differences are brought out more strikingly. During the controlled phase of rehabilitation, R1–R12, the recovery of fat lost in semi-starvation was faster than the recovery of body weight in all

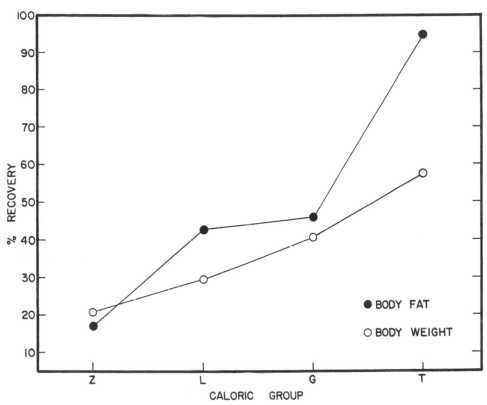

FIGURE 30. Recovery of Body Weight and Body Fat in the Minnesota Experiment at R12. The values express the amount of weight and fat regained as percentages of semi-starvation losses.

caloric groups except the lowest one (see Figure 30). In the extreme groups, Z and T, the recovery of weight at R12 corresponded to 20.8 per cent and 63.6 per cent, respectively, of the semi-starvation loss; for fat the corresponding values were 16.7 and 95.1 per cent.

Body Weight in Later Rehabilitation — Minnesota Experiment

The average weekly rates of recovery of body weight and the caloric intake during the period R13–R20 are indicated in Figure 31. The values pertain to 12 men who remained at the Laboratory for continued observations but who were released from strict adherence to a prescribed dietary regimen and were completely free over the weekends. The caloric intakes were estimated on the basis of careful records of the foods and amounts consumed.

During the first week of uncontrolled rehabilitation (R13) the individual intakes varied from 4400 Cal. to 5800 Cal., with the week's average of 5219 Cal. per man per day. Some of the men commented that they were still hungry at the end of the very large meals, even though they were unable to ingest any more food. All of the men ate snacks between meals and in the evenings.

The largest average caloric intake of 5801 Cal. daily was reached in R14, the

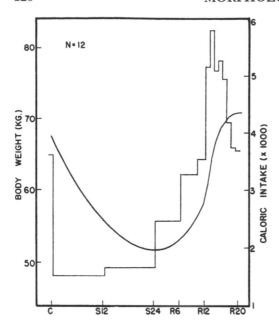

FIGURE 31. Average Weekly Values for Body Weight (Continuous Change) and Caloric Intake (Discontinuous Change), from R13 to R20. The data up to R12 are based on the average values for C, S12, S24, R6, and R12. N = 12. (Minnesota Experiment.)

second week of unrestricted feeding. From then on the caloric intake began to decrease, reaching by R20 the average value of 3683 Cal., with the range from 3200 to 4200 Cal.

Figures 32 and 33 present daily values of body weight and caloric intake obtained for 2 subjects during the first 4 weeks of uncontrolled rehabilitation (R13–R16); these illustrate the tendency of the men to "splurge" on the weekends, when intakes of 8000 and 10,000 Cal. were not uncommon, with occasional intakes as high as 11,000 Cal. in a day.

The effect of the dietary level in the first 12 weeks of rehabilitation was clearly evident in the food intake from R13 to R20. In these 8 weeks the 6 men who from R1 to R12 were placed in the 2 lower caloric groups (Z and L) ate 4860 Cal. per man per day while the 6 men in the upper caloric groups (G and T) ate 4660 Cal. per day. The difference amounts to 200 Cal. per day or 11,200 Cal. per 8 weeks. At the same time the L and Z men gained an average of 14.9 kg. (from 55.7 to 70.6) in this period as compared with a gain of 10.4 kg. (from 60.6 to 71.0) by the G and T groups.

Body Weight Gain and Protein and Vitamin Supplements — Minnesota Experiment

Within the range of supplements used in the Minnesota Experiment, the addition of vitamins (Figure 34), proteins (Figure 35), or both (Figure 36) did not result in a weight recovery more rapid than was obtained in the unsupplemented subjects. The lack of statistical significance of these differences contrasts sharply with the marked differentiation between the 2 upper (G and T) and the 2 lower (Z and L) caloric groups (see Table 43). Neither protein nor vitamin supplements increased the efficiency of caloric use.

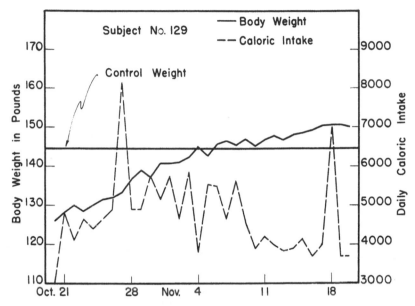

FIGURE 32. INDIVIDUAL DAILY FOOD INTAKES AND BODY WEIGHTS during the first 4 weeks of uncontrolled rehabilitation (R13 to R16). Subject No. 129 was in the lowest (Z) caloric group during the early phase of the rehabilitation period (R1 to R12).
(Minnesota Experiment.)

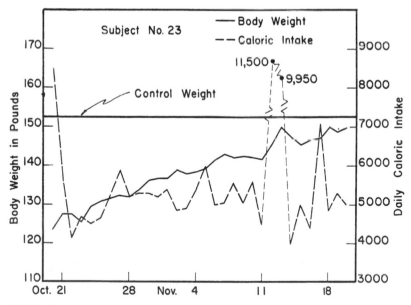

FIGURE 33. INDIVIDUAL DAILY FOOD INTAKES AND BODY WEIGHTS during the first 4 weeks of uncontrolled rehabilitation (R13 to R16). Subject No. 23 was in the low (L) caloric group during the early phase of the rehabilitation period (R1 to R12). (Minnesota Experiment.)

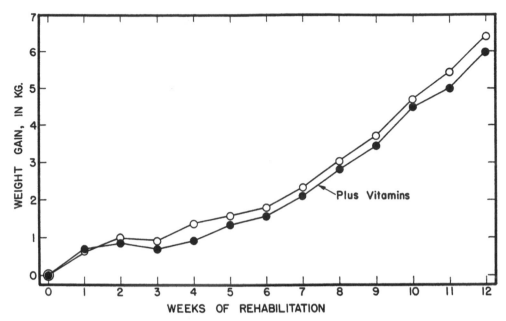

FIGURE 34. RECOVERY OF BODY WEIGHT in subjects with and without vitamin supplements (Minnesota Experiment).

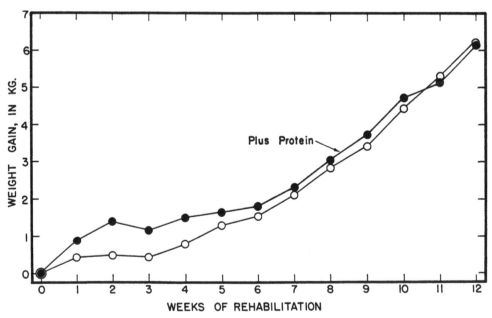

FIGURE 35. RECOVERY OF BODY WEIGHT in subjects with and without protein supplements (Minnesota Experiment).

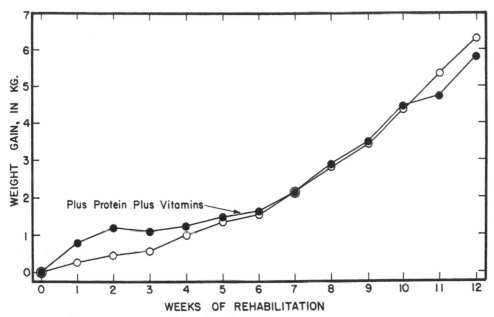

FIGURE 36. RECOVERY OF BODY WEIGHT in subjects receiving both protein and vitamin supplements, or neither (Minnesota Experiment).

TABLE 43

STATISTICAL SIGNIFICANCE OF THE DIFFERENCES IN WEIGHT GAINS, IN KG., FOR THE GROUPS RECEIVING PROTEIN AND/OR VITAMIN SUPPLEMENTS. At R6 and R12 the replicate variances were equal to 1.15 and 3.02, respectively. For comparison, the values for the combined caloric groups (Z + L) and (G + T) are included. (Minnesota Experiment.)

	Groups Compared			
	U vs. Y	P vs. H	UP vs. YH	(Z + L) vs. (G + T)
		R6		
V_{bGr}	0.66	0.36	0.02	21.45
F				18.65[**]
		R12		
V_{bGr}	0.04	1.44	1.00	109.55
F				36.27[**]

Recovery of Body Weight on Refeeding — Clinical and Field Data

In view of the great numbers of men who underwent nutritional rehabilitation during World War II, the data on weight changes in the course of recovery are surprisingly slim. Most of the available material is summarized in this section.

Detailed observations on changes in the weight of cachectic patients who received supplementary rations were made by Zimmer, Weill, and Dubois (1944). The loss of body weight was presumably 30 per cent or more below the postu-

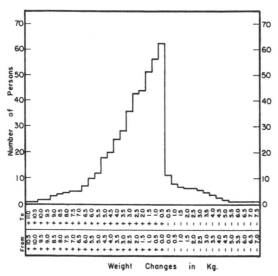

Weight Changes in Kg.

FIGURE 37. Weight Changes in Semi-Starvation Patients Provided with Supplementary Rations at Camp Rivesaltes in southern France, February 27 to May 11, 1942 (Zimmer et al., 1944). The caloric content of the supplements was not given. The basal diet supplied about 950 to 1100 Cal.

lated normal weight. A sample of 100 cachectic patients were given dietary treatment for about 3 months; during the first 2 weeks the weight increased in 32 per cent of the cases, remained stationary in 20 per cent, decreased somewhat (as a result of loss of edema fluid) in 40 per cent, and decreased seriously in 8 per cent. During the third and fourth weeks the percentage of patients gaining weight reached 60 per cent. In the fifth and sixth weeks 43 per cent continued to gain. During the seventh and eighth weeks 51 per cent of the patients gained weight, 19 per cent kept a stationary weight, and 30 per cent exhibited a loss of weight ascribed to the disappearance of edema.

The changes in body weight in Camp Rivesaltes after 6 weeks of refeeding are given in Figure 37. In another camp in which the refeeding was instituted in February 1942, by March the weight had increased in 60 per cent of the cases, remained unchanged in 14 per cent, and decreased in 27 per cent; a month later the percentages were 64, 18, and 18, respectively.

A rate of weight gain which was rather similar to that observed in the Minnesota Experiment was recorded for civilians rescued from Japanese prison camps (Butler et al., 1945). After their liberation in February 1945 the internees were placed for a short time on Army "hospital" and K rations. During the 22-day passage to the United States the average intake was estimated at 3164 Cal. In a random sample of 100 ambulatory cases examined on arrival in the west coast ports approximately 9 weeks after their rescue, the average weight gain was 24 lbs. or a recovery of 53 per cent of the loss sustained during imprisonment. In 25 returnees who needed hospitalization, the average weight loss was 55 lbs. and the gain on return was 23 lbs., or 42 per cent of the loss.

In 22 patients (Mitchell and Black, 1946) who lost on the average 35 lbs. during their internment by the Japanese, the average gain of weight was 5.5 lbs. during the first week of refeeding in the hospital and 5 lbs. during the second week. The picture was complicated in some cases by an initial decrease of weight due to the loss of edema fluid. In patients with extensive edema the weight decrements obtained within the first 10 days of maintenance on hospital diets were as large as 20 lbs.

Debray et al. (1946) frequently noted increases of 20 to 25 kg. of body weight in severely malnourished inmates of German concentration camps who

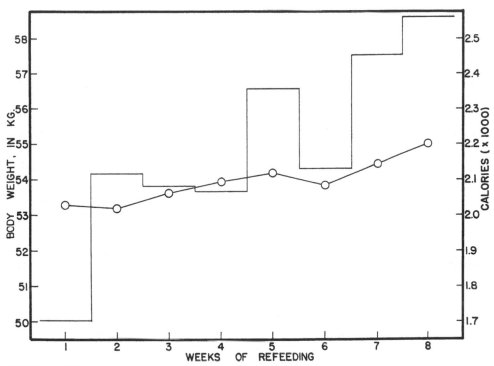

FIGURE 38. AVERAGE GAINS OF WEIGHT in 6 German prisoners studied in August and September 1946 (Beattie, 1947).

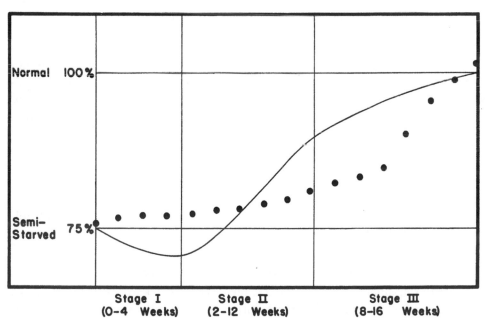

FIGURE 39. SCHEMATIC REPRESENTATION OF CHANGES IN BODY WEIGHT OF EDEMATOUS SEMI-STARVATION PATIENTS DURING SUCCESSIVE STAGES OF REHABILITATION (Walters *et al.*, 1947). The circles indicate actual values obtained for 32 men in the Minnesota Experiment.

had been transferred to Paris for treatment. The time course was not traced and no complete dietary intakes were given. It was noted that the patients received additional food from their families and would get up at night to eat.

Data for a group of German prisoners were obtained under controlled conditions by Beattie (1947). Figure 38 presents the results for a homogeneous group of 6 men. Their age varied from 21 to 25 years; their normal weight recorded in the summer of 1945 varied from 65 to 71 kg. with an average of 67.9 kg. The weight for the first week of refeeding was between 49 and 55 kg. with an average of 53.3 kg. There was an average decrement of 0.1 kg. during the second week, followed by steady but small gains during the next 6 weeks. The average weight for the eighth week of refeeding was only 55.0 kg.

Twelve semi-starved patients, Indian prisoners of war of the Japanese, had an estimated loss of 15 kg. from their previous normal weight of 61 kg. (Walters, Rossiter, and Lehmann, 1947b). In the course of rehabilitation the weight rose from 48.5 ± 4.1 kg. to 61.7 ± 5.7 kg. over a period of about 3 months.

Walters et al. (1947b) differentiated 3 phases of rehabilitation. During the first phase, which varied from 1 to 4 weeks of hospitalization, in many of the edematous patients the body weight actually decreased as the edema receded. In the second stage, from 2 to 12 weeks of rehabilitation, the weight began to increase rapidly, not infrequently by as much as 7 lbs. per week. In the third stage, from 8 to 16 weeks of rehabilitation, the rate of increase in body weight diminished as the values were approaching the normal. The three stages were not of the same duration for all the individuals, and there was an overlap between the successive stages of recovery. The actual course of recovery was not plotted but a schematic representation of changes in body weight was attempted (see Figure 39).

The hypothetical weight curve of Walters et al. may be compared with the data obtained in the Minnesota Experiment (see the circles in Figure 39). In obtaining the time-axis units, the total span of 16 weeks was divided into 16 equidistant units. Although there were individual weight losses in the early phase of rehabilitation, the Minnesota group as a whole registered weight gains starting with the first week of refeeding. The weight increases were slow during the first 6 weeks of rehabilitation, more rapid from R7 to R12, and very fast from R13 to R16. This course was directly related to the changing caloric intake and can by no means be regarded as a generally valid pattern. The coincidence of reaching the pre-starvation level of weight in both sets of data 16 weeks after the institution of the rehabilitation treatment is of interest. The data of Walters et al. followed the weight level only to the point where the previous normal weight was regained.

Tendency to Post-Starvation Obesity

The pre-starvation weight in the ZL men (66.9 kg.) and the GT men (68.2 kg.) was reached by both sub-groups during the sixteenth week of the rehabilitation period. At R20 their weight increased to 70.6 and 71.0 kg., respectively; in terms of the pre-starvation values this was 105.5 and 104.1 per cent.

The rise in weight continued still further. In 21 subjects who were weighed at R33–R35 the average weight was 75.8 kg. as compared to 69.36 kg. in the control period, or 109.2 per cent of the control value. The difference was statistically significant. The data obtained for a smaller group of 6 subjects on whom body fat determinations were made at R33 are instructive. The average deviation above the control value was 3.6 kg. for the body weight and 4.3 kg. for the body fat. Consequently, the "overweight" is fully accounted for on the basis of increased fat storage (see Table 44).

At R58 the body weight had returned close to the control value while the body fat, on the percentage basis, remained slightly elevated (see Table 45).

TABLE 44

Body Weight and Body Fat, in kg., in Later Part of Rehabilitation (R33), for 6 subjects (Minnesota Experiment).

	C	R12	R33	R33 — C
Body weight	68.8	58.4	72.4	3.6
Body fat	11.1	7.9	15.4	4.3

TABLE 45

Body Weight and Body Fat, in kg., in Later Part of Rehabilitation (R55), for 8 subjects (Minnesota Experiment).

	C	R12	R58	R58 — C
Body weight	68.4	59.2	69.8	1.4
Body fat	12.5	9.3	13.7	1.2

A tendency toward overeating following a period of reduced diet was observed by Benedict (1907, p. 526) in a group of subjects who were studied in a series of total fasts of short duration (2 days). After recovery all 5 subjects weighed more than at the start of the experiment, the average difference being +2.7 kg. with a range from +1.2 to +5.2 kg. In the undernutrition experiment carried out by Benedict et al. (1919, esp. pp. 229–30) the weight increased very rapidly on the termination of the reduced diet. Six weeks later the men not only regained all the weight lost (7.1 kg. on the average) but exceeded their pre-experimental weight by 3.1 kg. (minimum 0.7 kg., maximum 7.2 kg.).

The tendency toward overeating following a period of reduced food intake has been noted repeatedly under conditions of natural starvation when food suddenly became available again in large quantities. The experiences of semi-starved explorers document particularly well the surprising lack of self-control even in the face of clear knowledge that the overindulgence will certainly result in discomfort and may have fatal consequences (for details see Chapters 28 and 37). Excessive food consumption following religious fasts was noted in the days of czarist Russia by Pavlov (Benedict et al., 1919, p. 203).

In none of these cases, including the Minnesota Experiment, can the increase of body weight above the pre-starvation level be interpreted as a result of disturbed function of the organs of internal secretion. The overeating and the resulting overweight may be regarded with justification as psychogenic in nature —

that is, as a result of an excessive voluntary food intake not balanced by dissipation of energy in the form of physical work. The psychophysiology of appetite in man is only inadequately understood at the present time. We can do little beyond pointing out some factors which might have played a role in maintaining the appetite at a high level even after the body weight was restored to its normal, pre-starvation level. It may be that the subjects remained more conscious of hunger pangs than they were before participating in the experiment. At any rate, they had very low resistance to eating snacks between meals. Some men reported that at times they had a physical sensation of hunger even after they had eaten a large meal; subject No. 27 commented at the end of R13 on having "an odd sensation of being full yet still hungry." On the personality level there was anxiety, only rarely verbalized, that the food somehow would not last. For example, subject No. 29 decreased his intake only at R21 when he was no longer apprehensive that there would not be enough food.

Comment

The loss of weight is a universal accompaniment of starvation and famine and may serve as a gross index of the disturbed balance between caloric intake and caloric output. Whereas the biochemical characteristics of the "internal environment" can vary only within narrow limits, the range of variation in the relative amount of muscles and fat compatible with health, or at least with life, is truly amazing.

The total loss of body weight in the Minnesota Experiment reproduced the conditions of severe semi-starvation. It is evident that the course of weight reduction and weight gain depends on the magnitude and duration of the caloric deficit or surplus. A slowly increasing reduction of food intake will produce a quantitatively different picture not only of weight changes but of functional alterations as well. The duration of the various phases of the Minnesota Experiment and the rate of weight changes, both negative and positive, were dictated in part by the exigencies of the experimental program and by the effort to provide dependable information, limited as it might be, for the task of nutritional rehabilitation of the starvation victims of World War II in Europe and in Asia.

As a guide for evaluating the dangers present in future emergencies, it seems useful to attempt to make an estimate of the magnitude of weight losses, expressed as percentages of the original (normal, usual) body weights, which might result from subsistence on diets reduced to various degrees and applied for 3 to 12 months. Such estimates are given in Table 46. The values are based on data from many sources, including field reports, the Carnegie Laboratory undernutrition experiment (1919), and the Minnesota Experiment. They are average values, apply only to "normal" adults, and presume no extreme changes in occupational or other activity. It is believed that in general these estimates will be accurate within ±50 per cent of the indicated change.

The limitations of body weight in evaluating the degree of starvation and recovery deserve emphasis. Thus the differences in relative water content of the body present definite complications in the physiological interpretation of the weight changes. The hydration during semi-starvation, even when it does not as-

TABLE 46

Estimated General Magnitude of Weight Losses, as percentage of original body weight, resulting from subsistence on reduced diets with caloric values expressed as percentage of intake needed for maintenance of normal weight.

Duration (mo.)	Caloric Intake as Percentage of Normal Balance							
	90	80	70	60	50	40	30	20
3	5	8	10	12	15	20	25	30
6	8	12	15	20	25	30	35	45
12 or more.....	10	15	20	25	30	35	40	

sume the form of edema which can be diagnosed clinically, tends to mask the magnitude of the insult suffered by a starving organism. In the early phases of rehabilitation the loss of edema fluid may lead to an underestimate of the amount of tissue actually laid down. In refeeding, especially when the caloric intake is high, fat deposits tend to increase at a faster rate than the "active" tissues. This is another factor that limits the value of gross body weight as a criterion of the nutritional state.

CHAPTER 7

Physical Appearance and External Dimensions

IN PLANNING the Minnesota Experiment the application of the methods of physical anthropology to the study of physique appeared useful for a number of purposes. Thus it was desirable to characterize our sample of subjects during the pre-starvation (control) period in terms of single body dimensions and anthropometric indexes against the background of the general population. This aspect has been discussed in Chapter 4.

In this chapter we are concerned with the changes in body dimensions and body build produced by the semi-starvation and rehabilitation regimens. Although emaciation is one of the well-known and striking characteristics of the state of semi-starvation, information on the magnitude of changes in the soft tissues of the different parts of the body resulting from known decrements in the total body weight is very limited.

Some measurements which are regarded as dimensions of the bony framework of the body, such as the bi-cristal diameter, have also been included in the hope that we may obtain an indication of the magnitude of the error inherent in measuring the "bony" dimensions on starved populations, owing to the fact that such measurements on living persons include not only the true distance between the bony points but also the overlying soft tissues.

In addition to direct anthropometric measurements, individual photographs were taken during the control period and at intervals during the experiment. The subjects were photographed nude at a constant distance from the camera and from the screen, calibrated in 5 cm. units, against which the body was projected.

Somatoscopic Observations on the Effects of Semi-Starvation

Inspection of the subjects at the different stages of starvation and rehabilitation, as well as of the photographs, revealed marked changes in the physique, characteristically associated with prolonged inanition. Decreases in muscle tissues and subcutaneous fat gave the men a haggard, emaciated appearance. The faces became thin, with prominent cheekbones. Nicolaeff (1923), describing gross morphological changes taking place in starvation, commented on the large degree of atrophy affecting the facial musculature and attributed to this factor, at least in part, the apathetic, mask-like expression of the face in famine victims. At times and in some of our subjects edema masked the degree of emaciation and gave the face a bloated expression, with swollen eyelids and puffiness of the cheeks. The neck became thin, with a more marked reduction in the transversal than in the anteroposterior direction, and appeared long. The clavicular outline

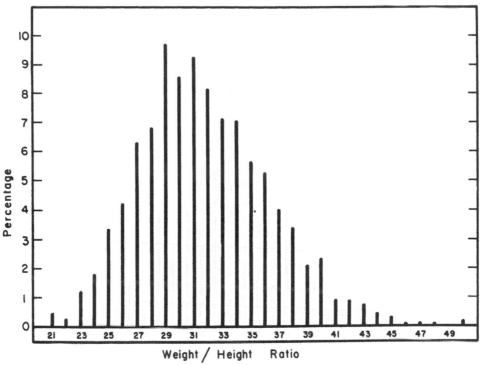

FIGURE 40. DISTRIBUTION OF WEIGHT/HEIGHT INDEX in female inmates of an internment camp in southern France (Zimmer *et al.*, 1944).

was sharp and the hollow deepened. The padding of the shoulder girdle was greatly reduced, with a marked decrease in the breadth of the shoulders.

The ribs became prominent, and there were the typical "winged" scapulae generally mentioned in field reports from regions in which undernutrition is prevalent (e.g. Adamson *et al.*, 1945). The pectoral area was flat. The vertebral column stood out because of the reduction in the dorsal muscle mass. The waist was narrow, "pinched." The iliac crests became prominent. The wasting of soft tissues was particularly marked in the region of the buttocks, which became thin and flat with the skin tending to hang in folds. The arms and legs were spindly. The photographs, taken with the feet in a standard orientation and at a constant distance apart, indicated a large increase in space between the thighs (see Figures 40 and 41 and Figures 151 through 158 in the Appendix).

Thus the diminution of soft tissues, including both the subcutaneous fat and the muscles, produced changes in physique characteristic of the "asthenic" body build. The bony framework remained the same, however. In a system of somatic typology which takes into account skeletal characteristics as well as fat and muscular development (Bullen and Hardy, 1946), one would expect both a shift in the direction of the "leptosome" (thin) habitus and an increase in the degree of dysplasia indicating the incongruence of the different characteristics of the body build.

The semi-starvation changes took place quite gradually and the staff as well

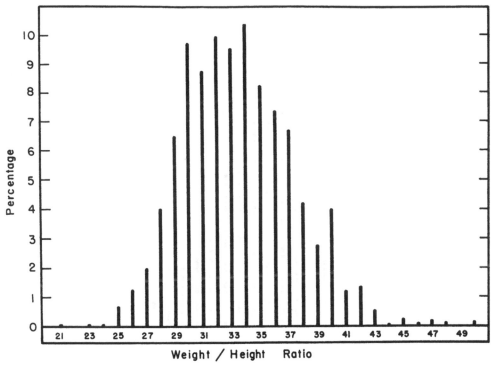

FIGURE 41. DISTRIBUTION OF WEIGHT/HEIGHT INDEX in male inmates of an internment camp in southern France (Zimmer *et al.*, 1944).

as the subjects tended to forget the original appearance of the men during the control period. The photographic records do not suffer from such a dulling of impressions and bring out vividly the degree of emaciation in starvation and the laying down of tissues in rehabilitation (Figures 151 to 158 in the Appendix).

Anthropometric Measurements in the Starvation Phase of the Minnesota Experiment

The linear anthropometric measurements were made by means of a steel measuring rod, calibrated in $\frac{1}{16}$ in., attached to a wall, and a large spreading caliper read to the nearest mm. The circumferences were measured by a flexible steel tape graduated in mm.

Care was taken to standardize rigorously the techniques of measurement, to determine quantitatively the landmarks when necessary (for example, in measuring the circumference of the upper arm), and to carry out the measurements during the morning to minimize effects of the daily variation in stature and in the distribution of edema fluid. All dimensions except the sitting height and the circumference of the extremities were made with the subjects standing. The definition of the dimensions and the procedures used are given in detail in the Appendix on methods. All determinations were made in duplicate and were repeated at a week's interval. The average of these four determinations has been used as the value for the individual at a given period.

Except for the bi-deltoid diameter, the linear measurements used in this study represent bony dimensions of the body. Actually no measurement on the living body is identical with a true skeletal dimension since the soft tissues enter in, even though pressure is exerted on the calipers to reduce this component of the measurement to a minimum. The magnitude of the error inherent in making body measurements based upon subcutaneous bony points was investigated by Todd (1925). The distance between the anterior superior iliac spines was selected as the dimension to be studied. In individuals with little subcutaneous tissue the bony landmarks are readily recognizable. In corpulent subjects it is more difficult to locate the bony points. In cadavers, and it is on this material that Todd made his observations, the difficulty is even increased because the fat tissues become solidified. The results of determining the bi-spinous diameter of the pelvis, first on the cadaver and subsequently on the torso after skin and other soft tissues had been removed and the bony points laid bare, are given in Table 47. The value obtained on the body before maceration is larger, the difference being 4.7 mm. for the male whites and 2.8 mm. for the male Negroes. The difference between the two racial groups suggests relatively less obesity for the examined Negro population as judged from the amount of subcutaneous adipose tissue. A mean value of about 5 mm. would be, then, the limit of shrinkage we might expect to obtain, on the average, in the most severe cachectic state. The individual differences will be large, of course, and it is unfortunate that Todd gave only the mean difference with no indication of the variability of the differences from individual to individual.

TABLE 47

Bi-Spinous Diameter of the Pelvis, in mm., determined on fresh cadaver and on the torso after stripping off the soft tissues
($N = 100$) (Todd, 1925).

	Male White		Male Negro	
	M	SD	M	SD
Cadaver	246.1	17.16	226.9	18.08
Torso	241.4	17.64	224.1	16.46
Difference	4.7		2.8	

In our subjects the breadth of the pelvis, determined as the bi-cristal diameter, was practically unchanged after 6 months of semi-starvation (see Table 48). This indicates that the error due to the inclusion of the soft tissues in the measurement of this diameter on the living is smaller than Todd's data would suggest. The change in the bi-trochanteric diameter is somewhat larger; in fleshy individuals considerable pressure has to be exerted to approach the bony landmarks. The decrease in the bi-acromial diameter when expressed in cm. is the largest of all bony dimensions. Under normal conditions this measurement is affected markedly by a change in posture. The fact that there was no progressive change after the twelfth week of the semi-starvation regimen may indicate that change in posture entered in, in spite of our efforts to maintain a standard posture while

TABLE 48

Linear Dimensions of the Body, in cm. M = mean of 32 individuals; SD = standard deviation of series; F = test of statistical significance of semi-starvation decrement; $d = M_C - M_{S24}$. For 1 and 31 degrees of freedom, $F_{0.05} = 4.16$, $F_{0.01} = 7.53$. (Minnesota Experiment.)

	C		S12			S24			d	$\dfrac{100d}{M_C}$	$\dfrac{d}{SD_C}$
	M	SD	M	SD	F	M	SD	F			
Standing height	178.65	5.77	178.51	5.71	5.43	178.33	5.73	22.97	−0.32	0.18	0.06
Sitting height	93.09	3.00	92.71	3.17	22.43	92.48	3.18	38.91	−0.61	0.66	0.20
Diameters											
Bi-deltoid	43.61	1.39	40.40	1.46	595.56	39.57	1.36	1017.25	−4.04	9.26	2.91
Bi-acromial	38.25	1.70	37.39	1.63	40.75	37.39	1.46	50.77	−0.86	2.25	0.51
Bi-cristal	27.97	1.36	27.92	1.63	2.00	27.92	1.61	2.21	−0.05	0.18	0.04
Bi-trochanteric	31.82	1.75	31.61	1.76	10.29	31.49	1.79	21.31	−0.33	1.04	0.19
Thorax, transverse	26.89	1.28	26.27	1.33	57.65	26.12	1.26	106.21	−0.77	2.86	0.60
Thorax, anteroposterior	18.59	1.02	18.50	1.20	1.53	18.28	1.26	14.13	−0.31	1.67	0.30

TABLE 49

Circumferential Dimensions of the Body, in cm. For legend see Table 48. (Minnesota Experiment.)

	C		S12			S24			d	$\dfrac{100d}{M_C}$	$\dfrac{d}{SD_C}$
	M	SD	M	SD	F	M	SD	F			
Thorax	89.28	3.33	83.69	3.03	314.38	82.61	2.79	423.83	−6.67	7.47	2.00
Abdomen	78.08	4.43	71.23	3.43	190.62	70.70	3.54	144.24	−7.38	9.45	1.66
Upper arm	28.49	1.78	23.37	1.32	1696.32	21.52	0.94	1238.11	−6.97	24.46	3.92
Thigh	47.11	2.85	40.30	2.26	745.06	37.96	1.90	886.71	−9.15	19.42	3.21
Calf	37.72	1.71	34.33	1.68	791.13	33.01	1.66	921.03	−4.71	12.49	2.75

TABLE 50

Changes in the Cross-Sectional Areas of the Limbs, in cm. (Minnesota Experiment).

	Upper Arm			Thigh			Calf		
	C	S12	S24	C	S12	S24	C	S12	S24
Limb									
Circumference	28.49	23.37	21.52	47.11	40.30	37.96	37.72	34.33	33.01
Diameter	9.07	7.44	6.85	15.00	12.83	12.08	12.01	10.93	10.51
Cross-sectional area	64.61	43.47	36.85	176.72	129.28	114.61	113.29	93.83	86.76
Area $d_s - c$		−21.14	−27.76		−47.44	−62.11		−19.46	−26.53
d as percentage of C		32.72	42.96		26.84	35.14		17.18	23.42

134

the anthropometric measurements were being taken. The changes in the thoracic diameters are small and probably reflect to a large extent a decrease in the subcutaneous tissues. There was a small decrease in standing height and a somewhat larger change in the sitting height. The change in the latter dimension was much less dramatic than the decrease in the size of the buttocks as seen in the photographs.

As one would expect, the decrements in the circumferences were more marked than were changes in the linear, predominantly bony dimensions, since the circumferences involved more soft tissues (see Table 49). On the absolute basis the change in the circumference of the thigh was the largest. On the percentage basis the decrease in the circumference of the upper arm was most striking and was parallel to the changes in body weight.

The changes are still more impressive when we deal with the cross-sectional areas of the limbs rather than with the circumferences. Assuming that the limbs at the levels where the measurements were made have a circular shape, we can readily compute the cross-sectional area. The values for control and after 12 and 24 weeks of semi-starvation are presented in Table 50. On the percentage basis, the decrements at the end of starvation varied from 23.42 per cent for the calf to 42.96 per cent for the upper arm.

In order to obtain an indication of the true changes in the soft tissues we should correct the gross values of the circumferences of the limbs for the bony component of the measurement. It has been possible to do so for the thigh. The dimensions of the femur were obtained from X-ray pictures of the leg, taken in a standard dorsoventral and mediolateral orientation and at a standard distance (30 in.) of the X-ray tube from the film. At a level corresponding approximately to the height of the leg at which the thigh circumference was determined, the average measured values were 32.4 mm. for the anteroposterior diameter of the femur and 29.7 mm. for the lateral diameter. These values were corrected for magnification resulting from the facts that the X-rays were not parallel and that the sagittal and coronal plane of the femur did not coincide with the plane of the film; in making the correction for magnification, both the focal-film distance and the object-film distance were taken into account (see Table 51). The percentage of distortion has been obtained by interpolation from Files' table (1945, p. 91).

By subtracting the value for the area of the femur from the gross cross section of the thigh we obtain the cross-sectional area of the soft tissues (see Table 52). The absolute differences between the control period and the successive stages of semi-starvation will remain, of course, the same. However, with a decrease in the denominator (C) the percentages are increased; the larger the cross-sectional area of the bone in relation to the surrounding soft tissues the larger will be this difference. The actual decrement in the cross-sectional area of the thigh, corrected for the bone, was 27.80 per cent at S12 and 36.40 per cent at S24. Using 3.8 cm.² as the estimated value for the cross-sectional area of the humerus at the approximate level at which the circumference of the arm was measured, the semi-starvation decrements were 34.8 per cent at S12 and 45.6 per cent of the control value at the end of the semi-starvation period (see Table 53).

TABLE 51

DIMENSIONS OF THE FEMUR at a level approximately corresponding to the thigh circumference measurements (Minnesota Experiment).

	Anteroposterior Diameter	Mediolateral Diameter
Raw value (mm.)	32.4	29.7
Focal-film distance (ft.)	2.5	2.5
Object-to-film distance (in.)	2.52	2.91
Magnification (%)	9.64	11.29
Magnification (mm.)	3.1	3.4
Corrected value (mm.)	29.3	26.3

TABLE 52

CROSS-SECTIONAL AREA OF THE THIGH, in cm.^{2}, corrected for bone (Minnesota Experiment).

	C	S12	S24
Limb cross-sectional area	176.72	129.28	114.61
Femur diameter (cm.)			2.78
Femur cross-sectional area			6.07
Muscle area — bone area	170.65	123.21	108.54
d_{s-c}		−47.44	−62.11
d as percentage of C		−27.80	−36.40

TABLE 53

CROSS-SECTIONAL AREA OF THE UPPER ARM, in cm.^{2}, corrected for bone (Minnesota Experiment).

	C	S12	S24
Limb cross-sectional area	64.61	43.47	36.85
Humerus, estimated area	3.80	3.80	3.80
Muscle area — bone area	60.81	39.67	33.05
d_{s-c}		−21.14	−27.76
d as percentage of C		−34.76	−45.65

Anthropometric Measurements in Starvation — Data from the Literature

When anthropometric measurements were carried out on a subject undergoing a 31-day voluntary fast, during which his weight decreased from 60.64 kg. to 47.39 kg., or 21.85 per cent of the original body weight (Benedict, 1915b), no change in the height was observed. The percentage change in the waist and thigh circumferences approached the magnitude of the total loss in weight (see Table 54).

Changes in body dimensions were studied by Benedict *et al.* (1919) in 2 groups of men participating in a semi-starvation experiment. The decreases in body weight were small. The relative changes in the various anthropometric measurements of the 2 groups are in general consistent, with the circumference

TABLE 54

ANTHROPOMETRIC MEASUREMENTS, in mm., of a subject undergoing a 31-day fast (Benedict, 1915, p. 68).

	Control	Day of Fast					Total Loss	Percentage Loss
		5th	12th	19th	25th	31st		
Standing height....	1707	1707	1707	1704	1707	1707	0	0
Girth								
Neck	376	371	368	361	338	335	41	10.9
Chest	879	871	856	825	805	800	79	9.0
Abdomen	800	785	742	757	686	681	119	14.9
Waist	780	749	696	673	648	627	153	19.6
Biceps	251	241	234	226	221	211	40	15.9
Thigh	488	465	450	432	427	394	94	19.3
Calf	335	335	323	310	305	300	35	10.4
Weight (kg.)	60.6	56.4	53.6	51.1	49.3	47.4	13.2	21.8

of the arms, thighs, and abdomen approaching or even exceeding the total percentage loss of body weight. The changes in height were small and not consistent (see Table 55).

The changes in some anthropometric dimensions and in weight of samples of individuals applying for life insurance during conditions of normal food supply and during the years of food shortage were reported for Germany by Fischer (1923). The weight average decreased from 1913 to 1918 by 10.8 kg., or 14.4 per cent of the prewar level, while the change in the circumference of the chest decreased by 6.2 cm. (6.9 per cent) and the abdominal circumference by 9.0 cm. (10.4 per cent). There was also a decrease in height amounting to 2.5 cm. (1.5 per cent). When the change of body weight was corrected for the simultaneous change of height, in 1918 the decrease below the normal as computed from the weight/height tables was given as —11.1 per cent. From 1918 up to the first months of 1922 there was a gradual return to the prewar levels (see Table 56).

From the internment camps in southern France (Zimmer, Weill, and Dubois, 1944) decreases in height were reported, caused by scoliosis due to the osteopathy of starvation. It is regrettable that no data are given, especially since the pre-internment weights and heights as recorded in the military identification papers of the men were available.

Most extensive anthropometric observations on starving populations were made by Ivanovsky (1923). The data were based on a large sample of adults (N = 2114 individuals: 1284 men, 830 women) including Great Russians, Ukrainians, and other national and racial groups of the Soviet Union (White Russians, Armenians, Georgian Gruzins, Tartars, Zyrians, Permiaks, Bashkirs, Kalmucks, and Kirghizs). We shall pay attention only to the first 2 groups, combining the data for the different provinces in which the observations were made. The author reported that the observations were repeated at 6-month intervals for 3 years. Unfortunately, only the figures for the periods "before the famine" and "after the famine" were given. The exact time of the investigations was not indicated, but it is presumed that the author and his unnamed colleagues who participated in

TABLE 55

AVERAGE ANTHROPOMETRIC DIMENSIONS of 2 groups of men participating in a semi-starvation experiment (Benedict *et al.*, 1919, p. 234).

	Squad A (N = 12)				Squad B (N = 11)			
	Sept. 29	Nov. 24	d	d%	Jan. 5	Jan. 27	d	d%
Standing height	171.07	171.26	+0.19	0.11	173.61	173.31	−0.30	0.17
Girths								
Arms at the axilla ...	30.91	27.87	−3.04	9.84	30.60	28.71	−1.89	6.18
Chest at nipples	90.09	84.72	−5.37	5.96	91.48	87.58	−3.90	4.26
Abdomen at umbilicus	75.47	69.06	−6.41	8.49	74.57	70.64	−3.93	5.27
Hips and buttocks ..	92.12	87.15	−4.97	5.40	94.13	90.37	−3.76	3.99
Thighs at perineum ..	55.42	50.85	−4.57	8.25	56.17	52.46	−3.71	6.60
Weight	66.95	60.78	−6.17	9.22	67.87	64.02	−3.85	5.67

TABLE 56

ANTHROPOMETRIC CHARACTERISTICS OF GROUPS OF INDIVIDUALS EXAMINED WHEN APPLYING FOR LIFE INSURANCE. Number of individuals per year = 984 (Fischer, 1923).

Year	Age (yrs.)	Weight (kg.)	Height (cm.)	Chest Girth (cm.)	Abdominal Girth (cm.)
1913	31.6	75.2	170.8	89.3	86.9
1916	30.0	68.2	170.2	85.0	81.1
1917	29.1	65.7	169.0	83.4	78.0
1918	30.7	64.4	168.3	83.1	77.9
1919	30.1	65.8	168.6	83.3	78.7
1920	30.5	67.1	170.3*	84.5	79.3
1921	31.6	68.0	170.3	85.0	82.0

* The value of 172.3 in the original table is considered to be a typographical error.

the anthropometric study collected their material in the years 1920–22. No information on the caloric intake is available.

The reduction in weight was a universal phenomenon, with the largest part of the population studied exhibiting a loss of 11 to 20 per cent of the original body weight (see Table 57). The changes in body dimensions are summarized in Table 58. It is unfortunate that important dimensions such as thoracic circumference and length of the extremities were given only as percentages of stature; the latter diminished markedly in the populations studied by Ivanovsky and can no longer serve as a stable and useful reference dimension. Most attention was paid by Ivanovsky and his co-workers to the head dimensions, which were not studied at all in the Minnesota Experiment and which, as a matter of fact, have little interest in relation to starvation. Thus the amount of useful material is relatively small.

Under the influence of inanition a marked degree of diminution of stature was observed in the population of the Soviet Union, the average decrease vary-

TABLE 57

DECREASE IN GROSS BODY WEIGHT AFTER 3 YEARS OF FAMINE. The tabular values indicate percentages of the population exhibiting a particular weight decrement. From Ivanovsky (1923, p. 349).

	Decrement as Percentage of Pre-Famine Weight			
	1–10	11–20	21–30	31–40
Great Russians				
Men (N = 313).........	28	43	26	3
Women (N = 229)	22	49	28	1
Ukrainians				
Men (N = 250).........	19	59	22	
Women (N = 200)	29	54	17	

TABLE 58

CHANGES IN BODY DIMENSIONS, in cm., after 3 years of famine (Ivanovsky, 1923).

	Great Russians		Ukrainians	
	Men	Women	Men	Women
Stature				
Before	165.9	154.5	165.8	156.1
After	161.2	151.0	161.6	152.0
Difference	−4.7	−3.5	−4.2	−4.1
Length of Trunk				
Before	52.0		52.1	
After	48.9		48.9	
Difference	−3.1		−3.2	
Chest Circumference				
Before	88.1		88.5	
After	81.4		82.8	
Difference	−6.7		−5.7	
Length of Arms				
Before	76.0	67.5	75.9	68.0
After	75.1	67.6	75.5	68.1
Difference	−0.9	+0.1	−0.4	+0.1
Length of Legs				
Before	85.3	79.5	86.1	81.6
After	85.1	79.8	85.6	81.0
Difference	−0.2	+0.3	−0.5	−0.6

ing in different parts of the country and for different racial and national segments from 3.8 to 6.1 cm. in men and from 3.6 to 4.8 cm. in women. Among the male populations the largest decrease was observed in the Crimean Tartars (6.1 cm.). As one would expect, in absolute units the tall individual diminished in stature more than the short person. Ivanovsky also commented that the stature decreased more in the intellectuals than in the laborers. In view of the emphasis on the effect of standing and physical work in decreasing the stature under normal conditions during the course of a day (and the possibility of a cumulative effect), this observation may appear paradoxical. In all probability this simply means that the intellectuals were taller at the start and showed a greater decrement as a result of the famine, a tendency which was previously noted for the population at large.

Individuals who had suffered from concomitant infectious diseases, particularly typhus, were reported to have had a larger decrease in stature than persons who were victims of uncomplicated semi-starvation. Also, the decrease in the older group (40–55 years) was greater than in the younger group (20–39 years), which might indicate a combined effect of starvation and the process of senescence.

In comparison with the diminution of stature during the Russian famine of 1920–22, the change in this dimension in the Minnesota Experiment was extremely small (—0.15 cm. at S12; —0.32 cm. at S24). This is particularly interesting in view of the fact that under conditions of natural starvation the largest part of the decrease in stature took place during the first 6 months of famine, with further noticeable decrease during the second half of the first year and essentially no change in the subsequent 2 years, in the face of continued famine. The difference may be due in part to the technique of measuring the stature. Ivanovsky noticed a tendency toward a bent posture, which we also observed. We insisted on a standard, erect posture while stature was measured; Ivanovsky did not specify his technique but probably a rod anthropometer was used in his field measurements.

Change in the length of the trunk, measured from the jugular notch to the pubic symphysis, appears to account for a large part of the decrease in stature. It may indicate a thinning of the cartilagenous intervertebral disks resulting in the decrease of the length of the vertebral column. However, postural factors (bending) may have affected this measurement as well.

The thoracic circumference and the length of the extremities were computed from Ivanovsky's data on the relative changes in these dimensions and the absolute values for stature. The decrease in the circumference of the chest was very close to that observed in the Minnesota Experiment (—6.7 cm. at S24). The changes in the length of the extremities were small and probably within the limits of the error of measurement.

Ivanovsky's data on the modification of the morphology of the head are summarized in Table 59. The greatest absolute decrement occurred in the horizontal circumference of the head. The changes in the other dimensions were small (mostly less than 1 cm.) and surprisingly uniform in magnitude. The anteroposterior diameter shortened somewhat less than the transverse diameter of the

TABLE 59

CHANGES IN MEASUREMENTS OF HEAD AND FACE after 3 years of famine
(Ivanovsky, 1923).

	Great Russians		Ukrainians	
	Men	Women	Men	Women
Horizontal Circumference of Head (cm.)				
Before	55.0	53.1	55.4	53.3
After	52.9	51.2	53.4	51.6
Difference	−2.1	−1.9	−2.0	−1.7
Length of Head, Vertex to Chin (cm.)				
Before	21.8	20.6	21.8	20.8
After	20.8	19.6	20.7	19.8
Difference	−1.0	−1.0	−1.1	−1.0
Anteroposterior Diameter of Head (cm.)				
Before	18.4	18.0	18.5	18.1
After	18.1	17.7	18.1	17.7
Difference	−0.3	−0.3	−0.4	−0.4
Transverse Diameter of Head (cm.)				
Before	15.2	14.8	15.4	15.2
After	14.5	14.2	14.7	14.5
Difference	−0.7	−0.6	−0.7	−0.7
Length of Face (cm.)				
Before	18.1	17.2	18.1	17.3
After	17.7	16.9	17.7	16.9
Difference	−0.4	−0.3	−0.4	−0.4
Breadth of Face (cm.)				
Before	13.9	13.1	13.9	13.4
After	13.4	12.6	13.4	13.0
Difference	−0.5	−0.5	−0.5	−0.4
Nasal Index				
Before	70.7	73.1	71.9	73.2
After	68.4	71.0	70.6	71.5
Difference	−2.3	−2.1	−1.3	−2.7

head, resulting in a slight lowering of the cephalic index (for example, from 82.4 to 80.2 for Great Russian men; from 82.4 to 79.7 for women). Modifications in the facial dimensions paralleled the changes in the head measurements. The height of the face, from the hair line to the chin, decreased less than the width; consequently, the facial index, obtained as the ratio of the two dimensions, also decreased. On the sample of the Great Russian population this change was from 76.8 to 75.6 in men and from 76.0 to 74.5 in women. The height of the nose was reported as unchanged or only slightly diminished (−1 to −2 mm.) while the width decreased, resulting in a diminution of the nasal index.

Photographic Analysis of Body Dimensions in the Starvation
Phase of the Minnesota Experiment

In addition to the measurements taken directly on the subjects at the time of the experiment, a photogrammetric analysis of the changes in bodily dimensions taking place during semi-starvation was carried out by Dr. Gabriel Lasker of the Department of Anatomy, Wayne University (Lasker, 1947).

The photographs were taken at the Laboratory of Physiological Hygiene by Mr. Jarrot Harkey at C, S12, S24, and R12. Complete measurements on photographs for the whole group were made only at C and S24. The details of taking the photographs and making the measurements are given in the Appendix on methods.

The subjects stood on a rotating stool 37.5 cm. in diameter. The three views — frontal, lateral, and dorsal — were obtained by turning the stool, without the subject's changing position. Fixed blocks at the top of the stool held the feet at a uniform angle and at a constant distance from the edge of the stool. Twenty-five cm. from the center of the pedestal was a background screen with a grid of lines 5 cm. apart. The location of the camera with respect to the grid screen was fixed.

The measurements were made by a pair of vernier calipers reading to 0.1 mm. All values for individual photographs were obtained as the average of two or more determinations, made at different times. The accuracy (repeatability) of the measurements and detailed definitions of the dimensions studied are discussed in the Appendix on methods. The selection of dimensions was made with the aim of obtaining metric data necessary for the characterization of somatic "types" according to Sheldon's system (1940). In this context we are concerned only with the changes resulting from semi-starvation, without reference to any typological scheme. The data for C and S24 are presented in Table 60. All measurements show an average decrease. Furthermore, with the exception of a few dimensions (ankle breadth, wrist thickness, and facial breadth) which were affected by edema, every individual showed a decrease in respect to all photographic measurements taken.

The values were expressed as percentages of the stature. The general tendency toward a decrease in all body dimensions is very striking. However, just as we have seen in the case of anthropometric measurements taken on living persons, a more detailed examination of the data disclosed marked differences in the degree of reduction for different parts of the body and in the extent of change in different individuals. In respect to every measurement some individuals changed at least twice as much as others, while the weight losses varied from 18.8 per cent to 29.3 per cent with an average of 24.1 per cent of control body weight.

The greatest decreases in size were observed in the upper thigh and upper arm diameters. The decrease in these measurements averaged 25 per cent of the original value. In these measurements and also in the neck breadth, the decrease was relatively constant for all individuals. A study of the photographs revealed especially marked loss of tissue in the gluteal region. However, the method used for measuring pelvic depth does not reflect the full extent of this change.

Areas which showed less decrease on the average and less regular decrease reflect several different processes. In the case of the head, for instance, the thin-

TABLE 60

BODILY DIMENSIONS ACCORDING TO PHOTOGRAPHIC MEASUREMENTS. All values are expressed as percentage of stature. (Minnesota Experiment.)

	Number of Subjects	C			S24			d = (S24 − C)			100d
		M	Min.	Max.	M	Min.	Max.	M	Min.	Max.	C
Upper facial breadth	32	9.07	8.4	10.0	8.72	8.0	9.7	−.35	+ .20	−1.19	−3.86
Lower facial breadth	32	7.51	6.9	8.2	7.16	6.6	7.7	−.35	0.00	−.76	−4.66
Neck depth	24	6.24	5.7	6.8	5.92	5.4	6.5	−.32	−.01	−.74	−5.13
Neck breadth	32	6.63	5.9	7.5	5.79	5.4	6.4	−.84	−.55	−1.32	−12.67
Upper trunk breadth	32	18.99	15.5	20.8	17.27	15.2	18.8	−1.72	−.28	−3.61	−9.06
Upper trunk depth	31	11.72	10.4	13.0	10.82	9.9	11.8	−.90	−.08	−1.70	−7.68
Waist breadth	32	15.06	12.8	16.7	13.28	12.0	14.6	−1.77	−.43	−2.64	−11.75
Upper arm thickness	31	6.93	5.7	8.3	5.07	3.9	6.2	−1.87	−1.16	−2.92	−26.98
Forearm thickness	31	5.88	5.3	6.4	5.16	4.6	5.7	−.72	.29	−1.36	−12.24
Wrist thickness	30	3.95	3.5	4.6	3.67	3.1	4.1	−.28	+ .26	−1.02	−7.09
Waist depth	31	11.17	9.5	12.4	9.70	8.8	11.0	−1.47	−.39	−2.49	−13.16
Hip breadth	32	19.09	16.4	22.0	18.03	15.4	19.7	−1.06	−.22	−2.29	−5.55
Pelvic depth	31	12.98	11.2	14.7	11.04	10.3	11.8	−1.94	−.74	−3.22	−14.95
Upper thigh depth	31	10.29	9.2	11.5	7.68	7.0	8.8	−2.61	−1.77	−3.56	−25.36
Lower thigh depth	31	7.32	6.6	8.0	6.44	5.9	7.1	−.87	−.20	−1.84	−11.89
Calf breadth	32	6.58	6.0	7.2	5.82	5.3	6.6	−.75	.21	−1.25	−11.40
Ankle breadth	32	3.30	2.9	3.7	3.20	2.7	3.8	−.10	+ .38	− .38	−3.03

143

ness of the tissues over the relatively immutable skull gives little room for shrinkage. The cartilages of the larynx and the spines of the vertebrae seem to have somewhat the same stabilizing effect on the depth of the neck; the breadth of the neck decreased more markedly. The decrease in the breadth and depth of the upper part of the trunk reflects a loss of fat in the subcutaneous tissues and also a decrease in the size and contour of the pectoral and dorsal muscles. Decrements in the forearm, lower thigh, and calf are considerable in all the individuals.

In the case of the wrist there is rather a large variability; two individuals actually showed an increase. This may be in part an artifact due to the difficulties

TABLE 61

DIRECT ANTHROPOMETRIC MEASUREMENTS expressed as percentages of stature. $N = 32$. (Minnesota Experiment.)

	C	S24	$d = S24 - C$	$\frac{100d}{C}$
Sitting height	52.11	51.86	−0.25	0.48
Diameters				
Bi-deltoid	24.41	22.19	−2.22	9.09
Bi-acromial	21.41	20.98	−0.44	2.06
Bi-cristal	15.66	15.66	0.00	0.00
Bi-trochanteric	17.81	17.66	−0.13	0.84
Thorax, transverse	15.05	14.65	−0.40	2.66
Thorax, anteroposterior .	10.40	10.25	−0.15	1.44
Circumferences				
Thorax	49.97	46.32	−3.65	7.30
Abdomen	43.71	39.65	−4.06	9.29
Upper arm	15.95	12.07	−3.88	24.40
Thigh	26.37	21.29	−5.08	19.26
Calf	21.11	18.51	−2.60	12.32

in exactly reproducing the position of the wrist, but on the other hand it may reflect the presence of edema. In ankle measurements 5 out of 32 individuals showed an increase in the diameter. The ankle, unlike the wrist, was very constantly oriented to and very accurately distanced from the camera, and the increases cannot be ascribed to inaccuracies in the technique of measurement. That the occasional increase in the size of the ankle was caused by nutritional edema appears clear from an inspection of the photographs of at least 2 of the 5 individuals for whom an increase in the dimension is reported. These 2 individuals showed an increase of more than 10 per cent, whereas 3 others showed an increase of less than 3 per cent, and all other individuals showed a decrease.

In order to provide anthropometric data in a form usable for comparison with the changes in the photographic diameters, the anthropometric averages were expressed as the percentage of the average stature at C and S24 (see Table 61). The two methods give similar results. In some measurements on living persons, such as the bi-trochanteric diameter, there is a compression of the soft parts; and since the form of the bone is probably little, if at all, affected by partial starvation in adult man, the photographic measurements tend to show greater change. The bi-deltoid diameter, in which the deltoid muscle as well as the bony framework is involved, decreased as much as the photographic upper trunk breadth. The chest circumference showed a decrease comparable to the diminution in photographic upper trunk measurements. The anteroposterior thoracic diameter, measuring the rib cage, showed little decrease in size; the photographic upper trunk depth measured also the decrease in fat and muscle of the chest and scapular regions and exhibited a greater decrement.

The bi-cristal diameter when expressed as a percentage of stature showed no tendency to decrease, and the bi-trochanteric diameter showed very little change. Both measurements involved pressure toward the bone and hence showed less decrease than the photographic hip breadth, because the changes in overlying soft tissues are eliminated to a considerable extent by the technique of anthropometric measurement. The abdominal circumference showed a considerable mean decrease but one not quite as marked as for photographic depth of the waist. Among the measurements made on living persons, the upper arm circumference decreased most; the loss here was closely proportional to that recorded on the photographs for upper arm thickness. Circumferences of the calf, and especially of the thigh, showed marked decreases which are comparable to those indicated by the photographic diameters of the respective regions.

In general, the measurements of various diameters on the photographs showed decreases which were closely parallel to but tended to exceed the degree of decrease recorded for the circumferences taken at approximately the same levels. Most trunk diameters taken on living persons were designed by anthropologists for measuring the bony frame (thorax, and pectoral and pelvic girdles). However, they do include to some extent the soft parts overlying these structures. The degree of decrease in these tissues may be expected to be roughly proportional to the amount of soft tissue overlying the skeletal support at the points where the measurements are made. Judging from the external configuration alone, the losses appear to have been in muscle tissue as well as in fat. They were very pronounced in the upper arm and thigh, the waist and buttocks. They also occurred over the thoracic wall, in the forearm and calf, in the neck and — at least to some small extent — everywhere, even over the skull.

Somatic Types in Starvation — Anthropometric Data

Scientific (and pseudoscientific) interest in human body build has been frequently motivated by a hypothesis that there is a significant correlation between the physical appearance, functional characteristics, temperament, and susceptibility to disease. In this context we are not concerned with the large and complex problem of the interrelationships between the structural, physiological, psychological, and pathological characteristics and tendencies of groups of subjects differing in body build, sex, age, ethnic origin, and other features which are generally considered "constitutional" factors. Human constitution, in the broad sense of the word, is regarded as being determined largely by heredity even though it is granted that the constitutional characteristics may be influenced, in varying degrees, by environmental factors (Montagu, 1945, p. 232). The smaller this environmental effect the more a given characteristic may be considered to depend on heredity and to be a stable, "constitutional" characteristic of the individual.

A large number of anthropometric indexes have been proposed for characterizing the gross "morphological constitution" (Tucker and Lessa, 1940, p. 413). In the Minnesota Experiment the following ratios were computed: weight/height, chest/height, abdomen/chest, and the Pignet index.* In all but the Pignet index, low numbers indicate a thin person and high numbers a stout one.

On inspection the individual body build in any large population will vary from emaciated through slender, sturdy, and plump to obese. Similarly, the index values obtained for a large number of individuals form a continuous series and

TABLE 62

Changes in Anthropometric Indexes of Body Build in Experimental Semi-Starvation. In all but Pignet's index, the low index values indicate a leptosomic habitus and the high values a eurysomic habitus. $N = 32$. (Minnesota Experiment.)

Index	C		S24		$d =$ (S24 − C)	$\dfrac{100d}{C}$	$\dfrac{d}{SD_C}$
	M	SD	M	SD			
$\dfrac{\text{Weight}}{\text{Height}}$	38.8	2.83	29.5	1.69	−9.3	24.0	3.3
$\dfrac{\text{Bi-cristal diameter}}{\text{Height}}$	15.6	0.65	15.6	0.63	0.0	0.0	0.0
$\dfrac{\text{Chest circumference}}{\text{Height}}$	50.0	2.49	46.3	2.08	−3.7	7.4	1.5
Pignet's index	20.0	8.39	43.1	5.79	+23.1	115.5	2.8

tend to follow a normal distribution curve. The concept of a "type," when derived from a one-dimensional continuum, represents necessarily an abstraction based on the extremes. The terms leptosomic, mesosomic, and eurysomic† may be used to indicate a subject's position along the thin-stout continuum for a given index; no reference to any typological system is involved.

To what extent are these indexes affected by variations in the nutritional state of an adult individual? The Minnesota Experiment provides data on one aspect of the question: the effect of a decrease in food intake resulting in the loss of 24 per cent of body weight.

The data for the 5 indexes obtained during the control period and at S24, and the differences, are given in Table 62. The bi-cristal/height index has an identical value at the control period and at the end of semi-starvation; also, the inter-individual differences indicated by the standard deviation of the series were not

* Weight/height index $= \dfrac{\text{Weight in kg.} \times 100}{\text{Stature in cm.}}$

   Chest/height index $= \dfrac{\text{Chest circumference in cm.} \times 100}{\text{Stature in cm.}}$

   Pelvis/height index $= \dfrac{\text{Bi-cristal diameter in cm.} \times 100}{\text{Stature in cm.}}$

   Abdomen/chest index $= \dfrac{\text{Circumference of the abdomen} \times 100}{\text{Circumference of the chest}}$

   Pignet's index $=$ Stature in cm. − Chest circumference in cm. + Body weight in kg.

† Greek soma = body; leptos = slender; mesos = middle; eurys = broad.

affected. The weight/height index decreased by 24 per cent of the control value and the men became a more uniform group, the standard deviation decreasing from 2.83 to 1.69. This reflects the fact that, in prescribing the weight losses at the start of the experiment, account was taken of the relative nutritional status of the individual, the overweight men losing larger percentages and the underweight men losing smaller percentages of the pre-starvation body weights. The chest/height index decreased by 7.4 per cent, the standard deviation decreasing from 2.49 to 2.08.

TABLE 63

FAMINE CHANGES IN THE PERCENTAGES OF THE POPULATION CLASSIFIED
ACCORDING TO PIGNET'S INDEX (Ivanovsky, 1923, p. 350). $N = 313$
Great Russians, 250 Ukrainians (men).

	Pignet's Index				
Below 10	11–15	16–20	21–25	26–30	31 and above
Great Russians					
Before 7	24	35	26	7	1
After 2	15	31	29	20	3
Ukrainians					
Before 6	24	24	33	12	1
After 2	15	16	39	24	4

One would expect that the external form of the body as characterized by indexes derived from anthropometric data would change as the result of loss of weight. There is very meager information available in the literature indicating the *relative* stability of the constitutional indexes in an adult population undergoing marked changes in nutritional status.

Valuable data on changes in Pignet's index in a population suffering from famine were presented by Ivanovsky (1923). The weight losses varied widely; the mode of the distribution of weight losses, expressed as the percentage of the pre-famine weight, fell within the range of 11 to 20 per cent of body weight lost. The distribution of the Pignet index, before and after 3 years of famine, is given in Table 63.

Before famine the index tended to be distributed "normally." The means have been estimated from Ivanovsky's data as 18.8 for the Great Russians and 19.7 for the Ukrainians, which would be regarded as indicating a "good" body build.* During famine the number of persons with a "weak" (that is, numerically high) physical index was uniformly augmented. The estimated mean index value increased to 21.5 and 22.5 respectively, not a very impressive change as compared with the changes in the Minnesota Experiment, in which the loss of body weight was larger on the average, and more uniform.

* Ivanovsky accepted the following qualitative gradations of the physical constitution in reference to Pignet's index: below 10, very strong; 10–15, strong; 16–20, good; 21–25, average; 26–30, weak; above 31, very weak.

The control values for our subjects fell in very much the same range as Ivan-ovsky's material, with an average of 20.0. After 6 months of semi-starvation the mean index value rose to 43.1 and there was a marked shift in the distribution toward the high values (see Table 64), with only a slight overlap of the distributions for the two periods, C and S24.

A procedure somewhat similar to the computation of the Pignet index was used by Fischer (1923), a physician for the Life Insurance Bank in Gotha, Germany. His "formula-number" ("F") was based on the finding that the sum of chest circumference ("C") and abdominal circumference ("A") at expiration in normal individuals tended to equal the sum of three fourths of the height (0.75 "H"), age in years ("Y"), and a complementary value of 15:

$$C + A = 0.75H + Y + 15$$

"F" was then defined as the difference of the two sides of the equation

$$F = (C + A) - (0.75H + Y + 15)$$

and would tend to approach zero. Negative values indicate a slender body build and positive values a stout body build. The author is aware of the mathematical weaknesses of his formula-number; he warns against the use of it as a constitutional index applicable in individual cases, stressing its usefulness for comparative studies on the nutritional status of large groups.

Fischer had an opportunity to examine on a number of occasions a group of clients of the insurance company who had applied for readmission. The average number of examinations per individual was 2.45 over a spread of 7 years. The average value of the formula-number for the 10 years preceding World War I in the insured population was zero, exhibiting a slight tendency toward positive

TABLE 64

CHANGES IN THE DISTRIBUTION OF PIGNET'S INDEX AFTER 24 WEEKS OF SEMI-STARVATION (Minnesota Experiment). $N = 32$ men. The tabular values represent percentage frequencies at each interval.

	Pignet's Index										
	Below 5	6–10	11–15	16–20	21–25	26–30	31–35	36–40	41–45	46–50	51 and above
C ..	3.1	9.4	21.9	18.8	15.6	28.1		3.1			
S24 .						3.1	6.2	21.9	43.8	21.9	3.1

TABLE 65

AVERAGE VALUES OF THE "FORMULA-NUMBER"* (normal adult average = 0) in a group of 717 individuals examined by an insurance company physician on a total of 1756 occasions (Fischer, 1923).

	1905–8	1910–14	1915	1916	1917	1918	1919	1920	1921
Formula-number ...	−0.8	+0.9	−5.0	−9.7	−9.7	−13.3	−10.0	−9.6	−8.6

* "F" = Chest circumference at expiration + Abdominal circumference − ¾ Height − Age in years − 15.

TABLE 66

AVERAGE VALUES OF THE "FORMULA-NUMBER"[*] (normal adult average = 0) in groups of individuals examined by an insurance company physician. Total number of persons per year = 984. From Fischer (1923).

Year	Age Group								Average	Average Difference from Prewar Level
	15-19	20-24	25-29	30-34	35-39	40-44	45-49	50+		
1907-13 ...	−4.0	−0.5	+ 1.9	+ 2.7	+ 2.0	+ 1.0	− 1.1	− 4.2	+ 1.2	
1915	−7.1	−3.3	− 1.7	− 1.1	− 2.6	− 4.0	− 6.5	−12.8	− 3.1	− 4.3
1916	−6.7	−3.9	− 4.7	− 4.7	− 4.5	− 5.1	− 6.6	−18.3	− 5.6	− 6.8
1917	−7.2	−5.6	− 9.3	− 9.4	− 9.8	−10.3	−15.1	−25.0	− 9.3	−10.5
1918	−7.8	−7.6	−10.2	−11.9	−14.2	−14.0	−17.6	−22.0	−13.3	−14.5
1919	−8.4	−6.6	− 8.9	− 9.8	−11.4	−13.4	−15.1	−17.5	−10.2	−11.4
1920	−6.4	−8.4	− 9.3	−10.8	−11.8	−11.9	−13.2	−19.7	−10.2	−11.4
1921	−5.9	−6.1	− 7.0	− 6.5	− 6.6	− 7.3	−10.9	−14.7	− 7.1	− 8.4

[*] "F" = Chest circumference at expiration + Abdominal circumference − ¾ Height − Age in years − 15.

149

values. In 1915 the formula-number began to fall, reaching the lowest level of
—13.3 in 1918 and improving from then on; the value for the first 2 months of
1922 was —6.0 (see Table 65). There was little shift in the age composition of
this group, the average before the war being 29.1 years, during the war 30.1
years, and after the war 34.1 years, with a general average of 33.2 years.

In addition, Fischer gives the figures for a much larger group of individuals
who were examined only once. Each year a random sample of 984 individuals
were examined — 492 cases during the first half of the year and 492 in the second
half. The values, broken down according to age groups, are given in Table 66.
It is evident that the effects of malnutrition were much more marked in the older
age group. This probably resulted from a difference in the dietary intake rather
than from a true age difference in response to caloric reduction. Fischer is more
inclined to affirm the latter possibility and points out that the food available during World War I was more easily digested by younger than by older people.

In the Minnesota Experiment the formula-number, computed from the average data for the group as a whole, was —7.3 at C and decreased to —21.4 at S24.

Zimmer, Weill, and Dubois (1944) obtained a weight/height index for the
semi-starved male and female population of the internment camps in southern
France. Computed on the basis of their Figures 1 and 2, the mean index value
for the men is 33.5 and for the women 31.8. No pre-starvation data are given, but
the value of 41.0 is postulated as the "physiological" normal. The distribution of
the index values in a section of the male and female population of one camp is
given in Figures 40 and 41. In the Minnesota Experiment the average value of
the weight/height index decreased from the control value of 38.8 to 29.5 at the
sixth month of semi-starvation.

Somatic Types in Starvation — Photogrammetric Data

It has been recognized that any single anthropometric index is an inadequate
characteristic of the body build. Attempts such as those of Pignet to combine a
few gross measurements go only halfway at best. In Italy, Viola (1936) developed an elaborate system for morphological classification of individuals according to the dimensions of the component parts of the body. In the United States
Sheldon approached the problem of physical constitution from a somewhat similar point of view. Sheldon's system has the advantage that the interindividual
variations in physique may be studied by means of measurements on photographs; the procedure is advocated as an exact and objective method of constitutional typology (Sheldon et al., 1940).

Sheldon's typological classification consists of assigning to the 5 regions (I,
head, face, and neck; II, thoracic trunk; III, arms, shoulders, and hands; IV, abdominal trunk; V, legs and feet), and subsequently to the body as a whole, a
number according to the intensity of the three "components" of physique: the
endomorphic, mesomorphic, and ectomorphic components. Less weighed with
theory and more useful appear the descriptive equivalents suggested by C. W.
Dupertuis: soft roundness (endomorphy), muscular solidity (mesomorphy),
and linearity-delicacy (ectomorphy) (Drapper, Dupertuis, and Caughey, 1944,
p. 132).

These components are conceived as representing the predominance of tissues and organs principally derived from the three embryonic layers (endoderm, mesoderm, and ectoderm).* In anthroposcopic somatotyping the values for the three components are obtained as estimates, varying from 1 (very low) through 4 (intermediate) to 7 (very high). The body build is characterized by a 3-digit number. Thus 7-1-1 would indicate an extremely endomorphic character of a body region or an individual; 4-4-4 a balanced middle form; and 1-1-7 an extreme ectomorphy. For photometric somatotyping Sheldon established mean values, empirically derived, for the various somatotypes.

Using Sheldon's tables it is possible to choose for any set of standard photographs of an individual (and for each region of the body separately) the type which shows the lowest mean-rank deviation from the measurements. In case of a tie, Sheldon recommends that one resort to anthroposcopic estimates. After typing the 5 regions of the body separately, Sheldon computes a general average to characterize the total physical constitution. In the pictures used for the present photogrammetric study, the head and neck were partly blocked out and hence unavailable for study. Furthermore, the arm is held in a somewhat different position than in the study on which Sheldon's standards were established. Therefore only the thoracic and abdominal trunk and the legs were available for somatotyping.

Concerning the permanence of the somatotype, Sheldon stated (p. 22): "The question of whether or not the somatotype can be modified during the lifetime of an individual can be answered with finality only after a few hundred physiques have been followed closely throughout the whole of a lifetime and photographed at regular intervals. This, of course, we have not yet done, but it has been possible to follow the development of several hundred individuals over a period of about a dozen years, and while many have shown sharp fluctuations in weight, we have discovered no case in which there has been a convincing change in the somatotype." He goes on to indicate that some of the measurements may change, but he believes the basic pattern can still be accurately determined by the trained worker.

A photogrammetric study of the material from the Minnesota Experiment, carried out by Dr. Gabriel Lasker, provided relevant information about the relative stability of the constitutional characteristics as determined by the photogrammetric method. Using the 3 available body regions and the 2 sets of pictures, obtained during the control period and at the end of 6 months of semistarvation, it was found that there was a mean decrease in endomorphy from 3.38 to 1.66 (on a scale from 1.00 to 7.00 for the total range of human body types);

---

* Sheldon et al. (1940, pp. 5–6) described the componental traits as follows: *Endomorphy* means relative predominance of soft roundness throughout the various regions of the body. When endomorphy is dominant the digestive viscera are massive and tend to dominate the bodily economy. *Mesomorphy* means relative predominance of muscle, bone, and connective tissue. The mesomorphic physique is normally heavy, hard, and rectangular in outline. Bone and muscle are prominent, and the skin is made thick by a heavy underlying connective tissue. *Ectomorphy* means relative predominance of linearity and fragility. In proportion to his mass, the ectomorph has the greatest surface area and hence the relatively greatest sensory exposure to the outside world. Relative to his mass he also has the largest brain and central nervous system.

mesomorphy decreased from 3.78 to 2.05; and ectomorphy increased from 3.28 to 5.98 (see Table 67). This represents a definite and marked shift in the average somatotype.

In order to determine whether the anthroposcopic criteria which are employed in the more usual method of somatotyping are also markedly affected by starvation, an attempt was made to have the photographs examined independently by several workers experienced in somatotyping. The task was undertaken by a group of somatotypists working at the Harvard Anthropometric Laboratory in the Peabody Museum in Cambridge, Massachusetts, directed by Professor E.

TABLE 67

SEMI-STARVATION CHANGES IN THE 3 "COMPONENTS" OF HUMAN PHYSIQUE (Sheldon's definitions), determined on the basis of measuring the photographs in the Minnesota Experiment.

	Endormorphy	Mesomorphy	Ectomorphy
Before (C)...............	3.38	3.78	3.28
After (S24)..............	1.66	2.05	5.98
Difference	−1.72	−1.73	+2.70

A. Hooton. We wish to express our gratitude to Dr. James M. Andrews, IV, who organized this part of the study, tabulated the data, and analyzed the findings.

The description of body build is made on the basis of inspection, not by measurement. The system follows essentially Sheldon's approach, with the "traits" (such as a thin long neck, characterizing the ectomorphic component of the body build) rated on a scale ranging from 1 to 7. The criteria by which somatotypes are determined in Professor Hooton's laboratory are given in detail in the Appendix on methods. The numerical values of separate traits under each component (endomorphy, mesomorphy, and ectomorphy) are averaged to the nearest whole number to obtain an index for each of the 5 regions into which the body has been divided. The regional indexes are then averaged to give the over-all type.

Every effort was made to ensure that the 5 observers did their work independently — without consultation with one another and without making any comparison of the control and semi-starvation photographs. Each observer somatotyped from 16 to 23 pairs of photographs. Each pair of photographs was somatotyped by 3 or 4 different observers.

As mentioned above, the head and neck region could not be somatotyped because this area was blocked out in the photographs used for this study. Consequently the total somatotypes were calculated by averaging the regional somatotypes assigned to regions II, III, IV, and V (that is, thoracic trunk, upper extremities, abdominal trunk, and lower extremities). Each of these regional somatotypes was calculated as an average of anthroposcopic assessments from 3 to 5 particular features within the body region. Agreement between the independent observers proved to be fairly good; in general, the over-all somatotypes assigned a given photograph by 3 or 4 observers did not disagree by more than one unit in any component. Regional somatotypes showed greater divergence. The analysis comprises a total of 105 assessments (performed quite independently) of control and semi-starvation photographs. The mean values (see Table 68) reveal a close parallelism to the data obtained by the metric analysis of the photographs.

All individuals dropped in endomorphy as a result of reduced diet, with the exception of a single case who was appraised by one observer as 2-5-4 before and 2-3-6 after. All but 5 individual assessments showed a drop in mesomorphy; these were appraised as 4 in mesomorphy before starvation and the same after starvation. All of the 105 assessments showed a clear gain in ectomorphy during starvation. However, Dr. Andrews believes that in general the partially starved individuals somehow look different from what he would consider a congenital or "habitual" ectomorphy.

TABLE 68

SEMI-STARVATION CHANGES IN THE 3 "COMPONENTS" OF HUMAN PHYSIQUE (Sheldon's definitions), determined on the basis of an inspectional analysis of the photographs in the Minnesota Experiment.

	Endomorphy	Mesomorphy	Ectomorphy
Before (C)...............	3.47	3.94	3.42
After (S24)	1.82	2.81	5.71
Difference	−1.65	−1.13	+2.29

The anthropological study of the semi-starvation series reveals marked changes in the external form of individuals undergoing starvation. These changes are evident in anthroposcopy, anthropometry, and photogrammetry. All techniques bring out the increasingly leptosomic character of the body build. The photogrammetric technique indicates somewhat greater changes in physique than do the values derived inspectionally.

Sheldon is aware of the influence of nutritional state on the traits which are regarded as indicative of somatotype: "In somatotyping physiques which show an obvious nutritional disturbance, we do not feel that the data are satisfactory, or the somatype scientifically certain, unless we have access to the previous physical history of the individual" (Sheldon et al., 1940, p. 223). However, his final conclusion about the stability of body form is affirmative: ". . . the evidence remains strong that, in adult life at least, no change occurs in the somatotype" (p. 225). The present study lends no support to this statement. Every criterion of somatotype which we have investigated is subject to change. In fact, the technique of somatotyping would appear to be more useful for determining the state of nutrition than for determining the inherent constitution.

Changes in Body Dimensions during the Rehabilitation
Phase of the Minnesota Experiment

In rehabilitation the changes observed in semi-starvation were reversed. However, this was not simply a return to "normal." It was apparent on the basis of somatoscopic observations that fat was being laid down more rapidly than muscle tissue. In the later stages of rehabilitation, especially in the interval from R20 to R33, "soft roundness" became the dominant characteristic, reflecting the tendency to a relative obesity.

As far as direct anthropometric determinations were concerned, in the period from R1 to R12 attention was focused on measurements which had shown marked changes in semi-starvation and which could be expected to respond differentially to the differences in the rehabilitation diets. The results are summarized in Table 69. One of the most striking features is the relationship of recovery to the caloric content of the diet. Not only did the lowest caloric group (Z) consistently exhibit the smallest degree of return to pre-starvation values while the highest caloric group (T) recovered most, but also the parallelism be-

TABLE 69

Body Dimensions in Rehabilitation, in cm. C = average value at control period; d = average decrement in starvation (S24 − C); $\triangle$ = average increment in rehabilitation (R12 − S24). (Minnesota Experiment.)

	Calories				Proteins		Vitamins	
	Z	L	G	T	U	Y	P	H
Bi-Humeral Diameter								
C	43.39	43.92	44.06	43.05	43.66	43.56	43.51	43.69
d	−3.15	−3.90	−4.10	−4.06	−4.16	−3.91	−4.00	−4.07
$\triangle$	+0.72	+1.08	+1.59	+2.40	+1.54	+1.36	+1.54	+1.35
$\frac{\triangle}{d} \times 100$	22.8	27.7	38.8	59.1	37.0	34.8	38.5	33.2
Upper Arm Circumference								
C	28.25	29.21	28.74	27.76	28.96	28.02	28.02	28.96
d	−6.86	−7.51	−7.21	−6.31	−7.28	−6.67	−6.68	−7.27
$\triangle$	+1.12	+1.94	+2.50	+2.86	+2.18	+2.03	+2.16	+2.06
$\frac{\triangle}{d} \times 100$	16.3	25.8	34.7	45.3	29.9	30.4	32.3	28.3
Thigh Circumference								
C	46.94	48.05	47.42	46.02	47.42	46.80	46.12	48.10
d	−9.25	−9.99	−9.19	−8.16	−9.48	−8.81	−8.45	−9.84
$\triangle$	+1.86	+3.12	+3.71	+4.44	+3.38	+3.23	+3.22	+3.39
$\frac{\triangle}{d} \times 100$	20.1	31.2	40.4	54.4	35.6	36.7	38.1	34.4
Calf Circumference								
C	37.99	37.49	37.89	37.51	37.72	37.72	37.23	38.21
d	−4.95	−5.05	−4.55	−4.29	−4.99	−4.43	−4.48	−4.94
$\triangle$	+1.02	+1.31	+1.39	+1.96	+1.36	+1.48	+1.29	+1.56
$\frac{\triangle}{d} \times 100$	20.6	25.9	30.5	45.7	27.2	33.4	28.8	31.6
Abdomen Circumference								
C	76.65	77.09	80.91	77.65	77.88	78.28	77.35	78.80
d	−7.58	−7.31	−7.99	−6.64	−7.70	−7.06	−6.38	−8.38
$\triangle$	+3.38	+3.69	+5.25	+6.68	+4.90	+4.59	+4.75	+4.74
$\frac{\triangle}{d} \times 100$	44.6	50.5	65.7	100.6	63.6	65.0	74.4	56.6

tween caloric intake and rehabilitation increments in the measurements held throughout the caloric range. The average values for the 16-man groups without (U) and with (Y) protein supplements, and without (P) and with (H) vitamin supplements, do not show any consistent trend, the supplemented groups being sometimes slightly superior, sometimes slightly inferior. The statistical evaluation of the significance of the differences in caloric groups is presented in Table 70.

TABLE 70

SIGNIFICANCE OF THE DIFFERENCES IN THE AVERAGE INCREMENTS IN BODY DIMENSIONS ($\triangle$ = R12 − S24) OF THE 8-MAN CALORIC GROUPS. V = between-group variance; V_{rep} = replicate variance used as the critical or "error" term, based on 16 pairs of subjects; F = the ratio of the between-group variance to the replicate variance. The F-tests were computed only for differences approximating or reaching significance. For 1 and 16 degrees of freedom $F_{0.05}$ = 4.49, $F_{0.01}$ = 8.53. (Minnesota Experiment.)

	Z vs. L	Z vs. G	Z vs. T	L vs. G	L vs. T	G vs. T	V_{rep}
			Bi-Humeral Diameter				
V	0.49	2.98	11.22	1.051	7.02	2.64	0.73
F		4.08	15.39[**]		9.63[**]		
			Upper Arm Circumference				
V	2.64	7.56	12.08	1.27	3.42	0.53	0.43
F	6.18[*]	17.69[**]	28.25[**]		8.01[*]		
			Thigh Circumference				
V	7.29	13.69	26.52	1.00	6.00	2.10	1.18
F	6.18[*]	11.60[**]	22.47[**]		5.08[*]		
			Calf Circumference				
V	0.33	0.53	3.51	0.03	1.69	1.32	0.61
F			5.78[*]				
			Abdomen Circumference				
V	0.39	14.06	43.56	9.78	35.70	8.12	4.55
F			9.58[*]		7.85[*]		

Comparison of the 8-man caloric groups reveals statistically significant (at the 5 per cent level) or highly significant (at the 1 per cent level) differences between the groups on the lowest and highest caloric intakes (Z vs. T). As would be expected, the differences are statistically less significant as the caloric differential between the groups decreases. The differences between L and G and between G and T never reach the level of statistical significance, those between Z and L infrequently.

The effectiveness of the protein and vitamin supplements at each of the 4 caloric levels can be expressed by calculating the ratio of the average percentage increments for the 4 men in each caloric group receiving and not receiving additional protein or vitamins. The ratios would be larger than 1.00 if the supplemented group (Y and H) had improved more than the unsupplemented group (U and P, respectively). The data in Table 71 do not show any consistent trend indicating a beneficial effect of either supplement.

TABLE 71

REHABILITATION INCREMENTS, expressed as percentages of semi-starvation loss, within caloric groups for the 4 men with (Y) and without (U) protein supplements, with (H) and without (P) vitamin supplements (Minnesota Experiment).

	Z	L	G	T		Z	L	G	T
			Bi-Humeral	Diameter					
Y	7.4	26.8	35.5	66.7	H	14.8	26.2	34.2	59.4
U	26.4	28.4	41.5	51.5	P	21.2	29.0	42.6	58.8
Y/U ...	0.28	0.94	0.86	1.30	H/P ...	0.70	0.90	0.80	1.01
			Upper Arm	Circumference					
Y	15.5	23.8	38.7	44.0	H	12.8	24.7	34.7	42.6
U	17.1	27.6	30.7	46.6	P	20.5	27.0	34.6	48.3
Y/U ...	0.91	0.86	1.26	0.94	H/P ...	0.62	0.91	1.00	0.88
			Thigh	Circumference					
Y	23.4	27.5	44.1	55.3	H	19.1	35.4	37.8	48.3
U	16.8	36.2	36.8	53.5	P	21.4	28.0	43.2	60.6
Y/U ...	1.39	0.75	1.20	1.03	H/P ...	0.89	1.26	0.88	0.80
			Calf	Circumference					
Y	26.1	24.1	36.0	53.1	H	21.1	29.1	35.2	43.4
U	15.2	27.8	25.5	40.5	P	20.2	22.5	25.0	48.0
Y/U ...	1.72	0.87	1.41	1.31	H/P ...	1.04	1.29	1.41	0.90
			Abdomen	Circumference					
Y	46.2	53.7	72.7	90.7	H	35.2	49.9	82.6	71.2
U	43.4	45.9	60.9	110.6	P	64.0	51.3	54.7	144.8
Y/U ...	1.06	1.17	1.19	0.82	H/P ...	0.55	0.97	1.51	0.49

In evaluating the effectiveness of these special protein and vitamin supplements, we may use means based on 16 men. The F-tests indicate that the mean recovery scores for the men who were given either protein (Y) or vitamin (H) supplements did not differ significantly from the means of the subjects who did not receive these special supplements. In contrast the differences between the means of the 16 men in the 2 lower caloric groups (Z and L) and the 2 upper caloric groups (G and T) are highly significant for all measurements in Table 72 except the calf circumference.

It is possible to make a hypothesis that the 3 diet treatments (calories, protein, vitamins) might be more effective in combination than they are singly. Statistically we speak of "interaction" of the treatments. Neither the first-order interactions (interaction of caloric and protein supplements, caloric and vitamin supplements, protein and vitamin supplements) nor the second-order interaction of caloric, protein, and vitamin supplements approached a level of statistical significance (see Table 73).

The bi-deltoid diameter and the limb circumferences were determined also 4 and 8 weeks after the release from rigorous dietary regimen (at R16 and R20). The average increments are given in Table 74, the values of the F-tests in Table 75. The 12 men measured at those times had belonged from R1 to R12 to the fol-

TABLE 72

SIGNIFICANCE OF THE DIFFERENCES IN THE AVERAGE INCREMENTS IN
BODY DIMENSIONS (Δ = R12 − S24) OF THE 16-MAN GROUPS. V = between-group variance; F = the ratio of the between-group variance to
the replicate variance, V_{rep}. (Minnesota Experiment.)

	ZL vs. GT	U vs. Y	P vs. H	V_{rep}
	Bi-Humeral Diameter			
V	9.57	0.26	0.30	0.73
F	13.13[**]			
	Upper Arm Circumference			
V	10.58	0.18	0.08	0.43
F	24.75[**]			
	Thigh Circumference			
V	18.91	0.18	0.21	1.18
F	15.34[**]			
	Calf Circumference			
V	2.05	0.11	0.58	0.61
F	3.36			
	Abdomen Circumference			
V	47.29	0.75	0.00	4.55
F	10.40[**]			

lowing groups: 4 to Z, 2 to L, 2 to G, and 4 to T. Taking the 6 men in the upper
and lower caloric categories singly or combining them, we observe a highly significant improvement from R12 to R20 for all the measurements in Table 71 for
all 3 groupings (Z + L; G + T; Z + L + G + T).

As we saw before, up to R12 the improvement in the Z + L men was slower
than in the G + T group. From R12 to R20 the trend was reversed as the men
were "catching up"; the difference between the recovery of the Z + L versus
G + T groups was significant for the bi-deltoid diameter, upper arm, and thigh
circumference.

Between R12 and R20 the body dimensions reflecting the magnitude of the
soft tissues not only returned to the pre-starvation state but exceeded the control
values in every instance. The changes in the abdominal circumference are particularly striking. In starvation the abdomen circumference decreased in the 12
men by 7.3 cm.; a large part was recovered by R12; by R20 the increment since
S24 was 13.9 cm., that is, the circumference was 6.6 cm. larger than at the start
of the experiment. This observation checks well with the increase of body fat as
estimated on the basis of the specific gravity of the body (see Chapter 8). At
R20 the dietary history of the men from R1 to R12 was no longer demonstrable.
The mean differences between the deviations from the control values (d =
R20 − C) for the Z + L and G + T groups were insignificant.

In contrast to the soft tissue measurements, the standing height continued to
decrease slightly even after refeeding had been instituted. With 178.65 cm. at C

TABLE 73

Interaction Variances and Replicate Variances for Rehabilitation Increments. The two lower (Z and L) and the two upper (G and T) caloric groups were combined. None of the F ratios were statistically significant. (Minnesota Experiment.)

| | Interaction Variances | | | | |
Dimension	Cal. × Prot.	Cal. × Vit.	Prot. × Vit.	Cal. × Prot. × Vit.	V_{rep}
Bi-humeral diameter............	0.48	0.11	0.09	0.00	0.73
Upper arm circumference........	0.45	0.03	0.01	0.50	0.43
Thigh circumference............	0.24	4.35	0.02	0.72	1.18
Calf circumference.............	0.00	0.04	0.43	0.00	0.61
Abdomen circumference........	5.36	8.30	0.07	8.30	4.55

TABLE 74

Mean Values for the Bi-Deltoid Diameter and the Limb Circumferences, in cm., at various periods up to R20. Six of the men were maintained from R1 to R12 on lower (groups Z and L) and 6 on higher caloric intake (groups G and T). (Minnesota Experiment.)

	C	(S24 − C)	(R12 − S24)	(R16 − S24)	(R20 − S24)	(R20 − R12)	(R20 − C)
			Bi-Humeral Diameter				
ZL	43.37	−4.08	0.97	4.15	4.73	3.77	0.65
GT	43.58	−4.12	2.23	4.18	4.48	2.25	0.37
ZLGT	43.48	−4.10	1.60	4.17	4.61	3.01	0.51
			Upper Arm Circumference				
ZL	27.63	−6.48	1.40	5.27	6.68	5.28	0.20
GT	27.42	−6.22	2.63	5.30	6.43	3.80	0.22
ZLGT	27.52	−6.35	2.02	5.28	6.56	4.54	0.21
			Thigh Circumference				
ZL	46.23	−8.80	2.53	8.47	9.42	6.88	0.62
GT	45.60	−7.65	3.95	7.62	8.75	4.80	1.10
ZLGT	45.92	−8.22	3.24	8.04	9.08	5.84	0.86
			Calf Circumference				
ZL	37.63	−4.45	1.22	4.50	4.97	3.75	0.52
GT	37.17	−4.22	1.63	3.82	4.52	2.88	0.30
ZLGT	37.40	−4.33	1.42	4.16	4.74	3.32	0.41
			Abdomen Circumference				
ZL	75.23	−6.28	4.10	12.50	13.72	9.62	7.43
GT	78.23	−8.32	6.70	14.85	13.98	7.28	5.67
ZLGT	76.73	−7.30	5.40	13.68	13.85	8.45	6.55

TABLE 75

F-Tests of Significance of the Differences between the Mean Values of the Bi-Humeral Diameter and Limb Circumferences measured at R20 as compared with the values at R12 and C. With 6 men in each of the two groups, (Z + L) and (G + T), the F-test for paired variates is associated with 1 and 5 degrees of freedom ($F_{0.05} = 6.61$, $F_{0.01} = 16.26$); combining (Z + L + G + T) the degrees of freedom increase to 1 and 11 ($F_{0.05} = 4.84$, $F_{0.01} = 9.65$). In testing the significance of the difference between the two groups (Z + L) and (G + T), the F-test for unpaired variates is associated with 1 and 10 degrees of freedom ($F_{0.05} = 4.96$, $F_{0.01} = 10.04$). (Minnesota Experiment.)

Group	Periods Compared	
	(R20 − R12)	(R20 − C)
Bi-Humeral Diameter		
(Z + L).............................	73.72[**]	2.96
(G + T).............................	36.20[**]	1.18
(Z + L + G + T)...................	70.82[**]	4.30
(Z + L) vs. (G + T)...............	6.95[*]	0.31
Upper Arm Circumference		
(Z + L).............................	280.23[**]	1.40
(G + T).............................	90.63[**]	0.10
(Z + L + G + T)...................	189.48[**]	0.39
(Z + L) vs. (G + T)...............	8.46[*]	0.00
Thigh Circumference		
(Z + L).............................	293.17[**]	2.35
(G + T).............................	49.37[**]	5.91
(Z + L + G + T)...................	141.34[**]	8.32[*]
(Z + L) vs. (G + T)...............	6.89[*]	0.63
Calf Circumference		
(Z + L).............................	150.94[**]	2.75
(G + T).............................	39.29[**]	1.48
(Z + L + G + T)...................	127.41[**]	4.52
(Z + L) vs. (G + T)	2.48	0.31
Abdomen Circumference		
(Z + L).............................	220.40[**]	214.90[**]
(G + T).............................	45.30[**]	13.68[*]
(Z + L + G + T)...................	147.15[**]	64.73[**]
(Z + L) vs. (G + T)...............	3.44	1.19

and 178.33 cm. at S24, or a mean decrement of −0.32 cm., the stature further decreased at R12 by −0.26 cm., or a total of −0.58 cm. There was no significant difference at R12 in the mean decrements (d = R12 − S24) of the various caloric groups. In 6 subjects measured at R33 this slight trend was still present, the average decrement from control being −0.71 cm. (see Table 76).

The change in standing height is probably a complex phenomenon. The continuation of the decrease in the apparent height in rehabilitation may be interpreted as resulting, in part, from a decrease in the edema of the feet. Where this factor is absent, the decrease in stature may be attributed to either a change in

TABLE 76

Standing Height, in cm., in a sample of subjects studied
at R33 (Minnesota Experiment).

Subject	C	S24	R12	R32	(R32 — C)
109	187.1	186.9	186.9	186.7	—0.4
112	176.0	175.4	175.3	175.0	—1.0
2	177.9	177.3	177.0	176.9	—1.0
22	177.5	177.5	176.9	177.2	—0.3
23	177.7	177.6	177.4	176.9	—0.8
127	180.1	180.2	179.8	179.3	—0.8
M	179.38	179.15	178.88	179.67	—0.71

the muscle tonus (and consequently in posture) or to a diminished thickness of
the intervertebral discs. An attempt was made to control the posture. In order to
explore the hypothesis that the main component of the decreased height was a
decrease in the thickness of the intervertebral discs, a section of the vertebral
column was measured. In 5 men, for whom clear radiographic pictures were
obtained in connection with the study of the gastric emptying time, the lengths
of the same 4 lumbar vertebrae were determined. An average decrement of 1
mm. was obtained when the values for the control (161.2 mm.) and for the end
of starvation (160.2 mm.) were compared. It is conceivable that these
changes may be irreversible and may essentially parallel the process of ageing in
which a decrease in stature has also been observed, even though longitudinal
studies of the decrease in stature in old age are needed before the results of investigations made on cross sections of the population may be safely interpreted.

Ivanovsky (1923, p. 335) reported that the stature of the population of the
Soviet Union, which decreased markedly during 3 years of famine, began to increase again when the diet improved until it reached the pre-famine values. It
is unfortunate that he did not give any figures to document this statement or to
characterize quantitatively the course of this process.

CHAPTER 8

Body Fat

IN POPULAR terms the most elementary difference between undernourished and well-fed persons is in their "fatness." The weight differences in these two states are commonly believed to be simply differences in fat. While this is a considerable oversimplification, the fact remains that changes in the relative amount of fat at different levels of nutrition are generally greater than changes in the other tissues of the body.

It is customary to differentiate between deposit, or depot, fat (adipose tissue and interstitial fat) and functional fat. The latter would represent the fatty compounds which are involved in the structure of active cells. While such a distinction has general validity, it defies precise definition in the terms of actual quantitative analyses which have any degree of practical application for human nutrition. Furthermore, the deposit fat does have definite functions in the total economy of the organism, including heat insulation, protection to more delicate tissues, and the maintenance of form, as well as acting as the principal fuel reserve of the body. The deposition (and mobilization) of fat in adipose tissue is an active process regulated by nervous and endocrine factors and involving continuous synthesis of new fatty acids from carbohydrates and transformation of fatty acids into one another (Wertheimer and Shapiro, 1948).

The amount of total fat can be determined by extraction and certain lipid fractions, such as phospholipids and unsaturated fatty acids, by more specific chemical methods. It is also possible to separate, mechanically, certain accumulations of fatty tissue in the fat depots, such as the perirenal deposits, and to weigh these. Quantitative studies on the total amount of fat in the human body are surprisingly few, and they form an inadequate basis for a discussion of the changes in starvation and subsequent rehabilitation. Direct measurement on one 35-year-old man in a fairly normal state of nutrition showed that 12.51 per cent of his body weight consisted of extractable fat (Mitchell et al., 1945). By indirect means the average fat content of the body was estimated as 13.6 per cent in a combined sample of 99 and 75 men in the U.S. Navy (Behnke, Feen, and Welham, 1942; Welham and Behnke, 1942). It seems likely that fat normally constitutes about 13 or 14 per cent of the body weight of young male adults in the United States and that this figure will be somewhat larger in females. However, there are no acceptable data on the averages or on the amount of individual variation in the percentage of total body fat for the various segments of the population, nor is it possible to state what proportion of the total fat is completely or partly expendable.

In the literature there are various measurements of the weights of separated adipose tissues but the data are insufficient for a quantitative characterization of the normal fat depots in man. The largest depot is undoubtedly the subcutaneous fat, which may account for half or more of the total adipose tissue of the human body as it does in the rat (Reed *et al.*, 1930). Other major fat deposits in man are those around the kidneys and those associated with the mesentery and omentum. Simple inspection of these deposits has been the chief basis for conclusions regarding the behavior of the body fat in starvation. This must be noted in the interpretation of the frequent statements in the literature to the effect that starved persons show a "complete absence of fat."

Students of human morphology are concerned with determining the components of the human body and its parts, including fat, in the effort to provide means for an intelligent interpretation of body weight and its changes in different physiological conditions (such as obesity and undernutrition), for the investigation of differential rates of growth and involution of the body tissues during the life cycle, for the characterization of the physical constitution and body types, and for the study of sex and race differences. The relationship of the fat content of the body to health and fitness (Kireilis and Cureton, 1947) is an important problem. Three general types of approach have been developed for the study of fat in living man: (1) measurement of the skin folds; (2) determination of the layer of subcutaneous tissue plus skin from roentgenograms; and (3) estimation of the body fat on the basis of the specific gravity of the body.

Thickness of Skin Folds as a Criterion of Subcutaneous Adipose Tissues

An approximate estimate of the nutritional state by pinching or picking up a fold of skin is a very old clinical procedure. During acute starvation the changes in the amount of subcutaneous fat were studied by Lehmann *et al.* (1893, esp. p. 97), using the thickness of skin folds as a criterion. Selected data, limited to the right side of the body and to the measurements of vertical skin folds, are presented in Table 77. No pre-fast values were given, so that the relative magnitude of these changes and thus the sensitivity of this criterion to starvation cannot be established on the basis of the available figures. The authors commented that both their subjects were thin. The subcutaneous fat present was not distributed in the same way in the two men, Cetti having thicker skin folds at the chest and abdomen, Brethaupt at the extremities. As would be expected, the decrease of the skin folds was directly related to the amount of fat available at the start.

The relationship between the measured thickness of skin folds and the clinically estimated leanness or fatness of an individual was studied by Oeder (1910). He measured the thickness of the abdominal fold in patients 26 years of age and older; out of the 1920 cases inspected, 936 were considered thin (undernourished), 607 normal (well-nourished), and 377 fat (overnourished). There was a general parallel between thickness of the abdominal fold and the relative fatness of the individual as appraised by general inspection. The average values of the fold, measured in the immediate neighborhood of the navel, were 1.1, 2.8, and 4.4 cm. for the 3 groups. The overlap indicated in Table 78 may reflect the crudity of the clinical impression of the nutritional status as much as anything

TABLE 77

DECREASE IN THE THICKNESS OF SKIN FOLDS DURING ACUTE EXPERIMENTAL STARVATION, for 2 subjects (Lehmann et al., 1893, p. 97).

	Cetti	Brethaupt
Length of fast (days)..................	10	6
Loss of weight (kg.)...................	−6.35	−3.62
Loss of weight as percentage of initial value..........................	−11.14	−6.00
Skin folds		
Chest (mm.).......................	−3.1	−0.5
Abdomen (mm.)....................	−2.1	−0.5
Thigh, lateral (mm.)...............	−1.5	−2.0

TABLE 78

PERCENTAGE DISTRIBUTION OF THE THICKNESS OF THE ABDOMINAL SKIN FOLD IN ADULTS classified on inspection as thin ($N = 936$), normal ($N = 607$), and fat ($N = 377$) (Oeder, 1910).

Thickness (cm.)	Nutritional Status		
	Thin	Normal	Fat
Below 0.9......................	39.0		
1.0–1.9	42.1	8.4	
2.0–2.9	18.6	66.6	11.4
3.0–3.9	0.3	23.7	42.2
4.0–4.9		1.3	28.4
5.0 and above			18.0

else. The patients were about equally divided between the sexes; in computing the averages sex was not taken into account.

Käding (1922) found low values for the thickness of the abdominal skin folds in undernourished children at Bonn. Käding's averages for boys and girls 13 to 15 years of age were approximately 5 and 9 mm., respectively, as compared with the normal values of 10.0 and 17.5 mm. reported by Kornfeld and Schüller (1930). Käding commented that this criterion corresponded closely to the clinical impression concerning the nutritional status of the individual but gave no data to substantiate his statement.

The thickness of the skin folds is a useful physical characteristic. However, the development of the adipose tissue in general, and of the skin fold thickness in particular, cannot be used as an exact measure of the nutritional status of an individual (Martin, 1928, p. 205).

Roentgenographic Measurements of the Thickness of Subcutaneous Tissues Plus the Skin

A technique for determining the thickness of the subcutaneous tissues plus the skin on the basis of roentgenograms was developed by Stuart, Hill, and Shaw (1940) at the Center for Research in Child Health and Development, Harvard School of Public Health. In essence the technique consists in the separation of the shadows on an X-ray film. After preliminary explorations of the parts of the body which would be suitable for study by this technique, the authors selected the lower leg, viewed in the anteroposterior direction. The shadows cast on the film were analyzed into 3 components: bone, muscle, and the subcutaneous tissue plus skin. The corresponding areas of the film were cut out and the segments weighed, the weight being proportional to the size of the particular area. Also, the widths of the 3 areas were measured.

The authors presented normative material for ages from 3 to 84 months, separated according to sex, and 19 case studies. The investigations were later extended to the age of 10 years (Stuart and Dwinell, 1942) with some modifications of the technique. The distance of the X-ray tube from the film was increased from 3 to 6 feet in order to reduce triangular distortion, and the sizes of the shadows corresponding to the 3 principal tissues were obtained by the use of a planimeter.

The method has the advantage that the values for the subcutaneous tissues can be related to the other two components, bones and muscles. However, the calf is by no means the area in which the fat in the adult tends to accumulate. The data for the subcutaneous fat in this region alone would be of only limited interest from the point of view of evaluating the over-all fatness or leanness of an individual, especially in cases of pathological shifts in fat distribution. This general method was developed for other parts of the body by Reynolds (1945) (see also Reynolds and Schoen, 1947, and Reynolds and Asakawa, 1948).

Estimation of Fat from the Specific Gravity of the Body

Measurements of the skin folds and roentgenographic determinations of the thickness of the subcutaneous adipose tissues are not adequate for the estimation of the total body fat. It is evident that they are limited to the surface layer and do not take into account the deeper fat deposits. Furthermore, the thickness of the subcutaneous fatty layer in different areas of the body is far from being uniform within the same individual and the localization of the maximum thickness in different individuals will also vary; the fatness of an individual must be approximated by averaging the values for a number of areas.

To eliminate the inadequacies of these techniques, a method was devised which makes it possible to estimate the amount of body fat on the basis of the specific gravity of the body. Attempts to determine the specific gravity of the human body go back to the work of Robertson (1757) and were reviewed by Boyd (1933). Most of the data are of little value because of the lack of standardization and the failure to make corrections for the effect of air in the lungs and in the air passages.

The method used in the Minnesota Experiment is described in detail in the Appendix. The volume of the body, needed for the calculation of specific gravity, was obtained by weighing the subjects under water. The formula developed by Rathbun and Pace (1945) relates the percentage of body fat to specific gravity (sp. gr.) as follows:

$$\% \text{ fat} = 100 \left[\frac{5.548}{\text{sp. gr.}} - 5.044 \right]$$

It has been pointed out in the discussion of the method in the Appendix that such alterations occurring in semi-starvation as increased body hydration and changes in the relative mass of bony tissues affect the validity of this formula. The effect of edema on the specific gravity of the body was dramatically demonstrated by Jamin and Müller (1903). In their studies they did not use true specific gravity but an approximation obtained as a quotient of the gross body weight to the volume of the body, excluding the volume of the head; no correction was made for the air in the respiratory system. It was possible to follow the value of their quotient in a patient with ascites before the release of 10 liters of the fluid which accumulated in the abdominal cavity, immediately after the puncture, and during the remission of ascites. The subject was a 56-year-old woman who, besides being extremely emaciated as a result of a carcinoma of the stomach, suffered from mitral insufficiency and developed edema in both legs in addition to the large amount of ascites. After the removal of the ascitic fluid (sp. gr. 1.012), there was a marked increase in the quotient of body weight to body volume, followed by a slow return to low values as the fluid again began to accumulate in the abdominal cavity. The details are given in Table 79.

TABLE 79

CHANGES IN THE QUOTIENT OF BODY WEIGHT TO BODY VOLUME (excluding the head and not corrected for air) in a patient with ascites (Jamin and Müller, 1903).

Condition	Body Weight	Body Weight / Partial Body Volume
Before paracentesis		
March 24	60.3	1.072
After paracentesis		
March 26	50.3	1.104
Return of ascites		
March 31	51.8	1.087
April 17	54.7	1.075

In Chapter 15 attempts are made to correct the estimated amount of body fat for increase in hydration and in the relative mass of the bones. That can be done with reasonable safety only to the averages. In this chapter we are concerned primarily with the responses to nutritional rehabilitation. The errors introduced in individual cases by using an average correction for the residual air are less serious when we are comparing the values obtained for the same individual measured at different periods of the experiment than when studying the interindividual differences.

Specific Gravity of the Body and of the Tissues in Undernutrition

Valid information on the specific gravity of the body is of considerable interest when evaluating the nutritional status of an individual because of the relation of this characteristic to the composition of the body and its changes in undernutrition and obesity.

Behnke, Feen, and Welham (1942) reported changes in the specific gravity of a man who was maintained on a restricted diet and an exercise regimen which produced a weight loss over a period of 7 months equal to 9.6 per cent of the initial value; the data are reproduced in Table 80. The specific gravity of the tissue involved in these weight changes, determined as the ratio of the change in body weight to the change in body volume, was 0.936. This value is practically identical with the specific gravity of body fat, which suggests that the decrease of body weight in the early stages of starvation may be due essentially to the loss of adipose tissue. The decrease in abdominal girth (89.4 cm. − 79.7 cm. = 9.7 cm. = 10.8 per cent of the initial value) was considered additional evidence that the weight loss was due chiefly to fat reduction. The specific gravity of the body rose from 1.056 to 1.071.

TABLE 80

SPECIFIC GRAVITY IN RELATION TO THE LOSS OF BODY WEIGHT IN ONE MAN (Behnke, Feen, and Welham, 1942).

	Specific Gravity	Weight in Air (kg.)	Net Weight in Water (kg.)	Circumference of Abdomen (cm.)
March 12	1.056	92.0	4.9	89.4
July 1	1.060	88.4	5.0	
August 13	1.066	85.0	5.3	83.8
October 9	1.071	83.2	5.5	79.7

However, not all the tissues and organs need follow the same pattern. This was confirmed by Tsai and Lin (1939), who investigated the variations in the density of tissues of animals under different physiological conditions, including acute starvation. In 3 rabbits, with water intake ad libitum, the loss of body weight after 9 days of acute starvation ranged from 26 to 36 per cent, with an average of 30 per cent. Whereas the densities of bone, kidney, stomach, small intestine, and brain remained within the normal range of variation, the density of skeletal muscles *decreased*. The average density of the thigh muscles was 1.0666 as compared with the control value of 1.0760. For the abdominal muscles the corresponding figures were 1.0614 (starvation) and 1.0711 (control). The decrease in the weight loss in these tissues was greater than the diminution in volume. A plausible explanation of these changes appears to be hydration of the tissues. The relative hydration of the body was studied in the Minnesota Experiment by measuring the volume of the extracellular fluid and that of the plasma (see Chapter 43).

Zook (1932) computed the specific gravity by means of water displacement in connection with his study of the physical growth of boys. The subjects were recruited from 2 groups: the University of Chicago High School (University group) and a settlement house in the stockyard district (Settlement group), representing well-nourished and poorly nourished children, respectively. The University group was taller and heavier and had a larger weight/height index; at the same time the specific gravity was also high, which may be regarded as an indication of better development of musculature and bones. Data for the ages 13–15 are given in Table 81. Although the absolute values of the specific gravity of the body are not correct since the residual air space was not taken into account, comparison of the 2 groups can be made. In the well-nourished group the average specific gravity was 1.053, in the poorly nourished group 1.045. This finding may be contrasted with the specific gravity in the Minnesota Experiment, where the values were higher at the end of the semi-starvation period than they were in the control or rehabilitation periods (1.0704 at C, 1.0838 at S12, 1.0887 at S24, and 1.0775 at R12). Zook's findings clearly indicate the need for caution in interpreting the specific gravity figures.

TABLE 81

MEAN VALUES FOR BODY CHARACTERISTICS, including specific gravity, in a group of poorly nourished (Settlement) and well-nourished (University) children. From Zook (1932).

	Settlement				University			
	Age Group			All Ages	Age Group			All Ages
	13	14	15		13	14	15	
Number	14	10	18	42	13	22	19	54
Weight (kg.)	36.7	42.7	46.0	42.1	44.1	48.8	55.8	50.1
Height (cm.)	128.0	132.5	134.7	131.9	136.2	140.2	147.5	141.8
100 × Weight/Height	28.7	32.2	34.1	31.9	32.4	34.8	37.8	35.3
Specific gravity	1.049	1.043	1.042	1.045	1.044	1.053	1.058	1.053

Body Fat of Undernourished or Starving Animals

The large extent to which the total body fat may decrease in undernutrition has been repeatedly demonstrated in various animals. Pfeiffer presented data for a rabbit that starved 13 days. The gross weight of the organs decreased from 1878 gm. to 1472 gm. (a total loss of 406 gm., or 21.6 per cent of the pre-starvation weight). The fat content of the subcutaneous, intermuscular, and abdominal adipose tissues decreased by 66, 51, and 75 per cent. The details are reproduced in Table 82. Of the total fat lost, 50 per cent was contributed by the tissues of the abdominal cavity, 30 per cent by the subcutaneous fat depot, and 11 per cent by the intermuscular tissues.

The figures calculated directly from Pfeiffer's figures disagree with his statement (1887, p. 365) that "The loss of fat of these three components (intermuscular connective tissues, adipose tissues of the abdominal cavity, and the subcutaneous tissues) . . . amount to 91 per cent." This figure was obtained by adding the data for the fat loss contributed by an organ, expressed as a percentage of the control fat content of the organ (last column of Table 82, section on fat content). This is clearly an erroneous method of arriving at the fat loss for these parts of the body; with the control fat content equal to 229.10 gm. and an absolute fat loss of 155.61 gm., the percentage loss is 67.9. Because of Pfeiffer's error in calculation we may view with suspicion the data quoted by him from Chossat and C. Voit, which we did not have an opportunity to check in the original

TABLE 82

DECREMENT IN THE WEIGHT OF ORGANS AND IN THEIR FAT CONTENT AFTER 13 DAYS OF STARVATION IN A RABBIT (Pfeiffer, 1887).

	Gross Weight (gm.)			
	Control	Starvation	d	d as % of Control Organ Weight
Blood	45.4	41.5	−3.9	−8.59
Heart	6.9	4.96	−1.94	−28.12
Liver	82.3	42.98	−39.32	−47.76
Muscles	854.6	703.0	−151.6	−17.74
Bones	173.8	171.2	−2.6	−1.50
Intermuscular connective tissues	61.2	34.6	−26.6	−43.46
Tissues of abdominal cavity ...	218.8	113.0	−105.8	−48.35
Skin	276.3	262.0	−14.3	−5.18
Subcutaneous tissues	115.3	51.0	−64.3	−55.77
Remaining tissues	43.1	47.7	+4.2	
Sum	1877.7	1471.5	406.2	−21.63

	Fat Content (gm.)					
	Control	Starvation	d	d as % of Control Organ Weight	d as % of Control Fat Content	Fat Loss % of Total
Blood	0.01	0.03	(+0.02)			
Heart	0.54	0.28	−0.26	−3.77	48.15	0.15
Liver	2.31	4.40	(+2.09)			
Muscles	37.68	24.78	−12.90	−1.51	65.76	7.60
Bones	16.96	16.10	−0.86	−0.50	5.07	0.51
Intermuscular connective tissues .	37.85	18.72	−19.13	−31.26	50.54	11.25
Tissues of abdominal cavity	113.50	28.60	−84.90	−38.80	74.80	49.94
Skin	7.27	3.09	−4.18	−1.51	57.50	2.46
Subcutaneous tissues..	77.75	26.17	−51.58	−44.74	66.34	30.34
Remaining tissues ...	2.34	3.98	(+1.64)			
Sum	296.21	126.15	170.06	−9.06	−57.59	100.0

TABLE 83

TOTAL FAT IN STEERS placed on different dietary regimens at the age of 11 months, as percentage of body weight (Moulton, 1920).

Dietary Regimen	Duration (mo.)	Condition	Fat (%)
Full-fed................	10	Very fat	32.6
Control.................	0	Fat	18.5
Gain of ½ lb. per day.....	6	Fat	20.2
Constant weight.........	6	Fat	19.7
Loss of ½ lb. per day.....	6	Thin	6.1
Loss of ½ lb. per day.....	10	Emaciated	1.9

publication. The cited figures indicate for starved pigeons a loss of fat tissues ("visible fat") corresponding to 93 per cent of the pre-starvation fat content, and a 97 per cent loss of fat in starved cats.

Moulton (1920) studied beef steers which were in a good state of nutrition until 11 months of age and then were fed at different caloric levels for 6 to 10 months. The carcass of the control animal contained 18.5 per cent of fat, whereas a fattened steer had nearly twice as much (32.6 per cent). This may be contrasted with an animal which lost 0.5 lb. per day for 10 months and became "very emaciated"; the fat content of the body decreased to 1.9 per cent. The data are given in greater detail in Table 83. It may be pointed out that while in the severely underfed steer the fat was almost completely resorbed, the weight loss was much less impressive. In terms of the control animal, the underfed steers lost at 6 months 67.0 per cent and at 10 months 89.7 per cent of the body fat present at the start of the underfeeding, while the decreases in the gross body weight were 14.6 per cent and 28.5 per cent of the pre-starvation value, respectively (Trowbridge, Moulton, and Haigh, 1918).

According to Jackson (1925, pp. 122–26), who reviewed the older literature on the effects of starvation on adipose tissue, the *tela subcutanea* in extreme starvation may decrease to 10 per cent or less of the normal weight. The changes in body fat in fasting rats were studied by Dible (1932). The fact that body fat decreased during a fast more rapidly than the gross body weight is indicated in Figure 42. In view of the large changes in the body fat shown by Dible, it is interesting to note that the nonsaponifiable lipoid material extractable from the carcass remained essentially unaltered; the values, expressed as per cent of initial body weight $\times 10^{-4}$, varied from 31.0 after 5 hours of fasting to 33.0 after 48 hours and 29.5 after 95 hours.

The changes in the testicular fat depot of fasting rats were described by

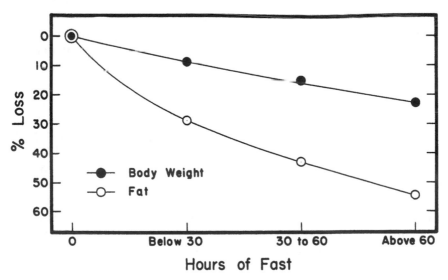

FIGURE 42. PERCENTAGE CHANGES IN BODY WEIGHT AND BODY FAT IN FASTING RATS (Dible, 1932).

Cremer (1939). This fat depot is anatomically well defined and the changes during starvation can be studied in terms of the weight of this adipose tissue. During acute starvation the body weight decreased by 35 per cent while the weight of the fat depot decreased by 92 per cent of the control values.

Human Body Fat in Starvation — Autopsy Material and Qualitative Observations

Donovan (1848) noted the disappearance of fat in the victims of the Irish famine of 1847. Formad and Birney (1891), in the autopsy of a woman who starved to death, found no subcutaneous or peritoneal fat.

Meyers (1917) made a thorough necropsy of a man who refused all food for 2 months and died on the sixty-third day. The loss of body weight was estimated as 41 per cent. The periorbital fat was absolutely depleted. There was only a trace of the popliteal and omental fat. Some fat was left near the root of the mesentery and in the mediastinum in the region of the thymus. The palmar and plantar skin lay in folds; in other areas the skin was loose but did not form folds. For the most part the adipose tissues contained no cells with fat globules and had the appearance of a fibrous web traversed by engorged capillaries. In some parts of the body (subpericardial fat) the boundaries of the fat cells were preserved but the cells were partly or completely filled with a granular protoplasm.

Schittenhelm and Schlecht (1918) noted a complete disappearance of the subcutaneous adipose tissue in their series of 200 undernourished patients with edema. Autopsies revealed that other fat depots in the body were also depleted. On microscopic examination a marked decrease in the fat content of the tissues was observed. Similarly, Park (1918) observed a "total absence of fat in the body" in 20 necropsies performed on cases of war edema. The fat deposits in the subcutaneous tissue, in the omentum and mesentery, and in the perirenal and pericardial space were replaced by a translucent, gelatinous substance. The patients exhibited an extreme muscular wasting. Jansen (1920, p. 152) also noted in autopsies of cases with undernutrition edema the disappearance of subcutaneous fat as well as a depletion of the pericardial and perirenal fat depots. Subcutaneous fat was reported absent on post-mortem examinations of malnourished Turkish prisoners of war (Enright, 1920a).

The autopsies of children who were the victims of starvation in Kharkov in 1922 showed a nearly complete disappearance of subcutaneous adipose tissue (Nicolaeff, 1923). This was also true of the omental, perirenal, and prevesical fat. Similarly, Tarassevitch (quoted by Ivanovsky, 1923) commented that in famine victims fat is the first tissue sacrificed and that the adipose tissue may almost completely disappear. According to Stefko (1928), even under conditions of a very large loss of weight (30–35 per cent) as observed in famine victims of the postwar years in Russia, fat may still be present in some areas (massa adiposa lumboglutealis, milk glands, and region of the symphysis) while other fat depots are totally depleted. He added that the fat in the region of the face and the *capsula adiposa* of the kidneys are the first to disappear during starvation. On the basis of observations made in German camps for the prisoners of war, Leyton (1946, p. 74) remarked that subcutaneous fat apparently disappeared before any gross decrease in the size of the muscles became noticeable. Dal (1946) commented on the striking decrease of body fat in the victims of starvation in Leningrad. The fat was mobilized from all depots but the subcutaneous adipose tissue was particularly depleted. On post-mortem examination of 7 fatal cases of starvation a complete absence of fat depots was noted by Picard (Simonart, 1948, p. 152).

All the reports cited present only qualitative descriptions of the semi-starvation changes and suffer from a lack of systematic approach. Some fat depots are mentioned in some reports while omitted by other investigators, which makes a critical synthesis of the literature impossible.

Relation of Fat to Starvation Resistance: Sex Differences in Fat Content

It has been universally observed that women withstand semi-starvation better than men. The illustrative data will be drawn from the material obtained during World War II. Valaoras (1946) presented figures on the effects of the 1941–42 famine in the Greek twin cities of Athens and Piraeus. The total number of deaths for men increased from 2794 in the third quarter of 1941 to 9108 in the fourth quarter of the same year and to 11,168 in the first quarter of 1942; this represents 100, 326, and 400 per cent. For women the data for comparable periods were 1999, 4379, and 6361, or 100, 219, and 318 per cent.

The fact that women have a greater resistance to semi-starvation was brought out also by the data collected during the German siege of Leningrad in 1941–42. The morbidity and mortality figures for the city as a whole are not available to us, but data on the death rate among individuals admitted to the First Leningrad Medical Institute from November 1941 to November 1942 indicate that the hunger deaths for men reached a peak in December 1941 (older men) and January 1942 (younger men); the death rate for women reached the maximal value in March and April of 1942, and even then the relative rise in the mortality curve was smaller than for men (Brožek, Wells, and Keys, 1946).

In Holland during the 1945 famine the mortality for males increased 169 per cent while the increase for females was only 72 per cent. These values are based on the differences in mortality between the first 6 months of the year 1945 and the corresponding period of 1944, recorded for 12 Dutch municipalities. The total number of deaths in the 2 periods was 13,155 and 29,122 (Dols and van Arcken, 1946).

One of the factors contributing to the better resistance of women to starvation may be the larger stores of body fat. The statement that in women fat represents a larger percentage of the body weight is generally accepted, although quantitative data are scanty, particularly as far as the total amount of adipose tissue is concerned. An indirect evidence of sex differences in the fat content of the body was obtained by Jamin and Müller (1903) in their studies on specific gravity. In approximating the value of specific gravity they measured a partial body volume by submersion up to the level of the head, averaging the values obtained at moderate inspiration and expiration. The volume of the head was not included and no correction was made for the air in the lungs and air passages. The ratio of the total body weight to the partial body volume was used as an index of the true specific gravity of the body. The average values in the age range comparable to that of subjects in the Minnesota Experiment (21–30 years) were higher for men than for women; this was also true of the grand averages based on 27 men (1.095) and 26 women (1.081). This fact may be interpreted as indicating that the women had a larger amount of fat, although the magnitude of the difference cannot be calculated from the above figures.

Sex differences were evident in the anatomical studies of Merselis and Texler (1925), who investigated the distribution of fat on the surface of the body. They used 36 cadavers, 20 women and 16 men, differing in nutritional status from normal to cachectic. The body surface was divided into a large number of small sectors and the thickness of the subcutaneous fat was measured on cuts made through the skin to the muscle or the bone. An extensive study in children was carried out with the same technique by Ogawara (1933). Out of Merselis and Texler's large but unsystematically presented material we have attempted to select data which would throw some light on the difference related to sex and the nutritional status. The latter was described only in qualitative terms; the heights of the individuals were given but not the weights so that a quantitative estimate of the deviation from the average-normal weight/height ratio cannot be obtained. Illustrative data are summarized in Table 84. The fat layer is considerably thinner for the 2 individuals in a poor nutritional state; also, it is thinner in men than in women.

Sex differences were brought out in an even more striking way by Traut

TABLE 84

Thickness of the Layer of Subcutaneous Fat, in mm., in 4 cadavers of different sex and nutritional status (Merselis and Texler, 1925).

Topographical Location	Poor Nutritional State		Good Nutritional State	
	Case No. 1, 17-Yr. Male	Case No. 5, 16-Yr. Female	Case No. 13, 20-Yr. Male	Case No. 18, 21-Yr. Female
Sternum	3	4	7	6
Breasts	5	8	11	16
Abdomen	4	7	13	11
Hips	5	18	10	29
Buttocks	16	21	23	36
Thigh	8	15	11	24

TABLE 85

Average Thickness of the Skin Folds in the Abdominal Region, in mm., of children at Vienna (Kornfeld and Schüller, 1930) and at Helsinki (Ruotsalainen, 1939).

Age (yrs.)	Vienna		Helsinki	
	Boys	Girls	Boys	Girls
6½–8½	7.5	9.5	8.0	9.0
8½–10½	8.0	11.5	8.0	10.5
10½–12½	9.0	14.5	9.0	13.5
12½–14½	10.0	17.5	11.0	17.0

(1926) who, using the technique of Merselis and Texler (1925), investigated the thickness of the subcutaneous fat in 42 individuals, 21 men and 21 women, in the older age group (50 years and up). These data are inadequate from the point of view of sampling, but they illustrate in a direct and quantitative way facts known from other sources of information, such as measurements of the thickness of skin folds on living individuals.

Sex differences in the thickness of skin folds, with higher values for girls, were noted by Kornfeld and Schüller (1930) in their study of Viennese children of normal nutritional status. The fact that the amount of body fat is larger in females than in males, as reflected in the greater thickness of the layer of subcutaneous tissue, was confirmed by Ruotsalainen (1939), who measured the thickness of the skin folds of 563 Finnish boys and 635 Finnish girls at Helsinki (see Table 85). Of the three regions investigated — abdomen, chest, and back — the abdominal fold was the thickest and showed the largest sex (and age) difference.

The observation that the average values for the thickness of skin folds are consistently higher in young adult girls (18 years) than in boys of the same age was confirmed in this country by the studies of Meredith (1935) and Boynton (1936). The differences were largest for the measurements made on the thorax and at the back of the upper arm. The detailed data are given in Table 86.

On the basis of the data available in the literature on the weight of the *tela subcutanea* in presumably normal adults (see Table 87), it appears that the sub-

TABLE 86

THICKNESS OF THE SKIN AND SUBCUTANEOUS TISSUES, in mm., obtained as one half of the thickness of the skin fold (Meredith, 1935; Boynton, 1936). The subjects were 18 years of age.

	Girls			Boys			M
	N	M	SD	N	M	SD	Difference
Upper arm							
Front	64	5.8	1.91	70	3.2	1.34	−2.6
Back	64	9.4	2.25	70	5.4	1.76	−4.0
Thorax							
Front	64	6.2	2.46	70	4.8	1.60	−1.4
Back	64	7.7	2.83	70	5.4	1.67	−2.3

TABLE 87

WEIGHT OF TELA SUBCUTANEA IN ADULT MEN AND WOMEN (Wilmer, 1940).

	Men	Women
Number	13	2
M age (yrs.).......................	37	39
M height (cm.).....................	168.8	159.5
M body weight (kg.)................	57.3	50.1
M skin weight (gm.)..................	3,289.7	3,000.0
M weight of tela subcutanea (gm.).......	6,604.0	11,897.0
Weight of tela subcutanea as percentage of body weight.............	11.5	23.7

cutaneous tissue in the average man forms about 10 per cent of the body weight while it amounts to about 20 per cent in the average mature woman (see also Scammon, 1942). The material from which these figures were derived is too limited to provide a reliable basis for establishing the degree of the sex difference in this characteristic. However, the striking differences between men and women with respect to the absolute and relative amount of subcutaneous tissue fall in line with the rest of the information on body fat.

Using their roentgenographic technique, Stuart, Hill, and Shaw (1940) observed a marked sex difference in the age range from 3 to 84 months. The linear width and the area of a shadow, cast on the roentgenogram by the subcutaneous and cutaneous tissues overlying laterally the muscles of the calf, were consistently higher for the girls than for the boys. The average values obtained by Stuart and Dwinell (1942) for Boston children in the age group from 6 to 10 years confirm the tendency of girls toward higher values in the subcutaneous fat. The data are given in Table 88.

TABLE 88

TOTAL WIDTH OF THE SHADOW OF THE SUBCUTANEOUS TISSUES AND SKIN ON BOTH SIDES OF THE LEG, in cm., measured on roentgenograms of the calf in Boston children (Stuart and Dwinell, 1942).

Age (yrs.)	Boys		Girls	
	N	M	N	M
6	59	1.06	58	1.35
7	73	1.04	63	1.32
8	64	1.06	52	1.35
9	45	1.06	37	1.33
10	25	1.11	24	1.38

The same observation was made in studies on children in Marseilles (Stuart and Kuhlman, 1942, esp. pp. 440–47); the thickness of subcutaneous tissue plus skin on both sides of the calf was 1.05 cm. in 35 6-year-old boys and 1.24 cm. in the 25 girls. In a similar study made in Madrid on children from 6 months to 9 years of age the average values were 0.90 cm. for 42 boys and 1.16 cm. for 42 girls; these values were calculated from the data of Robinson, Janney, and Grande (1942b).

In 49 boys and 58 girls 6 to 12 years old, studied with the roentgenographic technique by Reynolds (1944), the average values of subcutaneous tissue plus skin on both sides of the leg were 1.23 cm. for the boys and 1.38 cm. for the girls. The relative values of fat (fat thickness as a percentage of the total width of the shadow of the calf) were 14.0 for boys and 16.1 for girls. These figures represent averages weighted by the number of subjects measured in each age group.

In view of the cumulative evidence the larger fat content in the female can be regarded as an established constitutional characteristic. Since fat per unit of weight provides more energy than protein, muscular men are at a disadvantage as compared with women during periods of starvation.

Estimated Body Fat during the Semi-Starvation Phase
of the Minnesota Experiment

The areas of the body which normally contain a large amount of subcutaneous fat exhibited an early and marked diminution in size during semi-starvation. The buttocks decreased in a striking way. Loosely fitting belts were one of the early signs that the semi-starvation regimen required caloric supplements to the diet which had to be supplied by the body itself (see Chapter 7). The subcutaneous fat appeared to be easily mobilized for combustion and the skin folds became noticeably thinner. However, the decrease in the thickness of the skin folds was not studied quantitatively.

The total amount of adipose tissue in the Minnesota Experiment was estimated on the basis of specific gravity of the body (see Table 89). During the 6

TABLE 89

SPECIFIC GRAVITY AND ESTIMATED BODY FAT IN THE MINNESOTA EXPERIMENT. The values are not corrected for hydration and the relative increase in the bony mass. $N = 32$ subjects.

	C			S12		
	M	SD	Range	M	SD	Range
Specific gravity	1.0704	0.0104	1.046 to 1.086	1.0838	0.0117	1.052 to 1.112
Body weight (kg.)*	69.34			57.77		
Fat as % of body weight† .	13.9			7.5		
Fat (kg.)†.............	9.64			4.33		
Fat as % of the control value†	100.0			44.9		
Fat as % of body weight† .	13.96		26.0 to 6.5	7.75		23.0 to 0.0
Fat (kg.)‡.............	9.84	4.13	20.6 to 4.2	4.57	3.29	12.2 to 0.0
Fat as % of the control value‡	100.0			46.4		

	S24			R12	
	M	SD	Range	M	Range
Specific gravity	1.0887	0.0104	1.058 to 1.118	1.0775	1.052 to 1.103
Body weight (kg.)*.......	53.63			59.76	
Fat as % of body weight†...	5.2			10.5	
Fat (kg.)†...............	2.79			6.27	
Fat as % of the control value†	28.9			65.0	
Fat as % of body weight‡...	5.54		20.0 to 0.0	10.60	23.0 to 0.0
Fat (kg.)‡	3.05	2.51	11.9 to 0.0	6.40	15.3 to 0.0
Fat as % of the control value‡	31.0			65.0	

* The weight was obtained at the time of specific gravity determinations, made during the 2 weeks before the end of each period.
† The values were based on the average specific gravity and the average weight.
‡ The means were computed on the basis of individual values.

months of semi-starvation the average values for the estimated body fat showed a marked and progressive decline. The average values for the specific gravity of the group at C, S12, and S24 corresponded to 13.9, 7.5, and 5.2 per cent of the gross weight as body fat. On the basis of the average weights obtained at the time of specific gravity determinations, the specific gravity figures corresponded to 9.64 kg., 4.33 kg., and 2.79 kg. of body fat. In terms of the control value (C = 100 per cent), the fat content decreased to 45 per cent at S12 and 29 per cent at S24. Table 89 also contains averages and the range of the values obtained for the *individual* subjects. The fat value based on the average specific gravity is not exactly the same as the average of the individual values. The small differences between the 2 sets of figures at the control period were to be expected and are simply artifacts inherent in calculating the body fat on a percentage basis. Because the individual values for a specific gravity of 1.100 and higher were considered as equivalents of zero fat, the fat percentages averaged for the individuals at S12 and S24 are necessarily somewhat higher than the figures which correspond to the average values of specific gravity for the group.

Without correcting for hydration and other factors which affect the accuracy of the estimation of body fat, between C and S12 there was an average individual loss of 5.27 kg. of body fat, or 53.6 per cent of the control value, with the losses ranging from 0.8 kg. to 9.9 kg. The decrease was present in all individuals and the mean change was statistically highly significant (value of the F-test for paired variates was 200.14; for 1 and 31 degrees of freedom the F value at the 1 per cent level was 7.53). At S12 in 2 cases (subjects No. 8 and No. 122) the specific gravity reached or exceeded 1.100, which corresponds to zero per cent of body fat; in these individuals the true residual air volume was probably smaller than the applied correction of 1450 cc. On the other hand, in subject No. 109 the residual air volume was apparently much larger than the average correction, and the specific gravity (1.046 at C, 1.052 at S12) was too low, yielding a spuriously high value for body fat (26 per cent and 23 per cent of the body weight at C and S12, respectively).

At S24 the average individual loss of body fat amounted to 6.79 kg., or 69.0 per cent of the control value; the range of losses at this time was from 2.4 kg. to 14.3 kg. In 7 cases there was an apparent increase in the body fat from S12 to S24 (mean, 0.8 kg.; minimum, 0.1 kg.; maximum, 1.8 kg. of body fat) which may be explained as the effect of increased hydration and, possibly, less complete expulsion of air from the lungs. In the Minnesota Experiment the body fat decreased more markedly than the muscle mass, estimated from the average value of the cross-sectional area of the upper arm and the thigh, corrected for the bone. The gross body weight showed the least change (see Figure 43).

It has been repeatedly observed that fat is used preferentially as body fuel by animals deprived of all food. In semi-starvation, however, the body uses carbohydrates, proteins, and fats in proportions comparable to those in normal diets. The dramatic decrease in the fat content of the body observed in the Minnesota Experiment might tend to exaggerate the relative importance of lipids in the metabolic processes. Actually, during semi-starvation the lipids contributed to the total caloric expenditure at the same ratio as under normal conditions.

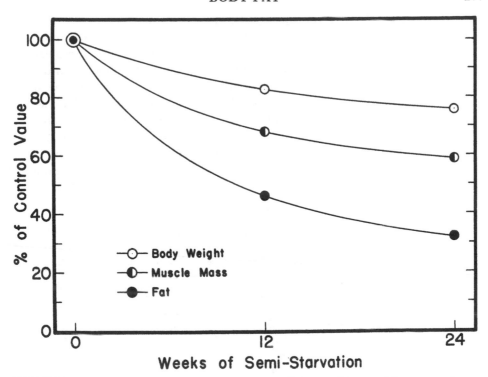

FIGURE 43. PERCENTAGE DECREMENTS OF BODY WEIGHT, MUSCLE MASS, AND BODY FAT in the Minnesota Experiment. $N = 32$.

Estimated Body Fat during the Rehabilitation Phase
of the Minnesota Experiment

In the first 12 weeks of the rehabilitation period the fat content of the subjects increased markedly, with the increases closely paralleling the caloric intake. The recovery of adipose tissue was more rapid than that of the muscles. Similarly, Debray *et al.* (1946, p. 865) noted that the early increase in the weight of individuals transferred from German concentration camps to a Paris hospital in a condition of severe starvation was largely due to the accumulation of fat rather than to the rebuilding of muscles. During the subsequent 8 weeks there was further increase in the body fat which brought the fat content of the body in all subjects well above the pre-starvation level. Thirty-three weeks after the end of the semi-starvation period the values for body fat showed a decrease and returned to a near normal level at 58 weeks. The data will be discussed in greater detail for each phase of the rehabilitation period.

In the first 3 months of rehabilitation in the Minnesota Experiment the mean individual gain was 3.35 kg., or 49.3 per cent of the loss sustained during semi-starvation. The maximum gain of body fat was 7.8 kg. At the other end of the distribution there was actually an apparent decrease of 1.9 kg. of body fat in subject No. 130, due to a clinically observable loss of edema. The edema fluid has a density lower than that of the body as a whole and its accumulation decreases the specific gravity of the body; the loss of edema fluid operates in the opposite direction, increasing the specific gravity and thus decreasing the estimated amount of body fat. In subject No. 130 the accumulation of edema, present at S12 and well marked at S24, was reflected in the change of specific gravity from 1.068 at S12 to 1.064 at S24; only at R12, after a return to a more normal state of hydration, did the true effects of starvation become manifest, with the specific gravity rising to 1.072. It should be noted that this subject was maintained during the rehabilitation period (R1–R12) on the lowest caloric intake (group Z).

At R12 the 4 caloric groups were well differentiated with respect to the amount of body fat regained, the percentages ranging from 17 per cent for the lowest caloric group, through 42 and 46 per cent for the intermediary groups, to 96 per cent for the highest caloric group (see Table 90 and Figure 44).

TABLE 90

Increase in Calculated Body Fat during the Period from R1 to R12. C = control value, in kg.; d = average decrement in starvation (S24 − C); △ = average increment in rehabilitation (R12 − S24). (Minnesota Experiment.)

	Calories				Protein		Vitamins	
	Z	L	G	T	U	Y	P	H
C	9.90	9.02	10.44	10.01	9.89	9.80	8.96	10.73
d	6.58	6.38	8.04	6.20	7.56	6.03	5.82	7.78
△	1.12	2.71	3.69	5.88	3.78	2.92	3.20	3.50
$\frac{100△}{d}$	17.0	42.5	45.9	94.8	50.0	48.4	55.0	45.0

TABLE 91

Significance of the Differences between the Average Increments of Estimated Body Fat ($△$ = R12 − S24) in the caloric groups. V_{bGr} = between-group variance; V_{rep} = replicate variance; F = the ratio V_{bGr}/V_{rep}. For 1 and 16 degrees of freedom, $F_{0.05}$ = 4.49, $F_{0.01}$ = 8.53; V_{rep} = 2.28. (Minnesota Experiment.)

	Groups Compared						
	Z vs. L	Z vs. G	Z vs. T	L vs. G	L vs. T	G vs. T	(ZL vs. GT)
V_{bGr}	10.08	26.27	90.25	3.80	40.01	19.14	65.55
F	4.42	11.52[**]	39.58[**]	1.67	17.55[**]	8.39[*]	28.75[**]

TABLE 92

Average Percentage Increments of Body Fat, 100 × (R12 − S24)/(C − S24), within caloric groups for the 4 men with (Y) and without (U) and with (H) and without (P) extra vitamins, and their ratios (Minnesota Experiment).

Special Supplements	Caloric Group			
	Z	L	G	T
Y (extra protein)	12.2	43.6	49.3	90.7
U (no extra protein)	21.1	41.7	43.3	98.2
Y/U	0.58	1.05	1.14	0.92
H (extra vitamins)..........	20.5	38.5	50.5	73.3
P (no extra vitamins)........	10.3	47.8	42.0	132.0
H/P	1.99	0.81	1.20	0.56

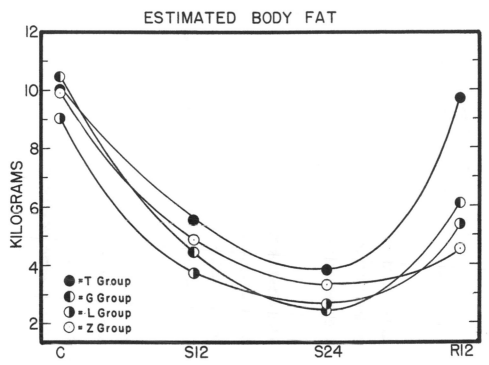

FIGURE 44. BODY FAT of the 8-man groups in the Minnesota Experiment.

All groups separated in their intake by more than 400 Cal. differed significantly in their average regain of body fat (see Table 91). As far as dietary factors other than calories are concerned, a comparison of the 16 supplemented with the 16 unsupplemented subjects indicates that neither the protein nor the vitamin supplements affected favorably the recovery of body fat (see Table 90). When we compute the ratios of the amounts of recovery within the 4 caloric groups, the protein and vitamin supplements do not affect the degree of gain of body fat in any consistent way (see Table 92).

It may be noted again that the body fat is more mobile than the gross body weight, which was regained at R12 by caloric groups Z, L, G, and T to the extent of 21, 30, 41, and 57 per cent of the semi-starvation loss. Anthropometric data support this conclusion and indicate that during rehabilitation the adipose tissue increased in size more rapidly than the muscles. In the highest caloric group (T) the circumferences of the upper arm, calf, and thigh showed an average recovery of 45, 46, and 54 per cent of the starvation decrement, whereas the abdominal circumference exhibited a recovery of 101 per cent. The recovery of the semi-starvation losses in body fat, abdominal circumference, and body weight is plotted for the 4 caloric groups in Figure 45.

The rapid recovery of body fat has also been observed by Kornfeld and Schüller (1931) in previously undernourished individuals undergoing treatment in the tuberculosis preventorium of the Viennese Children's Clinic. The thickness of the abdominal folds increased more rapidly than did the mass of other soft parts of the body.

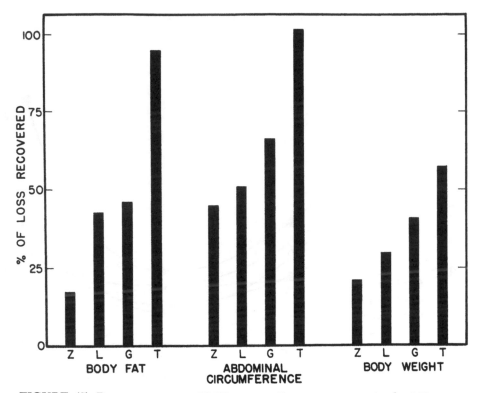

FIGURE 45. Recovery after 12 Weeks of Rehabilitation in the Minnesota
Experiment. The gains in body fat, abdominal circumference, and body weight
were expressed as percentages of the semi-starvation loss. $N = 8$ subjects in
each caloric group (Z, L, G, and T).

The fact that in convalescence, especially after a disease which results in
emaciation and a marked decrease of body weight, the fat increases more rapidly than the active tissues was pointed out by Jamin and Müller (1903). As indicated above, their quantitative criterion was a quotient of body weight to the
volume of the body, excluding the volume of the head. In a man who was normal
except for having scabies, the weight increased within a month from 70.3 kg. to
73.0 kg., while the quotient decreased from 1.110 to 1.081. This "equivalent" of
the specific gravity of the body decreased similarly in a woman who had severe
typhus. It went from 1.084 down to 1.071, while her body weight increased within 4½ weeks from 53.1 to 60.2 kg.

A marked increase in body fat, as indicated by the decrease of the specific
gravity during recuperation in a thin patient who had undergone an abdominal
operation, was reported by Rehn and Horsch (1934). While the body weights at
weekly intervals were 54.2, 55.2, and 55.8 kg. (showing an increase of 1.6 kg.),
the values for body volume — uncorrected for air content in the lungs and air
passages — were 50.2, 51.5, and 53.2 (an increase of 3.0 liters). The respective
ratios of the weight and volume were 1.080, 1.072, and 1.049.

The 12 men who remained at the Laboratory after the end of the first phase
of rehabilitation (R1–R12) continued to gain fat (see Table 93). At R12 the

TABLE 93

CALCULATED BODY FAT, in kg., for 12 men who continued to be studied in the second phase of the rehabilitation period (R12–R20) (Minnesota Experiment).

	C		S24		R12		R16		R20	
	M	Range	M	Range	M	Range	M	Range	M	Range
(Z + L) in kg..............	8.05	5.6 to 12.1	2.18	1.5 to 3.4	4.55	2.8 to 6.4	10.50	8.3 to 13.1	12.85	12.1 to 14.7
(Z + L) as percentage of C...........	100.0		27.1		56.5		130.4		159.6	
(G + T) in kg.	10.05	4.5 to 20.6	3.98	0.0 to 11.9	9.17	5.7 to 15.3	12.27	8.4 to 18.8	14.62	10.4 to 21.7
(G + T) as percentage of C...........	100.0		39.6		91.2		122.1		145.5	
(Z + L + G + T) in kg...........	9.05		3.08		6.86		11.38		13.73	
(Z + L + G + T) as percentage of C.	10.0		34.0		67.2		125.7		151.7	

181

body fat in the 6 men on low caloric intake (groups Z and L) was 56.5 per cent of the control value, in the other 6 men (groups G and T) 91.2 per cent. During the next 4 weeks (R12–R16) the men exhibited an extremely rapid increase in their fat content, with a gain of 5.95 kg. in the Z and L groups and 3.10 kg. in the G and T groups. At this time the total gains since the end of starvation were essentially identical in these two groups, 8.32 and 8.28 kg., and during the following 4 weeks both groups gained 2.35 kg. At R20 the average values for body fat were higher by 4.80 kg. and 4.57 kg., respectively, than in the control period, a difference which was statistically highly significant (see Table 94). In terms of percentages, the estimated body fat for the 12 men at R20 was 152 per cent of the control value, while the total body weight was 70.92 kg. compared with the control value of 67.53 kg., or 105 per cent of the control weight.

The 6 men examined 33 weeks after the end of starvation showed somewhat lower values for body fat than was the average at R20. The fat content of the body, expressed as a percentage of the pre-starvation value, was still higher than the body weight computed on the same basis; the values were 139 per cent for fat and 105 per cent for body weight (see Table 95). At R58 the return toward

TABLE 94

Differences between Average Values for Body Fat, in kg., obtained at various periods of the Minnesota Experiment. $N = 12$ subjects studied up to R20. For 1 and 11 degrees of freedom, $F_{0.01} = 9.65$.

	S24 − C	R12 − S24	R16 − S24	R20 − S24	R20 − C Δ	R20 − C F
(Z + L).........	−5.87	+2.37	+8.32	+10.67	+4.80	22.34[**]
(G + T).........	−6.07	+5.18	+8.28	+10.63	+4.57	16.84[**]
(Z + L + G +T)	−5.97	+3.78	+8.30	+10.65	+4.68	

TABLE 95

Average Calculated Body Fat for the 6 men examined at R33. For 1 and 5 degrees of freedom, $F_{0.05} = 6.61$, $F_{0.01} = 16.26$. (Minnesota Experiment.)

	C	S24	R12	R33	R33 − C Δ	R33 − C F
Fat (kg.)................	11.07	4.42	7.93	15.40	+4.33	15.39[*]
Fat as percentage of control..	100.0	39.9	71.6	139.1		

TABLE 96

Average Calculated Body Fat for 8 men examined at R58. For 1 and 7 degrees of freedom, $F_{0.05} = 5.59$. (Minnesota Experiment.)

	C	S24	R12	R58	R58 − C Δ	R58 − C F
Fat (kg.)	12.48	5.11	9.26	13.70	+1.22	2.71
Fat as percentage of control....	100.0	40.9	74.2	109.8		

pre-starvation values was nearly completed, with the average value of body fat equal to 110 per cent and the body weight 102 per cent of the control. The differences in the estimated body fat between C and R58 were not consistent and the F-test did not reach the value required for significance (see Table 96).

Comment

Ferguson (1905, p. 43) regarded the degeneration of the fat cells during starvation as a reversal of the process of fat cell formation. The fat cells were thought to arise from the connective tissue cells by deposit of fat droplets within the cytoplasm of the cells, until finally the cytoplasm becomes merely a limiting membrane, a cell wall. During starvation, as the fat is removed, the cytoplasm increases in amount. It does not return to the original state but assumes a peculiar fluid appearance.

Starvation in both animals and men produces a marked decrease in the fat content of the body. The rate of change of the different fat depots may vary considerably. In thyrogenic emaciation there may actually be an increase in the adipose tissues of the abdomen and the buttocks while the fat in the thoracic region disappears (Lauter and Terheddebrügge, 1937–38).

Attempts have been made in the past to estimate the fatness of individuals in different nutritional states on the basis of such characteristics as the thickness of the skin folds (which vary in proportion to the amount of subcutaneous fat), the width of the shadow on a roentgenogram corresponding to the subcutaneous tissues plus skin, the size of those parts of the body in which fat tends to accumulate (abdomen, buttocks), and the specific gravity of the whole body. Behnke and his collaborators (Morales, Rathbun, Smith, and Pace, 1945; Pace and Rathbun, 1945) postulated that fat, expressed as a percentage of the body weight, is the only major variable component of the body, and they developed a formula relating specific gravity to the fat content. Rathbun and Pace (1945) attempted to validate the use of specific gravity for estimating the fat content by experiments on guinea pigs.

Application of the technique to man appears promising although the method is not without drawbacks. The fact that the relative body hydration under different nutritional conditions may vary significantly complicates the picture. However, the primary difficulty inherent in the method as used today is the necessity for correcting the weight of the body fully immersed in water for the residual air. Variations in this quantity have a marked effect on the estimated value of the body fat. The methods for determinations of the residual air volume are time-consuming and present technical difficulties, especially if one attempts to determine the air volume at the time of immersion.

In spite of these handicaps the technique proved useful in the Minnesota Experiment. In view of the value of information on the fat content of the body for an intelligent appraisal of the nutritional status of an individual, techniques for a more effective determination of the body volume should be developed. Determination of the volume of the body on the basis of pressure changes in a closed-air system (Pfaundler, 1916; Bohnenkamp and Schmaeh, 1931; Joengbloed and Noyons, 1938) is a promising avenue of approach.

CHAPTER 9

Morphology of Some Organs and Tissues

THERE is an enormous literature on the effects of starvation on the size and appearance of the organs and tissues of the body. Several series of experiments on animals, quite respectable in both design and execution, were reported more than a century ago (Collard de Martigny, 1828; Chossat, 1843). With the development of semiquantitative methods in pathology, morphological studies on starved animals and occasional autopsy reports with measurements on starved human beings became especially numerous in the period from 1865 to 1915. Most of this older material was reviewed by Morgulis (1923) and Jackson (1925). More recent investigations are neither as numerous nor as illuminating as might be wished. Unfortunately, the great bulk of attention has been given to acute starvation, and there is relatively little information on famine victims or the cachexias which are almost always available in every large hospital. Perhaps the best gross morphological data from famine victims are still those of Porter (1889), but valuable data are provided in the monographs by Apfelbaum *et al.* (1946), Hottinger *et al.* (1948), and Lamy *et al.* (1948).

Though morphological studies on starvation sometimes report measurements, these are generally limited to simple organ weights — and usually even these are limited in significance because of the absence of real standards for normality. It may be presumed that the various investigators thought their data would have more than descriptive interest, but when the data are examined these many years later, the failure to arrive at functional or diagnostic significance is striking. In a huge literature, there is vastly more descriptive morphology than anything else.

In a general way, important morphological questions would seem to be: (1) How do the various organs and tissues participate, quantitatively, in the decrease of the body mass? (2) What happens to the individual cells of these organs and tissues? (3) What is the significance of these changes? The present chapter is a review of the literature on these points with regard to the muscles, brain, liver, spleen, lungs, kidneys, and mammary glands. Other organs and tissues are discussed in Chapters 8 and 10 through 15.

The Musculature

Marked atrophy of the skeletal musculature is a prominent characteristic of severe undernutrition and acute starvation in both men and animals. Since quantitative data are practically absent for the muscles in human autopsy material, reliance must be placed on the data from animal experiments. In general, the proportional loss of skeletal muscle mass is close to that for the body as a whole.

Jackson (1915) found an average loss of 31 per cent in the muscles in albino rats acutely starved until there was a body weight loss of 33 per cent; in chronic undernutrition with an average body weight loss of 36 per cent, the musculature was estimated to have lost 41 per cent. Cameron and Carmichael (1946) also starved albino rats and studied the anterior tibialis muscle; in 18 female rats the muscle weight loss was proportional to the body weight loss, but in 5 male rats this muscle appeared to lose only about two thirds as much, relatively, as the total body. With frogs, also, some sex differences in the atrophy of muscles in starvation have been reported, the females showing greater muscle atrophy in the early stages but retaining their muscles much better than males in extreme emaciation (Ott, 1924). Whether such a sex difference exists in man, and whether, indeed, it is a real phenomenon in other animals, remains a question.

The limitations of crude weight data are indicated in the fact that the percentage of dry matter in the muscles is decreased in starvation — that is, the muscles become waterlogged (Terroine, 1920; Moulton et al., 1922; Ott, 1924). It is uncertain whether this elevated water percentage represents intracellular edema to any extent or is only a reflection of the associated extracellular fluid. There are several descriptions of what seem to be intracellular vacuoles and separation of fibers from their sheaths (see Meyers, 1917).

There is agreement that the individual muscle fibers are decreased in size in undernutrition, the reduction in diameter generally being more or less in proportion to the loss of body weight (e.g. Morpurgo, 1889; Halban, 1894). Besides the reduction in fiber size, there is actual destruction of muscle cells in extreme emaciation (e.g. Meyers, 1917), but the resistance of the nuclei to degeneration is considerable (Beeli, 1908). Morpurgo reported occasional mitoses in adult rabbit muscles during refeeding after semi-starvation.

Histological studies on muscles taken from severely underfed animals often show little besides the marked reduction in fiber size, the attendant crowding of nuclei, and some loss of distinctness of striations (Morgulis et al., 1915; Meyers, 1917; Moulton, 1920). The numerous arguments about fatty, vitreous, waxy, granular, pigmentary, and primary degeneration are summarized by Jackson (1925, pp. 166–70). Though the muscles are a major source of fat and protein for general metabolic use in starvation, it appears that irreversible degeneration of the fibers does not occur except in the most extreme conditions of starvation.

In 359 autopsies on adults dead of "pure" starvation in the Warsaw Ghetto, well-preserved skeletal muscles were seen in only 2.7 per cent and severe atrophy was recorded for 61.0 per cent (Stein and Fenigstein, 1946). Microscopic studies in 6 cases revealed occasional loss of cross-striations and in some places there were amorphous accumulations of strongly colored material. Some of the fibers exhibited vitreous alteration or granular fragmentation. The number of nuclei per unit field was generally increased and occasionally there were accumulations of nuclei in groups. In 6 cases of pure famine Lamy, Lamotte, and Lamotte-Barrillon (1948) saw nuclear destruction and hyaline degeneration; in one case there was a granular deposit of calcium in the fibers and in 2 cases there were indications of an inflammatory reaction in the connective tissue of the skeletal muscles.

Cardiac musculature does not seem to differ from the skeletal musculature in response to starvation except, perhaps, that the relative decrease in mass may be slightly less. This latter difference might be expected in view of the fact that voluntary movements are ordinarily reduced to the absolute minimum in starvation, whereas the heart, though its work is much reduced, is still necessarily in fairly active "training." Starvation and disuse atrophy are both involved in skeletal muscles. In starved patients with tuberculosis, Chortis (1946) found cloudy swelling of the sarcoplasm and separation of the muscle fibers at their points of junction in the heart.

The smooth muscles apparently respond to starvation much as do the striated muscles. The muscle fibers in the stomach wall exhibit marked atrophy and various degrees of degeneration with vacuoles in occasional fibers (Gaglio, 1884; Miller, 1927). Indian famine victims examined by Porter (1889) frequently showed extreme atrophy of all layers of the intestinal wall, including the muscularis.

Weights of the Principal Organs

The simplest way of estimating the degree of atrophy of the various organs of the body is to compare their weights with normal standards. There are many data on organ weights in starvation but their significance is obscured because of the unsatisfactory character of the standards for normal states of nutrition. The most striking evidence of the failure of anatomy to become a quantitative science is the relative absence of good standards. The value of such data as those of Porter (1889) on famine victims is severely limited on this account. In animal experiments "control" data are frequently supplied, but these often fail to fulfill the needs for proper statistical analysis.

Data from the most uncomplicated cases of Porter (1889) are summarized in Table 97. These adults from the Madras area of India were, like the general population of that region, small in stature and slight in build. In the absence of any height/weight norms for India, the degree of body weight loss cannot be estimated precisely, but for rough calculations recourse might be had to Occidental height/weight standards for persons of slight build. By any method of estimation it would appear that the weight losses probably were between 25 and 45 per cent of the original weight. Obviously all the organ weights, except those of the brain, are markedly subnormal in Porter's series, but there is no proper basis for precise calculation of the deviation from normality. Porter himself attempted rather vaguely to apply the very primitive formulas of Quetelet, but besides the fact that figures derived from Europeans may not apply to Orientals, it is certain that Quetelet's values are far from precise even for Europeans. Porter included a few persons who died in the famine area from other causes than starvation and who appeared to be in a good state of nutrition. Calculations from these as controls will be referred to in subsequent sections of this chapter.

Uehlinger (1948) summarized the autopsy experience on victims of German concentration camps and prisons who died at Herisau, Switzerland; the data are given in Table 98. Much more significant material is available from the large number of autopsies in the Warsaw Ghetto for the starvation period of 1941–42.

TABLE 97

SUMMARIZED AUTOPSY DATA from the Madras famine of 1877–78. Uncomplicated cases selected from the protocols of Porter (1889). Organ weights are in gm. Kidney = the mean of the right and left kidneys; Males, A = cases of dysentery and atrophy without important edema; Males, B = cases of marked edema.

	30 Males, A		13 Males, B		19 Females	
	M	SD	M	SD	M	SD
Age (yrs.)	47.8		48.3		44.2	
Height (cm.) ...	164.6	5.4	161.8	6.7	153.4	6.1
Body weight (kg.)	36.5	6.3	47.0	9.4	29.5	5.2
Brain 1164		112	1188	118	1096	99
Heart	174.1	34.3	201.2	47.0	147.7	30.9
Liver	827	198	888	360	813	277
Spleen	81.8	52.5	117.5	97.0	68.8	66.1
Kidney	78.5	15.7	96.9	41.0	76.8	16.9
Pancreas	62.0	18.0	68.1	18.6	53.2	19.4

TABLE 98

GENERAL CHANGES IN ORGAN WEIGHTS IN PURE INANITION as summarized for victims of German concentration and prison camps who died at Herisau, Switzerland (by Uehlinger, 1948, p. 239).

	Average Weight (gm.)	Weight Loss (%)
Liver	1000–1200	30
Heart	200–220	25–30
Pancreas	70–90	10
Kidneys	265–280	10
Spleen	80–200	
Brain	1200–1300	10

TABLE 99

WEIGHTS OF ORGANS OF THE BODY IN UNCOMPLICATED CASES OF FAMINE DEATH, in gm., for persons from 20 to 60 years of age in the Warsaw Ghetto. Normal means are those cited by the Polish authors for comparable persons in a normal state of nutrition. Data from Stein and Fenigstein (1946).

	Brain	Heart	Liver	Kidneys		Spleen
				Right	Left	
Normal M	1310	275	1500–2000	150	155	150–250
Famine						
M	1390	220	865	112	114	103
Minimum	760	110	545	70	65	40
Maximum	1680	350	1600	190	190	310
M as percentage of normal	107	80	54	75	75	52

The findings are condensed in Table 99. They indicate very large losses in heart, liver, kidney, and spleen weights but no change in the brain. In these persons the gross body weight loss ranged from 30 to 50 per cent.

Krieger (1921) made a careful study of the autopsy records for extremely emaciated patients with various disorders and expressed all results as percentage deviations from normal. The summarized results given in Table 100 are instructive but they look more impressive than is warranted because the normal standards used, those of Vierordt (1893), are certainly far from perfect.

Rather indirect evidence on the effects of much smaller degrees of undernutrition on organ weights in man was provided by Weber (1921) from the autopsy records at Kiel, Germany, for the years 1914–15 and 1917–18. In the latter 2 years the population of Kiel was, in general, moderately undernourished, and it could be presumed that this condition also characterized the persons who came

TABLE 100

PERCENTAGE LOSSES OF WEIGHT OF THE BODY AND VARIOUS ORGANS IN AUTOPSIES ON EXTREMELY EMACIATED BODIES. Reference normal values were taken from the old tables of Vierordt (1893). Data from Krieger (1921).

	Cause of Death											
	Cachexia		Dysentery		Tumor		Chronic Infections		Tuber-culosis		Old Age	
	N	% Loss	N	% Loss	N	% Loss	N	% Loss	N	% Loss	N	% Loss
Body weight ..	11	38.8	7	48.4	27	38.0	31	43.9	40	43.0	19	35.8
Heart	11	34.7	7	45.0	25	33.2	31	30.7	39	31.9	20	29.6
Liver	10	42.1	6	43.5	21	32.8	15	28.0	29	27.7	17	26.0
Kidney	11	36.0	7	41.0	25	27.5	31	15.5	39	17.6	20	19.7
Spleen	8	46.6	5	36.0	22	27.6					14	36.0
Brain	10	4.6	5	4.1	19	3.4	23	2.8	25	4.0	12	3.1
Pancreas			5	44.8	6	33.0	26	30.5	17	38.6		
Thyroid			5	42.0	10	20.6	23	32.3	25	35.8		
Adrenals			5	17.2	9	21.5	23	+23.2	24	+6.6		
Testes			5	41.3	12	28.7	24	40.4	27	39.4		

TABLE 101

BODY LENGTH AND ORGAN WEIGHTS FROM ROUTINE POST-MORTEMS at Kiel, Germany, on adults, given as mean values for the 2 years 1914 and 1915 and for the 2 years 1917 and 1918. Only organs without evidence of pathology were included. Compiled from Weber (1921).

	1914–1915				1917–1918			
	Men		Women		Men		Women	
	N	M	N	M	N	M	N	M
Stature (cm.)	180	169.2	129	158.1	351	168.8	306	158.4
Brain (gm.)	110	1334	66	1222	36	1395	29	1270
Heart (gm.)	106	297.9	96	261.5	242	287.9	241	255.4
Liver (gm.)	93	1359	75	1272	142	1260	173	1213
Kidneys (gm.)	153	279.6	110	245.2	187	279.9	177	256.2
Spleen (gm.)	145	156.7	104	142.3	205	135.8	202	125.2

to autopsy in the hospital there. In any case the results showed that, though the body lengths were much the same in the two periods, there were distinct and consistent reductions in the weights of all the organs except the brain. For example, the liver, though it varies from individual to individual, decreased on the average 7 per cent in men and 5 per cent in women. The summarized data are given in Table 101.

Clearer indications of the losses of the different organs in starvation are seen in the results of animal experiments where control animals in a well-fed state have been used as a basis for estimating the pre-starvation values. Most of the studies were made with acute starvation but the results seem to be reasonably similar in chronic undernutrition. Acceptable data are summarized in Tables 102, 103, 104, and 105. These data are in general agreement with findings in less complete and systematic studies on other animal species and with autopsy records on

TABLE 102

TOTAL INANITION IN MALE FROGS. Percentage losses in weights of organs of the body at different levels of total body weight loss. Data from Ott (1924).

	Percentage Body Weight Loss					
	13.7	20.0	30.9	40.9	50.1	59.5
Skin	7	25	38	47	59	72
Muscles	19	25	40	52	63	80
Heart	6	15	31	40	54	52
Liver	30	40	56	74	83	87
Kidneys	16	18	25	39	59	69
Brain	7	6	3	13	15	22

TABLE 103

TOTAL INANITION IN GUINEA PIGS MAINTAINED WITHOUT FOOD. Losses in body weight and estimated corresponding weights of the body organs; average values for 10 animals in each group. The figures in parentheses for Group 1 give the average weights, in gm., for the organs which were used in computing the percentage losses for the other groups. Data from Lazareff (1895).

	Body Weight		Duration	Percentage Losses				
Group	Initial	Final	(hrs.)	Body	Lungs	Heart	Liver	Stomach
1	584	584	0		(3.22)	(1.86)	(22.08)	(2.62)
2	584	525	30.4	10.1	−0.9	4.8	18.0	−1.9
3	587	469	70.0	20.0	−0.3	9.1	23.5	6.5
4	580	405	121.4	30.2	0.6	21.0	31.0	6.5
5	580	374	164.2	35.5	5.0	33.3	35.1	11.8

	Percentage Losses							
Group	Intestines	Spleen	Pancreas	Kidneys	Skin	Cord	Brain	Femur
1	(16.32)	(0.48)	(1.50)	(3.91)	(76.76)	(1.32)	(3.97)	(1.41)
2	−0.5	0	3.3	−1.8	2.0	−1.1	1.5	−2.1
3	10.2	31.3	5.3	2.6	8.2	6.8	3.0	−2.1
4	10.5	37.1	24.7	10.2	12.7	6.8	5.5	3.6
5	25.8	43.8	39.3	11.0	17.9	6.8	6.1	2.8

TABLE 104

PERCENTAGE WEIGHT LOSSES OF THE ORGANS OF THE BODY IN ADULT WHITE RATS in acute starvation and in severe chronic undernutrition. Mean values of the differences from well-fed controls. Data from Jackson (1915; see also Jackson, 1925, p. 467).

	Acute	Chronic
Total body weight	−33	−36
Skeleton	0	+2
Musculature	−31	−41
Brain	−5	−7
Spinal cord	0	−4
Heart	−28	−33
Liver	−58	−43
Spleen	−51	−29
Kidneys	−26	−27
Lungs	−31	−40
Testes	−30	−40
Stomach and intestines	−57	−57

TABLE 105

PERCENTAGE WEIGHT LOSSES OF THE ORGANS OF ADULT WHITE RATS deprived of all food but allowed water ad lib. Mean values of the differences from the averages for well-fed controls, from Cameron and Carmichael (1946). Muscle = the right anterior tibialis muscles; body = the total gross body weight.

Sex	N	Body	Muscle	Liver	Heart	Kidney	Spleen	Gonads	Adrenals	Thyroid
F	9	−22.5	−22	−30	−13	−16	−23	−5	−2	−9
F	9	−30	−30	−50	−23	−24	−52	−16	+29	−30
M	5	−29	−19	−55	−26	−30	−37	−13	+9	−6*

* The thyroid in one animal was much enlarged.

individual human beings who have starved to death (e.g. Bright, 1877; Meyers, 1917).

Several generalizations are possible from the available information. The brain and spinal cord lose very little weight in starvation but the other soft tissues show large losses, the liver and intestines perhaps suffering to the greatest extent. The heart and kidneys tend to lose a little less weight than would be proportional to the body weight loss. There are discrepancies with regard to the lungs and some of the endocrine glands, but these may be, at least in part, reflections of differences in hydration and of technical difficulties in working with very small bits of wet tissue.

There are considerable differences in hydration of the tissues in various starvation conditions. In human cachexias the contrast between "dry" and dropsical states may be extreme. It is not possible to consider hibernating animals as the equivalent of cases of either acute or chronic starvation because of the attendant dehydration, which does not affect all organs and tissues to the same degree. Starvation edema extends to the parenchyma of all the viscera (Chortis, 1946). Studies on severe thirst show marked differential effects. Dehydration occurs most readily in the spleen, pancreas, lungs, and liver, but the kidneys, intestines, and eyeballs are relatively resistant and the brain seems to be even more resistant than the skeleton (Kudo, 1921).

The Brain and Nervous System

The fact that the brain loses little weight in any form of starvation has been noted above. In some instances it may be unusually heavy, as in the case reported by Meyers (1917) where the brain of an acutely starved man weighed 1600 gm.

It is incorrect, however, to conclude that the brain is "immune" to starvation. In Porter's (1889) series the mean brain weight of the greatly emaciated men was 9.8 per cent less than that of the "normal" (plump) men, and the difference in the women of these two categories was almost as large. Though these changes are larger than generally indicated in starvation studies, there seems to be no doubt that the brain does lose at least a little weight in starvation.

There are some histological alterations which suggest that the loss of brain tissue is greater than indicated by gross weight. All investigators seem to agree that the brain tissue of-starved animals, including man, shows increased pericellular spaces and frequent vacuoles. Part of the stability of the brain weight in starvation may be the result of replacement of tissue by water. However, it may be noted that Addis et al. (1936) found only small reductions in the total nitrogen content of the brains of starved rats.

Cytological evidences of degeneration in the brain are frequently but not always seen in starvation. They are not very marked and their significance is debatable. There is a good deal of evidence that the brain is edematous, and in most cases some degree of hyperemia is apparent (Chortis, 1946). Besides vacuolation, cloudy swelling, chromatolysis, and fibrolysis are reported occasionally in severe emaciation. As is the case in much cytological literature, there is a great deal of argument about the appearance and disappearance of various granules from which no clear picture emerges. In the Warsaw material the histology of the brain was not much altered, the most common peculiarities being accumulations of lymphocytes and rather abundant deposits of yellow-gold pigment in the perivascular lymph spaces (Stein and Fenigstein, 1946).

The changes in the spinal cord, both in gross morphology and in the finer structure, are very similar to those in the brain (Donaggio, 1906, 1907; Sundwall, 1917). There are few studies on the peripheral nervous system in starvation, and the limited information available was practically all covered by Jackson (1925, pp. 203–5). As a whole, the peripheral nerves show only small histological changes in starvation unless there is also pellagra or beriberi. The nerve fibers themselves are quite resistant, although slight atrophy of the myelin sheath of medullated fibers has been described. The nerve cells more often show degeneration, including vacuolation and chromatolysis, though these are by no means regular or very pronounced phenomena.

The Liver

In almost all cases starvation or chronic undernutrition produces a marked loss in weight of the liver, though the high degree of normal variability in this organ makes interpretation of individual data difficult. The relative loss of liver weight seems to exceed that of the body as a whole so that the liver weight is subnormal both in absolute and relative terms. Excluding patients with diseases directly affecting the liver itself, all forms of emaciation are accompanied by atrophy of the liver (Bean and Baker, 1919; Krieger, 1921; Stefko, 1923).

The cytological changes in the liver have been studied by scores of investigators (see Jackson, 1925, pp. 325–41; Lazarovich-Hrebeljanovich, 1936). Common findings are cloudy swelling, simple atrophy, hemosiderosis, vacuolation, and degenerative changes in the mitochondria. As to whether true fatty degeneration occurs there is no agreement. In some cases at least, there is fat deposition in the Kupffer cells and in the epithelium of some of the bile ducts (Okuneff, 1922). In 87 autopsies on starved men with tuberculosis, Larson (1946) found marked fatty infiltration of the liver in 54 cases (62 per cent); this change was most prominent in the most severely starved men. Parenchymatous or fatty degeneration of the liver cells was seen in 75 of the men in this group (86 per cent).

Since the liver more than any other organ combines active metabolic with storage functions, it is desirable to consider here the loss of stored materials. Terroine (1920) concluded that there is no important differential loss of fat from the liver in starvation and therefore this organ could not be considered a real storehouse of fat. Voelkel (1886) reported a fatty liver in a man who deliberately starved himself to death. But calculations are extremely difficult for an organ in which there is great lability of water, glycogen, and protein as well as fat. The weight loss of the liver is certainly not merely a change of water. Addis *et al.* (1936) found that the total protein content of the rat liver decreased by 40 per cent in a 7-day fast. In comparison, the kidney and heart lost, respectively, 20 and 18 per cent of their protein contents. Uehlinger (1948) estimated that five sixths of the protoplasm of the liver may be lost in some human cases.

Uehlinger (1948) likewise considered fatty degeneration of the liver to be one of the commonest changes in that organ in chronic undernutrition. In the peripheral lobules the cytoplasm of the liver cells was filled with fat droplets, mostly rather large in size, and the nuclei seemed to be displaced toward the outer walls of the cells on this account. In the most severe cases of starvation there were also large vacuoles which seemed to represent intracellular water droplets; they were not stained with carmine or fat-soluble stains. Those cells which contained little or no lipoid showed rather marked increases of brown pigment. Slight degrees of hemosiderosis were often seen in hepatic cells, star cells, and the portal connective tissue. In a total of 53 livers examined, fatty degeneration was seen in 39 cases, of which 7 were rated as severe and 21 as moderate. In the most extreme form it was estimated that fat represented 70 per cent of the total liver weight. Rather common was a slight lymphocytic infiltration of the periportal connective tissue. Since there were secondary infections in almost all the patients, these indications of inflammatory reactions are not surprising.

The picture of the liver in the material on uncomplicated starvation seen at Warsaw was somewhat different (Stein and Fenigstein, 1946). Four of the 6 livers studied in detail showed typical brown atrophy. The hepatic cells were small and often contained small, highly colored nuclei. In the interlobular spaces some of the blood vessels showed vitreous changes in the walls, and in places there were small groups of lymphocytes. In 2 out of 6 cases there were large amounts of hemosiderin, and this was especially marked in the hepatic cells in one case.

The Spleen

The spleen, even more than the liver, normally shows great variability among individuals, so quantitative analyses as to the effects of starvation are unusually difficult. In general, the spleen loses relatively more weight than the body as a whole, and individual apparent reductions of 60 per cent are not rare. But malaria, tuberculosis, and other complicating diseases have large opposing effects on the spleen size, so that the findings in famine, for example, are highly variable.

Porter (1889) concluded that the relative loss in spleen weight exceeded that of all other organs in the Madras famine. Krieger (1921) observed an average spleen weight of only 80 gm. in 8 cases of emaciation in insane persons who did not have chronic organic disease. Comparisons with animals are somewhat un-certain because there are considerable differences among species.

The structural changes in the spleen include partial obliteration or loss of the Malpighian bodies, general atrophy of lymphoid tissue, reduction in the splenic sinuses, and pigmentation. Because of the atrophy of the parenchyma, the tra-beculae and capsule are unusually prominent and may show signs of fibrosis and sclerosis (Stefko, 1923; Chortis, 1946). The rather constant appearance of hemo-siderosis is attributed to an increased destruction of erythrocytes. Details are given by Meyers (1917), Sundwall (1917), Inlow (1922), and Okuneff (1922, 1923).

In the Warsaw material Stein and Fenigstein (1946) reported prominent vitreous changes in the trabeculae and the membranes of the lymphatic fol-licles. In the red pulp there were large deposits of hemosiderin in rather large granules, and hemosiderin was also present in the cells and in the walls of the sinuses. In Uehlinger's (1948) material hemosiderosis was prominent in the spleen, being recorded in 41 out of 53 autopsies, and the degree of hemosiderosis was closely correlated with the degree of anemia. Atrophy of the spleen follicle, however, was not prominent in Uehlinger's material, and he concluded that in general the spleen is not very sensitive to starvation.

The Kidneys

The kidneys undergo atrophy in both acute and chronic starvation but the degree of weight loss is ordinarily somewhat less than that of the body as a whole. In human cases occasionally no important deviation from normal, except for some thinning of the cortex, is found on gross inspection (Lucas, 1826; Bright, 1877; Voelkel, 1886; Meyers, 1917; Nicolaeff, 1923). Several investigators have concluded that abnormalities in the gross morphology of the kidneys in emaciated infants and children are neither great nor constant (Mattei, 1914; Marfan, 1921). In Jackson's (1925, p. 375) summary of kidney weights for atroph-ic infants, the average was only slightly below normal for body length but was about 19 per cent below normal for age. In animal experiments a substantial loss in kidney weight has been recorded universally but the proportional reduction is of the order of half that of the body (see Jackson, 1925, pp. 375–83).

For 597 white males in general autopsy material (excluding renal disease), Bean and Baker (1919) recorded the average weight of both kidneys as 291 gm. in extreme emaciation and 379 gm. in obesity; similar values were found in white women and in Negroes of both sexes. From the material of Bean and Baker it might be estimated that the kidney loses something like half as much weight, proportionally, as the body as a whole. But the data of Krieger (1921) indicate a larger proportion of weight loss except in cases with tuberculosis and chronic infections (see Table 100).

There is some reason to believe that the kidney is relatively resistant to atrophy in slight to moderate degrees of undernutrition in man (Roessle, 1919; Weber, 1921). This is supported to some extent by animal experiments (Ochotin, 1886; Lazareff, 1895; Beeli, 1908; Cameron and Carmichael, 1946).

Stefani (1910) made experiments with starving dogs in which one kidney was removed at the beginning of the starvation. The result was that the remaining kidney underwent little or no atrophy, although in control dogs, with both kidneys intact, the kidneys showed a weight loss of 25 to 30 per cent. It was concluded that the tendency toward compensatory hypertrophy of a single kidney counteracts the normal tendency toward atrophy in starvation.

On section of the kidneys in starvation, more or less intense hyperemia of the cortex and medulla is usually found except where there is much edema; in the latter case the kidney section is often pale with only small contrast between the cortical and medullary zones (Nicolaeff, 1923). Under the microscope there is cloudy swelling of the cells of the tubules and Bowman's capsule is usually congested, with distortion of both the capsule and the tubules (Nicolaeff, 1923; Chortis, 1946). These same changes, as well as signs of more complete degeneration, are seen in animals (e.g. Sundwall, 1917; Asada, 1919).

Stein and Fenigstein (1946) observed, besides reduction in total size, only rather small changes in the kidneys of the victims of pure starvation in Warsaw. There was often some serous fluid in Bowman's capsule, and the tubules were rather contorted. The epithelium of the tubules appeared to be somewhat aplastic and showed vitreous changes in 5 out of 6 cases. Uehlinger (1948) observed even less alteration in the persons rescued from the Germans and brought to Herisau. He concluded that, with the exception of stasis albuminuria and ordinary age changes, the kidneys were substantially normal.

This is in great contrast to the report of Lamy, Lamotte, and Lamotte-Barrillon (1948), in which the renal changes were considered rather comparable to the situation in the crush syndrome. These latter workers suggested that the release of myoglobin from the degenerating muscles might account for their findings. In any case they reported cloudy swelling of the epithelium, desquamation of the tubular epithelium, and other changes which they interpreted as showing severe damage. Uehlinger (1948) sharply criticized this interpretation and stated that the published microphotographs did not support the French contentions about the findings at Mainau. This is the kind of debate that gives the investigator a sense of futility, impressing him with the limitations of classical pathology.

There are few measurements on the cells and the internal structure of the kidney. In white mice which had lost 28 per cent of their body weight, the kidney cells were reduced in size and the nuclei were estimated to have lost 23 per cent of their original volume (Lukjanow, 1898). Cesa-Bianchi (1909), however, claimed the nuclei did not change but that the thickness of the walls of the tubules is decreased by 15 to 20 per cent when there has been a 40 per cent loss in body weight.

The growth and development of the kidney in young rats was studied quantitatively by Kittelson (1920). Severe underfeeding retarded the growth of the medulla less than that of the cortex; the formation of new renal corpuscles may cease entirely but corpuscles in the process of development go on to completion. Refeeding after a period of severe undernutrition resulted in hypertrophy of the renal corpuscles and an increase in their number to normal or even beyond.

It should be noted that even in the most extreme starvation the gross function of the kidney is surprisingly well preserved; albuminuria does not appear, nor is there uremia (see Chapter 31).

The Mammary Glands

The data on the mammary glands in females are somewhat discordant. In severe famine the atrophy of the breasts is often described as extreme. For example, in Indian famine victims Porter (1889) stated that the position of the gland could be ascertained only from that of the nipple and that no glandular tissue could be felt. On the other hand, there is remarkable preservation of the breasts in some cases of anorexia nervosa. This was true in the 2 patients of Bruckner et al. (1938) whose body weights were close to 50 per cent below normal, and it is obvious from the photographs of the cases of Berkman et al. (1947) numbered 1, 2, 3, 4, 5, and 6. Schultzen (1863) reported a case of starvation in a 19-year-old girl with well-preserved breasts. Extreme atrophy of the breasts is common in tubercular patients who are much emaciated.

The evidence on milk secretion in moderately undernourished women is also variable. Decaisne (1871) studied 43 young mothers in the siege of Paris and found 12 of them able to nurse their infants fairly well, but in 16 women there was practically no milk secretion. During and after World War I variable findings were reported for Germany, but most observers agreed that there was some reduction in the volume of milk secreted but no change in its composition (Momm, 1920).

The Lungs

In the majority of cases, in both acute and chronic starvation, the changes in the lungs are minimal. This situation is to be contrasted with the apparent susceptibility of starved people to pulmonary infection, both tuberculosis and the pneumonias. Although there are few indications of primary atrophy and degeneration in the pulmonary tissues, there are substantial functional alterations.

At Warsaw the lungs were examined in detail in 5 patients representing uncomplicated starvation; the only positive finding was emphysema, and this was present in all 5 cases (Stein and Fenigstein, 1946). In a total of 370 autopsies at

Warsaw in which the lungs were examined there were 50 cases of emphysema, that is, 13.5 per cent. The character of the emphysema was like that seen in senile emphysema, and the Polish authors believed that it represented an involutional process. It was seen in Warsaw at all ages, 14 of the 50 cases being persons under 30 years of age. We have not found any other reports of this nature.

Another peculiarity of the lungs in chronic starvation seems to be an enhanced tendency toward cavity formation. Lamy *et al.* (1948) found non-tuberculous cavities in 5 out of 13 autopsies. Pulmonary infarction seems to lead readily to cavity formation in the starved. Uehlinger (1948) also remarked on the frequency of pleural reactions in the case of lung infections; pleuritis and pleural effusion developed in 3 out of 4 cases of lobar pneumonia. There is a general tendency toward reduced mucous formation in the bronchi and the respiratory passages. Pulmonary infection is seldom very productive but a dry type of chronic bronchitis is common. This was repeatedly observed in World War I and is evident in more recent studies. A mild but persistent bronchitis troubled a number of the men in the Minnesota Experiment.

Gross Composition of the Organs

The gross composition of the body, from the standpoint of major compartments, is discussed in Chapter 15. There are many indications that all organs of the body are edematous and unduly hydrated in both acute and chronic starvation, except perhaps in the terminal stage or when extreme dysentery tends to produce dehydration.

In the musculature the percentage of water rises progressively during fasting. The results of Mendel and Rose (1911) are typical. They found 75.42 per cent as the average for water in the muscles of well-fed rabbits, but for rabbits starved until they lost averages of 25, 36, and 45 per cent of their body weights the average percentages of water were, respectively, 77.50, 79.24, and 79.48. Similar results have been obtained with various species of fish, birds, and mammals (see Morgulis, 1923, pp. 100ff). Calculated on a fat-free basis, the change in water is equally striking. Data on the water in other organs are not numerous but in general the results are in agreement with those on muscle. In other words, the

TABLE 106

Relative Losses of Protein from Various Organs and Tissues of Albino Rats in a 7-day fast, as percentages of control values
(Addis, Poo, and Lew, 1936).

Organ or Tissue	Percentage Loss	Organ or Tissue	Percentage Loss
Liver	40	Heart	18
Prostate	29	Muscle, skin, and skeleton	8
Seminal vesicles	29	Brain	5
Gastrointestinal tract	28	Eyes	0
Kidneys	20	Testicles	0
Drawn blood	20	Adrenals	0

loss of weight of the body or of its organs is an underestimate of the actual loss in vital tissue.

In starvation the body as a whole loses fat, of course, and most of the organs appear to lose fat somewhat in proportion to their initial lipoid content. Exceptions are the brain and nervous tissues and, according to Terroine (1920), the kidney. The more directly intriguing question of nitrogen (protein) loss in starvation has been less well studied in the individual organs. The effects in albino rats of a 7-day fast were measured by Addis *et al.* (1936). In terms of percentage of the original protein content of the organ, the liver lost far more than the other organs studied. The findings are summarized in Table 106.

CHAPTER 10

Morphology of the Heart and Blood Vessels

TEXTBOOKS of cardiology, physiology, and nutrition in general state or imply that the heart is resistant to caloric undernutrition and does not undergo important atrophy or degeneration like the other tissues of the body. This idea has even been incorporated into a general principle proclaiming the wisdom of nature in protecting the "most vital organs" (heart and brain) in starvation.

However attractive this idea of immunity of the heart may be philosophically, or in terms of teleology, it is not supported by the actual facts, as is obvious from the data given in the tables in Chapter 9. The history of this error is discussed in a subsequent section of this chapter, and there are understandable reasons why it should have been so perpetuated. In general medical experience the disordered or abnormal heart has one or more of several outstanding characteristics: it is enlarged, it has a rapid rate, the heartbeat is irregular, or it has abnormal sounds. Moreover, the person with heart disease often complains of anginal pain or palpitation or dyspnea. None of these features is characteristic of starvation. The one item in starvation which suggests heart failure is edema, but even a superficial analysis seems to rule out the heart as the offender in the great majority of cases of famine edema.

Such negative evidence does not prove that the heart is protected in starvation or that it remains relatively normal, morphologically or functionally. Actually, as will be seen, both morphological and functional abnormalities of major degree develop regularly. Cardiac physiology is discussed in Chapters 28, 29, and 30. More purely morphological aspects are discussed below.

Heart Size in Starved Animals

As long ago as 1828 Collard de Martigny noted that the heart appeared atrophic and unusually small in dogs and rabbits deprived of all food and drink. The quantitative experimental studies of Chossat (1843) were surprisingly elaborate for the time. At death Chossat's starved pigeons had usually lost about 40 per cent of their body weight; comparison of their hearts with those taken from well-nourished birds indicated an average loss of about 45 per cent in heart weight. Since then a great many studies have been reported, mostly involving acute starvation, on rats, mice, marmots, guinea pigs, chickens, crows, pigeons, pigs, dogs, and cats (cf. Jackson, 1925, pp. 222ff). With few exceptions comparison of the starved animals with controls of the same species indicated that the proportional loss of heart weight in starvation averages something like 70 to 90 per cent of the body weight loss; there are a few reports that the heart weight loss was relatively greater than that of the body as a whole (e.g. Chossat, 1843;

Bourgeois, 1870; Pfeiffer, 1887; Beeli, 1908). In the most recent experiments 23 adult white rats lost an average of 26.9 per cent of the body weight in starvation and the heart weight loss was estimated to be 19.8 per cent (Cameron and Carmichael, 1946). This amounts to 73.6 per cent of the relative body weight loss.

The cat might seem to be a special case in view of the often quoted results of C. Voit (1866), in which the heart of one starved cat weighed 2.6 per cent less than that of one control animal, while the voluntary musculature weighed 30.5 per cent less. Unfortunately the heart is a highly variable part of the total body weight and, though Voit put no emphasis on his peculiar result, his data have been widely misapplied. That cats are, in fact, no different from other mammals in regard to the heart in starvation was observed as early as 1870 (Bourgeois).

Experiments with chronic undernutrition and semi-starvation in animals are less numerous than experiments with total lack of food, but the results on heart size are much the same in the two conditions. In rats Jackson (1915) found the average relative reduction in heart weight to be 85 per cent of the body weight loss in acute starvation, and in chronic undernutrition the proportion was 91 per cent; in both conditions about a third of the body weight was lost (see Table 104). In young steers the relation between heart weight and body weight is substantially the same irrespective of the plane of nutrition (Trowbridge et al., 1918, 1919; Moulton et al., 1922).

In very young animals starvation suppresses the growth of different organs to different degrees, and the heart weight tends to represent slightly more than the usual proportion of the total body weight in these cases. When rats are held nearly at birth weight for 14 to 16 days by severe underfeeding, the heart weight increases about 25 per cent (26 per cent according to Stewart, 1919; 24 per cent according to Jackson, 1932). This percentage increase of the heart ranked seventeenth in a list of 26 subdivisions of the body studied by Jackson (op. cit.); in other words, 16 other subdivisions of the body surpassed the heart in growth potential in the face of severe undernutrition. Bechterew (1895) had somewhat similar findings in newborn kittens and puppies.

It takes very severe undernutrition of the mother to interfere with the growth of the fetus, but if a degree of maternal starvation is enforced which does reduce the birth weight significantly, it is found that the heart weight reduction is slightly less than in proportion to the body as a whole. Barry (1920) reported that in proportion to body size the heart weight of rats delivered by very emaciated mothers averaged 8 per cent above normal.

Heart Size in Children at Autopsy

In children beyond infancy the relative degree of atrophy of the heart in starvation does not seem to differ greatly from that in adults. Stefko (1924) concluded that the heart size in starved children was roughly normal for the body weight but stated that the atrophy at puberty was more pronounced in females than in males. Nicolaeff (1923) did autopsies on children dead in the Russian famine after World War I. The heart weights were stated to be 20 to 40 per cent or more below the normal average for age; total body weights were also subnormal to about the same extent.

There is much information available from atrophic infants, though admittedly these cases do not represent pure and simple caloric undernutrition. In 571 autopsies on atrophic infants to 5 years of age, Bovaird and Nicoll (1906) generally found the heart weight to be subnormal to almost the same extent as the body weight. Jackson (1925) obtained data from several hundred autopsies and concluded (p. 227): "My own data confirm the principle that in malnourished infants the heart weight approximates the normal for corresponding body weight, although markedly below the normal for the previous maximum body weight, or the final body length, and especially retarded in comparison with the normal for corresponding age."

Heart Size in Adults at Autopsy

There are many individual case reports or autopsies on patients who died in a state of extreme cachexia; the smallness of the heart in such cases is usually a matter for comment or is evident from the reported measurements. For example, a man who died after 63 days of fasting weighed 80 pounds (36.3 kg.) and the body was 1.72 meters long; the heart weighed 178 gm. (Meyers, 1917). The large normal variability of heart weight makes it difficult to place much emphasis on such single cases.

The relation of heart size to the general degree of emaciation in miscellaneous cases at autopsy has been reported for several large series. Bean and Baker (1919) analyzed the data from several thousand autopsies in Baltimore and New Orleans. Excluding cases of heart disease, the average heart weight in very emaciated males averaged 73 per cent of that of the males classified as well nourished. The major findings are summarized in Table 107. Krieger (1921) reported heart weights in 133 cases of extreme emaciation and calculated that on the average the heart weight was 32.4 per cent below normal average. In Krieger's series the greatest apparent loss of heart weight, averaging 45.0 per cent, was found in 7 cases of chronic dysentery, and the least loss, averaging 29.6 per cent, was in 20 cases of senility with extreme emaciation (see Table 100).

TABLE 107

MEAN HEART WEIGHTS IN ROUTINE AUTOPSIES in Baltimore and New Orleans, excluding cases of heart disease. $N = 636$ white males, 385 white females, 279 Negro males, 234 Negro females. Data from Bean and Baker (1919).

	Very Emaciated	Emaciated	Thin	Normal	Fat	Obese
White male						
Weight	251	273	293	342	361	414
Percentage of normal	73.4	79.8	85.7	100	105.6	121.1
White female						
Weight	244	266	298	353	332	340
Percentage of normal	69.1	75.4	84.4	100	94.1	96.3
Negro male						
Weight	201	234	258	267	323	329
Percentage of normal	75.2	87.6	96.6	100	121.0	123.2
Negro female						
Weight	209	220	231	270	292	356
Percentage of normal	77.4	81.5	85.6	100	108.1	131.9

There are surprisingly few good data on heart size in famine victims. In spite of the great interest in famine edema in Germany in 1916–19, there were no sizable series of autopsies reported, though several authors cited "representative" cases, all of which had very small hearts (Paltauf, 1917; Oberndorfer, 1918). One of the largest series of measurements is that reported by Porter (1886, 1889) from 459 autopsies in relief camps set up in the Madras famine of 1877–78. For the men the heart averaged 0.51 per cent of the total body weight, while in the women the value was 0.55 per cent. Porter considered his results to indicate a disproportionate atrophy of the heart, but the absence of any normal standards for Indians makes it possible to conclude only that cardiac atrophy was severe and something like proportional to the whole body weight loss (see Table 97).

Roessle's (1919) data from World War I in Germany indicated that in moderate undernutrition the heart weight is reduced more or less in proportion to the total weight. At the time of the siege of Leningrad, 1941–42, the mean heart weight at autopsy of adults who died of uncomplicated semi-starvation was 168 gm., compared with a normal average of 300 gm. for the same region (Leningrad, 1946; Brožek, Chapman, and Keys, 1948).

The autopsy data from the Warsaw Ghetto are particularly valuable because of the large number of people involved and the rigid exclusion of cases with complications. In 492 adults who represented pure starvation the hearts were judged to be atrophic in 419, or 85 per cent; for the latter part of the famine period (1942) post-mortem measurements indicated the hearts to be definitely atrophic in 87 per cent of the bodies (Stein and Fenigstein, 1946). In the adults (approximately equal numbers of males and females between 20 and 60 years of age) in the Warsaw material, the heart weight ranged from 110 to 350 gm., with an average of 220 gm. as compared with an estimated normal average of 275 gm. (see Table 99). This would indicate a mean loss of 19.3 per cent in the heart weight. In this same material the heart weight averaged 15.8 per cent of the brain weight as contrasted with an estimated normal average of 21.0 per cent.

In 11 adult victims from the German concentration camps who died in Swiss hospitals in 1945, Uehlinger (1948) estimated losses of 13 to 117 gm. in heart weight, with a minimum absolute weight of 160 gm. In 5 cachectic females without clinical edema, who were between 23 and 50 years of age, the body weights ranged from 39 to 50 kg. and the heart weights from 200 to 240 gm. Lamy et al. (1948) noted that the heart weights in the concentration camp victims seen at Mainau were generally between 200 and 240 gm. (adult men), but in 2 men heart weights of 320 and 350 gm. were recorded.

The effect of chronic undernutrition on the heart size in persons with severe, long-standing cardiac disease is of interest. Occasionally at least the heart may be extraordinarily small under these conditions. For example, we recently saw the body of a 60-year-old man who died of gastric carcinoma in an extremely emaciated state. He had suffered from severe mitral disease for many years. The body length was 170 cm. (67 in.) and the heart weight was 115 gm. The mitral valve was practically destroyed and was of the "fish mouth" type, with an orifice of only 3 or 4 mm. In the concentration camp victims examined by Uehlinger

TABLE 108

Heart Weight in 8 Undernourished Persons with Severe Valvular Defects and Myocardial Damage. Data from Uehlinger (1948).

Age	Heart (gm.)	Cardiac Defect
	Men	
63	380	Valvular sclerosis
62	320	Myocardial fibrosis
69	350	Aortic endocarditis
72	270	Mitral endocarditis
	Women	
49	160	Mitral endorcarditis
70	380	Myocardial fibrosis
59	250	Mitral stenosis
55	250	Aortic and mitral stenosis

(1948) the heart weight was recorded for 8 patients with severe valvular and myocardial damage. The data are summarized in Table 108; the heart weights are far less than would be expected in well-nourished persons.

The History of an Error

For 122 years a large literature has been accumulating to show that cardiac atrophy is a regular result of undernutrition and starvation and that the degree of atrophy, estimated from the weight of the organ, is nearly, if not quite, proportional to the weight loss of the whole body. The only two general monographs on starvation, those of Morgulis (1923) and of Jackson (1925), abundantly documented the fact. The evidence from all forms of animals studied, including man, is in good agreement. Yet the curious fact remains that almost all textbooks uniformly state that the heart resists starvation and loses little or nothing of its substance while the rest of the body (except the brain and bony tissues) is wasting away (Bard, 1941; Best and Taylor, 1945; Evans, 1945; Fulton, 1946; Roaf, 1936; Wiggers, 1939; Winton and Bayliss, 1937; Zoethout and Tuttle, 1946; etc.). This erroneous teaching in the textbooks of physiology explains the total unconcern of cardiologists about the subject of the heart in undernutrition; Vaquez (1924, p. 290) stated the case flatly in his authoritative monograph on cardiology: "The conclusion is that inanition has no harmful effect on the heart."

When the textbooks provide an authority or data for their statement that the heart does not atrophy in starvation, it almost always turns out to be C. Voit (1866)— though E. Voit (1905a) is occasionally given as the reference. C. Voit, in a paper devoted to other problems, cited the weights of 2 cats, one starved and the other well fed, from which it appeared that the heart of the starved cat weighed only 2.6 per cent less than that of the other, though there was a difference of 30.5 per cent in the weight of the voluntary muscles. Voit was aware of the very different findings of Chossat (1843) and made no comment on the singularity of his one cat.

Voit was an outstanding investigator, his 1866 paper was widely read for other reasons, and his data were summarized in a table which is unusually clear and concise for his day. Just how this organ weight data first got into the textbooks is not certain, but, as a bar diagram still widely copied, they appeared in Waller's text (1896, p. 109). Some of the older texts, such as that of Ott (1907, p. 425), followed Waller in using Voit's data without comment, but an explanation was introduced by Landois (1904, p. 441), who stated with regard to the events in starvation, "Certain organs like the heart and the nerves, suffer slight loss, as they are able to maintain themselves on the decomposition products of other tissues" (see also Stewart, 1910, p. 530).

Starting with a single measurement, then, a spurious "fact" was set up as a general principle; an explanation, totally without basis, was provided; and it remained only to show a philosophical purpose in order to have a complete edifice of error. The final step occurred in the first edition of Starling's textbook (1912), from which comes the following statement as repeated in later editions (see Evans, 1945, p. 765): "It might be imagined that, since this loss of weight is determined by using up the tissues of the body for the production of energy, those organs which are most active would show also the greatest loss of weight. The very reverse of this is the case. . . . Those organs which are most necessary for life, the brain, the heart, the respiratory muscles such as the diaphragm, undergo very little loss of weight."

Starling's reference to the respiratory muscles is puzzling since there was no mention of these in the 1866 paper of C. Voit, which he clearly had in mind, nor could the statement be supported by the paper of E. Voit (1905), which he gave as authority. In 1894 C. Voit starved one dog for 22 days and compared it with a "control." The respiratory muscles, including the diaphragm, amounted to the same proportion of the total musculature in the starved dog (10.6 per cent) as in the well-fed dog (10.7 per cent). Moreover, in this work of C. Voit the atrophy of the heart was in the same proportion as in other organs. For the control and the starved dog, respectively, C. Voit found the heart/spleen ratios to be 0.46 and 0.49, the heart/lungs ratios 0.81 and 0.75, and the heart/liver ratios 0.298 and 0.297.

It has seemed worth while to devote some discussion to this error about the heart in starvation because it explains why neither cardiologists nor physiologists have properly studied the effects of starvation on the heart. Perhaps this documentation may stimulate the correction of an ingrained teaching. Finally, it should serve as a warning not to take the textbooks too seriously.

Heart Size in the Minnesota Experiment

Teleroentgenograms were taken regularly throughout the Minnesota Experiment, and roentgenkymograms were used to get measurements of the projection area in ventricular systole. From these it is possible to calculate cardiac volumes with considerable accuracy (Keys et al., 1940), but obviously all such measurements are for the heart plus contained blood at the end of ventricular systole. As starvation progresses and the heart becomes weaker, it is possible that systolic ejection becomes less complete so that the real shrinkage of the heart could be underestimated; an error in the reverse direction would seem unlikely. It could

hardly be suggested that the starved heart would be unusually efficient in emptying itself, especially in view of the maintenance of a distinct though small blood
plethora. The measurements are illustrated in Figure 46.

The values for the measurements of the heart before and during semi-starvation are summarized in Table 109. When the body weight had decreased by 17.4
per cent (at S12), the heart volume had decreased by 15.7 per cent. At the end
of semi-starvation, when the body weight had decreased by 24 per cent, the
heart volume had decreased by 17.1 per cent. This indicates a relative loss in
heart volume amounting to 71 per cent of the body weight loss. These values are
in reasonable agreement with the majority of animal experiments and autopsy
data in the literature.

The most common method of estimating heart size from X-ray films is by
measurement of the transverse diameter, though this is probably the worst
method that could be suggested (Hodges and Eyster, 1924; Beneditti, 1936; Lud-
wig, 1939; Keys et al., 1940). In the data of the Minnesota Experiment, calculation from the transverse diameter alone indicates a shrinkage of heart volume
considerably greater than in proportion to the loss in body weight. This overestimate results from the fact that in the starving man the length of the heart decreases less than the breadth and, furthermore, the heart assumes a more vertical
position in the chest. As a result the transverse diameter represents a different
proportion of the breadth. These points are obvious from the typical result
shown in Figure 46.

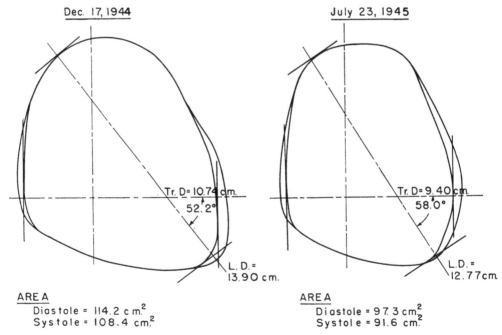

FIGURE 46. Representative Sample of the Measurement Tracings for Heart
Size before and at the end of semi-starvation in the Minnesota
Experiment, subject No. 129.

TABLE 109

Mean and Standard Deviations for Cardiac Dimensions, before and after 12 and 24 weeks of semi-starvation for 32 men in the Minnesota Experiment. Transverse and long diameters are in cm., for the heart in systole. Volume is in cc. for the heart in systole. Heart axis gives the angle of the long axis in degrees from the horizontal. The values are also expressed as percentages of the control value.

	Control		S12		S24	
	M	SD	M	SD	M	SD
Body weight (kg.)	70.0	6.20	57.8	5.14	53.2	4.68
Percentage of control...	100	8.85	82.6	8.89	76.0	8.80
Transverse diameter	11.71	0.87	10.44	0.89	10.20	0.91
Percentage of control...	100	7.4	89.2	8.5	87.1	8.9
Long diameter	14.16	0.90	13.71	1.00	13.62	0.92
Percentage of control...	100	6.3	96.8	7.3	96.2	6.7
Volume	620.4	86.4	536.2	93.4	514.1	84.0
Percentage of control...	100	13.9	84.3	17.4	82.9	16.3
Axis (degrees)	49.4	5.08	57.9	4.44	55.1	5.86

On the average in the Minnesota Experiment the long axis of the heart after 6 months of semi-starvation was 5.7° closer to the vertical than in the control period. When allowance is made for the changed anatomical axis and calculations are made from linear dimensions, the results indicate an average decrease in heart volume of about 20 per cent or, roughly, a little more than 80 per cent of the relative loss in body weight.

The relatively high degree of variability of heart size in different individuals is apparent from the standard deviations given in Table 109. In the control period the interindividual variability of heart volume was 57 per cent greater than the variability in body weight. In starvation the variability of the heart volume increased so that there was almost twice as much variation in the heart volume of the different men as there was in the body weight. Only a part of this variation can be ascribed to the technical imperfection of the measurement, so it must be concluded that the heart volume is normally quite variable among individuals and that this variability is increased in starvation. The first part of this conclusion is easily verified by such data as those of Kirch (1921) on normal hearts, and the second part would seem to be confirmed by Porter's (1889) autopsy data on famine victims.

During rehabilitation the heart size of the Minnesota subjects increased rather rapidly. The results from 12 men who were studied through 20 weeks of refeeding are summarized in Table 110. After 11 weeks, at which time these men were still on the average about 17 per cent under their control body weights, the heart volumes were within about 4 per cent of the control values. At R20 these men were about 4 per cent heavier than their control weights and their mean heart volume was likewise about 4 per cent larger than their control average; in other words, the normal heart size was regained after about 5 months of refeeding.

Whereas the heart volume loss in starvation may underestimate the shrinkage

TABLE 110

HEART SIZE in the control period, at the end of semi-starvation, and at 12 and 20 weeks of rehabilitation in the Minnesota Experiment. Mean values and standard deviations for 12 men.

	Control		S24		R12		R20	
	M	SD	M	SD	M	SD	M	SD
Body Weight (kg.)	67.5	5.1	51.7	3.3	58.2	3.8	70.8	4.4
Percentage of control	100	100	76.5	65	86.2	74	105	86
Transverse diameter (cm.) .	11.44	0.92	10.03	1.00	11.08	0.98	11.79	1.21
Percentage of control	100	100	88	109	97	107	103	131
Volume (cc.)	575	74	470	83	551	79	595	88
Percentage of control	100	100	82	112	96	107	104	119
Axis (degrees)	46.8	4.7	51.2	5.3	49.0	5.4	44.9	6.4

of heart tissue, the volume gain in rehabilitation may well overestimate the tissue restoration in some individuals. This was clearly the case with one man (not included in the 12 men summarized in Table 110) who had, in rehabilitation, a short period of heart failure with a sudden dilatation of the heart (see Chapter 28).

General Appearance and Histology

Reports of autopsies on famine victims are in general agreement that, besides being small, the heart in starvation is notable for its softness, paleness, and flabbiness; the gross appearance is that of anemia and edema. The coronary vessels are unusually conspicuous, partly because of the disappearance of the normal epicardial fat. Subepicardial edema is sometimes apparent (Porter, 1889; Nicolaeff, 1923). Oberndorfer (1918) remarked that the coronary vessels are much twisted, a fact which he ascribed to the shrinkage of the cardiac musculature. Paltauf (1917) noted that the left ventricle was usually contracted in the cases he saw. In 10 to 20 per cent of the cases autopsied by Porter (1889), the heart surface showed patches of a white gelatinous material which was easily detached. Porter referred to this as "soldier's spot"; perhaps it represents remnants of degenerating epicardial fat.

Brown atrophy of the myocardium is seen fairly frequently in starved persons (Stschastny, 1898; Krieger, 1921); it was observed in 72 out of 87 autopsies on tuberculous cases in famine (Larson, 1946). On microscopic examination the evidences of degeneration have been reported by many investigators following the first careful study in human starvation by Hayem (1877). Nicolaeff (1923) made histological studies on 19 starved children and remarked on the loss of cross-striations in the muscle fibers, an increased frequency of nuclei reflecting a reduction in the amount of sarcoplasm, and occasional vacuoles in the sarcoplasm. Besides cloudy swelling of the sarcoplasm of the muscle cells and serous impregnation of the interstitial material, Chortis (1946) also found separation of the muscle fibers at their point of junction in famine cases.

Stein and Fenigstein (1946) made careful histological studies on tissues obtained immediately after death from 6 adults who represented pure and prolonged starvation. In 4 of these the heart showed marked brown atrophy. The fibers were rather thin and had distinct striations. Around the nuclei there were fusiform accumulations of a finely granular brownish pigment. Around some blood vessels there were rather large amounts of connective tissues, which sometimes contained few lymphocytes; in places there were larger hyperchromic nuclei, often deformed in shape; in some places there were condensations of nuclei.

The autopsy material of Uehlinger (1948) was frequently complicated by severe infections, and in most cases death occurred after some weeks of refeeding. However, in all cases there was marked brown atrophy of the heart, and occasionally there was slight fatty infiltration of the myocardium. Lamy *et al.* (1948) saw no real anomalies of the cardiac fibers but in several instances remarked on a gelatinous atrophy of the fatty tissue of the heart; presumably this is the "soldier's spot" referred to in some older works. In 2 hearts there were large, white intracardiac thrombi.

There is much more detailed information on the histology of the heart in starved animals. Jackson (1925, pp. 230–31) reviewed the older literature, much of which debated whether or not fatty degeneration was involved; the question still seems to be unresolved. Recent studies on rats were reported by Lazarovich-Hrebeljanovich (1936). When attempts have been made to measure the size of the fibers in cardiac muscle, the results have all indicated a marked reduction (Morpurgo, 1889; Statkewitsch, 1894; Heitz, 1912). The cardiac ganglia also show degenerative changes, and vacuoles appear frequently in this tissue (Statkewitsch, 1894; Uspenski, 1896); this is reminiscent of the vacuoles described in the brains of starved persons and animals (see Chapter 9).

The Blood Vessels

The morphology of the blood vessels in starvation has received relatively little attention; the earlier literature, mostly from animal experiments, is fully reviewed by Jackson (1925). The possibility that abnormalities of the capillaries may be involved in famine edema has been discussed rather vaguely from time to time, but this whole problem can be dismissed with two comments: (1) Presently available histological methods are not capable of revealing the intimate structure of the vessels, or at least they offer no useful evidence on this question; the capillaries are not grossly abnormal in starvation. (2) The relative absence of protein in the edema fluid of famine edema seems to prove that the capillaries do not leak protein in this condition.

Active arteritis is not seen in famine, and the blood vessels do not show inflammatory reactions (Uehlinger, 1948; Lamy *et al.*, 1948). On the other hand, embolic phenomena are not infrequent. Uehlinger (1948) relates the frequency of venous thrombosis to the combined effects of decubitus, slow circulation, the high incidence of pyodermic infections, and possibly chemical peculiarities of the blood. The Swiss group reported 8 cases of thrombophlebitis, 6 of which did not involve tuberculosis. The emboli, which almost always develop in the lower extremities, resulted in pulmonary infarction in 4 of Uehlinger's patients. At autopsy Lamy *et al.* (1948) found large venous thrombi in a significant percentage of the famine victims; in some cases these were associated with atheromatous lesions.

Lamy *et al.* (1948) emphasized that the most significant effect of starvation on the blood vessels is an early appearance of changes ordinarily seen at more advanced ages. In one case a 17-year-old boy showed marked lipoid infiltration of the intima of the major blood vessels with edema of the adventitia and the subintimal connective tissue. Other patients also showed deposits and intimal changes of the atheromatous type.

CHAPTER 11

Morphology of the Endocrine Glands

THE results of semi-starvation in man in many ways bear a resemblance to the clinical symptoms of endocrine gland deficiency diseases. A low pulse rate, low blood pressure, and low basal metabolic rate are characteristic of both semi-starvation and hypothyroidism. The emaciation of semi-starvation is, superficially at least, suggestive of Simmonds' disease. The pigmentation of the skin in the semi-starved individual has some resemblance to the pigmentation of Addison's disease. The loss of the sex drive in males, cessation of menstruation in females, decreased incidence of hyperthyroidism and diabetes mellitus, and increased incidence of myxedema among the victims of severe famine are additional indications that endocrine gland function is disturbed by semi-starvation.

The endocrine glands have been implicated in the osteopathies of starvation (Naegeli, 1918; Fromme, 1919; Simon, 1919; Hirsch, 1920; Neuberger, 1921; De Gennes and Deltour, 1945). Bigland (1920a) stated that "most workers agree to a marked wasting of all organs, including endocrine organs," during severe caloric undernutrition. A general disturbance of all glandular activity was reported during the siege of Leningrad (Ryss, 1943). The clinical evidence of endocrine disturbances has been substantiated by a small number of histological observations on starved animals and men. The response of the endocrine glands to starvation is, no doubt, influenced by a variety of such factors as the length and degree of starvation, age, sex, activity, climatic conditions, and concomitant diseases which make both the clinical and morphological interpretations more difficult.

The discussion in this chapter will be limited to the adrenal glands, the thyroid, the pituitary, and the pancreas. The parathyroid glands are not included mainly because of the paucity of observations on adult human material in semi-starvation. The earlier literature on the effects of starvation on the parathyroid glands has been reviewed by Thompson (1907) and Jackson (1925). The clinical considerations of Simmonds' disease and diabetes mellitus are presented in Chapters 44 and 48, respectively. The effects of semi-starvation on the reproductive organs are discussed in Chapter 35.

Adrenal Gland Changes in Starvation

Studies on the endocrine gland changes induced by starvation in laboratory animals have the distinct advantage of being less complicated by other factors than are the human observations. In the adult albino rat, chronic inanition with about a 33 per cent loss of body weight was accompanied by an 8.9 per cent decrease in the weight of the adrenal glands (Jackson, 1915). The cortex of the adrenals was hyperemic, and histological changes in the cells of the outer and middle zones occurred; the inner zone appeared quite normal. The cell size in the middle zone was in many cases reduced about 20 per cent; the cytoplasm was decreased in amount and pycnotic nuclei were frequent (Jackson, 1919).

Mulinos and Pomerantz (1941) reported a progressive atrophy of the adrenals of rats during chronic starvation (see Table 111); the atrophy was more marked in the females than in the males. Noxious stimuli, such as injected foreign materials, disease processes, and drugs, caused the adrenals of chronically starved rats to become larger than normal. Pituitary transplants from normal rats into chronically starved rats resulted in normal sized adrenal glands; brain transplants had no effect. The authors concluded that "this indicates that the atrophy of the adrenal glands was not due directly to the malnutrition, but at least in part to an insufficient amount of adrenotropic hormone." Cameron and Carmichael (1946) observed no adrenal changes in rats with chronic starvation.

TABLE 111

PERCENTAGE LOSS OF BODY WEIGHT AND OF ENDOCRINE GLAND WEIGHT IN RATS during chronic inanition (from Mulinos and Pomerantz, 1941).

	Body Weight	Pituitary	Thyroid	Adrenals
Average	22	34	28	35
Range	12–38	20–58	10–45	17–50

The adrenal glands hypertrophy in acute starvation (Jackson, 1915; McCarrison, 1919; Stewart, 1917; Vincent and Hollenberg, 1920a, 1920b; Mulinos and Pomerantz, 1941; Cameron and Carmichael, 1946). At the time of death from acute starvation the adrenals in the rat were, according to Cameron and Carmichael (1946), large and red with the cortex friable and congested. The enlargement of the glands was the result of an increase in the water content of the cortical cells (hydropic degeneration) with a corresponding increase in cell size, not the result of an increase in actual tissue mass. The dry weight of the adrenal glands was not increased.

The normal adrenal cortex of animals and man contains considerable but variable amounts of fat (Babes and Jonesco, 1908a). In a study of factors influencing the fat content of the adrenals, Babes and Jonesco (1908b) observed that ordinarily in rabbits that had been starved to death the fat content of the adrenal glands was decreased, but that if the animals were starved in the cold the fat content of the glands was increased. The repeated injection of small amounts of adrenalin in normal rabbits produced a decrease in the fat content of the gland and an extreme hypertrophy. Babes and Jonesco (1908b) found the adrenal fat to be decreased or absent in 40 humans whose deaths were due to a variety of chronic infectious diseases with accompanying starvation.

The morphological appearance of the adrenal glands in infantile cachexia has been described by Lucien (1908a, 1908b). The adrenals in these infants were impressively small, often weighing only 1.05 gm. (the normal adrenal weight in infants is 3–5 gm.); actually the relative adrenal weight loss was greater than the relative loss in body weight. The glands were darker colored (café au lait) than normal, they had a firm consistency, and the surfaces were granular in appearance. On histological examination the capsule appeared thickened, the trabeculae were increased in size, and the cortex was sclerotic; the general appearance was that of hypoactivity. The cortical cells, especially in the *zona reticularis*, were often atrophic and devoid of fat. The capillaries in the *zona reticularis* were dilated and filled with erythrocytes. Lucien believed that the adrenal changes in the cachectic infants were similar to those reported by Bernard and Bipart in patients who died from tuberculosis. Marfan (1921) reported small adrenal glands in uncomplicated infantile athrepsia. In a series of 61 autopsies on starved Russian children, Nicolaeff (1923) found adrenal weights that were 5 to 37 per cent below normal; in some cases the glands were enlarged.

Meyers (1917) reported a 20 per cent decrease in the weight of the adrenals and a 36 per cent decrease in body weight in a man who ingested only water for 60 days prior to death. The atrophy of the glomerular cells was very marked, and many "shadowy cells" were seen in some portions of the glands. Meyers emphasized that these were not post-mortem changes but undoubtedly had occurred as a result of the prolonged period of starvation. Enright (1920a) stated that among the starved Turkish prisoners of war with edema "the suprarenals of all . . . were found to be atrophied." Pellegrini (1920) also found a small decrease in the size of the adrenal glands of Italian prisoners of war who died from undernutrition.

Quite the opposite findings have been reported by other observers. The 200 war edema cases reported by Schittenhelm and Schlecht (1918) had the usual signs of starvation — loss of body fat, wasting of muscles, bradycardia, low blood pressure, anemia, and weakness. On autopsy the adrenals were found to be enlarged but normal in general appearance. While a medical prisoner of war behind the German lines, Byrne (1919) made autopsies on several British and Allied soldiers who had died in the prison camps. All were poorly nourished and had been worked until they collapsed. "In a series of 8 cases of death from underfeeding the adrenals were definitely enlarged, almost to half as large again as normal, and the enlargement seemed, from naked-eye appearance, to be mostly in the cortex." Byrne explained the adrenal hypertrophy as an effort "to counteract the low blood pressure produced through insufficient nourishment." Krieger

TABLE 112

PERCENTAGE LOSS OF BODY WEIGHT AND OF ENDOCRINE GLAND WEIGHT IN SOLDIERS, with the number of autopsies. The deaths were due to a variety of causes involving emaciation (from Krieger, 1921).

	Body Weight		Pancreas		Thyroid		Adrenals	
	Cases	% Loss	Cases	% Loss	Cases	% Loss	Cases	% Loss
Chronic dysentery	7	48.4	5	44.8	5	42.0	5	17.2
Malignant tumors	27	38.0	6	33.0	10	20.6	9	+21.5
Chronic infections	31	43.9	26	30.5	23	32.3	23	+23.2
Tuberculosis	40	43.0	17	38.6	25	35.8	24	+6.6

(1921) likewise observed an increase in the weight of the adrenals in all the soldiers dying from a variety of chronic disorders with emaciation, except those with chronic dysentery (see Table 112).

Schilf (1922) concluded from a series of more than 2000 autopsies (804 of which were made during years of good nutrition) that the weight of the adrenal glands in humans was not influenced by either disease processes or nutritional status, but that the general level of physical activity did play a role. Soldiers, who were assumed to be very active physically, had adrenals that were about 20 per cent heavier than the average for civilians. Roessle (1919) found no relation between nutritional condition and the fat content of the adrenals in his series of autopsies on soldiers. The adrenal glands were large in both normal and malnourished Okinawans (Steiner, 1946). Stein and Fenigstein (1946) reported a diminution of fat in the adrenals of 49.2 per cent of the 285 persons studied in Warsaw.

Uehlinger (1948) found only relatively small histological changes in the adrenals of famine cases. The medullary tissue was normal. The cortex in some cases showed hydropic swelling, partial or total fatty degeneration, increased spongiocytes, and fibrosis of the capsule and *zona glomerulosa*. Partial cortical sclerosis was frequent. Partial fatty infiltration of the cortex was common but total infiltration was rare.

Lamy, Lamotte, and Lamotte-Barrillon (1948) observed a relationship between the lipoid content of the adrenal cortex and the length of time between liberation and death. In the cases that lived for about 4 months after liberation, the adrenal cortex was very rich in lipoids. When death occurred at about 2 months after liberation, the lipoid content of the cortex was normal. In 8 of 9 cases that died soon after liberation, the cortex contained little or no lipoids. The medullae were strongly pigmented, and in some cases there was cellular atrophy.

The evidence for degenerative changes in the adrenal glands during semistarvation is much less convincing for man than it is for animals. Some of the discrepancies in the observations on man are, no doubt, explainable on the basis of complicating factors other than undernutrition.

Thyroid Gland Changes in Starvation

The evidence is quite consistent that the human thyroid atrophies during starvation. In the one case of starvation described by Meyers (1917) the epithelium of the acini was often hardly recognizable. Areas of diminished colloid were present, but most of the gland was converted into a mass of colloid with distended acini. The thyroid weight was not recorded but "the picture is that of atrophy [and] exhaustion." Oberndorfer (1918) found thyroid glands in cases of war edema that were about 50 per cent smaller than normal. The follicles were small and the amount of colloid was markedly reduced, suggesting a decreased secretory activity. Thyroid glands of normal size were found among emaciated pellagrous cases in Egypt (Bigland, 1920b). Krieger (1921) observed reductions in thyroid weight of about the same degree as the body weight loss in a large series of autopsy cases in whom death was due to chronic disease accompanied by severe emaciation (see Table 112).

According to Hinz (1920) the adaptation or adjustment to a low caloric intake was accomplished by a change in the activity of the thyroid; the decreased thyroid activity was considered to be a major cause of the *Kriegsödem*. During the later part of World War I in Germany there were many cases of decreased thyroid activity with a symptomatology of both war edema and myxedema. These patients improved on thyroid therapy. Curschmann (1922) also observed that undernutrition exerted a depressant effect on thyroid function and that during the war there was an increase in myxedema (Kriegsmyxoedema) and a decrease in toxic hyperthyroidism. Tourniaire (1945) described a case of myxedema with cardiac enlargement in an emaciated patient; the symptoms were ascribed to starvation with thyroid degeneration.

In the Hamburg Clinic there were 217 cases of toxic hyperthyroidism during the 6-year period of 1909–14 and only 42 cases from 1915 to 1920. The incidence of toxic hyperthyroidism seen among private patients in Germany decreased from 2.5 per cent in 1912–13 to about 1.25 per cent in 1918–19 (Tallquist, 1922). Brull (1945) reported that in Belgium during the German occupation the average basal metabolic rate of hyperthyroid patients fell from +21.9 per cent in 1939 to +6.6 per cent in 1942. He concluded that the decrease in thyroid activity during periods of starvation was responsible for the small number of cases of severe Graves' disease observed during the war. Schur (1947) also found a decrease in the incidence of Graves' disease. In the acute phase of malnutrition in the Philippines in World War II, endocrine disturbances simulating hypothyroidism were frequently observed (Gottlieb, 1946). Thyroid glands that appeared hypofunctional were reported in 5 cases of death from starvation in France (Lamy et al., 1946, 1948). The vesicular epithelium was thin and the vesicles were distended with an abundance of homogeneous colloid.

On the other hand, Stein and Fenigstein (1946) saw no pathological changes in the thyroid gland; and Uehlinger (1948) found relatively little histological change in the thyroid, which led him to conclude that the decrease in B.M.R. in famine cases is primarily independent of thyroid changes. In Denmark, where semi-starvation was not present, there was a 4- to 8-fold increase in the incidence of hyperthyroidism in 1942 as compared with the period of 1932–41 (Meulengracht, 1945). The decrease in the basal metabolic rate that occurred in the Minnesota Experiment is discussed in Chapter 17.

In infantile athrepsia Thompson (1907) reported thyroid glands that weighed about 25 per cent of normal. The follicles were filled with detached epithelial cells, there was little colloid, and the interfollicular stroma was abundant. Sclerotic infiltration of the follicles was observed by Lucien (1908a). Marfan (1921) also found very small and sclerotic thyroid glands in infantile athrepsia. Among the children who died in the Russian famine of 1921–22, the thyroid gland weight was from 45 to 70 per cent of normal (Nicolaeff, 1923). On histological examination areas of aplasia were found with unequal coloration of the colloid, which was distributed unequally in the follicles. The glands appeared to have undergone a type of aplasia of the follicular cells with excess desquamation. Nicolaeff remarked, without giving any references, that hypothyroidism is common in starvation in both man and animals.

The thyroid gland atrophies in the rat during chronic inanition. With a body weight loss of 33 per cent, a thyroid weight loss of 21.8 per cent was recorded (Jackson, 1915). The histological changes included simple atrophy, vacuolization of the cytoplasm, granular degeneration, and hyperchromic and pycnotic nuclei (Editorial, 1917a). Mulinos and Pomerantz (1940) observed a 28 per cent loss in the weight of the thyroid of rats in which a body weight loss of about 20 per cent was produced by chronic starvation (see Table 111). The follicular cells were atrophic with dark shrunken nuclei.

Pituitary Gland Changes in Starvation

Data on the histological changes that occur in the pituitary gland of man during starvation are extremely limited. In the starved man reported by Meyers (1917) the anterior lobe of the pituitary was congested and hemorrhagic. The epithelium showed "great reduction and degeneration similar to that shown in the adrenals, and in some places [was] displaced completely by blood." Only a few masses of colloid were found, and the characteristic chromophile granulation was greatly reduced. Uehlinger (1947b) examined the pituitary glands of 36 inmates of German concentration and prisoner-of-war camps who had finally succumbed to slow starvation. Histologically the pituitary was normal in 10 cases. In the remaining 26 cases, 5 showed a dense central eosinophilia, 17 had an increase in the chief and stem cells which in places formed adenomas, and 4 showed both chief-cell proliferation and dense eosinophilia. In all cases the posterior pituitary appeared normal. Lamy, Lamotte, and Lamotte-Barrillon (1946c, 1948) reported eosinophilia in the pituitary of 5 out of 17 cases of death from starvation.

Stein and Fenigstein (1946) found the intermediate lobe of the pituitary underdeveloped in all cases; brown pigment cells and free pigment were seen in the posterior lobe in 2 cases, and in 2 cases there were more acidophilic cells than basophilic cells.

Neuberger (1921) incriminated the pituitary along with most of the other endocrine glands in the development of the osteopathies of starvation. The whole clinical syndrome of starvation was, according to Boenheim (1934), one of hypophyseal cachexia. Ryss (1943), from his observations during the siege of Leningrad, also concluded that the clinical symptoms pointed to a disturbance of function of the hypophysis. The pigmentation of the skin so frequently present among famine victims suggested to Musselman (1945) that the pituitary may be involved. Gottlieb (1946) suggested that some of the endocrine gland disturbances seen among the undernourished prisoners in the Philippines simulated the hypopituitary syndrome.

Lucien (1908a) found evidence of pituitary atrophy in infantile athrepsia. In the 15 cases of atrophic infants listed by Jackson (1925) the pituitary weight ranged from 55 to 110 mg., while in normal children of the same age the average weight was 130 mg. Jackson stated, "it therefore appears that the hypophysis undergoes a marked loss in weight during inanition."

A pituitary weight loss of 25 per cent occurred in adult rats during a period of chronic starvation in which the body weight loss was 36 per cent (Jackson,

1915). The loss in pituitary weight in chronic inanition was mainly of the paren-
chymal cells, which showed a decrease in size of both the cells and the nuclei
(Jackson, 1917). The nuclei became hyperchromic, and mitoses were depressed.
The cytoplasm was reduced in volume, frequently vacuolated, and exhibited a
marked tendency to lose its specific staining reactions so that strongly chromophilic cells became weakly chromophilic or even chromophobic. After a few
weeks of refeeding "the hypophysis has usually become nearly normal for the
most part, although more or less extensive atrophic areas may persist for indefi-
nite periods."

Definite evidence of pituitary hypofunction has been observed in chronically
underfed rats. Werner (1939) found the production of the gonadotropic principle reduced. In chronic starvation the loss of weight of the pituitary gland (34
per cent) was relatively greater than body weight loss (22 per cent) (see Table
111) and the cells showed evidence of degenerative changes (Mulinos and Pomerantz, 1940). From data obtained on more than 300 rats these authors concluded that "the effects of prolonged chronic inanition resemble those from hypophysectomy, especially upon the endocrine glands. The response of the endocrine organs of the rat to chronic inanition has been termed pseudo-hypophysec-
tomy not only because of the resemblance between the effects of inanition and
hypophysectomy but also because we believe that many of the effects of inanition are due to malnutrition of the hypophysis, resulting in a diminished secretion of hormones." When pituitary glands from normal rats were implanted into
chronically underfed rats, the adrenal glands and ovaries of the starved animals
became normal in size (Mulinos and Pomerantz, 1941), indicating "that chronic
inanition depresses the adrenotropic function of the pituitary glands as well as
the gonadotropic."

Maddock and Heller (1947) also observed that "the starved rats reacted as if
they had been hypophysectomized in that the ovaries, uteri and vaginal epithelium underwent marked atrophy. They did not differ from hypophysectomized
rats as concerns their potential capacity to respond to exogenously administered
gonadotrophins" provided the circulating gonadotrophins were low. However,
the gonadotrophin content of the pituitary of the starved rat was, on the basis of
potency per milligram of gland tissue, higher than in the normally fed rat. The
failure in release of gonadotrophin from the pituitary gland during acute starvation occurred along with at least some suppression of the hormone production.
The authors believed it was likely that the "release mechanism fails completely
and early in starvation and that the eventual content reflects minimal production." The mechanisms responsible for production and release of gonadotrophin
were not demonstrated by the experiment.

The Islets of Langerhans in Starvation

"The beneficial effects of temperate eating in diabetes were prominently illustrated during the siege of Paris (1870–71) as Bouchard tells us that sugar entirely disappeared from the urine of diabetics in whom up to that time it had
persisted, even though they had been living on a carefully regulated diet. The
diminution in the quantity of food, occasioned by its great scarcity during the

siege, effected that which alterations in quality had failed to accomplish" (Purdy, 1890, p. 99). It was apparent even at that time — and it has been fully substantiated since — that the level of caloric intake has a profound influence on the course of diabetes mellitus. The details of diabetes morbidity and mortality rates in times of famine and caloric undernutrition are discussed in Chapter 48.

In view of the decreased activity and atrophic changes of other endocrine glands during semi-starvation, it might be expected that the islet cells would react in much the same manner. If such were the case, it would raise the interesting point of a decreased incidence of diabetes with a decreased insulin production. Of course, with starvation there would be a decreased carbohydrate intake and consequently a decreased need for insulin. It is also conceivable that there would be little alteration in islet cell activity and few atrophic changes except in the extreme stages of starvation. In any event a decrease in the incidence of diabetes in semi-starvation is a recognized fact which is not necessarily inconsistent with the morphological changes that occur in the starved individual.

The victims of the Madras famine of 1877–78 had pancreas glands that were reduced in weight by about 40 per cent. Porter (1889) reported an average pancreas weight of 62 gm. in 30 adult males with "alvine flux and atrophy" while the average for those who were not emaciated was 108 gm. The loss in weight of the pancreas was less in the female (53.2 gm. in the emaciated and 64 gm. in the normal). Most of the difference in pancreatic weight between the emaciated and the normal persons was attributed to the loss of interlobular fat during starvation. No histological abnormalities were observed. A small pancreas with no morphological alterations was also observed in the starved human by Formad and Birney (1891).

The weight of the pancreas in Meyers' (1917) starved man was reduced 49 per cent while the body weight loss was 41 per cent. The island cells were small and degenerate and were difficult to recognize. All that remained of some of the islands was a fused mass of degenerated cells. While some of the cells were fairly well preserved, all were small and had shrunk away from the surrounding connective tissue. Meyers stated that "the histological condition of the pancreatic islands suggests that glycosuria should have been present, but this was not the case, [because] the need of the organism for glycogen was so great and the stored glycogen so long exhausted, that a diabetic condition did not arise." Krieger (1921) found the loss in pancreas weight during various chronic diseases with emaciation to be relatively about the same as the body weight loss (see Table 112). The glands were generally small, firm, and anemic. Roessle (1919) observed no consistency in the loss of weight of the pancreas in his large group of autopsies on soldiers who died from chronic emaciating diseases. Schittenhelm and Schlecht (1918) and Bigland (1920b) reported normal pancreatic glands in war edema and pellagrous cases. In the victims of the Russian famine of 1921–22 the weight of the pancreas was only 10 to 20 per cent below normal. The islets were sharply defined and appeared hypertrophic (Nicolaeff, 1923).

Smeeden (1946) reported histological observations on the pancreatic glands of a series of repatriated Allied prisoners of war who were severely malnourished. In most cases the immediate cause of death was tuberculosis and malnutrition.

All cases had been hospitalized for periods ranging from 1 to more than 85 days before death. At autopsy the pancreas was removed and weighed, and specimens were taken for histological examination. A total of 38 cases were studied histologically; 25 showed pancreatic changes and 13 were essentially normal. The range of weight of the pancreas was from 60 to 145 gm. No correlation was observed between the gross weight of the glands and the microscopic picture, or between the islet cell changes and the degree of generalized and pulmonary tuberculosis. The appearance of the islet cells bore a relationship to the length of hospitalization (see Table 113). The supporting stroma of the pancreas was

TABLE 113

FREQUENCY OF CHANGES IN THE PANCREATIC ISLETS as related to the period of hospitalization with refeeding after prolonged semi-starvation (from Smeeden, 1946).

Days Hospitalized before Death	Number of Cases	With Islet Changes	No Islet Changes
1 to 28	17	13	4
29 to 56	10	7	3
57 to 84	6	4	2
85 or more	5	1	4

edematous, but the acinar parenchyma was of normal architecture although some of the acinar cells were small. "In the islets there was a moderate to marked diminution in the number of cells. Those still remaining stained indistinctly and had pyknotic nuclei. Between the cells there were spaces in which no stainable material was present. The cytoplasm was small in volume, and homogenous and edematous in appearance." The changes in the islet cells appeared to be reversible and returned to normal in those cases who lived long enough to recover from most of the effects of starvation. It is interesting that none of the subjects with the severe islet cell damage had any glycosuria during life even though the "islet changes would certainly suggest diabetes."

Chakrabarty (1947a), in a study of famine victims in India, found the islet cells fewer in number than normal and foamy in structure. Blood sugar concentration was below 80 mg. per 100 ml. of blood in about 60 per cent of the cases.

CHAPTER 12

Bones and Teeth

THE major components of bone are minerals, water, fats, and proteins. Any changes that may occur in the bones during semi-starvation are dependent upon relative and absolute variations in these major components. It would be expected that the water, fats, and proteins in the bones would reflect the general change in these substances in the body as a whole; there is no reason to suppose that the protoplasmic constituents of bones are any more or any less resistant to caloric undernutrition than are, for example, those of muscles, blood, or skin (see Chapter 15). The question of changes specific to the bones and teeth in semi-starvation resolves itself mainly, then, into the problem of variations in mineral content.

Weight of the Skeleton in Starvation

It is generally believed that the skeleton loses less weight relatively during starvation than does the body as a whole or than do the blood, fat, muscle, and internal organs (Rokitansky, 1854; E. Voit, 1905b; Jackson, 1925). Whether the same conclusion holds for the conditions in semi-starvation and famine is not known. In spite of the abundance of potential human material that has existed for many centuries in times of famine, no analyses of adult bodies have been reported; there are, actually, few data on the weight and composition of even the normal adult skeleton. For lack of better evidence, it is necessary to assume that the changes that occur in the skeletons of animals during starvation are in general applicable to man.

In cachectic diseases the rate of weight loss is not the same in all the tissues and organs. Rokitansky (1854) listed the blood, fat, muscles, and visceral organs as being most labile in starvation; the bones were relatively more resistant and lost weight only when starvation was severe and of long duration. Marfan (1921) observed no changes in the bones of athreptic infants.

From animal experiments it is apparent that there is some loss of skeletal weight during starvation but that the loss is much less than for the body as a whole. In starved pigeons Chossat (1842) observed a 17 per cent decrease in the weight of the skeleton while the fat and skeletal muscles decreased 93 and 42 per cent, respectively. C. Voit's (1866) starved cat, compared with one normal cat, lost 33 per cent of its body weight and 14 per cent of its skeletal weight. Two cats that were starved until the body weight was reduced more than 50 per cent (see Table 114) had reductions in the weight of the bones of 6 and 19 per cent (Sedlmair, 1899). Wellman (1908) reported a decrease of 11.9 per cent in bone weight in cats in which a 40 per cent reduction in body weight had oc-

218

TABLE 114

SKELETON AND BODY WEIGHT LOSS IN STARVED CATS (from
C. Voit, 1866, and Sedlmair, 1899).

Body Weight			Skeleton Weight		
Before Starvation (gm.)	End of Starvation (gm.)	Loss (%)	Before Starvation (gm.)	End of Starvation (gm.)	Loss (%)
3105	2088	33	393	339	14
2988	1471	51	302	284	6
2969	1334	55	300	242	19

curred. In severely undernourished children whose body weights were about 40
per cent below that of normal children of comparable age and height, the fresh,
wet weight of the bones was reduced only 14 per cent (Ohlmüller, 1882). In
general the relation of body weight loss and fresh, wet bone weight loss is comparable for starved animals and atrophic children. Whether the same relationship holds for adult man in semi-starvation is not known.

A better concept of the changes that occur in the skeleton weight in starvation can probably be derived from data on the dry or dry, fat-free weights. On
the dry, fat-free basis the atrophic children reported by Ohlmüller (1882) had
skeletal weights that were 9.3 per cent below normal with actual body weights
40 per cent reduced. The three starved cats studied by Weiske (1897) had body
weights reduced 33 per cent, but the skeletons on the dry, fat-free basis changed
very little — 0.0, +2.4, and —4.7 per cent. The data for Sedlmair's (1899) two

TABLE 115

BODY WEIGHT AND DRY, FAT-FREE SKELETON WEIGHT before and
after starvation in 2 cats (Sedlmair, 1899).

Body Weight			Dry, Fat-Free Skeleton		
Before (gm.)	End (gm.)	Loss (%)	Before (gm.)	End (gm.)	Loss (%)
2988	1471	51	163	157	3.7
2969	1334	55	154	142	7.8

cats are presented in Table 115. Although the body weights of the two animals
had been reduced 51 and 55 per cent, the actual losses in the weights of the
skeletons after the water and fat had been removed were only 3.7 and 3.8 per
cent. Changes in dry, fat-free bone weights similar to those observed in atrophic
children by Ohlmüller (1882) have been reported in starved cats by Wellman
(1908). With body weight losses of 40 per cent, the weight of the skeletons was
reduced 16.0 per cent on the dry basis and 8.5 per cent on the dry, fat-free basis.
The fat-free organic components of the bones were reduced 12.2 per cent. Apparently the water, fat, and organic components are reduced more than the
minerals during starvation.

Mineral Changes in Bones and Teeth in Starvation

From Wellman's data (1908) there appeared to be, at least in starved cats, a real loss in the total bone minerals, but the loss was small in comparison to the loss of weight of the animal and of many of the organs. The percentage losses of the chief bone minerals, calculated from the total dry, fat-free bone weights, were 5.5 per cent for phosphorus, 7.8 per cent for calcium, and 8.3 per cent for magnesium. The data are particularly interesting because the percentage loss was almost identical for the three minerals even though the amount that each of the minerals contributes to total bone minerals is vastly different (see Tables 116 and 117).

TABLE 116

MINERAL CONTENT OF BONES AND TEETH IN NORMAL AND STARVED CATS, in per cent of dry, fat-free weight (after Weiske, 1897).

	Normal		Starved Cat #1		Starved Cat #2		Starved Cat #3	
	Bone	Teeth	Bone	Teeth	Bone	Teeth	Bone	Teeth
Minerals..	61.9	76.6	60.3	78.2	62.3	78.5	60.7	77.9
CaO	32.4	37.9	31.8	38.8	32.5	38.9	31.8	38.8
MgO	0.71	2.49	0.69	2.44	0.62	1.10	0.64	1.45
P_2O_5	24.2	33.9	24.0	34.2	24.7	34.5	24.2	34.2

TABLE 117

MINERAL CONTENT OF BONES IN NORMAL HUMANS AND IN PATIENTS WITH OSTEOMALACIA, as per cent of minerals in dry, fat-free bones (after Loll, 1923).

	CaO	MgO	P_2O_5	Others
Normal	51.8–52.0	0.78–0.85	38.7–38.9	8.55–8.66
Osteomalacia	56.8–76.8	0.74–0.80	16.2–37.3	4.04–8.59

The lack of any specific changes in the mineral composition of the bones of starved cats is substantiated by the observations reported by Weiske (1897). His data are presented in Table 116. There were no consistent differences between the normal and the starved cats for the mineral percentage of the bones and teeth, or for the percentage of any of the three major bone and teeth minerals. Whether there was a decrease in the total amount of the various minerals in the skeleton and teeth was not reported by Weiske; only one of his three animals had a decrease in the dry, fat-free skeletal weight (−4.7 per cent).

It could probably be assumed that the changes in bone mineral in man would be much the same as that in animals under similar degrees and conditions of starvation. Unfortunately, very few human data are available. It is well recognized, however, that in times of famine and semi-starvation there is an increase in the incidence of bone diseases (see the section below on osteopathies of starvation). It is generally believed that the osteopathies are due to, or are accompanied by, changes in bone mineral composition. The percentages of the three bone minerals in normal human bones and in bones of patients with osteomalacia have been determined by Loll (1923) and are summarized in Table 117. There was a relative increase in the calcium content of the bones in osteomalacia with a corresponding, but less regular, decrease in the phosphorus and no change in the magnesium. No data on absolute amounts of the minerals were given. The behavior of the bone minerals in osteomalacia appears to be quite different from that seen in animals during uncomplicated starvation.

Osteopathies of Starvation — Before World War II

Endemic bone disorders have long been recognized, but it was not until the food shortages developed in Central Europe in World War I that special emphasis was placed on the role of limited food intake in the development of these disorders. It was at that time that such designations as "hunger osteopathy," "hunger osteomalacia," and "hunger osteoporosis" appeared in the medical literature. The exact etiological role of semi-starvation was not satisfactorily proved, but that it was a factor cannot be doubted.

Gelpke (1891) attributed the endemic character of osteomalacia in certain regions of Switzerland mainly to the gradual development of a strain of people who have latent osteomalacia and whose bones are deficient in inorganic salts. The people are not poor, nor are their diets restricted in amount or variety of foods. The land in these regions, however, is quite low in phosphorus, and it might be assumed that food raised on the land would also be low in this mineral.

Januszewska (1910) emphasized the customs of the people instead of their diet. In a 10-year period she observed 3510 cases of osteomalacia in Bosnia. Only 12 of the patients were not Mohammedans, and the disease was limited almost exclusively to the large cities. Among the rural population there was not a single case of osteomalacia. According to Januszewska, the only factor that could explain this difference in incidence was that the urban people practiced seclusion (Purdah) while the rural population did not. It is conceivable, however, that the character of the diet, as well as the amount of sunlight (vitamin D), may have been quite different in the two groups.

The renewal of interest in the role of diet in osteopathies is apparent in the vast number of articles that appeared in the European medical literature during and following World War I, when near-famine conditions existed in many parts of the continent. Although the symptom complexes of the hunger osteopathies described had much in common, the disorders were designated by various authors as osteoporosis, late rickets, and osteomalacia — mainly on the basis of the age of the patient.

From March until late June 1919 Alwens (1919) saw 26 patients at the Frankfurt Medical Clinic with symptoms of spontaneous fractures, thinning of the bones, pain and tender bones, and difficulty in walking. In the group were 23 females and 3 males ranging in age from 19 to 72 years; 65 per cent of the women were at climacteric or post-climacteric age. The diet had for some months consisted chiefly of potatoes, turnips, watery soup, and war bread, with very little meat and no eggs, cheese, or butter. The disease was considered to be an osteoporosis, resembling senile osteoporosis, resulting from a diet inadequate in protein, calcium, and phosphorus; when a good mixed diet was provided the fractures and osteoporosis healed but the deformities remained.

Within a period of two months Porges and Wagner (1919) observed 20 cases of hunger osteoporosis. The patients were from 30 to 50 years of age, all had for some months been on a quantitatively and qualitatively poor diet and were emaciated, but none had pronounced edema or hypoproteinemia. Nothing specific was observed on physical examination except severe pains in the back, especially in the lumbar region, and ribs and vertebrae that were sensitive to pressure. On X-ray examination some of the bones appeared to have a decreased density. In a few cases the alveolar carbon dioxide tension was found to be as high as twice the normal values. The disorder was believed to be of dietary origin, mainly related to the low protein and mineral content of the wartime diets. Spontaneous fractures of the fingers, ribs, and femur were reported by Staunig (1919). Skeletal changes resembling the osteoporosis of senility were observed among the children during the Russian famine of 1921–22 (Nicolaeff, 1923).

Frequently the hunger osteopathies were reported as hunger osteomalacia or late rickets. Whether they differed fundamentally from hunger osteoporosis is not clear in most cases. The bone disorders described by Boehme (1919) were considered to be similar to rickets in children and much like the osteomalacia of adults. The 20 young women, 15 to 20 years of age, were malnourished and complained of pains in the legs, knees, and feet, especially on standing. The cortex of the long bones was thin, there was a delay in ossification, the epiphyseal border was diffuse, and the interspaces between the trabeculae of the metaphyses were enlarged. Edelmann's (1919) 19 cases and Hahn's (1919) 13 cases of osteomalacia were mostly undernourished old women.

Fromme (1919) observed 266 cases and Simon (1919) 40 cases, mainly adolescents, from regions of food shortages who had a disease that resembled rickets in children and osteomalacia in adults. Eisler (1919) and Looser (1920) insisted that late rickets in adolescents and hunger osteomalacia in adults were the same disease. Many cases of hunger osteomalacia were seen in Vienna during the winters of 1918–19 and 1919–20, mainly among the older people of both sexes. The incidence of the disease decreased in the spring and summer when fresh green vegetables were available. The chief complaints were pain in the bones of the legs and back and pain and difficulty in walking; in extreme cases the patients were bedridden (Dalyell and Chick, 1921). The simultaneous increase of rickets in the children, late rickets in the young adults, and osteomalacia in the older people was considered by these authors to be ample evidence that it was all the same disease.

Hume and Nirenstein (1921) found that hunger osteomalacia occurred mainly in people who had been on a quantitatively and qualitatively poor diet for some time. Cod liver oil in doses of 100 to 200 gm. per week cured most of the cases in 2 to 6 weeks while rapeseed oil plus phosphorus was not effective except in mild cases. They concluded that hunger osteomalacia was a vitamin A (*sic*) deficiency disease. This was, of course, before the discovery of vitamin D. Gribbon and Paton (1921) doubted whether malnutrition was a factor in the development of rickets in Vienna.

Schlesinger (1919a) considered hunger osteomalacia to be quite different from the osteomalacia of pregnancy mainly because hunger osteomalacia developed slowly, seldom affected the pelvis, seldom was accompanied by marked deformities, and was most frequent in the aged. Tetany may be present in hunger osteomalacia (Schlesinger, 1919b). Probably after having sat through innumerable medical meetings in which hunger osteopathies were the main topic, Strümpell (1919) presented five cases of osteomalacia in women who did not have unusually bad nutrition and in three of whom the disease was present before the war. He pointed out that more cases were being reported than usual because physicians were looking for them. Wassermann (1919), however, believed that these osteopathies were produced by inadequate nutrition and that they reflected a general labile nutritional state of the body which was, of course, made worse by the poor wartime diets.

Hirsch (1920) admitted that the pathogenesis of hunger osteopathy was not clear and that the state of knowledge was confused. He suggested dropping the term osteomalacia and using instead "osteomalacia symptom complex," and he questioned whether the cases described as hunged osteomalacia were actually all the same. Hirsch seriously considered hunger osteopathies to be the result of multiglandular defects in which diet was important because of its influence on endocrine gland activity. He classified the hunger osteopathies as late rickets in adolescents, osteomalacia in adults, and osteoporosis in the aged. From the data presented, it appeared that during 1917 to 1920 there was an increase in incidence and a spread of age range, but each category remained quite as distinct as it was during the years 1900 to 1917.

From observations during the famine in Poland in 1917–18, Chelmonski (1921) classified the starvation diseases of the bone in two main groups. "Osteoporosis alimentaria simplex," which is characterized by pain in the sacral bones, vertebrae, and ribs, by diminished calcification of the bones, and by spontaneous fractures, was considered to be the result of prolonged general undernutrition rather than the lack of any specific food item because it responded readily to a good diet. "Pseudo-osteomalacia alimentaria," on the other hand, was a progressive osteomalacia of old women and was characterized by pain in the legs, sacrum, joints, and thorax and by paresis and weakness of the legs and arms. It seems doubtful whether this classification would help much in clearing the confusion. Droese (1938) found no agreement in the literature on the classification of hunger osteopathies and concluded that, while hunger osteomalacia, late rickets, and rickets have much in common, hunger osteoporosis can generally be distinguished from the other forms.

The development of hunger osteopathies has been attributed by various authors to a variety of causes, although undernutrition has generally been considered a major cause (Alwens, 1919; Boehme, 1919; Fromme, 1919; Porges and Wagner, 1919; Wassermann, 1919; Chelmonski, 1921). The minerals, particularly calcium and phosphorus, have received surprisingly little consideration (Gelpke, 1891; Alwens, 1919; Fromme, 1919). Probably because there appear to be some age and sex differences in hunger osteopathies, the endocrine glands have attracted considerable attention (Naegeli, 1918; Edelmann, 1919; Fromme, 1919;

Sauer, 1921; Hirsch, 1920; Chelmonski, 1921; Neuberger, 1921; Simon, 1919; Droese, 1938; and others). The emphasis is placed on the general metabolic effect of the hormones rather than on a specific effect on the bone structure. Even though the action of vitamin D in bone metabolism was not known during World War I, cod liver oil especially was considered by some authors to have therapeutic value in many cases of hunger osteopathies (Dalyell and Chick, 1921; Hume and Nirenstein, 1921; Neuberger, 1921; Droese, 1938). In the same general category were the reports that stressed the importance of living conditions and, presumably, the production of vitamin D in the skin by the action of sunlight (Januszewska, 1910; Simon, 1919; Droese, 1938). Fromme (1919) included infections as another possible factor in hunger osteopathies.

Osteopathies of Starvation — World War II

There apparently was a sharp decrease in the incidence of bone disorders when food became more plentiful again in the early 1920s. But as famine and near-famine conditions developed in many regions during World War II, "hunger osteopathies" again were frequently observed. Reports on *l'ostéopathie de famine* started to appear in the French literature in 1942. Belger (1942) reported one 62-year-old female with a typical pseudofracture of the upper femur and pain in the bones. Dereux's (1943) case was a female 35 years of age who had pains in the hips and a pseudofracture of the left femur. She responded to vitamin D therapy. The three cases observed by Besançon (1942) were women 62 to 80 years of age. All were undernourished, complained of pains in the back, hips, and ribs, and had painful spontaneous, progressive, and symmetrical fractures of the ribs, forearms, and jaws. Some of the bones showed intense osteoporosis. Besançon considered the cases to be identical with those seen in Vienna in World War I. Zimmer, Weill, and Dubois (1944) and Besançon (1945) were convinced that the decreased food intake in some regions of France was the cause of the marked decalcification of the ribs and vertebrae and of the symmetrical and bilateral spontaneous fractures that were appearing in increasing numbers.

The number of cases of *ostéopathies de carence* seen in Paris were 3 in 1940, 5 in 1941, 8 in 1942, 24 in 1943, and 51 during the first 6 months of 1944 (Coste and Berget, 1945). The 20 cases reported by de Sèze, Ordonneau, and Godlewski (1946) were all women from 50 to 70 years of age who had lost 15 to 20 kg. of body weight. Their diet was mainly bread, vegetables, and a little meat, but no milk. They had pains in the bones of the legs, pelvis, and spine, and on X-ray examination the bones showed thinning of the cortex and fracture. The response to vitamin D and calcium therapy was prompt. Snapper (1949), impressed by the incidence of osteomalacia and hunger osteopathy among persons deprived of sunlight, stressed the importance of vitamin D in the etiology of these disorders.

Bone dystrophies with areas of translucency in the cortex of the long bones and narrow transverse fractures were seen in western Holland (Burger *et al.*, 1945; Stare, 1945). The dystrophies appeared to be of two different types. In emaciated patients there was a general osteoporosis with pain in the bones and a progressive kyphoscoliosis of the upper part of the trunk. The less emaciated patients complained of pain that was at first considered to be of rheumatic origin, and many had transparent areas in the cortex of the long bones. The cases with focal decalcification responded rapidly to vitamin D therapy, but the ones with generalized osteoporosis made a very slow recovery. Hunger osteopathy with transverse decalcification of the bones was also reported from the Belsen prison camp (Pollack, 1945). It is interesting that in the recent reports the bone disorders are referred to as osteopathies or dystrophies instead of by the more specific terms — osteoporosis, osteomalacia, and late rickets — which were used in the older literature.

Pompen *et al.* (1946) described 24 cases of typical hunger osteopathy in Holland during 1943–45. These authors were "inclined to regard hunger osteopathy as the result of a combined deficiency of multiple food factors, of which animal protein might be one and of which vitamin D is probably the most important. If untreated, hunger osteopathy develops progressively into typical osteomalacia." Twenty-two of the cases occurred in women but the peculiar sex distribution was attributed "to a difference of food habits between men and women rather than to hormonal" factors. The typical translucent notches and bands of decreased X-ray opacity in the bones were considered to be "the effect of mechanical forces upon a skeleton that, due to deficiency, is unable to resist and adapt itself in a normal way. Similar areas have been observed in normal bones under the influence of overtaxation through repeated pull and pressure. Thus hunger osteopathy is related to march-fractures and fatigue-fractures." In hunger osteopathy, however, the bones are weakened by the deficient diet to such an extent that the normal strain of everyday life produces the same effect as does repeated overstrain on normal bone. The response of the osteopathy to a full diet, rich in animal protein and milk with extra vitamin D, was rapid in all cases.

Lamy, Lamotte, and Lamotte-Barrillon (1948) made radiographic examinations of the bones of 28 starvation victims. In most cases many of the bones showed qualitative evidence of moderate to severe decalcification; no quantitative data on bone density were reported. The blood calcium level was normal or slightly elevated.

We have repeatedly observed that in times of food shortages the most vulnerable persons in the population are old people who live alone. Such persons tend to be most completely dependent on official rations and may obtain little else because of infirmity or lack of ingenuity or effort. Moreover, elderly recluses rarely receive the benefits of sunshine on the skin. It is perhaps to be expected, then, that some old people will develop hunger osteomalacia even in areas where food shortages are not extreme. Gsell (1945b) had 2 such patients in Switzerland; both were women, aged 77 and 75, who lived alone. Treatment consisted of dietary improvement and huge doses of vitamin D.

In the Netherlands East Indies hunger osteopathy appeared in the Japanese concentration camps, and again the victims were mainly older women, although some men and some younger persons were affected; the youngest was 36 years old (Netherlands Red Cross Feeding Team, 1948). Blood studies on 15 patients before treatment revealed nothing peculiar except a tendency to high serum phosphatase concentrations. Gsell (1945b) also recorded high serum phosphatase values in his 2 patients.

Both in the Netherlands and in the East Indies vitamin D was considered of much value in the treatment (Netherlands Red Cross Feeding Team, 1948). In the East Indies about 30 patients with hunger osteopathy were treated with injections of 600,000 units of vitamin D_3, usually repeated 2 or even 3 times. Some remission of the pains in the bones was reported within 3 or 4 days after the first injection, and evidence of improvement was obtained from calcium and roentgenological studies.

In the Warsaw Ghetto a tendency toward spontaneous fractures and delay in bone reunion was noted by Fliederbaum et al. (1946). Among the children, however, there was no evidence of bone abnormalities, and rickets was rarely present (Braude-Heller et al., 1946). Not a single case of late rickets was seen, and X-rays of the long bones were normal except for indications of some slight decalcification.

At the Pootung Civil Assembly Center, Shanghai, China, X-ray examinations frequently revealed a general absorption of the bone calcium (Graham, 1946). Some of the internees complained of pains in the bones and muscles but in only two cases was tetany observed. The calcium intakes were low, averaging 0.262 to 0.520 gm. per day. A varying degree of osteoporosis of the long bones was observed among the American prisoners of war in Japanese camps. Hibbs (1947) stated, "interestingly enough, about a dozen of us ended up at Gardner General Hospital in Chicago after our return to the states, and X-ray pictures there of our long bones revealed evidence of osteoporosis, after 3 to 4 months of good adequate rations." The prison diet was probably low in calcium and vitamin D, but all the men had ample exposure to the sun and fish bones were frequently boiled and eaten.

Dental Caries

The opinion is frequently expressed that bad nutrition results in an increased incidence of dental caries and that dietary improvements should reduce the tendency of teeth to decay. The general proposition is reasonable enough; the problem is to define what is a "bad" diet and what constitutes "improvements." Since the usual concept of dietary improvement has to do with increased intake of vitamins, proteins, and minerals, it might be thought that the same definition should apply to dental caries and that a low intake of these substances, or simply general undernutrition, would be associated with a high level of tooth decay. The facts, however, point to more complex or even quite different relationships.

During World War II there was moderate undernutrition, without specific vitamin deficiencies, in Northern and Western Europe; in the Japanese prison camps there was moderate to severe undernutrition with multiple vitamin deficiencies. In neither area was there an increase in caries incidence; the data actually point to a decrease. At the end of the war the teeth of the French and Belgian children were considered to be in good condition (Struthers, 1945; Sydenstricker, 1945). In Norway (Mellanby and Mellanby, 1948; Collett, 1946) and England (Oliver, 1946) there was a decreased incidence of caries in children which King (1947) interpreted as being coincident with improved wartime nutrition. From other viewpoints the wartime nutrition in Norway would be con-

sidered "bad," but there was a dramatic reduction in caries in children (Tov-
erud, 1945). Even in prison camps like Buchenwald and Dachau there was no
evidence that the extreme malnutrition had any deleterious effect on the teeth
(Dechaume, 1947).

Quantitative appraisals in Europe provide numerical values of interest. Steijling (1947) found a decrease in the incidence of carious teeth in the adult population of the large cities of Holland. The estimated incidence of 48 per cent in
1941 fell to 35 per cent in 1945. Steijling could find no explanation in dental hygiene or professional treatment and concluded that at least the poor nutrition of
the war period was not harmful to the teeth.

TABLE 118

INCIDENCE OF DENTAL CARIES IN CHILDREN OF PARIS during the
years 1942–45 (Dechaume, 1947).

Date	Number Examined	Without Caries	
		N	%
1942	1000	131	13.1
1943	500	131	26.2
1944	500	188	37.6
1945	500	224	44.8

TABLE 119

COMPARISON OF DENTAL DEFECTS IN MALNOURISHED ITALIANS AND IN
THE ORDINARY POPULATION OF THE UNITED STATES. Mean numbers, per
person, of decayed, missing, and filled teeth. Data
from Schour and Massler (1947).

Age 11–15		Age 51–60	
Italy	U.S.A.	Italy	U.S.A.
1.05	4.66	10.80	23.20

More striking changes were observed in the abandoned young children seen
at Saint Vincent de Paul in Paris (Dechaume, 1947). The steady decline in caries
incidence produced a reduction of caries in four years of almost 50 per cent (see
Table 118). Dechaume offered no explanation for the improvement but was obviously surprised to find this associated with a protracted state of chronic mild
undernutrition.

Schour and Massler (1947) studied 3905 persons in cities of southern Italy.
The prewar incidence of caries was not known but comparisons with experience
in the United States are of interest. The data are summarized in Table 119. The
Italian children between 11 and 15 years of age had less than a fourth as many
dental defects as children of the same age in the United States; in persons of 51
to 60 years of age the differential was almost half as great. The Italians were
relatively malnourished, and oral hygiene was not commonly practiced.

The data from the Japanese prison camps are less satisfactory but tend in the
same direction as in Europe. The teeth of American prisoners of war in Japan

were in remarkably good shape at the time of liberation (Carroll, 1945). Morgan, Wright, and Van Ravenswaay (1946) reported that among repatriated prisoners and inmates of concentration camps from the Far East "the number of carious teeth was less than that of the same age group living under normal conditions." Graham (1946) thought there was rapid tooth decay in some of the inmates of the Pootung Civil Assembly Center near Shanghai. With an average camp population of 1085 there were, in 18 months, 960 dental fillings and 580 extractions; this does not indicate a high incidence of caries if the dental work done represented most of that needed.

There was caloric undernutrition in all the groups studied abroad but the findings are rather similar to those in the United States in groups exhibiting vitamin deficiencies with a better state of caloric nutrition. In 124 patients with a clinical diagnosis of malnutrition there was an average per person of 4.54 decayed, missing, and filled teeth, compared with a figure of 14.94 for 99 control subjects (Mann *et al.*, 1947). The patients in the malnourished group were considered to show signs of vitamin A deficiency (9%), thiamine deficiency (22%), riboflavin deficiency (68%), nutritional macrocytic anemia (12%), subclinical pellagra (23%), definite pellagra (27%), and scurvy (14%). These findings confirm earlier studies on 42 malnourished patients (Kniesner *et al.*, 1942).

The complexity of the problem as to the relation between nutritional state and the incidence of caries is illustrated by an experiment on 72 children with clinical evidence of malnutrition and on 25 well-nourished "controls" (Dreizen *et al.*, 1947). Twenty-five of the malnourished children were given a supplement of a quart of milk daily for 18 months, while the other 47 children in that group continued to be malnourished. The data are summarized in Table 120; they suggest that the milk supplement raised the incidence of dental caries in the malnourished children to the level of the children who were habitually well nourished.

TABLE 120

DENTAL DEFECTS (average number of decayed, missing, and filled teeth) observed initially and 12 months later in 72 children initially malnourished (Groups 1 and 2) and in 25 well-nourished children (Group 3). The 25 children in Group 2 each received a quart of milk daily during the year of the experiment. Data from Dreizen *et al.* (1947).

	Group 1	Group 2	Group 3
Initial defects	4.52	2.80	7.34
New caries in 12 months	0.95	0.96	2.38
New caries as percentage of initial	21	34	32

Complete mouth X-rays of the teeth were made on all the semi-starvation subjects during the control period and at the end of 24 weeks of semi-starvation in the Minnesota Experiment. The X-rays were carefully examined at the Dental Clinic, University of Minnesota, for old caries present, new caries that developed, and the progression of old caries.

At the beginning of the semi-starvation regimen the subjects had, on the average, 5.6 caries per person. The condition of the gums and gingival tissue was in most cases considered to be satisfactory. At the end of the 24 weeks of semi-starvation an average of one new cavity had developed per subject, which was considered normal for any group of adults of similar age. The progression of old caries was not unusually rapid in any of the subjects. The condition of the gums and gingival tissue was neither better nor worse than during the control period.

Bone Density — Minnesota Experiment

Bone density studies by radiological methods were made in the Minnesota Experiment by Dr. Pauline Beery Mack of the Pennsylvania State College.

Teleroentgenograms were taken of the little finger of the left hand, the distal end of the left femur, and the left *os calcis* of the 32 subjects at 24 weeks of semi-starvation and after 6 and 12 weeks of rehabilitation. No control values were obtained on the 32-man group, but the measurements that were made on the same days on a group of 12 normal young men, who were in a good nutritional state, were assumed to represent normal bone densities and consequently could be used as representative of the semi-starvation group during the control period. Furthermore, a comparison of the normal group and the semi-starvation group at the end of semi-starvation and during rehabilitation could be used as an index of any change in the density of the bones that may have occurred in the semi-starvation subjects as a result of either the starvation or the rehabilitation diets.

The densitometric method used for the estimation of bone density in the Minnesota Experiment is described in detail in the Appendix on methods. The densities of the bone and the standard ivory ladder on the X-ray film are measured by means of a microphotometric scanning procedure. The density of the bone is expressed in terms of cm. of equivalent ivory thickness, and the concentration is the equivalent ivory thickness per sq. cm. of cross-sectional area. Referring the bone density to an ivory standard is justifiable because the X-ray absorption characteristics of ivory and bone are very similar even though the chemical composition of the two is not identical.

In determining the density of a bone a series of sections of the bone are scanned, the ivory equivalent density and concentration for each section is calculated, and the values for each section are integrated to give an integrated average value for the bone. The density in cm. of equivalent ivory for each of the 6 sections of the second phalange of the left little finger and the integrated averages for the 4 caloric groups and a normal group are presented in Table 121 for the X-rays taken at the end of the semi-starvation period. The average density for any one section of the second phalange of the little finger varied among groups by not more than 10 to 15 per cent. The variation among sections within the same group was greater, as would be expected, because the structure of the bone is not homogeneous nor is the diameter constant. A comparison of the integrated average densities and the standard deviation for any one group and between groups indicated a rather small interindividual variation in the bone density of the second phalange of the little finger. The coefficient of variation was 15.5, 13.6, 16.6, 11.5, and 13.2 per cent for groups Z, L, G, T, and N, respectively.

TABLE 121

Bone Density at the End of Semi-Starvation for the 4 caloric groups (Z, L, G, T) and the 12-man normal group (N), Minnesota Experiment. The values are given for each of the 6 sections of the second phalange of the left little finger, with integrated averages for the entire second phalange and standard deviations of the integrated averages. Density is expressed as equivalent ivory thickness in cm.

Section	Z	L	G	T	N
1	0.564	0.616	0.576	0.583	0.544
2	0.493	0.448	0.462	0.481	0.459
3	0.522	0.484	0.522	0.533	0.513
4	0.516	0.494	0.527	0.519	0.513
5	0.484	0.431	0.471	0.462	0.447
6	0.441	0.434	0.479	0.447	0.394
Integrated average ...	0.503	0.484	0.506	0.504	0.478
SD	0.078	0.066	0.084	0.058	0.063

TABLE 122

Bone Density at the end of semi-starvation (S24) and after 6 (R6) and 12 (R12) weeks of rehabilitation, Minnesota Experiment. Mean values and average standard deviations for the second phalange of the left little finger for each of the 8-man caloric groups (Z, L, G, T) and for a 12-man normal group (normals). For each subject 6 segments of the second phalange of the left little finger were measured and integrated on each occasion. "Density" is the integrated average density for the entire second phalange of the left little finger in terms of equivalent ivory thickness (cm.), and "Conc." (concentration) is the equivalent ivory thickness per unit (sq. cm.) of cross-sectional area.

	S24		R6		R12	
	Density	Conc.	Density	Conc.	Density	Conc.
Z	0.503	1.365	0.482	1.160	0.460	1.100
L	0.484	1.270	0.467	1.093	0.444	1.033
G	0.506	1.310	0.504	1.298	0.445	1.173
T	0.504	1.321	0.474	1.262	0.456	1.199
M	0.499	1.317	0.482	1.203	0.454	1.126
SD	0.072		0.066		0.055	
M of normals..	0.478	1.315	0.442	1.212	0.429	1.177
SD	0.063		0.055		0.050	

The densitometric data for the second phalange of the left little finger for the semi-starvation and rehabilitation periods are summarized in Table 122.

The mean integrated average density for the 4 caloric groups at S24 was substantially the same as for the normal group measured at the same time. During the 12 weeks' rehabilitation period there was a small apparent decrease of about 10 per cent in the bone density in both the normal and the semi-starvation groups. The mean bone density was actually 4.2 per cent and 5.5 per cent lower at S24 and R12, respectively, in the normal than in the semi-starvation subjects. The changes in concentration, which is density per unit of cross-sectional area, were similar to the density changes.

It is obvious that the 24 weeks of semi-starvation, with a 24 per cent body weight loss, did not produce a decrease in the density of the second phalange of the little finger. The small decrease that occurred in both the normal and semi-starvation subjects during the 12 weeks of rehabilitation is probably due to some systematic technical error rather than to actual loss of bone minerals.

The variation in the relative density and concentration of the *os calcis* in the semi-starvation and normal subjects was essentially the same as for the finger bone, with no significant difference between the two groups at any of the three test periods. The data are presented in Table 123. A small increase in the density of the *os calcis* occurred in all subject groups during the 12 weeks of rehabilitation instead of the small decrease in density observed in the finger bone. The differences in size and structure of the *os calcis* and the finger bone are apparent from the data in Tables 122 and 123. The density is dependent upon both the size of the bone and the compactness of its structure, while the concentration is primarily a measure of minerals present per unit area of the bone. As seen from the relative values, the finger bone is a more compact structure than is the *os calcis*.

TABLE 123

Bone Density at the end of semi-starvation (S24) and after 6 (R6) and 12 (R12) weeks of rehabilitation, Minnesota Experiment. Mean values and standard deviations for the *os calcis* for each of the 8-man caloric groups (Z, L, G, T) and for a 12-man normal group (normals). For each subject 10 segments of the *os calcis* were measured and integrated on each occasion. "Density" is the integrated average density for the entire *os calcis* in terms of equivalent ivory thickness (cm.), and "Conc." (concentration) is the equivalent ivory thickness per unit (sq. cm.) of cross-sectional area.

	S24		R6		R12	
	Density	Conc.	Density	Conc.	Density	Conc.
Z	1.703	1.103	1.631	1.048	1.709	1.081
L	1.788	1.145	1.771	1.116	1.781	1.111
G	1.738	1.140	1.700	1.117	1.768	1.127
T	1.734	1.153	1.681	1.076	1.730	1.112
M	1.738	1.135	1.700	1.089	1.747	1.108
SD	0.222	0.140	0.180	0.112	0.189	0.114
M of normals..	1.694	1.125	1.714	1.114	1.767	1.150
SD	0.424		0.426		0.428	

Data on the density of the distal end of the femur are available on the normal and semi-starvation subject groups at only R6 and R12. However, a comparison of the femur densities of the two groups can be made to indicate whether any serious demineralization had occurred in the semi-starvation subjects. Two regions on the distal end of the femur were measured. Region B was through the most distal tip of the femur, and region A was about 5 cm. proximal to region B. The average density in region A for the semi-starvation and normal subjects at R6 was 2.760 and 2.440, and at R12, 2.604 and 2.507, respectively;

for region B the density for the semi-starvation and normal subjects was 2.571 and 2.243, respectively, at R6 and 2.464 and 2.300 at R12. There was a slight decrease in density for both regions of the femur in the semi-starvation group from R6 to R12. In the normal subject group there was a slight increase in density. The changes were small and probably not significant in either group.

The densitometric data for the Minnesota Experiment is admittedly incomplete and does not include many bones, such as the ribs, pelvis, and proximal femur, in which interest would be greatest. The data are sufficient, however, to warrant the definite conclusion that a loss of 24 per cent of the body weight in a 6 months' period on the type of starvation diet used in the Minnesota Experiment is not accompanied by a generalized demineralization of the bones.

CHAPTER 13

Skin and Hair

BESIDES the obvious loss of subcutaneous adipose tissue, many clinical abnormalities of the skin have been observed among the underfed inhabitants of famine areas. Some of the skin changes are attributable to the starvation, and their severity and incidence are roughly proportional to the degree of undernutrition, but the breakdown of sanitary and medical facilities and the lack of sufficient clothing, shelter, warm water, and soap in times of famine no doubt play an important role in the development of many of the skin disorders. It is naive to assume, as many observers have done, that there is necessarily a direct relationship between nutritional status and all the clinical abnormalities of starved people.

Appearance and Texture of Skin

The pale and cold skin of semi-starvation (Maliwa, 1918; Burger *et al.*, 1945; Surveys, 1945) can probably best be explained on the basis of anemia (see Chapter 14) and peripheral vasoconstriction. The incidence of pallor in a random sample of people examined at street clinics in western Holland is presented in Table 124; 20.1 to 35.1 per cent of the individuals exhibited pallor, but that percentage represents only about half the number who had anemia. However, the degree of anemia was mild in most cases and, it might be assumed, not sufficiently severe to produce such obvious pallor.

The skin in semi-starvation has frequently been described as dry, scaly, thin, and inelastic. Petenyi (1918) reported that dry, scaly skin was common in Germany during World War I. The skin was rough and scaly in the group of French prisoners of war examined by Richet and Mignard (1919). The men had been doing hard work on 2000 Cal. per day for about 5 months and were obviously emaciated. In the Russian famine of 1921–22 the skin of the undernourished persons became dry and rough (Abel, 1923; Nicolaeff, 1923), lost its elasticity, became wrinkled, and resembled the skin of old age (Ivanovsky, 1923). The same general skin changes have been reported in cases of anorexia nervosa (Berkman, 1930). A thin, dry skin was observed at autopsy in children during the siege of Leningrad (Efimova and Elpersin, 1944; Garshin, 1943) and was present in the victims of the Travancore famine of 1943–44 (Sivaswamy *et al.*, 1945).

In the Western European area during World War II, Macrae (1944) found a dry, scaly skin in 78 per cent of the children examined in Belgium. The skin was described by Zimmer, Weill, and Dubois (1944) as loose, dry, scaly, and

MORPHOLOGY

TABLE 124

INCIDENCE OF PALLOR as percentages of people examined in street clinics during nutritional and medical surveys in 4 western Holland cities in 1945, grouped according to age and sex (Surveys, 1945).

| | Age Group (yrs.) | | | | | | | |
| | 1–12 | | 13–18 | | 19–59 | | Over 60 | |
	Male	Female	Male	Female	Male	Female	Male	Female
Amsterdam	26.0	22.5	16.4	15.3	28.9	18.2	14.8	27.7
Utrecht	11.2	12.9	11.0	8.1	23.1	17.5	21.4	9.5
Delft	46.0	36.0	31.0	44.0	39.7	56.7	46.7	56.7
Rotterdam	16.9	22.7	23.1	34.5	20.1	39.0	33.2	46.6
M	25.0	23.5	20.1	25.5	27.9	32.9	29.0	35.1

TABLE 125

INCIDENCE OF DRY SKIN as percentages of people examined in street clinics during nutritional and medical surveys in 4 western Holland cities in 1945, grouped according to age and sex (Surveys, 1945).

| | Age Group (yrs.) | | | | | | | |
| | 1–12 | | 13–18 | | 19–59 | | Over 60 | |
	Male	Female	Male	Female	Male	Female	Male	Female
Amsterdam ..	4.3	3.9	4.5	3.4	10.9	5.7	22.7	12.7
Utrecht	12.7	11.8	7.7	11.8	8.5	6.4	11.3	15.2
Delft	3.7	3.5	6.5	15.4	5.7	3.9	14.6	13.1
Rotterdam ..	1.0	0.4	2.0		1.8	2.3	5.8	8.4
M	5.4	4.9	5.2	10.2	6.7	4.6	13.6	12.4

atrophic among the internees in the camps in unoccupied France in 1941 and 1942. A dry and inelastic skin was reported from Holland. The incidence of dry skin observed during street clinic examinations in western Holland is summarized in Table 125 (see also Table 129). The quantitative data indicate a much lower incidence of dry skin than one would be led to believe from the more general reports.

Fliederbaum *et al.* (1946) described the skin of the Warsaw famine victims as dry, rough, and inelastic, with irregular cornification and an absence of the usual pilomotor reflexes. Among the children the skin was dry in all cases with some desquamation or, in severe cases, exfoliation (Braude-Heller *et al.*, 1946).

With only a single exception (Graham, 1946), the reports of conditions among the American and British internees and prisoners of war in the Japanese camps do not list dry, rough, scaly, and inelastic skin as prominent findings. It is not possible to determine now whether these skin changes actually did not develop in the Orient or whether they were simply not considered significant.

Some histological evidence is available in the literature to account for at least the thin, wrinkled, and inelastic appearance of the skin during semi-starvation. Formad and Birney (1891) found no subcutaneous adipose tissue present in a case of death from starvation. In Marfan's (1921) case of uncomplicated athrepsia, the subcutaneous fat was almost completely gone and in its place was a loose network of fibers. Degeneration of the adipose tissue was observed by Lubarsch (1921) in 13 cases of "pure" malnutrition. Nicolaeff (1923) reported histological findings from 19 typical cases of death from starvation in children during the Russian famine of 1921–22. The subcutaneous adipose tissue was gone in all cases, and the skin, which in normal children is from 1 to 1.5 mm. thick, had a thickness of from 0.5 to 1.0 mm.

Follicular Hyperkeratosis and Folliculosis

In the recent nutritional and medical survey reports reference has frequently been made to follicular hyperkeratosis and folliculosis as two of the skin changes associated with undernutrition. Follicular hyperkeratosis is recognized clinically as small, hard, elevated nodules around the hair follicles that give the skin a "nutmeg grater" texture. The hair follicles may be filled with keratotic plugs which can be squeezed out, leaving small holes in the skin. It is most often found on the extensor surfaces of the upper arms and thighs but may be present on the back and lateral surfaces of the trunk.

No description of "folliculosis" has been given in the nutritional and medical survey reports. It is probable that by folliculosis is meant a relative prominence of the hair follicle — "follicular pouting" or "permanent goose flesh" — due to the thinning of the epidermal, dermal, and subcutaneous layers of the skin. Actually it may represent an early and mild form of follicular hyperkeratosis. In some of the reports, however, the impression is given that folliculosis has been confused with folliculitis, which is defined as an "inflammation of the [hair] follicle or follicles" (Dorland, 1945). Unfortunately the literature on skin changes in semi-starvation contains no histological data pertaining to folliculosis.

In a series of 561 children and adults examined by Robinson, Janney, and Grande (1942b) in Madrid, Spain, during the summer of 1941 when food intakes were below normal, 72 (12.8 per cent) had follicular hyperkeratosis. The lesions were usually on the lateral surfaces of the arms and thighs and tended to occur more frequently in certain families. Unfortunately the relationship of the lesions to the state of vitamin A nutrition was not evaluated by laboratory methods; vitamin A intakes were estimated to average 3852 ± 273 I.U. per day. Of the 561 individuals, 13 complained of night-blindness, but the incidence of follicular hyperkeratosis among the 13 was not given.

Hyperkeratosis occurred frequently during the siege of Leningrad, and it was suggested that hypovitaminosis A may have been a causative factor even though clear-cut signs of vitamin A deficiency (xerophthalmia) were rarely seen (Garshin, 1943). A dry, horny, goose-fleshy type of dermatosis was reported by Youmans, Patton, and Kern (1943a) among the inhabitants of rural Tennessee. No relationship to vitamin A nutrition was established, however. Follicular hyperkeratosis was present in 32 per cent of the group of undernourished German prisoners of war studied by Davidson, Wilcke, and Reiner (1946). There were no eye changes suggestive of vitamin A deficiency; the degree of keratosis was roughly proportional to the weight loss.

Toward the end of World War II a 78 per cent incidence of follicular hyperkeratosis was reported among the children in Belgium (Macrae, 1944) and a 38
per cent incidence in the people of Rotterdam, Holland (Stare, 1945). These
figures appear extraordinarily high in view of the nutritional and medical survey
data from western Holland and the absence of severe undernutrition in Belgium. The incidence of follicular hyperkeratosis and folliculosis for 4 cities of
western Holland are presented in Tables 126 and 127.

TABLE 126

INCIDENCE OF FOLLICULAR HYPERKERATOSIS as percentages of people
examined in street clinics during nutritional and medical surveys in 4
western Holland cities in 1945, grouped according to age
and sex (Surveys, 1945).

| | Age Group (yrs.) | | | | | | | |
| | 1–12 | | 13–18 | | 19–59 | | Over 60 | |
	Male	Female	Male	Female	Male	Female	Male	Female
Amsterdam	1.3	5.5	4.7	18.7	0.5	3.4	0	0
Utrecht	2.5	1.4	3.8	16.3	0.8	6.5	0	0
Delft		1.4	6.7	6.7	1.1	1.2	2.7	9.1
Rotterdam	0.5				0.8	0.3	0.3	
M	1.4	2.8	5.1	16.8	0.8	2.9	0.8	3.0

TABLE 127

INCIDENCE OF FOLLICULOSIS as percentages of people examined in street
clinics during nutritional and medical surveys in 4 western Holland
cities in 1945, grouped according to age and sex (Surveys, 1945).

| | Age Group (yrs.) | | | | | | | |
| | 1–12 | | 13–18 | | 19–59 | | Over 60 | |
	Male	Female	Male	Female	Male	Female	Male	Female
Amsterdam	19.3	24.6	23.2	59.9	7.7	29.1	1.4	2.6
Utrecht	22.2	12.9	13.3	45.6	2.7	23.2	0	0
Delft	13.9	9.3	19.6	30.5	16.9	22.7	14.6	18.8
Rotterdam	12.0	13.2	26.0	32.7	12.2	20.8	7.7	2.9
M	16.9	15.0	20.5	42.2	9.9	24.0	5.9	6.1

Follicular hyperkeratosis was actually rather rare except in girls from 13 to
18 years of age (see Table 126) in the 4 large cities of western Holland. In the
6 smaller cities in Holland only 1.1 per cent of those examined had follicular
hyperkeratosis (see Table 129). Folliculosis was frequently present in Holland
(see Tables 127 and 129), and there appeared to be a sex and age relationship.
The incidence of folliculosis was more than twice as high in the females as in
the males in both the 13–18-year and the 19–59-year age groups. In both sexes
the incidence was considerably less among the adults than among the adolescents.

Apparently follicular hyperkeratosis and folliculosis did not occur among the internees and prisoners of war in Japanese camps. Butler *et al.* (1945) and Carroll (1945) specifically mention the absence of the two skin disorders while the numerous other reports contain no reference to the subject.

Although vitamin A is important in normal skin metabolism, its role in the development of follicular hyperkeratosis in semi-starvation is far from established. In fact, there is some doubt whether follicular hyperkeratosis is a clinical entity peculiar to semi-starvation. After a critical review of the evidence, Stannus (1945) concluded that follicular hyperkeratosis was probably identical with keratosis pilaris — a skin disorder commonly seen in childhood and adolescence even when food intakes are normal. Whatever may be the clinical status of follicular hyperkeratosis, "it seems questionable whether vitamin A deficiency should be looked upon as [the] specific cause, though under certain circumstances it may be a factor in [the] causation" (Stannus, 1945).

Pigmentation of the Skin

"Our skin was black like an oven because of the terrible famine," Lamentations of Jeremiah 5:10.

The presence of an abnormal pigmentation is a characteristic skin change in semi-starvation. The pigmentation is brownish in color — a deeper brown than ordinary sun tan — and is generally present around the mouth and eyes and on the malar prominences, although it may be more widely distributed to the hands, arms, and even the trunk.

During the Irish famine of 1847 the skin became pigmented and had a dirty appearance. Donovan (1848) suggested that the pigmentation was actually an abnormal secretion produced by the altered metabolism of the skin. A grayish-brown discoloration appeared on the exposed areas of the skin in the under-nourished people of Germany during World War I (Maliwa, 1918). Richet and Mignard (1919) reported that pigmentation of the hands and face, similar to the "melanodermia of vagabonds," was present in a group of French prisoners of war who for 5 months had been doing hard physical work on 2000 Cal. per day. It is interesting that these authors exhibited no surprise over the appearance of the pigmentation in the prisoners and that they had observed a similar pigmentation in vagrants who at best eke out a precarious existence with periods of poor food and with miserable living conditions. Small, strongly pigmented areas that were sharply differentiated from the surrounding paler skin were frequently observed on the scrotum of cases of war edema (Reach, 1918).

The pigmentation reported by Enright (1920) during the pellagra outbreak in Egypt was probably not the same type as that seen in uncomplicated semi-starvation. Boenheim (1934) found abnormal pigmentation of the skin in 77 patients in Berlin who had no specific disorders except loss of body weight. Daily caloric intakes ranged from 1483 to 2156 Cal. A brownish skin pigmentation developed in the children during the siege of Leningrad, when the food intake was low and the general living conditions were bad (Efimova and Elpersin, 1944).

At the Belsen prison camp, where many thousands were on the verge of death from starvation, some degree of pigmentation of the skin was present in most of the internees (Lipscomb, 1945; Pollack, 1945). The pigmentation became progressively more intense as the degree of starvation increased. In a group of German prisoners of war who had lost from 10 to 13 kg. of body weight within a few months, there was some degree of deep brown pigmentation on the exposed parts of the body; when the food intake of these prisoners was increased, the pigmented skin peeled off, leaving a light-colored skin underneath (Davidson, Wilcke, and Reiner, 1946). A different degree or kind of pigmentation was reported by Leyton (1946) in the Russian prisoners of war at Tost, Germany, who had been living on 1611 Cal. per day with forced labor. In general the skin was normal except for a yellowish discoloration which slowly faded during refeeding.

A dun-colored pigmentation, particularly of the face, was frequently observed in western Holland during the spring of 1945 (Burger et al., 1945; Surveys: Amsterdam, Rotterdam, Delft, and Utrecht, 1945). The incidence of the pigmentation, as found at street clinics in the 4 cities of western Holland and in 6 other smaller cities in Holland, is presented in Tables 128 and 129. Of the people examined at random, from 22.0 to 51.2 per cent had suborbital pigmentation. A sex and age trend in the incidence of pigmentation was noted. The incidence was higher in the females than in the males for all ages above 12 years. In the males there was a progressive decrease in incidence with increasing age, but for the females the incidence remained quite constant except in those over 60 years of age.

TABLE 128

INCIDENCE OF SUBORBITAL PIGMENTATION as percentages of people examined in street clinics during nutritional and medical surveys in 2 western Holland cities in 1945, grouped according to age and sex (Surveys, 1945).

| | Age Group (yrs.) | | | | | | | |
| | 1–12 | | 13–18 | | 19–59 | | Over 60 | |
	Male	Female	Male	Female	Male	Female	Male	Female
Amsterdam	48.0	48.3	48.1	42.2	34.8	39.9	26.8	33.7
Utrecht	51.7	42.9	43.3	60.2	28.8	61.5	17.2	34.9
M	49.9	45.6	45.7	51.2	31.8	50.7	22.0	34.3

TABLE 129

AVERAGE INCIDENCE OF VARIOUS SKIN ABNORMALITIES as percentages of people examined during the rapid nutritional and medical surveys in 6 cities in Holland in 1945. The cities were Haarlem, Nunspeet, Hilversum, Dordrecht, Alkmaar, and Amersfoort (Surveys, 1945).

| | Skin Disorder | | | | | |
	Pallor	Dry Skin	Cracked Skin	Folliculosis	Follicular Hyperkeratosis	Pigmentation
Incidence (%)	11.0	12.8	13.4	14.5	1.1	34.0

At the end of World War II Arzt (1947) saw 80 cases of skin pigmentation in undernourished persons in Vienna. The patients were mostly women between the ages of 35 and 50 years who were concerned about the cosmetic effect. The pigment appeared first in flecks, which later became confluent, and was most marked in exposed areas, notably the face. Arzt considered this to be the same kind of hunger pigmentation seen by Riehl (1917) in World War I. It was definitely not a pellagroid pigmentation. It is entirely possible that this was the same type of pigmentation frequently seen in Canadian troops imprisoned by the Japanese after the fall of Hong Kong (December 25, 1941) and termed "skin pellagra" (Crawford and Reid, 1947). An entirely different "skin pellagra," with red-purple localized coloration especially on the skin exposed to the sun (but without other signs of pellagra), was common among the civilian internees in the Singapore internment camp (Landor, 1948). The disorder responded readily to niacin therapy.

In the 492 cases of death from "pure" starvation in the Warsaw Ghetto Stein and Fenigstein (1946) found marked brownish pigmentation in 13.8 per cent of the edematous corpses and in 19.0 per cent of the dry cachexia cases. Histologically the skin showed a large amount of melanin and many chromatophores. A dirty brown pigmentation, somewhat like Addison's disease but of different distribution, was also reported by Fliederbaum et al. (1946). The pigmentation was believed not to be a result of the starvation alone but a combined result of starvation, trauma, ectoparasites, and skin infections. The majority of the starving Warsaw children also showed more or less extensive brown patches of pigmentation of the skin, particularly on the back, nape of the neck, abdomen, and on areas where the clothing had rubbed the skin; especialy striking was the pigmention around the scars. In a few cases the entire body was pigmented (Braude-Heller et al., 1946). The same tendency toward pigmentation was reported from the Dutch East Indies concentration camps (Netherlands Red Cross Feeding Team, 1948).

It is not known whether any etiological factors other than low caloric intake and poor living conditions are important in the development of the pigmentation. The pigmentation is not influenced by ascorbic acid and is helped only slightly by niacin. The general lack of signs of pellagra in western Holland, where the incidence of pigmentation was so high, would make it unlikely that the pigmentation was a pellagrous dermatitis; as a matter of fact many of the observers noted that it was quite unlike that seen in pellagra. Musselman (1945), however, observed that the pigmentation of the malar eminences of the prisoners in Cabanatuan responded to niacin and to yeast.

Skin Infections

Furunculosis, scabies, impetigo, skin ulcers, and sores are always prevalent in famine and war-torn areas (International Economic Conference, 1919; Richet and Mignard, 1919; Rubner, 1919; Besançon, 1945; Ellis, 1945; Letterman General Hospital, 1945; Debray et al., 1946; Gottlieb, 1946; Jokl and Kloppers, 1946; and others). In the prison camp at Tost, Germany, where housing, fuel, clothing, and washing facilities were satisfactory, the incidence of skin infections was not abnormally high even though caloric intakes were low (Leyton, 1946). Increases in the incidence of such skin infections have been attributed mainly to the lack of warm water and soap and inadequate shelter and clothing; they do not seem to be primarily nutritional skin disorders but are secondary to the general economic collapse and the breakdown of sanitary and medical facilities that always occur when famines and wars strike.

Acne and Sweat Vesicles

Other rather unusual skin changes have been occasionally observed in semi-starvation. Two reports from Japanese prison camps have indicated that during the periods of serious food restrictions most cases of acne disappeared (Carroll, 1945; Letterman General Hospital, 1945). After liberation, when the food intake was restored to normal, the acne returned to its pre-starvation form. It has frequently been assumed that the common acne of adolescence is in some way related to the type and quantity of food eaten and to a general shift in hormone balance.

The Netherlands Red Cross Feeding Team (1948) observed that in the Dutch East Indies concentration camps, as in Holland, both eczema and psoriasis were either improved or dormant during the period of undernutrition. After liberation and return to good food, these skin conditions often got worse. But in both the Netherlands and the East Indies semi-starvation was frequently productive of unexplained itching and perspiration, often in paroxysmal attacks.

Small vesicles filled with sweat on the shoulders and upper trunk were observed by Davidson, Wilcke, and Reiner (1945) in a group of undernourished German prisoners of war. The vesicles were easily broken but a new crop would soon appear if the men were working or the temperature was high. These authors concluded that the vesicles were probably due to faulty desquamation which resulted in a covering of the openings of the sweat glands by the outer layers of the skin.

Richet and Mignard (1919) found small vesicles filled with blood on the skin surface in a group of undernourished French prisoners of war. A slightly different condition was reported by Burger, Sandstead, and Drummond (1945) in western Holland. Petechiae and superficial hemorrhages occurred frequently in the skin of the hands and face. The hemorrhages slowly disappeared without leaving any discoloration.

Changes in the Hair

Qualitative and quantitative changes in the hair have occasionally been reported to occur in semi-starvation. Enright (1920) observed a thinning of the hair among semi-starved, pellagrous Turkish prisoners in Egypt. In the Russian famine of 1921–22, Ivanovsky (1923) reported that the hair grew slowly, fell out prematurely, and rapidly became gray. The growth of the nails was retarded. Dry hair has been noted in anorexia nervosa (Berkman, 1930). During the siege of Leningrad the hair became dry and rapidly fell out, and the nails were fissured (Garshin, 1943). Dry and staring hair was frequently observed in the nutritional survey in Newfoundland (Adamson et al., 1945). At Belsen there was a loss of the pubic hair, and the hair on the face became soft and silky (Pollack, 1945).

A different type of change in the hair during starvation was reported by Curran (1880). Without giving a reference, he quoted from a "paragraph in an American paper" relating a remarkable "false growth caused by decay and hunger" of body hair on the face and arms of starving children. Curran had never seen anything like this but was informed by Reverend Canon Bourke that he had seen hundreds of such cases in Ireland.

Among the Warsaw famine victims there was a pronounced development of a downy hair all over the body. The head hair became coarse and rough, and hair was lost from the head, axilla, and pubic areas (Fliederbaum et al., 1946). An increase in the hair on the face and nape of the neck (lanugo) was frequently noted in the children (Braude-Heller et al., 1946). Graying of the hair was noted among both males and females in some prison camps (Pevný, 1947).

Clinical Appearance of Skin and Hair — Minnesota Experiment

Observations on the appearance and texture of the skin were included in the clinical examinations conducted by Drs. V. P. Sydenstricker, R. M. Wilder, J. B. Youmans, M. B. Corlett, and the Staff of the Laboratory of Physiological Hygiene during the twenty-third week of semi-starvation and the sixth week of rehabilitation. Skin biopsies, which were taken at 24 weeks of semi-starvation (S24) and 10 weeks of rehabilitation (R10) on 6 of the subjects showing definite clinical changes, were examined for histological alteration by Drs. H. Montgomery and O. E. Okuly of the Department of Dermatology, Mayo Clinic, Rochester, Minnesota.

A dry and scaly skin, mostly representing only a mild degree of abnormality, was reported for 15 of 31 subjects during the twenty-third week of semi-starvation (see Table 130). It was not limited to the exposed parts of the body but was frequently observed on a large part of the trunk, legs, and arms. The degree of dryness and scaliness of the skin became more marked as semi-starvation progressed into the hot summer months of June and July. The skin was not chapped

TABLE 130

SKIN AND HAIR. The number and percentage of the 31 subjects in whom the skin and hair on clinical examination were classified as showing mild to moderate changes, questionable changes, or no comment at 23 weeks of semi-starvation (Minnesota Experiment).

	Mild to Moderate Changes		Questionable Changes		No Comment	
	N	%	N	%	N	%
Skin dry and scaly	15	48.4	1	3.2	15	48.4
Follicular Hyperkeratosis..	24	77.4	1	3.2	6	19.4
Folliculosis	4	12.9	2	6.5	25	80.6
Skin pigmentation	19	61.3	1	3.2	11	35.5
Dry and staring hair	20	64.5	3	9.7	8	25.8

or cracked as would be expected if the change has been produced by exposure to wind, cold, or sun. The general appearance was one of faulty desquamation with an accumulation of cornified epithelial cells on the outer surface of the skin. Brisk rubbing of the skin would dislodge some of the scales without appreciably altering the appearance of the skin. The dislodged white flaky scales looked like dandruff.

The dry and scaly skin was probably an expression of an altered skin metabolism as a result of a decreased blood flow through the skin. There was indirect evidence that the blood flow through the skin had been decreased during semi-starvation. The skin was cold to the touch, and in many of the subjects a mild degree of cyanosis of the lips and nail-beds was present. Even during the warm weather of July all the subjects complained of feeling cold and used two or three woolen blankets on their beds at night; they wore jackets during the day. As semi-starvation continued the peripheral veins decreased in size, and it became progressively more difficult to obtain blood samples (15 cc. of blood from the antecubital vein) without using prolonged stasis or massaging the muscles.

A mild to moderate degree of follicular hyperkeratosis was recorded for 24 of 31 subjects examined at S23 (see Table 130). It was limited chiefly to the extensor surfaces of the arms and legs, but in a few subjects it was present also on the lateral surfaces of the trunk and on the back. The skin immediately surrounding the hair follicles was slightly elevated and hard, producing a "permanent goose flesh" appearance and a "nutmeg grater" feel of the skin. In a few cases the hair follicles appeared to be plugged and were pigmented, but there was little reddening or other indication of infectious processes.

The etiology and the possible role of vitamin A in the follicular hyperkeratosis seen in famine areas have been discussed above. A calculation of the vitamin A content of the semi-starvation diets used in the Minnesota Experiment gave average daily vitamin A intakes of 1810 I.U., mainly in the form of carotene. This vitamin A intake must be considered low, particularly since the low fat content of the diet may have interfered with the absorption of all the carotene in the digestive tract. Plasma vitamin A content was determined on 18 subjects at the end of semi-starvation and gave average values of 171.1 I.U. per 100 cc. of blood as compared to 167.8 I.U. in the control period. The complete absence of any of the classical signs of vitamin A deficiency in the Minnesota Experiment, and in many of the famine areas, would seem to indicate that lack of vitamin A did not play a primary role in the development of the follicular hyperkeratosis seen in the Minnesota Experiment.

At 23 weeks of semi-starvation, 19 of the 31 subjects had developed a brownish patchy pigmentation of the skin (see Table 130). In most cases the pigmentation was chiefly around the mouth, under the eyes, and on the malar eminences. The pigmentation was a darker brown than ordinary sun tan and was easily recognized even in those subjects who were deeply sun-tanned. No information on the exact character of the pigmentation or on its mechanism of formation was obtained. It has been suggested that the pigmentation of semi-starvation is the result of a niacin deficiency, though most of the observers who have seen it insist that it is totally unlike the pigmentation in pellagra. The Minnesota Experiment semi-starvation diets contained an average of 20.7 mg. of niacin per day, which would seem to rule out niacin deficiency as a cause.

Mild "folliculosis" was observed in 4 of the 33 subjects (see Table 130). In none of the subjects was there any scabies, impetigo, or other infectious skin disorders. This is in sharp contrast to the experiences in famine and war-torn areas. The differences are, however, readily explainable. In the Minnesota Experiment living conditions were excellent; there was always ample hot water and soap, showers, clean linen and clothing, laundry and sanitary facilities. Personal cleanliness was maintained throughout the 6 months of semi-starvation.

Two of the subjects in the Minnesota Experiment had a moderate degree of acne which had been present for years. The acne progressively cleared up while the subjects were on the semi-starvation diet and had almost entirely disappeared by the end of the 24 weeks of semi-starvation. During the 12 weeks of rehabilitation the acne reappeared, much to the subjects' disappointment and disgust.

Toward the end of the semi-starvation period little vesicles filled with sweat were frequently observed under the outer skin layer of all the subjects whenever they did work that provoked sweating or were exposed to the hot sun. The vesicles, which appeared mainly on the back, chest, and upper arms, were 2 to 4 mm. in diameter and could be easily broken. They were never observed on the face or legs. Although no counts were made, it was estimated that more than a thousand of the vesicles may have been present at a time. The vesicles were probably caused by a plugging of the sweat duct openings and the prevention of the free flow of sweat to the surface as a result of the faulty desquamation of the outer skin layers.

The head hair was described as "dry, lusterless, unruly and staring" in 20 of the 31 subjects at 23 weeks of semi-starvation (see Table 130). About a third of the subjects remarked that their hair was falling out at an abnormally rapid rate. A comparison of the anthropometric photographs taken before and at the end of semi-starvation lends some credence to the alleged changes of the hair in some of the subjects. No increase in the rate of graying of the head hair or changes in the texture of the pubic hair were observed.

Histology of the Skin — Minnesota Experiment

The skin biopsies taken on 6 of the subjects disclosed less serious histological abnormalities than might have been expected from the clinical appearance. A brief résumé of the major histological changes reported by Drs. H. Montgomery and O. E. Okuly are presented for 6 subjects. Photomicrographs of representative regions in the biopsies are given in Figure 47.

SUBJECT 102 (Figure 47A). The biopsy specimen was taken from the extensor surface of the left arm where the skin clinically showed mild roughening and dryness without any signs of follicular hyperkeratosis. Histologically there was a relative and absolute hyperkeratosis with keratotic plugging of some of the hair follicles. A few of the basal and prickle cells showed intracellular edema result-

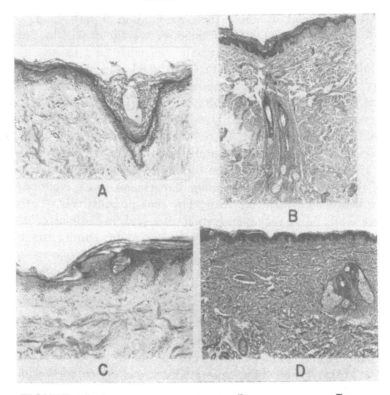

FIGURE 47. Photomicrographs of Representative Regions
of the Skin Obtained by Biopsy. A. Biopsy from subject No. 102
taken from the extensor surface of the left arm which showed
mild dryness or roughness at the end of semi-starvation. B. Biopsy
from subject No. 101 taken from the mid-back area over the
spine which exhibited grade 1 follicular hyperkeratosis at the end
of semi-starvation. C. Biopsy from subject No. 122 taken from
the extensor surface of the right arm where the skin showed grade
3 follicular hyperkeratosis and xerosis at the end of semi-starvation.
D. Biopsy from subject No. 122 taken at 10 weeks of rehabilitation showing relatively normal skin. (Minnesota Experiment.)

ing in vascuolization of the cells. The epidermis was atrophic in some areas with
flattening of the rete ridges. Some of the hair follicles were atrophic and the sebaceous glands were absent or small in size. There was dilatation of the capillaries and some edema in the cutis, with a resulting fragmentation of some of
the coarser elastic fibrils.

SUBJECT 101 (Figure 47B). The biopsy was taken from a mid-back area over
the spine that exhibited clinically grade 1 follicular hyperkeratosis. The sections
showed relative and absolute hyperkeratosis, keratotic plugging of the hair follicles, atrophy of the hair follicles, and the presence of only a few rudimentary
sebaceous glands. There was some vacuolization of the epidermal cells. The rete
ridges and papillary bodies were normal. The cutis was thickened and edematous and showed some fraying of the coarse elastic fibrils.

SUBJECT 122 (Figure 47C). The biopsy was taken from the extensor surface of the right arm where the skin clinically showed follicular hyperkeratosis and xerosis grade 3. Clinically this subject showed the most pronounced changes of the group. Histologically there was relative and absolute hyperkeratosis, keratotic plugging of the hair follicles, vacuolization of some of the epidermal cells, areas of atrophy of the epidermis with loss of the rete ridges, atrophy of the hair follicles, only remnants of the sebaceous glands, and atrophy of some of the sweat glands and ducts. The cutis showed edema, dilatation of the capillaries, and fragmentation of some of the coarse elastic fibrils.

Subject 108. The biopsy was taken from the flexural surface of the flank in an area of grade 2 follicular hyperkeratosis (in a most unusual location). The changes were much the same as those observed in the three preceding cases — relative and absolute hyperkeratosis, keratotic plugging of the hair follicles, some atrophy of the epidermis with loss of rete ridges, atrophy of some of the hair follicles and sebaceous glands, some changes in the sweat glands, edema of the cutis, and fraying of the coarse elastic fibrils.

Subject 111. The biopsy was taken from the extensor surface of the left arm where clinically there was no appreciable follicular hyperkeratosis or xerosis. The sections showed some atrophy of the epidermis and vacuolization of the basal and prickle cells. There was very slight relative hyperkeratosis. Hair follicles, sebaceous glands, sweat glands, and elastic tissue were normal. The sections would have easily passed for normal skin had the dermatologist not been forewarned.

Subject 20. The biopsy was taken from the extensor surface of the left shin where the skin was dry and scaly but not truly xerotic. The sections showed atrophy of the hair follicles, which probably was due to trauma from garters. There were many arrectores pilarum muscles without associated hair follicles. Sebaceous glands were absent. No other significant changes were observed.

The clinical and histological changes in the skin correlate quite well. The dry skin can be explained on the basis of the atrophic sebaceous glands. The absolute hyperkeratosis is expressed clinically as a scaly skin. The follicular hyperkeratosis (keratosis pilaris) is a keratotic plugging of the hair follicles. The atrophic changes seen in some of the hair follicles may account for the change in texture and loss of the hair. Of special interest is the fraying of the coarse elastic fibrils in the cutis because it confirms the impression that there was a loss of elasticity of the skin. The degenerative changes in the basal and prickle cells and the loss of the rete ridges were only minor but may be sufficient to cause some loss of regenerative capacity and to slow down normal skin growth.

Skin and Hair in Rehabilitation — Minnesota Experiment

A striking improvement in the clinical condition of the skin occurred during the first 6 weeks of rehabilitation (see Table 131). The skin became less dry and scaly in 93.3 per cent of the 15 who had shown definite changes during semistarvation. Follicular hyperkeratosis improved in 87.5 per cent; it was still present in a mild form in only 3 of the 24 subjects in whom it had been reported at S23. The skin pigmentation became much less apparent in 16 (84.2 per cent) of the 19 subjects in whom it was present at S23, with only 3 of the group still

TABLE 131

Skin and Hair. The number and percentage of the 32 subjects in whom the skin and hair on clinical examination were classified as showing mild changes, questionable changes, or no comment at 6 weeks of rehabilitation (Minnesota Experiment).

	Mild Changes		Questionable Changes		No Comment	
	N	%	N	%	N	%
Skin dry and scaly	1	3.1	5	15.6	26	81.3
Follicular hyperkeratosis ...	3	9.4	9	28.1	20	62.5
Folliculosis	1	3.1	4	12.5	27	84.4
Skin pigmentation	3	9.4	6	18.8	23	71.8
Dry and staring hair	4	12.5	11	34.3	17	53.2

having unmistakable signs of pigmentation. The condition of the hair was noticeably improved in 80.0 per cent; it had more sheen and was much less rebellious and staring.

The histological examination of second skin biopsies taken on the 6 subjects at 10 weeks of rehabilitation showed little that could be interpreted as abnormal (see Figure 47D). Subject 122 still had epidermal changes similar to those at the end of semi-starvation, but much less marked. In this case there was no evidence of regeneration of any of the dermal appendages, and the elastic tissue was absent around the atrophic hair follicles. In the other 5 biopsies the skin would, for all practical purposes, be considered normal.

TABLE 132

Skin and Hair. Number of times skin and hair changes were classified as mild or moderate (+) or questionable (?) during clinical examinations in the 4 caloric groups: Basal (Z), +400 (L), +800 (G), and +1200 (T) at the end of 23 weeks of semi-starvation (S23) and at 6 weeks of rehabilitation (R6) (Minnesota Experiment).

	S23		R6	
	+	?	+	?
Z	21	1	6	5
L	22	2	1	15
G	18	2	3	9
T	21	3	2	6

The level of caloric intake during the first 6 weeks of rehabilitation appeared to have some effect on the disappearance of the skin and hair changes (see Table 132). The clinical designation of the degree of severity of the skin changes was, of course, mainly subjective, so not too much reliance can be placed on the small differences.

Protein and vitamin supplementation did not exert any advantageous effects on the recovery of the skin and hair changes during the first 6 weeks of rehabilitation. The total number of skin changes still remaining at R6 and classified as mild or questionable, respectively, were 7 and 16 for the vitamin-supplemented group and 5 and 19 for the vitamin-unsupplemented group. In the group receiving extra proteins there remained 8 changes classified as mild and 17 as questionable, while in the group that received no extra protein there remained 4 changes classified as mild and 18 as questionable.

CHAPTER 14

Blood Morphology, Including Bone Marrow

PALLOR is as characteristic of starvation as is the loss of body weight, and the pale, sallow appearance of the starved is generally attributed to the presence of an anemia. Extensive observations on the hematological changes that occur during acute starvation and famines are often complicated by processes, such as shifts in body hydration and concomitant diseases, which in themselves may radically alter the composition of the blood. Variations in the degree of hydration will produce a change in the concentration of hemoglobin and the red blood cells even though the total amounts may remain constant. Many diseases and physiological processes are known to influence the total and differential leucocyte counts. Consequently, reports of alterations in blood composition during acute and semi-starvation must be viewed critically to eliminate, as far as possible, factors other than the restricted caloric intakes which may be responsible for the changes.

Erythrocytes in Acute Starvation — Qualitative Changes

Alterations in the appearance of the red blood cells have been recorded in some of the earlier observations during acute starvation in man. Curtis (1882) made a systematic and detailed study of the erythrocytes in one subject during a 40-day fast and reported the presence of macrocytes, microcytes, and crenated and distorted red blood cells. The appearance of numerous microcytes was observed by Senator (1887) in a case of almost complete starvation. Charteris (1907), however, observed no change in the size, shape, or staining qualities of the red blood cells in one subject during a 14-day fast. According to Dr. Wiles' report (cited by Ash, 1914) no anisocytosis or degenerative red blood cells were present in the blood of Dr. Gayer on either the eighteenth or the thirtieth day of his fast. Ash (1914) has attributed the appearance of abnormal red cells in both humans and animals during acute starvation to poor hematological techniques. No change in the corpuscular volume of the red blood cells occurred from the fourth to the twenty-sixth day of a 27-day fast in the professional faster studied by Berri and Weinberger (1927).

Erythrocytes in Acute Starvation — Quantitative Changes

Rather divergent results have been reported on the effects of acute starvation on the red cell number and hemoglobin content of the blood of man. Malassez (1874) recorded the case of a boy with an esophageal stricture in whom the red cell count had fallen to 2.6 million per cu. mm. of blood one week before death from starvation. In an almost identical case the red cell count was 4.9 million per cu. mm. of blood at the time of death (Brouardel, 1876). In neither of these cases was an indication of the state of body hydration given. Senator (1887) concluded that there might have been a slight decrease in red cell number in his patient who received only small amounts of wine and milk for 7 weeks. Only a slight, if any, over-all decrease in the red cell number and hemoglobin content was observed by Ash (1914) during a controlled 45-day fast. Berri and Weinberger (1927) reported red cell counts and hemoglobin values for the fourth and twenty-sixth days of a 27-day fast. The red cell count dropped from 4.9 to 3.8 million per cu. mm. of blood, and the hemoglobin decreased from 90 to 87 per cent. In view of the fact that there was no change in the corpuscular volume of the red blood cells, it is surprising that a decrease of 1.1 million in red cell count was accompanied by only a 3 per cent decrease in hemoglobin concentration.

If no change in body hydration takes place during acute starvation in man, the red cell number and hemoglobin content of the blood appear to remain quite constant (Hermann, 1888; Luciani, 1890; Renosse and van Wilder, 1903; Gordon, 1907; Ash, 1914). However, large day-to-day variations in red cell numbers have been reported. In Curtis' (1880) subject the red cell count ranged from 6.8 to 2.1 million per cu. mm. of blood during a 45-day fast. On the twentieth day the subject felt quite ill and had a red cell count of 6.5 million; he took an enema causing a bowel movement, and on the twenty-second day the red count was 2.1 million. On the twenty-third day the count had risen to 5.4 million. A 2½-hour excursion on the lake on the fortieth day of the fast resulted in a drop of 1 million in the red cell count. A daily red cell count variation of about 1 million was observed in Levanzin (Ash, 1914) even though the daily water intake (900 cc.) and the conditions for blood sampling were carefully controlled. After reviewing the literature on blood changes during acute starvation in man and animals, Ash (1914) concluded that "the red cells and hemoglobin are particularly resistant, though in long fasts there is no doubt a slight loss." The slight loss in red cells and hemoglobin may be masked in many cases by dehydration.

Erythrocytes in Semi-Starvation before World War II

The characteristic pale appearance of the people in famine areas has led both past and present observers to comment frequently on the high incidence of anemia. Anemia has, in fact, been listed as one of the 3 major contributing causes of death among the severely undernourished. The blood composition in famine and war starvation is often complicated by concomitant diseases. Particularly important in this respect is malaria, which in itself will produce a severe anemia regardless of the adequacy of the dietary intake.

Marked anemia was reported by Donovan (1848) in the Irish famine of 1847, and by Porter (1889) and Aykroyd (1939) in Indian famines. Hayem (1882) observed that in chronic cachexia the loss of total blood mass was greater in proportion than for the body as a whole. The loss in blood mass was often especially rapid in the later stages of the disorders. In chronic inanition in man Grawitz (1895) found a general decrease in the blood proportional to the loss of body weight. A progressive anemia with red blood cell counts below 3 million per cu. mm. of blood has been observed in cases of athrepsia (Thiercelin, 1904). Schlesinger (1903) concluded that the anemias in infantile atrophy were, except in severe cases, mainly due to plasma dilution.

The experiences in World War I indicate a consistent appearance of anemia among the people who were living on restricted caloric intakes. Knack and Neumann (1917) reported hemoglobin concentrations of 70 to 90 per cent of normal with corresponding decreases in red cell numbers (normal color index), and Maliwa (1918) frequently observed pallor and secondary anemia. In a prison camp where the food intake had been limited to 1300 Cal. per day for several weeks many of the internees were anemic (Castaldi, 1918). Red cell counts of 1 to 2 million below normal, with relatively normal color indexes, were common among the undernourished Turkish and German prisoners of war in 1917 (Bigland, 1920). Cases of moderate anemia with hemoglobin values of 70 to 80 per cent have also been reported by Strauss (1915) and Vandervelde and Cantineau (1919).

Lewy (1919) described several hundred Russian prisoners of war as being pale, with some of them livid; a fairly marked anemia was present in 40 to 60 per cent. Out of a group of 24 German soldiers with "hunger edema," the red blood cell number was below 5 million per cu. mm. in 17 (Wassermann, 1918). Maase and Zondek (1920) reported hemoglobin concentrations of 60 to 75 per cent of normal and red cell counts from 3.3 to 4.7 million, with color indexes of 0.73 to 0.96, in hunger edema patients in Germany. Red cell counts from 3.1 to 4.5 million per cu. mm. of blood were observed by Richet and Mignard (1919) among a group of French prisoners of war. The mild to moderate anemia so common among the undernourished in postwar Germany (Reiss, 1921; Lubarsch, 1921) was attributed by Lubarsch to an excessive rate of red cell destruction. The incidence of hemosiderin in the internal organs was high in cases of starvation deaths.

Blood studies were made by Jansen (1918) on 32 cases of war edema. The hemoglobin concentration ranged from 65 to 105 per cent, the red blood cell numbers from 1.5 to 4.5 million per cu. mm. of blood, and the color index from 1.0 to 1.9. The values are summarized in Table 133. On microscopic examination the red blood cells were substantially normal. There were no megaloblasts, no poikilocytes, only occasional anisocytes, and little evidence of polychromophilia. Although the general conclusion that a mild to severe anemia was present is undoubtedly correct, the few unusually low red cell counts and the astonishingly high color indexes cast some doubt on the accuracy of the determinations.

Jansen's (1918) observation that the morphology of the red blood cell remains normal in semi-starvation was substantiated by Rostoski (1929) in 165 Russian prisoners of war who during a hunger strike in 1918 lived on 200 gm. of bread per day for 24 days.

In Madrid, after the Spanish Civil War, 16 per cent of the males and 18 per cent of the females had a hemoglobin concentration of less than 12 gm. per 100 cc. of blood, and the red cell counts were below 4.1 million in one third of those examined. The anemia was a macrocytic hyperchromic type which definitely

TABLE 133

HEMOGLOBIN CONCENTRATIONS, RED BLOOD CELL NUMBER, AND COLOR INDEXES for 32 cases of war edema. The hemoglobin concentrations are given as percentage of normal and the red blood cells in millions per cu. mm. of blood (Jansen, 1918).

Number of Cases	Hemoglobin	Red Cells	Color Index
9	65–75	1.5–3.0	1.3–1.9
20	75–95	3.0–4.0	1.0–1.4
3	95–105	4.0–4.5	1.2–1.3

was not associated with an iron deficiency (Robinson et al., 1942a). The authors suggested that the anemia might have been due to the low protein diets used by the people.

Erythrocytes in Semi-Starvation in World War II — Far East

During World War II the concentration camps, the prisoner-of-war camps, and the occupied areas of Europe and the Far East furnished excellent opportunities to observe the effects of semi-starvation in man. Millions of people were forced, by either necessity or design, to subsist for long periods of time on vastly inadequate rations. In most cases the caloric intake and the time of subsistence on the diet are only approximately known, and little is known of specific diseases that may have contributed to the final clinical condition of the individual. In spite of the obvious inadequacies of the available data, there is a general consistency in the observation that anemia develops in man when the caloric intake is restricted.

The recent reports from the Far East indicate that anemia was common in the prison and concentration camps and in the famine areas. Malaria no doubt was a complicating factor since it was frequently present and since antimalaria therapy was generally inadequate under such conditions. Anemia was common among the victims of the Travancore famine of 1943–44 (Sivaswamy et al., 1945) and became impossible to cure when of long standing. The anemias were thought to be of nutritional origin and in most cases improved considerably on a diet including large amounts of iron, liver, proteins, and thiamine.

Morgan, Wright, and Van Ravenswaay (1946) made hematological examinations on a large group of repatriated American prisoners of war from the Far East and found a mild to severe anemia in 40.8 per cent of the cases. Hemoglobin concentrations were 8 to 12 gm. per 100 cc. of blood in 5 per cent of the cases and 12 to 14 gm. in 35.8 per cent. In a similar group of 750 American prisoners of war taken aboard the U.S.S. Rescue from prison camps in Japan, there were no severe anemias and hemoglobin concentrations were normal or only slightly depressed (Carroll, 1945). Hibbs (1946) reported that blood counts on the prisoners from Bataan routinely showed a secondary hypochromic anemia.

An extensive series of observations were made at Letterman General Hospital in San Francisco (1945) on 3204 American military personnel who had been Japanese prisoners of war. During the 3 weeks or more that had elapsed between release from the prison camps and the laboratory examinations, the men had been receiving a high calorie, high vitamin diet, and in some cases liver extract, iron therapy, chemotherapy, and vermifuges had been administered. The laboratory findings were no doubt affected by the regimen. In the first 1500 cases 52 per cent had a mild anemia with hemoglobin concentrations of 11 to 14 gm. per 100 cc. of blood. Hematocrit values ranged from 27 to 42 cc. per 100 cc. of blood with most of the values being between 34 and 40 cc. The mean corpuscular volume was more than 92 cubic microns with a maximum of 125 cubic microns in 73 per cent of the cases. This was accepted as evidence of a mild macrocytic anemia; the normal range of red cell volume is given as 72 to 110 cubic microns (Maximow and Bloom, 1931, p. 55) with an average of 87 $\pm$ 5 cubic microns (Simmons and Gentzkow, 1944, p. 113). Seven of the cases were classified as a normocytic normochromic anemia and 3 cases as a mycrocytic hyperchromic anemia. In general the red cells were well filled with hemoglobin, and there was a moderate polychromatophilia and a slight anisocytosis. An occasional nucleated red cell was present. The reticulocytes were 0.4 to 2.0 per cent in about half the cases and 2 to 6 per cent in the rest, indicating a good red blood cell regeneration in more than 50 per cent of the cases.

Anemia was common among the British prisoners of war and internees of Japanese prison camps (Gupta, 1946; Price, 1946). The anemias were from moderately severe to severe in degree, and many were macrocytic and hyperchromic with a color index as high as 1.5. Among the 10 cases presented in detail by Gupta, the hemoglobin concentration ranged from 5.0 to 11.0 gm. per 100 cc. of blood, the red cells from 1.7 to 3.7 million per cu. mm. of blood, and the color index from 0.80 to 1.45.

Little information is available at this time on the hematological changes that may have occurred among the Japanese in Allied prison camps. Kark (1946) reported observations on 29 Japanese soldiers captured in Burma who had evidently been isolated from their units and had been living off the land for some time. All showed evidence of weight loss but only 3 were emaciated. The average hemoglobin concentration in the group was 12.0 gm. per 100 cc. of blood, and "a considerable number were obviously anemic." Only 6 had a normal hemoglobin concentration while in 17 the hemoglobin was less than 80 per cent of normal. One of the prisoners, who was very emaciated and had edema of the feet and ankles, had a hemoglobin concentration of 8.9 gm. per 100 cc. of blood (56 per cent of normal).

Erythrocytes in Semi-Starvation in World War II — Europe

Caloric restrictions approaching famine proportions were present particularly in the larger cities of western Holland during the last months of the German occupation. Although food supplies were not abundant during most of the occupation, it was not until December 1944 that the situation became serious. The estimated average body weight loss during the period of acute food deprivation was about 20 to 30 lbs. Hematological examinations were made on random samples of the population by nutritional and medical survey teams that closely followed the liberating Allied armies into the cities. Burger, Sandstead, and Drummond (1945) found the inhabitants pale and anemic. They reported average hemoglobin values of about 11 gm. per 100 cc. of blood (69 per cent of normal). The anemias were mostly of the normochromic type with a color index of about 1. A few were mildly hyperchromic, however, with a color index of slightly above 1. Drummond (1946a) observed that the anemia was especially prevalent among the women and older children.

TABLE 134

AVERAGE HEMOGLOBIN CONCENTRATION in gm. per 100 cc. of blood, for 4 cities in western Holland, grouped according to age and sex (Surveys, 1945).

	Age Group (yrs.)							
	1–12		13–18		19–59		Over 60	
	Male	Female	Male	Female	Male	Female	Male	Female
Amsterdam	13.3	13.3	13.4	13.1	13.8	13.0	12.9	13.0
Rotterdam	12.7	12.2	12.6	12.5	13.8	12.2	12.7	12.0
Delft	12.4	12.2	12.9	12.6	13.4	12.9	13.0	12.3
Utrecht	12.4	12.7	13.1	12.8	13.6	12.5	12.6	13.0
M	12.7	12.6	13.0	12.8	13.7	12.7	12.8	12.6

The 1945 preliminary reports of the nutritional and medical survey teams listed average hemoglobin concentrations that were higher than those given by Burger, Sandstead, and Drummond (1945). The average hemoglobin concentrations in gm. per 100 cc. of blood, grouped according to age and sex, are given in Table 134 for 4 of the major cities in western Holland. The values were for individuals picked at random in street clinics and consequently did not include the severely starved or the inmates of institutions where food supplies were lower than for the general population. Actually the average degree of anemia was not very pronounced. Based on normal adult hemoglobin values of 16 ± 2 gm. per 100 cc. of blood for males and 14 ± 2 gm. for females (Simmons and Gentzkow, 1944, p. 113), the mean hemoglobin concentrations were, in the 19–59 year group, 90 and 85 per cent of normal for the females and males, respectively. Using mean values is, however, a misrepresentation of the degree of anemia in those who were anemic because not all the people had anemia. In the city of Hilversum, Holland, for example, of 102 examined the degree of anemia was recorded as mild in 55, moderate in 4, and severe in 2.

The general population of France fared quite well during World War II. It was only in isolated individual cases or in some internment camps that food intakes reached semi-starvation proportions. Among 11,000 individuals in the internment camps in unoccupied France studied by Zimmer, Weill, and Dubois (1944) the body weight was more than 22 per cent below normal in 32 per cent of the males and in 51.6 per cent of the females. In 63.2 per cent of the males and in 44.8 per cent of the females the body weight was less than 22 per cent below normal. Only a small percentage were overweight. Anemia was present in nearly all who showed evidence of gross body weight loss. The hemoglobin concentration was from 60 to 70 per cent of normal, and red cell counts of 3 million per cu. mm. of blood were common. Hemoglobin concentrations of more than 70 per cent of normal were found in only 5 per cent of the cases. The anemias were of the hyperchromic type in 43 per cent of those examined, with a color index as high as 1.23. A mild to moderate hyperchromic anemia with erythrocyte counts of 2.5 to 3.8 million and hemoglobin concentrations of 55 to 75 per cent was present among the internees at the Gurs prison camp in southern France (Schwarz, 1945).

TABLE 135

RED CELL NUMBER, HEMOGLOBIN, AND COLOR INDEX with group means for 6 cases of hunger edema. The red cells are given in millions per cu. mm. of blood and the hemoglobin in percentage of normal (Nicaud, Rouault, and Fuchs, 1942).

Case Number	Red Cells	Hemoglobin	Color Index
1	2.888	70	1.0
2	4.225	90	1.0
3	3.210	70	1.0
4	3.953	75	0.9
5	4.117	80	1.0
6	3.927	80	1.0
M	3.720	77.5	0.98

Hematological studies on several cases of hunger edema and hunger osteopathies have been reported from the hospitals of France. A red blood cell count of 2.9 million per cu. mm. of blood was reported by Belger (1942) in a 62-year-old woman with Milkman's disease. A mild to moderate anemia was present in the 6 cases of hunger edema, all females aged 48 to 64 years, reported by Nicaud, Rouault, and Fuchs (1942). The hematological data for the 6 cases are presented in Table 135. The anemia must be classified as the normochromic type. A moderate hypochromic anemia was present in the 771 repatriates from political concentration camps in Germany who were admitted to the Salpêtrière Hospital in Paris (Debray et al., 1946). Delafontaine, Guillaume, and Routier (1946), however, reported an increased incidence of hyperchromic anemia among the patients entering hospitals in Paris during the occupation. Lamy, Lamotte, and Lamotte-Barrillon (1948) found red blood cell counts below 3.3 million per cu. mm. in 25 of 28 starvation cases; of these 9 had values between 2.7 and 1.7 million. Hemoglobin concentration was also reduced, but the hemoglobin-cell value was about normal in all but 8 cases, in which it was elevated. These findings again point to the anemia of starvation being of the normochromic or hyperchromic type.

The underfed mine workers from the Island of Elba studied by Di Granati et al. (1947) had been living on a diet that furnished about 1500 Cal. with 70 gm. of protein per day. The estimated body weight loss was from 10 to 15 kg. with some having lost as much as 20 kg. Blood morphology data were given for

8 individuals for before and after a rehabilitation regimen. When first examined the red blood cell number ranged from 2.9 to 4.3 million per cu. mm. of whole blood with an average of 3.5 million. The average Sahli hemoglobin index was 97.0, and the hemoglobin–red cell ratio was 1.42. Polychromasia and anisocytosis were present in all. During the course of rehabilitation the red cell count increased 500,000 per cu. mm., and the hemoglobin–red cell ratio decreased to 1.23.

Anemia ranging from moderate to severe was frequently present in the patients at the Belsen prison camp at the time of the liberation. It was more severe in those with edema than in the "dry" forms; in the latter the anemia may have been masked by hemoconcentration. Lipscomb (1945) found a color index of about 1 and normal-appearing red cells without any evidence of iron deficiency or abnormal red cell maturation. In 21 emaciated men at Belsen the average hemoglobin concentration was 66.5 per cent of normal (Vaughan, 1945). Mollison (1946) made observations on 30 males and 23 females at Belsen about 4 weeks after the liberation. Body weight loss in the group averaged 39 per cent. All patients were anemic with average hemoglobin concentrations of 62.5 per cent and 57.3 per cent of normal in the males and females, respectively. The anemias were normochromic and normocytic, and in the uncomplicated cases recovery of the hemoglobin during refeeding was good without any special iron or liver therapy.

Hematological observations have been reported from other regions of Europe where semi-starvation occurred. Davidson, Wilcke, and Reiner (1946) reported a mild hypochromic anemia among a group of German prisoners of war who had been living for some months on intakes of less than 1000 Cal. per day. In Leningrad during the early months of the siege a microcytic hypochromic anemia developed. The few cases of hyperchromic macrocytic anemia that were seen in the spring and early summer of 1942 were believed by Drazdova (1945) to be of pellagrous origin (Brožek, Wells, and Keys, 1946). In an excellent description of the effects of slow starvation, Leyton (1946) reported a moderate anemia among Russian prisoners of war who were doing hard physical work on 1611 Cal. per day. The hemoglobin concentration ranged from 44 to 105 per cent of normal with an average of 82 per cent (13.1 gm. per 100 cc. of blood, using 16 gm. as normal). About 75 per cent of these men had hemoglobin concentrations below 85 per cent of normal (13.6 gm.).

The data from the Warsaw Ghetto also indicate a moderately severe to severe normochromic or hyperchromic anemia in starvation. Among the children the red cell number was in most cases between 3 and 3.5 million per cu. mm. of peripheral blood (Braude-Heller et al., 1946). Fliederbaum et al. (1946) found a pronounced anemia in all of the 80 starvation cases studied. In the moderately starved the erythrocyte count was around 3 million per cu. mm. of blood but in the severe starvation cases the number was from 1.2 to 1.5 million. Blood hemoglobin concentration was generally 60 to 70 per cent of normal with occasional values as low as 30 per cent. The color index was unity or below. Anisocytosis, microcytosis, and hypochromasia were the rule. The general characteristic of the blood was a hypochromic anemia with normal red cell regeneration, reticulocytes making up 8 to 10 per cent of the red cells in most cases.

In another report from the Warsaw Ghetto Szejnman (1946) recorded, for 32 starvation cases, erythrocyte numbers of 4 to 5 million in 6 cases, 3 to 4 milion in 10 cases, 2 to 3 million in 7 cases, 1 to 2 million in 7 cases, and less than 1 million (570,000 and 690,000) in 2 cases. The color index was above 0.9 in 28 of the 32 cases and higher than 1.0 in 18, again indicating a normochromic or hyperchromic anemia. The reticulocyte number was about normal in those cases where the red blood cell number was between 3 and 5 million per cu. mm. In the more severe anemia cases the reticulocytes were higher.

The evidence is overwhelming that anemia develops during prolonged periods of caloric restriction, and the degree of anemia appears to be related to the extent of starvation. The anemia may be either normochromic or slightly hyperchromic and is not of the iron deficiency type.

Leucocytes in Acute Starvation

Observations on the leucocyte response in man to acute fasts of from one to four weeks' duration present conflicting results. Luciani (1890) reported a leucopenia with a drop in leucocyte count from 14,536 on the first day to 861 on the seventh day of fasting. On the twenty-ninth day the count had risen to 1550. Benedict (1907) observed a progressive decrease in the number of white blood cells during the first week of fasting, after which there was a slight increase that did not reach normal values. An increase in leucocyte number from a control value of 5300 to 14,000 on the sixth day of fast was reported by Charteris (1907). In two subjects Howe and Hawk (1912) found an early increase in the polymorphonuclears followed by a slow decrease to below normal by the end of a 7-day fast. The lymphocyte response was in the opposite direction. Penny (1914) observed a slight leucocytosis during fasting with a relative increase in polymorphonuclear cells and a marked decrease in the lymphocytes.

Ash (1914) carefully followed the leucocyte response in Mr. Levanzin during a 31-day fast. Total and differential leucocyte counts were made about every 48 hours under rather carefully controlled conditions of activity, water intake, temperature, and time of day of taking samples. The total leucocyte count reached 12,400 on the third day of the fast, 8400 on the fourth day, and stayed at that level until the sixteenth day, when it returned to the control value of about 6000. The variations in total count were due entirely to changes in the number of polymorphonuclear cells. The lymphocytic cells remained constant.

In a 27-day fast on one subject Berri and Weinberger (1927) observed only a slight decrease in total leucocyte number but there was a progressive decrease in the proportion of polymorphonuclears from 80 per cent on the nineteenth day to 60 per cent on the twenty-fifth day. The lymphocytes increased from 17 per cent on the nineteenth day to 25 per cent on the twenty-fifth day. Unfortunately no data are given for the first 19 days of the fast. The values reported for the twenty-fifth day of the fast are within the normal range (Maximow and Bloom, 1931).

Leucocytes in Semi-Starvation

Hayem (1882) remarked that in chronic cachexia the leucocyte count varies with the cause of the cachexia. Schlesinger (1903) found normal or slightly decreased leucocyte counts with normal differentials in uncomplicated cases of infantile atrophy. Cabot (1904) reported 3000 leucocytes per cu. mm. of blood in undernourished patients. In anorexia nervosa leucocyte counts as low as 2400 per cu. mm. of blood have been observed (Morlock, 1939).

Benedict *et al.* (1919) concluded from their semi-starvation experiments that there was no effect on the number of leucocytes and that the histological appearance of the cells was normal. The normal proportion of lymphocytes and polymorphonuclear leucocytes was reported to be disturbed, with an average lymphocyte proportion of 36 per cent and a corresponding decrease in the polymorphonuclear leucocytes to 56 per cent. These values are, of course, at most only slightly beyond the normal and may have been influenced by the time of day that the blood samples were taken. The blood samples were taken in the early evenings (generally between 7 and 9 P.M.), and no mention was made whether it was before or after the evening meal.

In 32 cases of war edema Jansen (1918) observed leucocyte counts of less than 6000 per cu. mm. of blood in 24 persons, but no differential counts were made. The histological appearance of the cells was normal. A mild leucopenia with normal differential counts was present in a group of undernourished repatriated French prisoners of war (Richet and Mignard, 1919). Leucocyte counts ranging from 4300 to 8975 with a relative lymphocytosis (45 per cent) were reported among hunger edema patients in Germany (Maase and Zondek, 1920). A relative lymphocytosis with a normal total leucocyte count was reported by Bigland (1920) in 17 of 25 badly starved Turkish prisoners of war. In another group of Turkish prisoners of war the leucocyte count was 4000 per cu. mm. of blood with 36 per cent polymorphonuclear leucocytes, 48 per cent lymphocytes, and 16 per cent mononuclear leucocytes (Enright, 1920). In Russia during the famine of 1921–22 there was a moderate leucopenia with a relative decrease in the proportion of polymorphonuclear leucocytes (Abel, 1923). Boenheim (1934) found a total leucocyte count of 8800, with 23 per cent lymphocytes and 70 per cent polymorphonuclear leucocytes, in one severely undernourished patient who had no other specific disorders; these values are normal.

Only meager data are available on the leucocytic responses among the semi-starved people of Europe and the Orient during World War II. Burger, Sandstead, and Drummond (1945) mention a tendency to leucopenia in the urban population of western Holland. Leucocyte counts as low as 2700 per cu. mm. of blood were reported by Zimmer, Weill, and Dubois (1944) in the internment camps of unoccupied France. Leucocyte counts between 4000 and 5500 with a relative lymphocytosis were observed in the Gurs prison camp in southern France (Schwarz, 1945). Some patients admitted to the Salpêtrière Hospital in Paris from German concentration camps had leucopenia (Debray *et al.*, 1946).

There was no consistent trend in the leucocyte counts in hospital patients during the siege of Leningrad. Leucocyte counts ranging from 3000 to 16,000 per cu. mm. of blood were reported, but there was little relationship to the degree of starvation (Brožek, Wells, and Keys, 1946a). A leucopenia with a tendency toward a relative eosinophilia was found in the undernourished Russian prisoners of war at Tost, Germany (Leyton, 1946).

Starvation in the Warsaw Ghetto produced a slight decrease in the total leucocytes, a more marked decrease in the neutrophiles, a slight increase in the lymphocytes, and a decrease in the eosinophiles (Szejnman, 1946). There was no relationship between the decrease in the erythrocytes and the leucocytes in any of the cases studied. Among the children there was usually a moderate leucopenia with a relative lymphocytosis. The eosinophiles appeared to be completely absent except in the cases with intestinal parasites (Braude-Heller *et al.*, 1946).

Normal leucocyte numbers were reported by Di Granati *et al.* (1947) for the starved miners from the Island of Elba. The values ranged from 5100 to 7850 per cu. mm. of whole blood with an average of 6526; 57.9 per cent were neutrophiles and 22.9 per cent were lymphocytes. After a rehabilitation regimen the number of leucocytes remained essentially unchanged, the average number being 6677 per cu. mm. with 63.4 per cent neutrophiles and 21.9 per cent lymphocytes.

The Medical Survey Board report from Letterman General Hospital (1945) included some data on leucocyte counts. In general there was a leucocytosis with 65 per cent of the counts being above 10,000 per cu. mm. of blood and no counts falling below 5000 per cu. mm. The differential counts showed a slight increase in lymphocytes without any other changes except an extreme eosinophilia in many cases. Hibbs (1946) found a leucopenia with leucocyte counts of 3000 to 5000 per cu. mm. of blood in American Army personnel who had spent 3 years in Japanese prison camps.

Practically any conclusion one wishes to draw concerning the response of the leucocytes to semi-starvation can be substantiated by observations reported in the literature. But a trend does appear within the maze of conflicting reports; in relatively uncomplicated semi-starvation (as in urban western Holland in early 1945) the evidence indicates that a mild leucopenia with a relative lymphocytosis develops. The leucocyte response is, however, altered by diseased processes, by the state of body hydration, and by the extent of the starvation.

Bone Marrow in Starvation

Few references to the histological changes in the bone marrow of man during uncomplicated semi-starvation are available in the literature. Autopsy reports on a large number of deaths from chronic disease with accompanying cachexia include observations on the condition of the bone marrow. Whether or not the alterations in the histological appearance of the bone marrow in cachectic diseases are similar to those that may be present in semi-starvation is, of course, problematical. In some disorders, as in esophageal stricture, starvation is the prominent feature and the body changes should not differ greatly from those of uncomplicated starvation. The situation may conceivably be quite different in chronic infectious diseases, such as tuberculosis, even though extreme cachexia may be present at the time of death.

In a series of 35 autopsies in which death was due to various chronic disorders with cachexia, Blechmann (cited by Ricklin, 1879) observed normal-appearing bone marrow in the humeri of only 4 cases; three of these were tuberculosis and one was cancer of the esophagus. The bone marrow changes were apparent on gross examination and were mainly a gelatinous degeneration of the fatty marrow and a hyperplasia of the red marrow. The hyperplastic red marrow had an embryonic appearance. Blechmann concluded, as Newmann (1869) had earlier, that the hyperplasia was a compensatory reaction on the part of the bone marrow to alleviate the anemia resulting from the excessive red cell destruction caused by the cachectic diseases. Ricklin (1879) developed the thesis that the apparent hyperplasia of the red marrow in chronic cachectic diseases does not represent increased red cell production but, on the contrary, represents an increased red cell destruction in the bone marrow tissue.

A decrease in the amount of fatty marrow and an increase in the cellular marrow were reported by Litten and Orth (1877), Grohe (1884), and Geelmuyden (1886) as consistent findings in the long bones of about 750 autopsies where death was due to chronic cachectic diseases; the cases were mainly tuberculosis, cancer, and senility. Old age is usually accompanied by atrophic changes in muscles and internal organs quite similar to those produced by starvation. In a series of 100 males and females, aged 65 to 90 years, Reich, Swirsky, and Smith (1944) found the bone marrow, obtained by sternal puncture, to be normal in all respects. Grohe (1884) believed that the anemia so often seen in cachectic diseases was due to storage or phagocytosis of erythrocytes within the marrow.

Meyer (1917) reported bone marrow observations in the case of a man who died after a 60-day fast (water was ingested). The depletion of the bone marrow in the ribs and humerus was extreme; whole areas were devoid of hematopoietic cells. In some places small areas of lymphoid cells and red cells were present. The red cells were the more numerous but no nucleated red cells were present. According to Stefko (1923) the bone marrow of starved adults (50 autopsies) contained an abundance of red blood cells, and the capillaries in the marrow were often filled with blood cells. The bone marrow of the starved internees at Belsen prison camp showed normal maturation of the erythrocytes (Mollison, 1946). Mucoid degeneration of the fatty marrow, with absorption of the fat and metamorphosis of the fat cells, and a hyperemia of the red marrow, at least in some stages of starvation, have been observed in experimental starvation in animals (for a review of the literature, see Jackson, 1925).

Bone marrow biopsies were made on 5 cases of starvation by Szejnman (1946). In the majority of the cases the bone marrow was rich in cells, particularly of the myeloid series and normoblasts. In some areas that were poor in cells the lymphocytic forms were predominant. Histological analyses of the bone marrows from 4 of the cases are given in Table 136. Data for 2 areas of the bone marrow are included for cases 3 and 5. Areas 3a and 5a were poor in cells, areas 3b and 5b were rich. The author concluded that the anemia appeared to be related to a block in the migration of cells into the blood rather than to a decrease in the hematopoietic properties of the bone marrow.

Myelograms on 9 starvation patients were presented by Lamy, Lamotte, and Lamotte-Barrillon (1948). The essential data are included in Table 137. The authors believed there was evidence of some increased erythroblastic reaction with a corresponding decrease in myelocytes in some of the cases.

Undoubtedly marked changes occur in the bone marrow during starvation in

TABLE 136

ANALYSES OF THE BONE MARROW from 4 cases of starvation, showing percentages of cells of the various types. Two areas of bone marrow are included for cases 3 and 5 (Szejnman, 1946).

Cell Type	Cases					
	2	3a	3b	4	5a	5b
Proerythroblasts	0.6	0.0	0.8	0.4	0.5	0.8
Erythroblasts	0.4	0.4	3.0	2.2	0.4	1.6
Erythroblast basophils	4.2	2.8	9.6	22.4	7.2	21.4
Polychromatic normoblasts ..	1.2	5.0	6.6	6.4	8.8	7.4
Orthochromatic normoblasts.	5.4	4.0	4.4	5.6	9.2	6.6
Myeloblasts	0.6	0.4	0.2	0.0	0.0	0.0
Promyelocytes	2.2	0.8	1.8	1.0	3.0	0.6
Neutrophil myelocytes	32.2	6.8	16.4	16.4	6.4	5.8
Eosinophil myelocytes	2.6	0.8	0.6	2.0	0.4	0.6
Neutrophil metamyelocyte....	10.0	9.4	19.2	9.2	4.6	12.0
Compact neutrophil	1.4	6.0	16.6	5.8	12.4	13.4
Segmented neutrophil	5.6	2.3	4.4	13.4	25.4	7.0
Segmented eosinophil	2.2	0.4	0.2	1.0	2.0	0.2
Lymphocytes	10.0	36.6	11.8	8.4	15.8	21.0
Monocytes	2.0	2.4	0.4	1.8	1.4	0.8

TABLE 137

MYELOGRAMS from 9 cases of starvation in France (Lamy et al., 1948).

Cell Type	Cases								
	Voi	Pol	Pola	Co	Sa	La	Vo	Ba	Pe
Myelocytic neutrophils	21	16	21	25	16	18	23	15	28
Myelocytic eosinophils	1	1			1			1	0.5
Myelocytic basophils						1		1	0.5
Metamyelocytes	2	5	9	11	2	3	6	7	2.5
Polymyelocytic neutrophils	14	35	36	38	40	26	31	53	36
Polymyelocytic eosinophils	3	2				2			0.5
Polymyelocytic basophils	0.5								1.0
Lymphocytes	10	13	11	10	15	16	7	7	6
Monocytes		2	1	2	2	5	8	3	6
Plasmocytes	1	2	1	2	3	9	1	2	3
Erythroblasts	4.7	22	21	12	21	20	23	12	16

man. The loss of fat from the fatty marrow is to be expected, but the interpretation of the changes in the red or cellular marrow is more difficult. If the changes represent a compensatory hyperplasia in response to the anemia that is usually present in cachexia, it is necessary to explain the origin of the anemia. It is hard to accept the theory that the anemia is the result of the phagocytosis of the red cells by phagocytic cells in the bone marrow during starvation. The existence of anemia in the presence of active erythropoiesis in the bone marrow may be due, in these cases, either to failure of the marrow to deliver more or less adult erythrocytes to the peripheral blood (Szejnman, 1946) or to increased peripheral destruction of red cells. Lubarsch's (1921) observation that an intense hemosiderin deposit is found in the organs in cases of death from starvation would tend to substantiate the latter possibility. In any case there is insufficient information to prove either hypothesis.

Sedimentation Rate

In the Warsaw famine victims the sedimentation rate of the erythrocytes did not show any significant abnormality except in the presence of concurrent infection (Fliederbaum *et al.*, 1946). An increased sedimentation rate was a valuable indication, often very sensitive, of the presence of tuberculosis, but in some instances tuberculosis was present but the sedimentation rate was normal. Fliederbaum *et al.* remarked that in ordinary states of nutrition, especially in persons with liver disease, a high protein meal results in an accelerated sedimentation rate. In tests on these starved persons, however, a meal of eggs did not alter the sedimentation rate.

Bachet's (1943) observations in France agree with those made at Warsaw. In 8 men with famine edema but no signs of tuberculosis, syphilis, or other infections, the 60-minute sedimentation rate ranged from 5 to 13 mm., with a mean of 10 mm. Seven men in a similar state of severe undernutrition but with pulmonary tuberculosis showed 60-minute sedimentation rates of 25 to 80 mm., with a mean value of 55 mm. Another semi-starved man who had aortitis and positive serology for syphilis had a sedimentation rate of 40 mm. in 60 minutes. In 10 undernourished men Simonart (1948) recorded 60-minute sedimentation rates of 2 to 9 mm., with a mean of 5.2 mm. Di Granati *et al.* (1907) reported a higher sedimentation rate in 5 of 8 subjects when in a semi-starved condition as compared to the rate following a period of rehabilitation.

From the rather limited evidence it would appear that, in the absence of other complications, simple caloric undernutrition does not alter the sedimentation rate no matter how severe the emaciation may be. Furthermore, the response of the sedimentation rate to the presence of infection is substantially the same as in the well-nourished person.

Erythrocytes and Hemoglobin in Semi-Starvation — Minnesota Experiment

Fasting basal bloods for estimation of hemoglobin, red cell numbers, white cell numbers, and differential white cell counts were obtained by antecubital venapunctures in the mornings before the subjects had got out of bed. The blood was drawn into oiled syringes with a minimum of stasis and was used for the determination immediately without an anticoagulant. Hemoglobin was determined as oxyhemoglobin with an Evelyn colorimeter which was standardized against the oxygen capacity (Van Slyke and Neill, 1924). The red cell counts, white cell counts, and white cell differential counts were made according to the procedure described by Simmonds and Gentzkow (1944).

The usual laboratory and clinical practice of making hemoglobin determinations without any regard to time of day, previous meals, and activity is not applicable in any exact experimental investigation. Shifts in water distribution between the plasma and extracellular fluid occur rapidly with change in activity, posture, and circulation. For the most consistent results the blood should always be taken in the fasting subject before he has got out of bed in the morning. If venapunctures are used they should be made with a minimal amount of stasis, and if possible the blood should be used immediately without anticoagulants to eliminate the problem of complete mixing.

Blood examination made during the control period gave values which placed the subjects within the normal range to be expected in such a group of active young men. The mean hemoglobin concentration was 15.1 ± 0.88 gm. per 100 cc. of whole blood with a range of 13.3 to 17.3 gm. (see Table 138); the red cell number was 5,222,000 per cu. mm. of whole blood, and the color index was 0.98 (see Table 139). The red cell number and color index are for only 7 of the subjects who according to other tests were reasonably representative of the whole group and in whom the hemoglobin values were identical with the 32-subject group (15.1 gm. per 100 cc. of whole blood in both groups).

The hemoglobin concentration in the peripheral venous blood progressively decreased during the 24 weeks of semi-starvation. In the 32-subject group the

TABLE 138

HEMOGLOBIN VALUES in gm. per 100 cc. of whole blood with standard deviations and as percentages of control values for the 32-subject group at control (C), 12 and 24 weeks of semi-starvation (S12 and S24), and 6 and 12 weeks of rehabilitation (R6 and R12); for the 21-subject group at control (C), 24 weeks of semi-starvation (S24), and 12 and 33 weeks of rehabilitation (R12 and R33); and for 12 subjects at 20 weeks of rehabilitation (R20) (Minnesota Experiment).

	32-Subject Group		21-Subject Group		12-Subject Group	
	Gm.	%	Gm.	%	Gm.	%
C	15.1 ± 0.88	100	15.0 ± 1.22	100		
S12	12.6 ± 0.80	83.4				
S24	11.7 ± 0.80	77.4	11.8 ± 0.73	78.7		
R6	12.3 ± 0.93	81.5				
R12	12.8 ± 1.01	84.8	12.8 ± 1.09	85.3		
R20					14.6 ± 0.81	97.3
R33			15.3 ± 1.72	102.0		

TABLE 139

ABSOLUTE VALUES AND PERCENTAGE CHANGES FOR VARIOUS HEMATOLOGICAL DATA at control (C), 24 weeks of semi-starvation (S24), and 12 weeks of rehabilitation (R12) for 18 subjects; control values for erythrocyte count and color index are for 7 of the subjects (Minnesota Experiment).

Function	C	S24	R12	$\dfrac{dS24}{C} \times 100$	$\dfrac{\triangle R12}{dS24} \times 100$
Erythrocytes (million/cu. mm.)..	5.222	3.782	4.026	−27.6	16.9
Hemoglobin (gm./100 cc.)	15.1	11.7	12.8	−22.3	32.4
Color index	0.98	1.11	1.07	+14.3	30.8
Hematocrit (% cells)	46.78	36.38	40.19	−22.2	36.6
Blood volume (liters)	5.295	5.225	4.803	−1.3	
Total hemoglobin (gm.)	797.0	611.3	613.4	−23.4	1.1
Body weight (kg.)	68.0	51.8	58.1	−23.8	38.8
Hemoglobin (gm./kg.)	11.72	11.80	10.56	+0.68	
Total erythrocytes ($\times 10^{12}$)	27.8	19.7	19.3	−29.1	

hemoglobin concentration had fallen to 12.6 ± 0.80 gm. per 100 cc. of whole blood (a decrease of 16.3 per cent) at the twelfth week of semi-starvation (S12). During the second 12 weeks on the semi-starvation regimen the rate of hemoglobin loss was less rapid, reaching a final average low concentration of 11.7 ± 0.80 gm. per 100 cc. of whole blood; this indicates a total reduction in hemoglobin concentration of 22.6 per cent at S24 in the 32-subject group. The hemoglobin values at S24 and the percentage losses are approximately the same for the subject group as a whole (32 subjects), for the 21 subjects on whom hemoglobin values were obtained through 33 weeks of rehabilitation, and for the 18 subjects on whom red cell counts were also made through 12 weeks of rehabilitation (cf. Tables 138 and 139).

The number of red blood cells per cu. mm. of whole blood decreased from a control average of 5.222 to 3.782 million at the end of 24 weeks of semi-starvation, a decrease of 27.6 per cent (see Table 139). During the same period the color index increased 14.3 per cent from 0.98 to 1.11. For the 18-subject group the decreases in the hemoglobin concentration of the venous blood, the packed red blood cell volume (hematocrit), and the red blood count were 22.3, 22.2, and 27.6 per cent, respectively. The relatively larger decrease in the number of red blood cells per cu. mm. of blood than in hemoglobin concentration and hematocrit and the increase in the color index are definite evidence that the anemia that developed during semi-starvation was a macrocytic type. The total number of red blood cells in the body decreased 5.3 per cent more than did the total grams of hemoglobin (see Table 139). The relative increase in the volume of each red blood cell and the increase in amount of hemoglobin in each red blood cell can be calculated from the ratio of hematocrit to red blood cell number and the ratio of hemoglobin concentration to red blood cell number during the control period and at 24 weeks of semi-starvation. Such calculations indicate that there was approximately a 7 per cent increase in both the volume of each red blood cell and the amount of hemoglobin each cell contained.

The total amount of hemoglobin in the body, calculated from the total blood volume and the hemoglobin concentration, decreased from 797.0 gm. during the control period to 611.3 gm. at 24 weeks of semi-starvation, a decrease of 23.4 per cent in the group of 18 men (see Table 139). The percentage loss in total hemoglobin was only slightly less than the percentage loss of body weight. On the basis of grams of hemoglobin per kilogram of body weight the values were 11.72 for the control period and 11.80 at the end of semi-starvation. If allowance is made for the increased hydration — increase in plasma volume, extracellular fluids, and edema — that occurred during semi-starvation, the hemoglobin in grams per kilogram of normal hydrated body weight increased from 11.72 during the control period to 12.94 at the end of semi-starvation, an increase of 10.4 per cent.

According to the usual definition of anemia (a decrease in hemoglobin concentration and red blood cell number), a moderate macrocytic anemia developed during starvation. At the same time there were changes in body weight and body hydration. Consequently, if the hemoglobin is expressed in terms of normal hydrated body weight there was an increase in grams of hemoglobin per unit of body tissue. This does not, of course, imply that the oxygen supply to the active tissue per unit of time was increased or even normal during semi-starvation.

Red Blood Cells and Hemoglobin in Rehabilitation — Minnesota Experiment

The venous blood hemoglobin concentration and red blood cell number increased slowly during rehabilitation. At the sixth week of rehabilitation (R6) the hemoglobin concentration had increased 4.1 per cent from a semi-starvation low of 11.7 ± 0.80 gm. to 12.3 ± 0.93 gm. per 100 cc. of blood. In the second 6 weeks of rehabilitation the hemoglobin increased to 12.8 ± 1.01 gm. per 100 cc. of blood for a total 12 weeks' gain of 1.1 gm. per 100 cc. of blood (see Tables 138 and 139). During the same period the red cell number increased from a semi-starvation average of 3.782 to 4.026 million per cu. mm. and the color index decreased slightly from 1.11 to 1.07 (see Table 139). The hematocrit increased to 40.19 from a semi-starvation low of 36.38.

In the 8-week period from R12 to R20 when the subjects were eating an unrestricted but measured diet, the hemoglobin concentration increased rapidly. The average hemoglobin concentration at R20 was 14.6 ± 0.81 gm. per 100 cc. of blood, which was 97.3 per cent of the control value for the group. A further increase of 0.7 gm. occurred during the subsequent 13 weeks of unrestricted diet, so that at 33 weeks of rehabilitation (R33) the hemoglobin concentration had passed the control values with 15.3 ± 1.72 gm. per 100 cc. of blood in the 21-subject group (see Table 138).

The increase in hematocrit, hemoglobin concentration, and red blood cell number during the first 12 weeks of rehabilitation was accompanied by a decrease in blood volume (mainly plasma volume). As a result there was no increase in total grams of hemoglobin in the body or in total number of red blood cells (see Table 139). The hemoglobin expressed in grams per kilogram of body weight actually decreased 1.24 gm. or 10.5 per cent.

Data for total blood hemoglobin in the body are available for only 4 subjects for 20, 33, and 58 weeks of rehabilitation. The general response of the 4 subjects through R12 was similar to that of the 18-subject group (cf. Tables 139 and 140). During the 8 weeks from R12 to R20 the total blood hemoglobin increased to 835.3 gm., an increase of 150.8 gm. or 69.8 per cent of the total amount lost

TABLE 140

BLOOD VOLUME, HEMOGLOBIN CONCENTRATION, TOTAL HEMOGLOBIN, AND HEMO-GLOBIN PER KG. OF BODY WEIGHT for 4 subjects at control (C), 24 weeks of semi-starvation (S24), and 12, 20, 33, and 58 weeks of rehabilitation (R12, R20, R33, and R58) (Minnesota Experiment).

Function	C	S24	R12	R20	R33	R58
Blood volume (liters)	5.963	5.903	5.565	5.841	5.863	5.933
Hemoglobin (gm./%)	15.1	11.8	12.3	14.3	14.6	15,0
Body weight (kg.)	70.7	54.0	62.0	73.3	74.5	72.7
Total hemoglobin (gm.) ...	900.4	696.6	684.5	835.3	856.0	890.0
Hemoglobin (gm./kg.)	12.7	12.8	11.0	11.4	11.5	12.2

during the preceding months of semi-starvation and rehabilitation (see Table 140). The body weight increased rapidly from R12 to R20 so that there was only a small increase (0.4 gm.) in the grams of hemoglobin per kilogram of body weight. The increase in total blood hemoglobin amounted to only 20.7 gm. in the period of R20 to R33. On the whole the hemoglobin values were substantially back to normal except for the grams of hemoglobin per kilogram of body weight, which were still 1.2 gm. below the control value at 33 weeks of rehabilitation. This difference is mainly due to the higher than control body weights. At R58 blood volume, hemoglobin concentration, and total grams of hemoglobin were back to control values, with body weight 2 kg. above normal.

Effect of Calories, Vitamins, and Proteins on Hemoglobin — Minnesota Experiment

The effects of the level of caloric intake during rehabilitation on the rate of increase of hemoglobin in the venous blood are presented in Tables 141 and 142.

A comparison of the hemoglobin levels or the percentage regain of lost hemoglobin at 6 and 12 weeks of rehabilitation indicates that the caloric intake had no effect on the rate at which the hemoglobin concentration in the blood increased (see Tables 139 and 140). The greatest gain in hemoglobin both as

TABLE 141

EFFECT OF LEVEL OF CALORIC INTAKE ON HEMOGLOBIN. Hemoglobin values in gm. per 100 cc. of whole blood and as percentage of the control value for the 4 caloric groups (Basal, +400, +800, and +1200), at control (C), 12 and 24 weeks of semi-starvation (S12 and S24), and 6 and 12 weeks of rehabilitation (R6 and R12) (Minnesota Experiment).

	Basal		+400		+800		+1200	
	Gm.	%C	Gm.	%C	Gm.	%C	Gm.	%C
C	15.0	100	15.3	100	14.6	100	15.6	100
S12	12.7	84.2	12.9	84.5	12.1	82.5	12.8	82.5
S24	11.7	77.8	11.9	77.8	11.2	76.8	12.0	76.9
R6	12.1	80.6	12.3	80.2	12.0	82.0	12.4	79.4
R12	12.8	85.0	12.9	84.4	12.5	85.0	12.9	83.0

TABLE 142

EFFECT OF CALORIC INTAKE ON HEMOGLOBIN INCREASE DURING REHABILITATION. Hemoglobin values in grams lost or regained per 100 cc. of whole blood and as percentage of amount lost at 24 weeks of semi-starvation (S24) for the 4 caloric groups (Basal, +400, +800, and +1200) at 24 weeks of semi-starvation (S24) and 6 and 12 weeks of rehabilitation (R6 and R12) (Minnesota Experiment).

Hemoglobin	Basal		+400		+800		+1200	
	Gm.	% of dS24	Gm.	% of dS24	Gm.	% of dS24	Gm.	% of dS24
Lost at S24	3.33	100	3.39	100	3.40	100	3.69	100
Regained at R6	0.42	12.6	0.36	10.6	0.76	22.4	0.39	10.6
Regained at R12	1.08	32.4	1.01	30.0	1.21	35.6	0.94	25.5

gm. per 100 cc. of blood and as a percentage of the amount lost during semi-starvation occurred in the +800 caloric group at 6 and 12 weeks of rehabilitation. The least percentage gains at 6 weeks of rehabilitation were in the +400 and +1200 caloric groups with the +1200 group being lowest at 12 weeks of rehabilitation. The absolute gains during the 12 weeks of rehabilitation ranged from 0.94 to 1.21 gm. per 100 cc. of blood and the percentage gains from 25.5 to 35.6 per cent. Actually, as pointed out before, during the first 12 weeks of rehabilitation the increased blood hemoglobin concentration merely reflected the decrease in plasma content of the blood.

The design of the rehabilitation program was such that half the subjects in each caloric group received extra vitamins and half the subjects on each caloric and vitamin level received about 20 gm. of extra protein per day (see Chapter 4 for the factorial design of the experiment). In the 32-subject group, 16 subjects received the vitamin supplementation while for 16 subjects the vitamin intake was limited to that present in the food eaten, and 16 subjects received about 20 gm. of extra protein per day while for 16 subjects the protein intake was just that present in the natural foods eaten. Groups of 16 subjects can then be compared to determine the effects of extra vitamins or proteins irrespective of the level of caloric intake. To determine the interaction between calories and vitamins or proteins, the subjects in the two high caloric levels (+800 and +1200) and the subjects in the two low caloric levels (basal and +400) were combined to give 4 sub-groups of 8 subjects each.

The mean increase in blood hemoglobin during 12 weeks of rehabilitation was 0.86 gm. and 1.28 gm. per 100 cc. of blood for the vitamin-supplemented and vitamin-unsupplemented groups, respectively (see Table 143). The mean difference of 0.42 gm. for the 2 groups was not statistically significant, owing mainly to the large variability of individual responses in the 2 groups. There was no significant effect produced by the vitamins at any of the caloric levels even though the mean values for the unsupplemented groups were consistently higher than for the supplemented groups at all caloric levels. The macrocytic

TABLE 143

EFFECT OF VITAMIN AND PROTEIN SUPPLEMENTATION ON RECOVERY OF HEMOGLOBIN DURING REHABILITATION. Hemoglobin lost in gm. per 100 cc. of blood at 24 weeks of semi-starvation (S24) and regained at 3, 6, and 12 weeks of rehabilitation (R3, R6, and R12) for the 32-subject group (ZLGT), the combined two high caloric groups, +800 and +1200 (GT) and the combined two low caloric groups, Basal and +400 (ZL). H = vitamin-supplemented, P = no extra vitamins, Y = extra proteins, U = no extra proteins (Minnesota Experiment).

Gm. Hemoglobin per 100 cc. of Blood	Vitamins						Proteins					
	ZLGT		GT		ZL		ZLGT		GT		ZL	
	H	P	H	P	H	P	Y	U	Y	U	Y	U
Lost at S24.....	3.39	3.46	3.48	3.52	3.31	3.41	3.30	3.56	3.22	3.78	3.38	3.35
Regained at R3..	0.55	0.78	0.37	0.82	0.73	0.74	0.64	0.68	0.58	0.61	0.71	0.76
Regained at R6..	0.35	0.78	0.31	0.85	0.39	0.70	0.68	0.44	0.76	0.40	0.60	0.49
Regained at R12..	0.86	1.28	0.94	1.23	0.78	1.33	1.12	1.01	1.16	1.00	1.08	1.03

hyperchromic anemia reported in the hospital patients during the siege of Leningrad was thought to be of pellagrous origin (Drazdova, 1945). In the Minnesota Experiment the anemia was in no way benefited by a vitamin supplementation which included 10 mg. of niacin amide per day. This observation is consistent with reports from German-occupied western Holland, where supplementation with vitamins or iron during refeeding appeared to have no effect on the course of the anemia.

It has been suggested by Drummond (1946a) that the anemia of Western Europe may have been the result of the low protein intakes. The low protein, low calorie diets of urban western Holland in the spring of 1945 may not have contained sufficient amounts of the protein moiety necessary for the formation of new hemoglobin. If such was the cause of hunger anemia, it might be expected that hemoglobin formation after semi-starvation would take place more rapidly on a high protein diet.

In the Minnesota Experiment the mean increase in hemoglobin during rehabilitation, without reference to caloric intake, was 0.44 and 0.68 gm. per 100 cc. of blood for the protein-unsupplemented and the protein-supplemented groups, respectively, at 6 weeks of rehabilitation, and 1.01 and 1.12 gm., respectively, at 12 weeks of rehabilitation (see Table 143). The differences, however, were not statistically significant. It is possible, of course, that the mean difference of about 20 gm. in daily protein intake between the 2 groups was not large enough to allow any potential effect of the protein to be exhibited.

There was no significant effect of protein supplementation on hemoglobin formation within the caloric groups (see Table 143). The mean increase in hemoglobin concentration was 0.36 and 0.16 gm. per 100 cc. of blood greater at 6 and 12 weeks of rehabilitation in the subjects who received the extra proteins in the two combined higher caloric groups (G and T). In the two combined lower caloric groups (Z and L) the increase was 0.11 and 0.05 gm. greater in the protein-supplemented subjects for the same periods. Although the mean differences may suggest a possible advantage in hemoglobin formation obtained from the extra proteins, the statistical analysis of the data does not warrant even a suggestion.

White Blood Cells — Minnesota Experiment

Although leucocyte counts and/or differential counts were made on about half the subjects at various times during the experiment, the more complete data are available for only 7 subjects at control, 24 weeks of semi-starvation, and 12 weeks of rehabilitation. The leucocyte counts and differentials for the 7 subjects at the 3 major testing periods are presented in Table 144.

A definite leucopenia developed during semi-starvation; the leucocyte count, which was 6346 per cu. mm. of blood in the control period, dropped to 4129 at S24, a decrease of 34.9 per cent (see Table 144). During the same period the total plasma volume for the 7 subjects increased from 3174 cc. to 3628 cc., an increase of 14.3 per cent. If the leucocyte count is corrected for the increased plasma, the leucocyte count per cu. mm. of blood would be 4819 and the decrease that occurred during semi-starvation would be 24.1 per cent. The cause of the leuco-

TABLE 144

Total Leucocytes (WBC) per Cu. Mm. of Whole Blood, Percentage Lympho-
cytes (L), and Percentage Polymorphonuclear Leucocytes (PMN) at control
(C), 24 weeks of semi-starvation (S24), and 12 weeks of rehabilitation
(R12) for 7 subjects (Minnesota Experiment).

Subject	C			S24			R12		
	WBC	L	PMN	WBC	L	PMN	WBC	L	PMN
1	4525	32	67	4600	16	84	5150	25	74
122		50	49	3750	52	44	5020	42	55
29	5750			4750	34	64	4625	33	66
30	7950	22	77	3600	44	52	5075	35	64
123	6925	48	50	3650	44	56	5425	26	72
130	6350	28	72	4750	38	58	5500	20	70
11	6575	52	47	3800	48	52	5000	30	69
M	6346	38.7	60.3	4129	39.4	58.6	5114	30.1	67.1

penia — whether a decreased production, an increased destruction, or a storage
of the white cells in the tissues and organs — is not apparent. In any event the
picture was not complicated by disease processes as has so often been the case
in past observations.

The differential white blood cell counts revealed that no significant change
in the percentage of lymphocytes and polymorphonuclear leucocytes occurred
during starvation (see Table 144). This is in contrast to the observation of Benedict *et al.* (1919), Bigland (1920), and Abel (1923) that there is a relative lymphocytosis during semi-starvation. In the present group of 7 subjects, two had
an increase in the percentage of lymphocytes and one a decrease at the end of
the semi-starvation period. In the other 4 subjects the slight changes were well
within the limits of accuracy of the methods.

During the 12 weeks of rehabilitation the number of leucocytes increased
from the semi-starvation value of 4129 to 5114 per cu. mm. of blood, an increase
of 23.9 per cent and a recovery of 44.4 per cent of the loss that occurred during
semi-starvation (see Table 144). Only one of the subjects failed to have a higher
leucocyte count at the end of 12 weeks of rehabilitation. The small decrease in
plasma volume that occurred during the first 12 weeks of rehabilitation would
account for only a minor part of the increased leucocyte concentration. The regeneration of leucocytes during rehabilitation must have greatly exceeded the
rate of destruction or loss, or leucocytes which may have been held in the tissues
and organs during semi-starvation were again liberated into the blood.

Small changes occurred in the percentages of lymphocytes and polymorphonuclear leucocytes during the 12 weeks of rehabilitation (see Table 144). At
R12 the mean percentage of lymphocytes was 8.6 and 9.3 per cent lower and of
polymorphonuclear leucocytes 6.8 and 8.5 per cent higher than at control and 24
weeks of semi-starvation, respectively. Whether or not the differential distribution of the leucocytes was abnormal at any of the testing periods is problematical. The normal distribution for adults is given as 25–35 per cent lymphocytes
and 57–67 per cent polymorphonuclear cells (PMN) by Simmonds and Gentz-

kow (1944, p. 115). The percentage of PMN cells would then be within the normal range at control, S24, and R12, and the lymphocytes would be normal at R12 but slightly increased at control and S24. Only with caution can significance be attributed to small differences in leucocyte counts and differential distributions. As pointed out by Maximow and Bloom (1931, p. 62), "The number of leukocytes in the circulating blood . . . can change rapidly under . . . numerous conditions which are hard to control. Consequently, many of the leukocyte counts . . . have only a relative value."

Complete leucocyte data are available in too few of the subjects to determine whether differences in caloric intakes, vitamin supplementation, or extra proteins had any influence on the leucocyte counts.

Bone Marrow in Semi-Starvation — Minnesota Experiment

Sternal bone marrow was aspirated from the body of the sternum opposite the second intercostal space by the method routinely employed in the University of Minnesota Hospitals (Sundberg, 1946). With the technique used, there is a relatively constant ratio between the volumes of nucleated marrow cells and platelets (myeloid-erythroid volume) on the one hand and fat and particulate marrow on the other. Sternal bone marrow was aspirated from one subject (No. 123) at 6 weeks of rehabilitation and from 4 subjects at 10 weeks of rehabilitation; no aspirations were made during either the control or semi-starvation periods. The bone marrow specimens obtained probably reflected the general re-

TABLE 145

PERIPHERAL BLOOD VALUES of the 5 subjects from whom sternal bone marrow was aspirated: at the end of semi-starvation (S) and at the time during rehabilitation when the aspirations were made (R) (Minnesota Experiment).

	Subject				
	123	20	8	130	126
Hemoglobin (gm./%)					
S	10.1	12.1	10.4	10.9	12.0
R	11.1	12.9	12.5	13.7	10.4
Difference	1.0	0.8	2.1	2.8	−1.6
Erythrocytes (millions/cu. mm.)					
S	2.960	4.290	2.840	3.356*	3.860
R	3.400	3.525	3.785	4.320	3.435
Difference	0.440	−0.765	0.945	0.864	−0.425
Color Index					
S	1.02	0.95	1.09		0.93
R	0.98	1.23	0.99	0.95	0.90
Leucocytes (per cu. mm.)					
S	3,650	6,400	3,200	4,750	4,200
R	5,425	8,075	6,975	5,500	5,875
Difference	1,775	1,675	3,275	825	1,675
Neutrophils: lymphocytes					
S	56:44	48:50	44:56	58:38	54:46
R	65:30	64:27	67:25	68:20	67:20

* Erythrocyte value caculated on the basis of the hemoglobin concentration.

covery processes that occurred during rehabilitation but cannot be assumed to represent the condition of the bone marrow at the end of semi-starvation.

The changes in the peripheral blood which occurred between the end of semi-starvation and the time the bone marrow aspirations were made are summarized in Table 145. Erythrocyte counts were not available for subject 130 at the end of semi-starvation but from the relatively large increase in hemoglobin concentration (2.8 gm. per 100 cc.) during rehabilitation, it may be assumed that there was a corresponding increase in the number of erythrocytes. If that assumption is made, then the hemoglobin concentration increased in all but subject 126 and the number of erythrocytes increased in all but subjects 20 and 126. The decrease in the number of erythrocytes in subject 20 is difficult to explain in view of the fact that in all other respects his blood changes were similar to those of subjects 123, 8, and 130. Except for a very slight macrocytosis, there were no particular morphological changes in the erythrocytes in the peripheral blood; the blood platelets appeared normal in number and morphology.

There was a restoration of the total and differential leucocyte counts to relatively normal values in all the subjects at the time the bone marrow aspirations were made. The increase in leucocytes was greatest in subject 8, who had the

TABLE 146

DIFFERENTIAL STERNAL MARROW CELL COUNT (% of 1000 cells) for 5 subjects with normal values included in column N. V = variable in number, P = present, P+ = relatively prominent, A = absent, + = increased, and − = decreased (Minnesota Experiment).

	Subject					
	123	20	8	130	126	N
Myeloblasts	0.0	0.4	0.5	0.5	0.0	1.0
Leukoblasts	1.3	1.4	0.2	1.0	1.0	2.0
Neutrophils	52.0	51.7	37.1	36.5	61.6	63.0
Promyelocytes	2.0	3.0	2.1	1.0	3.0	4.0
Myelocytes	8.7	6.4	5.9	5.3	3.0	13.0
Metamyelocytes ..	19.0	10.6	11.1	15.0	7.1	15.0
Mature and band .	22.3	31.7	18.0	15.2	48.5	31.0
Eosinophils	3.0	2.9	3.5	2.9	1.3	3.0
Myelocytes	2.0	1.9	2.4	1.4	0.5	2.0
Mature	1.0	1.0	1.1	1.5	0.8	1.0
Basophils	0.9	0.5	0.2	0.3	1.6	0.5
Normoblasts	31.7	27.6	44.5	44.4	11.8	18.5
Pronormoblasts ...	0.0	0.6	2.0	0.4	0.0	0.5
Basophilic	1.3	2.4	5.0	4.0	0.3	2.0
Polychromatic	8.3	9.4	15.0	20.0	6.5	12.0
Orthochromatic ...	22.1	15.2	22.5	20.0	5.0	4.0
Lymphocytes	9.1	11.1	10.0	9.0	17.2	10.0
Plasma cells	0.5	0.6	2.0	1.5	0.0	1.0
Monocytes	1.5	3.8	2.0	3.9	5.5	1.0
Histiocytes	P	P	P+	P	P	P
Megakaryocytes	+	+	+	+	−	50–75[*]
Platelets	+	+	+	+	+	P
Stippling	P	P+	P+	P+	P	A
Pigment	P	P	P+	P+	P	V

[*] The number in the terminal 18 × 18 sq. mm. of the smear.

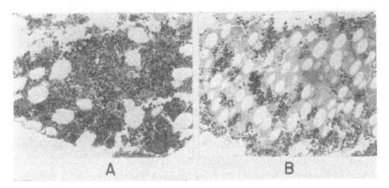

FIGURE 48. STERNAL BONE MARROW, Minnesota Experiment.
A. Section of sternal bone marrow from subject No. 130 showing
an area of representative cellularity of the marrow. Prominent aggregates of normoblasts and myeloid cells are present (×200).
B. Section of sternal bone marrow from subject No. 126 showing
gelatinous degeneration and only small foci of normoblasts and
other types of developing cells (×200).

lowest total leucocyte count at the end of semi-starvation. The peripheral blood
of these five subjects indicated an active regeneration of leucocytes.

The data from the sternal bone marrow studies are presented in Table 146.
The sternal bone marrow sections in all cases contained a relatively abundant
amount of adipose tissue and the intravascular and extravascular erythrocytes,
as well as cells of the megakaryocyteic series, were present in variable numbers;
they appeared to develop in patchy or nodular aggregates rather than diffusely
through the marrow. In subjects 20, 8, and 130 there were numerous islands of
normoblasts and developing granulocytes. In the latter two subjects the aggregates of normoblasts were more prominent than adipose tissue and contained
many mitoses; the aggregates were sufficiently large to constitute areas of definite normoblastic hyperplasia (see Figure 48A). This normoblastic hyperplasia
no doubt represented an accelerated erythropoieses which was reflected by an
increase in erythrocytes and hemoglobin in the peripheral blood (see Table
145). In subject 126 the normoblasts were sparse in both sections and smears.
Perivascular lymphocytes and plasma cells were increased in all cases and were
most prominent in subject 126. Areas of gelatinous degeneration were present in
all sections of the marrow from subject 126 (see Figure 48B); small and incon-
spicuous areas were occasionally present in the sections from subject 20. The absolute and relative reduction in the amount of erythropoietic tissue in subject
126 was accompanied by a relative increase in the percentages of lymphocytes
and monocytes in the bone marrow smears (see Table 146). Such a relative increase in lymphoid cells and a fairly normal relative percentage of granulocytes
are commonly encountered in hypoplastic marrows regardless of the etiologic
factors responsible for the hypoplasia.

Subject 126 was the only one of the five subjects in whom a definite absolute
normoblastic hypoplasia existed; he was also the only one who had a decrease in
the erythrocyte and hemoglobin levels during the early weeks of rehabilitation.

From the peripheral blood data, this subject was, even at the time of the bone marrow aspiration, fairly representative of the group at the end of semi-starvation, and it might be assumed, then, that a normoblastic hypoplasia of the bone marrow would have been found in the other subjects if the specimens had been obtained before refeeding had been instituted. The increase in normoblasts and the tendency toward a normoblastic hyperplasia in the other four bone marrow samples might well be regarded as an indication of an accelerated erythropoiesis in response to the rehabilitation diet.

All the bone marrow smears showed variable amounts of basophilic stippling (see Table 146), which would suggest that erythrocyte regeneration was not absolutely normal in any of the subjects. Pigment, both free and within the phagocytic cells, was more prominent in the smears and sections than would be expected in normal marrow. Cytoplasmic and nuclear changes in the normoblasts suggestive of degeneration were present in all the bone marrow smears but were most pronounced in subject 126. Evidence of phagocytosis of erythrocytes was minimal, but occasional macrophages containing erythrocytes were found in subject 130.

A shift to the more mature forms of both the neutrophil leucocytes and normoblast series was observed in the bone marrow (see Table 146). The shift in the cells of the neutrophil series was suggestive of that seen in the marrow of patients recovering from leucopenia. A shift to the right in the cells of the normoblast series may be found in a variety of conditions, including the posthemorrhagic state, following hemolysis, and following specific therapy for anemia (Sundberg, 1947).

One other finding which may be of some significance was the apparent increase in the basophils in the most hypoplastic of the marrows (subject 126). Similar minimal increases in the basophils have been observed in hypothyroidism (Thompson, 1941).

CHAPTER 15

Compartments of the Body

ALL parts of the body undergo changes in starvation. We have already discussed the morphology of the major organs in the ordinary terms of anatomy, but for the purposes of metabolic analysis, quantitative estimates of the changes in over-all compartments of the body are needed. In general the body is made up of the living cells and, in contrast, the less immediately vital materials such as interstitial fluid, fat, and bone. Moreover, the blood, with its two divisions of plasma and cells, needs quantitative estimation as a total mass or organ.

From the more ordinary anatomical evidence it is clear that in starvation much of the fat disappears, the body as a whole tends to become more hydrated, and the changes in the bones are relatively small. Heretofore there have been no quantitative estimates of these changes, except perhaps in the case of the bone minerals. But methods recently developed made it possible to get fairly detailed information on these points in the Minnesota Experiment.

The Bone Minerals

About one fourth of the bony skeleton of the body is made up of minerals which contribute extremely little to the total energy metabolism of the body. This part of the body is also relatively constant in starvation. The response of the bones and teeth to caloric undernutrition has been discussed in Chapter 12. It may be recalled that animal experiments have disclosed only trivial changes in the mineral compartment even in very severe undernutrition and that the mineral density studies in the Minnesota Experiment indicated that no significant changes in the bone minerals occurred in either the starvation or the rehabilitation phases of the experiment.

It is probable that in the Minnesota Experiment, as in other studies of severe starvation, there were changes in the amounts of water, fat, and protein in the bones. But these elements are, for the present analysis, included in other compartments of the body. It may be assumed that much of the water in the bones is accessible to thiocyanate and so would be included in that estimation. Changes in fat in the bones, as elsewhere in the body, would be reflected in changes in total density. Accordingly, we have first the problem of estimating the original mass of bone mineral.

It is generally stated that in the normal, well-nourished adult male body the bony and cartilaginous skeleton makes up from 15 to 20 per cent of the total body weight. The bony skeleton comprised 15.8 per cent in the 42-year-old man studied by Dursy (1863). Volkmann's (1873) cadaver was that of a 38-year-old man selected to represent normality; the bony skeleton comprised 15.8 per cent of the total weight. The fresh bony skeleton made up only 14.8 per cent of the body weight in the 35-year-old man analyzed by Mitchell *et al.* (1945); since this person died at this early age from cardiac decompensation, there may be some doubt as to the complete normality of the tissues. It seems possible that some degree of edema, perhaps occult, was present, in which case the percentage of the body weight represented by the bones would naturally be somewhat lower than normal. From a careful collation of the older literature Wilmer (1940) arrived at an average of 17.60 per cent for the total skeleton. For our present purposes no serious error will result from an assumption that on the average 16 per cent of the body mass is normally provided by the bones.

The mean mineral (ash) content for whole bony skeletons has rarely been measured in normal human cadavers. The data of Volkmann (1873) indicated 22.1 per cent ash and those of Mitchell *et al.* (1945), 28.9 per cent of the wet skeleton. A compromise at 25 per cent would result in an estimate that 4 per cent of the normal whole body is represented by the mineral ash of the bones. We have used this figure for the bone mineral compartment in the control period of the Minnesota Experiment. It will be observed that, even admitting a possible ±20 per cent error in this estimate, the effect on the relative accuracy of the calculations for the other body compartments would be very small.

It may be assumed, then, that about 4 per cent of the normal (control) body weight in the Minnesota Experiment was made up of bone mineral and that this mass remained substantially constant throughout the experiment. For the total group of 32 men the control body weight was 69.40 kg., so the average mineral ash compartment may be taken to be 2.78 kg. The 4 groups of 8 men each who received diets of different caloric content in rehabilitation had slightly different average control weights; the corresponding mineral ash compartment values for the Z, L, G, and T groups may be taken to average, respectively, 2.73, 2.78, 2.89, and 2.70 kg. Similarly, separate average estimates, adjusted to the initial body weights, can be provided for the other groupings, such as the men examined at later stages in rehabilitation — R20, R35, and R55. Since only the relative changes during the experiment are of present concern, the absolute accuracy of these estimates is of little consequence; they are at least relatively correct and provide a suitable identical basis for correction in the calculation of the other compartments of the body.

The Plasma and Total Blood Volume

Except for the Minnesota Experiment there are few satisfactorily controlled observations on the behavior of the blood or plasma volume in starvation. Some estimates were made on the inmates of the Belsen concentration camp at the time of their liberation (Mollison, 1946). The results indicated some reduction in the total blood volume but it was believed that this change was less than in proportion to the body weight loss. It was concluded that the total blood volume per kg. of body weight was unusually high. These persons were all anemic to a considerable degree, so that the plasma volume was definitely much elevated in proportion to the actual body weight. In 5 moderately underweight persons Perera (1946) reported that the plasma volume averaged 17.2 per cent above the normal average (from Gibson and Evans, 1937) when calculated per square meter of body surface. When calculated per kg. of body weight the same data averaged about 5 per cent above the value of 45 cc. of plasma per kg. suggested for the normal average by Gregerson (1944).

In Chapter 43 the data of Beattie, Herbert, and Bell (1948) are discussed in some detail. For 2 different groups of severely undernourished adults the estimated total plasma volumes corresponded closely to the Minnesota average for the normal (pre-starvation) body size; in other words, their rather extensive data indicated that semi-starvation produced little if any change in the absolute total plasma volume. Depending on the degree of anemia observed, then, the total blood volume would be estimated to have undergone a slight to moderate reduction. The general picture well substantiates that seen in the Minnesota Experiment.

From studies on some Indian ex-prisoners of war in a state of severe undernutrition Walters, Rossiter, and Lehmann (1947a) likewise concluded that there was little change in the total blood volume. When these men were brought under treatment, it was observed that there was often a great increase in plasma and blood volume during the first few weeks. It is possible that this latter phenomenon was the result of overzealous refeeding.

In the Minnesota Experiment the plasma volume was estimated at regular intervals with the dye dilution (T-1824) method. The estimates were made in the basal state and were repeated several times in some individuals at the same stage of the experiment so as to gauge the reliability (repeatability) of the results. At the time of each estimate the relative mass of the blood cells was measured by the hematocrit method; from this it was possible to estimate the total blood volume and that of the cellular compartment.

The principal results for the plasma and blood volumes in the control and semi-starvation periods and through the tenth week of rehabilitation are summarized in Table 147. Two independent trials in the control period gave averages, as cc. per kg. of body weight, of 45.01 and 45.57 for the plasma volume and 83.80 and 85.47 for the total blood volume. After 12 weeks of semi-starvation, when the body weight had declined slightly more than 15 per cent, the

TABLE 147

PLASMA, BLOOD CELL, AND TOTAL BLOOD VOLUMES, average values. In the control period (C), 2 determinations were made about a week apart. (Minnesota Experiment.)

Period	Number of Men	Body Weight Kg.	Body Weight % of C	Hematocrit %	Plasma Volume Liters	Plasma Volume Cc./kg.	Cell Volume Liters	Cell Volume Cc./kg.	Blood Volume Liters	Blood Volume Cc./kg.
C, 1	32	69.54	100*	46.29	3.130	45.01	2.698	38.80	5.828	83.80
C, 2	30	69.45	100*	46.68	3.165	45.57	2.771	39.90	5.936	85.47
S12	18	58.90	84.75	39.85	3.176	53.92	2.113	35.87	5.289	89.80
S23	32	53.45	76.91	36.60	3.410	63.80	1.989	37.21	5.379	100.64
R5	14	54.49	78.40	38.16	3.232	59.31	1.982	36.37	5.214	95.84
R11	16	57.95	83.38	39.51	3.090	53.32	2.007	34.63	5.097	88.04

* Mean value of 69.5 for body weight was taken to be 100 per cent.

figures had risen to 53.92 cc. of plasma per kg. and 89.8 cc. of total blood per kg. These relative volumes continued to rise as starvation progressed so that after 22 weeks (S22), when the body weights had fallen more than 23 per cent, the plasma volume per kg. of body weight was 40.9 per cent greater than in the normal (control) state. The change in the total blood volume was not so dramatic, being 19.6 per cent greater per kg. of body weight at S22 than the control average. Even in absolute terms the plasma volume increased in starvation (+8.3 per cent); the total blood volume, however, diminished by 8.6 per cent, reflecting the fact that there was a considerable degree of anemia.

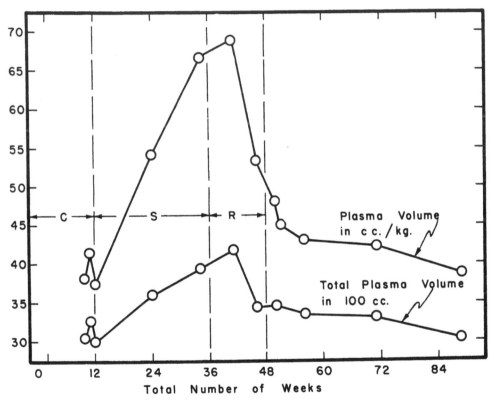

FIGURE 49. PLASMA VOLUME IN A TYPICAL SUBJECT IN THE MINNESOTA EXPERI-MENT. The period C is pre-starvation (control), S refers to the semi-starvation period, and R is the period of restricted rehabilitation. Plasma volume values in hundreds of cc. and cc. per kg. of total body weight.

During the first weeks of restricted rehabilitation the relative hydremia diminished only slightly; even after 10 weeks the plasma volume per unit of body weight was still 18 per cent above the previous normal. Data were obtained on only a few subjects in later rehabilitation. One example is given in Figure 49. In the case illustrated, it will be noted that normal levels for the plasma volume were approached within a few weeks after the return to a completely unrestricted diet but that values actually within the previous normal range for this subject were achieved only after the better part of a year of rehabilitation. In most of the other subjects, however, recovery was not so long delayed in this regard. For example, the 5 subjects examined at R20 were no longer hydremic. Some data for these men are summarized in Table 148.

TABLE 148

PLASMA AND CELL VOLUMES, per kg. of body weight, for 5 subjects at various periods of the Minnesota Experiment.

Subject Number	Control		S24		R11		R20		R35–R55	
	Plasma	Cells	Plasma	Cells	Plasma	Cells	Plasma	Cells	Plasma	Cells
2	46.5	42.9	74.2	43.9	72.3	52.3	42.4	34.8	47.5	39.1
104	44.8	37.7	60.7	38.6	45.5	34.2	39.0	30.7		
109	40.0	32.4	66.8	55.5	53.4	35.5	43.0	37.8	40.1	32.3
112	44.9	38.0	54.1	32.1	49.8	33.9	46.1	41.0	44.5	39.3
130	50.2	41.5	81.9	40.5	62.6	38.2	43.3	38.9	40.9	40.7
M	45.5	38.5	67.5	42.1	56.7	38.8	42.8	36.6	(43.3)	(37.9)

The Extracellular Fluid

The total mass of fluid in the body which is not contained within the cells — that is, the extracellular fluid — is a compartment of considerable interest, because it is presumably this compartment which undergoes the major changes when alterations in hydration occur. The development of methods for estimating this space has been proceeding for the past decade but the amount of critical work and the data reported for normal and diseased persons remain small. So far, however, it appears that at least a reliable relative estimate can be made by determining the dilution of a known amount of thiocyanate when this is injected into the body (Crandall and Anderson, 1934). Lavietes, Bourdillon, and Klinghoffer (1936) concluded that thiocyanate, sucrose, and inorganic sulfate are distributed through approximately the same fraction of the body fluids and that this fraction is, in all likelihood, the total extracellular fluid.

Although the distribution of thiocyanate in the body is not ideal or simple, its peculiarities do not prevent its use for a relative estimate of the extracellular fluid. Winkler, Elkington, and Eisenman (1943) concluded that the thiocyanate dilution method is a useful relative measure, but that it is not satisfactory as an absolute measure because in 3 dogs the thiocyanate space seemed to be larger than the space available for dilution of radioactive chloride and sodium. It should be noted, however, that Winkler and his colleagues obtained in their 3 dogs a thiocyanate space of about 36 per cent of the body weight, while Mellors et al. (1942) obtained an average of 25.9 per cent in one series of 9 dogs and an average of 28.7 per cent in another series of 9 dogs. These latter values are very close to what Winkler et al. concluded was the true extracellular space in their dogs from studies with Cl^{38} and Na^{24}. In any case, it is not certain that the distribution of these substances in dogs coincides with their distribution in human beings. It is clear that the percentage of the body represented by thiocyanate space is less in the normal man than in the normal dog (Molenaar and Roller, 1939; Stewart and Rourke, 1941).

In the Minnesota Experiment it was not possible to apply the thiocyanate method in the pre-starvation (control) period, but measurements were made at the end of semi-starvation and at various times during rehabilitation. For estimating the pre-starvation values we have several guides. Twenty-one control measurements were made on normal young men who lived with the starvation subjects but who were in a normal state of nutrition. The average thiocyanate space in these measurements was 23.64 per cent of the body weight ($SD =$ ±1.65). It may be noted that Molenaar and Roller (1939) found 20 to 25 per cent of the body weight for the thiocyanate space in normal persons. Finally, measurements of the thiocyanate space were made on a number of the starvation subjects late in rehabilitation, from 33 to 58 weeks after the end of semi-starvation. For 6 subjects studied at R33 the average was 23.33 per cent, range 21.3 to 25.3; for 7 subjects studied at R58 the average was 23.67, range 21.2 to 26.1. The grand average for these 13 measurements, which represent a post-experiment control, was 23.51 per cent of the body weight.

Near the end of semi-starvation (S22 to S24) measurements were made on 17 of the subjects. The thiocyanate space averaged 33.98 per cent of the body weight, range 30.3 to 38.7 ($SD =$ ±2.84). This large increase in the relative hydration of the body was only slowly removed in rehabilitation. After 5 weeks of refeeding the average had dropped to 28 per cent, but at R11 it was up to 30.5 per cent. With the subsequent institution of unlimited feeding, however, the thiocyanate space rapidly returned to the normal level; at R19 the average was 23.75 per cent of the body weight in the 7 men studied at that time. These data are summarized in Table 149.

TABLE 149

THIOCYANATE SPACE IN SEMI-STARVATION AND SUBSEQUENT REHABILITA-
TION. The control data are for men fully comparable to the
semi-starvation subjects. (Minnesota Experiment.)

	Control	S24	R5	R11	R19	Final
SCN space (% total weight)						
M	23.64	33.98	27.97	30.53	23.75	23.51
Maximum	26.3	38.7	32.5	34.3	25.8	26.1
Minimum	20.2	30.3	20.8	25.9	22.3	21.3
SD	1.65	2.84	2.98	2.37	1.29	1.43
Number of cases	21	17	15	16	7	13

The abnormality of the hydration in starvation is best indicated by calculation of the thiocyanate space which is in excess of the normal proportion; on the average, the normal may be taken to be 23.5 per cent of the body weight for the present material. If F is the fraction of the body made up by the thiocyanate space and W is the body weight, then the excess hydration in starvation could be estimated as:

(1) $$E = FW - 0.235W, \text{ or } E = W (F - 0.235)$$

But this calculation neglects the fact that the mineral mass of the skeleton, which makes up about 4 per cent of the body weight, is unchanged in starvation, so a more accurate formulation would be:

$$(2) \qquad\qquad E = FW - 0.235 \ (W - 0.04W_o)$$

where W_o is the body weight before starvation.

Equation (2) applied to the average data for S24, R5, R11, and R19 yields the results summarized in Table 150. At the end of semi-starvation these men had in their bodies an average of 6.25 liters (13.8 lbs.) of excess extracellular fluid, and a large part of this excess hydration persisted throughout the first 12 weeks of rehabilitation. But this excess hydration had almost completely disappeared by R19.

TABLE 150

Average Excess Hydration, expressed as liters and as percentage of the body weight at the time, at the end of semi-starvation and after 5, 11, and 19 weeks of rehabilitation. These estimates refer solely to extracellular fluid; there is no information on the water contained in the cells. (Minnesota Experiment.)

	S24	R5	R11	R19
Excess SCN space				
Liters	6.25	3.11	4.72	0.17
Percentage of weight	11.74	5.71	8.14	0.24

The data of Beattie, Herbert, and Bell (1948) confirm, so far as they go, the Minnesota findings on the thiocyanate space (see Chapter 43). The percentage of the total body mass represented by thiocyanate space in men who had been reduced to about 75 per cent of their normal weight was almost identical with that found in the Minnesota Experiment. The mean value for the thiocyanate space in the German subjects was 34.2 per cent ($SD = \pm 1.43$) of the body weight at the time; the mean value for the Minnesota subjects in the semi-starved state (at S24) was 33.98 per cent ($SD = \pm 2.84$). Beattie et al. also observed that a considerable part of the excess (relative) hydration persisted for as much as 56 days on a regimen of limited rehabilitation.

A recent study of a 54-year-old man who fasted 45 days indicates that the behavior of the extracellular fluid in acute starvation is much the same as in prolonged undernutrition (Sunderman, 1947). This man, who was previously close to the normal average weight for his height and age, lost about 40 lbs. (18 kg.), or roughly 30 per cent of his pre-starvation weight. At the end of his fast there was no visible clinical edema but both plasma volume and thiocyanate space were represented by abnormally large fractions of the total body weight at that time; the serum volume amounted to 57.5 cc. per kg., and the thiocyanate space accounted for 33.4 per cent of the total body weight. The absolute volumes, however, were not enlarged when expressed in terms of the pre-starvation weight, being 37.8 cc. of serum per kg. of pre-starvation weight and a thiocyanate space of 23.6 per cent of the original weight. After 43 days of rehabilitation the absolute volumes had not greatly changed from the values found at the end of the fast, but they then corresponded closely to the normal relationships, being 47 cc. of serum per kg. of actual weight and a thiocyanate space of 24 per cent of the total body weight.

Intravascular versus Extravascular Hydration

On the average, both extracellular fluid (thiocyanate space) and plasma volume, when calculated per unit of body weight, rise in starvation and fall in rehabilitation. What, then, is the correlation between the relative hydration of the blood and that of the tissues? Since the extracellular fluid space includes the plasma, it is necessary to subtract the volume of the latter from the former if two distinct body compartments are to be compared.

Thiocyanate is reputed to penetrate the red cells to some extent, so it would be proper to correct for this fact, but the exact relationship between the concentration of thiocyanate in the plasma and that in the red cells is not yet certain. This correction would be relatively small in any case, so no serious error would arise from the assumption that the thiocyanate space less plasma volume is a measure of the extravascular fluid, including interstitial fluid, and therefore is a general index of tissue hydration.

The total data in these terms from the Minnesota Experiment are summa-

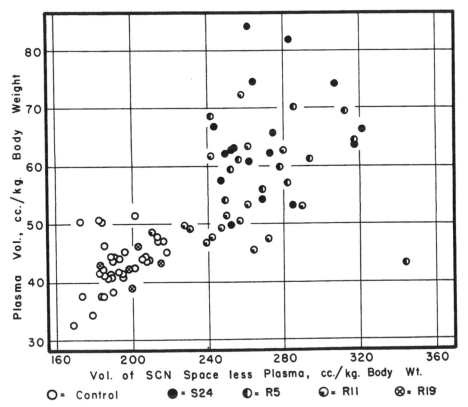

FIGURE 50. CORRELATION BETWEEN INTRAVASCULAR AND EXTRAVASCULAR HYDRATION in the Minnesota Experiment. Plasma volume plotted against the extravascular fluid, both in cc. per kg. of body weight.

rized in Figure 50. For the 34 pairs of observations at the normal nutritional state, the coefficient of correlation (r) between the plasma volume per kg. of body weight and the extravascular fluid volume per kg. of body weight is +0.421. For the 17 pairs of observations at S24, r = +0.215. For the entire data of 85 pairs of observations, including controls, starvation, and recovery, the coefficient of correlation is +0.706. It is clear that there is a strong general tendency for intravascular and extravascular hydration to change in the same direction and that the relation between these is definite but by no means exact for any one physiological state, but that over the whole range of states from normal through starvation to recovery the correlation is quite good.

There are few data in the literature on this point, the only relatively long series of normals being that of Stewart and Rourke (1941). Even their series of 27 pairs of observations is not ideal, since it is made up of persons ranging from 17 to 62 years of age and contains 10 males and 17 females. Such as it is, however, we have calculated from their whole series a correlation coefficient of −0.030 for the relation between plasma volume and extravascular fluid volume. The mixed nature of the Stewart and Rourke normal group is perhaps responsible for this discrepancy. In the mixed group of patients studied by Molenaar and Roller (1939) we have found a positive correlation between the plasma volume and the extravascular fluid volume, but the data were too few for proper statistical evaluation.

Obviously the relation between plasma volume and extravascular fluid volume must be variable depending upon the conditions, the degree of normality, and the type of pathology involved. The present evidence is that the two variables are fairly closely correlated in normal young men and in starvation and subsequent recovery.

Total Fat — The Specific Gravity Method

The estimation of the total fat of the body by means of specific gravity measurements (cf. Chapter 8) has been studied with what seem to be generally satisfactory results by Behnke, Feen, and Welham (1942) and Rathbun and Pace (1945). The major direct critical experiments have been made with guinea pigs but there are indirect indications that the calculations are also valid for man. In the Minnesota Experiment body weights in air and in water were measured periodically; from these data estimations of the total fat of the body might be made with this equation from Morales et al. (1945):

$$\text{(3)} \qquad \frac{M_f}{W} = \frac{5.548}{G} - 5.044$$

where M_f is the weight of the fat, W is the gross weight of the body in air, and G is the average density of the body.

Presumably the foregoing calculation would be satisfactory for the subjects before starvation and in late rehabilitation in the Minnesota Experiment. The details of the procedure and the question of the residual air in the lungs have been treated in the Appendix on methods. In the starvation and early rehabilitation phases of the Minnesota Experiment there were several complications to be considered. For example, there were very large changes in the hydration of the body at these times. In effect, body tissues were replaced to some extent by a lighter material, interstitial fluid, so that the constants for a normal body cannot apply without some correction.

The simplest way to correct for the abnormal hydration of the body would be to subtract the excess fluid, in terms of both weight and volume, in calculating the density of the body. The volume of the excess fluid can be estimated from the measurements of the extracellular fluid (thiocyanate space) as indicated in an earlier section of this chapter. The weight of this excess fluid requires data on its density. We have been unable to find useful data in the literature on the density of interstitial fluid or edema fluid from cases of famine edema. Buglia (1911) measured the specific gravity at 15° C. of the thoracic duct lymph of dogs; at minimal protein concentration the average for 5 animals was 1.0148. Lymph from the popliteal lymphatic near the node has a slightly lower protein concentration (about 3.5 per cent) and a correspondingly lower average specific gravity of 1.0128 (at 20° C.), as reported by Tsuji (1933).

Unquestionably these values are higher than those for true interstitial fluid or the edema fluid in cases of famine edema. The best value would probably be that for the actual edema fluid present in our own subjects. For this purpose we have made specific gravity measurements on edema fluid collected at the end of starvation. Only the clear fluid collected by a needle tap from the swollen ankles was used; it showed no trace of blood. This fluid had a protein content averaging 0.2 per cent or less and a specific gravity compared to water which averaged 1.0078 at 23° C.; measurements at 34° C., which were technically not quite as satisfactory, gave a specific gravity of 1.0089 for the edema fluid. The equation for the corrected body density would then be:

$$(4) \qquad\qquad G = \frac{W - 1.008E}{V - E}$$

where V is the volume displacement of the body and E is the volume of excess hydration.

But it is clear that the specific gravity method for calculating body fat must presume not only relatively normal hydration but also a relatively normal relation between the denser skeletal mass and the lighter material of the other nonfat tissues. In starvation, as we have seen, the skeletal mass, or more particularly its mineral component, remains practically constant so that it forms a larger fraction of the total mass than normal. Ideally, then, the proper procedure would be to correct both the mass and the volume of the total body for this "excess" skeletal fraction. Since the bony skeleton is normally about 16 per cent of the total body weight and has a density of about 1.4 (Tsai and Lin, 1939; Morales et al., 1945), it would seem possible to correct both weight and volume of the body for the excess skeleton in the starved state. But such a calculation would ignore the fact that the skeleton contains water, fat, and protein as well as minerals, and there may well be important changes in the proportions of these nonmineral components. A more satisfactory allowance for the skeletal disproportion in the starved man would be based on the bone minerals.

As we have seen, bone mineral normally represents about 4 per cent of the body weight. This bone mineral has an average density of 2.9 to 3 (Brekhus and Armstrong, 1935). For the purpose of calculating fat by the density method we must be sure that the density value used represents the body free from the disturbing influence on the density produced by excess water and bone or bone mineral. The proper equation is:

$$(5) \qquad G = \frac{W - 1.01E - 0.04\,[W_o - (W - 1.01E)]}{V - E - \dfrac{0.04\,[W_o - (W - 1.01E)]}{3}}$$

where W_o is the original body weight. This equation is more simply written:

$$(6) \qquad G = \frac{1.04W - 1.05E - 0.04W_o}{V + 0.013W - 1.013E - 0.013W_o}$$

Body Fat in the Minnesota Experiment

It should be noted that the value of G obtained in Equation (6) is for the specific gravity of the body less excess fluid and bone. The percentages of fat were recalculated using this figure. The average values for raw and corrected specific gravity and body fat are summarized in Table 151. Specific gravity was measured in the number of men indicated in the table; in general, the measurements of thiocyanate space involved a smaller number of men. But since thiocyanate space enters into the calculations as a percentage term and the individual variations in this are not great, it seemed justifiable to use averages even though they did not always refer to the same men. It will be observed that the "raw" and the uncorrected values for the specific gravity are quite similar; in general, the corrections for excess hydration and excess skeleton tend to cancel out. The mean corrected specific gravity of 1.0871 at the end of semi-starvation (S24) indicates 6 per cent fat in the "reduced ideal" body, or 5.2 per cent of the actual total weight. The uncorrected value of 1.0887 indicates exactly the same value for the fat percentage of the actual weight. At S12 the correction is of more consequence; the "raw" specific gravity value of 1.0838 corresponds to a fat content amounting to 7.5 per cent instead of 8.7 per cent of the actual body weight.

The sharp decline and subsequent even greater rise in the fat compartment of the body is clearly shown in Table 151. Perhaps the most significant feature is the excessive fat deposited in rehabilitation. This persisted for a very long time; even more than a year after the end of semi-starvation the average for 8 men was about 10 per cent above the pre-starvation control for the same men (see Table 153). It will be recalled that at this time the thiocyanate space was back to normal and had been so for months.

The Active Tissues

For our present purposes the "active" tissues can be defined as the difference between the body weight and the sum of the fat plus the thiocyanate space plus the bone minerals; this would include the red cells, or at least a considerable

TABLE 151

Body Fat in starvation and in rehabilitation. Mean values, for number of men as indicated, for actual body weight at time of specific gravity determination (mean weight), for body weight less excess fluid and bone mineral (reduced weight), uncorrected specific gravity (raw specific gravity), specific gravity as calculated from Equation (6) in the text (corrected specific gravity), calculated total body fat as percentage of the reduced body weight, fat as percentage of the actual total body weight, total fat in kilograms and as a percentage of the control value. Column headings indicate the periods. (Minnesota Experiment.)

	Control	S11	S23	R12	R16	R20	R32	R58
Number of men	32	32	32	32	12	12	6	8
Mean weight (kg.)	69.34	57.77	53.63	59.76	68.92	70.88	72.43	69.80
Reduced weight (kg.)	69.34	54.58	46.41	54.28	66.88	70.77	72.43	69.80
Raw specific gravity	1.0704	1.0838	1.0887	1.0775	1.0653	1.0595	1.0560	1.0594
Corrected specific gravity ...	1.0704	1.0803	1.0871	1.0765	1.0660	1.0602	1.0560	1.0594
Fat								
Percentage of reduced weight ...	13.9	9.2	6.0	11.0	16.3	18.9	21.0	19.3
Percentage of total weight	13.9	8.7	5.2	10.0	15.8	18.9	21.0	19.3
Total fat								
Kg.	9.64	5.02	2.78	5.97	10.90	13.38	15.21	13.47
Percentage of control	100	52.1	28.8	61.9	113.1	138.8	157.8	139.7

TABLE 152

Final Average Values for the Body Compartments, adjusted on a percentage basis to the averages for the 32 men at all periods. Values in kg. and as percentages of the total body weight for control, 12 and 24 weeks of starvation (S12, S24), and 12, 20, 33, and 58 weeks of rehabilitation (R12, R20, R33, R58). "Active" tissue is calculated as the total body weight less the sum of the fat, bone mineral, and thiocyanate space. Total body weights are the averages for the weeks indicated. (Minnesota Experiment.)

	Control		S12		S24		R12		R20		R33		R58	
	Kg.	% Wt.	Kg.	% Wt.	Kg.	% Wt.	Kg.	% Wt.	Kg.	% Wt.	Kg.	% Wt.	Kg.	% Wt.
Body weight	69.50	100.00	56.90	100.00	53.63	100.00	59.76	100.00	70.88	100.00	72.48	100.00	69.80	100.00
Bone mineral	2.78	4.00	2.78	4.89	2.78	5.29	2.78	4.65	2.78	3.92	2.78	3.84	2.78	3.98
Plasma	3.15	4.53	3.18	5.59	3.41	6.48	3.09	5.17	3.04	4.28	3.01	4.15	3.05	4.37
Red cells	2.74	3.94	2.11	3.71	1.99	3.78	2.01	3.36	2.60	3.66	2.55	3.52	2.66	3.81
SCN space	17.13	23.50	17.21	30.25	17.88	33.98	17.93	30.00	16.88	23.75	17.51	24.15	16.61	23.79
Fat	9.66	13.90	4.95	8.70	2.78	5.18	5.97	10.00	13.38	18.87	14.30	19.72	10.61	15.20
"Active" tissue	39.93	57.45	31.96	56.16	29.19	55.51	32.07	53.66	37.84	53.39	37.89	52.27	39.80	57.02

284

part of their mass, in the "active" compartment. The active tissues are capable of some alterations in hydration but no allowance for such changes is at present possible; we have discussed the question of intercellular edema in Chapter 44.

The mean values for the active tissues in the Minnesota Experiment are summarized in Table 152. Changes in the proportion of the active tissues to the total body weight were not very great; the gross loss of body weight in starvation only slightly underestimates the loss in the active tissues. This is of consequence, of course, in considering the metabolic changes. In rehabilitation there was a lag in the restoration of the active tissues. At R20, when the total body weight had surpassed the pre-starvation level, the mass of the active tissue was still some 8 per cent below the control value.

TABLE 153

BODY COMPARTMENTS, in kg., for each of 8 men studied in the control period before starvation and again 58 weeks following the end of semi-starvation (Final). "Active" tissue represents the difference between total body weight and the sum of the thiocyanate space, fat, and bone mineral. In all cases the latter value was taken to be 4 per cent of the control body weight and is not tabulated separately here. (Minnesota Experiment.)

Subject Number	Body Weight		SCN Space		Fat		"Active" Tissue	
	Control	Final	Control	Final	Control	Final	Control	Final
2	73.4	71.3	17.2	16.7	12.5	12.8	40.8	38.9
22	65.3	66.4	15.4	15.4	9.5	10.0	37.8	38.4
23	69.0	72.0	16.2	16.1	12.1	13.3	37.9	39.8
101	64.7	65.0	15.2	14.9	11.6	11.4	35.3	36.1
102	68.2	66.1	16.0	17.3	13.6	10.9	35.9	35.2
109	79.1	80.2	18.6	17.0	20.6	24.1	36.7	35.9
112	61.9	66.0	14.5	16.6	6.1	9.6	38.7	37.3
130	65.7	71.4	15.4	16.8	13.8	17.5	33.9	34.5
M	68.41	69.80	16.06	16.35	12.48	13.70	37.13	37.01

In the later stages of rehabilitation the computations are complicated by the fact that relatively few men were available for all the measurements needed. These men were a fair but not a perfect sample of the original "population." In order to preserve proper proportionality, the calculations for R33 and R58 were made in percentage terms and the results were then applied to the mean body weights for the entire group. The degree of consistency of the material is shown in Table 153, which gives the observed values on 8 men studied at control and at R58.

With all due allowances for these complications in sampling, it seems clear enough that by R58 the active tissues had been restored to values very close to those characterizing the men before starvation. It appears that at R33, however, the active tissues may not have been fully restored.

Biochemistry

"The most essential connection between the animal organism and the environment is that brought about by certain chemical substances which must continually enter into the composition of the given organism, i.e., the connection through food."

I. P. PAVLOV, in "Some General Facts about the Cerebral Centers," *Transactions of the Society of Russian Physicians*, 1909–10.

CHAPTER 16

Nature of the Biochemical Problems

THE basic defect in starvation and undernutrition is chemical; the supply of the materials for chemical energy is inadequate. This fact alters both the quantities of chemical reactants in the tissues of the body and the rates of their reactions. In fasting there are changes, notably those associated with ketosis, which may be termed qualitative, but these are presumably only the result or expression of unusually large and rapid quantitative changes in certain reactions. In ordinary undernutrition and semi-starvation the chemical characteristics of the body and its metabolism appear to be only quantitatively remarkable. As will be apparent in subsequent sections and chapters, there are no indications that fundamentally new or different metabolic processes are involved either in the catabolism of undernutrition or in the anabolism of rehabilitation. The biochemical problems of starvation, then, are specific aspects of the problems of metabolism in general, and their full solution cannot precede the solution of the problems of ordinary (non-starvation) metabolism.

It is not suggested, however, that the special biochemical problems of starvation cannot be attacked in the present stage of incomplete knowledge regarding ordinary metabolism. At the very least, biochemical comparisons between the starved and the ordinary states of the body should be made. How do they differ in terms of the amounts — and kinds — of chemical materials present in the body? The oldest approach, that of simple comparison between intake and outgo, is of limited value in indicating processes taking place in the body and is technically troublesome even for the estimation of resultant gross changes in body composition. The sheer bulk of materials and the structural relationships in which they exist are grossly suggested by the morphological studies which have been summarized in preceding chapters. Chemical morphology remains in a rudimentary state. One tissue, the blood, may be sampled and analyzed directly; besides its intrinsic interest, it reflects dimly and very incompletely the chemistry of the active tissues. The urine, in turn, reflects the blood and so brings third-hand information about the chemistry of the tissues. Direct studies on the composition of the tissues themselves in starvation have been very few, at least with modern methods.

At best, information on the composition of the body, its organs and tissues, by itself provides only a static picture. Comparisons between the normal and the starved states show the resultant changes. Comparisons over time indicate rates of change; this is the beginning of dynamic analysis. The next level of question has to do with mechanism. How are these changes wrought? Furthermore, important questions arise in regard to over-all function. How does the changed body composition affect the integrated biology of the individual? Theoretically this latter question has both morphological and physiological aspects.

Obviously, the most significant and most difficult questions are those which have to do with the rates of the various metabolic processes and their regulation and control. These determine the alterations in the bodily composition as well as the functional characteristics of the organism and its component parts. Elementary consideration is enough to show that in this area the approach of a restricted biochemistry is inadequate; the determinants are equally physiological. As a matter of fact, this imprecision of the demarcation between biochemistry and physiology, though sometimes operationally troublesome, must be accepted in favor of a larger biology, free from any hierarchy of disciplines, in which the classification of variables is adjusted to the analytical needs. The biochemical problems here are labeled as such only because the items of measurement are best expressed in chemical terms or, when some aspects of larger questions are conveniently segregated, temporarily as chemical problems.

Energetics and Metabolic Regulation

It might seem entirely reasonable that the energetic processes of the body diminish in intensity as the exogenous food supply (potential chemical energy) is reduced. Certainly this seems to be useful to the starving organism in that it minimizes the imbalance; it is reasonable in the sense that a wise man reduces his expenditures when his income is cut. In terms of machines, the analogy would be that of an automobile which would reduce its speed, but not stop, whenever the gasoline in the tank began to get low. The most interesting feature is that this favorable adjustment which occurs in starvation is not directly a result of the law of mass action, though it may depend on it in the long run.

In any heterogeneous system, such as the body, or a single organ, or even a single cell, the character and rate of the chemical reactions are determined both by the reactants and by their presentation to each other. Under ordinary conditions, the renewal of reactants in suitable juxtaposition, brought about both by catalysts, including enzymes and hormones, and mechanically by all forms of circulation and displacement, presumably exerts a more limiting control than the supply of the reactants themselves.

Minced or sliced tissues initially always have much higher rates of metabolism, that is of chemical reaction, than the tissues *in situ,* and only a drastic reduction in the concentration of the substrate reactants suffices to limit the reaction rates. The addition of easily metabolized materials like glucose or ketonic acids prolongs the high metabolic rate of normal excised tissues but does not elevate the initial rate. In other words, the metabolism of isolated tissues is relatively unaffected by considerable changes in the concentration of energy-rich substances in the substrate, the metabolic rate being determined by the enzyme concentrations.

The above statement is subject to the important qualification that the mechanical situation must be constant. High rates of metabolism in excised tissues are attained only when there is maximal renewal — circulation, agitation — of the reactants at the site of reaction. In the isolated but intact organ the metabolism is readily reduced by decreasing the rate of perfusion to a low level.

There are, then, at least 3 possible means whereby the rate of metabolic processes may be reduced in the starving organism. The first, simple fuel exhaustion, cannot be excluded on the basis of the present inadequate information about the composition of the tissues, but we can hazard the opinion that it is ordinarily not very important, except perhaps in the terminal state, where a condition like insulin shock may develop. The second, decrease in oxidative enzymes in the tissues, would seem likely to occur but the question requires investigation. The possibility of the third, reduction in circulation, is supported by the fact that the circulation is certainly reduced in starvation; the total blood circulation is apparently reduced more than in proportion to the total basal metabolism (see Chapter 28). But the question may be raised as to which is cause and which is effect. Finally there is the important question as to whether the metabolic rate per unit of metabolizing tissue is actually reduced. Calculations from the data of the Minnesota Experiment indicate that there is a real, though slight, reduction in the basal metabolism per unit of total active tissue, that is, excluding fat, extracellular water, and bone mineral (see Chapter 17).

If our calculations are approximately correct, there remains the question as to how much of the reduction in metabolism can be accounted for by the lowered body temperature of the starving organism. But this again raises a question. What is the mechanism of the reduction in body temperature?

The Composition of the Body

Chemical data on the composition of the body in severe undernutrition are surprisingly deficient. For man in particular, even data on the gross composition of the normal body are very few and of dubious applicability. The one man analyzed by Mitchell et al. (1945) died in heart failure and so probably was not in a normal state of hydration. It seems captious to complain about the absence of data on enzymes in the starved body when there is a dearth of good information about such elementary things as fat, glycogen, and water.

There seems to be no doubt that there is a relative superabundance of extracellular water in the starved body (see Chapter 15). More data on this point would be valuable but a more intriguing question remains regarding the intracellular water. There are histological suggestions that starvation produces intracellular edema, but this question, previously inaccessible to direct examination, should now be investigated with the new biochemical methods.

All experimental evidence is in agreement that there is very little loss of bone mineral in starvation. Yet in every area of famine it is possible to find cases of hunger osteomalacia. Is there a loss of total bone mineral in these cases and, if so, how much? In any event, what is the biochemical fault which produces this condition in a few people in a starving population?

Metabolism of Brain and Nervous Tissue

The morphologists are agreed that the brain and other nervous tissues lose very little weight in starvation and that, though histological changes can be discerned, these are usually not large (see Chapter 9). Functionally the brain seems to be little affected by quite severe degrees of undernutrition (see Chapters 39 and 42). In rats at least, gross alterations in total metabolism produced by experimental hyperthyroidism and hypothyroidism do not seem to alter the metabolic rate of the brain (Fazekas, 1946). The metabolism of the brain can be reduced by circulatory or oxygen deficiency but when large changes are so produced irreversible lesions result (Schmidt *et al.*, 1945). All this would suggest that the metabolism of the brain may continue with small change until late in starvation.

But if this is so, there is a quantitative problem. The metabolic rate of the brain is normally very high and in man at rest it is estimated to account for a fourth of the total oxygen metabolism of the body (Schmidt *et al.*, 1945; Kety and Schmidt, 1945, 1946). In severe undernutrition, as in the Minnesota Experiment, the total basal metabolism of the body frequently falls by 40 per cent, and there are no signs of impairment of function of the brain. On the basis of no changes in the metabolism of the brain, this would mean that the total basal metabolism of all the rest of the body must fall, not by 40 per cent, but by almost 50 per cent. This would indicate a very considerable decline in the metabolic rate of the other active tissues and would substantially alter the analysis of metabolic adjustment. Clearly these are interesting points for investigation.

The peculiarities of brain metabolism (cf. Folch, 1947) present other problems. Since the brain ordinarily derives all its energy from carbohydrate metabolism even in starvation ketosis (Mulder and Crandall, 1941), the conclusion is forced that: (1) the brain's use of carbohydrate is reduced in starvation, or (2) the respiratory quotient of the rest of the body is much lower than indicated by the total R.Q., or (3) the estimates of the large proportion of the total basal metabolism in man accounted for by the brain are in error. A sample calculation can be made. For the total basal metabolism of the body in the normal state we can take it that carbohydrate accounts for about 60 per cent; of this some 40 per cent, according to Kety and Schmidt (1945, 1946), would be used by the brain. Now in the severely undernourished state the total carbohydrate metabolism is reduced by something over 40 per cent. If the brain metabolism is not at all or only slightly affected, this means that the metabolism of the brain would account for some 70 per cent of the total carbohydrate used by the whole body. This may actually be the case, but these are challenging points for study.

The Nutritional Status of Fats

It is generally agreed at present that fats, except for minute amounts of certain fatty acids, are not essential in ordinary nutrition. Since fats are readily synthesized in the body from carbohydrate, this is understandable, and there might seem to be an obvious justification for rejecting claims from famine areas that their populations suffer from fat deficiency. But this may be an oversimplification.

In the body, fats are of two main types: (1) deposit fat, which is a useful energy reserve but otherwise has little metabolic significance, and (2) true tissue fat which is either an essential part of the cellular architecture or is otherwise directly involved in the vital machinery. When there is caloric inadequacy and carbohydrate is in great demand for energy metabolism, is it possible to maintain the true tissue fat without an exogenous fat supply? It is conceivable that from this standpoint it is not a matter of indifference whether fat or carbohydrate is provided in the diet when the total is calorically inadequate. There is no answer for this question at present, but it would seem to have both theoretical and practical importance.

Nutritional calculations are universally made on the basis of the calorie system; carbohydrate and fat calories are considered interchangeable. In heat units this is true, of course, but the efficiency of these calories for various physiological purposes does not seem to be identical. For example, in the production of human muscular work one fat calorie is apparently equivalent to only about 0.9 of a carbohydrate calorie (cf. Keys, 1945a). There are indications that under non-starvation conditions in man a high fat diet must be fed at a higher calorie level than a low fat diet to maintain the same body weight, at least over a period of some weeks (Mickelsen and Keys, 1948). This may be in part a result of a changed tendency in body hydration, but there are problems to explore.

Studies on the growing rat indicate that on isocaloric diets the efficiency of caloric utilization, nitrogen retention, and rate of growth all decrease as the percentage of fat in the diet is diminished below about 30 per cent (Forbes et al., 1946a, 1946c). In the mature rat the proportion of fat in a liberal isocaloric diet had little effect on nitrogen utilization but did influence the loss of energy in the feces, so that the fecal calories increased as the fat in the diet increased from 2 to 30 per cent (Forbes et al., 1946b, 1946d). Indirectly this seems to be in agreement with our own studies on well-fed men.

In experiments on rats with ad libitum feeding, it has been reported that liberal fat (20 to 40 per cent) in the diet produced better growth and general performance than did low fat diets and that different caloric intakes did not explain all of the difference (Deuel et al., 1947). With isocaloric feeding of rats at a severely restricted level, mortality was greatest on a low fat diet and recovery was best with a relatively high proportion of the calories of the rehabilitation diet in the form of fat (Scheer et al., 1947a, 1947b). These observations on rats led these workers to suggest that the participation of fat in the total metabolic machinery is more direct and important than previously supposed (Scheer et al., 1947b). Such evidence would seem to support popular and governmental insistence in food shortage areas that fats be a major item in relief supplies.

There are numerous problems concerning the details of fat metabolism in prolonged undernutrition. Those concerning certain particular lipoids are potentially of much importance. Cholesterol metabolism is a case in point. The ramifications into many fields, including the development of atherosclerosis, the production of steroid hormones, and so on, are obvious. As will be seen in Chapter 22, the evidence is fragmentary and confused.

The Use of Deposit Fat

The mechanism of the removal and utilization of deposit fat in semi-starvation presents interesting problems. Descriptively, fat is progressively withdrawn

and burned, but this is not necessarily uniform throughout the body. Fatty infiltration and degeneration in the liver, heart, and other organs is sometimes described in severely emaciated persons. In any case, how is the utilization of fat
brought about? It is not simply a case of exhaustion of carbohydrate with subsequent recourse to fat as a fuel of second choice. Although the glycogen reserves
are quickly reduced in severe undernutrition, they are not totally depleted, and
the blood sugar is seldom extremely low before the terminal stage.

There are indications that the pituitary gland is importantly involved in such
adaptive changes in the utilization of nutrients in the starving body. So far at
least, examination of the metabolites themselves does not indicate the mechanism, and the presumption is that there must be controlling hormone influences
(see Chapter 22 for further discussion). Beyond this point at present there is
only speculation. But these questions are not idle. The preferential use of fat as
fuel is imperative in starvation; without this adjustment the persistence of life
would be short indeed. Actually it seems clear that different individuals do differ markedly in their resistance to undernutrition, and they do not all collapse
or die at the same level of emaciation. It may be presumed that such differences
are, at least in part, dependent on differences in the hormonal regulation of the
nutrients metabolized. The difference in resistance of men and women may well
have some such basis.

The Products of Cellular Disintegration

Starvation atrophy involves reduction in cell size, degenerative changes in
some cells, and in extreme cases final dissolution of the cells. What happens to
the various substances which were present in the cell originally? Are some of
them retained preferentially or does the proportional composition of the cell remain unchanged in starvation? It is possible that in the course of degeneration
the contained enzymes, hormones, and vitamins are destroyed *in situ*, or alterna-
tively they may simply be released for use, or at least have effects, elsewhere.

Some of these points are discussed in Chapter 21, but it may be noted here
that direct evidence is absent. These questions are important. The difference
between the total cellular mass before and during starvation may be considered
to have represented an addition to the exogenous food supply during starvation.
The estimation of the vitamin needs in starvation would be quite different depending on whether the intracellular vitamins are or are not made available to
the body as a whole. Similar questions arise for other substances. One may ask,
In what form is the nitrogen of the disintegrating cell released? Are the cellular
proteins first destroyed? Such questions are particularly pertinent in the case of
the endocrine glands and their contained hormones but they have not been
studied, although there are many reports on changes in total mass and in the
histological appearance of the endocrine glands (see Chapter 11).

The properties of the cell in starvation degeneration have not been well
studied, so it is possible only to speculate on such matters as the permeability
of the cell wall and the mechanism of the progressive diminution in size and
eventual disintegration of the cell. Why does the nucleus retain its size so much
better than the total cell? It may well be, of course, that these questions in starva-

tion are no different than in other forms of degeneration, but in any case they are intriguing. The erythrocyte in starvation does not diminish in size, but instead increases, and there is a macrocytic anemia (see Chapter 14). What physico-chemical explanation is there for this, and why is iron ineffective in preventing it?

The Response to Infection

Some of the problems of infectious disease in relation to undernutrition are discussed in Chapters 46 and 47. The over-all results of undernutrition in terms of incidence, course of the diseases, and mortality are of immediate practical consequence and have their theoretical implications as well. But such evidence scarcely touches the biochemical problems.

Where the initial introduction of the infecting agent must be through barrier or limiting membranes, the chemical character of these membranes will deter-mine, at least in some measure, the amount and rate of penetration of the infect-ing organism and is therefore an important part of the complex which is sus-ceptibility. It is unthinkable that starvation has no influence on these barrier membranes, but the factual data are extremely few. Assuming the initial infec-tious exposure or inoculation is the same — that is, that a given number of equiva-lent pathogens have got into the blood stream or tissue in which multiplication may take place — then subsequent events depend on several factors of a chemical nature: (1) the suitability of the substrate to support growth and multiplication of the pathogens; (2) the capacity of the body to synthesize antibodies which can destroy or restrict the pathogens; (3) the capacity of the body to produce and mobilize other defenses, including phagocytes, encystment, and limiting cir-culation; (4) the capacity of the body to invoke thermal means to destroy some pathogens; (5) the capacity of the body to neutralize, detoxify, and excrete toxic products of the pathogens; and (6) the total vigor or margin of strength of the body to support its own ordinary demands as well as those made by the pathogens and the defensive processes. The possibility must also be considered that the body may suffer irreparable damage by overviolent response of the de-fensive mechanisms.

In each of the above categories of items which will influence the course of events, and therefore the total response and resistance, there clearly must be de-pendence on the composition of the tissues, the rate and course of the several metabolic processes, and the availability of ingredients to supply energy and to be used as constituent parts of influential chemical substances. In short, it is in-conceivable that any one of the determining items is independent of the nutri-tional state. But the direction of effect on the individual items, or on the net final outcome, cannot be predicted on a priori grounds. Here is a very large series of problems, different for different types of pathogens, which is almost unexplored. Only the question of antibody formation has received attention and this leaves much to be desired (see Chapter 46).

Cellular Rehabilitation

In rehabilitation the atrophic, partly degenerated cells must be restored to their normal or former states. Just what chemical problems are involved in this cannot be stated in the absence of detailed information on the normal and the starved cell. In any case, however, there are certainly large differences to be restored and the question may be asked, Are the processes in rehabilitation merely a reversal, a mirror image, of those of atrophy? There is reason to think that it is more complicated than this. At least the over-all composition of the body is not restored in quite this way, and the pattern of water, fat, and active tissue restoration is complicated (see Chapter 15).

Wound Healing and Growth

The biochemical problems of wound healing and growth in the undernourished animal are still more complicated than those of rehabilitation of the starved but otherwise ordinary adult cell. Again the over-all picture may be examined with profit even though a detailed analysis of the chemical processes has not yet been attempted. Some descriptive and statistical studies on growth have been made (see Chapter 45), and these are of great interest in demonstrating the tremendous growth potential of young tissues and young organisms during and after caloric inadequacy. Wound healing and regeneration in the adult presents problems which are related but far from identical, and the influence of caloric deficiency has been very little studied.

The Nitrogen "Requirement"

One of the commonest questions raised in connection with the deterioration observed in semi-starvation is whether the caloric deficiency or the protein deficiency is the more important. Similarly, in rehabilitation the question is asked whether emphasis should be placed on calories or on protein in the diet. These questions, which admit of no simple answer, are treated in some detail in Chapters 18 and 19. Here we will only note some of the broad problems involved.

In spite of many decades of attention to the problem of the nitrogen requirement at caloric balance, controversy continues. When the caloric intake is or has been reduced the problem is considerably more complicated. The metabolism of proteins and of calories cannot be separated in any strict sense except at very high food intakes, but evaluation of their relative effects in undernutrition is essential if only for practical reasons of food planning and distribution.

In the adult it is generally assumed that the minimum nitrogen requirement should be the lowest level of intake which will allow exact nitrogen balance. But different values are indicated with different proteins in the diet and questions arise about both digestibility and utilizability ("quality"). It would be impossibly tedious to arrive at exact balances by trial and error with all varieties of food proteins and in all combinations. The problem is only relatively simplified by using the extrapolation method, which is based on the linear relation between nitrogen balance and nitrogen intake at certain levels of nutrition (cf. Mitchell, 1948, p. 64).

If the limiting requirement of the body is considered to be set by the endogenous catabolism apart from any intake, then the starting point would be an attempt to measure this endogenous catabolism. The first suggestion is that this is represented simply by the nitrogen excretion at zero nitrogen intake.

Justification for this approach is offered by such statements as "the nitrogen output in the urine in previously well nourished men during the early days of fasting is remarkably constant" (Lusk, 1928, p. 91). But examination of the data offered in proof of this contention shows that in the first 2 or 3 days of fasting the nitrogen excretion reflects the previous diet and that thereafter the excretion tends to decline progressively. This is clear from the summarized data cited by Lusk, particularly when recalculated on a percentage basis as done in Table 154. As fasting is continued for protracted periods the nitrogen excretion falls to much lower levels. For example, at around 3 weeks of fasting the 24-hour urinary output of adults is commonly of the order of 3 to 4 gm. of nitrogen, or less than half the value in the first few days of fasting.

TABLE 154

TOTAL URINARY NITROGEN EXCRETION IN FASTING MEN. All values calculated as percentages of the values for the third day. "All" is the mean for the 5 men. Recalculated from Lusk (1928, p. 91).

	Cetti	Bre	Succi, 1	J.A.	Succi, 2	All
Day 3	100	100	100	100	100	100
Day 4	94	96	92	100	102	96.8
Day 5	82	89	92	84	106	90.6
Day 6	78	74	73		104	82.3

The obvious abnormality of the fasting man as an indicator of normal metabolism led to studies in which caloric intake was maintained on a nitrogen-free diet. But here again the nitrogen excretion slowly but progressively declined to very low levels (Smith, 1926; Deuel et al., 1928). A really fixed level of endogenous catabolism cannot be found with any nitrogen-free regimen, even when calories are supplied in abundance. Mitchell (1948) has argued that the decline in nitrogen excretion in these cases is analogous to the fall in the basic metabolism when the caloric intake is grossly deficient. This may be true, but it still leaves the question of how to estimate the intrinsic catabolism of the body both in the normal state and on inadequate intakes.

It may be that at least rough estimates for undernourished men can be made on the basis of the generalization offered by Terroine and Sorg-Matter (1928) that endogenous nitrogen metabolism is closely related to the basal energy metabolism. This relationship seems to hold fairly well in comparing different species of animals (Smuts, 1935) and is accepted by Brody (1945) and Mitchell (1948), who admit, however, that it is affected by age and other factors. The relationship suggested is that 2 mg. of urinary nitrogen represent endogenous catabolism for each calorie of basal heat. The examination of this theory in chronic starvation and in nutritional rehabilitation, where there are large changes in basal metabolism, should be of much interest.

Suitable data are not available to test whether or not the postulated relationship between basal nitrogen loss and basal energy metabolism holds in prolonged undernutrition in man. But it seems to be a reasonable proposition, and there is some indirect evidence from the Minnesota Experiment that is supporting as far as it goes. This is the demonstration that the active tissue mass changes roughly as does the basal metabolism in the starvation phase. On the other hand, similar evidence in the rehabilitation phase suggests that the relationship may not hold when tissue is being restored. In any event it seems worth while to calculate the potential effects of the operation of this principle, assuming that it holds in starvation.

To begin with, the basic nitrogen requirement before starvation must be estimated for the Minnesota men who then averaged 70 kg. in body weight. The exact value of this nitrogen requirement is of no consequence, but for the kind of diet used in semi-starvation and, for the men who received no supplementary protein, also in rehabilitation, 30 gm. of protein daily cannot be far from actuality, accepting the data of Lauter and Jenke (1925), Bricker, Mitchell, and Kinsman (1945), Hegsted *et al.* (1946), and Murlin *et al.* (1946a, 1946b). At that time — that is, before starvation — the basal energy use (Cal. per day) averaged 1570 Cal. But after 24 weeks of semi-starvation the basal energy use had fallen to an average of about 958 Cal. per day. If the basal nitrogen requirement fell in proportion, its value would then have been 18.3 gm. of protein daily. A protein intake which would have just met minimal requirements before starvation would, after semi-starvation, provide an excess of 11.7 gm. of protein.

Obviously, if this principle holds even approximately, nitrogen balance can be secured at extremely low protein intakes in starved individuals, and very moderate intakes should, if accompanied by sufficient calories, suffice to establish a substantial positive balance of nitrogen. What are the actual facts? In the Minnesota Experiment, as indicated in Chapter 19, special protein supplements in the rehabilitation period had little effect on recovery. It would be understandable that raising the protein intake from something like 5 times the basal balance requirement, as on the unsupplemented rehabilitation diet, to 6 times, as with the protein supplement, might have little or no effect on the speed of tissue restoration or any of the functions dependent on it.

Efficiency of Use of Body Protein

It is conceivable that not all the endogenous catabolism of protein observed on a nitrogen-free diet is obligatory in the sense of representing a necessary final destruction of protein as a part of the machinery of metabolism. If there is caloric balance, for example, why should there be any substantial nitrogen excretion? The "wear and tear" concept has never been clearly defined; it is presented largely as a philosophical explanation of a puzzling fact. But it is possible to offer a more specific theory on the basis of recent demonstrations of the dynamic nature of many if not all complex structures in the body.

We may conceive of the body proteins and amino acids as constantly in dynamic equilibrium with each other, with the several enzymes competing to produce synthesis and analysis at all times. The amino acids, however, are also susceptible to deamination, so that the over-all picture may be represented as follows: tissue protein $\leftrightarrows$ amino acids $\rightarrow$ urea $+$ glucose, etc. It is inconceivable that the whole protein synthesis is accomplished in a single step or that each of the several amino acids is built into the structure by an identical set of processes governed by identical mass action equations and facilitated to exactly the same degree by the appropriate enzymes. In other words, the incorporation of the various amino acids into the protein molecule may well be accomplished with differing degrees of difficulty, and the optimal concentrations of the several amino acids for synthesis may not correspond to the proportion of these amino acids in the completed protein molecule.

Such a concept, which seems eminently reasonably in theory, has far-reaching implications. From it could be deduced the proposition that a small intake of limiting amino acids could result in a smaller total nitrogen excretion than would obtain on a protein-free diet or even in the early stages of total fasting. Specifically, it may be invoked to explain findings such as those of Swanson (1946) regarding the effect of eggs in reducing nitrogen excretion.

The basic finding was that when a small amount of whole egg is fed to rats subsisting on a diet very low in protein, the urinary nitrogen excretion is decreased (Swanson, 1946). Subsequent studies showed that several individual amino acids possess some of the nitrogen-sparing action of eggs for rats and that methionine is particularly active in this way (Brush et al., 1947). The methionine effect was confirmed in dogs by Allison, Anderson, and Seeley (1947) and in both dogs and rats by Cox et al. (1947). Carefully conducted experiments on man, however, have failed to reveal any such effect of methionine. R. M. Johnson et al. (1947) studied young men on diets providing 600–2000 Cal. and 5–75 gm. of protein daily. Cox et al. (1947) studied surgical patients receiving intravenous alimentation, infants on a luxury diet, and young men on a protein maintenance diet and again after 21 days of protein depletion (see also Chapter 18).

The failure to obtain a favorable effect of methionine in man is disappointing and provides another proof that it is unsafe to rely on animal experiments for predicting nutritional phenomena in man. It must be noted, however, that the principle may still apply to man. The results simply indicate that methionine is a "critical" amino acid in the rat and the dog but not in man. Some other amino acid, or combination of amino acids, may be critical for man.

The Question of Protein Reserves

It is now generally agreed that, even though the body has no specific and discrete reservoir of protein, some portion of the total body protein can be used for emergencies without significant injury to vital tissues. Moreover, it is believed that this relatively free protein store can be altered by dietary means. If this is true, then the consequences of subsistence on a starvation regimen could differ according to the preceding state of protein nutrition, independent of the obvious effect of relative obesity.

This question has received little attention and seemingly could only be studied in controlled experiments. There are, in fact, two questions having to do respectively with quantity and with quality of protein in the preceding diet. Neither question has been investigated in situations involving prolonged undernutrition. Experiments with rats fasted for 24 hours have indicated some differences in the rate of weight loss dependent on the kind of protein in the preceding diet (Harte et al., 1948).

The Effect of Dietary Balance Preceding Undernutrition

Human dietaries in normal times, that is, in the absence of special food short-ages, differ widely in the proportions of the calories supplied as fat and as car-bohydrate. Do these affect the responses to subsequent dietary inadequacy? Conceivably there may be different metabolic patterns established by such vari-ations in energy sources. In acute starvation and in chronic starvation on ex-tremely inadequate calories, the metabolic diet is necessarily a high fat type and the peculiarity of this in comparison with the preceding period of normal feed-ing may be of consequence. Presumably the enzyme systems are established in accordance with the habitual diet and would be most efficient in metabolizing the customary mixture of food energy types. There would be, in effect, a prefer-ential food utilization corresponding to the character of the previous habitual diet.

There are a few studies on this question in acute starvation. In adult male rats forcibly fed a high fat diet for 3 to 6 weeks, sudden withdrawal of food re-sulted in more speedy and pronounced ketosis than in animals previously fed a high carbohydrate diet (Roberts and Samuels, 1943a). The fasting nitrogen ex-cretion was not affected in these rats. Such rats also showed different responses to administration of glucose and of insulin after a brief fast, the fat-fed animals exhibiting a more pronounced and prolonged glycocemia after sugar and a high-er insulin tolerance (Samuels et al., 1942; Roberts and Samuels, 1943b). From these and other evidences these workers concluded that "the foodstuff predom-inately burned by the extra-hepatic tissues during the early stages of fasting cor-responds to the major constituent of the previous diet" (Roberts et al., 1944, p. 644).

Unpublished experiments in the Minnesota Laboratory of Physiological Hy-giene failed to demonstrate similar effects in men who subsisted on high fat and high carbohydrate diets for periods of several weeks. Possibly a much longer period of subsistence on the special diets would show the phenomena in man. In any case, the possible influence of the character of the preceding normal diet on the course of events in partial starvation is pretty much a blank page.

The "Premortal Rise" of Nitrogen Excretion

According to common scientific belief the excretion of nitrogen from endoge-nous sources does not pursue a steady course in starvation. At an advanced stage of undernutrition there is supposed to occur an increased outpouring of nitrogen, associated presumably with the final dissolution of the degenerating cells. The idea is theoretically attractive and might seem to have practical utility. It would be useful to have a metabolic criterion for the severity of undernutri-tion, especially for a critical point in advanced emaciation. This so-called "pre-mortal rise" of nitrogen excretion might seem to be available for this purpose. The generality of the phenomenon is not questioned in monographs on starva-tion (e.g. Morgulis, 1923) and is reported without question in many textbooks.

It is hard to understand why this concept has been so universally accepted. As we shall see, the actual evidence is quite different from expectation. This seems to be another example of the perpetuation of error by the selection of individual data to suit a particular idea. Like the belief that the heart is resistant to starvation, this idea can be traced back more than 80 years to C. Voit (see Chapter 10).

The first experiment on nitrogen excretion in starvation was performed on a cat by Bidder and Schmidt (1852). The cat died after 18 days without food, and there was no rise in nitrogen excretion at any time. A year later Bidder fasted a rabbit for 12 days and again the nitrogen excretion did not rise. This rabbit did not die but the starvation was severe; the survival duration of rabbits deprived of food varies from 3 to 19 days. It is interesting that reports of a premortal rise of nitrogen excretion are by no means confined to experiments where the animals succumbed; the term *premortal* would be a misnomer in any case.

The classical experiment is that of Voit (1866) on one cat which survived 13 days of acute starvation. Voit was aware of the discrepancy between the earlier findings and the sudden late increase of nitrogen loss exhibited by his cat; the variability of the phenomenon was expressed in the title of his paper, "Ueber die Verschiedenheit der Eiweisszersetzung beim Hungern." Like Voit, later experimenters were aware of the irregularity of the phenomenon, although some tendency toward a late rise could be discerned in the majority of the experiments during this period. By 1901 the total material was still scanty, comprising 19 rabbits, 8 dogs, 2 cats, and 4 chickens, mostly studied in Voit's laboratory (Falck, 1877; Feder, 1878; Schimanski, 1879; Rubner, 1881; Kuckein, 1882; Schöndorf, 1897; Schulz, 1900, 1901; Kaufmann, 1901).

In spite of the variability of the results, the premortal rise was presented as an established and regular phenomenon as early as the influential textbooks of von Bunge (1898) and Hammarsten (1899). Apparently these textbook statements have not been really questioned for 50 years.

Voit explained the differences in the responses of individual animals as being related to the fat reserves of the body; the premortal rise might be expected more regularly in animals that were initially lean. Other investigators agreed on the importance of the initial fatness, but its functional significance was the subject of controversy (Schulz, 1900, 1901; Kaufmann, 1901). In spite of their controversy Schulz and Kaufmann agreed that the term *premortal rise* suggested by Voit was unfortunate because the rise in nitrogen excretion sometimes occurred rather early in starvation, mortal injury was not necessarily involved, and in some rare cases the nitrogen excretion even showed a "premortal fall."

Variability of results continues to characterize the more recent literature. Boy (1933) claimed that a premortal rise is rather regularly seen, but the results of Chambers, Chandler, and Barker (1939) were inconstant; of their 11 dogs, 7 showed a rise and 4 did not. The species seems to be of importance. In rabbits the premortal rise may be more prominent than in other species. Hogs are very resistant to hunger, and studies of this species failed to elicit the phenomenon (Terroine and Razafimahery, 1935). In none of the rather numerous studies of prolonged fasting in man was there any sign of a late rise of nitrogen excretion (cf. Benedict, 1915b).

There is no evidence that a premortal rise occurs in any species in chronic undernutrition or semi-starvation. In Morgulis' dog (1923), whose body weight dropped from 13.94 to 8.27 kg. in 9 weeks of undernutrition, the nitrogen excretion was elevated in the fifth and sixth weeks, when the diet was somewhat altered, but there was no trend in the succeeding 3 weeks. Experiments with a protein-deficient diet with adequate calories gave conflicting results (Schulz, 1900; Kaufmann, 1901). Severe protein depletion in dogs, resulting in losses up to 40 per cent of the body weight in 7 to 11 weeks, resulted in a decrease rather than in any premortal rise of nitrogen excretion (Whipple *et al.*, 1947). Studies on man in semi-starvation and famine give no support to the idea of a premortal rise.

It would not be surprising — and it would not really touch the fundamental idea — if there should be increased fecal nitrogen loss in famine victims with severe ulcerative lesions of the bowel, such as were seen in India by Porter (1889); such studies have not been made, however. Whipple, Miller, and Robscheit-Robbins (1947) suggested that a rise in nitrogen excretion shortly before death is an accidental phenomenon associated with a superimposed terminal infection.

The idea of a premortal rise can lead to ridiculous suggestions. Loewy observed a rather high nitrogen excretion on himself toward the end of World War I when there was a severe food shortage in Germany; it was suggested that this was a premortal rise in spite of the fact that Loewy, though underweight, was really quite fit (Zuntz and Loewy, 1918).

Severe debilitating diseases are often attended by a stubborn tendency toward a negative nitrogen balance. That hormonal influences are involved is strongly suggested by the effects of preparations like testosterone propionate in reducing or even reversing the imbalance. It is interesting to speculate whether the mechanisms may operate to produce changes of nitrogen excretion in acute starvation. It should be noted that in normal man fluctuations in nitrogen excretion from day to day and over periods of a few days are regularly seen but are as yet unexplained.

CHAPTER 17

Basal Metabolism

THE basal metabolic rate (B.M.R.) is probably the most important element in the total energy metabolism for several reasons. It constitutes, at a level of light or moderate physical activity, the major part of the total energy exchange — and a part that cannot be voluntarily changed. With the restriction of physical activity as observed in severe or moderate undernutrition, the energy deficit of starvation diets is very largely determined by the B.M.R. Because of the importance of the B.M.R. in influencing the dietary requirements under starvation conditions, all factors which modify it, such as age, race, sex, and climate, together with the distribution and variation of normal values, must be considered in order to evaluate any changes that may appear during a period of caloric insufficiency.

After World War I there was considerable discussion as to whether the decrease in B.M.R. which occurs in starved individuals is a physiological adaptation to the reduced caloric intake (Rubner, 1930; Brugsch, 1927, 1928; Benedict *et al.*, 1919). Of all possible metabolic adaptation mechanisms, a decrease in the B.M.R. would be the most important in its over-all significance.

To answer this question of metabolic adaptation, it is necessary to know whether the decrease of the B.M.R. corresponds to the loss of body tissue. The disagreement of different authors on this question is largely due to the various ways of expressing the B.M.R. and the various units to which the B.M.R. can be referred. For this reason, a discussion of the physiological significance of the various units seems to be appropriate. We believe that none of the units used before the Minnesota Experiment is adequate to decide the question of metabolic adaptation.

Definition and Methods of Expression

The basal metabolic rate (B.M.R.) is defined as the metabolic level obtained in a resting subject, in the supine position, 12 to 14 hours after a moderate meal. This is not the lowest rate which can be obtained; the rate may be lower during profound sleep. Also, the preceding level of nutrition has some influence; ample nutrition tends to increase the B.M.R. (Grafe, 1923). For this reason Krogh (1916, 1923) suggested a standardized diet several days before the B.M.R. determination. Krogh used the term *standard metabolism*, which appears to be more objective than the term *basal metabolic rate*, which was introduced by Plummer and Boothby (1924) and has been generally accepted in this country. Magnus-Levy (1895) used the term *Grundumsatz*, which has been used in the

German literature since that time. Magnus-Levy and Zuntz expressed the B.M.R. as cc. of oxygen consumption or calories per minute, and Lusk and DuBois expressed it as calories per hour, while Benedict preferred the expression as calories per 24 hours.

Before World War I B.M.R. determinations were made only for research purposes. The B.M.R. of the individual was usually compared with the values for a small group of "normal" persons of the same sex and of similar weight and age. The custom developed of expressing the results in terms of oxygen consumption or calories per unit of body weight. After the value of the B.M.R. for clinical purposes was recognized, the need for normal standards became urgent. The considerable bodies of data collected by Aub and DuBois (1917) and Harris and Benedict (1919) afforded material for empirical development and testing of many formulas.

In all the debates two criteria have been used almost exclusively: (1) the formula should eliminate or minimize the differences among normal individuals, and (2) it should be easy to apply and should involve a minimum of measurements. Occasionally a third requirement, that the formula be theoretically reasonable, has been emphasized (e.g. Kleiber, 1947). It is surprising, however, that so little thought has been given to the criterion of sensitivity for demonstrating abnormality. Presumably, Benedict (1938, p. 194) had this point in mind in his rather maladroit plea for "emphasizing the non-uniformity in heat production," which seems to have been misunderstood by Kleiber (1947).

The theory that the heat production of animals of different sizes is related to their surface rather than to their mass was vaguely suggested more than a century ago (Sarrus and Rameaux, 1839) but was more explicitly stated, with quantitative examples, by Rubner (1883, 1902) and Richet (1889). The table of Voit (1901a) impressively demonstrated that differences in oxygen consumption between various homeotherms largely disappear when expressed in units of body surface. Rubner originally conceived of the "surface law" as directly related to the heat exchange at the body surface but later abandoned this in the face of overwhelming evidence against such a simple explanation (cf. Wels, 1925).

Aub and DuBois (1917) used the body surface (square meters of skin) in calculating their standards. Harris and Benedict (1919) stated that their analysis was not based on preconceived ideas and arrived at their formulas: $H = 66.473 + 13.7516W + 5.003S - 6.755A$ for men, and $H = 655.0955 + 9.5634W + 1.8496S - 4.6756A$ for women. (H is the heat production (Cal.) for 24 hours, W is body weight in kg., S is height in cm., and A is age in years.) They claimed their formulas, tested on their material of 167 men and 103 women, were better — that is, showed less individual variation — than surface area relationships. The same data were used by Dreyer (1920) to derive a formula excluding stature which he considered equally good:

$$H = \frac{W^{0.5}}{K + A^{0.1333}}$$

in which the constant K is different for males and females. This formula was later modified (Stoner, 1923, 1924) by estimating weight from sitting height or chest circumference. Dreyer's formula has been much used in England in spite of its lack of theoretical elegance.

The rather large literature on the merits of the various standards does not need to be reviewed in detail; in general, mathematical comparisons have been made with considerable naïveté. Except in rare instances (DuBois and DuBois, 1915), body surface has not been estimated independently but has been calculated from weight, as in Meeh's (1879) formula, or from height and weight. The equation for surface area (A, in square meters) developed by DuBois and DuBois (1916) proved to have an average error of only 1.7 per cent and has been generally used in all recent work. The equation reads:

$$A = W^{0.425} \times H^{0.725} \times 71.84.$$

All earlier B.M.R. standards have been largely superseded by the Mayo Clinic standards of Boothby, Berkson, and Dunn (1936) and Berkson and Boothby (1936). These are based on 639 males and 828 females selected as normals from some 80,000 measurements of B.M.R. at the Mayo Clinic. The age range is from 5 to 70 years. The body surface was estimated from the equation of DuBois and DuBois (1916). The large material and the systematic and discriminating use of statistical procedures makes this the best available information on age and sex trends of the B.M.R. A most valuable feature of the Mayo standards is a nomogram indicating the probability of occurrence in a normal population of particular basal metabolic rates.

The actual B.M.R. values in the Mayo material refer to apparently normal persons who have undergone the test for the first time under the conditions prevailing at Rochester, Minnesota. As will be noted in more detail in the next section, these values are appreciably higher, perhaps by 10 to 12 per cent, than more strictly basal values.

As good as the Mayo standards are, they are not primarily concerned with the question of the best method of expression for the metabolic rate. As a matter of fact this question has no meaning unless "best" is precisely defined. Arguments about the virtues of this or that exponent which may be applied to the body weight in estimating the metabolism for an animal are mostly academic and have the undesirable effect of distracting attention from the more important problems of the mechanics and regulation of metabolism.

Experimental Limitations

Of the several methods for determination of the B.M.R., only the calculation of the caloric output from the oygxen consumption, by means of Zuntz' "indirect calorimetry," has attained general application. This method, to be precise, requires the determination of the fat and carbohydrate proportion of the non-protein respiratory metabolism by means of the respiratory quotient (R.Q.). Since the protein proportion is rather constant and usually not very large (about 10 per cent) and the protein R.Q. (about 0.83) is not far from the usual non-protein R.Q., only a slight error would occur if no correction for the nitrogen excretion were made. Consequently, the nitrogen excretion is very rarely determined for the purpose of B.M.R. measurement. In most closed circuit models used for the clinical B.M.R. determination, only the oxygen consumption is measured. Since no R.Q. is obtained, an average caloric value for oxygen corresponding to a respiratory quotient of about 0.85 must be assumed. This might involve a maximum error of about ±5 per cent.

The B.M.R. is usually determined in short periods of 5 to 10 minutes; this procedure gives lower values than longer periods of one or several hours (used in respiratory chambers), because it is difficult to maintain complete relaxation for so long a period. The procedures for short-run experiments, and consequently the B.M.R. values, vary; some authors use only one run, others 2 or 3 consecutive runs of 5 or 6 minutes each. Some authors use the lowest value out of 2 or 3 measurements, while others prefer the averages. DuBois (1936) suggested making 3 tests and if the 2 lowest agree within 5 per cent to use their average as the best estimation.

Of great importance is the factor of training, which probably means the degree of relaxation. The importance of training has long been recognized, and the discrepancy in the data of several authors may be largely due to this factor. DuBois (1936) estimated that repeated tests give values about 5 per cent below those obtained in untrained subjects, but this estimate may be too low. In the Minnesota Experiment group of 34 "normal" subjects, the control average for the B.M.R. was —11.8 per cent below the Mayo standards, and this is in agreement with results on a rather large number of other healthy, young subjects investigated in this Laboratory during the past 5 years. According to Vogelius (1945) the average training factor amounts to 8 per cent. The training factor depends very much, of course, on the setup and procedure in a given laboratory as well as on the emotional type of the subject. Boothby, Berkson, and Dunn (1936) were aware of the training factor, but because of the difficulty of a correct estimate and in view of the usual clinical procedure, only the results of first-run measurements were considered in their standards. This limitation is important for the correct interpretation of some results and the over-all estimate of the total energy expenditure.

In the usual clinical routine procedure, little attention is paid to the diet prior to the B.M.R. determination. Krogh (1923) and Kleitman (1926) emphasize the importance of this factor; in Kleitman's series, the B.M.R. per 24 hours varied between 1355 and 1914 Cal. depending upon the calorie content and especially the protein content given the day before the B.M.R. measurement. This effect, probably due to prolonged specific dynamic action (luxus consumption), might well account for differences in the average B.M.R. in limited samples of population. Deuel et al. (1928) found that a protein-free diet for a period of 63 days produced a definite decrease in the basal metabolism, and this is corroborated by observations on Eskimos living on a meat diet (Heinbecker, 1928) and on a mixed diet (Heinbecker, 1931). However, Benedict and Roth (1915) found no difference in the basal metabolic rate between vegetarians and non-vegetarians. In Boothby, Berkson, and Dunn's (1936) study no mention is made of the diet before the B.M.R. determination. Differences up to 5 or 10 per cent of the B.M.R. between a subject on a restricted or just sufficient diet and those making up the Mayo or other standards might be explained as due to the absence of the prolonged S.D.A. and need not necessarily indicate an actual decrease of the B.M.R.

In several studies the B.M.R. in a state of undernutrition was compared to that of small groups of "normal" subjects. Unfortunately, in most cases sufficient data were not given to make possible a recalculation in terms of real B.M.R. standards. That the B.M.R. of a subject in a famine area before starvation is known is the exception rather than the rule. How large the deviations must be in order to be safely attributed to undernutrition would depend on the range of distribution of the normal material.

For discussions of various factors which might affect the B.M.R., the reader may refer to DuBois' (1936) or Lusk's (1928) monographs.

Sex, Age, Climate, and Race

It is well known that the B.M.R. drops with age, but the rate of decline is not uniform and differs somewhat for males and females. The B.M.R. of women is consistently lower than that of men, but owing to the somewhat different age trend, the discrepancy is greatest between 18 and 22 years of age. Standards for children are discussed in a later section of this chapter.

TABLE 155

BASAL METABOLIC RATE of individuals of the same race living in different climates, expressed as percentage deviation from Aub and DuBois' or Harris and Benedict's standards. A deviation within ±5 per cent is regarded as not significantly different from these standards.

Race and Location	Men	Women	Author
White			
Batavia	0		Eykman (1896)
Batavia	0		Eykman (1921)
Java	−8		Radsma and Streef (1932)
Tropics	−10		Knipping (1923a, 1923b)
Brazil	−20		De Almeida (1920a, 1920b)
Brazil	−8		De Almeida (1924)
New Queensland	−22	−25	Sundstroem (1926, 1927)
Sydney	−9	−9	Hindmarsh (1927)
Australia	−7		Wardlaw, Davies, and Josephs (1934)
Madras		−10	Mason (1934)
Beirut		−6	Turner and Aboushadid (1930)
Beirut	0		Turner (1926)
China	−6	−6	Earle (1928)
Argentine	0		Mazzocco (1928)
New Orleans	−17	−17	Hafkespring and Borgstrom (1925)
Chinese			
America (U.S.)		−10	MacLeod, Crofts, and Benedict (1925)
America (U.S.)		−18	Benedict and Meyer (1933)
America (U.S.)		0	Wang (1934)
China	−8	−8	Earle (1928)
China	−8	−9	Necheles (1932)
Filipino			
New York	0		Basset, Holt, and Santos (1922)
Philippines	−7	−10	Ocampo, Cordero, and Concepción (1930)

TABLE 156

B.M.R. OF NATIVES IN TROPICAL, SEMITROPICAL, AND TEMPERATE ZONES, expressed as percentage deviation from Aub and DuBois' or Harris and Benedict's standards, a deviation of ±5 per cent being regarded as coinciding with the standards.

Race and Location	Men	Women	Author
Native New Queenslanders.	−22	−25	Sundstroem (1926)
Black			
Australia	−30	−30	Wardlaw and Horsley (1928)
Australia	−13	−16	Hicks, Matters, and Mitchell (1931)
Australia	−12		Wardlaw and Lawrence (1932)
Australia	−10		Wardlaw, Davies, and Josephs (1934)
Malay, Dutch East Indies.	0		Eykman (1896)
Malay, Dutch East Indies.	−10	−7	Van Berkhout and Teding (1929)
Malay	−15		Knipping (1923a, 1923b)
Various native races			
India		−17	Mason and Benedict (1931)
India	−13		Mukherjee and Gupta (1930–31)
India	−7		Banerji (1931)
India	−11	−17	Krishman (1932)
India	0	0	Bose and De (1934)
India		−17	Mason (1934)
Syria		−13	Turner and Aboushadid (1930)
Japan[*]	0	0	Okada, Sakurai, and Kameda (1926)
Japan[*]	0	0	Takeya (1929)
Japan[*]	0	0	Takahira (1925)
Japan[*]	0	0	Kisé and Ochi (1934)
China	−8	−8	Earle (1928)
China	−8	−9	Necheles (1932)
China, Szechwan	−2	−3	Kilborn and Benedict (1937a)
China, Miao race[*]	+12		Kilborn and Benedict (1937b)
Philippines	−7	−10	Ocampo, Cordero, and Concepción (1930)
South America (Brazil)			
Various native races	−14		De Almeida (1924)
Black	−20		De Almeida (1920b)
Cuba	0	0	Coro (1930)
Cuba	−16	−13	Montoro (1921–22)
Jamaica (brown)	0	0	Steggerda and Benedict (1928)
Chile,[*] Araucanian Indians	+10	+14	Pi Suñer (1933)
Yucatan, Maya Indians	+5		Williams and Benedict (1928)
Yucatan, Maya Indians	+6		Shattuck and Benedict (1931)
Yucatan, Maya Indians	+8		Steggerda and Benedict (1932)
Baffin Islands, Eskimos	0		Heinbecker (1931)

[*]Temperate zones.

The effect of climate, particularly tropical climate, is summarized in Table 155, based on observations on white subjects, Chinese and Filipinos in temperate and tropical zones. Although there are some discrepancies in the observations of several authors, it appears that the B.M.R. of white subjects is decreased in tropical and subtropical zones. The B.M.R. of Chinese is low in China as well as in the United States. Kilborn and Benedict (1937a) found normal values for the B.M.R. of Chinese in Szechwan, which was attributed to emotional strain due to civil war conditions (see Table 156). The material in Table 155 concerns people who had been living in the tropical zones for a year or more. A short time in a tropical climate increases the B.M.R. of white subjects according to Knipping

(1923a, 1923b), while Radsma and Streef (1932) obtained opposite results. According to unpublished experiments made in this Laboratory, a stay of 4½ days by 6 subjects in an air-conditioned suite at a room temperature of 49° C. (120° F.) and 25 per cent relative humidity decreased the B.M.R. by 15 per cent.

Table 156 summarizes the results obtained with natives in various countries The B.M.R. of natives in Australia, India, China, and South America (at sea level) seems to be substandard, while the B.M.R. of Japanese is within the accepted standards. It is interesting that the B.M.R. of Araucanian and Maya Indians tends to be somewhat above the ordinary standards. Since most of the data were obtained in comparatively small samples of the populations, it is hard to decide whether or not the basal metabolic rate of natives of tropical countries is actually lower than that of white people in these regions. The values are obviously overlapping. Sundstroem (1926, 1927), whose values for New Queenslanders are among the lowest reported, denied the existence of racial differences, and so did De Almeida (1920b) for the B.M.R. of white men and Negroes in Brazil. On the other hand, Mason and Benedict (1931) and Mason (1934) believed that there is a racial factor of difference between natives in India and white subjects living in India. Whether the reported slight differences are true racial differences or are due to the effect of a protein-poor diet or to a low level of physical activity is open to question.

The low basal metabolism of South Indian women does not seem to be explained by a greater ability to relax; Mason and Benedict (1934) found in 7 young women (from 19 to 22 years of age) with an average B.M.R. 20.7 per cent below the Harris-Benedict standards in waking condition an average additional decrease of −9.8 per cent during sleep. This decrease is of the same magnitude as observed in white Occidentals. An effect of the general level of physical activity on the B.M.R. has been observed by Takahira (1925) and Sundstroem (1926). This might explain the comparatively high values of the Araucanian Indians, a physically active tribe.

It has been mentioned that there is a definite training trend in repeated B.M.R. determinations, probably owing to a decrease of muscle tension. It may be that the training trend is less pronounced in the Indian and Chinese races than it is in the Caucasian race, so that the first determination gives a more correct value. However, in the mixed racial population of Szechwan, the comparatively high B.M.R., which was believed to be due to emotional strain (Kilborn and Benedict, 1937a), did not show any training trend in repeated experiments. No systematic study of the training trend in different races has yet been made.

The B.M.R. of Miao-Chinese is definitely higher than that of other Chinese races (Kilborn and Benedict, 1937b). The Miao live under primitive conditions at 7000 feet altitude in the mountainous regions of southern China. The normal metabolic rate of Eskimos in Heinbecker's study was found under the condition of subsistence on a mixed diet; the metabolic rate of Eskimos living on a meat diet might be higher (Heinbecker, 1928).

Some recent investigations support the idea that the basal metabolism is slightly depressed by continued residence in a warm climate. Galvao (1948a, 1948b) studied the basal metabolism of 100 healthy white men living in Sao

Paulo, Brazil; 50 of these men were lean, but not emaciated, and 50 of them were "well-proportioned." For both groups the basal metabolic rate was significantly lower than the accepted standards for countries of the north temperate zone. Galvao's conclusion that the B.M.R. is low in the warm climate of Brazil is not supported by Orsini's (1947) studies on 250 Brazilian girls. Using careful technique and referring to the Boothby-Sandiford standards, Orsini secured results that were rather closely distributed around the North American standards. There did not seem to be any important differences between Negro, mulatto, and white girls.

Normal Range and Variability

Boothby, Berkson, and Dunn (1936) calculated the distribution of their large material around the average age curve for men and women. Table 157, based on their nomogram, shows the percentage of B.M.R. deviation which can be expected to occur in between 10 and 0.1 per cent of a normal male population of 20 and 40 years. It can be seen that 20 years' age difference changes the normal distribution curve only slightly. A deviation of ±10 per cent from the standard value would be about the border line of statistical significance, while a deviation of ±14 per cent is statistically significant since it can be expected in only 1 per cent of the normal population. The expectancy as given in Table 157 refers to the value of a single individual compared to the standard material, and not to the comparison of groups.

TABLE 157

MAGNITUDE OF PERCENTAGE DEVIATION FROM THE MAYO B.M.R. STANDARDS which can be expected to occur in random samples in 10, 5, 1, and 0.1 per cent of normal men 20 and 40 years of age (Boothby et al., 1936).

Percentage Expectation in a Normal Population	Percentage Deviation from Normal Average B.M.R.	
	20 Years	40 Years
10	±7	±8
5	±9	±10
1	±13	±14
0.1	±17	±19

There is no significant trend of the B.M.R. during the day (Benedict, 1935); the average deviation in duplicate tests made on the same day was only 1.2 per cent in 5 subjects (Griffith et al., 1929). In the same series the average deviation of daily determinations was only 2.5 per cent from the monthly means and 3.4 per cent from the yearly means. The extreme variations for the same day, month, or year were 4.1, 8.1, and 9.7 per cent. Benedict's (1935) B.M.R. in 2 periods of 18 and 33 consecutive days varied between 216 to 242 and 223 to 237 cc. oxygen consumption per minute. The B.M.R. was not affected by variations of length and depth of sleep, the normal fluctuations of room temperature, or variations in the composition and amount of the meal the evening before. In contrast,

Krogh (1923) and Kleitman (1926) found an effect of the content of the meal the day before. There is general agreement about the constancy of the B.M.R., so that the above data may suffice as an example of a larger number of references (cf. DuBois, 1936).

Even over a period of many years the B.M.R. is remarkably constant (Benedict, 1915), although, of course, the age trend must be considered. For instance, in Carpenter's study (1933) the subject T. M. C. was 11.3 per cent above the average at the age of 30 and 15.9 per cent below at the age of 49. The B.M.R. of E. F. D. B. (DuBois, 1936), measured on 29 occasions from his thirtieth to his fifty-second year, varied between 33.2 and 38.4 Cal. per square meter per hour (average 36.1), with a tendency to decline with age (about 38 Cal. per square meter per hour at 30 years, about 34 to 35 Cal. at the age of 52).

Physiological Significance of Units

In the earlier literature the basal metabolic rate was expressed mostly as oxygen consumption or calorie expenditure per unit of body weight. It is surprising how long this procedure of calculation was continued after Voit (1901) and Rubner (1902) had shown that the basal metabolism is proportional to the surface area of an animal. For example, the horse and the mouse are much the same per unit of body surface, but per unit of weight the basal metabolism of the mouse is almost 20 times that of the horse.

Different formulas have been suggested, as we have noted, for standards of basal metabolic rate in man, the most important being Harris and Benedict's (1919) linear formula and Aub and DuBois' surface formula. Rubner's original working hypothesis, that the actual heat loss (which would follow Newton's surface law) determines the heat production, was quickly abandoned after demonstration that the surface relationship holds also at thermal neutrality. Rubner, however, maintained that the surface law was the result of a long-range climatic adaptation to the heat loss in the course of mammalian evolution. This, of course, is a conjecture, which was contradicted by Benedict (1915), who denied that the B.M.R. of the human body depends on the body surface. Harris and Benedict (1919) claimed that the data fitted their linear formula much better than they did the body surface formula. However, in most cases Harris and Benedict's and Aub and DuBois' standards agree rather closely. At this distance of time the controversy seems to have been largely a confusion in semantics. A large literature has accumulated, which cannot be reviewed in detail, on the agreement of various formulas with the actual B.M.R. of larger or smaller groups; the reader may be referred to Lusk (1928) and DuBois (1936). All investigators found good agreement between their material and the various standards proposed (Harris-Benedict, Aub-DuBois, Dreyer, Krogh), perhaps with a slight advantage for the Harris-Benedict formula in the majority of reports.

Berkson and Boothby (1936) recalculated Harris and Benedict's original data and denied that the linear formula fits better than the surface formula. The contrary conclusions of Harris and Benedict were due to the omission of the age factor in applying the surface area formula. In their own very large material, Berkson and Boothby failed to find any anthropometric relationships to which the data could be better fitted than to the surface area. However, there was no significant difference in the precision of prediction of heat production between Harris and Benedict's linear formula and the surface area formula. The reason for the good agreement of the two formulas is that the same variables (body weight, height, and age) are used, and any inaccuracy in either formula contributes only a comparatively small part to the total discrepancy between predicted and observed values (Berkson and Boothby, 1936). It appears that the question of the validity of the surface area for human metabolism cannot be decided by comparison of the different formulas, all of which in effect indirectly provide an estimate of the surface.

However, the surface formula is preferable for technical and biological reasons. There is reason to assume (Berkson and Boothby, 1936; DuBois, 1936) that the linear formula will be less reliable in more extreme variations of body build; the surface area formulation is relatively simpler, and the relationship of the B.M.R. to age can be studied independently of its relation to stature and weight. Biologically the use of the surface area formula brings the observations in line with the general surface law, which has been established for small and large animals beyond any doubt. It is probably because of the comparatively small range of body weight in human subjects (compared to the differences between the extremes of animal species) that the validity of the surface area formulation is not so obvious. This does not necessarily imply any causal relationship between skin surface and B.M.R. (Lusk, 1928; Simonson, 1928; DuBois, 1936); in fact, other areas have been suggested to which the heat production could possibly be related (cf. Simonson, 1928). However, the skin surface is the only one which has been exactly defined and measured.

The Active Tissue Mass

Rubner (1902) thought of a possible relationship between heat production and the mass of "active tissue," so he calculated a hypothetical "active cell surface area." He estimated the mass of active tissue in a man of 60 kg. body weight to be 37.8 kg., corresponding to an active cell surface area of 9014 square meters, or 150.2 square meters of such surface per 1 kg. of body weight, with an energy expenditure of 0.2 Cal. per day per square meter of active cell surface. In the mouse the energy expenditure, calculated in the same way, was 11 times as high per square meter of active cell surface. For this calculation an average size and shape of cells was assumed. Since the size of cells is not significantly different in small and large animals, the values are comparable. Rubner concluded that the body surface and not the active cell surface determines the heat production. His calculations are, of course, only rough estimates, and he made no allowance for other fundamental factors, such as enzyme concentrations. In this respect it is of interest that Potter (1947) found the succinoxidase concentration of kidney tissue of mice to be 10 times that of human kidney. This is in good agreement with the differences in metabolism calculated by Rubner but it may be only a coincidence. Perhaps Rubner's conception of "active tissue" is in principle a sound approach, although he himself was inclined to deny its importance.

We believe that the Minnesota method of experimental estimation of the mass of active tissue may represent some progress in this direction which might finally lead to the establishment of a more functional relationship. This method has been described in detail in Chapter 15; it consists, in principle, of the determination of the fat content and of the extracellular fluid, including blood plasma. This approach appears to be the more valuable since the validity of the formulas for metabolic standards which have been obtained in normal material is questionable in greatly abnormal states of emaciation or obesity.

B.M.R. at Minimal Normal Dietary Equilibrium

From time to time claims have been made that most people would not only get along with, but even benefit from, a smaller diet. Such claims have been based mostly on observations on a few individuals or on very limited groups of the population living under exceptional conditions. Larger groups have been studied by means of dietary surveys which are reviewed in Chapter 19. There is evidence that a warm climate tends to decrease the B.M.R. and, beyond that, to reduce the over-all energy expenditure through a reduction of physical activity (cf. Keys, 1949). However, in some of the surveys of the customary diet in southern states the calorie intake was so low that some doubt has been cast on the validity of our present metabolic standards. It is clear that the measurement of the B.M.R. is of fundamental importance in order to substantiate such claims, but such measurements have been made in comparatively few cases.

Because of the wide range of distribution of the B.M.R. in the normal population (see Table 157), it would be hazardous to generalize from observations on single individuals. Nevertheless, such claims have received widespread attention. Prominent exponents of this idea were Fletcher (1903) and Chittenden (1903, 1907).

The Cases of Fletcher and Kellogg

Metabolic studies were made with Fletcher and with Kellogg in Chittenden's, Benedict's, and Zuntz's laboratories. Fletcher was first studied by Chittenden over a period of 6 days in 1903; he was able to maintain a body weight of 74.8 kg. (at a height of 168 cm.) with a slightly positive nitrogen balance on an intake of only 1606 Cal. per day. He was 55 years of age at that time. However, Benedict and Milner (1903) found Fletcher's B.M.R. to be within normal limits in the same year. The results obtained by Benedict and Milner cannot easily be reconciled with those obtained by Chittenden. Fletcher was investigated again by Zuntz and Schirokich in February and March 1912, after he had been living for 3 months on a diet mainly of potatoes and butter. The diet during the first period of the experiments (February) amounted to 2750 Cal. and during the second period (March) to 2116 Cal., which is not particularly low. Fletcher was appreciably overweight at that time (76 kg.). His B.M.R., recalculated by us in terms of the Mayo standards, was −10 per cent. Three months later he was investigated again by Benedict, Emmes, Roth, and Smith under basal conditions; his body weight had increased to 82.1 kg. (17 kg. overweight) within the interval of 3 months. The B.M.R. was −5 per cent according to the Mayo standards — that is, completely within the normal range, the more so since Fletcher must be considered as a trained subject, so that the normal standards would be 5 to 10 per cent too high for him. While no direct information on his diet between March and May 1912 was available, it must have been quite liberal to account for a body weight increase of 6 kg. within so short a time.

T. H. Kellogg, who believed he subsisted on 1200 Cal. per day, was studied by Benedict and Carpenter (1910) in the respiration chamber when he was asleep, sitting, standing, and walking. His metabolic rate was found to be within the normal range, and his minimum food requirement (for a body weight of 56.1 kg.) was computed as probably exceeding 2000 Cal. per day.

Other Reports of Balance at Low Food Intakes

Winters and Leslie (1944) found, by means of a dietary survey, the daily average caloric intakes of a group of 12 wives of university professors, 4 faculty women, and 4 women students to be 1667, 1720, and 1920 Cal. per day, respectively. There was no evidence of undernutrition, and only one subject was underweight. The caloric intakes were definitely below accepted standards, but the B.M.R. of 13 of the subjects was within low normal limits (—5 to —15 per cent). The percentage excess of the caloric intake over the estimated B.M.R. ranged from 6 to 81 per cent in 11 subjects, while in 2 women the recorded caloric intakes were 2 and 13 per cent below the basal caloric requirements. It is obviously impossible that these 2 subjects were able to maintain their body weight. The results with these 2 women cast doubt on the reliability of the procedure of food sampling, which was left entirely to the subjects.

A careful analysis of food and excretions over a period of 7 days was made by Yukawa (1909) on 11 Japanese monks in a monastery, who were living on their customary diet with physical activity described as light. No determinations of the basal metabolism were made, but the gross caloric intake per square meter of body surface, calculated according to Miwa and Stoeltzner's (1898) formula, was, on the average, only 1129 Cal. for a group of 8 younger subjects (average age 26 years) and 1197 Cal. for a group of 3 older subjects (average age 64 years). Probably because of the vegetarian character of the diet, the food utilization was low, and the utilized caloric intake was only 1040 and 1106 Cal. per square meter of body surface per 24 hours. Stoeltzner and Miwa's formula gives values quite close to those obtained with the formula of DuBois and DuBois (Takahira, 1925). All except 2 younger subjects were in weight and nitrogen equilibrium; in fact, most subjects gained a little weight, and there was an average nitrogen balance of +0.79.

In this group of observations belongs also the experiment of Süsskind (1930), who attempted to live for 2 years according to Hindhede's recommendations on a protein-poor (33.3 gm. per day) but calorically sufficient diet (2300 Cal.). This diet produced "complete breakdown," which was not defined but was obviously not due to starvation. Süsskind was only 10 per cent underweight at that time. His B.M.R., measured in Rubner's (1930) laboratory, was within normal limits when his diet was increased to 2600 Cal. after his "breakdown." Obviously, a prolonged protein-poor diet at a barely sufficient caloric intake does not necessarily decrease the B.M.R.

The B.M.R. of Apparently Normal Underweight People

There is a rather large proportion of moderately underweight people in the normal population who do not have clinical signs, symptoms, or history of any pathology. The B.M.R. of such "thin" people is within normal limits, as has been shown in numerous investigations (Blunt and Bauer, 1922; Blunt et al., 1921, 1926; McKay, 1928; Coons, 1931; Strang and McCluggage, 1931; Stark, 1933; Strang et al., 1935). Strang (1935) reported a normal B.M.R. in 9 subjects who were approximately 25 per cent underweight from a persistently inadequate food intake.

The normal basal metabolic rate in underweight but apparently normal people is of great interest in view of the decreased basal metabolic rate produced by semi-starvation, as discussed in the following sections of this chapter. The degree of departure from normal weight is no explanation for this discrepancy, since in the Carnegie Experiment a definite decrease of the basal metabolic rate was observed after a body weight loss of 10 per cent. Many of the "thin" people investigated in the above experiments were more underweight than the subjects in the Carnegie Experiment.

Several explanations for this difference are possible, but no direct evidence to support any of these hypotheses is available. It is possible that a reduction in the metabolic rate is associated with the process of weight loss and that the normal metabolic rate is associated with caloric equilibrium. Neither the Carnegie Experiment nor the Minnesota Experiment was prolonged enough to provide adequate data on this point. An actual increase of the B.M.R. with prolonged undernutrition was observed by Zuntz and Loewy (1918).

B.M.R. in Self-Observations in World War I

Loewy's and Zuntz's self-observations (Loewy and Zuntz, 1916; Zuntz and Loewy, 1916, 1918) are of especial interest because they extend over a period of about 30 years, including the time of World War I. Zuntz's basal metabolic rate varied within the narrow limits of from 773 to 804 Cal. per square meter of body surface per 24 hours during a period of 22 years before the war (from 41 to 63 years of age). No age trend was apparent. In May 1916 his body weight had dropped from 67.5 kg. to 60.6 kg. and his B.M.R. to 708 Cal. per square meter of body surface.

Loewy's B.M.R. showed greater fluctuations (666 to 805 Cal. per square meter of body surface) from 1888 to 1914, corresponding to the period from his twenty-sixth to his fifty-second year. Again there was no age trend during that period. His body weight, which was 64 kg. in April 1914, had decreased to 61.2 kg. in April 1915, and thereafter it dropped gradually to 51.5 kg. by May 1917. His daily food intake at that time was between 1500 and 1800 Cal. with 50 to 60 gm. of protein. During the following year (until August 1918) his body weight remained at the low level of 51 kg. The B.M.R. in May of 1916 was 16 per cent lower than his normal average. However, in July 1917 the B.M.R. was again within Loewy's normal range. The authors assumed that the increase of the B.M.R. indicated a more progressive phase of malnutrition, comparable to the "premortal" increase of metabolism occasionally observed in starving animals.

While Loewy complained about weakness and fatigability, his condition, of course, was far from being *in extremis.*

The B.M.R. in Population Samples during World War I and World War II

Nitrogen balance experiments were made by Jansen (1917) in 13 normal subjects who had lost from 2 to 11 kg. of body weight owing to the war diet. The diet was not exactly known, but the experimental diet was set at 1600 Cal. for a period of 6 days. Probably the 1600 Cal. intake was lower for all subjects than their ordinary wartime diet. The nitrogen balance as well as the body weight change was negative (see Chapter 18). In 2 of Jansen's subjects the B.M.R. was determined on each of the 6 days; the average values were 33 and 31 Cal. per square meter per hour, or 17 and 23 per cent below the Mayo standards — that is, lower than could be expected in 99 or 99.9 per cent of the normal population.

The scarcity of information about the B.M.R. in Central Europe during World War I may be due to the fact that the measurement of the B.M.R. had not yet been introduced as a routine clinical procedure at that time. However, it is surprising that only a few data are available for World War II. Laroche, Bompard, and Trémolières (1941) found in 3 patients, who had weight losses of 30 to 40 per cent and edema, a B.M.R. of —15 per cent, —8 per cent, and +4 per cent. The percentage probably refers to normal standards, but no clear indication was given. Duvoir *et al.* (1942) reported the case of a man 43 years of age with a height of 183 cm., whose body weight dropped within 4 months from 106 to 54 kg. The B.M.R. was —27 per cent, probably with reference to normal standards. A similar decrease of the B.M.R. (from —20 to —30 per cent) was observed by Loeper, Varay, and Mende (1942), while Gounelle, Marche, and Bachet (1942c) found much more pronounced decreases (down to 52 per cent). Unfortunately, the basis of reference for calculation is not indicated in any of these papers. Govaerts (1947) reported on 39 cases of famine edema in adults, mostly older people. Using the standards (for body surface) of Dubois, the average B.M.R. was —23 per cent, with extremes of —60 and +10 per cent.

Beattie and Herbert (1948a) made observations on 11 German subjects who at the time of the first oxygen consumption determination were living on a ration similar to that on which they had lived for the preceding 3 months (about 1800 Cal. per day). The body weight loss in these German subjects ranged from 17 to 34 per cent. The percentage change of active tissue was estimated on the basis of the data obtained in the Minnesota Experiment. The B.M.R. changes were calculated per square meter of body surface, per kg. of body weight, and per kg. of body weight corrected for the change of percentage of active tissue. This correction was estimated to be small, however, amounting to zero for body weight losses up to 25 per cent and increasing to 10 per cent for weight losses up to 34 per cent.

The heat production per square meter of body surface was below normal (ranging between —5 and —29 per cent), while no uniform change in the rate of heat production per kg. of body weight was observed (ranging between +11 and —19 per cent). The decrease of the heat production per unit of body surface was associated with a decrease of the body weight per unit of surface area. It was concluded that the reference of the B.M.R. to the body weight (with or without active tissue correction) is the more correct procedure in semi-starvation. However, it should be noted that, in contrast to other observations and to the Minnesota Experiment, there was no consistent decrease of the Cal./kg. of body weight. Possibly this may be due to the longer duration of starvation; this question was discussed in an earlier part of this chapter.

In 70 "pure" famine cases at Warsaw a low basal metabolism was uniformly recorded, usually 30 to 40 per cent below normal in the more severe cases which made up the larger part of the group studied (Fliederbaum *et al.*, 1948). In 2 cases of extreme cachexia and dehydration, in the premortal phase, the B.M.R. was 60 per cent below normal. It is impossible to assess the exact significance of these data because of the lack of details as to methods, standards, and bases of calculation.

In these measurements at Warsaw, the respiratory minute volume was also low, usually between 2.5 and 3.5 liters, and in one patient only 1.5 liters. The respiratory quotient was also of interest in the Warsaw patients. After correcting the R.Q. for the protein metabolism estimated from the urinary nitrogen, the basal non-protein respiratory quotient was somewhat elevated in the majority of cases, several being between 0.95 and 0.98. But in the most extreme cases, where the body weight was about 50 per cent reduced, values of 0.72 to 0.74 were found.

Simonart (1948, p. 197) stated that in the undernourished men in a Belgian civilian prison the basal metabolic rate averaged —14.4 per cent and (p. 100) was as low as —55.8 per cent. in one man. For this material detailed figures are available (*ibid.*, pp. 221–23) which raise major questions as to the acceptability and meaning of the data. There were 21 men, ranging in age from 26 to 61 and averaging 45.2 years. For each test there were duplicate determinations of the oxygen used, and we have averaged these for each man. The results indicate a range of 35.2 to 71.8 Cal. per hour per square meter of body surface, with an average of 56.6 ($SD = \pm11.2$). For each man we have compared the reported B.M.R., as Cal./hr./m.^{2}, with the Mayo Clinic standards of Boothby and Berkson. This indicates that the Belgian prisoners had extraordinarily high metabolic rates, ranging from —9 to +94 per cent, with an average of +52 per cent ($SD = \pm23.7$ per cent).

We have been unable to discover the basis for Simonart's figure of —14.4 per cent. Simonart (1948) also gives the figures for 11 "normal" subjects who prove to have had, according to the published values, a mean basal metabolic rate, as Cal./hr./m.^{2}, of 81.4 ($SD = \pm9.3$). According to the Mayo standards the departure from normal averages was +98 per cent ($SD = \pm23.6$ per cent). These figures are absurd, of course, but the possibility must be considered that Simonart may have used them for his own normal standard. If this was the case, however, then his starved prisoners should have averaged —22.2 instead of —14.4 per cent.

In these patients of Simonart's the respiratory quotient was also estimated; the mean was 1.05 with a range in the individual means of 0.87 to 1.72. In the 11 "normal" men the mean R.Q. was 0.88, range 0.71 to 1.00. The significance of these figures is dubious in view of the indications that there were gross faults in the methods used. Simonart merely suggested that the high values for the R.Q. may have been due to a thiamine deficiency, though how and why are not indicated.

B.M.R. in Undernourished Patients without Metabolic Disease

Undernutrition may result from interference with food intake or digestion, in such conditions as mechanical obstruction of the digestive tract, peptic ulcer, and anorexia nervosa. In the absence of metabolic disorder (diabetes, hyperthyroidism, chronic infections) the resulting undernutrition is comparable in effect to that produced by dietary caloric insufficiency. The degree of emaciation in some of the patients that have been studied was rather extreme. Although no determinations of the B.M.R. were made in some of the cases, the control of the diet appears to have been adequate since the patients were kept in hospitals for long periods of time. Table 158 summarizes the results obtained in such patients. The maintenance diet is defined as that level of caloric intake where, over a period of several days, both the body weight and the nitrogen exchange were maintained in equilibrium or in slightly positive balance. Tuczek's case was included because the nitrogen balance was only slightly negative at the extremely low intake of 591 Cal. The caloric intake in these cases is so low, even for bedridden patients, that there must have been a considerable decrease of the B.M.R. In Müller's (1889) and Klemperer's patients it is safe to explain the reduction in B.M.R. on the basis of semi-starvation. Tuczek's patient with stupor recalls the experiments of Grafe (1911) on patients with stupor but without evidence of malnutrition. In 6 out of 18 patients with catatonic or paralytic stupor the B.M.R. was definitely subnormal, with a maximum decrease of 39 per cent below normal limits. There seemed to be a parallelism between the degree of stupor and the decrease of the B.M.R.

TABLE 158

MAINTENANCE DIET, controlled by nitrogen balance, in several emaciated patients without metabolic disease.

Sex	Age	Disease	Weight Loss (kg.)	Maintenance Weight (kg.)	Maintenance Diet (Cal.)	Author
Female	19	Esophageal stricture		31	765	Müller (1889)
Male	24	Esophageal stricture	15.5	45.9	620	Klemperer (1889)
Female	21	Anorexia nervosa	15.2	34.8	614	Klemperer (1889)
Female	38	Mental (stupor)		52.8	591*	Tuczek (1884)

*Slightly negative nitrogen balance.

Actual B.M.R. measurements in a mental case were made by Magnus-Levy (1906). The patient, a man 19 years of age, had lived on a calorically deficient diet for 9 months "in order to spare his sick stomach." He entered the hospital in a state of severe emaciation. Although his height was 160 cm., his body weight was only 36.2 kg. He was kept for several days on his previous diet of

700 to 800 Cal.; at this time his B.M.R. (26.6 Cal. per square meter per hour) was 37 per cent below the Mayo standards.

The B.M.R. has been fairly frequently studied in patients with anorexia nervosa; the results are quite uniform and are summarized in Table 159.

TABLE 159

B.M.R. IN PATIENTS WITH ANOREXIA NERVOSA.

Number of Patients	Nutritional State	B.M.R. as Percentage Change	Author
4...................	8.8 to 14.5 kg. weight loss	−18.1 to −31.6	Möller (1924)
8 out of 10...........	Prolonged malnutrition	−10 to −39	Labbé and Stévenin (1925)
5...................	21 to 40 per cent below normal	−15 to −21	Mason, Hill, and Charlton (1927)
117...................	0 to 45 kg. weight loss (Av. −13.5 kg.)	0 to −42 Av. −25	Berkman (1930)

Svenson (1901) investigated metabolic trends during the convalescence of 4 patients after typhus abdominalis. Immediately after the cessation of fever the basal metabolism was usually within normal limits, but it dropped to very low values during the first days of convalescence. For instance, in one patient the B.M.R. declined to 45 per cent and in another to 75 per cent below normal standards. This decrease was observed only after prolonged typhus with resulting emaciation.

In 52 children between 5 and 12 years of age who were convalescent after various diseases (acute infections, hepatic disease with icterus, brain tumor, and chorea or epilepsy under treatment with phenylethylhydantoin), Schick and Topper (1933) found a decrease of the B.M.R. and pulse rate, independent of the state of malnutrition. In recovery, both B.M.R. and pulse rate increased. The contrast to Svenson's report is marked.

Master, Jaffe, and Dack (1935) placed patients with coronary disease on diets of 800 Cal. and observed a drop in the B.M.R. of −20 to −30 per cent.

The material reviewed shows that caloric undernutrition, produced by various non-metabolic diseases or by treatment, decreases the B.M.R. One exceptional case (Richter, 1904) should be noted; in a 17-year-old girl who was estimated to be underweight by 30 kg. (esophageal stricture), the B.M.R. was within normal limits (+5 per cent according to the Mayo standards). The reason for this exception is not clear.

Controlled Experiments

Excepting the Carnegie Experiment, which has been separately reviewed in Chapter 3, there is only one prolonged, well-controlled experiment which has relevance here (Boothby and Bernhardt, 1931). The subject was 33 years of age with a height of 166 cm. and an initial body weight of 64.5 kg. The diet was restricted, but not controlled, in a preliminary period starting at the end of April

1930. From May 20 to June 1 the diet was reduced to 800 Cal. From June 2 to July 20 the calorie intake was gradually increased in 4 steps (periods I to IV, 2 to 3 weeks in duration). A period of uncontrolled diet followed. The B.M.R., weight, and nitrogen balance were determined each day. The subject's weight loss from May 20 until June 2 amounted to 2.6 kg. Table 160 summarizes the most important results.

TABLE 160

BERNHARDT's B.M.R., BODY WEIGHT, AND NITROGEN BALANCE during semi-starvation and refeeding (Boothby and Bernhardt, 1931). The tabular values are averages for each period.

Period	Food Intake		N Balance (gm.)	Body Weight (kg.)	B.M.R. (Cal./hr.)	R.Q.	Pulse Rate	Blood Pressure
	Cal.	N (gm.)						
I	849	9.4	−3.40	60.0	52.6	0.75	64	104/58
II	1296	10.1	−2.09	58.4	51.9	0.76	62	104/63
III	1620	11.6	−1.17	58.2	52.1	0.82	67	100/65
IV	3353	15.2	+2.00	60.4	60.6	0.88	77	113/68

The nitrogen balance was still negative in period III. In period IV the nitrogen balance became positive, together with a distinct increase of the B.M.R., blood pressure, and pulse rate; it must be assumed that these had been depressed during the period of semi-starvation. At the end of period IV (August 12) the body weight was 62.9 kg., only 1.6 kg. below the initial weight. The gain in body weight during period IV was 4.8 kg., or about 1.6 kg. per week.

Joffe, Poulton, and Ryffel (1919) studied a 47-year-old, 60 kg. man during and after a period of drastic reduction of food intake. The total duration of the experiment was 2 months. The experiment was poorly controlled; although the subject stayed in the hospital during the experimental period, he prepared his own food and the calculations of his food intake were based on his reports. On several occasions he was absent for a day or two and probably lived on a relaxed diet during these intervals. For the first 3 weeks his reported diet was only 400 to 500 Cal. a day, associated with a weight loss of 5.45 kg.; later the diet was increased to 1000 Cal., with the result that the body weight became constant after 1 week, at the level of 55 kg., although the nitrogen balance was still negative. The B.M.R. was 20 per cent below the Mayo standards; even so, it exceeded slightly the total calorie intake reported by the subject. On the basis of his level of activity and his actual B.M.R., the authors recalculated his requirement and arrived at a value of 1514 Cal. a day instead of his reported 1000 Cal. However, there is no doubt that the subject lived on a restricted diet — although substantially higher than the reported caloric intake — which produced a body weight loss of about 10 per cent and a definite decrease of the B.M.R.

Observations on Undernourished Infants

Howland (1911) compared the basal metabolism in sleep of a normal child of 7 months with a malnourished child of 6 months. The normal child (4.32 kg. body weight) had a metabolism of 1270 Cal. per square meter of body surface, or 83 Cal. per kg.; the malnourished child (3.05 kg. body weight) had only 737 Cal., or 62 Cal. per kg. Howland also investigated an 8-year-old atrophic child and found a similar low metabolism (733 Cal. per square meter), but this child was idiotic, his legs were paralyzed as a result of cerebrospinal meningitis at 18 months, and his body weight was only 6.6 kg. Obviously this child was abnormal in too many respects to be informative on the problem of uncomplicated nutritional atrophy.

In 15 infants aged from 11 to 41 weeks the calories per kg. of actual weight, plotted against percentage of expected or "normal" weight, increased with increasing emaciation (Fleming, 1920–21). Talbot (1921) confirmed Fleming's results. In 17 severely undernourished infants between 2 and 14 months of age he found an increase of the basal metabolism compared to that of normal infants when he plotted the actual calories per kg. against actual weight, or when he plotted the calories per square meter of body surface against actual weight or age. The calories per unit of expected weight, when plotted against the expected weight, were below the normal values.

Although the accumulation of data obtained from infants in a state of severe malnutrition is not large, it all points in the same direction; the calorie production per unit of actual weight seems to be increased. The only exception is Howland's case of a 6-month-old child, but this might possibly be explained by the presence of toxins which had depressed the metabolic rate. That toxins may have such effect in infants has been suggested by Stransky (1922).

Thus the effect of undernutrition on the basal metabolic rate in infants is opposite to the effect in adults. It is surprising that this discrepancy was not discussed by Talbot, although he must have known at least of Benedict's experiments. Talbot's suggestion that the increased metabolism in undernourished infants might be due to a loss of heat insulation from the disappearance of the subcutaneous fat layer does not hold, because this layer also disappears in starving adults. Technical explanations may not be entirely dismissed; it is hard to obtain reliable relaxation and a condition of rest in small children, and it is possible that malnourished children are even more restless than normal children. However, Fleming and Talbot were aware of this source of error, and care was taken to obtain reliable data.

For a possible explanation of the discrepancy in the B.M.R. in semi-starva-

TABLE 161

CHEMICAL COMPOSITION OF THE BODIES OF A NORMAL ADULT AND 1 NORMAL AND 3 ATROPHIC INFANTS. Data from Mitchell *et al.* (1945) and Sommerfeld (1900).

Condition	Age	Weight (kg.)	Water (%)	Fat (%)	Ash (%)	N (%)
Normal	35 years	70.55	67.85	12.51	4.81	2.33
Normal	3 months	4.34	70.1	13.4	2.73	2.27
Atrophic	3 months	2.63	79.9	1.45	2.73	2.32
Atrophic	2.5 months	1.96	82.3	1.80	3.20	2.13
Atrophic	3.8 months	3.19	79.9	1.90	3.34	2.51

tion between small children and adults, the data from chemical analysis of the bodies of a normal adult (Mitchell *et al.*, 1945), a normal infant (Sommerfeld, 1900), and 3 atrophic infants are of interest (see Table 161).

In spite of limitations, some interesting conclusions may be drawn. The fat and water content of the 3-month-old normal infant is somewhat higher and the ash content lower than that of the adult man, and the differences are in the direction expected from studies on animals. In the 3 atrophic children the fat content is drastically reduced, and the water content is much higher than normal. However, the sum of fat plus water is about the same in the 3 atrophic infants and in the normal infant. It appears that the inactive material (water + fat) was diminished in the same proportion as the body weight; that is, quantitatively speaking, water replaced the fat. This only indicates the over-all changes and does not necessarily mean that such replacement actually occurs in the same proportion in the various tissues. However, there is some old histological evidence in the case of serous atrophy (Flemming, 1871a, 1871b, 1876) that fat replacement by water actually does occur. It is interesting that the nitrogen concentration was about the same in the normal and in the 3 atrophic children — that is, that the active cellular material was diminished in the same proportion as the body weight. This was not the case in the young adult subjects in the Minnesota Experiment. During 24 weeks of semi-starvation the body weight decreased by 24 per cent, while the mass of active tissue decreased by about 27 per cent. It seems that the rapidly growing organism maintains its active cellular material with greater tenacity than does the adult organism. This is in agreement with Fleming's data (1920–21) and with the fact that young animals continue to grow even when severely undernourished (Jackson, 1925).

Since the young organism tends to conserve protein during starvation, it might be assumed that fat is more readily sacrificed. This is indeed the case. In the young adult subjects of the Minnesota Experiment the fat content decreased from 13.9 per cent during the control period to 5.21 per cent at the end of semi-starvation, a decrease of 63 per cent (in the percentage content). In contrast, the percentage fat content in the atrophic children was decreased by about 85 per cent. It could be objected that the 3 atrophic children were in a more advanced state of semi-starvation than were the Minnesota subjects. This is probably correct. However, there is agreement in autopsy reports that in adult people who died from starvation even the macroscopically visible fat had not always disappeared completely.

It appears, therefore, that the response of the young, growing organism to starvation is fundamentally different from that of adults, and this would explain the discrepancies between adults and infants in the response of the metabolic rate to semi-starvation. We advance this hypothesis with reservations because of the limitations discussed above, but at the present time it seems to be the most likely explanation for the reported phenomena.

Studies on Undernourished Older Children

Blunt, Nelson, and Oleson (1921) measured the basal metabolism of 26 undernourished children from 7 to 12 years of age. According to the norms for their age and height, these children were from 3 to 27 per cent underweight. From 1 to 4 measurements were made on each child, and the activity was carefully controlled. All the children had elevated metabolic rates calculated in terms either of calories per unit of body weight or of calories per unit of body surface. The average of excess calories per square meter of surface was 19.5 per cent; there was no consistent relationship between the extent to which the children were underweight and the elevation of their metabolism. The basal metabolisms of 2 overweight children studied at the same time were +5 and +7 per cent.

Wang *et al.* (1926) compared the basal metabolism of 9 normal children with that of 32 undernourished children; all were from 5 to 10 years of age. The division between the two groups was arbitrarily made at 5 per cent below average weight. The activity during the experiment was usually recorded graphically. The basal metabolism of all the children was within the normal limits, but several trends were obvious. When calculated as calories per kg. of body weight, the metabolism was higher in the underweight children, although there was no consistent relationship between the increase of the metabolism per kg. and the extent to which they were underweight.

Wang, Kern, and Kaucher (1929) also found the basal metabolic rate to be within normal limits in another group of underweight children (between 6 and 10 years of age). During dietary treatment (rehabilitation), the percentage of nitrogen retention was greater than in children of normal weight. In some cases the basal metabolic rate decreased as the child gained weight, probably owing to a gain in inactive tissue (fat).

Topper and Mulier (1929) doubted the validity of the previous reports of elevated metabolism in underweight children where the B.M.R. was calculated on the basis of body weight or surface area. They maintained that the sitting height is the best basis for calculating basal metabolism in children. On this basis, the metabolic rate of 35 boys and 35 girls from 5 to 15 years of age, who were from 10 to 33 per cent underweight, was within normal limits in 70 per cent of the children, with a tendency toward a subnormal rate. In those children who were more than 25 per cent underweight the metabolism was at least 10 per cent below the normal average calculated from the seated height. Topper and Mulier are justified in their criticism that a calculation of the basal metabolism of undernourished children on the basis of the height, weight, or surface relationships of normal children might be improper, but the same may be said about their own choice of a method of comparison.

Since the relative metabolism in undernourished infants is increased and that of adults is decreased, one would assume that there must be an age range of older children in which the metabolic rate is not greatly abnormal in the state of undernutrition. And this seems to be the case in a large proportion of the reported material.

The Metabolic Rate of Obese Subjects on Reduction Diets

In the United States caloric inadequacy is encountered most frequently in the use of reducing diets to control obesity. Such practical efforts to lose weight are extremely common, but careful metabolic studies are rarely made. Unfortunately, the subject of obesity is beset with spurious "science" and fantastic claims. This problem and its investigation do have an amusing side, as witness the study by McCaskey (1927). The results of this study, in itself certainly not remarkable, led the author to the startling conclusion that fat consists mainly of water, while its other constituents are gas and heat.

The general subject of obesity has been reviewed frequently, notably by Grafe (1933) and Newburgh (1942). There is no doubt that in simple obesity the basal metabolism is generally within normal limits when calculated per unit of body surface. In 91 obese patients Boothby and Sandiford (1922) found 81 per cent of the group to be within 10 per cent of normal B.M.R.; 3 patients were definitely low (−15 to −20 per cent), and 1 patient had a rather high rate (+16 per cent). Strouse, Wang, and Dye (1924) reported no consistent differences in the basal metabolic rate per unit of body surface between underweight, normal, and overweight persons. Grafe (1933) found only 3 persons out of 180 extremely obese patients who had a definitely low metabolic rate. Strang and Evans (1928) found no abnormality in the B.M.R. in careful studies of 5 obese women. It would seem, therefore, that findings in obese persons on reducing diets would have some pertinence to the general problem of undernutrition. However, as we shall see, there are serious questions about the proper method of calculating the metabolism in very fat people.

Evans and Strang (1931) kept 187 obese patients for an average of 8.7 weeks on a diet of from 400 to 600 Cal. a day. The diet consisted mainly of proteins. The caloric deficits ranged from 1500 to 2500 Cal. a day. The B.M.R., calculated on the basis of normal standards, was within the normal range or slightly decreased. This type of calculation is obviously questionable with obese patients since it disregards the large proportion of inactive fat tissue. When the basal metabolic rate of the obese patients was calculated from the ideal weight, it was 25 per cent above normal. This form of calculation is an attempt to reduce the excess fat to a normal proportion. The result would mean that the metabolic rate of the active tissue is higher than normal. But calculation from the ideal weight disregards the fact that some small but definite oxygen consumption occurs in fat tissues; this factor might account for a sizable proportion of the total metabolism if the total weight of fat is as high as 50 kg. Moreover, the existence of such large amounts of fat adds substantially to the load of the circulatory system and, to a lesser degree, of the respiratory system, increasing their oxygen requirement in turn. Thus, the estimate of a 25 per cent increase in the B.M.R. in obese patients may be somewhat too high. In any case, during the reduction diet the B.M.R., calculated with the same procedure, declined to a point only 6 per cent above normal. The rate at which the B.M.R. dropped was greater than the rate of weight reduction.

Keeton and Bone (1935) gave 10 obese patients a diet that provided an average of approximately 55 per cent of their actual calorie requirements. The metabolic rate did not change during the period of diet reduction.

In spite of the difficulty of calculating the B.M.R. in obese patients as a strict percentage of normal standards, there is no question that a drop in the basal metabolism such as is observed in semi-starvation does not occur. Of course, an obese patient on a reduction diet is not in a state of undernutrition as long as he is still overweight. The rate of weight and nitrogen loss will be discussed in Chapter 20.

The Metabolism of Animals in Experimental Semi-Starvation

Morgulis (1923) placed a dog on a diet of approximately one third of the maintenance requirements for a period of 9 weeks. Table 162 gives a condensation of the results. The metabolic response of this dog to semi-starvation shows marked differences from the response of man. In man no transitory plateau of the B.M.R. has been observed, although the diet reduction in the reported experiments was far less drastic. In the later part of semi-starvation, the calories per kg. of body weight increased in Morgulis' dog, while in human semi-starvation there is a marked decrease.

TABLE 162

METABOLIC RATE OF A DOG DURING SEMI-STARVATION
(Morgulis, 1923, p. 266).

Time	Body Weight		Caloric Intake	B.M.R./Day	
	Kg.	% Decrease		Cal.	% Decrease
Control					
10 days	13.94		987	546	
Semi-starvation					
1st week	13.59	2.5	349	493	9.7
2nd week	12.79	8.2	349	458	15.8
7th week	9.70	30.4	310	454	16.7
8th week	8.92	36.0	313	417	23.6
9th week	8.27	40.7	333	360	34.1

TABLE 163

WEIGHT OF STEERS, before and at the end of 140 days of reduced
diet (Benedict and Ritzman, 1923).

Group	Number of Steers	Restricted Diet, as Percentage of Maintenance Diet	Initial Weight (kg.)	Weight Loss	
				Kg.	%
II........	5	50	527	120	22.7
III........	4	66 (later 40)	468	111	23.9
IV........	2	60	586	128	21.7

Benedict and Ritzman (1923) studied the effect of reducing the diet to 40–66 per cent of the maintenance level in 11 steers over a period of 135 to 140 days, compared to a control group of 3 steers (group I) kept on a diet which maintained the body weight constant. Table 163 shows the degree of diet reduction and the average weight changes. The effect of the diet reduction on the metabolic rate was complicated by a seasonal drop of the metabolic rate, shared by the experimental animals with the control group (see Figure 51). The decline of the metabolic rate in the experimental groups was only slightly more pronounced

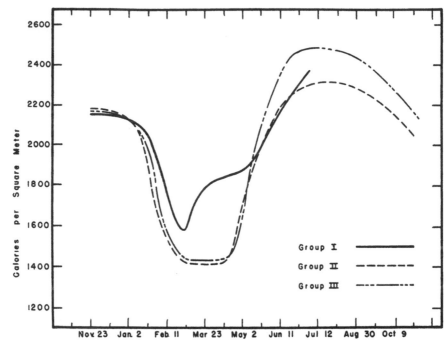

FIGURE 51. CHANGE OF THE METABOLIC RATE OF STEERS (calories per square
meter of surface and 24 hours) during semi-starvation and subsequent refeeding (groups II and III), compared to a control group (I) kept on a maintenance diet (Benedict and Ritzman, 1923).

than in the control animals (group I). Subsequently the metabolic rate exceeded
the initial level in both the control and the experimental groups.

The B.M.R. during Nutritional Rehabilitation

Prior to the Minnesota Experiment, the only controlled rehabilitation experiment was that made by Boothby and Bernhardt (1931), which has been summarized in an earlier section. In the last experimental period of that study, on a
diet of 3353 Cal. a day, the basal metabolism increased from a semi-starvation
level of 52 Cal. per square meter per hour to 60.6 Cal., but no control value of
the B.M.R. before semi-starvation is available.

In several patients B.M.R. determinations have been made during nutritional
rehabilitation. Strang, McCluggage, and Brownlee (1935) produced an average
weight gain of 5.1 kg. in 18 undernourished patients during treatment with a
high caloric diet (between 2800 and 5000 Cal.). There was no significant trend
of the basal metabolism during that period except in 6 patients who made very
large weight gains (averaging 9.3 kg., or 22 per cent of the initial weight); in
them the B.M.R. increase was parallel to the increase in the body surface (+8
per cent). There was no relation between caloric intake and basal metabolism.

In Magnus-Levy's case the B.M.R. was 26.6 Cal. per square meter per hour
(33 per cent below the Mayo standards) at a body weight of 36.2 kg. Five
months of dietary treatment associated with a body weight gain of 13.9 kg. re-

sulted in a B.M.R. of 40.5 Cal. in the same units ($= +2$ per cent). Svenson (1901) found that during nutritional rehabilitation in convalescence after typhoid fever the B.M.R. increased rapidly from a subnormal value and reached a maximum within 3 or 4 weeks, at which time the rate exceeded the normal rate, in some cases by as much as 50 per cent.

The effect of dietary rehabilitation was studied by Beattie and Herbert (1948b) in a group of 11 severely emaciated patients in Holland in May and June 1945 and in another group of 11 inmates of a German civil prison. The intake of the Dutch group, estimated at about 1600 Cal. per day for 6 months prior to the experiment, in 4 of the subjects dropped to 1000 Cal. per day for the last 3 months preceding the experiment. The diet of the German subjects was about 1800 Cal. per day for the preceding year. The weight losses of the Dutch group ranged between 12 and 31 kg., and the German subjects were from 11.7 to 24.1 kg. under their normal weight. The basal metabolic rate was expressed as Cal./kg. of body weight per 24 hours, with a correction for the active tissue based on the data of the Minnesota Experiment.

In the Dutch group the first B.M.R. determinations were made within 3 days after a measured diet was started; this diet was higher than the previous starvation intake. Consequently, the determinations may not represent the B.M.R. at the end of starvation. It is amazing, however, that only 2 of the 11 subjects had a subnormal heat production per kg. (6 and 15 per cent below normal) while in the other subjects the first value was from 2 to 84 per cent above normal. For comparison, it may be noted that in the Minnesota Experiment even in the high caloric group the B.M.R. was still below the control value after 6 weeks of rehabilitation. The caloric intake in the Dutch group ranged from 27 to 79 Cal./kg. of body weight, compared to 38.8, 46.0, 50.6, and 60.3 Cal./kg. of body weight in the 4 caloric groups Z, L, G, and T of the Minnesota Experiment during the first 6 weeks of rehabilitation. The caloric intake in the Minnesota Experiment is within the range of Beattie and Herbert's Dutch series.

In the group of German subjects the first B.M.R. value was significantly decreased in only 1 subject and was within the normal range in the other 10. The initial caloric intake ranged within the narrow limits of from 31.3 to 42.8 Cal. per kg. per 24 hours and was increased to 42.5–48.7 Cal. per kg. of body weight within periods of 30 to 56 days. The smaller increase of the B.M.R. in the German group compared with the Dutch group might be explained by the higher level of food intake during the starvation period and the smaller increase in food intake during rehabilitation. Still, the discrepancy between Beattie and Herbert's subjects and the Minnesota subjects remains in that in Beattie and Herbert's data there is no evidence of any consistent decrease of the oxygen consumption per unit of body weight at the end of semi-starvation (Beattie and Herbert, 1948a) or in the early phase of rehabilitation (ibid., 1948b). When the first values (calories per kg.) in Beattie and Herbert's material were plotted against the caloric intake, a correlation between them was apparent. At a caloric intake level of 35 Cal. per kg. per day, the heat production was at the normal level. There was no direct relationship between nitrogen intake and heat production, the caloric intake being the determining factor.

The high rate of heat production per kg. of body weight in the Dutch group during the early phases of recovery was associated with only very small weight gains, even on high intakes of calories or nitrogen. A small weight gain in the early phase of rehabilitation was also an outstanding feature in the Minnesota Experiment. When the dietetic treatment was continued for a period of 3 to 4 weeks, the level of heat production ceased to be related to the level of caloric intake. When the intake was raised, the heat production tended to increase, but when the intake remained constant the heat production tended to fall. The tendency of the heat production to fall during dietary treatment was most marked in the German series, so that with only one exception the final level of heat production was either at or below the initial level in spite of an increase of the caloric intake by 800 Cal. At an intake between 43.8 and 48.7 Cal. per kg. the heat production tended to fall about 10 per cent below the normal level. Beattie and Herbert suggest that this decrease might be due to the conversion of carbohydrates into fat, which would reduce the demand for external oxygen. There was evidence of an increase of the subcutaneous fat, in spite of the low fat intake of a maximum 20 gm. However, no actual estimations of the body fat content were made.

Basal Metabolism in Control and Semi-Starvation — Minnesota Experiment

In the control period of the Minnesota Experiment the basal metabolic rate was measured several times on each man. Preliminary practice sessions were provided so that all measurements may be considered as referring to trained subjects. Details are given in the Appendix on methods, but it should be noted here that all measurements in the control period and throughout the Experiment were made shortly after the men rose from bed in the morning before any actual work was done. In comparison with the Mayo Clinic standards the control values on these men were significantly low, as would be expected under the conditions used. For the entire group the control average, as a percentage of the Mayo Clinic normal averages, was −11.78 ($SD = \pm5.32$).

It is generally found in this Laboratory that average rates of metabolism on normal persons are lower than the Mayo Clinic averages, even when the subjects do not sleep in the Laboratory the night before the measurement. This reflects the unusual degree of relaxation customarily attained here and the practice of having at least one training session before the actual measurement. For comparison there are results on 33 other "normal" young men recently studied in this Laboratory; the average of these, again as a percentage of the Mayo Clinic normal averages, was −10.39 ($SD = \pm4.60$).

During semi-starvation the basal oxygen consumption promptly and steadily declined. Figure 52 shows the average results calculated per man, per square meter of body surface and per kg. of body weight, all expressed as percentages of the control value. At the end of semi-starvation the mean oxygen consumption per man had decreased 38.89 per cent, per square meter of body surface 31.26 per cent, and per kg. of body weight 19.46 per cent. At this time (S24) the rate of change in the B.M.R. was close to zero; that is, the metabolic rate change roughly corresponded to the rate of weight loss.

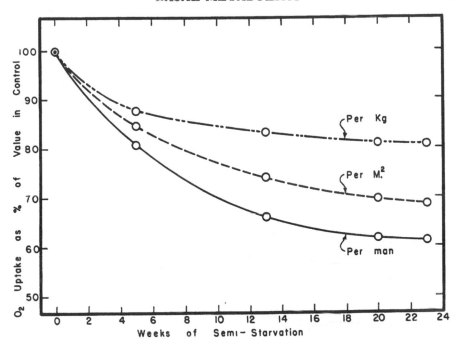

FIGURE 52. MEAN BASAL METABOLISM FOR 32 MEN before and during 24 weeks of semi-starvation. All values are expressed as percentage of the pre-starvation (control) values for the oxygen uptake per man, per square meter of body surface, and per kg. of body weight. (Minnesota Experiment.)

Obviously the B.M.R. was more constant per unit of body weight than per unit of body surface. Calculation of the B.M.R. per unit of active tissue (cf. Chapter 15) might be a better method of expressing the results. In these terms the mean oxygen consumption for the 32 men at C was 5.64 cc. per minute per kg. of active tissue and 4.77 cc. at S24; this amounts to a decline of 15.5 per cent.

With such calculation it should not be implied that the "active tissue" as defined by our procedure (see Chapter 15) is an entity. It is probable that the decrease of the metabolic rate differs for various tissues and organs during starvation, as has been shown by Kleiber (1947a) for rats.

Basal Metabolism in Early Rehabilitation — Minnesota Experiment

The fact that the subjects were divided into 4 groups receiving different amounts of food in the first 12 weeks of rehabilitation allows a more detailed examination of the recovery of the basal metabolism as related to "relief" feeding. The average changes in the B.M.R. during semi-starvation of these 8-man groups were very similar. For example, in terms of total cc. of oxygen consumption, the averages for the Z, L, G, and T groups declined, respectively, by 38.7, 39.0, 39.8, and 38.5 per cent. The recoveries in B.M.R. per unit of body surface, expressed as percentages of the losses in semi-starvation, are summarized for the first 12 weeks of rehabilitation in Figure 53.

The greatest average recovery in 12 weeks, 69.9 per cent of the loss in B.M.R.

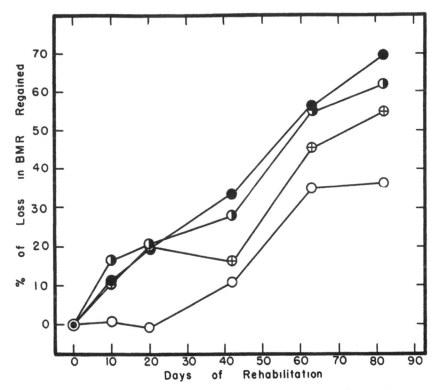

FIGURE 53. BASAL METABOLISM (OXYGEN CONSUMPTION) PER SQUARE METER OF BODY SURFACE during the first 12 weeks of rehabilitation, expressed as percentage of the reduction that occurred during semi-starvation. Mean values for the caloric groups: group Z = open circles; group L = crossed circles; group G = half-moons; group T = solid circles. (Minnesota Experiment.)

per square meter, was achieved by the best fed group (group T), and the recovery in the other caloric groups was more or less proportional to the differences in caloric intake. There was a tendency, however, for the lowest caloric group to lag behind at all times; this group showed no rise in the B.M.R. per unit of body surface in the first 3 weeks of rehabilitation.

At R12 the recovery of the body weight, as a percentage of the loss in semi-starvation, averaged 20.9, 29.4, 40.4, and 57.1 per cent, respectively, for the Z, L, G, and T groups. Obviously the rise in basal metabolism per square meter in this period exceeded the proportional recovery of lost weight. This is true also if we consider the gross oxygen consumption per man. In these terms the recovery at R12, as a percentage of the decline in semi-starvation, averaged 30.4, 45.8, 55.2, and 65.2 per cent for the Z, L, G, and T groups, respectively.

Again reference may be made to the changes in the active tissue mass. In semi-starvation it was calculated that there was an average loss, for the 32 men, of 10.73 kg., or 26.9 per cent, in the active tissue mass. At R12, 3.01 kg., or 28.1 per cent, of the loss had been regained. This is to be compared with an average recovery of 49.2 per cent of the total oxygen consumption. The rate of increase of the oxygen consumption was proceeding considerably faster than the restoration of the mass of the tissues themselves, although the absolute rate of metabolism at R12 — per man, per unit of surface, per unit of weight, or per unit of active tissue weight — was still a good deal below the normal control level.

Protein Supplementation — Minnesota Experiment

During the first 12 weeks of rehabilitation 16 men (group Y) received a daily average of 20.74 gm. of protein more than the 16 men in group U, although the average caloric and vitamin intakes of the two groups were substantially identical. These 2 groups of men showed exactly the same average decline in B.M.R. per square meter of body surface in semi-starvation. After 6 weeks of rehabilitation group U had regained, on the average, 22.03 per cent of their semi-starvation loss in basal metabolic rate, while group Y had regained 25.03 per cent. At R12 the corresponding average percentage recoveries were 60.15 and 53.23, respectively. These small average differences between the U and Y groups are not statistically significant.

TABLE 164

BASAL METABOLIC RATE, as cc. of oxygen per minute per kg. of gross body weight, for the 4 caloric groups and for the sub-groups, regarding protein and vitamin supplementation. Values for C (control period) are in cc. of O_2 per kg.; values for the end of semi-starvation (S24) and for 6 and 12 weeks of rehabilitation (R6, R12) are in percentages of the values at C. Note that there were 4 men in each sub-group; the tabulated values are the averages for these men. (Minnesota Experiment.)

Protein and Vitamin Groups	Z (Zero) Caloric Group				L (Low) Caloric Group			
	C	S24	R6	R12	C	S24	R6	R12
Extra protein (Y)	3.278	77.4	82.1	86.5	3.198	80.3	83.7	92.0
No extra protein (U) ..	3.318	83.9	87.8	93.5	3.282	82.7	89.4	101.0
Extra vitamins (H) ...	3.228	85.7	86.2	90.6	3.188	82.6	89.6	96.8
No extra vitamins (P) ..	3.368	75.8	83.8	89.4	3.292	80.5	83.6	96.4

Protein and Vitamin Groups	G (Good) Caloric Group				T (Top) Caloric Group			
	C	S24	R6	R12	C	S24	R6	R12
Extra protein (Y)	3.230	76.7	88.2	96.8	3.372	85.2	90.9	95.0
No extra protein (U) ..	3.378	84.2	89.3	95.9	3.382	73.6	86.0	96.2
Extra vitamins (H) ...	3.192	76.4	87.8	94.9	3.248	83.5	93.1	99.4
No extra vitamins (P) ..	3.415	84.3	89.7	97.7	3.508	75.5	84.0	92.0

It might be suggested that, although these 16-man groups did not show any significant difference in the metabolic rate response, there could have been differences of more consequence related to the protein intake at high or at low caloric levels. There were some larger differences at particular caloric levels. For example, at R12 in caloric group L (basal + 400 Cal.) the 4 protein-supplemented men (LY) regained only 46.6 per cent as compared with an average recovery in B.M.R. of 67.6 per cent for the 4 men who received no extra protein (LU). But there was no consistent trend, and none of the differences in these sub-groups was significant. Much the same is true when the basal metabolism changes are analyzed in terms of rates per kg. of body weight (cf. Table 164). The failure to discover any significant effects of the protein supplementation on the B.M.R. may have been due to the small numbers of men involved, but it is clear enough that 20 gm. of extra protein per day has no great effect under these conditions.

Vitamin Supplementation — Minnesota Experiment

The method of analysis of the possible effect of supplementary vitamins on the B.M.R. is similar to that for supplementary proteins. On the average, the 16 men who received vitamin supplements (group H) had, in comparison with the 16 men in the non-supplemented group (P), 44 per cent more vitamin A, 68 per cent more thiamine, 56 per cent more riboflavin, 64 per cent more niacin, and 22 per cent more ascorbic acid.

The H and P groups were very similar but not quite identical in their declines in B.M.R. in semi-starvation. In cc. of oxygen per square meter of body surface per hour, the H group showed an average decline of 87.5 in semi-starvation, compared with a drop of 90.3 for the P group. After 6 weeks of rehabilitation (with supplementary vitamins) the average B.M.R. recovery for the 16 men in the P group was 16.95 per cent of the loss in semi-starvation; the corresponding average for the H group was 18.48. At R12 the average recovery for the P group was 51.30 per cent compared with 46.85 for the H group. These differences between the P and H groups are not statistically significant.

Again, as in the analysis of the protein supplementation, the possible effect of the vitamin supplementation may be considered at each of the 4 caloric levels. There was a tendency, at the lower caloric levels, for the men who received vitamin supplements (group H) to show a larger total rise at R12 in B.M.R. per unit of body surface than those who had only placebos. The average R12 recoveries, as percentages of the B.M.R. decline in starvation, were 22.4, 43.5, 55.0, and 66.5 for the groups ZP, LP, CP, and TP, respectively; these are to be compared with 37.8, 48.2, 55.2, and 64.0 per cent, respectively, for the vitamin-supplemented groups ZH, LH, GH, and TH. This trend did not reach acceptable statistical significance, however. The analysis of the changes in oxygen consumption per kg. of body weight did not yield any significant differences that could be specifically ascribed to the differences in vitamin intake (see Table 164).

Caloric Equivalent of Oxygen — Minnesota Experiment

The primary purpose of measuring the oxygen consumption is, of course, for metabolic analysis in terms of energy or calories. But the caloric equivalent of oxygen consumed is variable according to the nutrients being burned. The usual factors applied for calories per liter of oxygen are 4.49 for protein, 4.69 for fat, and 5.05 for carbohydrate. For rough calculations a constant factor of say 4.8 Cal. per liter might be suggested, but for more precise calculation it is necessary to have information on the respiratory quotient (R.Q.) and the nitrogen excretion. For almost all purposes of simple energy calculation the latter correction for protein metabolism is so small that it may be disregarded.

In the starvation phase of the Minnesota Experiment the R.Q. was studied systematically in exercise but not in basal rest, the only regular measurements in the latter state being made near the end of semi-starvation (S23). At that time the mean gross R.Q. in basal rest was 0.836, with a range from 0.79 to 0.89. In rehabilitation the R.Q. was only slightly higher than in semi-starvation. At R14, when the food intake and weight gain were highest, the mean basal R.Q. was 0.853 (range 0.83 to 0.88) in the few cases studied. The proportion of the oxygen consumption derived from protein catabolism averaged about 25 per cent in semi-starvation and about 12 per cent in rehabilitation. These data indicate that in both starvation and rehabilitation it would be proper to use a factor of 4.8 Cal. per liter of total oxygen consumption.

Another approach to this question may be made from calculations on the nutrients burned during the period of semi-starvation. The total food intake in terms of fat, protein, and carbohydrate is known for the whole period of semi-starvation. During this time the indicated average change of composition of the body amounted to losses of 6.72 kg. of fat and 10.11 kg. of active tissue (see Table 166). The latter, which is free of extracellular water, represents roughly 2.5 kg. of protein. Finally, there was obviously a small loss of carbohydrate (glycogen) from the body in this period; we may suggest that this loss amounted to about 0.2 kg. of carbohydrate. The combustion of all these materials, including both exogenous and endogenous sources, would entail a gross average R.Q. of 0.880 and 4.856 Cal. per liter of oxygen. This is the value for the total metabolism including all activities.

Basal Metabolism in Later Rehabilitation — Minnesota Experiment

After 12 weeks of controlled rehabilitation on restricted food intakes, the men in the Minnesota Experiment were allowed to eat what they would. The result was a large increase in food intake, a sudden gain in weight, and an abrupt rise in B.M.R. In some cases the B.M.R. change was startling. For example, subject 126 used 153 cc. of oxygen per minute in basal rest on October 17 and 265 cc. on October 24. October 20 was the last day of restricted feeding, so the rise from 153 to 265 cc. of oxygen per minute occurred in 3 days. In this period of 3 days the subject gained 6 pounds and the basal pulse rate rose from 36 to 60.

The principal results on the 12 men who remained to be studied in detail from R12 through R20 are summarized in Table 165. These are presented as the averages of 2 groups of 6 men each, representing the higher and the lower caloric groups in the period from S24 through R12. A salient feature of Table 165 is the abrupt rise in the basal metabolism which took place between R12 and R14. This was striking in both groups but was particularly prominent in group II, comprising the men who had been the least well fed from S24 through R12. The basal oxygen consumption of this latter group rose, on the average, 49 per cent per man in 2 weeks; even in terms of the B.M.R. per square meter of body surface it rose 42 per cent in this period. This was, of course, associated with a sudden rise in food intake, although the diet for several weeks immediately preceding had been fairly generous, averaging 4096 Cal. for the men in group I and 3205 Cal. for the men in group II.

By R20 the men from caloric groups T and G had returned to, or very close

TABLE 165

BASAL METABOLISM, in cc. of oxygen per minute, calculated per man, per square meter of body surface, and per kg. of body weight. Mean values for 2 groups of 6 men each in the control period, after 24 weeks of semi-starvation, and at 5 stages of rehabilitation. Group I comprises 4 men from caloric group T and 2 men from group G; group II comprises 2 men from caloric group L and 4 men from group Z. (Minnesota Experiment.)

	C	S24	R6	R12	R14	R16	R20
			Group I				
Per man	225.2	145.7	163.8	197.3	246.0	254.2	231.2
Percentage of C	100	64.7	72.7	87.6	109.2	112.9	102.7
Per square meter	122.3	88.0	97.7	111.8	135.5	136.7	122.7
Percentage of C	100	71.9	79.9	91.4	110.8	111.8	100.3
Per kg.	3.30	2.79	3.02	3.25	3.75	3.66	3.25
Percentage of C	100	84.5	91.5	98.5	113.6	110.9	98.5
Body weight in kg.	68.18	52.18	54.32	60.62	65.58	69.37	71.05
			Group II				
Per man	230.5	144.8	152.0	175.0	260.5	252.5	246.8
Percentage of C	100	62.8	65.9	75.9	113.0	109.5	107.1
Per square meter	123.2	87.3	91.5	101.8	144.5	135.3	130.0
Percentage of C	100	70.9	74.3	82.6	117.3	109.8	105.5
Per kg.	3.45	2.83	2.94	3.14	4.11	3.73	3.49
Percentage of C	100	82.0	85.2	91.0	119.1	108.1	101.2
Body weight in kg.	66.88	51.20	51.65	55.70	63.35	67.72	70.63

to, their pre-starvation values in metabolism. The men from caloric groups Z and L were nearly but not quite back to normal, their B.M.R. values, by any method of calculation, being slightly high. At R34 the average for the 6 men from groups Z and L was 121.5 cc. of oxygen per minute per square meter of body surface; this may be compared with their control average of 123.2 cc.

The very high metabolic rates at R14 and R16 were associated with a high rate of increase in body weight. Group I gained weight at an average rate of 2.19 kg. per week from R12 to R16; for the same period of 4 weeks, group II gained at the rate of 3.01 kg.

Metabolic Cost of Weight Gain — Minnesota Experiment

During rehabilitation there were at least 3 major processes involving energy metabolism. In the first place, the maintenance requirements must have been changing; they would have to change to return to normal. Secondly, fat was being laid down at a rapid rate. Finally, active tissues were being restored or rebuilt. The present data provide no proper basis for exact calculation of each of these several components of the total metabolism, but certain over-all calculations are possible. For this purpose the averages for the entire 12 men studied from R12 to R20 are most suitable. The essential data are given in Table 166.

Before starvation these men were in metabolic balance and the total basal metabolism can be considered as solely maintenance of the active tissues at this time. This amounted to an average of 227.9 cc. of oxygen per minute for 38.80 kg. of active tissue (body weight less the sum of fat, bone mineral, and extra-

TABLE 166

Energy Cost of Rehabilitation, mean values for 12 men. Active tissue is the difference between gross body weight and the sum of fat, extracellular fluid, and bone mineral. Line 5 is calculated on the basis that at C (pre-starvation) and at S24 (end of starvation) the total basal oxygen consumption was used for maintenance of the active tissues and that in rehabilitation this maintenance cost per kg. of active tissue is proportional to the total active tissue as indicated by the values at C and S24. Line 6 is the difference between line 4 and the product of line 2 times line 5. (Minnesota Experiment.)

	C	S24	R12	R14	R16	R20
1. Gross body weight (kg.)	67.53	51.69	58.16	64.47	68.55	70.85
2. Active tissue (kg.)	38.80	28.69	31.21	(34.50)	(36.63)	37.82
3. Fat (kg.)	9.39	2.67	5.82	(9.58)	(12.01)	13.37
4. Basal O_2 (cc./min.)	227.9	145.3	186.2	253.3	253.4	239.0
5. Maintenance O_2 (cc./min./kg. active)	5.87	5.06	5.26	5.52	5.69	5.79
6. Excess O_2 (cc./min.)			22.0	52.9	45.0	40.0

cellular fluid). This amounts to 5.87 cc. of oxygen per minute for each kg. of active tissue. At the end of starvation (S24) the men were not laying down fat or building active tissues and were close to metabolic balance. At this time an average of 28.69 kg. of active tissue was being maintained with 145.3 cc. of oxygen per minute in the basal state. This amounts to 5.06 cc. of oxygen per minute for each kg. of active tissue. Between these two limits of active tissue mass it seems reasonable to consider the metabolic maintenance requirement per kg. of active tissue as being roughly proportional to the active tissue mass.

For the present data, and between the limits of 38.8 and 28.7 kg. of active tissue, the relation holds: $Y = 0.0801X + 2.76$, where Y is the cc. of oxygen per minute required for basal maintenance of 1 kg. of active tissue, and X is the mass of active tissue in kg. For these men the active tissue mass as well as the fat mass is also known for R12 and R20, and the value of Y for these periods can therefore be calculated. As a first approximation we can use the average of Y for R12 and R20 to characterize this period of 8 weeks during which the average mass of active tissue in these men rose from 31.21 kg. to 37.82 kg. The average value of Y ($= 5.53$) times the average value of X ($= 34.52$) gives 190.9 cc. per minute as the indicated average maintenance requirement over this period.

The total basal oxygen consumption was measured on each man on 4 occasions from R12 to R20. The average of these measurements, integrated over the 8-week period, was 241.4 cc. of oxygen per minute. The difference between this total basal oxygen consumption and that calculated for maintenance alone amounts to 50.5 cc. of oxygen per minute, or 4072 liters of oxygen for the entire 8 weeks. At an average of 4.85 Cal. per liter, corresponding to a non-protein R.Q. of 0.84, this amounts to 19,784 Cal.

The foregoing calculation may be relatively crude because of the assumption that the active tissue mass changed in a strictly linear fashion between R12 and R20. No direct estimates of the active tissue are available between these times. This assumption of strict linearity would appear to be less acceptable than the assumption that the increase of active tissue between R12 and R20 proceeded in proportion to the increase in body weight. On this basis we have calculated the most probable values for the average active tissue mass at R14 and R16 to have been 34.50 and 36.63 kg., respectively. The corresponding values for Y are 5.52 and 5.69 cc. of oxygen per minute for each kg. of active tissue. The integrated average value for the maintenance of the active tissue mass over R12, R14, R16, and R20 is 200.7 cc. of oxygen per minute. This indicates an excess metabolism for laying down fat and addition to the active tissue mass amounting to 40.7 cc. of oxygen per minute, or 15,918 Cal. for the entire 8 weeks.

It is clear that something like 16,000 to 19,000 Cal. may be assigned to the metabolic cost of an average increase of 12.68 kg. of total body weight involving a gain of 6.61 kg. of active tissue and 7.55 kg. of fat and a loss of 0.48 kg. of extracellular water. Per kilogram of gross body weight gained, the total metabolic cost was of the order of 1200 to 1500 Cal. if it may be presumed, as seems proper, that the entire metabolic cost is represented by the excess of basal metabolism over the maintenance requirement.

This indicates a highly efficient mechanism if only the caloric values are considered. The fat gain alone amounted to the equivalent of close to 68,000 Cal. The caloric value of the active tissue gained may be calculated as that of lean meat (without extracellular water) and would be equivalent to about 10,000 Cal. Disregarding nonassimilation in the gut and the caloric equivalent of materials excreted in the urine, the growth, conversion, and storage cost amounted to about 15 to 20 per cent, and the process had an indicated efficiency of 80 to 85 per cent.

It must be realized, of course, that these calculations are limited in several ways. The fundamental assumption made above is that the excess oxygen consumption in the basal state over bare maintenance requirements accounts for the total energy cost of laying down fat and rebuilding active tissues. In turn this presumes that these latter processes were proceeding at a constant rate during the day. It is possible, however, that such "growth" is not regular and might even be at a low point at the time the basal metabolism measurement is made, that is, some 12 hours after the last meal. The ordinary specific dynamic action following a meal could conceivably be a reflection of "growth," in the sense used here, so that a proper calculation of the energy cost of regaining weight should allow for any increase in specific dynamic action as well as an increase in basal metabolism. Specific dynamic action was not measured in this study. The present estimates for the metabolic cost of growth may be too low; allowance for a rise in specific dynamic action might well add 10 or even 20 per cent to this "growth" cost. Accordingly, it may be estimated that the maximum efficiency was between 80 and 85 per cent; the minimum estimate would be more like 70 or 75 per cent.

There are no really comparable data in the literature. The measurements and calculations made regarding the growth and fattening of livestock have serious limitations because all ordinary physical activity of the animals is included in the metabolic balance calculations (cf. Brody, 1945). Calculations on the net energy cost of producing milk and eggs are more nearly comparable. In the case of milk production the net efficiency is of the order of 60 to 65 per cent — that is, the caloric value of the milk produced is from 60 to 65 per cent of the metabolized calories over and above tissue maintenance (Brody and Procter, 1935; Gaines, 1938). In the case of egg production the net efficiency is of the same order of magnitude, or even higher — 77 per cent according to Brody, Funk, and Kempster (1938).

Units and Interpretation — Minnesota Experiment

The basal metabolism data from the Minnesota Experiment again call attention to the question of the mode of expression for the basal metabolism and the relationship of this to interpretation. We have discussed these controversial points earlier in this chapter. Should we compare values for oxygen consumption per unit of time: (1) per individual, (2) per square meter of body surface, (3) per kg. of body weight, or (4) per unit of active tissue? The results are not identical for these various methods of calculation, and important points of interpretation might be decided differently according to the method used to express the basic data. The virtue of the customary expression in terms of body surface is that it renders individuals comparable in a way which seems to minimize ordinary variability. It by no means follows that this is the best method of expressing the results when there are considerable changes in body weight to be evaluated.

From our present results it appears that the basal metabolism, in changing from the normal to the undernourished state, is least variable when expressed per unit weight of the active tissues, but nearly the same result is obtained with the simple gross weight. In rehabilitation it again appears that computation per unit of body weight makes for greater apparent stability of the basal metabolism than does the surface calculation. For example, using the data in Table 165, the oxygen consumption per square meter rose in group I to a maximum (at R16) of 55.4 per cent above the value at S24; the calculation per kg. indicates a maximum rise (at R14) of only 34.4 per cent. In group II the same calculations give, respectively, 65.6 and 45.2 per cent.

It is rather academic to adhere to calculation of the B.M.R. per unit of body surface in a situation where, as in the present case, we are concerned with analysis of the meaning and effect of the responses of the organism to changes in dietary intake. It is now agreed that, even though the "surface law" is useful in comparing individuals of different sizes, the surface area does not immediately and directly determine or limit the basal energy exchanges. The oxygen consumption or calorie production per unit of surface area has no particular theoretical significance in describing or understanding the changes which take place in the individual. The total calorie production or demand of the individual obviously has great practical significance, particularly to the individual concerned. And the calorie production per unit mass of the metabolizing tissue has significance because it is a direct expression of the intensity or "concentration" of the metabolic processes.

It is clear enough that in starvation and recovery the rate of oxygen consumption shows the same direction of change in any terms. In semi-starvation it goes down very significantly; in rehabilitation it may almost as significantly surpass the normal pre-starvation levels. The latter phenomenon we may vaguely ascribe to "luxus consumption" or to the "cost of growth" — without necessarily explaining it — and the former we may just as vaguely explain as "adaptation."

Adaptation in the Basal Metabolic Rate

By adaptation we generally mean simply a useful adjustment to altered circumstances. When the total basal metabolic rate decreases in starvation, as it indubitably does, it is certainly a favorable change in that it reduces the calorie deficit as compared with what it would be in the absence of such change in the basal metabolism. To the starving individual the reduced metabolic rate means that, at a given food intake, his rate of loss of strength and endurance is diminished and that, to carry it to the limit, he will survive longer. In the Minnesota Experiment the total rate of basal metabolism at the end of semi-starvation was almost 40 per cent less than in the control period. This is equivalent to a gain of almost 600 Cal. per day.

If the metabolic saving in starvation were achieved solely by reducing the size of the individual, it would not be proper perhaps to speak of metabolic adaptation. But, as we have seen, the metabolic rate is reduced more than would be expected from the reduction in body size. The intensity of the metabolism per unit of tissue mass is also reduced. This seems to fulfill the requirements of a true adaptive process in the metabolic rate.

There have been 2 sources of confusion about adaptation. In the first place, there is a tendency — to which we object — to consider that adaptation must be purposeful in that the change should come about *in order to produce a better adjustment*. In the second place, it is necessary to distinguish between adaptation and *perfect adaptation*, the latter meaning such adjustment to the new circumstances that complete biological freedom of the organism is preserved. Rubner and others have argued that the reduction of metabolism in starvation could not be considered adaptation because, though it effects economies and is attended by advantageous adjustments, it does not produce a new adjustment which preserves the full biological freedom of the organism. The starving man is weak, for example. But there is no proof that he is weak because of his lowered basal metabolism, and it is certain that he would be weaker if his basal metabolism had not diminished. The sole detriment that can be ascribed to the reduced metabolism itself is his increased sensitivity to cold. The metabolic adjustment to starvation is by no means a perfect adaptation, but it is a substantial adaptation.

The Mechanism of B.M.R. Reduction

There are many possible mechanisms which might reduce the basal metabolism. The reduction in tissue mass has been mentioned; this by itself affects only the total metabolism and does not necessarily reduce the rate per unit of tissue. In Chapter 29 we have noted the reduction in cardiac work. The work output of the heart per minute is reduced about 50 per cent; this is far more than in proportion to the reduction in tissue mass and accordingly can account for a part of the reduced metabolism per unit mass of the total body. By all criteria of ordinary manual examination there seems to be a reduction in the resting tension or tonus in the voluntary muscles, and it is reasonable to suppose that a similar change occurs in the involuntary muscles as well. This might account for some reduction in metabolism beyond that resulting from the simple reduction in muscle mass. There is a small decrease in body temperature; this would result in a slight decline in the rate of all the chemical reactions in the body. Finally, we may suggest that the concentration in the tissues of the enzymes and substrates concerned with metabolic reactions may be reduced. There is no direct evidence on this point, but it is indirectly suggested by the small decline in the blood sugar level.

CHAPTER 18

Energy and Protein Requirements of Normal Persons

SINCE 1881 when Voit made his recommendations that a 70 kg. man doing moderate work needed 118 gm. of protein, 56 gm. of fat, and 500 gm. of carbohydrates daily, many reports and estimates on this subject have appeared. As early as 1902, Neumann reviewed 128 papers. The protein level has been discussed much more often than the energy requirement.

The early literature revolved around the question of whether or not Voit's recommendation of 118 gm. of protein was too high. Voit himself was much less dogmatic than his followers; he was well aware that nitrogen equilibrium could be maintained at a much lower level of protein intake. This was shown by numerous nitrogen balance experiments performed on small groups of individuals for comparatively short periods of time (some of these will be discussed later in this chapter). Voit had in mind, however, the food intake necessary for the long-time maintenance of health and working capacity; he particularly inferred this high protein recommendation from estimates of the voluntary intake of Germans who were in "good health." It is clear that well-controlled experiments of short duration cannot answer this question. More prolonged experiments covering a period of several months or more can be performed only on a few individuals, and the results of such experiments cannot be applied to the population at large without reservations.

It may be inferred that the average diet on which Voit's figures were based is sufficient for maintenance of health and working capacity for the majority of the population, and that one could not go far wrong in accepting these averages as recommendations. This does not mean that the average food consumption constitutes a minimum requirement, but this misinterpretation has been made since the moment Voit's work was published. The large literature reviewed by Neumann (1902) revolves mainly around the question of whether or not Voit's protein recommendation of 118 gm. is a minimum requirement. Rubner (1930), who started his career in Voit's laboratory and later succeeded him as *the* authority on nutrition in Germany, interpreted the German literature on semi-starvation during World War I as confirmation that Voit's figures were minimum requirements. Since that time this confusion in definition and terminology has caused endless trouble. For instance, quite recently the National Research Council dietary allowances were treated, even in the Council's bulletins (1943a), as if they were minimum requirements. For this reason, we define the various expressions as they are used in this chapter.

Requirement is used as "minimum requirement," defined as the minimal amount of food intake necessary for maintenance of health under normal conditions of activity and living. *Allowance* is defined as the suggested level of food intake proposed by various authorities; these levels include provision for an arbitrary safety factor. *Recommendations* or *nutritional standards* are practically synonymous with allowances; they are sometimes based on the actual food intake of individuals or groups but more often are derived from theoretical considerations plus a "margin of safety."

Methods

Dietary surveys have been used extensively to estimate the average energy and protein intakes and thus to arrive at nutritional standards. In order to do so, the assumption is made that most of the people surveyed are in weight and nitrogen equilibrium. These surveys might be subdivided into several types: (1) analysis of freely chosen diets on the basis of individual or family questionnaires; (2) analysis of over-all food consumption and waste in institutions such as army or labor camps, monasteries, mental hospitals, etc., where the food intake is reasonably well controlled; (3) analysis of the over-all food consumption of geographical areas such as cities, provinces, or entire countries, on the basis of agricultural production, exports, and imports, with a correction for waste and spoilage.

It is clear that each of these 3 types has both advantages and limitations. Analysis of individual questionnaires has the advantage of detailed breakdown into the food items actually consumed, with the possibility of physiologically or clinically characterizing the given individual. One disadvantage is uncertainty as to the reliability of the individual reports. Most of the reports are likely to give a low rather than a high estimate of the food consumed. This seems to have been the case even in surveys where all precautions were made to obtain reliable results. Usually the caloric and protein contents of the diets have been calculated from tables of food composition. This procedure is questionable for small groups and may be very unreliable for single individuals in view of the variability in the composition and preparation of foods. Analysis of the food consumption of larger groups is more reliable in this respect, but the results so secured refer to the entire sample. An evaluation of the results would depend on the age and sex distribution as well as the occupations of the persons investigated. For this reason, recommendations from such material must be calculated on the basis of the consumer unit. Lichtenfelt (1905) was one of the first to use this approach in his survey of the food consumption in 5 provinces of southern Italy.

Another way to arrive at dietary recommendations was suggested by Lusk (1918). He maintained that the basal energy requirements can be known with absolute accuracy and that the energy expenditure for work can be estimated with approximate accuracy. He made the further assumption that the energy expenditure for work does not depend on the body weight or other constitutional factors. This may be true for some but not for all types of industrial work; it does not apply to any activity that requires walking or running. Lusk calculated the basal metabolism according to DuBois' formula for an average man and

woman of a given height and age. He secured his data from the Medico-Actuarial Mortality Investigation of 1912, which included several hundred thousand
individuals. Although the weight included the clothes, the basal metabolic rate
was calculated on the assumption that the weight with clothes was the weight
of the naked individual. The B.M.R. values are somewhat above the normal average, which is due partly to this error and partly to the influence of training
discussed in Chapter 17.

Lusk calculated the B.M.R. on the basis of the energy expenditure per unit
of body surface, but when he plotted the 24-hour basal metabolism against body
weight he obtained a straight line relationship. According to the surface area
formula, where the height is constant, as was the case in Lusk's study, the above
relationship should be a curve, especially when plotted for the wide range of
body weights used (128 to 201 lbs.). To these values were added (1) 150 to 200
Cal. per day for minor uncontrolled movements such as dressing, etc.; (2) 150 to
200 Cal. for every hour of walking at the rate of 3 m.p.h.; and (3) the estimated
energy expenditure for specific occupations, based on the few investigations that
were available at that time (i.e., tailoring, shoemaking, carpentering, metal-
working, filing, and hammering). Lusk added to his graph, parallel to the B.M.R.
line, lines that related the energy expenditure to the body weight for various occupations. This was done on the assumption that the energy expenditure for
work was independent of the body weight.

All the assumptions on which this approach is based are open to serious criticism. Nevertheless, the principles of factorial computation of the total energy expenditure were adopted by the Technical Commission on Nutrition of the
League of Nations' Health Organization (1936a). The basal standard of energy
expenditure per day for an average 70 kg. man or woman engaged in sedentary
work was established at 2400 Cal. The assumption of sex equality is certainly
not justifiable. It was concluded that to this basic caloric requirement 35, 75, 150,
and 300 Cal. per hour should be added for light, moderate, hard, and very hard
work, respectively. This scale provided 3000 Cal. for 8 hours of moderate work,
which is identical with that recommended by various other authors or commit-
tees. The main difference lies in the fact that in the other recommendations 3000
Cal. is taken as the basic unit, while in the recommendations of the League of
Nations' Commission 2400 Cal. is the unit. The adoption of the latter unit would
necessitate a new scale of conversion units (Stiebeling and Phipard, 1939).

TABLE 167

FOOD-REQUIREMENT COEFFICIENTS FOR INDIVIDUALS OF DIFFERENT AGE
AND SEX, as proposed by a committee of experts of the League of
Nations (after Stiebeling and Phipard, 1939, p. 43).

Age and Sex	Coefficient	Age and Sex	Coefficient
Under 2 years	0.2	10–11 years	0.7
2–3 years	0.3	12–13 years	0.8
4–5 years	0.4	14–59 years, male	1.0
6–7 years	0.5	14–59 years, female	0.8
8–9 years	0.6	Over 60 years	0.8

Dietary recommendations are primarily devised for and used with population groups. Regardless of arguments about the absolute caloric levels to be recommended, the factor of age must be considered. Relative coefficients for this purpose have been proposed as indicated in Table 167.

Accepted Recommendations

In Table 168 several of the "accepted" dietary standards for caloric and protein intake are given. In general, the recommendations for the caloric intake, obtained from different groups studied over a period of 60 years, agree surprisingly well. The discrepancies are probably due mainly to the estimate of physical activity, which is arbitrary to a large extent. Rubner's, Atwater's, Cathcart's *et al.*, and the National Research Council recommendations refer to 3 levels of work, while in the reports of the League of Nations and of the U.S. Bureau of Home Economics sedentary work and light work are separated. The League of Nations' commission made recommendations for both hard work (3600 Cal.) and very hard work (4500 Cal.). It should be noted, however, that a good share of the consistency in the accepted recommendations arises from the fact that they start from the same assumptions, use much the same method of calculation, and in every instance were influenced by previous recommendations.

Caloric Intake

The total energy requirement consists of the basal requirement, the energy for the specific dynamic action, and the energy cost of work. In cool climates, some energy is used for thermoregulation, but this fraction is small. The indirect effect of climate on the energy expenditure by modification of the general level of activity and of appetite is more important and will be discussed later in this chapter. The specific dynamic action is relatively unimportant; it does not exceed 10 per cent of the total food intake and may well be closer to 5 per cent (cf. Chapter 19). The contribution of the basal metabolism is generally overestimated by approximately 10 per cent through neglect of the training factor (see Chapter 17). In addition to this a correction should be made for the decrease of the metabolism during sleep. Since the B.M.R. with these minor corrections is constant for the same individual, the degree of muscular activity is the most important item in determining the actual energy requirements.

We are not able to agree with Lusk (1918, 1919) that the energy requirements can be estimated with reasonable accuracy. Although only a few measurements of the energy expenditure in industry have been made, they show how difficult it is to arrive at an estimate of physical activity. This may be illustrated by the results obtained in 3000 experiments on 50 workers in 4 different steel mills (Simonson and Dobrin, 1933). The excess energy expenditure (above the B.M.R.) calculated for the whole day from repeated determinations of the oxygen consumption during the day varied between 335 and 1346 Cal. for different workers engaged in work at the furnaces or in the mills. The variation depended largely on the working place and on the size of the steel or iron blocks, but other factors were important too. The tremendous spread in the energy expenditure within one job category makes any attempt to classify energy requirements hazardous, or even, for a given individual, impossible.

Nonetheless, in dietary surveys the workers in a certain job category are lumped together, and the considerable differences existing within that group are not considered. Work in steel mills is generally considered to be very strenuous, yet the excess energy expenditure of all workers investigated in the mills was, on an average, 845 Cal. This is far below the estimate of 1200 Cal. for this type of activity made by the League of Nations' commission. Obviously, work is considered "hard" not only because of the energy requirements but also because of the physical stresses it imposes and the condition of comfort under which it is performed. A similar study has shown that the energy expenditure of miners is only moderate (Berkowitsch and Simonson, 1935), although mining is often regarded as very hard work. Our knowledge about the energy expenditure in various occupations is too fragmentary to permit a reliable classification of jobs from this standpoint. Errors probably will occur in both directions. Technological differences produce great differences in energy expenditure for the same occupation in different areas or in the same area at different periods of technological development.

This explains why the estimates in Table 168 show the greatest discrepancies for hard work; at light work, differences in the energy expenditure, although probably as great on a percentage basis as those observed in heavy work, will affect the over-all energy requirement only slightly. In occupations classified as light or moderate work, muscular activity during the time off the job is of considerable importance. It is not unusual for workers to spend as much, if not more, energy on their hobbies — gardening, sports, etc. — as they do during their working hours.

TABLE 168

STANDARD DIETARIES FOR A MAN OF 70 KG.

Author	Physical Activity							
	Sedentary		Light		Moderate		Hard	
	Cal.	Protein (gm.)	Cal.	Protein (gm.)	Cal.	Protein (gm.)	Cal.	Protein (gm.)
Voit (1881)					3055	118	3574	145
Rubner (1902)	2445	123			2868	127	3362	165
Atwater (1891)			2700	100	3400	125	4150	150
Cathcart et al. (1934)..			2600–3000	37[°]	3000–3400	50[°]	3400–4000	80–100[°]
League of Nations (1936a)	2400		2700		3000		3600–4500	
Stiebeling and Phipard (1939)	2400	67	2700	67	3000	67	4500	67
National Research Council (1943b, 1945)†	2500	70			3000	70	4500	70
National Research Council (1948)†	2400	70			3000	70	4500	70

[°] First-class protein.

† No "light activity" category; in the 1948 table "moderately active" was replaced by "physically active."

Unfortunately, this is a factor beyond control in most cases. Therefore, all recommendations for caloric intake are only rough approximations at best. Perhaps they can be used as rough general standards for an average population living under average conditions at a particular time and place. They cannot be used for single individuals or even for limited groups. Although the maximum energy expenditure for very hard work is given as 4500 Cal. a day, it is well known that the energy expenditure in certain occupations may exceed 8000 Cal. (see Table 169).

TABLE 169

Examples of High Energy Requirements for Several Occupations. In these cases the energy utilization was calculated to be about 92 per cent. (Woods and Mansfield, 1904.)

	Protein (gm.)	Calories
Maine lumbermen		
Chopping and yarding	206	8140
Drawing logs to landing	173	6888
River driving	152	5035
Bavarian lumbermen	130	6015
New England marble workers	254	7551
New England brickmakers	180	8569

Keys (1945a) has re-emphasized the fact that the above calculations may have to be modified for differences in the physiological utilization of nutrients. One instance where such a factor enters is in the relative efficiency of carbohydrates and fats for mechanical work. Krogh and Lindhard (1920) showed that even though these foodstuffs are equivalent in maintaining body temperature, carbohydrates are about 12 per cent more efficient than fats in supplying energy for work on the bicycle ergometer. This fact has been confirmed on the treadmill in the Minnesota Laboratory of Physiological Hygiene.

Any rigid application of the recommendations in Table 168 is obviously impossible. It is quite likely that diets observed in certain groups of a normal population may not agree with the "standards." An illustration of this is the report of Wiehl (1944), who found that the caloric intake was below the "requirement" in 55.3 per cent of 272 pupils. In spite of this caloric "deficiency," only 11.1 per cent of the pupils were underweight, whereas 47.7 per cent were overweight. This does not necessarily mean that the standards are wrong, for the activity of the special group may be far above or below that of the average person. The over-all requirements depend to a great extent on uncontrolled activity, which certainly will differ widely in different communities. A considerable saving of energy may be obtained by curtailing all types of muscular activity, including productive work, during periods of food shortages (Kraut and Müller, 1946). The concomitant decrease in the basal metabolic rate in famine acts in the same direction.

Protein Intake

The protein intakes in Voit's, Rubner's, Atwater's, and the British recommendations (Cathcart et al., 1934) increased with the work load. The dietaries on which these recommendations are based show an increasing protein consumption with increasing energy expenditure. This can be seen also in Table 169, where the increase in caloric intakes is associated with concomitant increases in protein content. The fraction of the total calories supplied by protein in the very high caloric diets is about the same (between 10 and 14 per cent) as that observed at lower caloric levels (light or moderate work). The protein consumption of persons occupied in hard manual work in different countries is listed in Table 170. The protein intake is equal to, if not higher than, the recommendations.

TABLE 170

PROTEIN CONSUMPTION OF INDIVIDUALS ENGAGED IN VARIOUS TYPES OF HARD PHYSICAL ACTIVITY (Jones, 1939).

Country	Activity	Protein (gm./day)
Sweden	Laborers, hard work	189
Russia	Laborers, hard work	132
Italy	Laborers, hard work	115
France	Laborers, hard work	135
England	Laborers, hard work	151
Germany	Soldiers	145
United States	Lumbermen	160–270
United States	College athletes	160–270

Many athletes appear to labor under the misconception that hard physical exercise imposes an undue strain upon the muscles which can be compensated by an increased protein intake. This is borne out by a dietary study made by Schenk (1936) on 4700 athletes participating in the eleventh Olympic games. The meat consumption was especially high among the athletes taking part in the more strenuous sports. These men usually ate something like 1 kg. of meat a day as well as several eggs. In addition, they consumed 1 to 2.5 liters of milk a day. Similar dietary habits can be observed at many of the training tables maintained for college football players.

Schenk (1936) claimed that daily intakes of 320 gm. of protein and 7300 Cal., as observed among the Olympic athletes, are essential for hard manual work or strenuous exercise. Such a statement is surprising since Pettenkofer and Voit as long ago as 1866 showed that when sufficient carbohydrate and fat are supplied to meet the caloric requirements, hard physical exercise has no significant influence on the urinary nitrogen excretion.

Another misconception of some athletes is illustrated by a long-distance runner who consumed large amounts of olive oil just before a race (Jokl, 1936). The resulting indigestion forced the runner to drop out of the race which he otherwise had a good chance of winning.

The theory that physical work increases the protein metabolism dies hard and is resurrected again and again. In a recent instance (Peters and Van Slyke, 1946, Vol. I, p. 664), emphasis was placed on the changes in the blood non-protein nitrogen during a 25-mile marathon race (Levine, Gordon, and Derick, 1924). The blood non-protein nitrogen increased in 5 athletes from an average control value of 26.6 to 44.2 mg. per 100 cc., with only a very small part of this increase being due to uric acid. No interpretation can be made of such findings in the absence of data on renal excretion, blood volume, state of hydration, and so on.

Much better evidence has been offered in other studies. Work for 8 hours on the bicycle ergometer, totaling about 360,000 kilogrammeters, increased the nitrogen excretion in a long-distance cyclist who was kept on a diet of approximately 4100 Cal. The protein intake varied between 39 and 213 gm. (Wilson, 1934; Wishart, 1934). The nitrogen excretion was greatest the day after the work. The work load was extremely heavy; at a 20 per cent mechanical efficiency, the caloric requirement for the work would be 4218 Cal. When the resting requirements (at least 2000 Cal.) are added to this, it is apparent that the subject was in severe caloric deficiency during the work day, which was repaid on the following rest days. The caloric deficiency certainly must have contributed to the increased nitrogen excretion; protein was simply being burnt as fuel. Taken all in all, Wilson's and Wishart's experiments do not support the assumption that a high protein intake is needed for hard muscular work, which is in line with other observations (Cathcart, 1925).

In contrast to the prevailing opinion, Kraut and Lehmann (1948) hold that the minimum nitrogen requirement for the maintenance of working capacity is affected by the level of physical activity but their own data are inconclusive. They studied work performance and nitrogen balance of 3 miners over a period of 32 weeks at different levels of nitrogen intake and activity. The caloric intake was sufficient throughout the experiment, while the nitrogen intake was gradually diminished until the balance became negative. The minimum nitrogen requirement for nitrogen equilibrium was the same (7.0, 7.5, and 8.0 gm. of nitrogen per day for the 3 subjects, respectively) at the 2 levels of activity investigated, light work and heavy work (mining). Also the endurance in static work, muscle strength, and performance with Mosso's finger ergograph was independent of the nitrogen intake. The rate of work that could be performed with the bicycle ergometer at a pulse rate between 110 and 120 beats per minute remained the same when the low nitrogen diet was given during 4 weeks of light work, but it dropped sharply when the subjects were working in the mines. Also, their daily output in the mines decreased during the time of the low nitrogen diet; however, no data were given.

It is not easy to reconcile Kraut and Lehmann's findings with other observations — for instance, those of Darling et al. (1944) or those reported by Johnson and Kark (1947b, p. 14). A body weight loss of 8 lbs. and a marked deterioration in the military efficiency occurred during a period of 60 days on a diet which was deficient in calories (the estimated caloric deficit was between 300 and 500 Cal. a day) but rather high in proteins (132 gm. per day). This is just the opposite result from that observed by Kraut and Lehmann. Material in which the work performance has been controlled and measured at different levels of nitrogen intake is still scanty and the problem deserves further study.

The normal fact of increasing protein consumption with increasing food intake is primarily accidental since natural diets providing more than 4000 Cal. would be much too bulky if composed only of carbohydrate foods. Excluding

pure fats, those foods that are high in calories are also high in protein. Therefore, as the caloric consumption increases, the protein intake rises also. At the
other extreme, it is almost impossible to devise low calorie diets containing only
natural foods without including at least a fair amount of protein. Most diets
used in Europe or America which contain 2000 or more calories are very likely
to provide at least 50 gm. of protein. Even foods which are considered to be composed primarily of carbohydrates still supply some protein.

The recommendations of Stiebeling and Phipard are based on Sherman's
(1920) investigations of 29 men and 8 women. Sherman found 0.5 gm. of protein per kg. of body weight to be sufficient for maintenance of health and working capacity for several months. The suggested recommendation of 1 gm. of protein per kg. of body weight provides a 100 per cent margin of safety. The recommendations of the National Research Council are similar. No increase in the protein intake is suggested for increasing severity of work, which is consistent with
the accepted physiological data. Diets unusually high in protein have no obvious
disadvantage. Eskimos live on a diet rich in animal protein without any apparent untoward consequences (Krogh and Krogh, 1915; Heinbecker, 1928; Thomas,
1927). The ample meat quota is explained by its availability; physiologically,
there is no reason known at present why a great part of the Eskimo diet could
not be replaced by carbohydrates, and a change in this direction is actually occurring in Greenland and Alaska.

That a low protein intake is adequate even for a fair amount of physical
work over short periods was indicated by the investigation of Darling et al.
(1944). They studied 3 well-matched groups, each containing 8 subjects; the
groups received, respectively, a balanced diet with 95–113 gm. of protein, a high
protein diet of 157–92 gm., and a low protein diet of 39–57 gm. The average
body weight in each group was about 75 kg. The subjects were engaged in
manual work requiring, on an average, 3300 Cal. The work, however, was not
standardized. No beneficial or harmful effects of the high protein or the low protein diet could be seen during a period of 2 months. This was verified by measurements of endurance, nitrogen balance studies, and examinations of blood
constituents. Thus it seems that a diet supplying about 50 gm. of proteins is sufficient for maintenance of working capacity and health in young men for a period
of at least 2 months and that no improvement is obtained by increasing the protein intake.

Protein requirements or recommendations are usually estimated in gm. per
kg. of body weight. Whether this procedure is correct is open to question.
Smuts (1935) compared the B.M.R. and the protein metabolism in animals (including man) ranging in weight from 0.1 to 681 kg. Among warm-blooded animals, the maintenance protein requirement varied more nearly with the body
surface than with the body weight, and the relationship between the B.M.R.
and the nitrogen excretion may be estimated from the body surface. This indicates that the protein requirement should be related to body surface rather than
to body weight. Recommendations based on body weight would tend to give
more protein than needed by heavier subjects and less protein than needed by
lighter subjects. Brody (1945) also supported the view that both B.M.R. and

nitrogen metabolism depend on the body surface. Hegsted *et al.* (1946) found a correlation coefficient of 0.71 between nitrogen balance and nitrogen intake when the intake per unit of body surface was used, whereas it was only 0.61 when the intake on a body weight basis was used.

Effects of Income, Geographical Region, and Race on Energy and Protein Intakes

The caloric and protein contents of natural diets ordinarily increase with increasing expenditure for food, as indicated in Table 171. At both levels of food expenditure the diet of the Negroes was larger than that of the whites. The interpretation of such data is highly debatable; the attempt to use them for the estimation of nutritional status is to be deplored. The implication of Table 171 as it stands is that a considerable proportion of the people in low income groups in the United States are calorically underfed but that they fare better in regard to protein; a different set of standards, which could be defended equally well, would produce a very different picture. For the present purposes, at least, perhaps the most instructive fact to be derived from Table 171 is that the protein provided per 1000 Cal. is much the same in all groups, in spite of the large differences in total calories and dollar expenditure.

Since the diet improves (increases) with an increase in the economic status, one would expect a deterioration of the average diet during a depression. It is surprising that this is not the case according to the material collected by the

TABLE 171

AVERAGE FOOD ENERGY AND PROTEIN INTAKE OF U.S. POPULATION per day and requirement unit (Bureau of Home Economics Scale), according to level of food expenditure per capita, region, and race, December 1934–February 1937 (Stiebeling and Phipard, 1939).

Region and Race	Calories	Protein (gm.)	Percentage of Diets Furnishing: Calories		Percentage of Diets Furnishing: Protein	
			Less Than 2400	3000 or More	Under 45 Gm.	100 Gm. or More
Weekly Food Expenditure, $1.25–1.87						
White						
North Atlantic	2530	64	38	7	5	0
East-North Central ...	2650	64	33	12	1	1
South Atlantic	2740	62	35	15	0	0
East-South Central ...	3050	65	11	35	2	1
Pacific	2570	63	38	4	5	0
Negro, South	3460	70	10	53	4	5
Weekly Food Expenditure, $2.50–3.12						
White						
North Atlantic	3320	88	2	54	0	18
East-North Central ...	3710	94	2	68	0	33
South Atlantic	3660	90	0	65	0	18
East-South Central ...	3980	93	0	87	0	12
Pacific	3660	90	0	70	0	20
Negro, South	4880	111	0	94	0	78

League of Nations (1936b). Table 172 shows no change in the consumption of important food items during the depression years 1930–34, compared to the years of normal production 1925–29. Most of the countries in Table 172 were highly industrialized and were severely hit by the economic depression. The values for the different countries are not comparable because of differences in the collection of material and variations in the age and occupational distribution of the inhabitants. However, comparison within the same country appears to be valid. The data of Table 172 do not allow any calculation of energy or protein consumption, but use of the food items included in the table, which are known to be consumed in greatly differing amounts in various income classes, seems not to have been affected by the depression. Obviously, food prices are adjusted to the reduced income, and a greater part of the reduced income is spent for food. The data about the nutritional state of the population during the depression are scanty, but the report includes a statement that no change occurred in the general nutritional state of the British population during that time.

TABLE 172

CONSUMPTION OF IMPORTANT FOOD ITEMS per capita per annum during 1925–29 (normal) and 1930–34 (depression). The meat column does not include game, poultry, or fish. (League of Nations, 1936b.)

Country	Butter (lbs.)		Meat (lbs.)		Eggs (N per capita)		Sugar (lbs.)		Cheese (lbs.)	
	Norm.	Depr.	Norm.	Depr.	Norm.	Depr.	Norm.	Depr.	Norm.	Depr.
Belgium	17	21	85	90	93	98	57	62	6	6
Denmark	13	18	93	125	241	236	116	120	11	12
France	10	13	72	72			53	57	11	12
Germany	14	16	107	110	123	129	55	52	11	13
Italy	2	2	44	35	112	119	21	18	10	11
Netherlands	12	16	85	91			72	68	12	14
England	16	22	138	140	149	172	104	110	8	9
United States	18	18	137	135	197	199	114	103	5	4
Australia	30	29	237	202			117	107	5	4

The absence of dietary differences between the North and the South in the United States is surprising (see Table 171). That the climate has a profound effect on appetite and level of physical activity is very well known; Lusk (1928) discussed a few typical examples of this. Murlin and Hildebrandt (1919) found from analyses of the food in 427 messes in different army camps that the average American soldier (average weight 66.6 kg.) consumed more food in the winter than in the summer months; the average food intake from June to September varied between 3500 and 3650 Cal., from November to February between 3700 and 4150 Cal. Important material on this question has been recently published by Johnson and Kark (1947a). The average caloric and protein intakes of healthy young North American soldiers were compared in temperate, arctic, and tropical areas. Figure 54 represents the average caloric intakes for groups of from 50 to 200 men fully acclimatized to the climate in which they were living. The food available was ample and of wide variety. The data show a close correlation be-

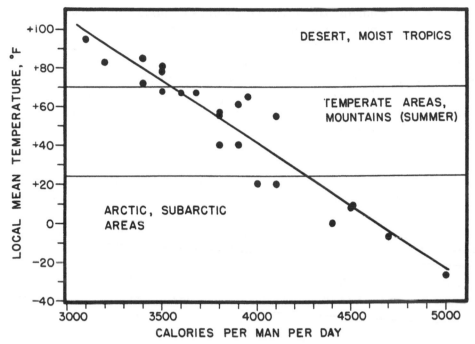

FIGURE 54. ESTIMATED VOLUNTARY CALORIC INTAKES OF NORTH AMERICAN TROOPS in hot, moderate, and cold climatic zones; averages for groups of 50 or more men (Johnson and Kark, 1947a).

tween the voluntary daily caloric intake and the mean environmental temperature, ranging from 3100 Cal. in the desert (92° F.) to 4900 Cal. a day in the Arctic (- 30° F.). The work (job) activity was believed to be about the same and could not account for the great differences; the hobbling effect of arctic clothing and equipment undoubtedly played a role but was believed to account for no more than a part of the variation. Probably the greater part must be explained by the energy requirement necessary to maintain a normal body temperature in a cold climate and the automatic adjustments of voluntary activity on this account. No generalization on civilian populations can be made.

Minimum Energy and Protein Requirements

As we have already pointed out, average energy and protein intakes or allowances are not necessarily applicable to limited groups of the population. The number of dietary surveys reporting caloric and protein intakes below the accepted dietary standards is quite large. Such discrepancies, if not too great, may be explained on the basis of a lower level of activity than that of the average population in which the standards were obtained. If this is not so, then the conclusion must be: (1) that the original standards are too high and that a lower caloric intake is sufficient to maintain health and activity, or (2) assuming the original standards to be correct, that the nutritional state of the people living on the lower caloric intakes is poor as a result of the substandard diet. Both conclusions have been made.

Often the standards have been too rigorously applied, and diets below the standards have been considered inadequate. For instance, it was estimated (Stiebeling, 1942) that less than one fifth of the families in the United States had "adequate" diets in 1936. The adequacy was defined by arbitrary standards, not only in regard to energy and protein content, but also as far as certain vitamins and minerals were concerned. The percentage of adequate diets in any country depends upon the economic group or groups being surveyed. There are many intangible factors involved in the calculation of such data, and this fact calls for a large measure of individual judgment on the part of the statistician. The difficulties involved are illustrated by the fact that in a later publication Stiebeling (1943) made another estimate using the same original data but now indicating that approximately one fourth of the families had an adequate diet. The conclusion that the majority of the working population in the United States live on diets inadequate in one or more dietary factors seems hard to believe. This conclusion, however, was accepted by the National Research Council's Committee on Diagnosis and Pathology of Nutritional Deficiencies (1943a) when their own standards were used as the criterion of adequacy. Since the average diet in the United States is better than that in many European countries (when compared to prewar conditions), one would have to conclude that the diet of perhaps close to 100 per cent of the population in those countries is inadequate. Such a conclusion seems to be absurd, but the definition of adequacy is arbitrary.

In the same report of the National Research Council (1943a), the results of several dietary surveys were compared. Table 173 has been prepared from this material. It can be seen that the caloric intake was held to be substandard in most of the surveys, and the same is true of the protein intake in 5 out of 7 cases. A substantial percentage of the various groups surveyed (between 4 and 40 per cent) were reported as receiving only 50 per cent of the "standard" caloric in-

TABLE 173

INADEQUACY OF DIETS, expressed as percentages of persons (adults) in the groups surveyed who did *not* receive diets that were 100, 75, 50, or 25 per cent of the National Research Council's recommended calorie or protein allowances (National Research Council, 1943a).

Author	Subjects and Region	Diet in Percentage of N.R.C. Standards							
		100		75		50		25	
		Cal.	Protein	Cal.	Protein	Cal.	Protein	Cal.	Protein
Smith, J. M., et al. (1943)...	38 white, Pa.	87	8	21	0	5	0	0	0
Milam (1942)....	110 white, N.C.		83		42		10		1
Youmans et al. (1943a).......	457 white, Tenn.	82	51	50	24	12	7	3	0
Youmans et al. (1943a).......	194 colored, Tenn.	95	74	77	45	40	18	12	0
Wiehl (1942) ...	250 workers, Cal.	78	15	38	3	4	0	0	0
Milam (1944) ...	56 white, N.C.		55		14		2		0
Milam (1944) ...	39 colored, N.C.		9		21		10		0

take, and, curiously enough, it was indicated that in one group 12 per cent of the population were living on a diet containing only 25 per cent of the caloric intake recommended for that age group. These reports seem fantastic. The only explanation for such conclusions is that the individual reports on which the survey was based are unreliable. A subject living for any length of time on a diet providing 50 or 25 per cent of the standard caloric intakes would soon become very emaciated. Apparently no such decrease in body weight was observed, as is shown by the comparison of the absolute caloric intake and the body weight in the material given by Youmans et al. (1943a, 1943b) (see Table 174).

TABLE 174

AVERAGE CALORIC INTAKE OF ADULTS IN TENNESSEE, calculated from a 3-day record of food intake (Youmans et al., 1943a).

Subjects		Caloric Intake	Body Weight (lbs.)	Normal Body Weight (lbs.)
Sex and Race	N			
Males				
White	228	2522	157	170
Negro	98	2005	152	163
Females				
White	229	1736	139	138
Negro	96	1315	145	138

In this study the body weight of Negroes, in spite of the low caloric content of their diet, was about the same as that of the white population, and in both cases the body weights were not essentially different from the expected (normal) weight. The difference in diet between the white population and Negroes was explained by the difference in income. In males (both Negroes and white) there was about an equal number of overweight and underweight individuals (judged by clinical estimation of the subcutaneous fat), while in women the number of overweight subjects was even greater than the number of underweight subjects.

In order to explore further the discrepancies between weight and caloric intake, the body weight expressed as a percentage of normal weight was plotted against the food intake expressed as the percentage of deviation from the basic caloric requirement. This was done for a representative sample (10 per cent) of the population selected at random. The basic caloric requirement was calculated on the basis of body surface, age, and sex and was not actually determined. There was no correlation between body weight and caloric intake. As many as 22 out of 43 adult men and 15 out of 18 women recorded intakes less than 40 per cent above basal requirement. Four men (9 per cent) and 11 women (60 per cent) recorded intakes below their basal requirements, which, of course, casts serious doubt on the reliability of the reported food intakes. The estimate of activity was very uncertain, but even assuming the women to be very inactive, their caloric intakes were far below recommended levels.

In several surveys Milam and his co-workers (Milam, 1942; Milam and Anderson, 1944; and Milam and Darby, 1945) found the caloric intake in North Carolina communities definitely below the standard requirement, which was estimated to be 3000 Cal. for men (see Table 175). The protein intake in Milam and Anderson's survey amounted to 61 and 54 gm. for whites and Negroes, respectively. Again, the level of protein intake was more nearly adequate than the level of caloric intake. Milam and his associates also failed to find any evidence of serious malnutrition.

TABLE 175

AVERAGE CALORIC INTAKES OF ADULTS IN SEVERAL NORTH CAROLINA COMMUNITIES. Calculations based on 7-day diet records. The individuals included in the studies lived either on farms or in small communities. They were all moderately active.

Author	Subjects	Calories
Milam (1942)	400, white	2000
Milam and Anderson (1944)	253, white	1954
	207, Negroes	1744
Milam and Darby (1945)	Male white	2500
	Male Negro	2200
	Female white	1850
	Female Negro	1700

TABLE 176

CALORIC INTAKES OF SMALL GROUPS OF WOMEN IN AUSTIN, TEXAS, classified either by race or occupation (Winters and Leslie, 1943, 1944).

Number of Subjects	Race or Occupation	Income Group	Caloric Intake
	Year 1943		
11	Anglo-American	Low	1248
7	Latin-American	Low	1249
6	Negro	Low	938
	Year 1944		
12	Professors' wives	High	1667
4	Faculty women	High	1720
4	Students	High	1920

Winters and Leslie (1943, 1944) studied the caloric intakes of comparatively small groups of women in Texas. The period during which the diet was studied was longer than in most surveys (from 7 to 21 days) and analyses were actually made for the caloric and protein contents of the actual diets consumed by the subjects. The diet was the customary one of the subjects. In a few subjects the basal metabolic rate was determined. The results are summarized in Table 176. In the first group of 24 women (1943) no body weights were reported, except for the general statement that very few were underweight and several were considerably overweight. Their activity was characterized as sedentary. The caloric intake of the 20 subjects in the higher income group was greater than that of the lower income group, although definitely below standard requirements. One subject of this series (1944) was underweight, two were slightly overweight, and the others were within the normal range of body weight. There was no correlation between body weight and caloric intake in the group of 12 professors' wives. The basal metabolic rate was within low normal limits. The calorie content of the collected food of a few subjects was below the basal requirements as calculated from the B.M.R. This discrepancy can only be explained by a failure to collect food samples identical to those actually eaten. A repetition of these findings under rigidly controlled laboratory conditions is desirable.

In the study by Wiehl (1942) of California aircraft workers (some results of which were included in Table 173), the basal requirement was estimated on the basis of Boothby, Berkson, and Dunn's nomogram (1936). Nearly 40 per cent of the 250 men reported diets which provided less than 1.4 times the basal metabolic needs, and yet there was no sign to indicate that this level was insufficient to maintain health and working capacity. It should be kept in mind, however, that the data of Boothby, Berkson, and Dunn were designed for clinical use rather than for calculation of dietary requirements. The estimate of basic requirements will be too high because of the neglect of the training factor and the decreased requirement during sleep (cf. Chapter 17). Unfortunately, the weight of the workers, which would be the best criterion for dietary sufficiency, was not given in Wiehl's study. The median of the calories reported was about 2675, which agrees closely with the average of 2540 Cal. for 93 male wage earners in Toronto (Patterson and McHenry, 1941) and the average of 2622 Cal. for male workers in 82 families in Halifax (Young, 1941). Of all the nutrients analyzed by Wiehl, protein was most often obtained in "adequate" amounts; 85 per cent of the workers had a protein intake equaling or exceeding the allowance of 70 gm. This is in harmony with the results of the other studies discussed above. In contrast to this, Ashe and Mosenthal (1937) reported a protein intake of 42 gm. or less in 348 out of 593 "normal" adult subjects in New York City. No data of the caloric content of the diets were given, but it was probably equally low.

In England, Widdowson and McCance (1936) found an average caloric intake of 2039 per day in 63 middle-class women and 1871 Cal. per day in 6 wives of unemployed men. A similarly low figure of 1998 Cal. per day was observed in another group of 57 middle-class women (Widdowson and Alington, 1941).

In all the above surveys there were no signs of any definite caloric undernutrition; the people surveyed could be considered as representative of a normal working population. Although the accuracy of food sampling or the reliability of the reports is open to serious doubt in several cases, as a whole the results cannot be ignored. Of great interest in this respect is Fleisch's (1947) report on the nutrition in Switzerland during World War II. Switzerland imported about 50 per cent of her food in peacetime and was compelled to introduce a system of rationing very early in the war (1942). The rationed food was always available, and vegetables and potatoes were not rationed. Black markets played a negligible role, so that from the rationed foods, agricultural production, and imports the over-all consumption could be estimated with considerable accuracy.

As in the League of Nation's scheme, Fleisch chose as a basic unit a man of 70 kg. doing light work. Additional calories were added according to 3 classes of occupation (moderate, heavy, and very heavy). However, the unit for the average adult, 2160 Cal., was deliberately set below that of the League of Nations. This low average food intake did not decrease the body weight or productive capacity or result in deterioration of health. However, in 1945, when the food intake was reduced to 1800 Cal., the body weight and hemoglobin dropped. This trend was quickly reversed in the winter of 1945–46, when the rations were increased again, although they were still below the prewar level. Fleisch concluded that the League of Nations' standards are higher than necessary for maintenance of health and working capacity over a prolonged period. It seems that the limit must be close to 2000 or 2100 Cal. per consumer unit (for light muscular work). This would agree with Neumann's self-observations and other material reviewed in this chapter.

The accumulation of recent reports that sedentary adults can maintain their weight and carry on their normal activity at a very low caloric level would indicate the desirability of re-evaluating the standard caloric requirements. Since most of the gross caloric deficiencies have been reported as occurring among women, it would appear mandatory to include them in any such study.

A number of suggestions have been made that the dietary habits of the American people have drastically changed during the past century. Mendel (1932) supports this contention by quotations from cookbooks, hotel menus, and the like. These findings, however, undoubtedly do not apply to the large majority of the people. More pertinent in this case is the report of Stiebeling (1943) on the change in the average consumption of various foodstuffs for the period 1910 to 1942. In all these cases, it must be realized that eating habits have only an incidental and very minor relationship to caloric requirements.

It is interesting to compare the recent American dietary surveys with Lichtenfelt's (1905) study of the diet in southern Italy. The average caloric intake (per man of 70 kg. body weight) in 5 southern Italian provinces was 2601 Cal. per day, compared with an average diet of 3683 Cal. in northern and central Italy; that is, the diet in southern Italy was 29 per cent below the northern Italian diet. The average protein intake of the people in southern Italy was 67 gm., of which 50 gm. were estimated to be utilizable. The caloric intake in southern Italy was similar to that reported in recent surveys of United States population samples, yet Lichtenfelt concluded that the diet of the Italians was inadequate. It was claimed, without supporting evidence, that the average person in southern Italy was inferior in all respects (economic, moral, political, intellectual, and physical) to the average person in northern Italy, and it was implied that this inferiority was referable to the difference in the diets used in the 2 regions.

From the surveys discussed, it can be seen that it is difficult to arrive at any definite statement about normal or minimum requirements. The standard requirements in current use are obviously not minimum requirements, but how low the minimum requirements may be without interference with health or activity cannot easily be indicated by surveys alone.

Minimum Energy and Protein Requirements — Experiments on Individuals

The question of minimum energy and especially minimum protein requirements has also been approached in small experimental groups and in single individuals by an analysis of nitrogen excretion or balance data.

Chittenden (1903), who suffered from persistent rheumatism, restricted his daily diet to about 1600 Cal. and 38 gm. of protein. At that time he weighed 57 kg. On this intake he was able to maintain nitrogen equilibrium for 9 months, with what he considered to be beneficial results to his health. Later he investigated small groups of teachers, university athletes, and soldiers. These groups had average nitrogen excretions equivalent to 46, 58, and 44 gm. of protein per day and average body weights of 64, 70, and 61 kg., respectively. All groups maintained "perfect" health and strength during a period of 5 months. Lusk (1928) felt that Chittenden's work showed that the allowance of protein necessary for maintenance of health and activity may be reduced for many months to half or less of that in the customary diet.

TABLE 177

NITROGEN BALANCE ON PROTEIN-POOR DIETS, at different caloric levels, from the experiments of various authors. 0 = the subjects were in nitrogen equilibrium.

| Body Weight (kg.) | Intake per 70 Kg. | | Nitrogen Balance | Author |
	Cal.	Protein (gm.)		
64	5520	26	0	Klemperer (1889)
70	3000–5000	32	0	Hindhede (1913)
69* ...	4570	49	+	Caspari and Glässner (1903)
73	3320	41	0	Hirschfeld (1888)
57* ...	3290	66	0	E. Voit and Constantinidi (1889)
58* ...	3280	40	+	Caspari and Glässner (1903)
75	3250	29	0	Peschel (1891)
59	2940	46	0	Sivén (1900)
62	2935	30–36	0	Sivén (1901)
66	2905	60	0	Lapicque (1894)
41* ...	2870	46	0	Caspari (1905)
73	2640	28	—	Hirschfeld (1888)
37.5* .	2620	63	0	Albu (1899)
†	2200–2700	59–72†	—	Albertoni and Rossi (1908)
65	2480	38	—	Ritter (1893)
74	2460	45	—	E. Voit and Constantinidi (1889)
86	2250	45	—	Ritter (1893)

* Vegetarians.

† Farmers in southern Italy, living on their customary diet, mainly vegetarian; nitrogen utilization about 80 per cent.

The protein requirement depends, of course, on the caloric intake; a part of the protein requirement can be replaced by carbohydrates. In Table 177 the results of different workers who studied persons on comparatively low protein intakes are given. The results are arranged in a decreasing order of caloric intake; both caloric and protein intakes were calculated for an individual weighing 70

TABLE 178

BALANCE OF FOOD INTAKE AND EXCRETIONS IN 8 ITALIAN SUBJECTS (Manfredi, 1893).

Subject	Sex	Age	Weight (kg.)	Nitrogen (gm.)		Caloric Intake for 24 Hrs.			Calories Utilized	Activity
				Intake	Balance	Absolute	Per Kg.	Per Square Meter		
1	M	34	55	11.46	+0.30	1997	25.42	1103	1927	Light
2	M	18	47	12.65	−0.10	2423	36.09	1496	2243	Light
3	F	70	38	9.31	+0.30	1793	33.01	1272	1694	Light
4	F	40	62	15.00	+0.32	2854	32.20	1457	2728	Light to moderate
5	M	40	48	10.06	+0.08	1848	26.97	1120	1664	Moderate
6	M	29	55	11.28	+0.27	2155	27.42	1191	1977	Moderate
7	F	25	50	10.45	+0.20	1982	27.75	1161	1828	Light
8	F	20	52	9.72	+0.13	1727	23.26	993	1601	Light
M			51	11.24	+0.13	2098	30.25	1219	1750	

kg. The table shows individual discrepancies in the ability to maintain nitrogen equilibrium. In general, at diets of 3000 Cal. or more, nitrogen equilibrium can be maintained at an intake of 40 gm. of protein or less. At diets below 3000 Cal., nitrogen balance is harder to obtain, even at higher nitrogen intakes. All results were obtained in short-term experiments. For most of the subjects the diet was an experimental one, while for the vegetarians it corresponded more nearly to their customary diet.

A mainly vegetarian diet providing a caloric intake between 1600 and 1800 Cal. and a protein intake between 30 and 40 gm. was sufficient to maintain nitrogen equilibrium, health, and working capacity (including military service, medical practice, and physical exercise) in one subject over a period of 7 years. During the first year of this diet, the B.M.R. was decreased to approximately 20 per cent below standards, but it increased later to normal values (Abelin and Rhyn, 1942).

Nitrogen Balance in Group Investigations

Manfredi (1893) and Yukawa (1909) observed adults living on low caloric intakes. Eight Italians from the poorer section of Naples were studied by Manfredi, and Japanese monks living in a monastery were studied by Yukawa. In both cases, the subjects lived on their customary diets and did their usual work.

The diet of both groups was mainly vegetarian. Manfredi's results are summarized in Table 178. Food intakes and excretions were analyzed over periods of 3 to 7 days. The subjects were quartered in the hospital but followed their ordinary occupations (tailoring, carpentering, household work, etc.). Their activity was observed but not standardized. The estimate of the physical activity appears to be fairly reliable. All subjects were in nitrogen equilibrium. This series demonstrates again the large individual variability in the ease with which nitrogen equilibrium can be established. The caloric intakes of subjects 1, 5, 6, 7, and 8 were low and agree fairly well with the intakes reported in some recent surveys in the United States, as discussed above. However, poor utilization due to the vegetarian character of the diet should be taken into account (see column in Table 178 showing calories utilized).

The results of Yukawa's careful investigations of Japanese monks are shown in Tables 179 and 180. The body surface was calculated according to Miwa and Stoeltzner's formula (1898). The food utilization was poor as would be expected with the vegetarian type of diet. The caloric intake appeared to be adequate for

TABLE 179

Caloric Intake and Food Utilization of Japanese Monks living on their customary diet; average values for 2 age groups (Yukawa, 1909).

Number of Subjects	Age	Calories per Day			Calories Utilized	Percentage Utilization		
		Absolute	Per 70 Kg.	Per Square Meter		Protein	Fat	CHO
8	26	1804	2854	1129	1659	66.6	69.5	99.0
3	64	2020	2732	1197	1872	68.1	59.1	99.0

TABLE 180

AVERAGE BODY WEIGHT AND NITROGEN BALANCE OF JAPANESE
MONKS (Yukawa, 1909).

Number of Subjects	Age	Days of Experiment	Weight in Kg.		Nitrogen in Gm.	
			Original	Change	Intake	Balance
8 26		9	44.5	+0.2	9.06	+0.78
3 64		7	51.8	+0.4	9.34	+0.79

the type of activity (sedentary or light), even when the poor utilization is considered. The protein content of the diet was low, but still it exceeded Sherman's (1920) suggested minimum requirement. While Sherman (1920) and Darling et al. (1944) concluded that normal people could maintain health and activity on a similar or even somewhat lower protein intake for a period of several months, Yukawa's data suggest that this might be true for much longer periods. The average for both groups of monks shows a positive nitrogen balance and a small weight gain.

Hetler (1932) determined urinary nitrogen excretion, basal metabolism, and protein and caloric intakes of 85 women students from 19 to 37 years of age at the University of Illinois. The caloric intake was estimated from dietary reports for periods of from 3 to 7 days. Calculated for a person weighing 70 kg., the average daily nitrogen excretion was 9.5 gm., which corresponds to a consumption of 66 gm. of protein. This was calculated from the urinary excretion on the assumption that 10 per cent of the nitrogen was lost in the feces. The B.M.R. was 7.1 per cent below the Harris-Benedict standards and amounted to 1260 Cal. per day, while the total food intake was 1700 Cal. as shown by the dietary reports. The average body weight of the group was 56.8 kg. Recalculated for a body weight of 70 kg., the B.M.R. amounts to 1554 Cal. and the total caloric intake to 2095 Cal. This is one of the lowest intake figures reported for a group of working people, checked by means of the nitrogen excretion. Curiously enough, the author does not comment upon the low caloric intake of this group.

De Venanzi (1947) claimed that the Venezuelans attending the Health Certificate Service lived on a very low protein intake. To verify this, he collected 24-hour urine samples from 118 women and 76 men. The average urinary nitrogen excretion in these cases was equivalent to a daily protein intake of 47 and 56 gm. These figures would be increased when corrected for the fecal nitrogen. Again it is obvious that a population living under relatively poor conditions is probably receiving a protein intake adequate to maintain nitrogen balance, providing the people are in caloric equilibrium.

Neumann's Self-Observation

The most prolonged and carefully controlled self-experiment on a low maintenance diet was carried out by Neumann (1902). It lasted over a period of almost 2 years. The protein, fat, and carbohydrate contents of food and excretions were analyzed. From these results the caloric content of the food was calculated.

His average body weight was 66.5 kg., and his height was 165 cm. The following data from his experiments have been recalculated for a body weight of 70 kg.

During the first period of 305 days Neumann maintained body weight and nitrogen equilibrium on an average diet of 69.1 gm. of protein and 2427 Cal. During that time, as well as during the subsequent 2 experimental periods, he continued his laboratory work. His muscular activity was probably light. During the second period of 66 days the caloric and protein contents of his diet were increased in 6 intervals of 5 to 15 days' duration each. The diet for the first interval supplied 51 gm. of protein and 1535 Cal., whereas the diet during the last interval consisted of 79.5 gm. of protein and 2777 Cal. During the first 4 intervals of the second period, when the protein intake was increased from 51 to 79 gm. and the calories from 1535 to 1937, the nitrogen balance was negative (between −2.11 and −3.11 gm. daily). There was no relationship between the nitrogen balance and the caloric intake. During the first 10 days at the lowest caloric level the body weight remained constant in spite of an average nitrogen loss of 2.81 gm. a day. Throughout the first 4 intervals (35 days) the loss in body weight was small (0.8 kg.). This was obviously due to water retention. Nitrogen equilibrium, with a daily average gain of 0.22 gm., was attained in the fifth interval at a level of 2659 Cal. and 76 gm. of protein. During the last part of the experiment the body weight increased by 1.3 kg., so that the final weight slightly exceeded the initial weight.

From these results one might conclude that the minimum caloric requirement necessary to maintain the body weight of this subject must have been between 2400 and 2700 Cal., but Neumann lived for a subsequent period of 240 days on a diet of 74 gm. of protein, 106 gm. of fat, 164 gm. of carbohydrates, and 5 gm. of alcohol, or a total of 2000 Cal., without losing any body weight or showing any deterioration of working capacity or health.

The level of this maintenance diet is surprisingly low, even when only light activity is assumed. There are some differences between the first maintenance diet (2427 Cal.) and the last one (2000 Cal.) in that the percentages as well as the absolute amounts of protein and fat are higher, and those of carbohydrates and alcohol are lower, in the 2000 Cal. diet. The decrease of alcohol accounts for about two thirds of the caloric differences. Whether the different composition of the diet explains the different maintenance level is questionable, but at present it seems to be the only valid explanation. Another possibility would be a change of muscular activity, but it is implied that no such change occurred.

Neumann's experiment shows that maintenance of body weight, nitrogen equilibrium, ordinary (light) activity, and health is possible on a rather low dietary level. This is, indeed, the only experiment other than the recent work by Hegsted *et al.* (1946) in which the minimum requirement level was actually determined by producing a negative balance and then increasing the diet until the maintenance level was reached. Therefore Neumann's experiment is of great interest. But it is questionable whether any general application can be made from it because the basal requirements show such a wide range (cf. Chapter 17). Neumann may have had a basal metabolic rate below the normal average; unfortunately, his B.M.R. was not determined.

Hegsted *et al.* (1946) studied the nitrogen balance in 26 normal subjects aged 19 to 50 years who were fed on diets low in protein; the effects of altering the diet by the use of meat, soy flour, wheat germ, and white bread were investigated in periods long enough to assure a reasonably constant nitrogen excretion. The daily requirement for nitrogen equilibrium was 2.9 gm. of nitrogen (18 gm. of protein) per square meter of body surface, equivalent to 30–40 gm. of protein intake for an average 70 kg. man. Replacement of one third of the vegetable protein by meat decreased the protein requirement by 17 per cent. The low protein requirement of a mixed diet necessary for maintenance of nitrogen equilibrium in this carefully controlled series substantiates the earlier claims (for instance, Sivén, 1901) based on shorter experiments which were not so well controlled. Hegsted *et al.* calculated the protein requirements for the basic vegetable, meat, bread, soy flour, and wheat germ diets as 33.6, 29.4, 31.6, 34.3, and 27.2 gm., respectively. The authors concluded that the National Research Council recommendations of 70 gm. of protein per day are overly generous and might be reduced to 50 gm., still leaving an ample margin of safety.

Sufficient evidence has been presented to indicate that there is little likelihood that protein starvation will occur unless the caloric intake is grossly inadequate. There is one factor that may modify this statement somewhat: the digestibility of the proteins. In the Far East where the percentage digestibility of proteins has been reported as being very low (McKay, 1910; Basu and Basak, 1939; Basu *et al.*, 1941), the requirement is increased correspondingly.

Some Physiological Considerations

Of the authors who found a comparatively low protein intake sufficient for maintenance, only Chittenden (1903) and Hindhede (1913) claimed that such an intake was beneficial. However, Chittenden's and Hindhede's claims that a low protein intake improves health and performance have not been confirmed (Darling *et al.*, 1944).

Süsskind (1930) attempted to live according to Hindhede's recommendations on a diet of 33.3 gm. of protein and a total of 2300 Cal. After being on this diet for 25 months, he had a "complete breakdown." The type of breakdown was not described. At that time he increased his diet from 2300 to 2600 Cal. and on the higher intake he claimed he recovered slowly but not completely. Shortly thereafter he entered Rubner's laboratory, where the same diet was controlled for a period of 18 days by means of balance studies. The first measurements of any kind were made when Süsskind entered Rubner's laboratory. The body weight at that time was 53 kg., and the height was 160 cm. According to German standard height/weight tables, Süsskind was 10 per cent underweight, yet his B.M.R. was within normal limits (Rubner, 1930). The caloric intake preceding his entrance into the laboratory should have been adequate for light physical work. In Rubner's laboratory he was able to maintain both weight and nitrogen equilibrium at a protein intake of 36 gm. and a level of 2600 Cal. When the protein intake was increased to 52.5 gm. per day in a diet supplying 3200 Cal., there was, for 4 days, an average daily protein retention of 15 gm. without any change in body weight.

Rubner (1930) used Süsskind's experiment to show that the older recommendations of a high protein diet were correct. He based this statement on the fact the Süsskind's capacity for work was low and that the extra protein during the last 4 days of the experiment was largely retained. To him this indicated that the fact that nitrogen equilibrium could be maintained at a low level of protein intake did not mean that the low level was physiologically adequate. It must be pointed out that these conclusions of Rubner's were based on the nitrogen retention at the higher protein intake, which was maintained for a period of only 4 days, during which time there was no change in the body weight.

There is a possibility, as Rubner suggested, that an individual in nitrogen equilibrium may not receive sufficient protein for maximum physical performance. The nitrogen equilibrium has the advantage of clear-cut definition and measurement. Physical fitness is a complex which is more vaguely defined and which can be approached at best to only a limited extent by means of a multiple battery of tests (cf. Taylor and Brožek, 1944; Keys, 1947). If, as Rubner implies, the nitrogen equilibrium (and even more so, the body weight equilibrium) is an inadequate index of dietary requirements, then the term *safety factor* for arriving at protein recommendations by adding 50 per cent or more to the level of nitrogen required for equilibrium is misleading. In practice, by means of this arbitrary procedure one might arrive at the physiological level of protein intake. Actually one does not have to refer at all, then, to the minimum requirement levels. There is no doubt, however, that the earlier protein recommendations of 100 gm. or more are excessive and are not based on any known physiological needs. The physiological needs for all nutrients are at present not fully known; we know the importance of the over-all caloric intake, but individual variability is large.

The minimum nitrogen requirement, as discussed above, refers to protein as it occurs in an ordinary mixed diet. The actual nitrogen requirement is determined primarily by the intake of the essential amino acids with the nonessential amino acids functioning mainly as sources of energy. The actual amino acid requirement of humans is only in the initial stages of study. There are some indications (Holt and Albanese, 1944) that less than the 10 amino acids required by the growing rat are needed to maintain nitrogen equilibrium for short periods in male adults. Adult rats can be kept in nitrogen equilibrium in the absence of lysine, leucine, histidine, arginine, and phenylalanine, all of which are required for growth in the rat (Burroughs, Burroughs, and Mitchell, 1940). In various foods the nitrogen content of the 10 amino acids essential for the growth of rats (arginine, histidine, isoleucine, lysine, methionine, phenylalanine, threonine, tryptophane, and valine) amounts to approximately 50 per cent of the total nitrogen on a weight basis (Edwards et al., 1946). Accordingly, the minimum nitrogen requirement, when supplied as the essential amino acids, would be approximately one half that of the accepted minimum protein requirement. If we accept the minimum protein requirement as 35 gm., this then is equivalent to 5.6 gm. of nitrogen per day. On the basis of the essential amino acids, only half of this, or 2.8 gm. of nitrogen per day, would be sufficient.

However, certain unknown factors may influence this absolute minimum nitrogen requirement. In the first place, we have no idea whether the nitrogen present in the nonessential amino acids can be dispensed with completely in human beings as was done in the above calculations. Secondly, it is not known definitely whether amino acids alone will support adequate growth even in animals, much less in human beings. Recent work with protein hydrolyzates indicates the presence of a peptide-like substance that must be added to an amino acid mixture in order to secure adequate growth in mice (Woolley, 1945). It is possible that the nonessential amino acids have functions other than purely energetic.

The results of Murlin *et al.* (1946a) lend support to the idea that the presence of factors in proteins other than the amino acids is necessary. These workers found that when a mixture of the 10 essential amino acids replaced the protein, the retention of nitrogen was less on the amino acid diet than on the natural protein diet. This was true even though the nitrogen intake was the same in both cases. In this instance, the difference was later shown (Murlin *et al.*, 1946b) to be due to the presence of unnatural isomeric forms of the amino acids in the mixture.

The effect of the intake level of individual amino acids on nitrogen retention is far from being properly understood. Brush, Willman, and Swanson (1947) were able to show that eggs had a nitrogen-sparing effect in rats when added to a nitrogen-low ration. The urinary nitrogen reduction following the addition of small quantities of egg amounted to one-third of the urinary excretion during the week when the nitrogen-low ration was fed alone. When methionine replaced eggs in the nitrogen-low ration at a comparable level of nitrogen, the same effect was produced. The influence of the other essential amino acids on the urinary nitrogen excretion was considerably less than that of methionine. Results similar to this were secured by Cox *et al.* (1947) with both rats and dogs.

But further studies show the limitations of animal experimentation when considering the protein nutrition of human beings. This was emphasized by the work of Johnson *et al.* (1947), who found entirely different results with human subjects. The addition of methionine to the diets of their subjects had no influence on the nitrogen excretion. The addition of methionine had no influence on a pre-existing positive or negative nitrogen balance produced by the diets containing 75 or 14 gm. of protein, respectively. The sulfur in the 3 gm. of dl-methionine given to each subject per day was completely recovered in the urine during the period when this amino acid was fed. Somewhat similar findings were reported by Cox *et al.* (1947), who found that the addition of methionine to a casein hydrolyzate had no influence on nitrogen retention in man. This was true when the casein hydrolyzate was given intravenously to surgical patients as the sole source of nitrogen, when children were fed a high level of nitrogen, and when normal adults were maintained on a maintenance nitrogen level.

CHAPTER 19

Total Energy Exchange and Nitrogen Balance

AT ALL levels of undernutrition the nitrogen balance and the total energy balance seem to be inextricably related. The total intakes of energy (calories) and nitrogen, and the nitrogen output, are readily measured. On the other hand, estimation of the total energy output is difficult.

The total energy expenditure is determined by the basal metabolism, the specific dynamic action of food eaten, the level and efficiency of physical activity, and, indirectly, the environmental temperature. If all these components were known, a complete energy equation could be written. As a matter of fact, however, even in the best controlled studies the information is incomplete, so that only a rough estimate of energy exchange can be made.

The basal metabolic rate during undernutrition and rehabilitation has been reviewed in Chapter 17. There is general agreement that the B.M.R. decreases during semi-starvation, and the decrease can amount to a considerable saving of energy. In the Minnesota Experiment the decrease in the B.M.R. amounted to 40 per cent of the control values. However, most of the B.M.R. observations in the literature as well as in the Minnesota Experiment pertain to comparatively short periods of undernutrition, as a rule not exceeding one year, and it is possible the situation is different after very prolonged periods of undernutrition. It has been reported that in persons who are chronically underweight (Blunt and Bauer, 1922; Blunt et al., 1926; McKay, 1928; Coons, 1931; Strang and McCluggage, 1931; Stark, 1933; Strang et al., 1935) and in those who were emaciated at the end of World War II in Holland or Germany (Beattie and Herbert, 1948a) the B.M.R. approached normal values when expressed on the basis of surface area. Similar findings were reported by Zuntz and Loewy (1918). This means that no definite prediction of the state of undernutrition can be made from the B.M.R. without further definition.

The specific dynamic action of food plays a comparatively minor role, probably not exceeding 5 to 10 per cent of the total energy expenditure. Even if there were a decrease during starvation, the total effect would be insignificant. However, this component of the total energy balance should not be completely ignored, and it is reviewed further in a separate section of this chapter.

Besides the B.M.R. the most important factor determining the total level of energy expenditure in the normal condition as well as in undernutrition is the level of physical activity. It is the only component which may be changed voluntarily, and spontaneous reduction of physical activity is one of the most prominent features of semi-starvation. This is certainly the natural response of any starved individual, but in World War II (in German and Japanese prison camps) muscular work was often enforced in spite of advanced semi-starvation. A considerable energy expenditure is often involved in the quest for food in famine areas. Unfortunately, there is practically no precise information on the level of muscular activity in any of the literature. However, in a few experiments the level of physical activity was varied and the effects on nitrogen balance and body weight were studied. These data will be reviewed in this chapter.

Information about the effect of climate on the energy exchange in semi-starvation is equally deficient. It is known that the undernourished individual is sensitive to low environmental temperatures, and it seems possible that cold temperatures may increase the total metabolism more than in normal subjects. However, no actual experimental data are available.

In view of this situation, the only criterion for estimating the energy exchange is the change in body weight or the nitrogen balance. It may be assumed that a subject in weight and nitrogen equilibrium is also in energy equilibrium; in this state the total food intake would equal the total energy expenditure. If the B.M.R. is known, a breakdown into the three main components — B.M.R., specific dynamic action of food, and physical activity — may be attempted. However, the assumption of energy equilibrium in a person who is maintaining body weight and nitrogen equilibrium cannot be made without reservation. The composition or energy content of the body might change without a corresponding change in body weight. A change in the water content of the body is the most important source of error, but there is also the possible conversion of carbohydrates into fat, or vice versa, to be considered. It is well known that obese persons on reducing diets often show false equilibriums for periods of a week or more.

During undernutrition or rehabilitation the changes in body composition are pronounced. Owing to changes in water content, there is often no parallelism between body weight and nitrogen loss or gain. This was recognized by Voit in 1881 and has been confirmed in many subsequent observations. The water retention during starvation (see Chapter 15) is not enough to overshadow the loss of body weight. However, in some situations it may produce the paradoxical phenomenon of weight gain during short periods of semi-starvation on an intake of only a few hundred calories even though the subject is in negative nitrogen balance (Mühlmann, 1899). There is no question that the nitrogen balance is a better criterion of the true changes in body composition than the body weight; for this reason, we will consider in this chapter primarily those reports in which nitrogen determinations were made. The general question of body weight changes in semi-starvation is discussed in Chapter 6.

Nitrogen loss is of great importance in semi-starvation since it is practically equivalent to loss of active tissue mass. The fat loss, although greater on a percentage basis, is functionally much less consequential. Therefore, the accumulated nitrogen loss would be an excellent, if not the best, criterion for the degree of undernutrition. Complete data in human semi-starvation are available only in the Carnegie Experiment (see Chapter 3), but valuable data were obtained by Moulton (1920) in steers and by Morgulis (1923) on a dog.

There are limitations to the use of nitrogen balance as a criterion for energy exchanges. The nitrogen metabolism follows dietary changes with considerable inertia. One of the best illustrations is the slow adaptation of the daily nitrogen excretion to acute starvation. Figure 55 shows this in one of the experiments made on Succi (Freund and Freund, 1901); other experiments on the same subject show essentially the same picture (tabulated by Benedict, 1915b, p. 249). Adaptation of the nitrogen excretion to a given new dietary level requires at least one week, but most experimental periods in the literature have been shorter. In a state of nitrogen or energy imbalance, trends are apparent which may make it difficult to correlate a given level of food intake with a definite level of nitrogen excretion.

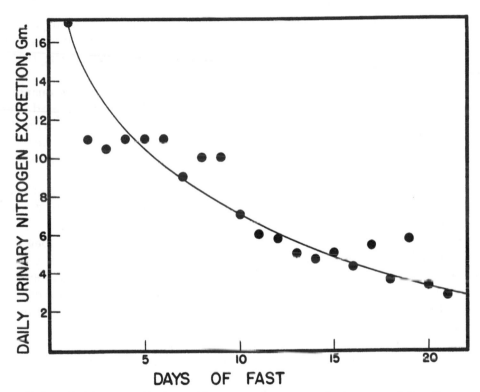

FIGURE 55. DAILY URINARY NITROGEN EXCRETION during one of Succi's prolonged fasts (Freund and Freund, 1901).

Even a superficial survey of the available literature reveals tremendous differences. The trend changes in body weight or nitrogen loss are due mainly to the length of the starvation period, as well as to the level of intake. The responses of nitrogen excretion to different levels of caloric intake are complicated. At levels of moderate caloric insufficiency the nitrogen excretion decreases as a rule, but at still lower levels the nitrogen excretion may rise because tissue protein is being used as fuel. The level at which this occurs depends upon the dietary history and seems to vary widely; the data available from different studies are not comparable. Besides divergent experimental conditions, there are large individual differences in the response to undernutrition, as revealed by Loewy and Brahm (1919) in investigations on women of the poorer population of Berlin and by Kohn (1919) in Vienna.

Most of the literature from starvation areas is concerned with rehabilitation rather than with the process of semi-starvation. In starvation areas, as well as in the case of emaciated patients, the most urgent practical problem is always speedy dietary rehabilitation.

Specific Dynamic Action

The specific dynamic action of nutrients (S.D.A.) is important for two reasons: (1) it represents a source of energy loss which must be allowed for in the over-all energy balance, and (2) it is a biological response which could possibly be used as a criterion of the nutritional state of the organism. The large general literature on S.D.A. is covered in numerous reviews (Lusk, 1928, 1931a, 1931b; Brody, 1934; Grafe, 1934a, 1934b).

In general, proteins produce a higher S.D.A. than do fats, but the quantitative values reported by different workers show wide discrepancies, partly because of differences in procedure. All agree that the S.D.A. increases with the amount of food given. The bulk of the information concerns single nutrients, but Hamilton (1935) reviewed the literature to show that the values for the single nutrients are not simply additive.

With large mixed meals (containing 2142 to 4378 Cal.) Benedict and Carpenter (1918) found the total S.D.A. (per 24 hours) to be from 5.2 to 7.4 per cent of the ingested calories. Richardson and Mason (1923) fed diabetics diets containing the same proportions of protein, fat, and carbohydrate that they calculated their subjects metabolized under basal conditions. When these diets were fed in small portions at 2-hour intervals no elevation in metabolism was found — that is to say, the S.D.A. was zero. Lusk (1931a) concluded from these experiments that the S.D.A. of a mixed diet may be negligible if the diet conforms in quantity and proportion to the exact needs of the tissues.

Recent experiments with rats clearly show the difference between the S.D.A. for single nutrients and that for mixed meals. Hamilton (1939), on diets ranging from 4 to 54 per cent protein, found a definite minimum for the S.D.A. at a level of 18 per cent protein in the diet. Ring (1942) and Forbes and Swift (1944) found the S.D.A. of all food combinations to be smaller than could be calculated for the sum of the single nutrients.

Specific Dynamic Action in Undernutrition

The literature may be cited to indicate that the specific dynamic action in undernutrition is less than, equal to, or greater than that in a normal state of nutrition. These differences may be explained partly by the difficulty in accurately characterizing the level of nutrition and, more important, by the absence of a standardized technique for determining the S.D.A.

Carbohydrate ingestion (75 gm. of glucose) failed to increase the metabolism of a fasting man (Johansson, 1908) and the same was true for fat ingestion in an emaciated diabetic patient (Richardson and Mason, 1923). These reports indicate that the S.D.A. is reduced or absent in undernutrition.

In contrast, there are the somewhat better studies which indicate that undernutrition does not alter the S.D.A. McCann (1920) found that a fast of one week's duration did not alter the S.D.A. in one man. Boothby and Bernhardt (1931) measured the S.D.A. in a 65 kg. man who lost 7 kg. during a 4-week period of undernutrition. Throughout this time the S.D.A. of small amounts of glycocoll (5 to 30 gm.) was not affected. It should be noted that glycocoll has a powerful S.D.A.

Finally, there are data which indicate that the S.D.A. may be considerably increased in undernutrition. Mason, Hill, and Charlton (1927) studied the S.D.A. of fat in normal subjects and in patients who were 21 to 40 per cent underweight. While the maximum S.D.A. in the 5 normal subjects was an increase of the metabolism amounting to 15.9 per cent above the basal rate, the undernourished patients exhibited rises of 27, 20, 48, 27, 15, 26, and 31 per cent. There was no correlation between the excess S.D.A. and the extent of weight loss in this small group.

Specific Dynamic Action in Rehabilitation

Long ago Rubner (1902) stated that no S.D.A. appears when ingested protein is deposited in the form of new tissues. In any case, there is reason to believe that the S.D.A. of protein, and perhaps of other nutrients as well, is dependent on the physiological destination of the nutrient.

In the experiment of Boothby and Bernhardt referred to above, the subject was subsequently given a diet of 3353 Cal.; at this time the S.D.A. of the glycocoll became negative. Occasionally a decrease in 'the metabolism has been observed in normal persons after fat ingestion (Grafe, 1934a, 1934b).

Allowance for Specific Dynamic Action in Calculating Total Energetics

The available information is certainly not adequate for making exact allowances for the S.D.A. in the total daily metabolism of normal persons, let alone in starvation and recovery. The best present estimate might be that the S.D.A. is not importantly changed in undernutrition. In rehabilitation it may be reasonable to suggest that the specific dynamic action is somewhat reduced. Finally, it will be noted that in all practical situations we are concerned with the effect of mixed meals, and therefore smaller allowances for S.D.A. should be made than would correspond to that of the single nutrients.

Physical Work — The Mechanical Efficiency

The mechanical efficiency of muscular work is not changed in semi-starvation (see Chapter 34; also Joffe et al., 1919). The claim of an increased efficiency made by Benedict et al. (1919) was unfounded, as is shown by recalculation of their data (see Chapter 3). However, fatigue trends seem to be more prominent in undernutrition as shown by Zuntz and Loewy (1918). While Loewy's energy expenditure in 1917 when walking on the treadmill was the same as before the war for very short periods of exercise, there was a pronounced increase (up to

32 per cent) during walking on a steep grade for 20 minutes or more. An increase of the energy expenditure in walking occurs in normal subjects only under much more strenuous conditions. For instance, Zuntz (1902) found only a small increase (about 10 per cent) in 2 healthy subjects after 5.5 hours of walking while carrying a load of 10 kg.; with a smaller load the increase was negligible. The early breakdown of the steady state, as shown in Loewy's observations, undoubtedly indicates reduced endurance and corresponds well with Loewy's subjective sense of weakness and fatigability.

In Boothby and Bernhardt's (1931) subject, the excess energy expenditure during short periods of standard exercise (bicycle ergometer) showed only minor differences at caloric intakes of 849, 1296, 1620, and 3353 Cal. a day. However, in the later phases of recovery there was a drop below the resting level after exercise on the 1620 and 3353 Cal. diets, which was not apparent on the preceding starvation diets of 849 and 1296 Cal.

The effect of changing the level of physical activity on weight and nitrogen balance in undernourished persons is of much practical importance, but there are few data of value. Jansen (1917) compared the effects of heavy and light work in 3 men who had lost from 3.5 to 7.5 kg. on the restricted diets of World War I in Germany. Before the work load studies began, these men were standardized for 6 days on daily intakes of 1630 Cal. and 9.5 gm. of nitrogen (60 gm. of protein). Days of light activity were alternated with days of hard work.

Two of the 3 subjects had a markedly higher nitrogen loss during hard work, but there was no real change in the rate of loss of body weight. It is interesting that the B.M.R. was 25 to 33 per cent higher the morning after the walk. Another subject in Jansen's series was put on the same experimental diet of 1600 Cal. for 6 days and then brought back to his initial (pre-experiment) weight. He was studied during bed rest for another 6 days on a diet of 1600 Cal. There was no significant difference in the weight loss in bed rest, but the nitrogen loss fell to an average of 1.19 gm. daily as compared with 2.37 gm. when the subject was up and about. The diets and nitrogen intake were not precisely matched during the two periods of this experiment; however, the difference was small. It is improbable that the very slight differences in the diet can explain the drastic reduction in nitrogen loss during bed rest.

Even in bed rest the nitrogen balance was negative at an intake level of 1600 Cal. These observations may be compared with the results in this Laboratory (Miller *et al.*, 1945) on the effect of prolonged bed rest in normal young men. We found that during a 3-week period of bed rest the nitrogen balance was negative even at the relatively high intake of 2600 Cal. Obviously, the effect of bed rest is somewhat different in normal people and in severely undernourished persons.

Level of Over-All Physical Activity in Semi-Starvation

Reduction of physical activity is a prominent feature of undernutrition. For a working population, living on their earnings, one might expect occupational activity to be maintained at the expense of leisure activity — the more so since during periods of semi-starvation wages fall far behind the cost of food. While this may actually be the case in jobs requiring only slight to moderate energy expenditure, the situation is different in more demanding work.

According to Kraut and Müller (1946) the daily coal output in tons per miner in the Ruhr district declined gradually from 1939 to 1944, more or less in parallel with a decline of the daily food intake from 4200 to 3600 Cal. This relationship was observed for the whole Ruhr district but not for single plants. Since the actual energy expenditure in mining, contrary to general opinion, is not extreme (Berkowitsch and Simonson, 1935), the lowest caloric intake in the Ruhr district during World War II was still above the undernutrition level. For this reason, undernutrition alone cannot be considered the cause of the decreased work output. The drop in output may have been related to the increasing proportion of slave labor in the Ruhr mines and to the general disruption of transportation and supply.

The data of the Minnesota Experiment can be used to arrive at a general estimate of the reduction of physical activity on the basis of available work calories (see Table 181). Such estimates can be readily made if there is energy equilibrium. In the Minnesota Experiment the body weight approached equilibrium at the end of starvation, and the data on nitrogen balance also indicated at least a close approach to equilibrium. In these calculations the caloric intake and the B.M.R. were known and the S.D.A. was estimated to be 7.5 per cent of the caloric intake; any error involved in this estimate cannot be great.

TABLE 181

ESTIMATE OF CALORIES EXPENDED IN PHYSICAL ACTIVITY in the Minnesota Experiment.

	Control	S24 Absolute	% C
Caloric intake	3468	1570	45
B.M.R.	1595	964	60
S.D.A.	260	118	45
Physical activity			
As calories	1613	488	30
As percentage of caloric intake	46.5	31.0	
Body weight (kg.)	70	53.2	76

Table 181 shows an over-all reduction of 70 per cent in the total calories available for work at the end of the starvation period. A breakdown of the overall activity into various types could not be made. However, it is obvious that the reduction of work calories far exceeded the reduction of body weight (24 per cent), and the reduction of activity was at least 50 per cent.

Nitrogen Balance at Low Levels of Caloric Intake

It would be of the greatest practical value in times of food shortage to know the lowest combinations of calories and of protein in the diet which can maintain nitrogen equilibrium, and this question is also of considerable theoretical interest. Many efforts to provide an answer have been made but none of the

TABLE 182

NITROGEN EQUILIBRIUM observed at intakes of 1800 Cal. or less per day.

Subject		Degree of Undernutrition	Caloric Intake	Protein Intake (gm.)	Author
N	Sex				
1	M .... moderate		1600	25	Caspari (1905)
1	M .... slight		1500–1800	50–60	Zuntz and Loewy (1918)
1	F .... advanced		1684–1849	50–55	Loewy and Brahm (1919)
1	F .... advanced		1300	50	Loewy and Brahm (1919)
3	F .... moderate		1400	58	Kestner (1919)
6	F .... moderate to advanced		1000 (av.)	35	Kohn (1919)
10	M .... advanced		1744	62	Hoesslin (1919)

TABLE 183

NEGATIVE NITROGEN BALANCE observed at intakes of 1800 Cal. or less per day.

Subject		Degree of Undernutrition	Caloric Intake	Protein Intake (gm.)	Author
N	Sex				
1	M .... slight		1620	72.5	Boothby and Bernhardt (1931)
1	M .... slight		1500–1800	50–60	Zuntz and Loewy (1918)
13	M .... slight to moderate		1600	60.5	Jansen (1917)
7	F .... advanced		1400 (av.)	40 (av.)	Loewy and Brahm (1919)
10	F .... moderate to advanced		1000 (av.)	35 (av.)	Kohn (1919)
7	M .... advanced		1350	41	Brull (1943)

results have more than suggestive value. The extraordinary difficulty of obtaining exact nitrogen equilibrium at the lowest possible level of intake is not generally appreciated.

We have attempted to analyze all cases reported in the literature where undernourished people on low caloric diets have been studied with a view to finding the lowest nitrogen intake which can maintain nitrogen balance. The results are set down in Tables 182 and 183. Table 182 summarizes the observations of positive nitrogen balances at caloric intakes of 1800 or less per day, and Table 183 summarizes the observations of negative nitrogen balances. The level of 1800 Cal. was arbitrarily taken as the upper limit because maintenance at such levels has been reported in normal population samples (see Chapter 18). In most of the cases there were no data, or only inadequate data, on basal metabolism, body size, and sex, and the precision of the estimate of food intake is often questionable. In all cases the physical activity may be characterized as sedentary to light.

These data, such as they are, indicate a considerable degree of individual variability, but in general they strongly suggest that the protein requirement for maintaining nitrogen balance is not reduced in undernourished people on low caloric intakes. In other words, such metabolic adaptation as may result from continued subsistence on a low caloric intake may be insufficient to allow adults to maintain balance on 50 gm. or less of protein daily when the caloric intake is of the order of 1500 Cal. or less. In none of the studies was there a proper separation of the variables of calories and protein, although not infrequently the authors' conclusions disregard this point. Loewy and Brahm (1919) insisted that their findings showed that the nitrogen balance is broadly independent of caloric or protein intake in the range they studied, but this might be due to the large individual differences observed in their series. In passing it may be noted that dogs on a low caloric intake (25 Cal. daily per kg.) may be maintained in positive nitrogen balance if the protein in the diet is very high but not if the proportion of protein to carbohydrate is reduced to 1 to 4 (Elman *et al.*, 1944).

The studies of Kohn (1919) on 16 adult women in Vienna at the end of 1917 merit some examination (see Table 184). Eleven of these women were hospital convalescents and all had been on the severely restricted food intakes imposed by the rationing system of the time. Although these studies have many technical flaws, they are interesting because of the indications of individual variations and the fact that 2 of the women were apparently in positive nitrogen balance on caloric intakes of 30 Cal. or less per kg. A distinct but crude relationship between caloric intake and the nitrogen balance appears in Kohn's data, but this is more clearly evident in the studies of Beattie, Herbert, and Bell (1948).

Nitrogen equilibrium has been observed at extremely low levels of food in-

TABLE 184

NITROGEN BALANCE ON LOW CALORIC INTAKES of 16 women in Vienna, December 1917 (Kohn, 1919).

Diet (Cal./kg./day)	Number of Subjects	Nitrogen Balance			
		Positive		Negative	
		N	Range (gm. daily)	N	Range (gm. daily)
38.0	2	2	2.36–8.39	0	
35.1–38.0	6	2	6.27–6.98	4	0.49–4.47
30.1–35.0	2	0		2	0.33–0.71
25.1–30.0	1	1	1.12	0	
20.1–25.0	3	1	2.18	2	0.59–2.28
20.1	2	0		2	1.53–3.18

TABLE 185

NITROGEN EQUILIBRIUM IN EMACIATED HOSPITAL PATIENTS at low caloric intakes.

Subject		Disease	Caloric Intake		Protein (gm.)	Author
Age	Sex		Total	Per Kg.		
19	F	esophageal stricture	765	24.7	44.8	F. Müller (1889)
24	M	esophageal stricture	620	13.5	48.8	Klemperer (1889)
21	F	anorexia nervosa	614	18.0	48.2	Klemperer (1889)
38	F	mental disease	591	11.2	60.0*	Tuczek (1884)
28	M	tuberculosis	1625	56	67.0	Albu (1899)

* Slightly negative nitrogen balance.

take in emaciated hospital patients (see Table 185). The nitrogen balance was only slightly negative in Tuczek's (1884) case; this woman retained 17 gm. of protein per day when the diet subsequently was increased to 1691 Cal. per day. Albu's (1899) patient is included because of the high positive nitrogen balance achieved, although the level of intake he used is not particularly low. This patient was severely emaciated (body weight 29 kg.); on 56 Cal. daily per kg. he retained 106 gm. of protein over a period of 2 weeks.

Caloric Intake in Germany after World War II

Most of the opinions concerning Central Europe in 1947 indicated widespread undernutrition and semi-starvation among large segments of the population. Much weight has been given to statements of various individuals who were keenly impressed by their visits to the different countries, but there are unfortunately very few objective details on the actual amount or extent of the undernutrition. In order to secure some independent information we requested a German physician[*] in the British Zone to determine the caloric intakes and body weights of 10 "representative" college students over a 4-week period in February 1947, when the food supply was reported as being very low.

TABLE 186

BODY WEIGHTS OF GERMAN COLLEGE STUDENTS IN 1947. The reported caloric intakes of the first 7 men averaged 1560 Cal. per day for each of the 4 weeks of the experiment; the intakes of the last 3 men averaged 1900, 2700, and 1775 Cal. per day with considerable week-to-week variation in each case. The degree of underweight was calculated from the height/weight tables secured by Dr. Josef Brožek from Czechoslovakian workers in Zlín (unpublished).

Subject	Age	Original Height (m.)	Original Weight (kg.)	Underweight (%)	Weekly Weight Change from Original Weight, in Kg. 1st	2d	3d	4th
1	23	1.70	65.5	−4.4	0.0	+1.0	+1.0	+1.6
2	27	1.75	64.0	−5.5	−0.5	+1.0	+4.0	+3.0
3	25	1.82	62.0	−14.2	+1.5	+1.5	0.0	+1.0
4	21	1.75	58.0	−14.5	+0.5	+0.1	+0.5	+1.1
5	23	1.70	50.9	−20.6	+1.8	+2.1	+2.4	+2.3
6	20	1.71	49.0	−24.6	+0.5	+1.5	+1.7	+2.1
7	22	1.84	51.5	−30.2	−2.5	−0.5	+1.8	−0.5
8	24	1.68	56.0	−9.2	+1.5	+2.1	−1.2	−0.7
9	25	1.81	68.6	−5.1	−0.8	−0.5	0.0	+0.4
10	24	1.85	64.0	−14.5	+5.5	+1.4	+2.8	+3.3
Average	23.4	1.761	58.95	−14.28	+0.75	+0.97	+0.94	+1.36

The original body weights in this group showed a great deal of variation in the degree of emaciation (see Table 186). On the basis of height/weight tables 3 of the 10 students were 20 per cent or more underweight, while 4 of them were less than 10 per cent underweight; none was overweight. No mention was made

[*] The scientist requested that his name and institution be omitted in any report on his findings because he was afraid of possible reprisals. We can vouch for the reliability of his findings.

of any edema in these subjects either during the experiment or at any time preceding it. These students continued on their regular diets at their usual dining place for the 4 weeks of the experiment.

Six of the 7 students who received an average of only 1560 Cal. per day showed small but significant weight increases; the one person in this group who showed a consistent weight loss was the one who was underweight to the greatest extent at the start. The other 3 subjects, whose diet showed greater fluctuations, showed also greater variations of body weight; in general, the correlation between body weight changes and the caloric intakes was poor.

It should be noted that all reports of the military governor of the American Zone in 1947 indicated a very low food intake and only an inconsequential weight loss. The more detailed study of these 10 students is entirely consistent. These findings can be explained on one of two bases: (1) for the students, as well as in the military reports, the entire food intake was not recorded, or (2) the physiological adaptation to a very restricted caloric intake is much greater than has previously been thought possible. There are other reports, both from the United States and from other parts of the world, which tend to emphasize the second possibility. More research is needed, particularly well-controlled experiments, on persons with widely differing nutritional histories over a period of some years.

The reports of the military governments of the occupied zones of Germany might seem to be valuable sources of data, but they are acceptable only to a limited degree. Where definite measurements were recorded, as in body weights, they seem to be reliable, but the estimates of food consumption are far less satisfactory. For example, in June 1947 it was estimated that the total average food consumptions in Würtemberg-Baden, Hessen, and Bavaria amounted to 1150, 1350, and 1050 Cal. daily, respectively. It is impossible to reconcile these figures with the comparatively slight weight losses actually recorded.

Reports from German sources on the state of nutrition in Germany from 1945 to 1947 have been made available through Dr. Gunther Lehman, director of the Institut für Arbeitsphysiologie, Dortmund.* The reports from the various regions are not uniform, but taken as a whole they seem to agree with the reports of the occupation authorities about the degree of weight loss.

The best estimate of the actual caloric intake of a rather large group of the working population in Germany was found in a report of Dr. Wille, Gewerbehygienische Untersuchunsstelle der Medizinischen Klinik of the Kruppsche Krankenanstalten, Essen, August 1946. Five physicians and 5 nurses made a thorough clinical examination of 1059 workers. At the same time, the subjects were questioned as to their activities, complaints, observations, and various sources of food procurement. Physicians and workers had known one another for many years so that a very favorable background was established for obtaining confidential information. The following sources of non-rationed food were checked: (1) support from relatives in the country; (2) black market; (3) trips to the country; (4) small gardens; (5) keeping of small animals such as chickens and

* The reports were sent by Commander H. I. Alvis, Medical Intelligence Section, EUCOM Headquarters.

rabbits; and (6) parcels. The caloric value of the food from each source was estimated. The estimated caloric expenditure necessary to obtain the additional calories from gardening or trips to the country was subtracted in order to estimate the net calories obtained from that particular source. On this basis, the over-all consumption per non-working member of the family was estimated to be 1500 Cal. a day, including the 1000 Cal. of rationed foods. For the average workingman the over-all caloric intake was estimated to be 2000 Cal., and the caloric deficit for both the working and the non-working population was assumed to be about 500 Cal. a day.

Only 3.6 per cent of the group was overweight; 19.4 per cent were slightly underweight (up to 10 per cent); 48.2 per cent ranged between 11 and 20 per cent underweight; and in 29.1 per cent the degree of underweight exceeded 20 per cent. However, the distribution of the weight loss according to age showed great differences; the largest weight losses (average about 16 kg.) occurred in people over 40 years of age, while the weight loss in the younger age groups (up to 30 years) was much smaller (about 10 kg.). The weight loss of the manual workers (average 15.5 kg.) was smaller in all age groups than that of the white collar workers (average 20.5 kg.), probably because the higher food requirement for the manual workers was more nearly met by the larger food ration supplied them. The weight loss of female workers was smaller than that of male workers.

Nitrogen and Weight Balance at Different Levels of Deficiency

Different levels of caloric deficiency have been compared in very few studies. The efforts in this direction in the Carnegie Experiment have been discussed in Chapter 3.

In Boothby and Bernhardt's (1931) experiment, the subject lost 2.6 kg. of body weight. During the 4 dietary periods that followed (see Table 187), progressive increases of the food intake up to 1620 Cal. produced no increase in body weight; in fact, the body weight still declined slightly. It was only when the food intake was raised to 3350 Cal. for 24 days that any substantial increase in body weight occurred and that a positive nitrogen balance was established. The last period was also marked by a large increase in the B.M.R. (see Table 160). The R.Q. increased throughout the period of refeeding. The caloric and

TABLE 187

Food Intake, Nitrogen Balance, and Body Weight during a Period of Refeeding following a short period of semi-starvation (Boothby and Bernhardt, 1931).

Period	Date	Food Intake		Nitrogen Balance (gm.)	Change of Body Weight (kg.)
		Cal.	N (gm.)		
I	June 2–15	894	9.4	−3.40	−2.9
II	June 16–July 1	1296	10.1	−2.09	−0.3
III	July 2–July 19	1620	11.6	−1.17	−0.2
IV	July 20–August 12	3353	15.2	+2.00	+2.3

TABLE 188

INFLUENCE OF A SEMI-STARVATION DIET ON BODY WEIGHT AND NITROGEN BALANCE. Average results of Jansen's (1917) experiments. Caloric intakes and nitrogen values are expressed on the basis of 24 hours.

Subject	Sex	Age	Height (cm.)	Weight			Caloric Intake	Nitrogen		
				Before War Diet (kg.)	Before Experiment (kg.)	Daily Loss in Experiment (kg.)		Intake (gm.)	Excretion (gm.)	Balance (gm.)
E.J.	M	32	173	62	57.6	−0.25	1632	9.72	11.97	−2.25
F.E.	M	26	174	63	57.5	−0.23	1630	9.61	11.83	−2.22
P.L.	M	23	174	62	58.5	−0.28	1630	9.61	11.55	−1.94
E.K.	M	23	182	67	60.1	−0.25	1630	9.61	11.74	−2.13
C.K.	M	22	178	68	64.6	−0.40	1625	9.72	11.29	−1.58
M.G.	M	22	171	59	54.1	−0.25	1619	10.02	11.34	−1.32
B.	M	57	175	74	66.5	−0.21	1578	9.89	11.97	−2.08
H.	M	36	165	70	59	−0.25	1523	9.10	12.98	−3.88
G.B.	F	22	165	52.5	61.3	−0.50	1630	9.61	11.81	−2.20
H.F.	F	35	168	58	60.4	−0.38	1630	9.61	9.93	−0.33
B.	M	33	164	68	61.3	−0.20	1578	9.89	12.26	−2.37
R.H.	M	21	181	65	63.5	−0.35	1617	9.89	12.63	−2.74
W.J.	M	30	169	79	75.6	−0.26	1639	9.81	12.24	−2.42
M		28.6	172.2	65.2	61.54	−0.293	1612.4	9.699	11.811	−2.112

nitrogen intakes that were necessary to obtain nitrogen equilibrium were not determined.

Nitrogen balance experiments were made by Jansen (1917) on 13 normal subjects who had lost from 2 to 11 kg. on the wartime diet. Their previous diet was not exactly known, but the experimental diet was set at 1600 Cal. The food, as well as the urine and feces, was analyzed for nitrogen. The caloric content of the meals was calculated according to tables, but some control measurements were made by Rubner. As a rule each experiment lasted 6 days. In 2 subjects the 1600 Cal. diet was given over a period of 28 and 31 days, with analyses performed during the first 6 and the last 6 or 9 days. In one subject the values for the two periods were essentially identical; in the other subject the weight and nitrogen loss was significantly less in the later period. Table 188 is the summary of the data in Jansen's Tables II to XIV.

Jansen's subjects G.B. and H.F. were exceptions in that they had increased their body weights during the war and were, at the beginning of the experiment, in a good nutritional state. Their body weight loss during the experiment was larger than the average. The data on the other subjects were rather uniform as far as nitrogen deficit and body weight loss were concerned (from —1.32 to —3.88 gm. per day and from —0.2 to 0.5 kg., respectively). The average nitrogen intake corresponded to a protein intake of 60.5 gm. per day; the average nitrogen loss corresponded to a loss of 11.8 gm. of protein per day. All the subjects performed light work during the experiment.

In 3 of Jansen's subjects the protein content of the diet was increased to 67.6 gm. (10.59 gm. of nitrogen intake) while the caloric content was reduced to 736

TABLE 189

DAILY WEIGHT AND NITROGEN LOSSES at caloric intakes of 1630 and 736 Cal. (Jansen, 1917).

Subject	Weight Loss (kg.)		Nitrogen Loss (gm.)	
	1630 Cal.	736 Cal.	1630 Cal.	736 Cal.
E.K.	0.4	0.2	1.58	4.15
G.B.	0.5	0.35	2.20	3.19
H.F.	0.38	0.42	0.33	2.83

TABLE 190

NITROGEN EQUILIBRIUM AND WEIGHT CHANGES IN GERMAN CIVILIANS during World War I who were maintained on 2 different caloric intakes, 1630 and 2100 Cal. The higher diet was given for 6 days, the lower one for 5 days. (Jansen, 1917.)

Subject	Weight Change (kg.)		Nitrogen Balance (gm.)	
	1630 Cal.	2100 Cal.	1630 Cal.	2100 Cal.
E.K.	—0.25	+0.2	—2.13	—0.21
E.J.	—0.25	0.0	—2.25	—0.56
R.H.	—0.36	+0.28	—2.33	—0.14

Cal. for a period of 4 days. Table 189 shows the comparison of weight and nitrogen loss for the 1630 and 736 Cal. diets. While the nitrogen loss was much greater, the weight loss was the same or even less with the 736 Cal. diet.

In 3 other subjects the diet was increased to 2100 Cal. for a period of 5 days without any change in the nitrogen intake (9.6 gm. daily). Table 190 indicates that 2100 Cal. were sufficient to maintain the body weight and produce nitrogen equilibrium in Jansen's subjects doing light physical work. The slightly negative nitrogen balances are not significant; in subjects E.J. and R.H. the nitrogen balance was positive during the last 2 days of the 5-day experiment.

Reduction Diet in Obesity

There are many reports of metabolic studies on obese persons while on reducing regimens. The literature on the B.M.R. in obese patients has been reviewed in Chapter 17.

It is generally believed, particularly in clinical practice, that the loss of weight in obese persons on reducing diets is accounted for precisely by a loss of fat. The main justification for this is the success of calculation methods whereby the weight loss is predicted from the caloric intake on the assumption that the entire weight change must represent pure fat. The usual procedure is to estimate the caloric "requirement" for "ideal" body weight. The difference between this value and the actual food fed is taken to be the "caloric deficit," and the latter value, divided by the caloric value of a unit weight of fat, is the predicted weight loss. Sometimes an adjustment factor, derived from empirical experience, is included. Surprisingly enough, such formulas may work rather well in spite of the fact that some of the assumptions involved are probably wrong. But such results have little or no value as evidence for the actual metabolic processes or the changes in body composition.

The subject of the quantitative responses of obese persons to reducing diets is a large one and cannot be treated in detail here. But some of the outstanding studies which involved actual nitrogen balances are instructive and will be examined.

Evans and Strang (1931) kept 187 obese patients for an average of 8.7 weeks on a diet of 400 to 600 Cal. a day. The diet consisted mainly of proteins. The estimated caloric deficits ranged from 1500 to 2500 Cal. a day. In the beginning the weight loss curve was quite steep but later the slope became considerably less. The average weekly weight loss was 1.6 kg. (3.5 lbs.). During the first weeks stored carbohydrates and body nitrogen were lost as well as fat. Later nitrogen equilibrium was established, and only fats were used in quantities from 150 to 300 gm. a day.

The creatinine excretion did not change in these patients during weight loss. This was interpreted to mean that there was no destruction of "vital" tissues. The subjective changes were remarkable, and in sharp contrast to those which would occur in normal people on such a low diet. There was no feeling of hunger, nor any disagreeable sensations; on the contrary, there was an outspoken feeling of well-being, decreased fatigability, and return of vigor, and minor ailments such as headaches and skin disorders disappeared. This general improvement was noted after a loss of only 2.2 to 4.5 kg. In patients with increased blood pressure, a pronounced drop was one of the commonest and earliest signs. The same favorable response was obtained in both men and women, but on the whole the men responded more rapidly. Age (ranging from 12 to 67 years) was without any influence on the response. In several female patients irregularities of menstruation were restored to normal.

A more detailed study (Evans and Strang, 1931) was performed on 5 patients with an average initial weight of 157.5 kg. They received first a maintenance diet of 2413 Cal. containing 69 gm. of protein, which was sufficient for nitrogen equilibrium. On the reducing diet, which contained 59 gm. of protein, 10 gm. of carbohydrate, and 7 gm. of fat, totaling 335 Cal., a nitrogen loss occurred, but this was amazingly small (2 gm. of nitrogen in 5.5 weeks). The addition of only 20 gm. of carbohydrate (increasing the total caloric intake to 445 Cal. per day) brought these patients into nitrogen equilibrium at an excretion level of 0.15 gm. of nitrogen per kg. of "ideal" weight. Nitrogen equilibrium was maintained in one patient for 260 days, during which time there was a total weight loss of 67.1 kg.

Keeton and Bone (1935) gave 10 obese patients a diet that provided an average of approximately 55 per cent of their actual caloric requirements and 52 to 70 per cent of their basal metabolic needs. The patients were confined to the hospital but were allowed the freedom of the ward. The protein content was changed during the experiment from high (90 gm.) to low (13 gm.), and in some patients this change was repeated 2 or 3 times; each high or low protein period lasted from 30 to 40 days, with a total duration of the experiment ranging from 8 to 53 weeks. The weight dropped linearly with time, and no effect of the high or low protein diet on the rate of weight loss was apparent. Figure 56 summarizes 2 typical experiments. No break in the linear weight loss occurred on the low protein diets (broken lines). All these patients of Keeton and Bone gained nitrogen during the period when they were on the high protein diet but did not lose nitrogen on the low protein intake. The metabolic rate, apparently calculated as deviation from normal standards, did not change.

Sainsbury and Smith (1937) kept 12 obese women for 6 days on a daily diet of 1 quart of skimmed milk and 6 bananas, with an energy content between 600

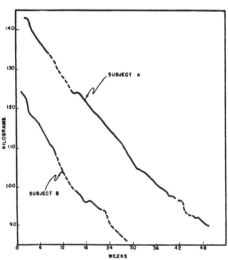

FIGURE 56. Diet Experiments on Obese Patients. Subject A received 70 per cent of his basal calorie requirement and lost 52.9 kg. Subject B received 58 per cent of his basal calorie requirement and lost 38.5 kg. In both cases the protein intake was maintained at about 90 gm. daily except where indicated by broken lines when the protein intake was about 13 gm. daily. (Keeton and Bone, 1935.)

and 1037 Cal., depending on the size of bananas eaten. The nitrogen intake varied from 4.36 to 6.85 gm. per day. The body weight initially ranged from 62 to 91 kg. (137 to 201 lbs.) and decreased from 2.0 to 4.5 kg. (4.5 to 10 lbs.) in the 6-day test. The nitrogen balance was negative in 10 subjects (daily nitrogen losses ranged from 0.69 to 3.02 gm.), but there was no correlation between the individual nitrogen loss and the nitrogen or caloric intake. In one subject with only a moderate degree of obesity (153 lbs.) a positive nitrogen balance (+1.52 gm.) was observed at the low caloric intake of 767 Cal. with 5.62 gm. of nitrogen. Another subject who weighed 75 kg. was in nitrogen equilibrium at only 600 Cal. and 3.86 gm. of nitrogen.

In another experiment involving 6 subjects, part of the bananas were replaced by beef, increasing the nitrogen intake to 7.8–13.6 gm. and the caloric intake to 737–1269 Cal. The initial body weight ranged from 63 to 93 kg. (138 to 205 lbs.). The weight loss during the 6 days of observation was from 1.6 to 3.2 kg. (3.5 to 7.0 lbs.). One patient had a negative nitrogen balance (−3.2 gm.) on an intake of 1104 Cal. and 7.8 gm. of nitrogen, but all the others had positive nitrogen balances (+3.2 to +7.0 gm.) which showed good correlations with the nitrogen intake.

The discrepancy in the nitrogen metabolism in these studies as compared with Keeton and Bone's (1935) observations may possibly be explained by different degrees of obesity. Both series of studies show that obese subjects can maintain nitrogen equilibrium on very low intakes of calories and nitrogen. In general the results obtained in obese patients on a very low reduction diet are strikingly different from those obtained in normal subjects during semi-starvation. It seems that as long as ample fat reserves are available the organism is protected against protein loss. The loss of fat produces no untoward symptoms such as accompany the weight loss of normal subjects during semi-starvation. Furthermore, no decrease of the B.M.R. below normal standards has been reported in obese patients during the period of weight reduction.

Nitrogen Excretion in Starvation — Minnesota Experiment

The nitrogen excretion in the urine was determined in all 32 subjects during the control period and at S12 and S24. Complete nitrogen balances for 3-day periods at these times were made on 10 men. The diets during the control and semi-starvation periods were constant for some weeks prior to the balance estimations, so the 3-day periods, with fecal excretions marked off with carmine, should be acceptable for defining the mean balances. Tables 191, 192, and 193 give the essential data for these 10 men at C, S12, and S24.

The mean urinary nitrogen excretions for the 10 men were 13.8, 8.1, and 7.5 gm. per day at C, S12, and S24, respectively; the corresponding means for the total group of 32 men were 13.2, 8.1, and 7.4 gm. Obviously, the smaller group of 10 men may be regarded as representative of the total group of 32 men.

The total caloric intake of the 10 men decreased from an average of 3470 Cal. in the control period to 1638 Cal. at S12 and 1533 Cal. at S24. Although the absolute as well as the caloric intake per kg. decreased somewhat from S12 to S24, the nitrogen balance was definitely less negative at S23 than at S12. At S12,

TABLE 191

NITROGEN BALANCE AND BODY WEIGHT of 10 men in the control period. The total
nitrogen excretion is the sum of that in the urine and feces.
(Minnesota Experiment.)

Subject	Nitrogen Excretion (gm.)		Nitrogen Intake (gm.)	Nitrogen Balance (gm.)	Body Weight
	Urine	Total			
5	12.2	14.2	16.1	+1.9	80.8
9	14.9	16.6	16.1	−0.5	72.9
23	13.7	15.7	16.1	+0.4	69.2
26	14.0	16.4	16.1	−0.3	71.3
30	14.9	16.7	16.1	−0.6	68.1
105	13.4	15.6	16.1	+0.5	68.7
112	14.4	16.2	16.1	−0.1	61.7
119	13.1	14.9	16.1	+1.2	66.3
106	14.0	15.5	16.1	+0.6	83.6
130	13.0	14.3	16.1	+1.8	66.0
M	13.8	15.6	16.1	+0.8	70.9

9 out of the 10 subjects were in a definitely negative nitrogen balance; at S23, 2
subjects showed some nitrogen gain, 2 were about in nitrogen equilibrium, and
only 5 (out of 9) were in negative balance. It will be recalled that the body
weight approached a plateau at S24.

There was little or no correlation between the individual nitrogen balances
at S12 and at S23, but it will be noted that the relative intakes of the different
individuals were not precisely the same at the two periods. When the intakes of
calories and nitrogen per kg. of body weight were plotted against the nitrogen
balance (Figures 57, 58, 59, 60), there were no evident correlations at S12 but
correlation trends appeared at S23. This may be explained in part by the general fact that individual variability is more pronounced when metabolic changes
are rapidly progressing than when relative stability is achieved. Further points
of significance are the facts that the net caloric deficiency was much greater at
S12 than at S23 and that individual differences in fat reserves were much reduced at S23.

The cumulative nitrogen loss can be indirectly estimated on a different basis
in the Minnesota Experiment. The active tissue mass (see Chapter 15) was estimated in 18 of the subjects, including 4 of those in the 10-man nitrogen balance
group, and these 18 men may be considered representative, on the average, of
the entire group of 32 men. The data in Table 194 indicate a total protein loss
of 1990 gm. for the first 12 weeks and 690 gm. for the second 12 weeks of semi-
starvation. These values correspond to average daily losses of 3.8 gm. in the
first 12 weeks and 1.3 gm. for the second 12 weeks. This indicates a progressive
reduction in the negativity of the nitrogen balance as semi-starvation proceeded.
If the course of the nitrogen loss for the entire 24 weeks is reconstructed from
these values, assuming a mathematical form of the curve similar to that for body
weight, the indicated values at S12 and at S23 are in good agreement with the
nitrogen balances measured at those times.

TABLE 192

Nitrogen Balance, Body Weight Loss, and Caloric Intake of 10 men after 12 weeks of semi-starvation (S12). The total nitrogen excretion is the sum of that in the urine and feces. (Minnesota Experiment.)

Subject	Nitrogen Excretion (gm.)		Nitrogen Intake (gm.)		Nitrogen Balance			Caloric Intake (per kg.)	Percentage Weight Loss
	Urine	Total	Absolute	Per Kg.	Absolute (gm.)	Mg. per Kg.			
5	6.1	7.6	7.0	.11	−0.6	−10.0		26.6	22.7
9	7.4	8.9	8.0	.14	−0.9	−15.3		27.7	19.3
23	8.6	10.1	9.8	.18	−0.3	−4.6		30.7	21.8
26	9.2	11.0	7.0	.12	−4.0	−71.7		29.4	20.3
30	7.9	8.9	7.0	.12	−1.9	−33.5		29.5	17.4
105	10.4	11.9	7.0	.13	−4.9	−88.7		29.9	18.7
112	6.3	7.8	7.0	.14	−0.8	−16.3		32.8	18.2
119	8.6	10.1	8.8	.16	−1.3	−24.8		31.0	19.1
106	7.8	9.3	5.9	.09	−3.4	−49.4		24.3	18.5
130	9.1	10.6	8.8	.15	−1.8	−31.8		29.4	13.9
M	8.1	9.7	7.6	.13	−2.0	−34.6		29.1	19.0

TABLE 193

Nitrogen Balance, Caloric Intake, and Body Weight Loss in 10 men after 23 weeks of semi-starvation. The total nitrogen excretion is the sum of that in the urine and feces. (Minnesota Experiment.)

Subject	Nitrogen Excretion (gm.)		Nitrogen Intake (gm.)		Nitrogen Balance			Caloric Intake (per kg.)	Percentage Weight Loss
	Urine	Total	Absolute	Per Kg.	Absolute (gm.)	Mg. per Kg.			
5	8.0	9.8	8.7	.15	−1.1	−18.5		29.0	29.1
9	8.8	10.1	8.7	.15	−1.4	−23.6		29.0	21.5
23	8.0	9.2	10.2	.20	+1.0	+18.9		37.6	26.3
26	7.2	8.4	5.9	.11	−2.5	−47.3		28.9	25.5
30	8.1	9.4	5.9	.11	−3.5	−65.5		21.6	22.1
105	7.3	8.9	8.7	.17	−0.2	−3.5		32.3	25.3
112	6.5	7.1						21.9	20.5
119	8.3	9.1	8.7	.17	−0.4	−8.2		33.4	24.6
106	6.9	8.2	5.9	.10	−2.3	−38.3		21.1	26.5
130	5.5	6.3	8.0	.15	+1.7	+29.4		30.3	18.2
M	7.5	8.6	(7.8)	(.15)	(−1.0)	(−17.4)		28.5	24.0

TABLE 194

Changes in Body Fat and Protein, and their caloric equivalent; from the averages for 18 men in the Minnesota Experiment. These calculations are based on data summarized in Table 152, Chapter 15.

	Body Fat		Active Tissue			
	Kg.	Cal.	Kg.	Protein (gm.)	N (gm.)	Cal.
Control	9.66		39.93			
S12	4.95		31.96			
C–S12, total	4.71	42,300	7.97	1990	320	7970
C–S12, per day		504		23.7	3.8	107
S24	2.74		29.20			
S12–S24, total	2.21	19,800	2.76	690	110	2760
S12–S24, per day		136		8.2	1.3	33

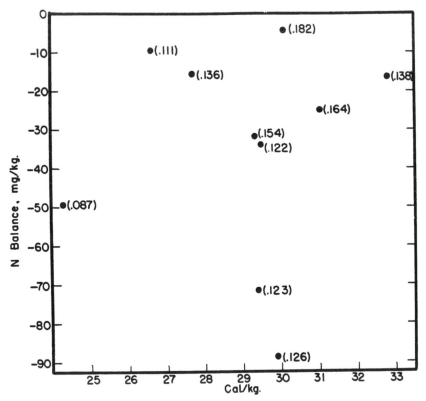

FIGURE 57. Lack of Correlation between Caloric Intake (Cal./Kg.) and Nitrogen Balance (Mg./Kg.) after 12 weeks of semi-starvation (S12) in the Minnesota Experiment. The figures in parentheses indicate the nitrogen intake (gm./kg.).

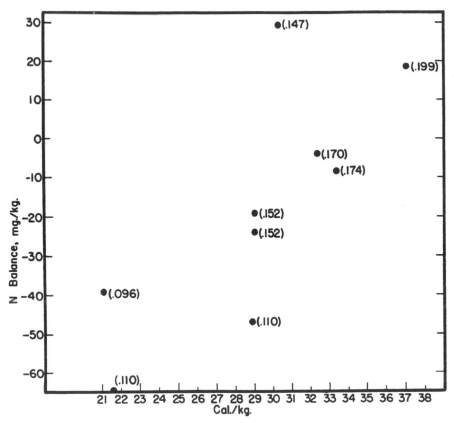

FIGURE 58. CORRELATION BETWEEN CALORIC INTAKE (CAL./KG.) AND NITRO-
GEN BALANCE (MG./KG.) at the end of semi-starvation in the Minnesota Experiment. The figures in parentheses indicate the nitrogen intake (gm./kg.).

From these data it is possible to calculate the average daily destruction of
body fat and of body protein for the first and second halves of the semi-starvation period in the Minnesota Experiment. It is indicated that the average daily
catabolism of body fat amounted to 504 Cal. for the first 12 weeks and 136 Cal.
during the second 12 weeks; in the second 12 weeks the body fat destruction was
only 27 per cent as great as in the first 12 weeks. For the body protein the average daily destruction amounted to 107 Cal. during the first 12 weeks; during the
final 12 weeks this average was 33 Cal., that is, only 32 per cent of the value
for the first half of the total period of semi-starvation.

Protein and Body Weight Gain during Rehabilitation

In most reports from starvation areas interest has centered on the changes
occurring during nutritional rehabilitation rather than on the process of semi-
starvation. The protein (active tissue) gain would appear to be the best criterion for nutritional rehabilitation; changes in body weight may reflect mainly
alterations of water or fat which are of smaller functional consequence. For this
reason, emphasis is placed on nitrogen balances rather than on body weight

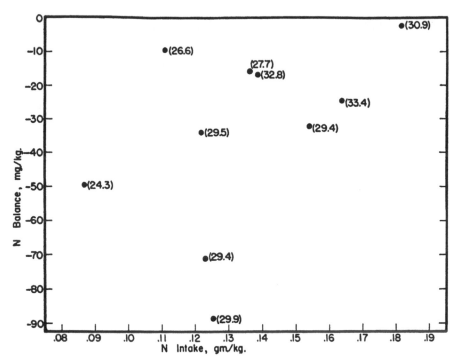

FIGURE 59. LACK OF CORRELATION BETWEEN NITROGEN INTAKE (GM./KG.) AND NITROGEN BALANCE (MG./KG.) after 12 weeks of semi-starvation in the Minnesota Experiment. The figures in parentheses indicate the caloric intake (Cal./kg.).

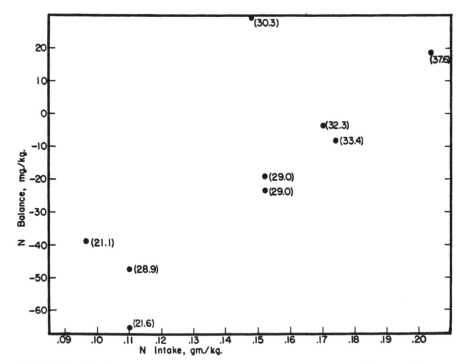

FIGURE 60. CORRELATION BETWEEN NITROGEN INTAKE (GM./KG.) AND NITROGEN BALANCE (MG./KG.) at the end of semi-starvation in the Minnesota Experiment. The figures in parentheses indicate the caloric intake (Cal./kg.).

changes in rehabilitation, although the latter cannot be disregarded since it is the only evidence available in some studies.

The most interesting, and from a practical standpoint the most important, question is the relative efficiency of caloric versus protein intake in effecting nutritional rehabilitation. Unfortunately, in most studies the two intakes were changed simultaneously, although not necessarily in the same proportion.

Some of the shortcomings in previous studies should be pointed out before their results are discussed. As a rule, the dietary periods used during rehabilitation were too short or followed one another in an unsystematic sequence, so that the effect of any particular level of caloric or nitrogen intake can be surmised only approximately. In other studies (Kestner, 1919; Loewy and Zuntz, 1918; Loewy and Strauss, 1919; Boothby and Bernhardt, 1931; Strang, McCluggage, and Brownlee, 1935; Brull, 1945) the increase in the caloric intake was so large that it is impossible to estimate the minimum caloric and nitrogen intake necessary to initiate rehabilitation.

Rehabilitation of Undernourished Patients

The very low level of caloric intake at which nitrogen equilibrium can be maintained in emaciated patients was discussed in an earlier section of this chapter. Table 195 shows the effect of rehabilitation diets on emaciated patients. In the first 5 patients a positive nitrogen balance was obtained at very low levels of caloric or protein intake. This could be expected from the very low level at which nitrogen equilibrium was maintained in these patients before the rehabilitation program (see Table 185); any slight excess of nitrogen beyond the maintenance level is readily stored. In Richter's patient the caloric and especially the protein intakes were relatively high but the rate of nitrogen storage was extremely high.

Clinically there has been more interest in the level of food intake that produces the most rapid rehabilitation than in the efficiency of the rehabilitation process. The latter is of primary concern in starvation areas, where even in the period of relief the food supply is likely to be critical. Kissling (1913) studied the rehabilitation of 450 cases of undernutrition from various causes. Diets be-

TABLE 195

DIETARY REHABILITATION IN EMACIATED PATIENTS. All values are given on a daily basis.

| Subject | | Caloric Intake | | Nitrogen Intake (gm.) | Days of Experiment | Gain in Gm. | | Author |
Age	Sex	Total	Per Kg.			Body Weight	N	
19	F	765	24.7	7.6	5	200	1.66	F. Müller (1889)
		881	27.6	9.0	8	110	1.95	
24	M.....	1550	33.7	13.8	14	170	0.27	Klemperer (1889)
21	F	614	18.0	7.7	3	0	0.86	Klemperer (1889)
		2090	83.4	18.7	8	412	4.50	
38	F	626	11.8	3.9	27	−92	2.84	Tuczek (1884)
28	M.....	1625	56.0	11.2	14		1.21	Albu (1889)
17	F	2560	62.5	27	56		12.0	Richter (1904)

tween 3430 and 4812 Cal. were given over periods of from 19 to 58 days. The weight gains were tremendous (daily averages between 160 and 430 gm.).

More informative are Strang, McCluggage, and Brownlee's investigations (1935) with a smaller but better controlled group. They studied the metabolism, nitrogen balance, and body weight in 18 undernourished patients before and during treatment with a high caloric diet of between 2800 and 5000 Cal. for 2 to 12 weeks. The part of their work concerning changes in the B.M.R. has been reviewed in Chapter 17. The daily nitrogen intake averaged 10.96 gm. and the total nitrogen excretion 8.70 gm., with an average daily nitrogen storage of 2.26 gm. in the 18 patients. The rate of storage was not directly related to the level of the nitrogen intake; the amounts stored varied widely between 5 and 41 per cent of the intake. The caloric intake had some influence on the nitrogen storage, but this relationship could not be expressed in simple terms; the amounts of nitrogen stored both per calorie and per calorie in excess of the basal requirement varied widely for the different individuals, with no apparent correlation to either total or excess calories. Furthermore, the rate of nitrogen storage was not parallel to the size of the patient or to the degree of undernutrition.

The results of the nitrogen balances made for these patients showed that the body tissue stored during this period contained very little protein. The 8 patients who showed an average gain of 8 kg. of body weight retained an average of 11 gm. of nitrogen per kg. of body weight gain. This is considerably lower than the 30 gm. of nitrogen per kg. of tissue stated by Lusk (1928) to be normal. From these figures, Strang, McCluggage, and Brownlee calculated that the solids gained by their patients consisted of approximately 75 per cent fat and 25 per cent protein.

Rehabilitation Observations in World War I

Loewy and Brahm (1919), Kestner (1919), and Kohn (1919) confirmed the theory that in undernutrition nitrogen retention occurs with the slightest increase of the nitrogen intake over the level of the starvation diet. All of the nitrogen in 110 to 120 gm. of meat was retained since this addition to the diet did not increase the nitrogen excretion (Kestner, 1919). This, of course, is true only for very short periods and with fairly reasonable caloric intakes. In general, most of the observations of these authors were exploratory rather than analytical.

Hoesslin's Studies on Refeeding

The most thorough research during World War I was performed by Hoesslin (1919) on more than 200 undernourished people. The semi-starvation was due to wartime restrictions of the diet; the subjects were free from specific diseases. The nitrogen analyses of food, urine, and feces were performed in Rubner's laboratory. Hoesslin's results are summarized in Tables 196, 197, 198, and 199.

The weight changes in Hoesslin's patients were quite irregular in the several periods and in different individuals and groups. For the entire 7 weeks of series 1 (see Table 196) the 210 men lost an average of 0.76 kg. on an average daily intake of 2284 Cal., including 72.7 gm. of protein. This can only be interpreted to mean that water shifts dominated the entire picture of weight change in this period. The average results for Hoesslin's groups A and B are similar to those obtained in the Z and L groups during early rehabilitation in the Minnesota Experiment, where comparable caloric and protein intakes produced substantially no gain in gross body weight in the first 6 weeks.

TABLE 196

EFFECTS OF VARIATIONS OF CALORIES AND OF PROTEINS IN REHABILITA-
TION DIETS ON THE WEIGHT GAINS OF 210 UNDERNOURISHED MEN. Food
intakes refer to the daily values for each man. Mean values for 70 men
in each of 3 groups: A = mean height 166 cm., weight 40 to 50 kg.;
B = mean height 172 cm., weight 50 to 60 kg.; C = mean height 170
cm., weight 60 to 70 kg. Data from series 1 of Hoesslin (1919).

| Week | Food Intake | | | Weight Change, in Kg. | | |
	Protein (gm.)	Total Cal.	Cal./Kg.	A	B	C
1	70	2392	45			
2	68	2065	40	0.0	−0.5	−0.5
3	68	1977	37	−0.1	−0.1	−0.2
4	92	2191	42	+1.1	+0.9	−0.4
5	65	2256	43	−0.8	−1.3	−1.4
6	73	2225	42	−0.1	0.0	0.0
7	73	2493	47	+0.1	+0.3	+0.7
8	67	2111	41		+0.2	−0.4
M*	72.7	2284	42.3	+0.03	−0.11	−0.30

* Excluding week 8.

In Hoesslin's subjects the tendency to gain weight on a given dietary intake was related to the degree of undernutrition. Hoesslin's group C was initially far less undernourished than the men in the Minnesota Experiment, as can be seen by comparing heights and weights. The results suggest that for this group C a daily intake of 2284 Cal. was actually inadequate.

The results in Hoesslin's series 2a and 2b (see Tables 197 and 198) are in general agreement with his series 1. As a whole these measurements suggest that in the early stages of rehabilitation high protein intakes may be useful in promoting early weight gain and that relatively high intakes of both proteins and calories are indicated for rehabilitation. But no definite conclusions can be drawn because of the unknown factor of water shifts and the confusion produced by too many dietary changes in the short periods of study. Note that in period 5 of series 2b there was practically no weight gain on a daily intake of more than 4000 Cal., including 127 gm. of proteins.

Hoesslin studied the nitrogen balance in some of his patients. In one group of 5 men he was able to maintain nitrogen equilibrium on intakes of 5.1 gm. of nitrogen (32 gm. of protein) with 2316 Cal., and of 4.2 gm. of nitrogen (26 gm. of protein) with 2618 Cal.; the addition of 300 Cal. lowered the protein requirement by 6 gm. A positive nitrogen balance was readily achieved with 2196 Cal., including 67.5 gm. of protein. But a rapid response in body weight was obtained only when the daily protein intake was 90 gm. or more.

TABLE 197

Effects of Variations of Calories and of Proteins in Rehabilitation Diets on the Weight Gains of 5 Undernourished Men averaging, initially, 51.3 kg. in weight. Mean values computed from series 2a of Hoesslin (1919).

| Period | Days | Food Intake | | | Weight Change (kg.) |
		Protein (gm.)	Total Cal.	Cal./Kg.	
1	6	68	2353	46	−0.3
2	9	97	2381	47	+1.1
3	9	91	1936	37	+0.8
4	7	90	3229	61	0.0
5	6	150	2952	55	+1.5
6	9	68	2196	40	−0.7

TABLE 198

Effects of Variations of Calories and of Proteins in Rehabilitation Diets on the Weight Gains of 5 Undernourished Men averaging, initially, 50.4 kg. in weight. Mean values computed from series 2b of Hoesslin (1919).

| Period | Days | Food Intake | | | Weight Change (kg.) |
		Protein (gm.)	Total Cal.	Cal./Kg.	
1	6	68	2353	47	−1.3
2	9	97	2381	48	+1.2
3	9	96	2976	61	−0.3
4	7	122	3316	66	+2.6
5	6	127	4071	80	+0.1
6	9	68	2196	43	−1.4
7	4	32	2316	46	−1.8
8	4	26	2618	52	−0.1

In 10 men (see series 3, Table 199) Hoesslin obtained nitrogen retention at all levels of intake tested, but the nitrogen balance was unrelated to the weight gain in the test periods of a week each. In another series (4) of 5 men, both nitrogen and body weight gains were obtained in 5 out of 6 weeks, but there again the 2 items were not correlated.

In general, Hoesslin's data indicate that substantial nitrogen retention occurs in the first weeks of rehabilitation of undernourished men on any diet providing more than about 60 gm. of protein and about 1800 Cal. Larger nitrogen retentions result from increasing the intake. The efficiency of nitrogen use seems to be low at all levels of intake, amounting to something like a storage of 10 per cent of the nitrogen supplied over and above the minimum requirement for equilibrium. In all cases there is at best a very poor correlation between nitrogen storage and weight gain in the early weeks of rehabilitation.

Nitrogen retention data for Hoesslin's series 3 and 4 are summarized in Figures 61 and 62. The significance of these correlations is questionable because

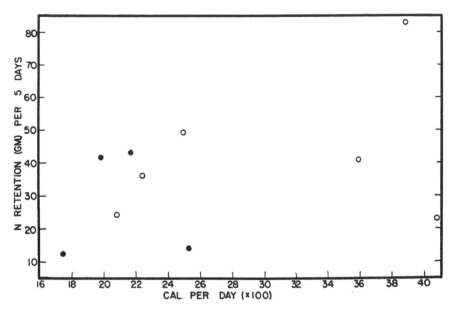

FIGURE 61. Correlation between Caloric Intake and Nitrogen Retention, plotted from experiments by Hoesslin (1919). Solid circles = series 3;
open circles = series 4.

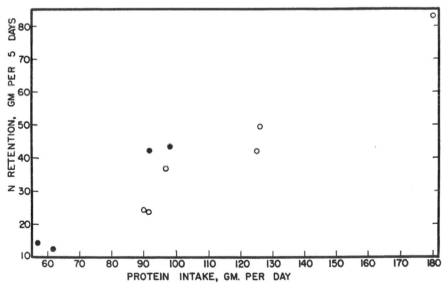

FIGURE 62. Correlation between Protein Intake and Nitrogen Retention, plotted from experiments by Hoesslin (1919). Solid circles = series 3;
open circles = series 4.

TABLE 199

EFFECTS OF VARIATIONS OF CALORIC AND PROTEIN INTAKE ON BODY WEIGHT AND NITROGEN BALANCE IN UNDERNOURISHED MEN. Series 3 = mean values for 10 men with an average initial weight of 45 kg.; series 4 = mean values for 5 men with an average weight of 50.3 kg. Calorie values are daily averages for the week shown Data from Hoesslin (1919).

| Week | Food Intake | | | Weight Change (gm.) | Nitrogen Balance (gm.) |
	Protein (gm.)	Cal.	Cal./Kg.		
		Series 3			
1	62	1744	39	−29	+1.8
2	57	2529	56	+43	+2.0
3	92	1928	42	+214	+6.0
4	98	2161	47	−43	+6.2
5	81	2571	56	−29	
		Series 4			
1	97	2247	46	+43	+5.2
2	126	2484	51	+170	+7.1
3	125	3539	72	+270	+5.9
4	183	3891	75	+170	+11.8
5	90	2084	41	−170	+3.5
6	91	4080	78	+86	+3.3

the protein and calorie variables are not properly separated and the time sequence is such as to render the whole analysis dubious. As rehabilitation progresses there is a rapid change in the level of endogenous catabolism (maintenance requirement) as well as in the avidity of the tissues to store nitrogen.

Hoesslin concluded from his work that continuous nitrogen storage in severely undernourished people demands a protein intake of 120 gm. or more daily. His data warrant no such conclusion.

Rehabilitation Data from World War II

We know of only one major study from World War II on nitrogen balance during recovery from severe undernutrition. Beattie, Herbert, and Bell (1948a) studied 6 emaciated Dutch patients and 11 Germans at various levels of protein and caloric intake. The experimental periods varied from 10 to 16 days. In the Dutch subjects the protein intakes were varied from 24 to 50 gm. of nitrogen per day in a total of about 3000 Cal. However, the different levels of protein intake were compared in different individuals; only one of the subjects received both the "low" and the "high" protein intake. In this subject the nitrogen retention was 4.5 gm. a day at an intake of 25 gm. of nitrogen per day and increased to 10.2 gm. a day at an intake of 50 gm. of nitrogen a day. In 4 subjects the average daily nitrogen retention was 6.1 gm. on a nitrogen intake of 25 gm. daily. One subject receiving a mixed diet with 40 gm. of nitrogen had a daily nitrogen retention of 7.8 gm., and 2 subjects stored an average of 10 gm. of nitrogen per day at the high intake level of 50 gm. of nitrogen per day.

The purpose of the investigation on the Dutch subjects was to determine the calorie and protein intake that would secure the most rapid rehabilitation and retention. Consequently, the level of protein intake was very high, even exceeding that in Hoesslin's series. The nitrogen storage, expressed as a percentage of the intake, was 28.6, 19.6, and 20 per cent at intake levels of 25, 40, and 50 gm., respectively. The economy of nitrogen storage seemed to be best at the lowest nitrogen intake, that is, at 25 gm., but the total storage was greater at 50 gm. Owing to the small number of individuals in the series all conclusions are somewhat questionable, but as a whole the results seem to confirm the findings in Hoesslin's series.

The studies of Beattie, Herbert, and Bell on the German subjects had a different purpose — to find the caloric and protein intake necessary to secure nitrogen *equilibrium* in emaciated persons. The lowest level provided 1700 Cal. and 9 gm. of nitrogen, and the highest level was 2500 Cal. and 19 gm. of nitrogen in the last experimental period. Different dietary levels were studied in 7 of the 11 subjects. The fact that the caloric and protein intakes paralleled each other makes the differentiation of the relative importance of nitrogen versus caloric intake very difficult, if not impossible.

Beattie and his colleagues attempted to overcome this difficulty by means of graphic analysis of their results. For both the German and the Dutch group they plotted the nitrogen retention against both the caloric intake (see Figure

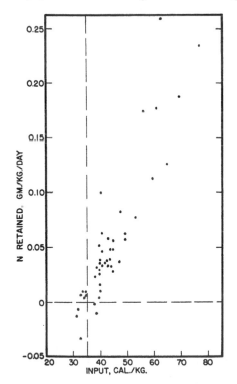

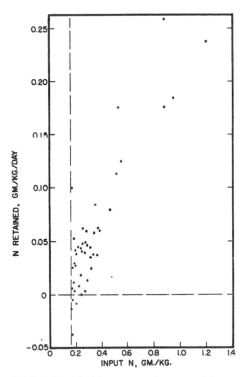

FIGURE 63. CALORIC INTAKE (CAL./ KG./DAY) VS. NITROGEN RETENTION (GM./KG./DAY) in experiments by Beattie, Herbert, and Bell (1948).

FIGURE 64. NITROGEN INTAKE (GM./ KG.) VS. NITROGEN RETENTION (GM./ KG./DAY) in experiments by Beattie, Herbert, and Bell (1948).

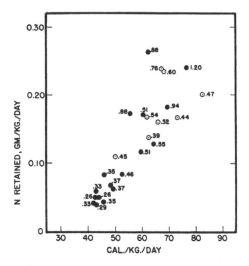

FIGURE 65. CALORIC INTAKE BETWEEN 40 AND 80 CAL./KG./DAY VS. NITROGEN RETENTION (GM./KG./DAY). The values attached to the circles indicate the nitrogen intake (gm./kg./day). (Beattie, Herbert, and Bell, 1948.)

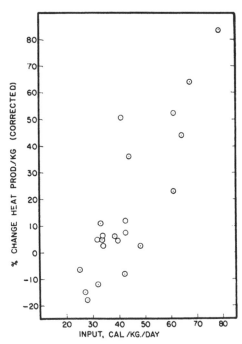

FIGURE 66. RELATIONSHIP BETWEEN CALORIC INTAKE (CAL./KG./DAY) AND BASAL METABOLIC RATE (Beattie, Herbert, and Bell, 1948).

63) and the nitrogen intake (see Figure 64). Both graphs show approximately the same degree of correlation. The authors were inclined to interpret a slight shift to the right in the nitrogen retention at higher nitrogen intakes (Figure 64) as an indication that retention at such levels is not proportional to the nitrogen input. The scatter is probably too great, however, to permit any definite conclusions. The lowest caloric intake at which nitrogen retention occurred was 35 Cal. per kg. of body weight (Figure 63), while the nitrogen intake at which retention occurred was 0.15 to 0.2 gm. per kg. of body weight (Figure 64). These values agree fairly well with Hoesslin's results. It would appear that it is possible to obtain significant positive nitrogen balances at nitrogen intakes as low as 0.17 gm. per kg. of body weight, provided the caloric intake exceeds 35 Cal./kg.

In order to examine the relationship between caloric and nitrogen intake more carefully, two graphs were prepared for a limited range of the caloric intakes. Figure 65 shows the nitrogen retentions of the Dutch group for caloric levels above 40 Cal./kg. Short balance periods (4 to 6 days) were added (open circles) to make the information more complete. The authors interpret this figure as evidence that the caloric intake was more important than the protein intake for nitrogen retention. At certain retention levels there was a wide variation in the nitrogen intake associated with only a small difference in the caloric level. For instance, between 67 and 77 Cal. per kg. the nitrogen retention was constant at 0.23 to 0.24 gm. per kg. even though the nitrogen intake varied from 0.6 to 1.2 gm. per kg. per day. On the other hand, at the same level of caloric intake (60 to 62.5 Cal./kg.)

there was a tremendous variation in the nitrogen retention ranging from 0.118 to 0.264 gm./kg. In view of such variations, it is impossible to draw from this work any final conclusion as to the relative importance of caloric versus protein intake in enhancing nitrogen retention.

Beattie, Herbert, and Bell also investigated the effect of the proportion of non-protein to protein calories on nitrogen storage. Again the results were not conclusive, probably because too many unseparated factors were involved. There was a definite correlation between the caloric intake and the increase in the basal metabolic rate (see Figure 66). An increase in basal heat production does not interfere with protein retention, as might be expected from an energetic equation. On the contrary, increased basal metabolic rate and protein storage seem to be associated.

Nitrogen Balance in Rehabilitation — Minnesota Experiment

Two-day urine samples were collected from all 32 subjects in the sixth and twelfth weeks of rehabilitation (R6 and R12). This might appear too short a time for a true evaluation of the nitrogen excretion, but the subjects had been on a constant diet for a long time preceding the collection and the number of subjects was large enough to minimize random individual errors and differences. The caloric and nitrogen contents of the food were determined by direct analysis and are shown in columns 1 and 2 (R6) and 6 and 7 (R12) of Table 200. The urinary nitrogen excretions (columns 3 and 8), as well as the differences between nitrogen intake and urinary nitrogen excretion (columns 4 and 9), increase with the calorie and nitrogen intakes. In the absence of fecal nitrogen determinations, these differences may be used as a rough index of nitrogen retention, since the urinary excretion constitutes by far the larger part of the total nitrogen excretion. The difference between nitrogen intake and urinary nitrogen excretion in percentage of the N intake (columns 5 and 10) increased with the calorie and nitrogen intakes from group Z to T. However, it changed little from R6 to R12 within the same caloric group, in spite of the considerable increase in both calorie and nitrogen intakes. This might indicate that the percentage nitrogen retention, as expressed in columns 5 and 10, is determined in the early part of rehabilitation, although the absolute retention increased with calorie and nitrogen intake.

The above statement can be made only with reservations since no fecal nitrogen determinations were made, but it seems to be reasonably sound bcause the differences in nitrogen retention of the caloric groups are quite large and are present both at R6 and R12.

The situation is different for the subdivisions of protein-supplemented (Y) and non-supplemented (U) groups. Individual differences, which are probably evened out for a caloric group of 8 men, might constitute a rather large source of error in the comparison of sub-groups of 4 subjects. Table 201 shows the breakdown of the results into low protein (U) and high protein (Y) groups.

The presentation of results in Table 201 corresponds to that of Table 200 except that the urinary nitrogen excretion was not calculated as a percentage of the nitrogen intake. The urinary nitrogen excretion in the high protein groups

TABLE 200

NITROGEN INTAKE AND URINARY NITROGEN EXCRETION in the sixth and twelfth weeks of rehabilitation in the Minnesota Experiment. Mean values, in gm. of nitrogen per man per day for the caloric groups of 8 men each. Cal. = mean daily calorie intake for the week indicated.

	R6				
	Intake		Urinary	Nitrogen Differences	
	Cal.	N	Nitrogen	Absolute	% of
Caloric		(gm.)	(gm.)	(2 — 3)	N Intake
Group	1	2	3	4	5
Z	1981	10.73	8.85	1.88	17.6
L	2333	12.02	9.20	2.82	23.4
G	2700	13.34	9.90	3.44	25.8
T	3032	14.84	10.33	4.51	30.4

	R12				
	Intake		Urinary	Nitrogen Differences	
	Cal.	N	Nitrogen	Absolute	% of
Caloric		(gm.)	(gm.)	(7 — 8)	N Intake
Group	6	7	8	9	10
Z	3049	17.14	13.70	3.44	19.5
L	3401	18.43	13.88	4.55	24.6
G	3769	19.75	14.83	4.92	24.9
T	4101	21.27	14.95	6.32	29.7

TABLE 201

NITROGEN INTAKE, URINARY NITROGEN EXCRETION, AND ESTIMATED NITROGEN BALANCE at the sixth and twelfth weeks of rehabilitation for the protein-supplemented (Y) and unsupplemented (U) subjects (Minnesota Experiment).

	R6				R12			
	Intake		Urinary		Intake		Urinary	
	Cal.	N	Nitrogen	(2 — 3)	Cal.	N	Nitrogen	(6 — 7)
Group		(gm.)	(gm.)			(gm.)	(gm.)	
	1	2	3	4	5	6	7	8
ZU	1977	9.76	7.45	2.31	3042	14.05	10.10	3.95
ZY	1984	11.70	10.25	1.45	3056	20.22	17.30	2.92
LU	2329	11.05	8.05	3.00	3304	15.35	10.55	4.80
LY	2336	12.99	10.35	2.64	3408	21.50	17.20	4.30
GU	2696	12.37	9.35	3.02	3762	16.69	11.50	5.19
GY	2704	14.31	10.45	3.86	3776	22.80	18.15	4.65
TU	3029	13.87	8.50	5.37	4094	18.18	11.40	6.78
TY	3035	15.81	12.15	3.66	4108	24.35	18.50	5.85

(Y) was both absolutely and relatively higher than in the low protein groups (U) at all caloric levels. Consequently, the differences between nitrogen intake and urinary nitrogen excretion (balance) are, with the single exception of group G at R6, larger for the U than for the Y groups. This result is surprising, but no definite conclusions in regard to the total nitrogen balance can be made. The differences between the Y and U groups in columns 4 and 8 are rather small, so that the fecal nitrogen excretion cannot be ignored in a final evaluation of such findings. In fact, there is reason to believe that the utilization coefficient of the protein supplement was substantially higher than that of the protein in the basic diet. The protein supplements used, casein and soy protein concentrate, have a true digestibility of close to 99 per cent, but the proteins in the basic diet were supplied by items such as whole wheat, potatoes, and cabbage which have much lower percentages of digestibility (cf. Mitchell, 1948, p. 53). It is probable, therefore, that the fecal nitrogen excretion represented a greater percentage of the total nitrogen intake in the U groups than in the Y groups, so that the differences between the U and Y groups in regard in nitrogen utilization, as expressed in columns 4 and 8, could disappear or even be reversed if full data were at hand.

With all possible allowances for these uncertainties the fact seems to be that the protein supplementation used here had no favorable effect on nitrogen balance or utilization. It should be noted that the lowest nitrogen intakes in this period were 9.76 gm. from R1 to R6 and 14.05 gm. from R6 to R12; this means a minimal average of 74.5 gm. of protein daily. With basal diets providing less protein quite different results might be obtained.

CHAPTER 20

Nitrogen Metabolism

NITROGEN metabolism has occupied a very prominent place in almost all studies on undernutrition and starvation. A net loss of nitrogen in the body obviously means a destruction or wasting of living tissue if we accept the customary theory that the body has little or no protein reservoir. The minimal level of protein intake which allows nitrogen equilibrium is of paramount importance in a period of food deprivation, the more so since at such times the protein-rich foods are apt to be the first to disappear from the diet. The influence of caloric intake on the nitrogen loss becomes increasingly greater as the caloric intake is reduced, until at very low levels all the protein ingested may be combusted forthwith. This aspect of nitrogen metabolism was considered further in Chapter 19.

For both practical and theoretical reasons it is highly desirable to be able to estimate the rate at which tissues are breaking down in starvation and the extent to which this process has already occurred. In many experiments this has been attempted by means of studies on nitrogen balance. This type of measurement is both tedious and difficult; small systematic errors become very important when extrapolated over any protracted period. Furthermore, the nitrogen balance in starvation gives no indication of the nitrogen balance prior to starvation. For these reasons many investigators have emphasized the level of protein in the blood plasma as a reflection of the bodily stores of nitrogen. Cayer and Nabors (1947) go so far as to state that "Since the plasma proteins of the blood are maintained by the tissue reserve stores and the dietary intake, the laboratory finding of an actual decrease in the serum proteins of the blood is always indicative of tissue protein depletion." Similar statements have been made by others (e.g. Bieler *et al.*, 1947; Bruckman *et al.*, 1930; Peters and Eisenman, 1932; Weech, 1936; Peters, 1946).

Factors Influencing the Apparent Level of Plasma Protein

Part of the emphasis on the plasma proteins stems from observations on animals that have been subjected to a protein-low ration (Frisch *et al.*, 1929; Weech *et al.*, 1935, 1937), part from studies on patients suffering from various diseases which have resulted in a restricted food intake (Peters and Eisenman, 1932), and part from examinations made on individuals subjected to prolonged periods of starvation (Weech, 1936). Before any of these data can be critically evaluated, it is necessary to consider briefly the following factors which may influence the absolute level of protein in the plasma:

(1) Many analyses for protein have been made on both the plasma and the serum without discrimination but the results are not identical nor can the anti-coagulant used in the separation of plasma be entirely neglected (Boyd and Murray, 1937; Leichsenring et al., 1939; Lange, 1946). According to Lange (1946) heparin is the anticoagulant of choice when plasma is to be used, but Kagan (1942) claims that serum is to be preferred for all such analyses. Any differences that may exist between serum and plasma protein values are usually less than those reported as the result of dietary disturbances. Here the terms plasma and serum will be used without any implication as to the differences in their protein contents.

(2) From the standpoint of the body's reserve, the total circulating proteins are of greater consequence than the percentage in the plasma. The percentage of protein in the plasma has received most attention partly because so much emphasis has been placed on the level required to prevent edema (see Chapter 43) and partly because blood volume estimations are infrequently available.

(3) The normal variation in plasma protein levels must be known before conclusions can be arrived at concerning the significance of so-called abnormal values. Most workers have tacitly assumed that the plasma protein concentration of the individual is normally constant. Denz (1947) stated that "the constancy of the plasma-protein in the normal subject is usually accepted without question." That this is not necessarily the case has been shown by Lange (1946), who made repeated determinations of the plasma proteins in normal subjects over periods as long as 10 months. The maximum difference between the extreme values in any one subject was as much as 15 per cent of the mean of all determinations made on that subject. Simultaneous estimations af the hematocrit in these cases showed parallel variations. Assuming that the volume of the circulating erythrocytes is constant in normal persons, Lange (1946) concluded that there was no essential change in the amount of circulating plasma protein over the period studied; the assumption, and hence the conclusion, may be challenged. It is important to bear in mind that (a) there may be considerable day-to-day fluctuations in the absolute plasma protein levels for any individual, and (b) practically all values in the literature give only the absolute plasma protein level with no indication as to the hematocrit or plasma volume.

(4) Posture may exert a marked influence on the percentage of plasma protein. Lange (1946) states that the difference in the concentration of plasma proteins may be as great as 0.7 gm. per 100 cc. for successive blood samples secured from the same individual in the upright and the recumbent position. This difference is especially prominent after the subject has rested for a half-hour.

(5) Other factors that may possibly influence the plasma protein level include season, atmospheric temperature, diurnal fluctuation and diet. According to Lange (1946) these factors are of minor significance in the evaluation of the changes in protein levels ordinarily encountered in physiological experiments.

(6) The methods used for estimating the plasma proteins include the refractive index, the viscosity, the specific gravity, the amount of tyrosine-like material by colorimetric techniques, the determination of the nitrogen either by Nesslerization or Kjeldahl techniques, and the turbidimetric estimation of the

amount of protein following the addition of a nonspecific protein precipitating agent (Looney and Walsh, 1939) or a specific precipitin produced by immunization reactions (Chow, 1947). Most of the older methods are described and discussed by Starlinger and Hartl (1925). The extent to which these methods are exactly comparable is questionable. Studies in which different methods have been compared do not always give reassuring results (Rowe, 1916; Howe, 1925; Tuckman and Sobotka, 1932; Myers and Muntwyler, 1940; Sunderman, 1944).

The specific gravity of the plasma did not agree with the protein values as determined by the micro-Kjeldahl procedure when rabbits were subjected to gravity shock (Cole et al., 1943). Even the Kjeldahl method has been the object of much discussion (Chibnall et al., 1943; Miller and Houghton, 1945).

Perhaps of greater importance as far as this discussion is concerned are the methods for determining the plasma protein fractions. Much of this work has been done by means of the salting-out technique proposed by Howe (1921). It is common experience that there is a great deal of variation in the results thus secured. Part of this can be accounted for by the adsorption of the albumin on the filter paper, which in some cases may produce a considerable decrease in the albumin value (Robinson et al., 1937, 1938a). Problems of centrifugation and of temperature continue to be studied and discussed (Robinson et al., 1938b; Kingsley, 1940; Maclay and Osterberg, 1940).

The use of sodium sulfate in fractionating the plasma proteins does not give a clear-cut separation of albumin and globulin. Electrophoretic analysis of the albumin fraction thus secured has indicated the presence of small amounts of $alpha_1$ and $alpha_2$ globulins in the albumin fraction. Petermann et al. (1947) noted that in normal subjects the difference was slight when the albumin values were determined both by the Howe (1921) and the electrophoretic techniques. The work of Dole (1944), on the other hand, showed that the A/G ratio determined by electrophoretic analysis was roughly two thirds that found by chemical fractionation of the same sample. Other methods have been proposed for the fractionation of the plasma proteins but have been little used (Goettsch and Kendall, 1935; Chow, 1945; Pillemer and Hutchinson, 1945).

Level of Plasma Proteins in Normal Individuals

Since 1831, when Berzelius and Marcett independently determined the concentration of protein in plasma, there have been many reports on "normal" values. The most important of these reports are summarized in Table 202. There is a great deal of variation in the values given. Part of this represents intraindividual variability, but it is very likely that a fair share of the variation can be attributed to the factors discussed above, especially to the analytical techniques used and the absence of a standard procedure for securing the blood samples.

The values for total proteins in normal adults seem to range from well below 6 per cent (Linder et al., 1924; Kark et al., 1947) up to 8.86 per cent (Lange, 1946). The mean value for total plasma proteins secured from the above reports is 6.89 per cent. It is noteworthy that many of these reports indicate the mean protein level of normal adults to be greater than 7 per cent. There is some evidence that this may not be the case for young men in good physical and nutri-

TABLE 202

PLASMA PROTEIN VALUES FOR NORMAL INDIVIDUALS REPORTED IN THE LITERATURE. The values in parentheses are averages for the series studied. M = males; F = females. All values for protein are given in gm. per 100 cc.

Author	Subjects		Total Protein	Albumin	Globulin	Method
	N	Description				
Rowe (1916)	22	adults	6.5–8.2 (7.5)	4.6–6.7 (5.6)	1.2–2.4 (1.9)	refractometric
Linder, Lundsgaard, and Van Slyke (1924)	7	adults	5.62–7.45 (6.73)	3.62–4.9 (4.11)	2.45–2.89 (2.61)	micro-Kjeldahl
Salvesen (1926)	16M	adults	6.53–7.96 (7.00)	3.95–5.24 (4.44)	1.96–3.16 (2.58)	micro-Kjeldahl
Salvesen (1926)	16F	adults	6.34–7.96 (7.02)	3.77–4.80 (4.55)	2.18–3.55 (2.68)	micro-Kjeldahl
Wiener and Wiener (1930) ..	20	adults		4.20–5.00 (4.60)	1.50–1.90 (1.70)	micro-Kjeldahl
Moore and Van Slyke (1930)..	9	adults		4.0–4.5 (4.3)	(2.8)	micro Kjeldahl
Bruckman, D'Esopo, and Peters (1930)	13M	adults	(6.93)	4.37–5.65 (5.06)	1.52–2.91 (1.89)	micro-Kjeldahl
Bruckman, D'Esopo, and Peters (1930)	8F	adults		4.71–5.71 (4.98)	2.02–3.22 (2.62)	micro-Kjeldahl
Kumpf (1931)	8	adults	7.06–7.78 (7.36)	4.86–5.93 (5.20)	1.65–2.92 (2.16)	refractometric
Kumpf (1931)	8	adults	6.65–7.54 (7.17)	4.20–5.44 (4.82)	1.73–3.31 (2.34)	micro-Kjeldahl
Rennie (1935)	22	infants (3–23 mos.)	6.04–8.0 (7.08)	4.12–5.91 (4.95)	1.13–2.82 (2.13)	micro-Kjeldahl
Rennie (1935)	12F	children (2–11 yrs.)	6.88–8.04 (7.32)	4.15–5.94 (5.11)	1.52–2.79 (2.21)	micro-Kjeldahl
Rennie (1935)	12M	children (5–10 yrs.)	6.84–8.24 (7.54)	4.72–5.92 (5.20)	1.61–2.94 (2.34)	micro-Kjeldahl
Gutman et al. (1941)	26	adults	6.5–7.9 (7.17)	4.7–5.7 (5.24)	1.3–2.5 (1.94)	micro-Kjeldahl
Trevorrow et al. (1942)	45M	infants (0–30 days)	4.3–6.6 (5.3)	2.8–4.4 (3.8)	0.9–2.3 (1.7)	micro-Kjeldahl
Trevorrow et al. (1942)	48F	infants (0–20 days)	4.5–7.0 (5.3)	3.0–4.6 (3.8)	1.0–2.4 (1.7)	micro-Kjeldahl
Trevorrow et al. (1942)	41M	infants (1–6 mos.)	4.6–6.3 (5.7)	3.2–4.7 (4.2)	1.0–1.8 (1.3)	micro-Kjeldahl

TABLE 202 *Continued*

Author	N	Subjects Description	Total Protein	Albumin	Globulin	Method
Trevorrow et al. (1942)	26F	infants (1–6 mos.)	4.7–6.4 (5.7)	3.4–4.8 (4.2)	1.0–1.9 (1.3)	micro-Kjeldahl
Trevorrow et al. (1942)	46M	children (6 mos.–4 yrs.)	5.6–6.9 (6.4)	3.8–5.3 (4.7)	1.2–2.7 (1.6)	micro-Kjeldahl
Trevorrow et al. (1942)	40F	children (6 mos.–4 yrs.)	5.3–6.7 (6.4)	3.7–5.3 (4.7)	1.1–2.1 (1.6)	micro-Kjeldahl
Trevorrow et al. (1942)	89M	children and adults (4–40 yrs.)	6.0–8.1 (6.9)	4.0–5.4 (4.7)	1.3–3.0 (2.0)	micro-Kjeldahl
Trevorrow et al. (1942)	80F	children and adults (4–40 yrs.)	6.0–8.0 (6.9)	3.7–5.5 (4.7)	1.5–3.1 (2.0)	micro-Kjeldahl
Youmans et al. (1943b)	41	whites (1–3 yrs.)	(6.98)	(4.98)	(2.06)	micro-Kjeldahl
Youmans et al. (1943b)	34	Negroes (1–3 yrs.)	(6.78)	(4.73)	(2.12)	micro-Kjeldahl
Youmans et al. (1943b)	48	whites (4–6 yrs.)	(6.01)	(4.81)	(2.15)	micro-Kjeldahl
Youmans et al. (1943b)	24	Negroes (4–6 yrs.)	(6.96)	(4.78)	(2.15)	micro-Kjeldahl
Youmans et al. (1943b)	687	whites (over 7)	(6.95)	(4.70)	(2.22)	micro-Kjeldahl
Youmans et al. (1943b)	327	Negroes (over 7)	(6.96)	(4.53)	(2.42)	micro-Kjeldahl
Milam (1946)	1153	whites (all ages)	6.29–7.98 (7.13)	3.89–5.31 (4.60)	1.71–3.37 (2.54)	micro-Kjeldahl
Milam (1946)	426	Negroes (all ages)	6.46–8.30 (7.38)	3.77–5.03 (4.40)	2.15–3.83 (2.99)	micro-Kjeldahl
Lange (1946)	115	adults	6.30–8.86 (7.45)			micro-Kjeldahl
Dyson (1945)	353	adults	5.56–7.65 (6.56)			micro-Kjeldahl
Johnson (1945)	730	soldiers in U.S.	5.3–7.7 (6.46)			specific gravity
Johnson (1945)	149	soldiers in desert	6.0–8.7 (7.3)			specific gravity

tional condition. The Gurkha soldiers in the Indian Army had a mean plasma protein level of 5.3 per cent (Kark *et al.*, 1947). These men were in as good or better condition than American or British soldiers stationed in their respective countries as shown by various tests of physical fitness. Johnson (1945) found a mean plasma protein value of 6.3 per cent for 730 American soldiers who were stationed in moderate or cold climates. In the Minnesota Laboratory of Physiological Hygiene similar low values have been found for young men who were well trained physically and in excellent nutritional condition. There is no clear explanation for the difference between these low normal values and the large number of reports of levels over 7 per cent. Both the values secured by Johnson (personal communication) and those in the Laboratory of Physiological Hygiene were obtained while the men were in the fasting state (12 to 15 hours after the last meal) and after they had rested for at least 15 to 30 minutes.

The values in Table 202 indicate that there is no essential difference between the total plasma protein levels in males and females. This has been verified by the extensive studies of Trevorrow *et al.* (1942), Dyson (1945), Lange (1946), and Milam (1946).

The plasma protein level in children has been reported as reaching a low point at about 4 weeks of age, when it ranges from 4.3 to 6.0 per cent with a mean value of about 5.3. The total proteins in the plasma increase progressively thereafter until the adult value is reached in 4-year-old children (Trevorrow *et al.*, 1942). Youmans *et al.* (1943b) found that the protein level in their 1-to-3-year-old group was very similar to that in their adults. They did, however, find a slightly lower mean value in the white 4-to-6-year-old group which was not present among the Negroes of the same age. These differences in children of various age groups have been denied by Rennie (1935), who could find no difference between the values for infants and older children when compared with those in adults. Milam (1946) has suggested that there is a slight difference in the total plasma protein level of Negroes and whites. This difference is so small that it is probably of no biological significance. On the basis of these reports, it would appear that there is no trend in the plasma protein level with age aside from a possible lower value in infants.

Plasma Proteins in Starvation — Animal Experiments

Panum (1864) found only a very slight decrease in the total serum proteins in one dog that had gone for 9 days without food and a greater decrease in another dog that was starved for 13 days. This variability in the response of the total plasma proteins to starvation was emphasized by Burkhardt (1883), who found that when dogs were denied food for 5 to 6 days, some showed a slight decrease in plasma protein levels whereas others showed a slight increase. By means of salt precipitation he found a consistent decrease in the serum albumin level in all 7 dogs he studied. Lewinski (1903), however, was unable to confirm Burkhardt's findings. Lewinski concluded that there was no relationship between the plasma protein levels and the nutritional condition of the animal (see Table 203).

Githens (1904) determined the amount of food necessary to maintain the

TABLE 203

CHANGE IN PLASMA PROTEINS RESULTING FROM A 3-TO-4-DAY FAST IN DOGS, compared with the level in the subsequent refeeding period. All values have been converted to protein by multiplying the nitrogen values by 6.25. The second starvation experiment on dog III was made 8 days after the end of the first experiment. (Lewinski, 1903.)

	Total Protein	Albumin	Globulin	Fibrinogen
Dog I				
Starved	5.84	3.88	1.61	0.35
Fed	5.20	3.20	1.50	0.50
Dog II				
Starved	5.75	1.95	3.40	0.40
Fed	6.65	3.22	2.65	0.78
Dog III				
Starved	6.31	2.92	2.81	0.58
Fed	6.11	2.97	2.52	0.62
Starved	6.85	3.46	2.77	0.62
Dog IV				
Starved	6.56	3.35	2.02	1.19
Fed	5.47	3.38	1.55	0.54

weight of his dogs. For 2 or 3 days he removed all food from them and then for a 3-week period he gave them sufficient calories to supply one half of their maintenance requirement. Under these circumstances the plasma albumin decreased whereas the globulin increased.

Robertson (1912–13), by means of his refractometric technique, studied the plasma protein changes in both rabbits and rats during acute starvation. When all food was removed from rabbits for 5 days, the albumin and the total protein increased considerably whereas the globulin decreased. In rats that were fasted for 36 hours there was no change in the albumin but a marked increase in the globulin.

Following World War I emphasis was put on the role of dietary proteins in influencing the development and the course of edema. Kohman's experiments in 1920 showed that rats maintained on a low protein ration developed edema which she thought was similar to the hunger or famine edema so prevalent in Europe a few years earlier. This condition could be cured or prevented by adding casein to the ration. In 1929 Frisch, Mendel, and Peters repeated Kohman's work and included a study of the changes in the total plasma protein during their experimental periods. They found that the plasma proteins decreased considerably in some of the animals as the period on the low protein diet was extended. Edema developed in some of the animals after varying periods on the low protein ration, and the mean plasma protein level in these rats was lower than that for the normals. In individual animals, however, the appearance of edema was not predictable from the protein level.

Weech, Goettsch, and Reeves (1935) fed dogs on a low protein ration composed of carrots, rice, lard, and sugar. On this regimen the dogs lost weight and showed a marked change in plasma proteins. During the various experiments, which in some cases were carried on for 100 days, the total protein and the plasma albumin decreased throughout the entire period. By means of nitrogen balance studies, they were able to show that the change in the plasma albumin could account for only 3 to 4 per cent of the total protein lost during the feeding program. The globulin fraction remained very constant throughout the entire experiment. The standard deviation of the globulin values around the mean decreased from 0.77 in the control period to 0.46 gm. per 100 cc. on the fiftieth day of the experiment.

When these experiments were continued, Weech, Wollstein, and Goettsch (1937) found that there were marked changes in both the plasma and the red cell volumes of their dogs on the low protein rations. Utilizing these changes in calculating the loss of plasma proteins from the body, they found that after 80 days on the low protein ration the dogs had lost some 70 per cent of their plasma albumin. The decrease in the albumin fraction was far greater and at a more rapid rate than was the change in body weight. Because of edema the body weight changes did not accurately reflect the true loss of body tissue.

In some of their experiments (Weech, Goettsch, and Reeves, 1935) nitrogen balances were determined. By this means it is possible to construct a curve showing the rate of loss of body tissue in relation to body weight and plasma

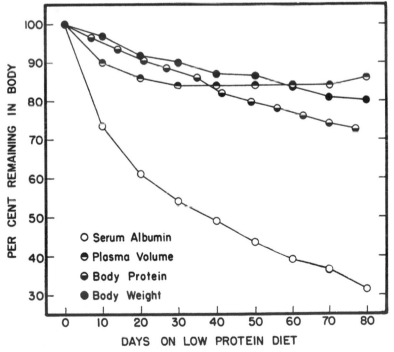

FIGURE 67. CHANGES IN BODY WEIGHT, PLASMA VOLUME, BODY PROTEIN, AND SERUM ALBUMIN IN DOGS FED A LOW PROTEIN RATION. The curves for body weight and serum albumin are from Weech, Wollstein, and Goettsch (1937); the curve for plasma volume is calculated from data in the same paper; and the curve for body protein is calculated from data in Weech, Goettsch, and Reeves (1935).

albumin changes (see Figure 67). The percentage loss of body protein was plotted on the basis of the average daily nitrogen losses of 2 dogs. It was assumed that initially the protein in the dogs amounted to 15 per cent of their body weights. No essential difference in the curve would result if slightly different values were used for the original protein content of the animals. The curves in Figure 67 indicate quite clearly that if proper corrections could be made for the extra water which accumulates during prolonged periods of semi-starvation, then the body weight changes would reflect the loss of body protein more accurately than would the plasma albumin. The curve for the percentage change in body weight and that for the change in body protein agree very well, especially for the first half of the experiment. It was only during the latter part of the period when the development of edema interfered with the weight determinations that these two curves separated. Even then the weight curve was much closer to the body protein curve than was the serum albumin.

In an attempt to find a means of quantitatively measuring the loss of protein from the body, Sachar, Horvitz, and Elman (1942) performed some similar experiments. They assumed that throughout their 3-week experiments 5 per cent of the body weight represented the blood volume. Using values for the total circulating plasma albumin calculated in this manner, they found that on the average the loss of plasma albumin represented about 3 per cent of the total tissue protein loss. On this basis, they derived a formula for computing the magnitude of body protein loss from the plasma albumin concentration alone. They supported their claims by reference to work with animals which indicated that the loss of plasma albumin was 4 per cent of the total protein loss.

After making various calculations on the ratio of serum albumin to the muscle mass in human beings, Sachar, Horvitz, and Elman (1942) could "see no reason to believe that a similar relationship does not hold true for human beings." This work suffers from the same limitations as that of Weech, Goettsch, and Reeves (1935). No determinations were made of the plasma volume, and the assumption that this changes in direct proportion to the body weight is probably in error. In the young men of the Minnesota Experiment there was a slight absolute *increase* in the plasma volume even though there was a reduction in the body weight equal to 24 per cent of the original weight (see Chapters 14, 15). Furthermore, the actual decrease in the plasma albumin of Sachar, Horvitz, and Elman's (1942) dogs represented from 2.0 to 4.7 per cent of the total protein loss; they selected 4.0 as the mean value for their formula. Clearly the errors introduced by the use of this formula may be very considerable.

Animal experimentation, especially with dogs, has indicated a decrease in the plasma protein level during periods of semi-starvation. The albumin fraction makes up the largest portion of the protein thus lost. The albumin loss, however, is not directly proportional to the total protein lost by the body. Under some circumstances the globulin level may decrease somewhat (Allison *et al.*, 1946), or it may not change at all (Weech *et al.*, 1935). The difference in this behavior has been related to the original globulin level. If the level of plasma globulin was initially below normal, the percentage of globulin may increase during periods of restricted feeding (Weech *et al.*, 1935).

The reduction in plasma proteins that occurs in animals during a period of acute starvation is rapidly restored on a liberal feeding program. Elman, Brown, and Wolf (1942) noted this in dogs that had been without food for 3 weeks. One week of ad libitum feeding with horse meat produced a prompt restoration of the albumin fraction (see Table 204) as well as a continued rise in the globulin.

TABLE 204

CHANGES IN PLASMA PROTEINS OF DOGS SUBJECTED TO A 3-WEEK FAST, and the subsequent change during a 1-week period of ad libitum feeding with meat. T.P. = total proteins. (Elman, Brown, and Wolf, 1942.)

Dog	Initial			After Fasting			After 1 Week's Feeding		
	Wt. (kg.)	T.P. (%)	Alb. (%)	Wt. (kg.)	T.P. (%)	Alb. (%)	Wt. (kg.)	T.P. (%)	Alb. (%)
E6	5.6	5.40	2.65	4.3	5.24	2.23	5.0	5.67	2.56
E2	7.4	6.42	3.64	5.4	5.63	3.07	7.1	5.98	3.54
F3	8.5	6.29	3.39	6.8	5.52	2.94	8.7	5.83	3.22
59	8.0	5.43	3.26	6.4	5.28	2.76	6.7	6.17	3.21
52	13.4	6.66	3.14	10.3	6.56	2.44	12.4	6.35	2.99
Average ..	8.6	6.04	3.22	6.8	5.65	2.69	8.0	6.10	3.10

Plasma Proteins in Acute Starvation — Human Beings

In most of the reports on acute fasting experiments (deprivation of all food) no mention has been made of changes in the plasma protein levels. The only work of this nature including such data is that of Eeg-Olofson (1941) with epileptics. His 4 men went without food for 14 days. During that time they showed no essential change in the total protein or in the albumin or globulin concentrations in the plasma, though there were considerable random variations during both the control and the starvation period. This finding is of interest, but in the absence of data on the blood volume it does not answer the question of the total amount of protein in the plasma compartment of the body. It is likely, however, that the blood volume usually decreases during a prolonged fast, and thus, even though there is no change in the percentage of protein, there may be a considerable reduction in the amount of circulating protein. Such a suggestion was made by Sachar, Horvitz, and Elman (1942) for dogs deprived of food. Unfortunately, in the latter experiments, the blood volume was determined only at the end of the starvation period and was compared with calculated values for the control period.

Fasting (acute starvation) and severe undernutrition (semi-starvation) differ in many important aspects, including, in all probability, the behavior of the plasma proteins. It is not possible to adjust those differences simply by comparing equal total caloric or nitrogen deficits produced in these two states of poor nutrition. The appearance of edema, which is often suggested to reflect the nitrogen deficit, emphasizes the difference. Edema does not appear in fasting though it is common in semi-starvation. A 40-day fast, such as that undertaken by Succi, certainly produces a large nitrogen deficit, but edema does not result (Anon., 1890). In the Minnesota Experiment edema appeared in several men after 2 months of semi-starvation although the nitrogen deficit was almost certainly less than that sustained by Succi. From the discussion of the edema problem and the material presented in Chapter 43, it is obvious that the explanation will not reside in any difference in the plasma protein levels.

Plasma Proteins in Semi-Starvation before 1920

One of the earliest investigations of the changes in the composition of the blood during semi-starvation was made by Grawitz in 1895. One of his experiments was on a 21-year-old medical student who, over a period of 8 days, showed a nitrogen loss of 17 gm. The serum protein was determined by the macro-Kjeldahl procedure but was uncorrected for non-protein nitrogen. During this time it changed from 9.0 to 8.6 per cent. After 3 days on a diet rich in protein the serum protein level decreased still farther to 8 per cent. Grawitz explained this latter decrease on the basis of blood dilution. In all 4 cases studied by Grawitz the serum protein levels decreased slightly during the period of restricted food intake. Grawitz concluded that a poor protein diet produces a corresponding decrease in the plasma protein level.

During World War I when the question of edema aroused so much interest and speculation, the European investigators made only slight reference to the possible role of the plasma proteins in the development of this condition. Some of the reports which appeared at that time did show a tendency toward a low plasma protein level in persons with edema. In most of these cases the decrease in the plasma protein level was ascribed to the hydremia which the investigators

TABLE 205

SERUM PROTEIN VALUES (gm. per 100 cc.) in cases of starvation (*Ödemkrankheit*) reported by Schittenhelm and Schlecht (1919c). Protein determined by refractometric method.

Patients with Edema							
Name	Serum Protein	Name	Serum Protein	Name	Serum Protein	Name	Serum Protein
Kop	3.94	Pon	5.20	Scha	5.81	Do	6.66
Per	4.08	Swan	5.20	Gon	5.90	Schka	6.77
Must	4.10	F	5.23	Rev	5.91	Deg	6.81
Kort	4.20	Bol	5.29	Ber	5.97	Lor	6.94
Mur	4.60	Sch	5.39	Lu	6.06	Ro	6.98
B	4.80	Ab	5.47	Ant	6.08	Bor	7.04
Mat	4.81	Story	5.47	Rub	6.08	La	7.09
Mur II	4.81	Sin	5.56	Gol	6.25	Rot	7.15
L	4.90	Rus	5.64	Al	6.34	Dol	7.29
Wen	4.90	Rom	5.68	Rob	6.34	Kost	7.29
Schma	5.03	Kug	5.72	Du	6.45	Kond	7.59
Zub	5.18	Pact	5.77	Mesch	6.61	On	7.63

Patients without Edema									
Name	Serum Protein	Name	Serum Protein	Name	Serum Protein	Name	Serum Protein	Name	Serum Protein
Bu	4.85	Jer	5.34	Mosch	5.47	Cem	5.93	Pi	6.10
K	5.18	Mal	5.47	Pup	5.81	Iw	6.10	Lup	6.94

believed accompanied the edema (Knack and Neumann, 1917; Schiff, 1917; Maase and Zondek, 1920).

The only investigators who reported their individual values for the plasma protein levels in any large series were Schittenhelm and Schlecht (1919c). Their results, reproduced in Table 205, show a range for total plasma protein from 3.94 to 7.63 per cent. Although Schittenhelm and Schlecht found that a large number of their subjects with edema had serum protein levels below the usual normal values, 35 per cent of them had levels within their accepted normal range (i.e., above 6 per cent). The patients who had no visible edema showed a plasma protein distribution very similar to that of the edematous group. These workers used the refractometric technique in determining their serum protein levels. They made the statement that their results were checked by comparison with the values secured by the Kjeldahl method. Unfortunately no data were given to show the agreement between these two sets of values. Also regrettable is the fact that Schittenhelm and Schlecht gave no values for strictly normal persons and only made reference to such studies by other investigators which indicated that normal values ranged from 6.4 to 9.0 per cent.

TABLE 206

Total Serum Protein Concentrations (gm. per 100 cc.) found among German civilians with edema (Jansen, 1918).

Number of Cases	Range of Plasma Protein
5	8.5–6.5
5	6.4–6.0
9	5.9–5.0
3	4.7–4.0

The only investigator of this period who emphasized the changes in plasma protein levels was Jansen (1918, 1920). He determined the total nitrogen in the serum as well as the non-protein nitrogen (the exact procedure is not given for either substance) and from the difference calculated the serum protein content. All of his 22 analyses were made on serum from patients with edema (see Table 206). He maintained that with his procedure the range of serum protein in normal persons was 6.5 to 8.5 per cent. He was impressed with the fact that 17 of his 22 edematous subjects had lower than normal serum proteins and 3 others were in the lower normal range. The low protein values, according to Jansen, were not the result of a simple hydremia since the color index was above one in most cases. Hypoproteinemia and low red and white cell counts were all indications of a general protein impoverishment in the body cells, tissues, and fluids. Jansen held that this deficiency resulted primarily from an insufficient caloric intake. He did not carry his argument beyond the point of trying to show that the edema which was so prevalent in Central Europe during World War I was due to a dietary deficiency. This deficiency, according to Jansen (1920), was one primarily of calories, since in many cases where edema developed while the subjects were under close observation, the protein intake was not considered to be very low.

Plasma Proteins in Semi-Starvation after World War I

The belief that low plasma protein values result from an insufficient protein intake developed to a large extent out of the animal experiments described earlier in this chapter. At the same time there were reports such as that by Ling (1931a) on the presence of low plasma protein levels in individuals who had subsisted for long periods of time on a deficient food intake. The people Ling studied were refugees from famine districts in the northern and northwestern parts of China. All the subjects had visible edema of varying duration. The percentage of total plasma protein in all cases was very low, the lowest being 2.32 per cent and the highest 4.91 (see Table 207). The albumin fraction showed a much higher proportional decline; the plasma globulin in most of the cases was within normal limits. When these persons were given a hospital diet providing 2050 Cal. per day, the concentration of their plasma proteins gradually increased until the total protein values approached, and in some cases exceeded, 6 per cent. The albumin levels in some cases increased threefold, whereas the globulin level showed a relatively small increase. In some cases the plasma protein fractions were determined by Howe's method in which the Kjeldahl procedure was used.

Weech and Ling (1931) observed similar low plasma protein levels among edematous patients admitted to a Chinese hospital. It was felt that the edema in

TABLE 207

PLASMA PROTEIN IN FAMINE VICTIMS IN CHINA. The protein values are
given as gm. per 100 cc. and the non-protein nitrogen as
mg. per 100 cc. of plasma. (Ling, 1931a.)

Subject	Total Protein	Albumin	Globulin	A/G	Non-Protein Nitrogen
1	3.38	1.28	2.10	0.61	20
2	3.62	1.21	2.41	0.50	
3	4.91	1.99	2.92	0.68	25
4	4.93	1.62	3.31	0.49	20
5	4.75	1.9	2.84	0.67	20
6	4.57	1.68	2.89	0.58	25
7	2.32	1.07	1.25	0.86	
8	4.50	2.21	2.29	0.96	25
9	4.35	2.54	1.81	1.40	
10	3.68	2.06	1.62	1.27	
11	3.92	1.85	2.07	0.89	
12	4.01	1.88	2.13	0.88	
13	3.97	1.27	2.70	0.47	24
14	4.23	1.25	2.98	0.42	20
15	4.09	1.63	2.46	0.66	32
16	3.40	1.44	1.96	0.74	32
17	2.76	1.03	1.73	0.60	24
18	2.70	1.18	1.52	0.78	31
19	3.25	1.64	1.61	1.02	27
20	3.02	1.69	1.33	1.27	21
21	3.46	1.43	2.03	0.70	20
22	3.77	1.33	2.44	0.55	20
23	3.98	1.84	2.14	0.86	22
24	3.06	1.25	1.81	0.69	
Average	3.78	1.60	2.18	0.73	23

all cases was the result of a poor diet. One of the two detailed reports given by these workers indicated that the plasma protein level in a 9-year-old boy was 3.98 per cent on admission. After 9 days on a liberal diet which supplied 71 gm. of protein per day, the plasma proteins increased to 6.93 per cent. Again the greater increase occurred in the albumin fraction. In this patient, as in the other one described in some detail by Weech and Ling, there was no essential change in the plasma protein level as long as the protein intake was relatively low (50 gm. or less per day).

Ling (1931a) studied the plasma protein changes in a group of Chinese students who had been on starvation rations for some months preceding their admission to the hospital. The initial plasma protein determinations all showed very low albumin levels and relatively normal values for globulin (see Table 208). When these subjects were put on a diet providing 90 gm. of protein and 2000 Cal. per day, there were prompt and substantial increases in both the albumin and globulin levels, with the albumin showing the more pronounced increase. Neither this nor the preceding report permits one to assess the relative role of caloric and protein intake in changing the plasma protein level.

The role of plasma albumin as an indicator of the body's stores of protein was re-emphasized by Bruckman, D'Esopo, and Peters (1930) in a study of mal-

TABLE 208

CHANGES IN THE PLASMA PROTEIN LEVELS OF STARVATION VICTIMS while being rehabilitated on a hospital diet composed of 330 gm. of carbohydrate, 41 gm. of fat, and 90 gm. of protein of which 20 gm. were of animal origin. All values for protein are in gm. per 100 cc. of plasma. (Ling, 1931a.)

Date	Total Protein	Albumin	Globulin	A/G	Date	Total Protein	Albumin	Globulin	A/G
	Subject 13					Subject 16			
9/26	3.97	1.27	2.70	.47	12/5.....	4.77	2.72	2.05	1.33
10/14	5.44	1.73	3.71	.47	12/27....	5.43	2.93	2.50	1.17
10/31	5.90	2.62	3.28	.80	1/7.....	5.99	3.59	2.40	1.50
12/2	5.81	2.67	3.14	.85					
						Subject 17			
	Subject 14				9/28,...	2.76	1.03	1.73	.60
9/26	4.23	1.25	2.98	.42	10/9.....	3.55	1.84	1.71	1.08
10/14	5.07	2.50	2.57	.97	10/22....	3.44	1.55	1.89	.82
10/20	4.77	2.34	2.43	.96	10/31....	5.10	2.35	2.75	.85
12/2	5.88	2.78	3.10	.90	12/2.....	5.38	2.95	2.43	1.21
	Subject 15					Subject 18			
9/23	4.09	1.63	2.46	.66	10/28....	2.70	1.18	1.52	.78
10/23	6.30	2.33	5.97	.39	11/19....	4.69	2.94	1.75	1.68
11/18	6.56	2.75	3.81	.72	12/11....	5.60	3.23	2.37	1.36
12/3	5.90	3.31	2.59	1.28	12/30....	5.14	2.63	2.51	1.05
12/28	6.60	3.54	3.06	1.16	1/9.....	6.33	3.66	2.67	1.37
	Subject 16					Subject 19			
9/23	3.40	1.44	1.96	.74	9/30....	3.25	1.64	1.61	1.02
9/28	2.61	1.10	1.51	.73	10/14....	6.23	3.36	2.87	1.17
10/23	4.34	2.52	1.82	1.38	10/22....	5.55	3.19	2.36	1.35
11/19	4.42	2.75	1.67	1.65	10/31....	6.57	3.92	2.65	1.48

nourished patients. They claimed that the condition of their subjects was not due to any specific disease but resulted primarily from an inadequate diet. The total plasma protein in these cases was usually below 7 per cent, which the investigators accepted as near the lower limit of normal. The greatest decrease was attributed to the albumin fraction. Again, when these individuals were fed diets rich in protein, the plasma albumin level increased.

Somewhat similar findings were observed by Youmans et al. (1932, 1933) in a study of the plasma protein levels in 31 adults in Tennessee. These persons, mainly women, had no complaint other than edema of the extremities which appeared in the late winter and persisted, in some cases, through the early part of the summer. The only explanation for the edema was the low caloric and low protein intakes revealed by a dietary survey. Here, too, the lower than normal plasma albumin levels were believed to be the result of the poor diet.

There are a number of reasons why it is difficult to accept wholeheartedly the thesis that the level of plasma protein is a reliable index of the body's nitrogen stores. In the first place, careful examination of the data presented by Bruckman, D'Esopo, and Peters (1930) and Youmans et al. (1933) indicates that their plasma protein values, even for the albumin fraction, are generally within the levels reported as normal by some investigators. This is especially true when their values are compared with those for American soldiers stationed in moderate or cold climates. The range of values for the total protein level in 730 soldiers was 5.3 to 7.1 per cent with a mean value of 6.4 (Johnson, 1945). These values are as low or lower than many reported by the above investigators as indicative of a deficient protein intake. All the soldiers in the group reported were in excellent nutritional condition. Dietary surveys indicated that such soldiers received an average of 125 gm. of protein per day in a diet providing from 3700 to 4000 Cal. (Howe and Berryman, 1945).

In the second place, some of the controlled observations on human beings indicate that the plasma protein changes do not reflect the fluctuations in the body's protein reserves as accurately as the Bruckman and Youmans reports imply. This was shown by the studies of Liu et al. (1932) on the relationship between nitrogen balance and plasma protein levels in two severely undernourished boys from a Chinese orphanage. One boy 11 years old weighed 30.7 kg., and the other boy, aged 20, weighed 27.3 kg. During the 85-day experimental period the dietary intake of protein was varied but the daily caloric intake was kept constant at about 1400 Cal. The total plasma proteins increased initially when the protein intake was raised, but even after 20 days on a diet providing 110 gm. of protein per day and an additional 16 days on 200 gm. the levels were not significantly greater than those observed during the first period (see Table 209). During this 36-day period the 2 subjects stored 340 and 348 gm. of protein respectively. The discrepancy between the plasma proteins and the amount of protein in the body was equally great in the last 20-day period when both subjects were on a low nitrogen intake and in negative nitrogen balance but showed no change in the level of either the total plasma proteins or the albumin fraction. The maximum changes in the level of plasma globulin on both an absolute (1.3 gm. in both cases) and a relative basis (45 per cent in both cases) were as great as the changes in the albumin level

TABLE 209

RELATIONSHIP BETWEEN NITROGEN BALANCE AND PLASMA PROTEIN LEVELS. The caloric intakes in the different periods were constant. (Liu *et al.*, 1932.)

	Case 1				Case 2			
	Nitrogen in Gm.		Plasma Protein (%)		Nitrogen in Gm.		Plasma Protein (%)	
Dates	Intake	Balance	Alb.	Total	Intake	Balance	Alb.	Total
Oct. 19–22	14.96	+1.26	2.18	5.13	18.29	−1.59	2.17	4.01
23–27	18.70	+0.18	2.30	4.69	22.00	−0.52	2.46	5.05
28–31	14.96	+0.94	2.67	5.56	17.60	−0.43	2.61	5.96
Nov. 1–4	1.22	−7.58			1.22	−10.58		
5–8	1.17	−6.39	2.52	4.89	1.17	−7.79	2.49	5.37
9–12	1.02	−7.27	2.14	4.17	1.02	−6.74	2.53	5.04
13–16	1.02	−6.36	2.18	4.27	1.02	−5.80	2.24	4.98
17–20	17.92	+7.52	2.08	3.73	17.92	+8.58	2.55	5.06
21–24	17.92	+6.39	2.10	4.07	17.92	+5.86	2.73	5.88
25–28	17.92	+5.36			17.92	+4.48		
29–2	17.92	+3.73	2.53	5.08	17.92	+6.38	2.97	6.08
Dec. 3–6	17.92	+1.85			17.92	+4.43		
7–10	31.92	+7.85	2.31	4.59	31.92	+7.74	2.83	5.64
11–14	31.92	+6.58	2.85	5.16	31.92	+5.38	2.98	6.02
15–18	31.92	+7.53	3.24	5.61	31.92	+6.68		
19–22	31.92	+7.50			31.92	+6.13	2.84	5.65
23–26	1.36	−6.48	2.89	5.31	1.36	−9.62		
27–30	1.36	−6.17			1.36	−5.65	2.73	5.10
Jan. 1–3	6.28	−3.09	2.52	4.55	6.28	−0.77		
4–7	6.28	−2.56	2.67	5.04	6.28	−3.02	2.76	5.65
8–11	6.28	−0.63	3.11	5.57	6.28	−0.68	2.87	5.48

(1.2 and 0.8 gm. and 38 and 27 per cent for the absolute and relative changes in the 2 subjects).

In the Minnesota Experiment, even though normal young men were maintained on a calorically inadequate diet until they lost 24 per cent of their original body weight, there was no marked decrease in the plasma protein levels or in the circulating plasma albumin.

Finally, some patients with anorexia nervosa whose body weights are approximately 50 per cent of the normal (on the basis of height/weight tables) have normal serum protein concentrations. Berkman, Weir, and Kepler (1947) showed that out of 23 such cases 14 had serum protein levels of 6.4 per cent or more; in one of these the concentration was as high as 7.7. The other 9 women had levels ranging as low as 4.4 per cent. Body weight increases in some of these women were associated with increases in serum proteins, especially when the original level was below 6 per cent. In most cases, however, there was no change in the protein level in spite of a marked increase in body weight under treatment. Bruckner, Wies, and Lavietes (1938) also found normal serum protein levels in 2 severe cases of the same disturbance.

These findings can lead only to the following conclusion: the plasma protein level, as such, is of dubious merit in evaluating a person's nutritional status, at least in so far as that pertains to the protein stores in the body. A low plasma protein level, especially one in which the decrease is due primarily to a change in the albumin fraction, may indicate that the person has suffered from a dietary protein deficiency or an insufficient caloric intake. Liberal feeding will increase the plasma protein level in these cases but usually at a slow rate. There are, however, many normal persons in good nutritional states who may have protein levels comparable to those exhibited by many semi-starved persons. It is doubtful whether the ingestion of large amounts of proteins will increase a low plasma protein concentration in normal individuals. Although there is no experimental evidence, it is highly unlikely that hyperproteinemia would ever result from such a regimen even when continued for a long period. In fact the presumption is that hyperproteinemia is seen only in a few disease conditions (Jeghers and Selesnick, 1937; Kagan, 1943).

Plasma Proteins in Semi-Starvation — World War II

Shortly after the German invasion of France and the Lowlands in 1940, a number of reports about famine edema appeared from these countries. The reports from France are difficult to evaluate since they usually do not indicate the method of analysis or else refer to a gravimetric procedure in which the separated protein was dried to constant weight after being washed with alcohol and ether (Kayser, 1930; Macheboeuf and Wohl, 1931). Plasma protein values observed by some of the French investigators in adults with famine edema are listed in Table 210. Some of these subjects were reported to have lost as much as 42 per cent of their original weight, presumably since the start of the war (Laroche et al., 1941).

The French investigators are unanimous in maintaining that there was a decrease in plasma protein among their starvation cases. The figures reported in support of such claims are a little difficult to interpret. Very few if any of the values for total protein are below normal according to most standards. Apparently

TABLE 210

SERUM PROTEIN LEVELS in gm. per 100 cc. From cases of starvation
edema in France during World War II.

Author	Total Protein	Albumin	Globulin	A/G
Laroche, Bompard, and Trémolières (1941)	5.68	2.54	3.14	0.8
	6.34	3.29	3.04	1.08
	5.76	2.73	3.04	0.9
	6.28	2.98	3.31	0.89
	6.64	3.32	3.32	1.0
	7.14	2.64	4.50	0.58
	7.66	4.72	2.89	1.6
Nicaud, Rouault, and Fuchs (1942)	6.80	3.66	3.14	1.17
	7.10	3.80	3.30	1.15
	7.90	4.20	3.70	1.14
	9.30	6.00	3.30	1.82
	6.70	4.40	2.30	1.92
	8.00	5.70	2.30	2.48
Warembourg, Poiteau, and Biserte (1942)	8.60	5.70	2.90	1.96
	7.00	3.60	3.40	1.06
	8.40	4.30	4.10	1.05
	6.50	4.20	2.30	1.82
Vallery-Radot, Loeper, and Tabone (1943)	6.50	3.57	2.93	1.21
	6.02	3.28	2.74	1.16
	6.72			
	5.40	3.40	2.00	1.7
	5.0	2.9	2.1	1.38
Raynaud and Laroche (1943) .	5.39[*]	3.2[*]	2.17[*]	

[*] These are the averages of 4 subjects.

allowance was made for this factor in the French reports since in the few cases
where normal values for total proteins are given they range from 8 to 9 per cent
(Raynaud and Laroche, 1943).

The plasma albumin considered by the French as normal agrees fairly well
with the majority of the values given in Table 202, whereas their normal globulin
level is much higher (Raynaud and Laroche, 1943). If the normal globulin values
used by the French investigators are too high, this might explain some of their
reports which indicate that the globulin fraction of the plasma protein decreases
before the albumin (Gounelle et al., 1945).

Gounelle and his co-workers (Gounelle, Marche, and Bachet, 1942a, 1942b,
1942c; Gounelle, Sassier, and Delbarre, 1945) maintained that in the early stages
of undernutrition, before any edema is evident, there is a decrease in the plasma
globulin level (see Table 211), with the plasma albumin remaining in the normal
range at this stage. As soon as there is any detectable edema, the plasma globulin
may show no change over the pre-edematous condition or may return toward a
normal value. The plasma albumin, on the other hand, decreases at this time, and
the magnitude of this drop determines the severity of the edema. According to this

TABLE 211

Changes in the Composition of Plasma Proteins before and after the Development of Edema, as reported by Gounelle, Marche, and Bachet (1942a). All values for the level of protein in the plasma are given as gm. per 100 cc.

Date	Edema	Total Protein	Albumin	Globulin	A/G
Subject L					
5/2/41	0	5.62	4.22	1.4	3.01
5/20/41	+	5.09	2.62	2.47	1.06
Subject J					
4/10/41	0	5.93	4.36	1.57	2.77
10/1/41	+	5.44	3.68	1.76	2.08
Subject C					
2/9/42	0	6.95	5.25	1.7	3.08
4/30/42	+	5.39	3.66	1.73	2.11
10/28/41	0	7.9	4.85	3.05	1.59
12/13/41	+	6.44	3.49	2.95	1.18
Subject E					
4/10/41	0	6.82	5.04	1.78	2.83
4/10/42	+	5.0	2.8	2.2	1.27
Subject D					
4/10/41	0	6.7	4.96	1.74	2.85
12/4/41	+	5.23	3.66	1.57	2.33
Subject A					
8/27/41	0	7.88	5.24	2.64	1.98
4/21/42	+	7.71	4.29	3.42	1.25

group (Gounelle et al., 1942b) severe edema is characterized by a plasma albumin concentration between 2.5 and 3.0 per cent, with moderate edema at levels of 3.5 to 4.0 per cent, and minimal edema at levels of 4.0 to 4.5 per cent. This is discussed further in Chapter 43.

In the early part of the German occupation of Belgium, Brull and Dumont (1945) in Liége made nitrogen balance studies on 7 men who had lost 20 per cent or more of their original weight (see Table 212). The serum proteins in all but 2 of these cases were within the range reported for normal adults. In fact, in 2 cases the serum protein values were in the upper normal range and in only 2 were they particularly low. Both the refractometric and Kjeldahl methods were used in these analyses (Neuprez, 1945).

An extensive investigation of the plasma protein levels was carried out at Liége during the German occupation (Neuprez, 1945). Between December 1942 and May 1944, 3137 analyses for total protein in the plasma were made by means of the refractometric technique. A correction factor for this procedure was established by comparing the results with those from the Kjeldahl method. In spite of an average weight loss of 12 to 13 per cent during this period, the mean of all plasma protein determinations was 6.99 gm. per 100 cc. There was no significant variation in the values throughout the entire period. No indication is given as to the range of the values or their deviation, but it is implied that the distribution about the mean was normal. One of the tables (*ibid.*, p. 29), giving the mean protein values for the different age groups, indicates a slight decrease for the older people. Whether this is of any biological significance is questionable. The distribution of total serum protein values in 100 adolescents in the same city is also given; this indicated that the mean value for the group was 6.42 per cent, with all but 10 of the children within the range 5.6 to 6.9.

TABLE 212

Total Serum Proteins in Belgians during World War II. The percentage weight loss represents that resulting from the wartime diet. (Brull and Dumont, 1945.)

Age	Height (m.)	Weight (kg.)	Percentage Weight Loss	Percentage Serum Protein	A/G
42	1.74	58		7.88	
41	1.77	61.8		6.75	1.16
41	1.75	66.9	31	8.37	1.9
58	1.71	55.2		6.4	
54	1.74	54	40	5.1	0.97
48	1.71	45.5	40	6.35	2.1
64	1.64	53.5	25	5.3	

According to the official ration, the diet of the Belgians at that time provided only 30 gm. of meat per day, which meant that the larger fraction of the nitrogen intake was from plant sources (22 gm. of a total of 30 gm. of protein). No indication is given of the extent to which this ration was supplemented by non-rationed foods. This undoubtedly varied with both the locality and the individual. Since the official ration provided only 1300 Cal. per day, it is certain that considerable supplementation occurred. Both cheese and skim milk were unrationed, at least during the period covered by the above study (December 1942 through May 1944) (Piersotte, 1945). It appears reasonable to assume that about three fourths or more of the protein intake was from vegetable sources. Yet in spite of the small amount of animal proteins in the diet, the total plasma protein levels were essentially normal. It has been shown that plant foods, when practically the only source of protein, can maintain the plasma protein level in adults for long periods of time. Abelin and Rhyn (1942) reported plasma protein concentrations of 6.85 and 7.50 per cent in 1937 and 7.61 per cent 4 years later in a practicing physician who was a vegetarian; the only animal food in his diet was 80 cc. of milk per day and occasionally small amounts of cheese.

Brull ignored the findings of other members in his group (Neuprez, 1945) when he stated that "A serum protein deficiency of 10 to 20 and sometimes 50% was a common finding even in the absence of decrease of weight" (Brull, 1945, p. 178). This is certainly not reflected in the mean values they report, and unless more data showing the range of values are given, it will have to be assumed that

Brull was referring to hospital patients whose very low protein levels were due to factors other than uncomplicated starvation.

With the end of the war in Europe and the release of the inhabitants of concentration and prisoner-of-war camps, a number of reports appeared indicating the presence of low levels of plasma protein among those persons who had been starved for a long time. Leyton (1946) found that 75 per cent of the 150 Russian prisoners examined in a large German camp had plasma protein levels below 6 per cent. The plasma protein level in only 5 of the 51 British soldiers examined in the same camp was below this level. Although no indication is given of the actual changes in body weight, it is stated that the Russians were markedly underweight and that many of them showed famine edema. The British soldiers had received supplemental food (Red Cross parcels) and showed few signs of starvation.

More severe cases of starvation were observed among the civilian inmates of the concentration camps in Germany. Mollison (1946) found that at Belsen the total serum protein in 37 adults varied from 3.7 to 6.9 per cent. The average value for 18 males was 5.2 and that for 19 females was 5.0 per cent. There was no relationship between the extent of starvation as shown by body weight loss and the serum protein level. One person who had lost 36 per cent of his original weight had the highest protein level, whereas another who showed the same percentage of weight loss had the lowest protein level (see Table 213). In 12 of the cases blood volume determinations were made together with the serum protein. From these data it was calculated that the average value for the total circulating proteins was 123 gm. per person (see Table 214).

Similar results for both the level of serum protein and the total circulating proteins were secured by Walters, Rossiter, and Lehmann (1947a) in their studies on Indian soldiers who had been interned in Japan. In the case of the Indian prisoners

TABLE 213

RELATIONSHIP BETWEEN THE LOSS OF BODY WEIGHT AND THE SERUM PROTEIN CONCENTRATION. Detailed data are given for only 11 of the 37 subjects studied. All these cases were either free of edema or at most had only a trace of it. (Mollison, 1946.)

Sex	Body Weight, in Kg.			Percentage Loss	Percentage Serum Protein
	Now	Original	Loss		
M	35	80	45	56	4.9
F	25	50	25	50	5.0
F	34	63	29	45	5.0
M	39	67	28	42	4.8
F	35	56	21	37	5.6
F	40	62	22	36	3.7
M	32	50	18	36	6.9
F	31	48	17	35	4.5
F	33	48	15	31	5.0
F	33	47	14	30	5.8
F	33°	45°	18	29	6.0

° These weights were obviously reversed in the original publication.

TABLE 214

TOTAL CIRCULATING PLASMA PROTEINS AMONG CIVILIANS RELEASED
FROM BELSEN, A GERMAN CONCENTRATION CAMP (Mollison, 1946).

Case	Weight (kg.)	Edema	Plasma Volume (liters)	Percentage Serum Protein	Circulating Proteins (gm.)
M 23	40	0	3.4	5.5	187
M 45	45	++	3.2	5.3	170
F 5		++	2.2	5.2	114
F 100	44	++	2.0	5.7	114
F 110	59	+++	2.5	4.5	112
F 110	47	++	2.5	5.7	143
F 111	31	0	2.3	4.6	115
F 114	35	0	2.2	5.6	123
F 116	25	+	1.7	5.0	85
F 117	36	0	2.4	5.5	132
F 122	36	++	2.3	4.0	92
F 124	33	0	1.9	5.0	90
M	39.2		2.38	5.13	123.1

of war, the albumin fraction of the blood was reduced to a value about 60 per cent of normal. On the basis of their observations, Walters and his colleagues set up a schedule of recovery from a protein deficiency in which the changes that might be expected during the rehabilitation period are listed. The restoration of physiological functions in their schedule is predicated on the assumption that the changes they observed among the Indian prisoners of war occur in all starvation cases. That such is not the case is evident from the preceding discussion of variations in plasma protein changes.

Reductions in the plasma albumin levels were also observed among some of the hospitalized starvation cases during the Bengal famine of 1943. These cases were so advanced that they could not tolerate ordinary gruel (Bose et al., 1946). The albumin level was reduced in many cases to approximately half the value found among normal controls. The globulin concentration in some cases was normal but in many it was doubled. This finding may have been due to intercurrent diseases such as malaria, dysentery, and pneumonia, which complicated all these studies. The total plasma proteins were lower in the cases that had edema, and in all cases the values were below normal.

Plasma Proteins in the Minnesota Experiment

Routinely, plasma protein samples were those that were used for the determination of the plasma volume. Heparin was the anticoagulant. The protein content was determined by means of the micro-Kjeldahl method (Keys, 1940). The albumin fraction was separated according to Howe's (1922) procedure. Globulin was calculated as the difference between the total protein and the albumin fraction.

The data for the individual subjects throughout the experiment are given in the Appendix Tables. At S24, after the men had lost 24 per cent of their original body weight, the total plasma protein concentration averaged 5.98 gm. per

100 cc. compared to a control period value of 6.67 (see Table 215). The decrease in the albumin fraction amounted to 10 per cent of the original albumin level while the globulin decrease was 11 per cent.

Although the actual concentration of plasma proteins decreased in most of the subjects during the starvation period, it increased in 2 of the men. Subject No. 119 showed an increase from the control value of 6.37 to 6.41 per cent at S24. The plasma volume decreased in this subject with a consequent slight decrease in the amount of circulating plasma protein. The level of plasma protein in subject No. 4 changed from 6.86 at the end of the control period to 7.08 at S24. The increase in the plasma volume accounted for an increase in the circulating plasma protein from 181 gm. to 211 gm. Here was one case at least where the amount of plasma protein at the end of the starvation period was greater than that during the control period.

TABLE 215

Plasma Protein Levels during Starvation and in the Subsequent Recovery Period. All protein values, in gm. per 100 cc. of plasma, are the means of the 32 subjects. (Minnesota Experiment.)

	Control	S12	S24	dS24	R6	R12
Total protein	6.67	6.39	5.98	−0.69	6.11	6.45
Albumin	4.28		3.86	−0.42		4.18
Globulin	2.39		2.12	−0.27		2.27
A/G ratio	1.89		1.95	0.06		1.92

When the changes in the total circulating plasma protein are considered, it becomes obvious that neither the total protein nor the albumin fraction in any way reflects the changes in the body's protein stores. Even though there was an average 10 per cent decrease in the concentration of these 2 plasma constituents, the average increase in the plasma volume during the starvation period just balanced this, with the net result that there was no significant change in either the amount of circulating total protein or albumin (see Table 216). In the Minnesota Experiment at least, it would appear that the decrease in concentration of proteins in the plasma at the end of the starvation period could be explained entirely by the increase in the plasma volume. Actually the loss of active tissue was reflected far more accurately in the hemoglobin changes (see Table 216). At S24 the hemoglobin concentration in the blood had decreased by about 23 per cent of its original value. Allowing for the measured decrease in the total blood volume, the circulating hemoglobin actually decreased by 29 per cent. This agrees very well with the 27 per cent reduction in active tissue calculated independently.

On this basis it would appear that the hemoglobin level reflected the changes in body tissue as accurately as did the body weight and that both of these were much closer in this respect than was the decrease in total plasma protein or any of its components. A similar conclusion was reached by Foy et al. (1946) from a study of the inhabitants in northern Greece made shortly after the end of World War II.

TABLE 216

Comparison of the Changes in the Circulating Proteins with the Changes in Active Tissue (Muscle Mass) during Starvation. All values are the means of the 32 men in the Minnesota Experiment.

	Control	S24	Difference	Difference as Percentage of C
Body weight (kg.)	69.50	52.60	—16.90	—24.3
Plasma volume (liters) ...	3.13	3.40	+0.27	+8.6
Total plasma protein (%)	6.67	5.98	—0.69	—10.3
Circulating total proteins (gm.)	208	204	—4	—1.9
Plasma albumin (%)	4.28	3.86	—0.42	—10.1
Circulating plasma albumin (gm.)	134	131	—3	—2.2
Hemoglobin (%)	15.12	11.70	—3.43	—22.7
Total plasma volume (liters)	5.89	5.40	—0.49	—8.3
Circulating hemoglobin (gm.)	890	631	—259	—29.0
Active tissue (kg.)	39.93	29.20	—10.73	—27.3

TABLE 217

Influence of the Dietary Protein Intake on the Plasma Protein Level during the Recovery Period. The protein supplement gave each man in the supplemented group 20 gm. of protein more than received by those in the basal group. (Minnesota Experiment.)

Protein Group	Plasma Protein Level, as Gm. per 100 Cc.			
	Control	S24	R6	R12
Caloric Group Z				
Basal	6.55	6.17	6.34	6.81
Supplemented	6.59	5.94	6.03	6.34
M	6.57	6.06	6.19	6.58
Caloric Group L				
Basal	6.63	5.77	5.82	6.18
Supplemented	6.61	5.76	6.02	6.17
M	6.62	5.77	5.93	6.18
Caloric Group G				
Basal	6.62	6.41	6.12	6.12
Supplemented	6.61	5.68	5.78	6.18
M	6.62	6.04	5.95	6.15
Caloric Group T				
Basal	7.04	6.03	6.36	7.04
Supplemented	6.68	6.12	6.44	6.72
M	6.86	6.07	6.40	6.88

During the recovery period each of 4 men in each caloric group received 20 gm. more protein per day than did the other men in the group. These men on extra protein showed the same general response in the level of plasma proteins as did the basal dietary groups (see Table 217). These results also indicate that the different caloric levels had no appreciable influence on the plasma protein levels. The lowest and highest caloric groups showed the same changes in plasma protein concentrations, with the control levels restored by R12. In the 2 intermediate caloric groups, the protein levels were not quite restored by R12.

Physical Chemistry of Plasma Proteins

The physicochemical properties of the plasma proteins in starvation are of major interest in connection with the problems of famine edema and the exchanges of fluid between blood and the tissue spaces. The nature of these problems is discussed in Chapter 43. It is enough here to note that the point of immediate concern is the effective osmotic pressure exerted by the plasma proteins at the limiting membrane of the capillary. It seems clear that in starvation the capillaries do not become permeable to these proteins, and therefore the limiting characteristics which determine the osmotic activity are the size of the protein molecules or particles and the electrical charges on them. These properties may be estimated by ultracentrifugation, or osmometry with collodion and similar membranes, according to the principles developed, respectively, by Svedberg, Tiselius, and Adair.

In normal man, and in at least some diseased states, the colloid osmotic pressure of the blood serum or plasma can be calculated with fair accuracy from the concentrations of the albumin and globulin fractions as estimated by the usual salting-out method. For colloid osmotic pressure at $0°$ C., measured in mm. of water, the equation applies: C.O.P. $= f_c (45.2A + 18.8G)$, where A and G are, respectively, the concentrations of albumin and globulin in gm. per 100 cc. of serum (Keys, 1938; Greenberg, 1943). The factor f_c varies with the total concentration of proteins, being 0.90 at 1.5, 0.95 at 2.5, 1.00 at 3.5, 1.06 at 4.5, 1.12 at 5.5, 1.22 at 6.5, and 1.35 at 7.5 gm. per 100 cc.

The foregoing equation is based on the assumption that Dalton's law of partial pressures applies and utilizes the universally accepted values for the molecular weights of the major plasma proteins. The factor for the effect of total concentration on the osmotic activity was obtained empirically; it agrees with many independent investigations. The fact that it conformed to the direct measurements of the colloid osmotic pressure in a considerable variety of samples tested, including diluted samples and bloods from patients with liver and kidney disease, indicates that in general the calculation of the C.O.P. from the results of careful chemical analyses is reasonably safe. Such calculations, however, would be inapplicable with serum in which a substantial part of the protein was of some aberrant particle size.

This equation has not been carefully checked with bloods from persons suffering from severe and protracted undernutrition. Govaerts (1947) produced data which he believed indicated that the osmotic behavior of the plasma proteins may be abnormal in such cases. His comparison of osmometric results with the chemical findings, however, is by no means unquestionable, and his calculation method is certainly unacceptable (see Chapter 43). More impressive evidence for a physical peculiarity of the plasma proteins in starvation was produced by Lamy, Lamotte, and Lamotte-Barrillon (1946). These workers studied 38 edematous patients recovered from German concentration camps, most of whom had a normal concentration of total protein in the serum and a normal ratio of albumin to globulin. In 7 of these patients the C.O.P. was estimated with a collodion membrane method which gave pressures from 11.5 to 23.5 mm. of mercury. These unexpectedly low values led to studies on the serums of two of the patients with the ultracentrifugation method of Svedberg. By this means it was found that these samples contained appreciable amounts of proteins with very high molecular weight (up to 1 million). These heavy proteins could not be found in the edema fluid.

The ultracentrifugation findings explain part of the discrepancy between the C.O.P. and the chemical data on these serums. Since only a small part of the total plasma protein appeared to be in the form of these large molecules, the effect on the C.O.P. would be small. Lamy and his colleagues (1946) concluded that the abnormality of particle size distribution in the serum could play a part but could not fully explain the genesis of edema in starvation. They suggested that the protein of high molecular weight is derived from the muscle tissues, and they believed they found histological evidence pointing to this origin.

Obviously, further work is needed on these questions, beginning with systematic osmometric and gravitational measurements on serums from starved persons. The publication of full details, which are lacking in the paper by Lamy, Lamotte, and Lamotte-Barrillon (1946), would aid in evaluation. Until further evidence is at hand it seems unwarranted to conclude that starvation regularly results in substantial physicochemical alterations in the plasma proteins, but the possibility must be seriously entertained.

The physicochemical events in the destruction of the protein-containing tissues of the starving man have so far received no real attention. Presumably, the assumption has been that the released proteins are metabolized and used for energy purposes directly. Just how this would come about is questionable, and in any event there has been no search for the facts. Are the proteins of the degenerating cell degraded to the constituent amino acids fully and immediately? Are the amino acids resulting from such a process in turn oxidized locally? What happens to the enzyme systems? Considerations like these would have important effects on the osmotic and fluid balance as well as on the more purely chemical events in the body.

Plasma Protein Fractions and Resistance to Disease

It is frequently stated that protein depletion results in a lowered resistance to infection. This question, along with the related problem of disease resistance in simple caloric deficiency, is discussed further in Chapter 46. It is of interest to note here that the starvation undergone by the subjects in the Minnesota Experiment resulted in only a minimal reduction in the circulating gamma globulins. While there is much evidence to indicate that the majority of the circulating antibodies are found in the gamma globulin fraction (Enders, 1944), it has not been shown that these are the only proteins measured by this peak in the Tiselius pattern. Consequently, the small loss of gamma globulin in semi-starvation in man does not necessarily represent a depletion of the circulating antibodies.

That resistance to infectious diseases is a complex phenomenon has been brought out by Schneider (1946) and more recently by Metcoff *et al.* (1948). The latter group found that the bacteriologic and immune responses resulting from infection and secondary antigenic stimulation were essentially unaltered by severe dietary protein deficiency. No significant difference in the incidence of bacteremia was noted between infected rats on the 2 per cent casein diet and those on one of 18 per cent. Despite relative quantitative depletion of circulating gamma globulin, protein-deficient rats attained antibody titers equivalent to those of the well-nourished animals in response to infection. Following specific secondary antigenic stimulation, the rise in humoral antibody titer was more marked in the deficient than in the well-nourished rat. Despite rise in titer, no significant change in circulating gamma globulin was observed in either group.

Like many other nutritional problems, the question of the relationship between the plasma proteins and infectious disease is far more complex than appears at first sight. From the limited evidence so far obtained, it appears that several negative statements may be applied to the condition in ordinary caloric undernutrition such as occurs in European famine and in anorexia nervosa: (1) the ability of the body to manufacture protein antibodies is not abolished or, perhaps, even severely diminished; (2) antibody proteins already in the blood plasma are not selectively destroyed or removed.

Blood Non-Protein Nitrogen — Acute Starvation

The concentration of the non-protein nitrogenous compounds in the blood plasma has generally been used clinically as an indication of kidney function. For some time it has been recognized that other factors might be operative in determining the levels within the relatively broad range designated as normal. Recent experimental work with normal young men would suggest that the level of blood urea increases with the amount of protein in the diet. Addis *et al.* (1947) observed an increase in the blood concentration of urea at the end of a high protein diet served for 6 days. Whether these differences would have been maintained over longer dietary periods is still an unanswered question.

The other extreme of protein intake, acute starvation, produces a variable change in the blood urea level. Lennox, O'Connor, and Bellinger (1926) determined this in one subject who fasted for 17 days. During the first week of the fast there was an increase in urea; this was followed by a return to the pre-starvation control level. The urea level followed quite closely the changes in the non-protein nitrogen (N.P.N.). This is what one would expect since in most conditions urea comprises the largest fraction of the blood N.P.N. Using the N.P.N. as an indication of the urea changes, the data of Lennox and his colleagues (1926) on 10 subjects who fasted from 12 to 21 days showed that in some cases there was no change throughout the fast, whereas in others there was an increase in the first week followed by a decline to normal. The decline in some cases continued through a part of the refeeding period and then returned to normal during the subsequent 2 weeks.

The only constituent of the blood N.P.N. that shows a consistent change during acute starvation is uric acid. Hoeffel and Moriarty (1924) and Lennox (1924) independently observed that the uric acid in the blood increased during the first few days of a fast and was maintained at high levels throughout the fast. Lennox (1924), who studied this problem in 22 starvation experiments, found some indication that the magnitude of the uric acid increase was proportional to the duration of the fast. The highest level he observed was 285 per cent of the pre-starvation value. The resumption of feeding produced an immediate decrease in the uric acid level. The variations in the blood levels were reflected by inverse changes in the urinary excretion of this compound, as shown in Figure 68 (Lennox, 1925). Since there were no comparable changes in other blood constituents, it would appear that kidney function was not grossly abnormal. This was further confirmed by means of the phenolsulfonphthalein test of kidney function and by the absence of albumin and casts in the urine (Lennox, 1924). The excretion of uric acid by non-starved individuals is linearly related to the blood level even above the highest concentrations observed in starvation (Brøchner-Mortensen, 1937).

The amino acid level in the blood during acute starvation remains surprisingly constant both in human beings (Lennox et al., 1926) and in dogs (Van Slyke and Meyer, 1913–14). On first thought it would seem that the breakdown of body tissue represented by urinary nitrogen excretions of 15 gm. per day during starvation should involve rather marked changes in the amino acid levels of the blood. Calculations, however, show that even the highest nitrogen excretion would involve transport of only 15 mg. of amino nitrogen per hour through the blood if the catabolism of the body tissue occurred at a uniform rate throughout the day.

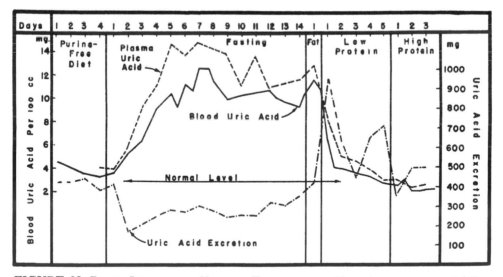

FIGURE 68. BLOOD LEVELS AND URINARY EXCRETION OF URIC ACID DURING A 14-DAY STARVATION. The ordinates for the uric acid excretion are so arranged that the excretion of 100 mg. represents a loss of the same amount from the total blood volume. In this case a change of 1 mg. in the blood uric acid level was equivalent to an excretion of 100 mg. (Lennox, 1925.)

Blood Non-Protein Nitrogen — Semi-Starvation

The blood constituents in semi-starved men have been examined by a number of investigators. The non-protein nitrogen appears to be within the normal range according to most of the reports (Jansen, 1918; Schittenhelm and Schlecht, 1919c; Ling, 1931; Weech, 1936; Youmans et al., 1932; Butler et al., 1945).

In severely emaciated cases of anorexia nervosa (Bruckner et al., 1938) and in the advanced starvation cases seen during the Bengal famine of 1943 (Bose et al., 1946), a slight increase in the blood N.P.N. was found. Since the kidney under such conditions shows no alterations (see Chapter 31), the elevated N.P.N. levels have been attributed to alterations in the circulation (Bruckner et al., 1938).

Maas and Zondek (1917) found among 27 men with varying degrees of edema only 7 who had blood N.P.N. values above 35 mg. per 100 cc. (up to 76 mg.). The mean value for all 27 men was 32 mg., which is within normal limits (Hawk et al., 1947). These investigators claimed that in the same cases the blood creatine level was elevated, with the concentration greater than 12 mg. per 100 cc. in 22 per cent of the cases. This is an exceedingly high value (Hawk et al., 1947), but since no details are given either about method or analytical results, it is difficult to interpret the finding. The urea and amino acid fractions were normal in most cases. This was also true of the blood uric acid and ammonia.

In view of the dramatic increase in the blood uric acid in acute starvation, analyses for this compound were made on a number of subjects in the Minnesota Experiment. Plasma and urine samples were analyzed for uric acid by a modification of Folin's method (1933, 1934). All but 3 of the subjects showed a decrease in the plasma uric acid levels when the control values were compared with those at S24 (see Table 218). A slight increase was seen in 2 subjects. For the third subject only one sample was available. The mean value for the control period was 3.90 mg. per 100 cc. of plasma, whereas at S24 it was 2.80. The significance of this

TABLE 218

PLASMA URIC ACID LEVELS AND 24-HOUR URINARY EXCRETION VALUES
in the Minnesota Experiment.

Subject	Plasma (mg./100 cc.)		Urine (mg./day)
	C	S24	S24
1	4.48	2.75	398
4	3.22	2.40	449
5	3.40		476
8	3.05	1.85	427
12	2.49	3.79	1318
23	4.59	3.56	368
27	4.60	2.60	256
111	3.10	1.55	249
122	5.45	3.28	
123	2.75	2.84	904
126	4.79	2.41	426
130	4.40	3.72	352
M	3.90	2.80	511

decrease in the uric acid concentration is unknown, but it does point out once more the important difference in the physiological response to acute starvation and to semi-starvation.

Urinary Nitrogen Partition — Acute Starvation

The excretion of total nitrogen, in both the urine and the feces, has been considered in Chapter 19. The presence of albumin in the urine from starvation cases is discussed in Chapter 31.

Practically all the early work on the distribution of the nitrogen compounds in the urine during periods of starvation was limited to experiments involving the complete absence of food. Weber (1902) compiled most of these early reports in a thorough review.

If the starvation is continued long enough, there is a decrease in the total nitrogen excretion. This is associated with a change in the relative proportion of the different urinary nitrogenous compounds. Although the creatinine excretion normally shows too much variation to permit its use in estimating changes in body protein, it does, nevertheless, appear in the urine at a fairly constant level during starvation. The urea fraction of the urine decreases throughout the starvation period. In normal urine about 90 per cent of the nitrogen is excreted as urea. This value decreases during the first week of starvation to about 70 per cent, where it may remain for two weeks or more (Weber, 1902; Cathcart, 1907; Benedict, 1915). The decrease in the percentage of nitrogen present as urea is associated with an increase in the ammonia fraction. The excretion of ammonia increases as ketosis develops, but no quantitative data on the relationship between the two are available. The value of such a study is questionable, because as ketosis progresses there is an increase in the acidity of the urine and a change in the excretion of cations.

A decrease in the excretion of uric acid during a fast was observed as early as 1914 (Watanabe and Sassa, 1914), but no attention was paid to this observation until Hoeffel and Moriarty and Lennox in 1925 independently showed that this was associated with a marked increase in the blood level of uric acid.

In the past considerable attention has been given to the excretion of creatine and creatinine during acute starvation. These substances are considered elsewhere in this chapter.

Urinary Nitrogen Partition — Semi-Starvation

There have been relatively few studies on the excretion of the various nitrogen compounds in semi-starvation. Perhaps the most extensive report is that of Knack and Neumann (1917), but unfortunately only ranges are given for each constituent and there is no indication of the number of subjects studied or of their nutritional status. The urea in the urine of the German civilians examined by Knack and Neumann (1917) accounted for "74 to 80 per cent of the total nitrogen with most of the values below 75." During rehabilitation this figure increased to an average of 88. All the individuals examined presumably showed some signs of famine edema.

Kohn (1920), who studied the nitrogen excretion of a laboratory attendant, also observed a high proportion of urea in the urine. Over a period of a year the attendant, who weighed 47 kg., was stated to have received an average of only 825 Cal. per day. The total daily nitrogen excretion was 5.25 gm.; 84 per cent of this was urea.

In the study by Knack and Neumann (1917), the ammonia nitrogen amounted to 7.0 to 7.5 per cent of the total nitrogen; this figure was 8.6 per cent in Kohn's (1920) subject. Since the total nitrogen excretion was so low in the latter experiment, the amount of ammonia excreted (0.55 gm. per day) was still below the mean (0.7 gm.) for normal individuals (Hawk *et al.*, 1947).

These two reports agree that there was a slight increase in the uric acid excretion. Knack and Neumann (1917) found a range between 1.0 and 1.6 gm. compared to their normal value of 0.5 to 0.8, while Kohn (1920) found a value of 1.2 gm. for his subject.

These results are in contrast to the values secured in the Minnesota Experiment. Urine samples taken at the end of the semi-starvation period were analyzed for uric acid by a modification of Folin's method (1934). When analyzed by this procedure the urine of normal adults contains approximately 0.7 gm. of uric acid per day (Hawk *et al.*, 1947). All but 2 of the 11 subjects on whom analyses were made excreted less than the average adult value (see Table 218). The mean excretion value for S24 was 511 mg. This reduction in uric acid excretion may be only a reflection of the lowered plasma values. It is noteworthy that the 2 subjects (No. 12 and No. 123) whose uric acid excretions were higher than normal were the same 2 men in whom the plasma uric acid concentrations increased during the starvation period.

The amino acid nitrogen made up about 2.5 per cent of the total nitrogen according to Knack and Neumann, whereas it amounted to only 0.3 per cent in Kohn's subject.

Of the changes in the distribution of the nitrogenous compounds in the urine during both acute and semi-starvation, none is so outstanding as the decrease in the amount of uric acid in acute starvation. After considering various possible explanations for his observations, Lennox (1926) was forced to conclude that "the exact mechanism involved in retention of uric acid, must await other investigators, and new knowledge."

Creatinine Coefficient as an Index of Muscle Mass

Firmly entrenched in the scientific literature is the concept that the urinary creatinine excretion reflects the amount of muscle tissue in the subject. Even the most recent monographs on biochemistry contain statements such as "For most normal men, the creatinine coefficient will vary between 18 and 32 with an average of about 25, and in women the normal range is between 9 and 26 with an average of about 18. Children have lower values. In general, the better the muscular development, the higher the creatinine coefficient. Consequently, obese individuals are likely to have low coefficients because much of their weight is not muscle. This indicates that every individual has a characteristic creatine-creatinine turnover dependent in a general way on the amount of functioning muscle tissue" (Kleiner, 1945).

No attempt will be made to review the plethora of literature in this field.

Twenty years ago an extensive survey was made by Hunter (1928) which led him to accept the theories held by most investigators. In going back over the more important papers that have been used to support the original contentions, one is impressed by the pertinency of Folin's (1905) statement about the early work on the composition of urine from normal men. He concluded that ". . . throughout the voluminous literature on this subject there is to be noted a strong general *desire* or tendency to corroborate the findings of Voit. In most investigations of certain nationalities, as, for example, the Japanese, or of certain groups of people, as, for example, vegetarians and some rural populations, there are to be noted more or less strained attempts to show either that such people come nearer Voit's standards than had been supposed, or that they are not properly nourished." This quotation, with only slight alterations, can be applied also to the subsequent creatinine work offered to substantiate Folin's original claims.

The quantitative studies in this field previous to Folin's colorimetric method for the determination of creatinine (1904) are of dubious value (van Hoogenhuyze and Verploegh, 1905). With this method, Folin studied the urinary creatinine excretion of normal men on various protein intakes. Among other things, he found that "the absolute quantity of kreatinin eliminated in the urine on a meat-free diet is a constant quantity different for different individuals, but wholly independent of quantitative changes in the total amount of nitrogen eliminated. . . . The chief factor determining the amount of kreatinin eliminated appears to be the weight of the person. The proportion between the body-weight and the amount of kreatinin in the urine is, however, not very constant. Fat or corpulent persons yield less kreatinin per unit of body-weight than lean ones. . . . The analytical data . . . indicate that moderately corpulent persons eliminate per twenty-four hours about 20 mg. kreatinin per kilo of body-weight, while lean persons yield about 25 mg. per kilo" (Folin, 1905, pp. 34–35). It was also suggested that the creatinine excretion would serve as a check on the completeness of 24-hour urine collections, a suggestion which found wide acceptance.

Folin's data (1905), however, show more variation in the daily creatinine excretion than his conclusions indicate. For instance, the data on Dr. E.V.S. over a period of 10 days show a range in daily creatinine excretion from 1.02 to 1.28 gm. with a mean of 1.17 ± 0.089. The difference between the extremes was 22 per cent of the mean. A similar range in the creatinine values occurred among the other normal men studied by Folin. The body weights of these subjects are given only at irregular intervals. Where no change in the body weight was indicated, the variation in the creatinine excretion was equally great. These variations occurred in spite of the fact that the men were on a low nitrogen diet. The simplicity and reliability of the creatinine method itself is such as to eliminate analytical error as the source of these variations.

Similar results appear in the data of Closson (1906). He studied 2 normal men who for months had been on diets devoid of meat, eggs, and fish. Even so, the daily creatinine excretion for one man ranged from 0.89 to 1.26 gm., a variation equal to 40 per cent of the mean. For the other subject the range was from 0.98 to 1.31 gm., corresponding to 28 per cent of the mean. According to the records kept on these men by Chittenden (1905), no changes in body weight occurred in either subject throughout the experiment. Entirely similar results were obtained by Albanese and Wangerin (1944) with 30 normal subjects who were studied for 38 to 60 days.

Shaffer elaborated on Folin's proposal (1905) that the creatinine excretion was related to the body weight and put forth the term *creatinine coefficient*. In his first report (1906) this was defined as the mg. of creatinine excreted in 24 hours per kg. of body weight. Later (in 1908) he suggested that the mg. of creatinine nitrogen be used in place of the actual weight of creatinine.

To explain the interindividual variations in this coefficient, the following explanations have been advanced (taken from Hunter, 1928):

(1) *Variations in body weight.* Folin (1905) indicated that the amount of adipose tissue had to be considered in the evaluation of body weight. This was interpreted by Shaffer (1908) to mean that the creatinine excretion was proportional to the active protoplasmic mass.

(2) *Variations in work capacity.* Shaffer (1908) was led to conclude that for "the muscular tissues . . . the amount of kreatinin excretion is an index of their efficiency, not the amount of work which the muscles are doing at the time, but the amount of work which they are capable of doing." The individuals "with the better muscular development and capable of the greater amount of muscular work have the higher kreatinin coefficients and *vice versa.*"

(3) *Variations in muscle creatine content.* From their studies on the muscles of rabbits, cats, dogs, and men, Myers and Fine (1913) concluded that "the creatinine elimination appears to bear a distinct relation to the percentage content of muscle creatine in a given species." Furthermore, "the constancy in the content of muscle creatine offers a satisfactory explanation for the constancy in the daily elimination of creatinine, first noted by Folin and subsequently confirmed by many workers."

(4) *Variations in the amount of muscle.* Bürger (1919) determined the creatinine excretion of a number of fever-free patients. He claimed he found a high degree of relationship when the muscle development of these subjects as judged by clinical examination was compared with their creatinine coefficient.

Although Hunter (1928) distinguished between Folin's (1905) suggestion that the body weight limited the creatinine excretion and Bürger's (1919) statement that the muscle mass was the important factor, the apparent difference is due to Bürger's misunderstanding of the earlier work. Folin (1905) determined the creatinine excretion per kg. of body weight for a lean and an obese individual. He specifically stated "that moderately corpulent persons eliminate per twenty-four hours about 20 mg. kreatinin per kilo. of body weight, while lean persons yield about 25 mg. per kilo."

Regardless of whether Folin's or Bürger's position is taken, it is still impossible to rationalize the large day-to-day variations in the creatinine excretion. These fluctuations for persons in weight equilibrium cannot be explained on the basis of changes in muscle mass ranging up to 25 per cent. To maintain that the constancy of the creatinine coefficient can be shown by a series of average creatinine excretions appears to beg the question. It is hard to see why the coefficient should not be constant every day in adults if the creatinine coefficient is related only to the mass of protoplasm. Furthermore, even a cursory survey of the reports on creatinine excretion verifies Myers' (1932) conclusion that "the absolute constancy of the creatinine excretion appears to be more marked in some individuals than in others."

Shaffer's (1908) argument that the creatinine coefficient is related to muscular efficiency can also be disposed of. Probably Shaffer's use of the term *efficiency* did not correspond with more recent usage. There is surprisingly little difference in the muscular *efficiency* of different individuals. Erickson *et al.* (1946) found that the oxygen consumption per kg. of body weight was practically the same for all 47 of the normal young men who were studied on the treadmill. Although there was a wide range in the physical condition of these men as measured by various "fitness" tests, the energy expenditure (oxygen utilization) per unit of external work done was closely similar for all of them. The interindividual standard deviation of the oxygen consumption was only 4 per cent of the mean.

Another prop that has been used to support Shaffer's contention is the finding that women have a lower creatinine coefficient than men (Tracy and Clark, 1914). Hodgson and Lewis (1928–29) found that women who were physically well developed and active in sports had creatinine coefficients approaching those for men. This appeared to strengthen the suggestion that the muscles of women ordinarily are not as efficient as those of men. Recent work has shown that the energy expenditure by women is very nearly the same as by men in a task such as walking. The slight sex differences in oxygen utilization observed by Metheny *et al.* (1942) cannot account for the much larger differences in the creatinine coefficients. The creatinine coefficients of men are about twice those of women.

It might be added that any attempt to explain the sex differences in the creatinine coefficients on the basis of body fat can neither be denied nor confirmed. There are no reliable data on the actual percentage of fat in women and only one acceptable analysis for a man (see Chapter 8). In normal men the mass of protein tissue is 3 to 4 times as large as the fat (Mitchell *et al.*, 1945). To explain the difference between men and women on the basis of active protoplasmic mass would require the assumption that this mass is only about half as great in a woman as in a man of the same total body weight.

The association of creatinine excretion with the muscles appeared reasonable since there was suggestive evidence at first (for references see Hunter, 1928), and then conclusive evidence (Bloch and Schoenheimer, 1939), for the conversion of creatine to creatinine by the body. It has been known since the time of Liebig that the highest concentration of creatine occurs in muscle. Myers and Fine (1913) seemed to integrate these observations when they claimed that the creatinine excretion of different species of animals was related to the muscle creatine content which was characteristic of the particular species. Chanutin and Kinard (1932–33) determined the creatine, moisture, fat, and nitrogen contents in the muscles from a large series of animals. Although there were differences in the mean concentrations of creatine in the thigh muscles of rabbits, dogs, rats, and guinea pigs, there was no significant correlation between the creatinine excretion and the creatine content of the muscles when this was expressed either on a wet weight basis or on a dry, ash, and fat-free basis.

It is surprising that so little reference has been made in the creatinine literature to the role of creatine in muscular contraction. Shortly after the isolation of phosphocreatine by Fiske and Subbarrow (1927), it was shown that most of the creatine in the muscles occurs in that form (Fiske and Subbarrow, 1929; Borsook and Dubnoff, 1947). More recent work indicates that this compound functions in the resynthesis of adenosinetriphosphate, which is one of the more important factors in both carbohydrate metabolism and muscular contraction (Kalchar, 1944; Meyerhof, 1942; Potter, 1944).

A priori, it might be assumed that if the creatinine excretion is an index of the participation of creatine (as phosphocreatine) in physiological reactions, then its excretion should be proportional to the rate of creatine utilization (or turnover) in the muscles. Some of the earlier workers in this field (Bürger, 1919; Schulz, 1921) claimed that physical work produces an increased excretion of creatinine, while others contradicted this (van Hoogenhuyze and Verploegh, 1905). Most of the recent work indicates that even severe exercise has no influence on the creatinine excretion; this is especially true when allowance is made for the normal variations that occur in control periods (Krüger, 1938; Hobson, 1939; Norris and Weiser, 1937).

It would appear from experiments with animals that physical training changes neither the creatine (Ovcharenko, 1939) nor the phosphocreatine (Klimenko and Kashpar, 1939) concentration of the muscles in rest. This would imply that training, among other things, increases the size of muscle fibers but, as far as creatine is concerned, does not change the percentage composition of the muscle. This finding was anticipated by Folin and Denis (1914b), who suggested that an addition of creatine to the body occurs only when the tissues are increasing in size.

If creatinine excretion is not directly related to the size of the protein tissue in the body, then what is its significance? There is sufficient proof that the level of excretion is not influenced by the protein intake, because on protein intakes ranging from 14 gm. of nitrogen per day down to 1.5 gm. there was no change in the urinary creatinine level greater than the day-to-day variations (Albanese and Wangerin, 1944). Even the serum creatinine levels are uninfluenced by a fivefold increase in protein intake (Addis et al., 1947; Barrett and Addis, 1947).

In some respects the urinary excretion of creatinine behaves in the same way as that of the 17-ketosteroids. The latter substances, by and large, are excreted in greater amounts by men than women (Venning and Kazmin, 1946; Barnett et al., 1946), and their excretion by both sexes increases to a certain extent throughout childhood and adolescence (Nathanson et al., 1939, 1941b; Hain, 1947). Their excretion is also a characteristic of the individual, as shown by studies on the same persons over periods as long as 3 years (Miller et al., 1948). A correlation between the creatinine and 17-ketosteroid excretion has been observed by Nathanson, Towne, and Aub (1941a).

Creatinine Excretion in Starvation

The many studies made on fasting subjects at the turn of the century revealed only variable and relatively minor changes in the urinary excretion of creatinine during acute starvation (van Hoogenhuyze and Verploegh, 1905; Benedict, 1907;

TABLE 219

THE URINARY EXCRETION OF CREATININE, preformed (A) and total (B) during acute starvation. Compiled by Hunter (1928).

Days	Van Hoogenhuyze and Verploegh (1905) A	Cathcart (1907) A	Cathcart (1907) B	Howe, Mattill, and Hawk (1911) (1) A	Howe, Mattill, and Hawk (1911) (1) B	Howe, Mattill, and Hawk (1911) (2) A	Howe, Mattill, and Hawk (1911) (2) B	Benedict (1915) A	Benedict (1915) B
Preceding fast									
1		1.40	1.40	1.78		1.71			
2		1.29	1.37	1.88		.1.74			
3	1.09	1.40	1.42	1.83		1.72			
Fast									
1	0.90	1.13	1.18	0.94	1.66	1.47	1.50	1.37	1.29
2	0.58	1.05	1.34	1.53	1.73	1.56	1.66	1.23	1.23
3	0.58	0.91	1.16	1.54	1.78	1.49	1.67	1.23	1.47
4	0.63	0.94	1.40	1.45	1.03	1.45	1.69	1.14	1.45
5	0.60			1.51	1.60	1.40	1.50	1.09	1.37
6	0.59	0.89	1.16	1.45	1.51	1.49	1.63	1.05	1.40
7	0.47*	0.91	1.13	1.55	1.56	1.37	1.46	1.03	1.31
8	0.69	0.86	1.16					1.03	1.35
9	0.72							1.00	1.35
10	0.60	0.78	1.00					1.00	1.31
11	0.45	0.81	1.02					1.00	1.31
12	0.57	0.81	1.05					1.00	1.31
13	0.55							0.94	1.28
14	0.42	0.65	0.91					0.89	1.19
15								0.81	1.03
16								0.86	1.14
17								0.83	1.07
18								0.91	1.09
19								0.81	1.03
20								0.83	1.01
21								0.83	1.01
22								0.83	0.96
23								0.91	0.98
24								0.81	0.92
25								0.75	0.94
26								0.78	0.96
27								0.78	0.95
28								0.75	0.91
29								0.78	0.04
30								0.78	0.89
31								0.81	0.86
After fast									
1	0.72	1.02	1.08	1.59		0.95	1.51	0.94	1.00
2	1.03	1.05	1.02	1.59	1.70	0.94	1.67	0.91	0.91
3		1.08	1.10	2.05		1.64	1.71		
4		1.05	1.10	2.04		1.65	1.74		
5		0.97	1.02	1.85		1.74			

*On this day a certain amount of dumbbell exercise was performed.

Cathcart, 1907; Watanabe and Sassa, 1914; Junkersdorf and Lisenfeld, 1926) (see Table 219).

It has been suggested that changes in the creatinine excretion during starvation are compensated by opposite trends in creatine with the result that there is

TABLE 220

URINARY CREATINE, CREATININE, AND NITROGEN EXCRETION OF A WOMAN
DURING STARVATION. The "calculated" creatine excretions have been based
on the assumption that during starvation 1 gm. of urinary nitrogen is
associated with 0.061 gm. of creatine (see text). All values are in gm.
per day. (Benedict and Diefendorf, 1907.)

Days	Total Nitrogen	Creatinine	Creatine	
			Found	Calculated
Control				
1	6.49	0.61	0.05	
2	6.65	0.60	0.03	
3	6.99	0.61	0.06	
4	7.45	0.65	0.03	
Fasting				
1	4.19	0.61	0.04	0.26
2	6.05	0.57	0.09	0.37
3	6.38	0.54	0.07	0.38
4	6.93	0.44	0.06	0.42
5	6.16	0.49	0.16	0.38
6	4.41	0.34	0.15	0.27
No food				
1	3.04		0.11	0.18
2	4.87		0.11	0.30
3	3.17		0.05	0.19

TABLE 221

URINARY CREATINE, CREATININE, AND NITROGEN EXCRETION OF A MAN
DURING A 2-WEEK STARVATION PERIOD. The "calculated" creatine excretions have been determined as in Table 218. All values are in gm.
per day. (Cathcart, 1907.)

Days	Total Nitrogen	Creatinine	Creatine		Creatine + Creatinine
			Found	Calculated	
Control 16.4		1.63	0.03		1.69
Starvation					
1 ....10.5		1.31	0.06	0.64	1.37
2 ....14.4		1.22	0.34	0.88	1.56
3 ....13.7		1.06	0.28	0.84	1.34
4 ....13.7		1.40	0.53	0.84	1.62
6 ....10.8		1.03	0.31	0.66	1.34
7 .... 9.7		1.06	0.25	0.59	1.31
8 .... 9.5		1.00	0.34	0.57	1.34
10 .... 8.4		0.91	0.25	0.51	1.15
11 .... 8.5		0.94	0.25	0.52	1.19
12 .... 8.8		0.94	0.28	0.54	1.22
14 .... 7.8		0.75	0.31	0.48	1.06
Refeeding					
1 .... 7.1		1.19	0.06		1.25
2 .... 3.6		1.22	0.03		1.19
3 .... 2.8		1.25	0.03		1.28

no marked change in the total creatine plus creatinine excretion (see Table 220). Partial confirmation of this was obtained in the study by Cathcart (1907), who found a decrease in creatinine excretion and an increase in creatine during starvation. There was, however, a decrease in the excretion of creatine plus creatinine (see Table 221). The generalization has been challenged by Watanabe and Sassa (1926), who found an increase in the excretion of these 2 compounds by a man who starved for 2 weeks. The change was from a control level of 1.26 gm. per day (expressed as creatinine for the sum of creatine plus creatinine) to a mean of 1.54 during the starvation period, with a tendency toward a peak of 1.86 gm. on the fourth day of starvation followed by a gradual decline to 1.23 on the fourteenth day.

Knack and Neumann (1917) reported the excretion of creatinine in semi-starvation for their edema cases. They simply stated that in the early phases of hunger edema the values for the daily urinary creatinine excretion ranged from 1.7 to 2.4 gm. During convalescence these same patients showed values of 1.4 to 1.7 gm. On the basis of this evidence, and from the data on the excretion of other nitrogenous compounds, they concluded that the metabolic disturbance in hunger edema was the same as that in acute starvation. At Warsaw the actively starving inmates of the ghetto excreted a reduced amount of creatinine in the urine, the general range being 200 to 400 mg. per day instead of the "normal" expectation of 600 mg. (Fliederbaum et al., 1946).

The creatinine excretion in the Minnesota Experiment was determined at S12, S24, R6, and R12. The analyses were made by the modification of Folin's method described in Peters and Van Slyke's manual (1932). From studies on well-nourished subjects in the Laboratory of Physiological Hygiene who were comparable in age and weight to those in the Minnesota Experiment, a mean creatinine excretion of 1.6 gm. per day may be assumed for the control value. Throughout the course of semi-starvation there was a decrease in the creatinine excretion that was greater, on a percentage basis, than the decrease in the body weight (see Table 222). Accordingly, the creatinine coefficient also decreased during semi-starvation. The average decreases in creatinine excretion, in body weight, and in the creatinine coefficient, as percentages of the control values, were 37.5, 24.3, and 18.3, respectively.

During the rehabilitation period the recovery of body weight was, to a certain

TABLE 222

CREATININE EXCRETION AND THE CREATININE COEFFICIENTS. The coefficients have been calculated on the basis of both mg. of creatinine nitrogen (1) and mg. of creatinine (2). All values are expressed as the means of 32 men. (Minnesota Experiment.)

	C	S12	S24	R6	R12
Urinary creatinine (gm.)	1.6	1.19	0.99	1.02	1.40
Body weights (kg.)	69.5	57.2	52.6	54.2	58.8
Creatinine coefficients (1)	8.56	7.72	6.99	7.25	8.84
Creatinine coefficients (2)	23.0	20.8	18.8	18.8	23.8
Active tissue (kg.)	39.9		29.2		32.2

extent, proportional to the caloric intake. The creatinine excretion of the different caloric groups did not in all cases reflect the changes in body weight (see Table 223). Although there was an average increase of 3 kg. in the weight of the 8 men in the T group during the first 6 weeks, the mean creatinine excretion decreased from 1.07 to 1.04 gm. Similar discrepancies in the creatinine excretion during a period of weight increase by markedly underweight persons have been observed by McCluggage, Booth, and Evans (1931).

TABLE 223

CREATININE EXCRETION of the caloric groups in the Minnesota Experiment and the creatinine coefficients calculated on the basis of mg. of creatinine. Each value is the mean of 8 subjects.

Caloric Group	Creatinine (gm.)	Body Weight (kg.)	Creatinine Coefficient
	S24		
Z	0.96	52.0	18.5
L	0.89	51.7	17.2
G	1.03	54.5	18.9
T	1.07	52.1	20.5
	R6		
Z	0.95	52.0	18.2
L	1.00	53.4	18.7
G	1.07	56.4	19.0
T	1.04	55.0	18.9
	R12		
Z	1.27	55.4	22.9
L	1.36	57.0	23.8
G	1.52	61.7	24.6
T	1.46	60.9	24.0

Creatine Excretion in Starvation

The significance of the creatine excretion has been subject to a misunderstanding comparable to that of the creatinine excretion. Neither time nor space permits a critical evaluation of the many papers that have attempted to interpret physiological activities on the basis of changes in the excretion of creatine. Many workers in this field have made a number of tacit assumptions which should be carefully considered before anyone ventures into this caldron of confusion and controversy.

Hunter (1928), Rose (1933, 1935), and Beard (1943) have reviewed the literature in this field. They have presented the conflicting claims for the relationship between creatine excretion and such factors as growth, sex hormones, carbohydrate reserves, and muscular dystrophies and myopathies, as well as starvation. Our attention will be confined to the last condition, and that only long enough to show how superficial have been many of the past pronouncements in this field.

Benedict did much of the earlier work on the excretion of creatine by starving men and women. In one of these experiments (Benedict and Diefendorf, 1907) the creatine and creatinine excretion of a fasting woman was followed. From the description of the experiment it is difficult to determine the food intake in the different periods. The woman presumably ate some food during the control period preceding the 6 "fasting" days, but this was followed by 3 days of "no food." If we accept as true starvation only the last part of the experiment, then the creatine excretion on those days was 0.11, 0.11, and 0.05 gm., respectively. On the first day part of the urine was lost, so the workers did not consider that figure reliable. These excretion values should be compared with the controls, which averaged 0.04 gm. per day.

On the basis of these findings, it was stated that "The most noticeable feature of the creatine observations was the marked increase in the preformed creatine excreted during the fast. . . . The creatine excretion during fasting appears to be the result of the disintegration of flesh, and the hypothesis . . . advanced [is] that as the flesh was katabolized or broken down, the creatine which existed in the flesh as such was excreted by the body as such and not converted to creatinine." The validity of this statement is generally accepted, but it appears that no one has either considered its implications or carefully scrutinized the loose foundation on which it rests.

It has been calculated that a 70 kg. man contains 140 gm. of creatine (Hunter, 1928, as quoted by Beard, 1941) and that this same man contains 2300 gm. of nitrogen (Mitchell et al., 1945). This means that on the average each gm. of nitrogen in the body is associated with 0.061 gm. of creatine. On this basis one should find an increase in the creatine excretion equal to 0.061 gm. for each gm. of nitrogen in excess of the intake. The woman in the experiment summarized in Table 220 lost more than 3 gm. of nitrogen each day that she received no food. This should have been reflected in creatine excretions greater than 0.18 gm. even without an allowance for the "control" creatine excretion. On each of her "no food" days, the creatine excretion calculated on the above basis was considerably higher than that actually observed.

Additional proof that a loss of body tissue is not necessarily associated with any rise in the urinary excretion of either creatine or creatinine is provided by the experiment of Deuel et al. (1928). A normal young man lived on a protein-free diet for 54 days, during which time he lost 9 kg. of body weight and 197 gm. of nitrogen in the urine. Using Lusk's (1928) figure of 30 gm. of nitrogen as the equivalent of 1 kg. of body mass, the subject lost 6.6 kg. of active body tissue (on the basis of the urinary nitrogen). In spite of this considerable breakdown of protoplasm, the creatinine excretion ranged from 0.56 to 0.58 gm. per day, with the latter figure appearing only once. On no occasion was there any indication of creatine in the urine.

A few studies on creatine excretion were made on victims of the German prison camps who were hospitalized in Switzerland. It was noted that some very high values were obtained with patients who were convalescing on diets high in calories and rich in proteins and that when such patients had attacks of diarrhea the creatinuria very sharply diminished (Gsell, 1948). Maximum urine creatine values in 4 men, aged 20, 26, 42, and 51 years, were, respectively, 203, 239, 111, and 573 mg. per 100 cc.

There are a number of problems that must be considered before a final report on the creatine excretion in starvation can be made. In the first place, the method most commonly used for the analysis of creatine involves its conversion to creatinine by heating in an acid solution (Folin, 1904). Recently it has been suggested that the conversion is not quantitative and that small amounts of creatinine are destroyed in the process (Albanese and Wangerin, 1944). The amount of creatinine destroyed in normal urine was reported to be just about the same as the creatine originally present therein. This compensatory action made it appear as though there was no creatine in urine from normal adult men. By making correction factors for both disturbances, Albanese and Wangerin (1944) claimed that urine from men normally contains some creatine. The report of Albanese and Wangerin (1944) has been challenged by Caspe, Davidson, and Truhlar (1947).

In addition to the above technical factor, the validity of the color reaction has been discussed by many workers (Behre and Benedict, 1937; Danielsen, 1936; Miller and Dubos, 1937). The use of a specific enzymatic method (Miller and Dubos, 1937) has shown that the Jaffe color reaction (adapted by Folin, 1904) applied to blood gives highly reliable figures but that when it is used in tissue analyses, the "true" creatine ranges from 98 down to 26 per cent of the apparent (Baker and Miller, 1939). So far the enzymatic technique has not been applied to starvation urines.

The other factor that may influence the starvation creatine excretion values is the presence of acetone bodies in the urine. It has been reported that both acetone and glucose interfere with the determination as proposed by Folin (1904), but this does not seem to have been considered by many workers in this field (see Hobson, 1939). It appears desirable to check this with the more specific enzymatic method for creatine and creatinine (Miller et al., 1939).

Electrophoretic Analysis of Serum — Minnesota Experiment

Serums from a random group of subjects in the Minnesota Experiment were analyzed in the moving boundary electrophoresis apparatus of Tiselius (1937) by the procedure of Longsworth (1942). The serum was diluted with 2 parts of veronal buffer which was at pH8.6 and of ionic strength of 0.1. The details of the procedure are given in the Appendix on methods.

The results of the electrophoretic analysis of the serum proteins from two subjects are given in Table 224. At S12 there was a small but definite increase in the percentage distribution of albumin. This was accompanied by an increase in the concentration of total proteins which meant that the albumin in the serum increased, on an absolute basis, from 4.63 per cent to 4.95 for subject 122 and from 4.45 to 5.06 for subject 123. By S24 the albumin level was slightly below the control value. These changes in the albumin were reflected in the A/G ratios, which, like the albumin, increased and then returned to the control value by S24. There was a small decrease in the gamma globulin which was apparent at 12 weeks and showed no consistent change at 24 weeks. While there was a small increase in the concentration of gamma globulins during recovery, it should be noted that this fraction had not returned to its control level at the end of recovery.

Results similar to these were secured for the serums from 6 other men (see Table 225), but the data are incomplete because of a refrigeration accident in-

TABLE 224

ELECTROPHORETIC ANALYSES OF THE SERUMS OF 2 MEN. The analyses are presented for
the control period, after 12 and 24 weeks of semi-starvation (S12 and S24), and 6, 12,
and 32 weeks of rehabilitation (R6, R12, and R32). (Minnesota Experiment.)

	Total Proteins (gm./100 cc.)	A/G	Percentage of Total Proteins				
			Albumin	Alpha $_1$	Alpha $_2$	Beta	Gamma
			Subject 122				
Control	6.89	1.70	62.8	4.9	8.4	11.8	12.1
S12	7.10	2.30	69.6	3.5	7.7	10.5	8.7
S24	6.88	1.70	63.0	4.7	9.7	12.9	9.7
R6	6.78	2.33	70.0	3.3	7.1	11.6	8.0
R12	7.10	1.95	66.2	4.2	8.0	13.0	8.6
R32	6.18	1.65	62.2	4.3	8.4	15.2	9.9
			Subject 123				
Control	6.99	1.76	63.7	4.4	7.6	11.2	13.1
S12	7.18	2.40	70.6	3.2	6.4	9.9	9.9
S24	6.16	1.88	65.4	3.2	7.5	13.7	10.2
R6	6.43	2.00	66.7	3.3	7.0	13.2	9.8
R12	7.24	1.91	65.5	3.3	7.2	13.1	10.9
R32	6.60	1.58	61.2	4.1	9.2	15.0	10.5

TABLE 225

MEAN VALUES FROM ELECTROPHORETIC ANALYSES OF BLOOD SERUM. Results from the
same 6 men are averaged for control, after 12 weeks of semi-starvation (S12), and
after 6 and 12 weeks of rehabilitation (R6 and R12). Values for 4 men are averaged
for the periods R20 and R34 (after 20 and 34 weeks of rehabilitation). For normal
comparison, mean values and their standard deviations are given for 30 normal men
of the same age. (Minnesota Experiment.)

	Total Proteins (gm./100 cc.)	A/G	Percentage of Total Proteins					
			Albumin	Globulin	Alpha $_1$	Alpha $_2$	Beta	Gamma
Control ..	6.90	1.84	64.7	35.3	4.4	7.7	10.9	12.3
S12	7.15	2.31	69.5	30.5	3.3	6.8	10.2	10.2
R6	6.58	2.13	67.7	32.3	3.6	7.1	11.9	9.7
R12	7.14	2.02	66.6	33.4	3.5	7.3	12.2	10.4
R20	6.58	1.71	62.8	37.2	4.3	8.0	14.1	10.9
R34	6.47	1.62	61.6	38.4	4.1	8.2	14.9	11.2
30 normals	7.01	1.72	63.0	37.0	4.1	8.0	12.0	12.9
SD	0.35	0.22	2.81		0.55	0.96	1.72	2.02

volving samples for S24. For purposes of comparison, the data from 30 normal
young men studied in the Minnesota Laboratory of Physiological Hygiene are
included. At S12 there were slight decreases in all the globulin components. The
magnitude of these decreases varied not only for the different fractions but among
the subjects as well. On a percentage basis the greatest drop occurred in the alpha$_1$
globulin, the concentration of which changed by one fourth of its control value.
These data also show a slight decrease in the gamma globulin. The finding of a
practically normal distribution of the various protein fractions after 24 weeks of

semi-starvation is similar to the experience of Bieler, Ecker, and Spies (1947), who found that the patterns of 8 underweight individuals who were suffering from anemia and hypoproteinemia did not differ materially from the patterns found in their healthy controls.

Evidence has been presented by Lamy, Lamotte, and Lamotte-Barrillon (1946b) for the presence of a globulin of high molecular weight in the serum from semi-starvation patients. The absence of any new peaks in the electrophoretic diagrams of the serums from the Minnesota Experiment is evidence neither for nor against this conclusion. It should be remembered that the antipneumonococcus antibodies have molecular weights in excess of the normal serum globulins. In spite of this, these heavy proteins migrate at so nearly the same speed as normal gamma globulin that they can be detected electrophoretically only by the increase in the gamma fraction (Tiselius and Kabat, 1939).

The absence of significant decreases in the albumin fraction in the serums of men who lost 24 per cent of their original weight is interesting in view of the current theories which consider albumin a sensitive indicator of the body's protein reserves. Elsewhere in this chapter it is reported that the serum albumin concentration, determined by the usual sodium sulfate precipitation techniques, showed a 10 per cent decrease, but when allowance was made for the change in plasma volume, the amount of circulating albumin at S24 was exactly the same as that in the control period (see Table 226).

Figures for the total circulating protein fractions are presented in Tables 226 and 227. There was an increase in the circulating albumin during the first 12 weeks which was probably reduced at 24 weeks. The principal decreases in circulating globulin, which were small, occurred in the alpha$_1$ and gamma fractions, and there was a definite increase in the beta fraction.

The use of the electrophoretic technique in the study of the plasma proteins from protein-depleted dogs has yielded results showing a decrease in albumin similar to that found by other workers (Weech et al., 1937) using the chemical method of fractionation. Chow et al. (1945), by means of electrophoretic analysis, found a marked reduction in both the concentration of serum albumin and the total circulating albumin in dogs that had been depleted of protein by means of plasmapheresis and protein-free rations. The circulating alpha globulin remained unchanged while the gamma globulin decreased slightly. Somewhat similar results were secured by Zeldis et al. (1945a) for the electrophoretic patterns of the serum proteins of 4 dogs which were depleted by a 9-week period on a ration that was practically free of protein. Like Chow et al. (1945), they found that the concentration of albumin determined by the electrophoretic technique was smaller than by the sodium sulfate salting-out procedure. This was due to the contamination of globulins in the albumin fraction. The differences between the two methods were still greater when the animals were placed on high protein diets. The albumin concentration determined by electrophoresis required twice as long to return to normal as when determined chemically. It should be emphasized, however, that these findings on dogs severely depleted in protein do not necessarily bear any relation to ordinary semi-starvation.

TABLE 226

The Plasma Volumes, in cc., and the Circulating Fractions of the Serum Proteins as determined by the electrophoretic technique, in gm., for 2 men before (Control) and after 24 weeks of semi-starvation (S24). For comparison, the values of the same variables found in a group of 17 normal men are presented along with their standard deviations. (Minnesota Experiment.)

	Plasma Volume	Circulating Protein Fractions					
		Albumin	Globulin	Alpha $_1$	Alpha $_2$	Beta	Gamma
Subject 122							
Control	2746	118.9	70.3	9.3	15.9	22.4	22.8
S24	3206	138.8	81.8	10.2	21.5	28.5	21.5
Subject 123							
Control	3059	136.1	77.7	9.5	16.2	23.8	28.1
S24	3404	137.2	72.5	6.8	15.6	28.5	21.4
17 normals	3249	140.7	79.1	9.4	17.2	24.9	27.7
SD	305	13.5	10.0	1.6	2.5	3.3	5.6

TABLE 227

The Plasma Volumes, in cc., and the Circulating Fractions of the Serum Proteins as determined by the electrophoretic technique, in gm., for 4 men before (Control) and after 12 weeks of semi-starvation (S12). (Minnesota Experiment.)

Subject	Plasma Volume	Circulating Protein Fractions					
		Albumin	Globulin	Alpha $_1$	Alpha $_2$	Beta	Gamma
		Control					
123	3059	136.1	77.6	9.5	15.9	23.9	28.1
119	2599	116.2	59.5	7.0	15.6	19.5	17.4
129	3316	142.6	67.6	9.0	16.2	19.9	22.5
26	3201	147.2	83.5	9.9	18.2	25.6	29.8
M	3044	135.5	72.1	8.8	16.5	22.2	24.3
		S12					
123	2936	148.5	62.2	6.8	13.5	20.9	21.1
119	2504	118.9	56.8	5.8	15.5	18.8	16.8
129	3508	185.2	65.3	9.1	17.5	22.5	16.1
26	3381	174.5	86.6	7.8	13.9	28.7	36.2
M	3082	156.8	67.7	7.3	15.1	22.7	22.5

Electrophoretic Analysis of Serum in Rehabilitation — Minnesota Experiment

During the rehabilitation period there was considerable fluctuation in the level of the various protein components. The albumin was at the original (control) level at R6, increased slightly by R12, and then dropped progressively through R34, when it was down to an absolute level of 3.98 per cent in the serum (see Table 225). This was considerably below the control value of 4.55 per cent and the peak at S12, which was 4.97. Of the globulins, the beta fraction was the first to reach its original level shortly after R6, but then it continued to increase until at R34 it was present in the serum at a concentration of 0.96 per cent, whereas in the control period there was only 0.76 per cent beta globulin. By R20 the alpha$_1$

and alpha$_2$ fractions were still considerably below their control values. The gamma globulin, which had decreased only slightly, had not returned to normal levels at R34.

Data for calculating the circulating grams of protein components as identified by electrophoresis were available for only a few men during the first 12 weeks of rehabilitation. In these men, who were in the group receiving the highest caloric intake, the amount of the various circulating protein fractions showed only a slight change between the sixth and twelfth weeks of rehabilitation (see Table 228). The concentration of plasma albumin was back to the control value in the men in the T caloric group by R12, so it is quite likely that even at R6 they had their normal quota of plasma proteins. The regeneration of the plasma protein appears to take place much before the completion of the recovery in most other functions. One would have to conclude, then, that the plasma protein is no better indicator of the general state of recovery than it is of the stage of debilitation.

TABLE 228

THE PLASMA VOLUME, IN CC., AND THE CIRCULATING FRACTIONS OF THE SERUM PROTEINS as determined by the electrophoretic technique, in gm., for 3 men after 6 and 12 weeks of recovery from semi-starvation (R6 and R12) (Minnesota Experiment).

Subject	Plasma Volume	Circulating Protein Fractions					
		Albumin	Globulin	Alpha $_1$	Alpha $_2$	Beta	Gamma
			R6				
104	2972	126.6	64.8	6.2	18.4	23.2	17.0
109	4182	173.9	98.3	10.0	21.3	33.9	33.0
112	2695	121.8	77.1	7.0	18.3	26.9	24.8
M	3283	140.7	80.1	7.8	19.3	28.0	24.9
			R12				
104	2700	121.0	79.9	9.2	25.1	25.7	19.7
109	3442	148.0	84.0	8.3	15.8	31.7	28.2
112	2907	123.0	86.3	7.3	21.2	32.0	25.9
M	3016	131.0	83.3	8.3	20.7	29.8	24.6

The unusual behavior of the beta fraction during the rehabilitation period may be related to the findings of Blix, Tiselius, and Svensson (1941). They observed that this component was associated with a considerable amount of lipoidal material which could be extracted with alcohol or ether. Recently, Zeldis et al. (1945b) showed that abnormally large beta peaks regularly occur in the presence of elevated lipids. No plasma lipid analyses were made on the men during the rehabilitation period. Consequently, there is a possibility that the increased beta peaks at R20 and R34 may be associated with increased plasma lipid levels.

During the depletion period there was an apparent increase in the concentration of the globulins, particularly of the alpha fraction, in the dogs used by Zeldis et al. (1945a). Further work on these plasmas by Zeldis et al. (1945b) indicated that the major part of the increase in the alpha globulins was attributable to alcohol-soluble material. They found that there was no change in the amount of the alcohol-extracted globulins during the low protein diets. This finding poses an intriguing problem for the interpretation of results secured by electrophoresis. The lipids present in the protein fractions are low in nitrogen and as such will have little influence on the results of the micro-Kjeldahl analysis. They can markedly alter the electrophoretic pattern, indicating a considerable increase or decrease in components of intermediate mobility which is due not to changes in the protein fractions but to changes in the lipo-protein complex. Such a condition may occur quite regularly during periods of protein depletion in dogs.

The difference in the behavior of the serum albumin as seen in the Minnesota Experiment and that reported for protein-depleted dogs demands at least passing notice. This may be considered purely a species difference. However, practically all the dogs on which plasma protein studies have been made were depleted by such extreme means as protein-free rations and/or plasmapheresis. A protein-free diet has no counterpart in actual famine experience, and plasmapheresis is an entirely artificial procedure. The special situation is emphasized by the work of Chow et al. (1945), in which marked reductions in plasma albumin were seen after short periods of protein depletion. None of their dogs, however, showed any appreciable change in weight.

There are other differences between the condition of these dogs and simple undernutrition in man. In the first place, dogs on protein-free rations show a decrease in plasma volume, together with a reduction in albumin concentration which begins early in the course of protein depletion (Weech et al., 1937; Zeldis et al., 1945a; Allison et al., 1946), and an increase in some of the globulin fractions (Zeldis et al., 1945a). The latter is associated with a hypercholesteremia (Zeldis et al., 1945b).

In contrast to the above, human beings who have been starved show no significant change or a slight increase in plasma volume even after marked loss of body tissue (see Chapters 14 and 15). When the condition is associated with significant reduction in plasma proteins, the blood volume, if anything, is increased (Gollan, 1948). Such change as does occur in the plasma concentration occurs only late in the semi-starvation period. Finally, hypercholesteremia does not occur in men subjected to semi-starvation (see Chapter 22). For these reasons, great care should be exercised in applying to human beings data obtained with protein-depleted dogs.

Serum Protein Patterns (Electrophoretic) in Starvation and Disease

It is frequently stated that protein depletion results in a lowered resistance to infection (Cannon, 1945). It is of interest, therefore, that the protein depletion produced in the Minnesota Experiment resulted in only a minimal reduction of circulating gamma globulins. While there is much evidence to indicate that the majority of circulating antibodies are found in the gamma globulin fraction (Enders, 1944), it has not been shown that gamma globulin is made up entirely of antibodies. Therefore, the small losses observed in semi-starvation in man do not necessarily represent depletion of circulating antibodies.

Disease or injury produces changes in the serum proteins that have been compared to the effects of undernutrition. Many diseases are characterized by a decreased serum albumin level and an increase in one or both of the alpha glob-ulins (Leutcher, 1947). Those chronic diseases which are accompanied by poor nutrition, such as tuberculosis (Chow, 1947; Siebert *et al.*, 1947), cancer (Siebert *et al.*, 1947), and ulcers (Chow, 1947), exhibit the characteristic decrease in serum albumin and increase in the alpha globulins.

An electrophoretic pattern similar to this is found in the protein-depleted dog (Chow *et al.*, 1945; Zeldis *et al.*, 1945a). This led Chow (1947) to wonder whether the serum protein changes in diseased conditions are due to the infective process or to the protein depletion which accompanies it.

The results of the Minnesota Experiment strongly suggest that the protein depletion seen in simple undernutrition does not produce the changes in the electrophoretic serum protein pattern characteristically observed in various chronic diseases. The subjects in the Minnesota Experiment suffered only from under-nutrition uncomplicated by any other factor, and they showed none of the altera-tions in serum protein associated with various chronic diseases.

A similar argument can be put forth to show that the alterations in the protein patterns seen in acute infectious disease are not due simply to undernutrition. The albumin is decreased and one (or both) of the alpha globulins is increased in pneumonia (Leutcher, 1941), scarlet fever (Dole *et al.*, 1945), rheumatic fever (Dole *et al.*, 1945), tonsillitis (Malmros and Blix, 1946), and malaria (Minnesota Laboratory of Physiological Hygiene). Since most infectious diseases are accom-panied by a negative nitrogen balance (Grossman *et al.*, 1945), it was reasonable to suggest that the nitrogen depletion in these conditions was responsible for the altered electrophoretic patterns (Chow, 1947), but this now seems untenable.

It has been suggested that there is a special form of nitrogen loss in acute infectious diseases in the so-called catabolic phase (Peters, 1946). There is sup-posedly a "toxic destruction of protein" (*ibid.*), defined as a negative nitrogen balance which will not respond to the addition of calories and/or protein.

Certain experimental findings have raised some question about the existence of the catabolic phase of nitrogen loss. It has been known for some time that the negative nitrogen balance in tuberculosis can be easily corrected by dietary means (McCann, 1922), and this is one of the diseases characterized by a decrease in serum albumin and an increase in the alpha globulins. Peters (1946) has presented evidence indicating that the serum albumin determined by the sodium sulfate technique increased in spite of a marked nitrogen loss. He noticed that in surgical patients the albumin concentration decreased before the negative nitrogen balance became apparent.

This discussion should not be misconstrued as questioning the fact of de-creased serum albumin concentration and increased alpha globulin in acute and chronic disease. This response to injury is independent of the magnitude of the nitrogen loss. Furthermore, there is no question about the very low plasma albumin and total protein levels seen among the extreme starvation victims in China. It may be suggested that such low plasma protein values result from prolonged inanition or extreme protein deficiency, or both. Our main contention is that there is a wide and significant range of tissue protein depletion that is not reflected by alterations in the electrophoretic pattern of the serum proteins.

CHAPTER 21

Vitamins and Vitamin Metabolism

WHENEVER food shortages occur and famine conditions arise there is not only a diminution in the amount of food eaten but also a change in the kinds of food available. Depending on the local circumstances, the vitamin content of the famine diet may be either larger or smaller than that of the ordinary diet. How the people will be affected by any change in the vitamin intake during starvation periods will depend upon their previous state of vitamin nutrition and bodily reserves. The quality of the diet used before a famine as well as that during the famine is of considerable importance in determining whether, and when, any signs of vitamin deficiency will appear. These facts probably go far to explain the great variation in the reports on the occurrence of vitamin deficiencies in different famine areas.

In the older literature on semi-starvation there are suggestions indicating the presence of vitamin deficiencies, since such terms as *scurvy* and *pellagra* were frequently used in describing the conditions observed. The significance that can be attached to these earlier reports is probably very small. The older definitions of these words were extremely vague. Conditions of undernutrition and nutritional defects in general were frequently called "scurvies" on little or no basis that would be accepted today (Hirsch, 1885). It is also necessary to realize that in the past 200 years there have been profound changes in the staple foods as far as the Western world is concerned. This is true not only for normal times but also when famine conditions arise. A starving population in Europe today eats a very different diet — and gets a different vitamin supply in it — than was true of the same area during the famines of the Middle Ages. The present importance of the white potato, which was unknown a few hundred years ago, is only one example. There is little information on the ordinary diets used in the past and still less about the food used during periods of shortages. This means that there is not much basis for estimates of the actual frequency, importance, and character of vitamin deficiencies in most of the famines of the past. In any case, it is clear that few generalizations can be made except for particular periods and places.

In modern times there is reason to doubt that vitamin deficiencies are usually of great importance in the picture of starvation. There are general opinions in the literature to the effect that signs and symptoms of vitamin deficiencies are rare in persons who have undergone prolonged periods of starvation (e.g. Hopkins, 1919–20; Bruckner *et al.*, 1938; Wilder, 1945). The evidence for such opinions has been drawn partly from observations on acute starvation — which is a very special case — and partly from observations in famine areas which were not very detailed until the last few years. Recently a number of investigators have called specific attention to the absence of vitamin deficiencies during starvation ·both in Europe (Bigwood, 1944; Leyton, 1946; French and Stare, 1947) and in the Orient (Chari, 1944). At the time of the Bengal famine in 1943–44 large numbers of people went to Calcutta, where diseases like keratomalacia, beriberi, and scurvy were absent among the famine victims (Bhattacharya and Sen, 1945).

In spite of the fact that nearly twenty different compounds have been isolated and shown to have a vitamin-like activity for different animals, only a few of these have been shown to be required by human beings. There is valid evidence that vitamin A, thiamine, niacin, vitamin C, vitamin D, and vitamin K are required for normal metabolism at some or all stages of human development. Unless there is some physiological disturbance, the requirement for vitamin K is provided by the bacteria in the lower part of the intestine (Brinkhous, 1940; Dam, 1943). Since there have been no indications of any change in this condition in famine areas (Walters *et al.*, 1947b), no further mention of vitamin K will be made. Although there is not a great deal of valid evidence that human beings require riboflavin, there are a number of recent claims from famine areas that certain disturbances in vision or skin changes were due to a deficiency of this substance. For this reason, the following discussion will be limited to riboflavin and the above vitamins which have been shown to be required by human beings.

Metabolic Rate, Vitamin Needs, and Adaptation

Prolonged undernutrition and starvation always result in a decrease in energy expenditure. Not only is the basal metabolism reduced, but there is an inevitable reduction in physical activity which may be relatively greater than the change in the B.M.R. A laborer who normally expends 4000 Cal. per day may be able, after prolonged severe undernutrition, to maintain caloric balance at a level of about 1500 Cal.

At this low level of energy balance, what are his vitamin requirements? It is now commonly taken for granted that the energy expenditure of the organism is one of the more important factors influencing the requirement for at least some members of the vitamin B complex (National Research Council, 1945). The direct evidence for this belief is astonishingly small, and even the few experiments on which it is based have been made with animals. But if there should be a direct, linear relationship between energy metabolism and the need for certain vitamins, then we could predict a reduction in the vitamins required of 50 per cent or more during prolonged starvation.

What may be the case with vitamins other than thiamine, riboflavin, and niacin is even more open to speculation. It is conceivable that all the vitamins are required in proportion to the total metabolism of the body or to the metabolism of certain nutrients. In the various tables of vitamin "requirements" or "recommended allowances," it is apparently assumed that the plane of activity has no influence on vitamin A, ascorbic acid, and vitamin D, since the same intake is suggested regardless of the caloric level (National Research Council, 1945).

A similar degree of uncertainty exists regarding the question of possible adaptation to a lower or higher level of vitamin intake. There are indications of the existence of a mechanism of adaptation to low intakes of thiamine, both in animals (Lanczos, 1939) and in man (Meyers, 1941). In a sense, the starving man adapts to his reduced caloric intake; it is not unreasonable to suggest that a similar change in his vitamin usage and needs may occur.

Vitamin Release from Tissues

One factor that cannot be overlooked in the starving man is the fate of the vitamins in his own tissues. As starvation progresses the body tissues themselves are used as fuel. The vitamins contained in these tissues are presumably available for meeting the needs of the body. The utilization of these vitamins may be singularly efficient since digestive and absorptive processes are not involved. Again, it must be emphasized that the fate of the released vitamins is unknown. The vitamin studies in the Minnesota Experiment do not contribute very much toward clearing up this point. The results thereof are at best suggestive and are not consistent when, for example, thiamine and riboflavin excretions are compared.

TABLE 229

URINARY THIAMINE AND PYRAMIN EXCRETION during the semi-starvation phase of the Minnesota Experiment. All values are in micrograms per 24 hours. Pyramin is expressed as 2-methyl-4-amino-5-ethoxymethyl pyrimidine hydrochloride. The expected excretion values are from Mickelsen, Caster, and Keys (1947). The thiamine excretion calculated from the loss of body tissue was determined as described in the text.

	Thiamine		Pyramin	
	M	SD	M	SD
Observed excretion				
S12	121	60	135	18
S24	132	71	135	27
Expected excretion	107	35	195	30
Difference	+19		−65	
Calculated from body loss	37			

During the semi-starvation period of the Minnesota Experiment the urinary thiamine excretion was greater than normal when related to the intake. Although this difference was not significant, there was an interesting implication in its magnitude. The amount of "liberated" thiamine was calculated from the changes in the active tissue. This agreed quite closely with the thiamine excretion, which was above the amount that would be predicted from the intake alone (see Table 229). This, however, was not the case for riboflavin. Here the excretion was about the same as that which would be expected of young men maintained at their normal body weight on the same vitamin intake. The riboflavin thus "liberated" by the breakdown of tissues would have added considerably to the daily excretion if it had behaved as thiamine apparently did.

Unfortunately, there is no agreement as to the physiological significance of the vitamins excreted in the urine. This is true for both normal and starving individuals. Even the interpretation of urinary excretion values is debatable. Some investigators maintain that the body's requirement for a vitamin is not met until the urine indicates that the body has been saturated with the vitamin. At the other extreme, a few investigators argue that the excretion of even traces of the vitamin is proof that the body's needs have been met (Holt, 1943).

The Diagnosis of Vitamin Deficiencies

Except in "classical" cases, the diagnosis of vitamin deficiencies is still highly controversial. Laboratory tests may give objective results for the amounts of the vitamin in blood, urine, or tissues, but the significance of the levels found is questionable (Mickelsen, 1943). So-called load, saturation, and tolerance tests are open to the same difficulties. It is possible, of course, to use laboratory procedures to define the relative blood or urinary levels of an individual or a special group in comparison with the general population or some other special group. But reliable and precise differentiation between "good" and "bad," or "adequate" and "inadequate," vitamin status may still escape us. In the case of starvation where metabolic derangement and tissue breakdown are added, the situation is still more obscure.

During the last few years there have been marked changes in the significance attached to some of the clinical signs that had been proposed as pathognomonic of specific deficiencies (Machella, 1942; Stannus, 1945; Leitner and Moore, 1946; Berg *et al.*, 1947). Even in a normal state of caloric nutrition and in the absence of metabolic peculiarities, it is clear that gingivitis, night blindness, corneal vascularity, and so on are at best only suggestive of certain vitamin deficiencies. In the presence of marked caloric deficiency and tissue breakdown, all such signs and symptoms become much more difficult to interpret. As we shall see in subsequent sections of this discussion, there is reason to believe that the metabolic and morphological consequences of starvation itself may produce changes similar to some of those which have long been considered indicative of vitamin deficiency.

Vitamin A — General Considerations

Signs of vitamin A deficiency usually do not occur among individuals subjected to starvation. When the normal food supply becomes restricted, greater use is made of green plants and vegetables. Since these are high in carotene, they are likely to increase the vitamin A intake above that in the ordinary diet of the individual. During World War II the British and Russians in a typical German prisoner-of-war camp received a diet which provided about 3000 I.U. of vitamin A per day (Leyton, 1946). Although this is somewhat below the presently recommended allowance (National Research Council, 1945), it is likely to provide an adult with more than his minimum requirement. When the food shortage becomes acute, carotene may still be secured from the herbs and grasses used in such extreme cases. The Russians at Leningrad noted a number of cases of carotenemia, which were supposedly due to the ingestion of weeds and grasses during the German siege of 1941–42 (Levin, 1944). This condition has been reported as common in other parts of the world during acute shortages of food (Musselman, 1945; Grant, 1946).

Even in those regions where less stringent measures are taken to supply the food requirements, the change in the diet may be of such a character that the need for vitamin A is reduced. This is suggested by the work of Mellanby (1944), who found that when potatoes replaced the bread or oatmeal in a ration deficient in vitamin A, it was impossible to produce the incoordination which he considered to be one of the early symptoms of a vitamin A deficiency in his dogs. If a similar relationship holds true for human beings, then it would decrease still further the possibility of a vitamin A deficiency. In most of the countries of Northern Europe food shortages during both World War I and World War II produced a marked increase in the use of potatoes (Maver, 1920; Drummond, 1946a).

Unless the starvation regimen extends over a number of years, it is likely that adults will be protected to a certain extent by their body stores of vitamin A. Recent experiments with young men showed that even on a diet providing practically no vitamin A, there were no signs of a deficiency during the first 18 months (Vitamin A Subcommittee, 1945). Children may be subjected much earlier than this to rather severe, and sometimes irreparable, damage, as shown by the experience in Denmark during World War I (Bloch, 1920–21).

Vitamin A — Keratomalacia

Keratomalacia has for a long time been thought of as a manifestation of semi-starvation (Isaacs et al., 1938). In 1892 Spicer indicated that this condition occurred among nurslings in countries where the mothers practiced long religious fasts. A more specific report of such cases was made by Schiele (1906), who saw 33 cases of keratomalacia during a 3-year period in Kursk, Russia. All these cases were in infants and occurred during the fasts of the Russian lenten season. Schiele treated one of the children with cod liver oil with such remarkable success that he suggested it as a special cure for the cases of keratomalacia that had not progressed too far. A case he had treated previously was more advanced in its development and went on to blindness in spite of the cod liver oil treatment.

A similar condition was reported by Wright (1922), who saw many cases among the people in southern India, especially during famines. It occurred most frequently in children under 5 years of age but was also found in older children and even in adults, particularly those whose emaciation was aggravated by gastrointestinal disturbances. Wright also found that cod liver oil cleared up the condition if it was not too severe.

There were a few cases of keratomalacia among infants in Denmark before World War I, and with the outbreak of the war there was a marked increase in the number of cases (Bloch, 1920–21). The eyes of many children were in such an advanced stage when first examined that they went on to blindness in spite of the various treatments tried. At that time most of the butter was being exported, and the main foods used for infants were skimmed milk and vegetable margarine. The German blockade effectively stopped the export of butter. A rationing system was immediately instituted whereby each person was assured of 250 gm. of butter per week at a reasonable price. This led to an almost complete eradication of keratomalacia in 1917. At the same time, Bloch became convinced from his own work with infants that cod liver oil was a cure for the condition. The fact that both butter and cod liver oil could be used to cure keratomalacia led Bloch to conclude that this condition was due to a deficiency of the recently discovered fat-soluble vitamin which was later identified as vitamin A. This finding showed conclusively that the visual defects among the Danish children were due to a vitamin A deficiency which was in no way associated with starvation.

Vitamin A and Night Blindness

Night blindness is another condition reported as occurring extensively among the inhabitants of famine areas. Wosika (1943), in a review of the literature on night blindness during wars, stated that since the time of the Crusades this ophthalmological difficulty has been associated with practically every major military campaign. Aykroyd (1939), in discussing the famine situation in India, stated that "night blindness and xerophthalmia are invariably observed in populations suffering from famine" (see also Aykroyd and Krishnan, 1936).

From the very beginning of World War I reports appeared on the presence of night blindness in various parts of Europe. Some of the first descriptions of hunger edema from the eastern front in Poland indicated that edema and night blindness frequently appeared simultaneously (Strauss, 1915). One of the first to mention the presence of night blindness within Germany at that time was Braunschweig (1915). As a criterion for the existence of night blindness, he measured the distance over which the person could see a luminous watch in the dark. Such a qualitative test, together with subjective complaints, was used by many German clinicians in their investigations at that time. The extent of this work can be judged by the fact that Jess (1922) mentioned more than 100 papers on the occurrence of night blindness in Central Europe during World War I.

The fact that the incidence of night blindness decreased very sharply in May when fresh foods again became available led Wietfeldt (1915) to suggest that the condition was due to a vitamin deficiency. The proof for this theory was not forthcoming until 1925, when Fridericia and Holm showed that on a restricted intake of vitamin A the amount of visual purple in the eyes of rats was markedly reduced. On the basis of their data the emphasis on night blindness shifted from the field of starvation to that of vitamin A deficiency. This led to the development of various techniques for quantitatively determining the rate of regeneration of the visual purple following exposure to a bright light. It was believed that this was an exact method for measuring the rate of dark adaptation in a subject.

Many of the early reports indicated an intimate relationship between the degree of night blindness as measured by various instruments and the vitamin A status (e.g. Jeans and Zentmire, 1934). Most of this work was based primarily on the improvement produced by large doses of vitamin A in those subjects who initially had poor dark adaptation. In these studies only passing attention was given to the dietary vitamin A intake of the subjects.

Investigations of night blindness are handicapped by a number of factors, some of which are worthy of emphasis. (1) At present there is no valid quantitative means of describing various states in the development of the condition. Unfortunately, there is no clear relationship between night blindness and dark adaptation as measured by any of the presently available instruments (Wittkower et al., 1941; Youmans et al., 1944). (2) There is, in much of this work, an incomplete understanding of the mechanism and theoretical limitations involved in the adjustment of the eye to dim light. This has been emphasized by Simonson, Blankstein, and Carey (1946). (3) The incompleteness of many of the field reports on night blindness limits their value. This is especially true in regard to the restriction of food intake and the presence of emotional or psychological factors which might have considerable influence on the subject's vision in the dark.

(4) A considerable number of factors may produce night blindness. Anomalies of vision resulting from congenital, hereditary, and neurasthenic factors may also be responsible for night blindness (Duke-Elder, 1937). The importance of non-dietary factors in the production of night blindness has been stressed by Witt-kower and his associates (1941). They studied 42 British soldiers in whom night blindness was so severe that on a number of nights they had either injured themselves or damaged army vehicles. In only one of the cases were there any physical findings that might explain the impairment of night vision. A complete psychiatric examination showed that 9 of the men had been night blind from childhood, that 13 had noticed the condition at the start of the war, and that 14 had developed it during the course of the war, usually following some trying emotional experience. All but 3 of these people were psychologically abnormal. A similar suggestion was made during World War I, but no substantiating data accompanied it (Paul, 1915).

(5) There is no strict relationship between dark adaptation and vitamin A intake. This has been emphasized by some of the more recent work in this field (see Steininger and Roberts, 1939, for references). In one of the longest controlled dietary experiments so far recorded, no change was observed in the dark adaptation of 8 out of 11 subjects; the other 3 showed only slight changes in this function. These young men had been on a diet practically devoid of vitamin A for 28 months (Vitamin A Subcommitte, 1945; Marrack, 1947b). There is sufficient evidence at present to indicate that night vision is a complex phenomenon influenced not only by dietary, hereditary, congenital, anatomical, physiological, and pathological factors, but also by various psychological and neurological changes (Heinsius, 1941).

One of the interesting findings about night blindness as it occurs in starvation areas is the marked difference in its incidence in the various age groups and in the two sexes. In a study of 330 such cases Birnbacher (1928) found that the incidence was ten times greater among Viennese males aged 5 to 25 years than among females of the same age. The incidence was very low among women of childbearing age, though there were many cases among the men in the same age group. This study was made at the time of the severe economic depression following World War I, and Birnbacher implied that all his cases were suffering from starvation.

Joachimoglu and Logaras (1933) observed a similar sex difference in the incidence of night blindness among the economically lower classes in Greece. Kirwan, Sen, and Bose (1943) found that the incidence of night blindness was 11 times more prevalent among men than among women in India. The only explanation that has been offered for this difference is that men are subjected to severe glare more frequently than women (Kirwan et al., 1943). If this were the only factor involved, it would be difficult to explain the fact that the incidence of night blindness is much higher among children than among adults (Birnbacher, 1928; Joachimoglu and Logaras, 1933; Kirwan *et al.*, 1943).

At present it is impossible to distinguish adequately between the effect of a vitamin A deficiency and that of a simple caloric deficiency in producing night blindness. The reports on the occurrence of this condition during World War II have emphasized the dietary deficiency of vitamin A. The Spanish refugees interned in southern France lost considerable weight, and many of them complained of poor night vision (Zimmer *et al.*, 1944). The investigators attributed this to a vitamin deficiency without presenting any evidence. A similar interpretation was given for the presence of night blindness among the residents of the poorer section of Madrid a short while after the end of the Spanish Civil War (Robinson *et al.*, 1942a, 1942b).

A report from Paris showed that during the German occupation the level of vitamin A in the plasma decreased and with this the dark adaptation deteriorated (Chevallier, 1947). No changes were seen in the plasma vitamin A level in Liége when Belgium was occupied, nor were any signs of night blindness seen (Lambrechts *et al.*, 1945). Leyton's (1946) study of some 15,000 Russian prisoners at Tost, Germany, indicated that even though there was extensive weight loss in most of the men, there were practically no signs of night blindness; these men had a fairly reasonable intake of vitamin A.

Many of the World War II reports from the Orient indicate the presence of an appreciable number of cases of night blindness and other difficulties of vision which, very probably, were in no way related either to semi-starvation or to a vitamin A deficiency (Hibbs, 1946). Among the Americans rescued from the concentration camps in the Philippines, 16 per cent were reported to have had night blindness during their internment and about a third of them had poor vision (Butler *et al.*, 1945). Poor vision during the day was found among the British and Americans interned in a number of the camps in China (Adolph *et al.*, 1944), but no mention was made of night blindness. Most of these persons received liberal amounts of fish liver oils.

It is very likely that whenever night blindness was observed during protracted periods of food restriction, some factor other than a simple caloric deficiency was the cause. The cases that occurred may well have been due to a drastic alteration of the diet which excluded most of the vitamin A and/or carotene. In other cases the trying emotional and psychological factors associated with the disrupted life in a concentration camp may have brought out an incipient night blindness (Pillat, 1940).

Vitamin A and the Skin

Many reports from areas of endemic starvation have mentioned a "folliculosis" which has been considered one of the early signs of vitamin A deficiency (Frazier and Hu, 1931). In an extensive review of the skin changes that might possibly be related to a deficiency of vitamin A, Stannus (1945) concluded that most of the changes observed during World War II were not a specific manifestation of any one vitamin deficiency. It was suggested that such factors as exposure of the skin to cold, the use of coarse clothing, the absence of hygienic facilities, the presence of sympathetic nervous stimulations, and an unbalanced diet might be far more important than vitamin A in the development of this condition. Besançon (1945) noticed a similar skin condition among the French during World War II and attributed it to a lack of personal hygiene and laundry facilities (see Chapter 13 for a further discussion of this subject).

Famine and Thiamine Deficiency

During World War I there were apparently few people in Europe who suffered from a deficiency of thiamine. True, a few clinicians did some speculating as to whether famine edema was not in some way related to a vitamin B deficiency (see the section in this chapter on beriberi and famine edema). The only report of beriberi among Europeans during World War I was that of Hehir (1922). A contingent of British soldiers was besieged at Kut-al-Amara in Iran. These men were cut off from their supplies for about 5 months and lived primarily on white flour. Of these British soldiers 155 developed beriberi, whereas 1050 Indians in the same army who ate primarily lentils developed scurvy. A similar distinction in the deficiency diseases among the British and Indian soldiers during another siege in the Middle East was observed by Willcox (1920).

The great awareness of vitamins during World War II may account for some of the references to various manifestations of thiamine deficiency at that time. It was reported that there were cases of athiaminosis among the Russians in besieged Leningrad (1941–42). The symptoms were those of the combined sensory-motor type, especially marked in the periphery, with hyperesthesia, paresthesia, decreased reflex response, aches, decreased muscular strength, and trophic and vegetative disturbances. Many of these signs and symptoms may well be due to factors other than a deficiency of thiamine (Denny-Brown, 1947). The skin on the extremities of these individuals was very yellow, and there were large amounts of carotene in the plasma. Those who were thus affected had eaten large amounts of weeds and grasses. The Russian workers (Levin, 1944) attributed the polyneuritic symptoms to the excess carotene in the body which had escaped detoxification by the liver. The identification of this condition with a thiamine deficiency is open to considerable question, but it is still more doubtful that the symptoms were due to the excess carotene in the body. Carotenemia has never been associated with any neurological symptoms (Boeck and Yater, 1929; Curtis and Kleinschmidt, 1932; Rabinowitch, 1928, 1930). Furthermore, no such symptoms were mentioned by others who observed starving people eating grasses and leaves (Musselman, 1945).

Aside from the description of the supposed thiamine deficiency in Leningrad, most of the reports from Europe are in agreement that few, if any, cases of unmistakable vitamin deficiency occurred in association with severe caloric restriction (cf. e.g. Robinson et al., 1942a; Besançon, 1945; Lipscomb, 1945; Lamy et al., 1946a; French and Stare, 1947). The high extraction flour and the large amounts of potatoes in the diet were the primary reasons for there being so few cases of vitamin deficiency in the European theater (Drummond, 1946b).

In the Orient at this time, several reports indicated the absence of beriberi among the advanced cases of inanition in both Calcutta and Barisal (Chari, 1944; Bahattacharya and Sen, 1945). These were victims of the Bengal famine of 1943. There were, however, a number of references to beriberi among Occidental intern- ees and prisoners of war (Adolph *et al.*, 1944; Butler *et al.*, 1945; Musselman, 1945). The diagnosis of beriberi was based primarily on clinical observations. Unfortunately, most of these reports do not describe in any detail the symptoms that were used as criteria in the diagnosis. In some of the cases where thiamine was tried as a cure for this condition, it proved ineffective (Hibbs, 1946). This would suggest that some of the cases called beriberi were more probably the result of a deficiency of something besides thiamine.

Beriberi and Famine Edema

It is noteworthy that during World War II practically all the emphasis in Europe, as far as deficiency diseases are concerned, was put on starvation edema, whereas most of the disturbances associated with a restricted food intake in the Orient were designated as beriberi. The differentiation of beriberi and starvation edema may be difficult if reliance is placed only on casual clinical observations. The two conditions have many of the same signs and symptoms. For instance, in both disturbances a loss of strength is one of the earliest signs to appear, followed by "fatigue and sensations of heaviness and stiffness of the legs and perhaps inability to walk long distances" (Williams and Spies, 1939, p. 19).

Edema is a most striking item in wet beriberi and in starvation. In both cases it originates in the lower extremities and is accentuated by activity. Bed rest, even overnight, redistributes the edema fluid so that an amelioration appears to have occurred. From the published reports of conditions among prisoners of war in the Orient, it appears that there was some edema among more than half the population of 7 different camps (Butler *et al.*, 1945; Adolph *et al.*, 1944; McDaniel *et al.*, 1946). Many of these cases were presumably diagnosed as beriberi (Butler *et al.*, 1945). There are statements that the edema in these cases showed no re- sponse to thiamine but did clear up with extra protein rations (McDaniel *et al.*, 1946). This suggests that the actual incidence of athiaminosis among this group was much smaller than originally supposed.

Of the other symptoms used in the diagnosis of beriberi, insomnia, nervous- ness, dizziness, low blood pressure, absence of tendon reflexes, and low voltage in electrocardiograms are common to both athiaminosis and starvation. There are no quantitative data on the electrocardiographic changes occurring in classical beriberi; consequently, no direct comparisons with semi-starvation can be made. The similarity between the electrocardiographic changes observed in some of the Americans liberated from German prison camps at the end of World War II and the changes observed in beriberi have been emphasized by Ellis (1946). The changes in the electrocardiogram during starvation are discussed in Chapter 30.

Most clinicians report that beriberi is associated with a tachycardia, but accord- ing to Williams and Spies (1939, p. 20) this develops only late in the course of the disease. There is no information available on the heart rate in the early stages of beriberi. Semi-starvation is associated with a decrease in heart rate which may be very pronounced when the condition becomes severe (see Chapter 28). In both beriberi and semi-starvation the changes in heart rate are progressive. This, together with the great variation in the heart rate of normal individuals, makes it difficult to differentiate between the two conditions in their early stages.

A similar situation exists as far as the size of the heart is concerned. The Minnesota Experiment has shown that in semi-starvation the heart size as measured by teleroentgenkymograms decreases markedly. The later stages of beriberi are associated with an enlargement of the heart, which occurs more frequently in the wet type of the disease than in the dry (see Williams and Spies, 1939, for references). Nothing is known about the heart size in the earlier stages of beriberi. It is doubtful whether a change from normal in either the rate or the size of the heart will permit a differential diagnosis of the two conditions. Hyperesthesia, paresthesia, and pains in the lower extremities differ quantitatively, and perhaps also qualitatively, in beriberi and semi-starvation.

More elaborate tests as a means of differentiating these two conditions hold some promise. As far as the basal metabolic rate is concerned, no reports are available as to its value in classical beriberi, but from the fact that it has not been mentioned, we may infer that there is little obvious change. In starvation there is a marked decrease in the basal metabolic rate.

A number of reports in the literature indicate that one prominent difference between beriberi and semi-starvation is the response of the pulse rate to exercise. Keefer (1930) stated, and Shimazono has been quoted (Aalsmeer and Wenckebach, 1928–29) as saying, that the pulse rate in beriberi following exercise increases to an extent much greater than would normally be expected. Each of these reports is apparently based only on clinical impressions. There are no reports on the pulse rates following exercise in cases of semi-starvation other than that from the Minnesota Experiment (see Chapter 34). In this experiment starvation produced no change in the pulse rate response to a standard work test.

To complete this comparison, it would be advisable to consider ship beriberi, which presents many symptoms similar to the beriberi of the Orient and to epidemic dropsy. Since ship beriberi practically disappeared before the advent of the pure vitamins, it is impossible now to say whether this disease was due to a deficiency of thiamine or of vitamin C or to a combination of the two (Williams and Spies, 1939; Holst, 1907; Rolt, 1930; Bassett-Smith, 1911–12). The symptoms of epidemic dropsy ("Bengal disease") have certain features in common with beriberi (Ray, 1927), but a great deal of emphasis has recently been put on the influence of the oil from *Argemone mexicanus* as a cause of the disease (Lal *et al.*, 1941). This seed is a contaminant of commercial mustard seeds marketed in India. The disease has been reported only from India, where it occurs when economic conditions become severe. The evidence that a toxic substance is the cause of epidemic dropsy rules out this disease as a condition due to starvation per se.

An attempt has been made here to point out the difficulties inherent in distinguishing beriberi from semi-starvation, without entering into a discussion of the role played by starvation in modifying and controlling the symptoms of beriberi. These difficulties were apparent shortly after the end of World War I, when there was some confusion about the relationship between hunger edema and beriberi. The extreme views on this matter were held by Schiff (1917b), Madsen

(1918), Maase and Zondek (1920), and Zambrzycki (1920), who maintained that the two conditions were the same, whereas Schittenhelm and Schlecht (1918, 1919e) stated that there was no relationship between them. That there are important differences between the two conditions is shown by the fact that the injection of thiamine into edema cases was without effect (Besançon, 1945; Dumont and Lambrechts, 1945).

Although it should not be difficult under present conditions to distinguish advanced beriberi from severe famine edema, it may be much more difficult to differentiate the two conditions in their early stages. The situation is probably not as hopeless as Whitfield (1947) implied when, as a result of his experience among the prisoners of war released from Japan, he stated, "I never felt capable of differentiating between the oedema of beriberi and that of hypoproteinemia (if there are two kinds). . . ." Beriberi, when it occurs in adults with "an enlarged heart, peripheral neuritis, edema, and tenderness and atrophy of the muscles, is easily diagnosed. The mild case is much more common than the classical one, but the manifestations vary so greatly that the disease may be recognized only with difficulty" (Williams and Spies, 1939, pp. 22–23). During World War II a large number of cases of mild beriberi were probably incorrectly diagnosed. It is likely that many of these cases were famine edema. This interpretation is strengthened by the suggestion that the geographical occurrence of the condition is important in diagnosing beriberi, especially if only a clinical examination is made (Vedder, 1944; Butler *et al.*, 1945; McDaniel *et al.*, 1946).

Famine and Pellagra

The occurrence of pellagra was reported from a number of areas during World War I. A considerable number of the Turkish prisoners who were transferred from Syria to Egypt showed dermatological signs of pellagra (Bigland, 1920a; Enright, 1920a). In many of these cases the disease started before the men were captured. Pellagra became so widespread among these prisoners while they were in Egypt that a medical commission was appointed to investigate the condition (Committee of Inquiry regarding . . . Pellagra, 1919). A similar condition developed among the German prisoners interned in Egypt (Bigland, 1920b; Enright, 1920b). At that time it was not firmly established that pellagra was due to a vitamin deficiency, so considerable emphasis in these reports was given to the absence of significant bacteriological findings. The similarity of the disturbance to classical pellagra is indicated by the fact that all the investigators implied that the diets served the prisoners contained large amounts of corn, but no mention was made of the actual foods served.

Only a few reports on the occurrence of pellagra were made during World War II. Vaguely described skin and tongue changes among the English soldiers who were interned in Germany were attributed to a deficiency of niacin (Edge, 1945). A similar report appeared concerning German soldiers interned in prisoner-of-war camps in the Middle East (Spillane and Scott, 1945). Besançon (1945) stated that mental, digestive, and dermatological signs of pellagra appeared in France late in 1940. Ordinarily pellagra is not seen in France; it is difficult to understand how pellagra could have developed there since, according to the same author, the people lived primarily on whole-wheat bread. This food should supply a fairly large amount of niacin (U.S. Bureau of Human Nutrition, 1945). In a study of 40 cases of uncomplicated undernutrition among French prisoners of war released from Germany, Lamy, Lamotte, and Lamotte-Barrillon (1946a) specifically noted that all signs of pellagra were absent.

Pellagra was stated to have occurred among the inhabitants of Leningrad during the early part of 1942 and to have increased in extent throughout the year (Khvilitskaia, 1943). Even though these cases showed the classical signs of pellagra, semi-starvation was present in less than half of them. Here also it is impossible to decide whether the condition which developed was due to the use of a diet which was markedly different from that on which the people had been living previously or whether only a reduction in the amount of food occurred.

Apparently the only report of a widespread occurrence of pellagra was from the Philippine Islands (Lewis and Musselman, 1946); 500 American soldiers captured at Bataan or Corregidor were reported to have developed most of the symptoms associated with this deficiency disease. The dermatitis used as a criterion in the diagnosis developed about 5 months after the Americans surrendered to the Japanese. Before this they had been on restricted rations for 3 or more months. These men were living primarily on polished rice, vegetables, flour, and oil. Not only was the caloric content of the diet low (about 1500 Cal. per day), but it was grossly deficient in niacin. The latter factor, and not the insufficient food intake, was probably responsible for the development of the pellagra. Even this seemingly authentic occurrence of pellagra is open to some question since the condition was not observed among the Philippine scouts who were captured at the same time as the Americans (Concepción, 1948); these soldiers presumably received the same food as the Americans.

Skin changes similar to those in pellagra were reported among the British a few months after their capture at Hong Kong (Clarke and Sneddon, 1946). Here again it is difficult to arrive at any decision on the validity of the implied vitamin deficiency.

There is at present no indication that the development of pellagra is accelerated by a restriction of food intake; in fact, most of the evidence from Europe during World War II indicates that almost all signs of pellagra were absent in those areas where starvation was acute (Lipscomb, 1945; Robinson et al., 1942a).

There have been many reports on the occurrence of a nonpathogenic type of diarrhea among the inmates of some of the concentration camps during World War II (see Chapter 26). Various remedies have been suggested for this condition. In one of the concentration camps in China, doses of niacin starting at 150 mg. per day, followed by maintenance doses of 50 mg., produced striking cures of the diarrhea. This response occurred in spite of the absence of any other signs of a niacin deficiency (Adolph et al., 1944). Aykroyd and Gopalan (1945) reported similar cures among emaciated adults in India, some of whom had edema that was apparently of nutritional origin. In the European theater the administration of large amounts of niacin (300 mg. per day) had no influence either on the diarrhea or on the pigmentation apparent in some of the internees (Lipscomb, 1945).

Burning Feet Syndrome

Burning feet and severe pains in the legs have been reported as occurring among individuals in various famine areas. From time to time suggestions have been made that the diet was the primary cause of the syndrome, which has been recognized in certain parts of the Orient for more than a century, most frequently in prisons. A few reports on the occurrence of the same, or a very similar, condition in France, Belgium, and Italy also appeared in the early part of the 19th century (Denny-Brown, 1947).

Interest in this condition was renewed at the end of World War II when it was discovered among fairly large numbers of American and British soldiers interned by the Japanese. Cruickshank (1946) observed 500 cases among the British at a camp outside of Singapore. This condition did not develop until nearly half a year after internment. It usually started after periods of standing or extended marches. The pain came on at night, especially when the men went to bed, but as the syndrome progressed, it became continuous throughout both day and night. Exercise, cold foot baths, and massage offered only moderate and, at the most, temporary relief. From the report it appears that large losses in body weight did not occur until the pain became so severe that it interfered markedly with sleep. Simultaneously with the pain, there appeared in a large percentage of the cases stomatitis, glossitis, scrotitis, and defective vision. These concomitant symptoms were also observed by Simpson (1946) among the British prisoners of war on the Malay Peninsula who had severe pains in their feet. In both instances the feet did not improve permanently until there was a sizable increase in caloric intake.

Although both authors believed that a deficiency of the vitamin B complex was responsible for the condition, their evidence is not convincing. Furthermore, Harrison's report (1946) of a similar condition among the British internees in Hong Kong and Hibbs' report (1946) of the same thing among the Americans from Corregidor and Bataan indicate that additions of yeast, thiamine, riboflavin, and nicotinic acid were of no avail in curing the disturbance. It has been suggested that pantothenic acid in doses of 30–40 mg. per day would cure this condition (Gopalan, 1946), but no other statements on the efficacy of this treatment are available.

A milder form of aches and pains in both hands and feet was mentioned as occurring among the inhabitants of Madrid during the Spanish Civil War (Peraita, 1946) and among the Dutch in western Holland during the later part of the German occupation (Burger et al., 1945). In the latter case the observers were impressed by the absence of all signs of vitamin deficiency. It is hard to determine whether this condition is a manifestation of semi-starvation, a sign of vitamin deficiency, or the result of poor footwear — as has been suggested by Hibbs (1946). Denny-Brown (1947) gives a more complete discussion of this condition. Its neurological manifestations are considered in Chapter 33.

Famine and Scurvy

Scurvy has long been associated with famine and starvation. Magnus, writing in 1555, noted the epidemic occurrence of scurvy in the Scandinavian countries, especially in times of food shortage. Since then many reports of a similar nature have appeared from all parts of the world (Curran, 1847; Hirsch, 1885; Hess, 1920). In most of the areas where scurvy has developed there has been a marked change in the type of food available to the inhabitants. Fresh fruits, vegetables, and meat have been replaced by other foods or dehydrated products with a consequent sharp reduction in the vitamin C intake. The marked change in the character of the diet, and not the reduction in calories, has been the cause of the scurvy observed in most famine areas.

The change in the character of the diet during periods of famine was emphasized by Curran (1847) in his discussion of the Irish famine of 1846. As a result of the failure of the potato crop, many people developed scurvy. The shortage of potatoes produced a change from a diet consisting mainly of potatoes to one "of Indian meal, or more frequently bread and tea" (Curran, 1847, p. 129). Under these conditions, the thiamine intake was maintained at a fairly high level by the coarse flour used in the bread, but the ascorbic acid was reduced to such a point that scurvy developed within a few months of the crop failure (as observed by Shapter, 1847).

Cases of scurvy following crop failures and during sieges and other military operations have been reviewed by Curran (1847), Hirsch (1885), and Hess (1920). Under such conditions the incidence of scurvy is always highest among those individuals who are confined to institutions or who, because of physical limitations, are unable to supplement their official food allotment (Turnbull, 1848; Delpech, 1871). These persons also show the most severe signs of starvation (Dols, 1946).

During World War I many cases of scurvy were reported from the eastern part of Europe (Hess, 1920). The largest number of these cases occurred among the soldiers in Russia and Austria from 1916 to about 1920 (Hoerschelmann, 1917). There were a few cases of scurvy among the Indian soldiers in the British Army engaged in the Middle East campaigns (Hehir, 1922; Willcox, 1020).

In World War II no cases of scurvy were reported from Europe outside of Russia (Besançon, 1945; Lipscomb, 1945; Lambrechts et al., 1945; Collins, 1947). Scurvy appeared among the inhabitants of Leningrad in the latter part of 1941 (Chernorutskii, 1944; Brožek, Wells, and Keys, 1946), and the number of cases increased considerably during the ensuing winter and spring. The disease was brought under control and finally eliminated by the administration of an extract made from pine needles which was rich in vitamin C.

Measures such as this and other prophylactic techniques undoubtedly went far to hold scurvy in abeyance during the period 1940–45. Practically all the packaged rations supplied to the American, Canadian, and British troops during World War II were fortified with vitamin C, as were many of the German rations also (Brenner, 1947). The medical corps of all the armies were well equipped with vitamin pills with which to treat any suspected deficiency disorders. Urgent requests for large shipments of crystalline vitamins came from the Russians in the beginning of lend-lease activities (1942). The civilians in Central Europe at that time used potatoes so extensively in their diet that an adequate supply of vitamin C was assured. Spicknall (1943) states that potatoes were served at practically every meal in Germany during the early part of the war and that the German government supplied miners, infants, mothers, and women past the sixth month of pregnancy with vitamin C tablets.

Famine and Rickets

It might logically be assumed that the incidence of rickets would increase during famines. At such times fats are one of the first food substances to be severely restricted. This is especially true of animal fats, such as those present in butter, milk, eggs, and meat. In recent times the fish liver oils have also been in short supply at such times. Since these are the principal dietary sources of vitamin D, famine should a priori be associated with a marked increase in the number of cases of rickets.

Prior to 1920 the situation was complicated by the fact that the incidence of rickets was so high in normal times that it was difficult to determine whether famine really produced any change in the number of cases. Hess (1929), without giving any statistics, stated that toward the end of World War I there was a tremendous increase in the incidence of rickets, hunger osteomalacia, and war osteopathy throughout Germany, Austria, and Poland. A similar suggestion was made by Simon (1919). These statements appear to ignore the high incidence of rickets in those areas before the start of the war. According to Gribbon and Paton (1921), when the normal incidence of rickets is taken into account "the evidence . . . does not indicate that there has been the marked increase in the disease since the war which is commonly stated to have occurred. It would rather seem that attention being directed to the condition has led to its diagnosis in children in whom it would not have attracted attention in former times." The true situation probably lies somewhere between these extremes, but there are no data available to aid in determining the facts.

There is some evidence to indicate that malnutrition exerts a sparing action on the skeleton with a concomitant decrease in the incidence of rickets. This is based to some extent on the statements of pediatricians who claim that children growing at a slow rate show no signs of this disease (Simon, 1919; Hess, 1929; Eliot and Park, 1945). Simon (1919) attempted to explain this by suggesting that such children were so weak they had to be carried by their parents, which took the strain off their bones and thus retarded or overcame the development of rickets.

The work of McCollum et al. (1922) on starvation and rickets has been cited as proof of the fact that starvation reduces the requirement for vitamin D. This work indicated that when rachitic rats were starved for 3 to 5 days the bones showed evidence of healing. Other factors than vitamin D may have been important in these experiments. The liberation of phosphorus from the body tissue during starvation may have been responsible for the initiation of the cure; McCollum et al. called attention to this point. The rachitogenic ration in these experiments was low in phosphorus and high in calcium. Under such conditions an improvement in the calcium-phosphorus ratio of the rat's diet leads to normal mineral deposition in the bones even though there is an inadequate supply of vitamin D.

Starvation in children is first manifested by a reduction in the rate of weight increase. If the period of food restriction is not too prolonged, the growth of the skeleton may continue at a normal rate. However, in prolonged starvation the size of the skeleton is smaller than normal (Valaoras, 1946; see also Chapter 45). No data are available on the mineral content of the bones of emaciated children except of those less than 4 months old (Korenshevsky, 1922). This makes it difficult to determine whether the slight retardation in height noted in starvation areas is associated with a change in the bones such as occurs in hunger osteomalacia among older people.

As far as human beings are concerned there is no valid evidence to indicate that the requirement for vitamin D is reduced during periods of starvation. It is likely that under such conditions the reserve of vitamin D in the body is able to maintain a normal rate of skeletal growth for a time. The length of this period would depend upon the magnitude of the vitamin D reserve, the amount of calcium in the diet, and the exposure of the children to sunlight. The latter factor has been prominently mentioned in some of the reports from civilian concentration camps during World War II. The Occidentals interned in a number of the Japanese camps were heavily tanned, and no signs of vitamin D deficiency were seen among them (Adolph et al., 1944; Butler et al., 1945). This, together with the fact that the children received whatever calcium was available in the form of milk (Salmon, 1946) and powdered egg shells (Adolph, 1944; Kramer, 1944), may have contributed toward the absence of rickets.

The reports from Europe confirm the absence of any increase in the cases of rickets during World War II (Robinson et al., 1942b; French and Stare, 1947). The only exception to this is the report of a large number of cases among the Spanish Loyalists interned in southern France (Zimmer et al., 1944). In most of these children, rickets was not severe but the proportion of scoliosis was large. Unfortunately, this report gives no indication of the degree of starvation among the rachitic children.

The behavior of the skeleton in animals during starvation is different from that in children (see Chapter 45). Experiments with growing animals indicate that the bony structure continues to grow at a nearly normal rate even though there is no appreciable change in body weight (Aron, 1911; Smith, 1931). When the food of pups was restricted to such an extent that they showed no increase in body weight, their skeletons were similar in many respects to those of litter mate controls fed ad libitum (Aron, 1911). At the end of one experiment the restricted dog weighed only one half as much as the fed dog; at that time there was only a slight difference in the two skeletons on the basis of length as well as weight (see Table 230). The bones in the two dogs contained practically the same amount of mineral. The most outstanding difference was in the fat content.

A somewhat similar experiment was performed with cattle by Trowbridge, Moulton, and Haigh (1918). Two yearling steers were fed at such a level that they lost ½ pound a day, two others were maintained at their original weights, while a fifth was liberally fed. At the end of the experiment (6 months for some and 12 months for the other animals), the weights of the skeletons varied surprisingly little. Excepting steer 592, which lost 28.5 per cent of its original body weight, the percentage of fat and water in the skeleton varied little among the

TABLE 230

INFLUENCE OF SEMI-STARVATION ON THE SKELETON. The dogs were litter mates; dog V was fed ad libitum whereas dog VI was given only enough food to maintain his original weight. Duration of the experiment was 206 days. (Aron, 1911.)

	Dog V	Dog VI
Weight (kg.)	5.9	2.8
Nose to *os occipitale* (cm.)	15.5	15.0
Nose to tip of tail (cm.)	53.0	49.0
Length of foreleg (cm.)	35.0	34.0
Length of hind leg (cm.)	38.5	37.5
Circumference of head (cm.)	26.5	25.5
Circumference of chest (cm.)	37.5	30.0
Skeleton,° fresh (gm.)	127.0	119.3
Skeleton,° dry (gm.)	72.8	47.7
Ash in bones (gm.)	131.0	126.3
Protein in bones (gm.)	138.0	110.0
Fat in bones (gm.)	15.7	1.1
Water in bones (gm.)	54.3	71.6

° The same 9 representative bones were used in both cases.

TABLE 231

COMPOSITION OF THE SKELETON OF BEEF CATTLE under various nutritional conditions (Trowbridge *et al.*, 1918).

	Maintenance		Submaintenance		Super-maintenance
	Steer No.		Steer No.		Steer No.
	597	595	591	592	593
Body weight					
At start of experiment (kg.)	336	276	260	301	317
After 6½ months (kg.)	335	292	218	261	347
After 12 months (kg.)		266		213	
Skeleton weight	51.5	45.2	40.2	49.9	51.3
Percentage water	34.8	35.9	37.0	52.6	35.4
Percentage fat	19.3	16.4	17.6	2.9	17.3
Percentage protein	19.4	20.1	19.5	19.1	19.7
Percentage ash	24.6	26.5	23.8	22.8	25.3
Percentage ash (dry basis)	35.9	32.5	35.4	31.9	33.2
Skeletal ash					
As percentage of final body weight	3.8	4.5	4.4	5.3	3.7
As percentage of original body weight	3.8	4.3	3.7	3.8	4.1
Skeleton as percentage of body weight	15.3	17.0	18.5	23.4	14.8

animals (see Table 231). Although steer 591 lost 17 per cent of its original weight, the fat and moisture content was essentially the same in its bones as in those of the full-fed animal. The skeleton of a control animal killed at the start of the experiment weighed 59 kg., which is somewhat above that of the other animals. Differences in the initial weights of the steers make it difficult to determine the changes that may have occurred in the skeletons. For this purpose it would seem appropriate to express the skeleton as a percentage of the body weight. When this was done for the skeleton and the skeletal ash, the animals that lost the most weight showed the highest values. Even these percentages were reduced to the level of the others when the values were referred to the original body weights (3.7 and 3.8 for steers 591 and 592, respectively).

Similar experiments in which the supply of vitamin D and bone-forming mineral were controlled have not been reported. For this reason it is impossible to determine the influence of starvation on the requirement for these factors.

Famine and Other Bone Disorders

Besides rickets in children, various disturbances of the skeleton in adults have been reported as occurring in starvation areas. Hunger or famine osteomalacia, hunger osteoporosis, and hunger osteopathy have all been used in describing the condition. A complete discussion of the differences between these abnormalities is beyond the scope of this review. The term *osteopathy* has been used more frequently in recent reports in order to avoid any implication as to the etiology of the condition. The distinction between osteomalacia and osteoporosis can be made primarily on the basis of biochemical studies (Bauer, 1944). In osteomalacia the serum calcium is normal or low, the level of phosphorus low, and the phosphatase elevated, whereas no such changes are seen in osteoporosis. There are only a few reports of analyses for these substances in the serum of starved individuals, and they indicate a normal condition (Gounelle *et al.*, 1941; Walters *et al.*, 1947b; van Buchem, 1948). In none of the patients so studied was there any indication of bone changes.

It has been suggested that the principal disturbance in osteomalacia is the inadequate deposition of mineral by the osteoblasts (Bauer, 1944). This is thought to be due to an insufficient amount of vitamin D. Osteoporosis, on the other hand, appears to result from the overactivity of the osteoclasts in resorbing bone mineral. The therapy for osteoporosis is still complex, and the results of any measures are none too conclusive (Bauer, 1944).

On the basis of this discussion, it would appear that many of the skeletal disturbances observed during periods of famine were due to osteomalacia (Dalyell and Chick, 1921; Hume and Nirenstein, 1921; Crawford and Cuthbertson, 1934). It is still difficult to determine whether this type of osteomalacia is the same as that occurring in pregnancy. Both men and women were susceptible to the osteomalacia that was common in Germany and Austria shortly after World War I. At that time people of middle age and beyond were affected to the greatest extent, while there was no increase in the incidence of pregnancy osteomalacia (Dalyell and Chick, 1921; see also Chapter 12).

Although there were a number of reports on the presence of osteopathies and spontaneous fractures among adults in various countries affected by the war (Hibbs, 1947a; van Buchem, 1948), there is not a great deal of evidence to show that these were directly attributable to starvation (see Chapter 12 for references). In fact, the evidence from China indicates that osteomalacia may occur among women who appear to be well nourished (Maxwell, 1925). The same problems arise here as in the case of rickets when one attempts to determine whether the adult osteopathies reported during famine periods are in excess of those occurring normally.

The results of the Minnesota Experiment confirm the fact that when normal adults lose 25 per cent of their original weight on a diet similar to that used by the people of North Central Europe, no bone disturbances are likely to occur during the first 6 months. The average calcium intake from the 3 menus used was 0.79 gm. per day. This was combined with an average daily phosphorus intake of 1.24 gm. Not only were the calcium and phosphorus contents of these diets quite high, but for the last 2 months of the starvation period most of the men spent a considerable amount of time sun-bathing. The resulting tan covered a large portion of the body. The combination of the high mineral intake and the exposure to sunlight may explain why there were no indications of decalcification of the bones on radiographic examination (see Chapter 12 for further details).

The problem of the influence of starvation on the skeleton of adults is rather complex and involves a number of factors. The more important of these are: (1) The question as to whether adults require any vitamin D. No specific answer to this question is available. The work in this field led Stearns (1943) to conclude that the "attempts to determine the vitamin D need of 'non-encumbered' adults have as yet been unsuccessful. It seems established, however, that vitamin D does not decrease the minimum requirement for calcium and phosphorus, and that the average adult is more likely to need additional mineral than vitamin D." (2) The supply of calcium and phosphorus in the diet. (3) The amount of fat in the diet which may influence the absorption of vitamin D as well as the minerals. (4) The changes in the physical strains and stresses on the bones resulting from the decrease in weight and occasionally from the development of a new gait. (5) The influence of inactivity upon the bone structure. This is believed to be of minor importance in hospital patients. (6) The influence of starvation upon the crystal structure of the bones.

Vitamin A — Minnesota Experiment

No attempt was made at any time to determine the rate of dark adaptation of the subjects. No signs indicative of night blindness were reported during the experiment. Each man kept a careful record of his complaints, but difficulty in seeing in the dark was never mentioned. On many occasions the subjects walked around the sleeping quarters at night when the illumination was very low. This undoubtedly offered ample opportunity for the discovery of night blindness had it been present.

The average daily intake of vitamin A during the semi-starvation period was 1600 I.U. This level of intake, by itself, might not be low enough to produce any abnormalities of vision — as shown by the work of Brenner and Roberts (1943) and the Sheffield Experiment in England (Vitamin A Subcommittee, 1945). In both these experiments very low intakes of vitamin A for periods of one to two years produced little change in the eye, even when measured with the best available instruments.

Plasma samples for vitamin A analysis were secured at the time of the blood volume determinations. They were kept frozen at —20° C. until used. These analyses were deemed necessary when the "nutmeg grater" appearance of the skin of many of the subjects at the end of the starvation period suggested a deficiency of this vitamin.

Blood samples from both the control period and the end of the starvation period were used in these studies. Samples from 6 subjects were analyzed for carotene and vitamin A by Dr. J. B. Wilkie of the Food and Drug Administration. Carotene was determined photocolorimetrically on the petroleum ether eluate secured from the material absorbed on aluminum oxide, while the vitamin A values were obtained from the unchromatographed samples by means of the antimony trichloride procedure. The latter results were in good agreement with those secured for vitamin A from the absorption values. Similar samples from 12 subjects were analyzed by Dr. Otto A. Bessey for vitamin A and carotene according to the micromethod developed by Bessey et al. (1946). In this procedure the vitamin A is destroyed by irradiation in order to get a better estimate of the true vitamin A content.

The results of both sets of analyses are given in Table 232. The two methods gave concordant results indicating the absence of any significant change in either

TABLE 232

PLASMA VITAMIN A AND CAROTENE LEVELS in the Minnesota Experiment. The first 6 samples were analyzed by Dr. J. B. Wilkie, the last 12 by Dr. Otto A. Bessey. All values are expressed as I.U. per 100 cc. of plasma. The conversion factors used are: 1 microgram of vitamin A = 3.5 I.U. and 1 microgram of B carotene = 1.67 I.U.

Subject	Vitamin A		Carotene		Total	
	C	S24	C	S24	C	S24
22	160	141	15	25	175	166
26	218	285	20	20	238	305
102	116	125	19	7	135	132
104	88	27	12	16	100	43
108	39	100	14	10	53	110
122	338	169	20	24	358	193
M	159.8	141.2	16.7	17.0	176.5	158.2
SD	97.4	77.8	3.0	6.7	99.6	80.7
1	126	178	14	72	140	250
9	206	221	112	82	218	303
12	171	175	28	50	199	225
104	136	206	20	60	156	266
109	161	154	14	33	175	187
111	206	192	25	40	231	232
119	168	140	15	85	183	225
120	231	175	62	107	293	282
123	171	178	15	60	186	238
127	161	182	17	60	178	242
129	189	217	80	38	269	255
130	136	196	23	100	159	296
M	171.8	184.5	35.4	65.6	198.9	250.1
SD	30.5	22.8	30.6	23.0	44.3	31.5

plasma vitamin A or carotene during the starvation period. Although the mean values in some cases appear to be slightly different when the control values are compared with those at S24, the magnitude of the standard deviations reduces the significance of any apparent change. Throughout the experiment the subjects had a lower plasma carotene level than that observed by other workers (Bessey et al., 1946), but the meaning of this difference is unknown.

The starvation diets provided an average of 1600 I.U. of vitamin A per day with practically all of this in the form of carotene. These diets were very low in fat, the average intake being about 30 gm. per day. Under such circumstances the absorption of vitamin A and carotene is at its poorest (Wilson et al., 1937; Eekelen and Pannevis, 1938; Virtanen and Kreula, 1941). It is impossible to determine whether the plasma levels were maintained by means of an efficient absorptive mechanism or at the expense of the liver stores. Liver biopsy would give some indication of the change in concentration of vitamin A in that organ, but this, by itself, would be of only minor consequence since there is substantial evidence that the size of the liver decreases in starvation (see Chapter 9). The reserves of vitamin A in the liver are sufficiently large to maintain the plasma level for a period considerably longer than that of the Minnesota Experiment (Vitamin A Subcommittee, 1945). These results do indicate, however, that the skin changes observed in the subjects at S24 and originally referred to a deficiency of vitamin A were undoubtedly due to some other factor or factors.

A similar conclusion was reached by Krause and Pierce (1947), who found no relationship between the serum level of vitmain A and the presence of keratosis in a large number of school children. For some unknown reason the serum carotene values were lower in the children who had keratosis than in those who did not. In a smaller group normal vitamin A levels in cases of keratosis were also observed by Carleton and Steven (1943), and Leitner and Moore (1946) found no relationship between the plasma level of vitamin A and the presence of skin disturbances.

Results similar to these have been reported from Belgium. There the plasma levels of vitamin A and carotene were normal during the period of severest food restriction following the German occupation in 1941 (Lambrechts et al., 1945). The manual workers, medical students, and tuberculous and diabetic patients examined in that study were within the limits observed for similar groups before the war. All these people had lost some weight; in certain groups the weight losses ranged from 5 to 15 kg.

The only report that does not agree with the above is from the Netherlands. Blood samples were secured from some of the people admitted to one of the Amsterdam hospitals shortly after the liberation in 1945 (Hoogland, 1947). The admissions were ostensibly because of famine edema. The level of vitamin A in the plasma was in general very low, and in 9 of the 34 patients it was reported as being zero. The carotene level on admission in practically all patients was near the normal level. It is possible that some intercurrent disease which was not mentioned in the report may have been responsible for the low vitamin A blood levels.

Thiamine in Starvation — Minnesota Experiment

No definite information is available about the diets consumed by the subjects just before the start of the Minnesota Experiment. It can be assumed that the food intake was fairly good, comparable to that of persons in the middle economic bracket in the United States. During the 3 months of standardization in the Laboratory the only source of vitamins was the diet, which supplied daily between 1 and 2 mg. of thiamine per man. This diet was slightly better than that of the average person in the United States, being more nearly comparable in composition to that served at the established army bases (for typical diets served during this period see the Appendix Tables). It is therefore likely that at the start of the semi-starvation regimen the Minnesota subjects were slightly above average in their body stores of all vitamins.

Three-day composite urine samples were collected at S12 and again at S24. These were representative 72-hour samples which were preserved with glacial acetic acid (enough to bring the final pH to about 4) and 5 ml. toluol. All samples were stored in the refrigerator as soon as the collections were completed. They were analyzed for thiamine by a modification of the thiochrome procedure (Mickelsen, Condiff, and Keys, 1945) and for pyramin by a modification of the fermentation method used by Schultz, Atkin, and Frey (1942) for the determination of thiamine.

The average daily urinary thiamine excretions for the 34 men after 12 and 24 weeks of semi-starvation were 121 and 132 micrograms, respectively. On both these occasions the average pyramin excretion, expressed as 2-methyl-4-amino-5-ethoxymethyl pyrimidine hydrochloride, was 135 micrograms per day. There was considerable variation in the thiamine excretion values, with a range from 42 to 273 micrograms at S12 and from 41 to 305 at S24. The pyramin excretion was much more constant on both occasions, the range being from 112 to 168 at S12 and from 83 to 200 at S24. Similar individual variability in the thiamine excretion values is characteristic of normal young men on an ordinary diet and maintaining normal activity (Mickelsen et al., 1947).

The urinary thiamine excretion during the starvation phase of the Minnesota Experiment was slightly higher than that predicted from normal intake-excretion relationships. The average thiamine content of the menus used during this period was 1.3 mg. per day. Other work has shown that for such an intake an average daily excretion of 107 micrograms of thiamine would be expected (Mickelsen et al., 1947). The difference between the observed and the theoretical values is statistically nonsignificant, as shown by the t-test, because of the great variation in the thiamine excretion values.

Although the preceding differences are not statistically significant, they may be of some theoretical interest. The absence of any signs or symptoms of vitamin deficiencies during periods of starvation has, by implication, been attributed to the availability of the vitamins released as a result of the breakdown of body tissue. All the men in the Minnesota Experiment lost a considerable amount of active body tissue. From the amount of muscle tissue lost the amount of thiamine that might have been made available can be calculated with a fair degree of certainty.

The average loss of active body tissue for the 32 subjects during the starvation period was 10.7 kg. (see Table 152). This tissue originally contained about 5.4 mg. of thiamine, on the basis of the findings of Ferrebee et al. (1942) that normal human muscle contains 0.5 mg. of thiamine per kg. The thiamine contributed by the liver tissue during the same period amounted to 0.6 mg. This calculation was based on the assumption that in starvation the change in size of the liver approximates, on a percentage basis, that of the body as a whole (for references, see Chapter 9). According to Mitchell et al. (1945) the liver amounts to 3.4 per cent of the body weight in normal young men, and for the average subject in the Minnesota Experiment the liver weighed 2.36 kg. at the end of the control period. Since the body weight decreased by 24 per cent, the loss of liver tissue should have approximated 0.57 kg. by the twenty-fourth week of starvation. Ferrebee et al. (1942) reported that normal liver tissue contains 1.0 mg. of thiamine per kg. Over the 6-month starvation period a total of 6.0 mg., or about 37 micrograms, of thiamine per day could have been mobilized from the body tissue on the basis of these calculations. This figure is about twice the difference between the observed and the theoretical thiamine excretion values.

At present it is impossible to indicate how much, if any, of the thiamine "liberated" from the tissues as a result of semi-starvation may appear in the urine. No studies of a similar type have ever been made either with animals or with human beings. The closest approaches to this problem are a few reports on vitamin excretion during the complete absence of food. Under these circumstances the urinary excretion of thiamine decreased to zero within a few days (Caster et al., 1945; Perlzweig et al., 1944; Wollenberger and Linton, 1947). Even though these experiments were accompanied by a loss of body tissue, none of the liberated thiamine appeared as such in the urine.

There is some indication that the excretion of pyramin is more intimately related to the metabolism of thiamine than is the excretion of thiamine itself (Mickelsen et al., 1947). With increasing thiamine intakes the urinary thiamine excretion is linearly related to the intake even at elevated levels, whereas the pyramin excretion levels off at thiamine intakes of between 3 and 5 mg. per day.

If it could be assumed that the utilization of thiamine is directly related to physical activity (but see Wang and Yudkin, 1940), then the decreased activity of the subjects during the starvation period should be reflected in a lower than normal excretion of pyramin. Such was actually the case. The observed average pyramin excretion both at S12 and S24 was 135 micrograms per day. A comparable group of young men on a normal diet would show a pyramin excretion of 195 micrograms for the same thiamine intake. The differences between this theoretical excretion value and those observed at both S12 and S24 are significant beyond the 0.01 level as shown by the t-test.

The level of pyruvic acid in blood secured from an individual at least 12 hours after the last meal has been advocated as a measure of the thiamine status of the body (Platt and Lu, 1939). More recently the ratio of lactate to pyruvate in such a blood sample has been suggested as a better measure of the same thing (Stotz and Bessey, 1942).

In the Minnesota Experiment fasting blood samples were secured from the subjects 5 minutes after a short walk. The pyruvate level in these samples was determined by the method of Lu (1939a and 1939b) as modified by Friedemann and Haugen (1942, 1943), while the lactate was estimated by the method of Friedemann, Cotonio, and Shaffer as modified by Edwards (1938). The level of pyruvate and lactate in the blood of normal young men after this mild exercise has never shown any increase over the resting values. The pyruvate concentration in the blood of the starvation subjects increased progressively from the control period to S24 (see Table 233). The average value found at S24 (1.4 mg. per 100 cc.) would be considered by some investigators as an indication of a thiamine deficiency (Platt and Lu, 1939).

TABLE 233

Level of Pyruvate and Lactate and Lactate/Pyruvate Ratio in the Blood. All blood samples were taken from the antecubital vein 5 minutes after the subject completed a 20-minute walk on a motor-driven treadmill moving at a rate of 3.5 miles per hour and inclined at an angle of 10 per cent. The work was done before the men had breakfast. Each value is the average of the same 7 men, and in all periods except S24 the experiment was performed on two or more occasions. All values are in mg. per 100 cc. of blood. The pyruvic acid analyses were made by means of Lu's method (1939) as modified by Friedemann and Haugen (1942, 1943). The blood lactate was estimated by the Friedemann, Cotonio, and Shaffer method as modified by Edwards (1938). (Minnesota Experiment.)

	Control	S12	S24	R6	R12
Pyruvate	1.1±0.23	1.3±0.39	1.4±0.25	1.3±0.26	1.3±0.25
Lactate	9.0±1.54	11.0±3.05	9.8±2.71	11.5±2.65	9.5±2.61
Lactate/Py- ruvate	7.94±0.70	8.20±0.77	6.80±1.32	9.00±1.12	7.53±1.08

All other observations on these subjects — clinical, physiological, biochemical, and psychological — confirmed the absence of any signs or symptoms of a thiamine deficiency. This indicates that the clinical interpretation of the vitamin status of hospital patients should be tempered with caution if such estimation is based primarily on the pyruvate level in the blood. Many patients present some signs indicating that they have suffered from an insufficient caloric intake for a considerable length of time. It is probable that in some instances the increase in pyruvate level of the blood could be attributed to a non-vitamin factor. The unreliability of the blood pyruvate and lactate values, as well as of their ratio, in the diagnosis of a mild thiamine deficiency has been emphasized by Henschel et al. (1945).

In the Minnesota Experiment the pyruvate level in the blood decreased only slightly during the rehabilitation period, and even after 12 weeks on a fairly high thiamine intake the pyruvate levels were still slightly above the control values.

The lactate concentration in the blood, however, showed no consistent change throughout the experiment (Table 233). The variations observed are similar to those that are routinely seen in the Minnesota Laboratory of Physiological Hygiene as day-to-day trial variations. The lactate/pyruvate ratios also showed no significant changes throughout the experiment. These ratios (Table 233) are comparable to the intraindividual variations observed in normal young men.

Devis and Simonart (1947) noted what they considered marked increases in the pyruvic acid level of civilian prisoners in Belgium during World War II. They were led to believe that this was a manifestation of a thiamine deficiency since so many other findings among these prisoners reminded them of beriberi. It seems more probable that the elevated pyruvic acid levels in their subjects were due to the same unknown factors operative in producing this condition among the subjects of the Minnesota Experiment.

During the 6 months of semi-starvation in the Minnesota Experiment there was a complete absence of signs or symptoms that could be attributed to a deficiency of thiamine. In countries subjected to a restricted food intake where unrefined foods are likely to form the largest part of the diet, it is highly improbable that any substantial change in the inhabitants can be directly attributed to a thiamine deficiency. Disruptions in the sanitation facilities, with a consequent increase in certain communicable diseases, or the psychological stresses resulting from the dislocation of daily activities may be more important as the causative factors underlying many of the vague neurological changes observed during World War II. These symptoms have been reviewed in some detail for the Asiatic theater by Denny-Brown (1947).

Thiamine in Rehabilitation — Minnesota Experiment

During the rehabilitation period, one half of the men in each caloric group received a Hexavitamin pill every day, whereas the others received placebos. The

TABLE 234

THIAMINE INTAKE AND URINARY EXCRETION during the rehabilitation period. All values for thiamine are expressed as mg. per day. The % columns represent the fraction of the thiamine intake excreted in the urine. The P columns are the averages for the subjects on the placebos; the H columns are the averages for those receiving the Hexavitamin pills. (Minnesota Experiment.)

Group	Values for R6					
	Intake		Excretion		%	
	P	H	P	H	P	H
Z	1.12	2.12	.191	.627	17.0	29.6
L	1.30	2.30	.298	.822	22.9	35.7
G	1.48	2.48	.336	.907	22.7	36.6
T	1.67	2.67	.362	.916	21.7	34.3

Group	Values for R12					
	Intake		Excretion		%	
	P	H	P	H	P	H
Z	1.33	2.33	.363	.764	27.5	32.9
L	1.51	2.51	.432	.961	28.8	38.4
G	1.70	2.70	.410	1.021	24.1	37.8
T	1.88	2.88	.450	.985	23.9	34.2

vitamin pills contained, among other things, 1.0 mg. of thiamine. The thiamine intake of each caloric group varied in proportion to the basic food allotments and ranged from 1.12 mg. for the unsupplemented men in the lowest caloric group at R6 to 2.88 mg. for the supplemented men in the highest caloric group at R12. Three-day urine samples were collected at R6 and R12.

When the average urinary thiamine excretion values of all groups for both R6 and R12 (see Table 234) were plotted against the corresponding vitamin intakes, a straight line was secured (see Figure 69). This line was significantly higher than that obtained from other young men maintained in weight equilibrium at comparable intakes of thiamine. For instance, the daily thiamine excretion of the men in the Minnesota Experiment who received 1.3 mg. of this vitamin per day was 298 micrograms, whereas that of normal men at the same intake was 107 micrograms (Figure 69).

In the Minnesota Experiment the fraction of thiamine recovered in the urine at both R6 and R12 ranged from 17 to 38 per cent (Table 234). These values are considerably higher than those found in normal young men. In caloric equilibrium the recovery of urinary thiamine increases with the intake until it equals an

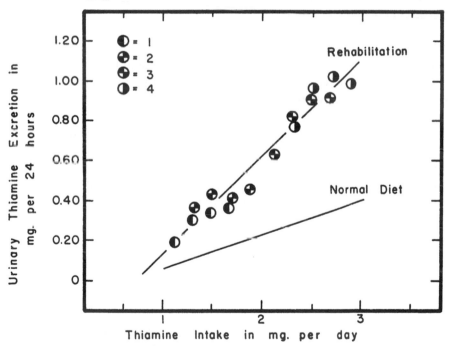

FIGURE 69. RELATIONSHIP BETWEEN THE THIAMINE INTAKE AND THE URINARY EXCRETION OF THIAMINE DURING THE REHABILITATION PERIOD. The thiamine intake for the rehabilitation phase of the Minnesota Experiment is the sum of the dietary content and that in the Hexavitamin pills wherever they were given. The circles for groups 1 and 2 represent the excretions at R6 and R12, respectively, of the 4 caloric sub-groups receiving placebos; the circles for groups 3 and 4 are the excretions at R6 and R12, respectively, of the 4 caloric sub-groups receiving Hexavitamin pills. The curve for the normal diet is from Mickelsen, Caster, and Keys (1947).

average of 13 per cent of that in the diet. There it remains even for intakes of 10 to 15 mg. of thiamine per day (Mickelsen *et al.*, 1947). It is difficult to reconcile the findings during the rehabilitation period with the recognized interpretation of urinary excretion data. Although few workers have rationalized their observations, most of them have implied that the urinary thiamine excretion represents only that fraction of the dietary intake which is in excess of the body's needs (Melnick, 1942). If this were so, it would be reasonable to assume that the thiamine excretion during periods of rapid tissue formation would be lower than normal as a result of the deposition of the vitamin in the new tissue. Just the opposite results were observed in this case during the rehabilitation period.

The trend in the urinary thiamine excretion throughout the Minnesota Experiment is shown in Figure 70. This indicates the relative excretion values for a thiamine intake of 1.3 mg. per day. For the control period the thiamine excretion value was secured from the data on normal young men (Mickelsen *et al.*, 1947). The excretion values for the semi-starvation period were averaged. At R6 the men on placebos in the L caloric group were on a daily thiamine intake of 1.3 mg. Their excretion values were used for the rehabilitation period. This was justifiable since the intake-excretion relationship was the same throughout the rehabilitation period (Figure 69).

There was no significant difference in the urinary thiamine excretion when the starvation period was compared with the control period. The urinary thiamine excretion during the rehabilitation period was considerably higher than that in the other two periods. All attempts to explain the thiamine excretion results of the rehabilitation period have proved futile. There is, however, a suggestion that a similar phenomenon occurs in children. The few experiments using recognized analytical procedures have shown that the recovery of urinary thiamine is never less than 20 per cent of the intake in normal children. This was shown by Benson and his group (1941) to be the case among 22 normal children. On an average calculated thiamine intake of 0.99 mg. per day, the average of 30 daily excretions for each child was 0.268 mg., which amounted to a recovery of 27 per cent. Oldham *et al.* (1944) confirmed this finding with two 5-year-old boys in whom the

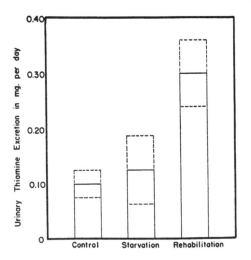

FIGURE 70. TREND OF URINARY THIAMINE EXCRETION DURING THE MINNESOTA EXPERIMENT on the basis of a constant thiamine intake of 1.3 mg. per day. The value for the control period is from Mickelsen, Caster, and Keys (1947); the values for the starvation and rehabilitation periods are the actual excretions at those times. The dotted lines indicate the limits of the standard deviations.

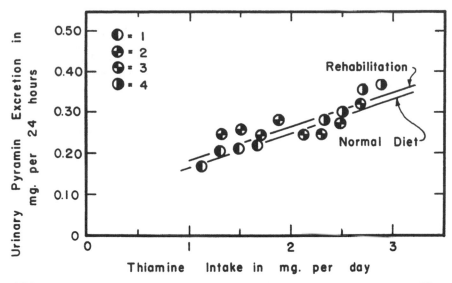

FIGURE 71. RELATIONSHIP BETWEEN THE THIAMINE INTAKE AND THE URI-
NARY PYRAMIN EXCRETION DURING THE REHABILITATION PERIOD. See the legend
to Figure 69 for an explanation of symbols.

urinary recoveries averaged 26 and 32 per cent when each boy received approximately 1 mg. of thiamine per day. This contrasts with a recovery in normal adults
of about 7 per cent for the same thiamine intake (Mickelsen *et al.*, 1947).

Here are two conditions — growth in children and recovery from starvation —
which are characterized by an increase in body tissue and an active storage of
thiamine. In spite of this, they are associated with a urinary thiamine excretion
higher than that observed among adults who are in weight equilibrium. More
work will have to be done on this problem, but the solution to this apparent
paradox may offer a better insight into the significance of urinary thiamine excretions.

During the rehabilitation period the pyramin excretion values were exactly
what would have been expected from the thiamine intakes. When the thiamine
intakes and the corresponding pyramin excretion levels were plotted against each
other, the line was practically identical with that previously secured for normal
young men (see Figure 71).

Riboflavin in Starvation — Minnesota Experiment

During the control period the diets supplied an average of 2.52 mg. of riboflavin per day, or 0.72 mg. per 1000 Cal. These values were calculated for the
diets served during the last 3 weeks of that period and are representative of those
used throughout the control period. Urinary excretion studies were not made
at that time since there was no reason to think they would be other than normal.

During the semi-starvation period 3-day composite urine samples were collected at S12 and S24. These samples were the same ones on which the thiamine
analyses were made. The riboflavin in the urine was determined by a modification of the Conner and Straub (1941) fluorometric method (see the Appendix on

TABLE 235

24-Hour Urinary Excretion of Riboflavin during the semi-starvation period (in micrograms). The calculated excretion values of young men on an adequate caloric intake are also given. The "calculated body loss" of riboflavin was determined as described in the text and represents the amount that might have been released from the tissues each day of the period. (Minnesota Experiment.)

	Riboflavin		Urine Volumes (ml.)
	M	SD	
Observed excretion			
S12	56.3	34.7	1826
S24	60.9	36.8	2449
Calculated excretion	70		1200
Calculated body loss	240		

methods for the modifications). The riboflavin content of the diet was determined by the microbiological method of Snell and Strong (1939).

The average daily riboflavin excretion was 56 micrograms at S12 and 61 at S24 (see Table 235). On both occasions there was considerable variation in the excretion values, as shown by standard deviations of 35 and 37 micrograms for the two periods. The individual excretion values are given in the Appendix Tables. The dietary intake of riboflavin throughout the starvation period was approximately 0.6 mg. per day, or 0.36 mg. per 1000 Cal. An average of about 9 per cent of the riboflavin intake was recovered in the urine. These recovery values are similar to those found for adults who are in weight equilibrium. For instance, Williams et al. (1943) found that 6 women maintained on a daily intake of 0.6 mg. of riboflavin for 60 or more days showed an average urinary riboflavin excretion of 77 micrograms per day. Similar results were secured with a group of 14 college women who were on a riboflavin intake of 0.79 mg. per day (Brewer et al., 1946). The comparison of the urinary excretions of the subjects in the Minnesota Experiment with those of women is valid since there is no evidence for any sex difference in this intake-excretion relationship.

Even though the riboflavin intake per 1000 Cal. was reduced from 0.72 mg. during the control period to 0.36 mg. in the starvation period, the excretion during the latter period still fell on the curve showing the relationship between intake and excretion values for normal persons maintained in weight equilibrium (see Figure 72). Hathaway and Lobb (1946) have also reported that the riboflavin excretion is more intimately related to the total intake than to the intake per 1000 Cal.

The normal excretion of riboflavin occurred in spite of a considerable loss of body tissue which could theoretically serve as a source of fairly large amounts of riboflavin. Calculations similar to those made for thiamine indicate that the active tissue which disappeared during the 168 days of semi-starvation "liberated" about 41 mg. of riboflavin, or an average of 240 micrograms, per day. These calculations are based on the value of 2.9 mg. of riboflavin per kg. of muscle (Axelrod et al.,

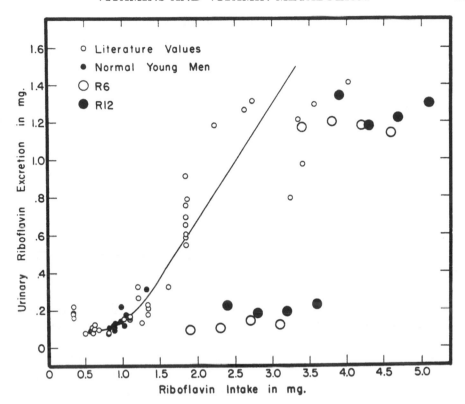

FIGURE 72. Relationship between the Intake and Urinary Excretion of Riboflavin in the Minnesota Experiment Compared to the Intake-Excretion Curve Secured from Normal Subjects. The following reports have been used in preparing the normal curve, and the values therein are represented by the smaller open circles: Brewer *et al.* (1946), Cogswell *et al.* (1946), Davis *et al.* (1946), Denko *et al.* (1946a, 1946b), Hathaway *et al.* (1946), Keys *et al.* (1944), Sebrell *et al.* (1941), and Williams *et al.* (1943). The smaller closed circles are the values previously secured for normal men in the Minnesota Laboratory (Keys *et al.*, 1944). The larger open circles are the excretion values for R6 in the Minnesota Experiment, the larger closed circles for R12. The placebo values are the lower group of 8 circles and the vitamin-supplemented values are the upper group of 8.

1941) and 17 mg. per kg. of liver (Taylor *et al.*, 1942). From the urinary riboflavin excretion data of the Minnesota Experiment there is no indication that any of the riboflavin released during the breakdown of body tissues appeared in the urine. In this respect the excretion of riboflavin during semi-starvation differs from that of thiamine.

The marked increase in the urine volume during semi-starvation had no apparent influence on the excretion of riboflavin — or, for that matter, on the excretion of thiamine and pyramin. This indicates that there was no "washing out" of these water-soluble vitamins.

The interpretation of the above findings is complicated by observations on riboflavin excretion under other dietary regimens. When all food was withheld from normal young men, the riboflavin excretion showed a very marked increase over that of the preceding control period (Mickelsen, Doeden, and Keys, 1945). During a 3-to-4-day fast the daily urinary riboflavin excretion in some of the subjects exceeded the preceding vitamin *intake* by 2 to 3 times. This is in marked contrast to the behavior of thiamine under similar conditions; it decreased within a few days to zero and remained at that level throughout the period of starvation. The riboflavin excretion of normal young men also increased on a diet deficient only in thiamine (Mickelsen, Doeden, and Keys, 1945). In the preceding experiments, both the acute starvation and the thiamine deficiency were associated with marked losses of body tissue. Originally the explanation of these urinary riboflavin findings appeared to involve the riboflavin "liberated" from the tissues during periods of weight loss; as the body tissue was metabolized to make up the energy deficits, the riboflavin thus released was poured into the urine. This explanation did not hold when similar subjects were put on a diet deficient in both riboflavin and thiamine. Under these conditions the riboflavin excretion decreased, even though there was a weight loss comparable to that in the above vitamin deficiency experiment (Mickelsen, Doeden, and Keys, 1945).

Not only do dietary factors influence the urinary excretion of riboflavin but so does physical activity. During periods of hard work, young men maintained in caloric balance showed a riboflavin excretion that decreased very markedly during the first 7 days of work. In contrast to this, extreme inactivity such as complete bed rest produced an increase in the riboflavin excretion (Mickelsen, Doeden, and Keys, 1945). In all these cases the increase or decrease in riboflavin excretion was based on the excretion expected from the dietary intake.

The only study made of the riboflavin concentration in the body organs during semi-starvation is that of Flinn *et al.* (1946). These workers used 2 groups of 3-week-old rats. One group was fed ad libitum while the other was given only enough food to maintain body weight. One week later all the animals were killed. In the restricted rats the liver was not quite one half the size of that in the controls. The concentration of riboflavin in the liver was the reverse of this, with the restricted rats showing a much higher value than the controls. Similar changes occurred in the kidney. Thus it would appear that these two organs retained most of the riboflavin liberated during the starvation period.

Riboflavin in Rehabilitation — Minnesota Experiment

Throughout the rehabilitation period of the Minnesota Experiment half of the men in each caloric group received a daily supplement of 1.5 mg. of synthetic riboflavin, whereas the others received placebos. Three-day composite urine samples were collected at R6 and R12. The results for the individual subjects are given in the Appendix Tables. On both occasions the interindividual variations in the riboflavin excretion values of the men in each intake group were much larger than in the case of thiamine. This was especially true in the 2 placebo groups, whereas the range of values for the vitamin-supplemented groups was of the same order of magnitude for both thiamine and riboflavin.

There was a marked difference in the amount of dietary riboflavin excreted by the placebo and the vitamin-supplemented groups (see Table 236). The

TABLE 236

Riboflavin Intake and Urinary Excretion during the rehabilitation period. All values for riboflavin are expressed as mg. per day. The % columns represent the fraction of the intake excreted in the urine. The P columns are for the subjects on the placebos; the H columns are for those receiving the Hexavitamin pills. (Minnesota Experiment.)

| Group | Values for R6 | | | | | |
| | Intake | | Excretion | | % | |
	P	H	P	H	P	H
Z	1.9	3.4	0.09	1.17	4.7	34.4
L	2.3	3.8	0.10	1.20	4.3	31.6
G	2.7	4.2	0.14	1.18	5.2	28.1
T	3.1	4.6	0.12	1.14	3.9	24.8

| Group | Values for R12 | | | | | |
| | Intake | | Excretion | | % | |
	P	H	P	H	P	H
Z	2.4	3.9	0.22	1.34	9.2	34.4
L	2.8	4.3	.18	1.18	6.4	27.4
G	3.2	4.7	0.19	1.22	5.9	25.9
T	3.6	5.1	.23	1.30	6.4	25.5

placebo groups excreted from 3.9 to 5.2 per cent of their intake at R6 whereas the other group excreted from 24.8 to 34.4 per cent. Although the differences between the 2 groups at R12 were not that great, they were still highly significant. These differences between the 2 groups persisted even in the area where the riboflavin intakes overlapped (Figure 72).

Similar differences in urinary excretion when riboflavin was fed as a synthetic compound or in natural foods have not been reported for normal individuals maintained in weight equilibrium. From the work of Sebrell et al. (1941), it appears that women receiving a large fraction of their riboflavin intake from the synthetic substance showed a riboflavin excretion which fitted the intake-excretion curve of Figure 72. Hagedorn et al. (1945) made a more direct comparison of the urinary riboflavin excretion when comparable amounts of milk, yeast, and synthetic riboflavin were fed to men. Twenty to 30 per cent of the riboflavin supplement fed as a synthetic compound was excreted, whereas 80 to 84 per cent of the riboflavin in milk appeared in the urine. The values for yeast were in between the other two.

The magnitude of the differences observed in the urinary excretion of the 2 vitamin groups in the Minnesota Experiment implicated a number of factors that had to be considered before any explanation of the difference could be suggested. The riboflavin content of the vitamin pills was checked by fluorometric analysis. The mean value for the 4 pills analyzed was found to be 1.55 mg. This compares well with the rated potency of 1.5 mg. per pill. The correctness of the urinary analyses was checked by means of 2 different analytical procedures. In the first the riboflavin in the samples was destroyed by irradiation, and in the other the

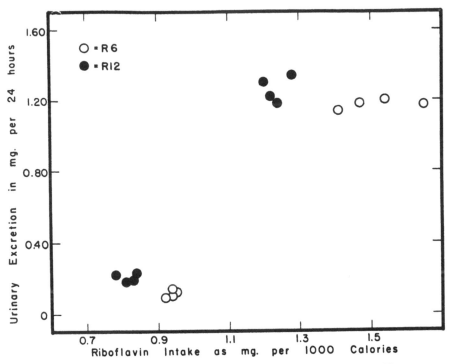

FIGURE 73. RELATIONSHIP BETWEEN THE URINARY RIBOFLAVIN EXCRETION DURING THE REHABILITATION PERIOD AND THE INTAKE EXPRESSED ON THE BASIS OF CALORIES. The circles represent the excretion values. The placebo values are the lower group of 8 circles, and the vitamin-supplemented values are the upper group of 8. (Minnesota Experiment.)

riboflavin was destroyed by heating the diluted urine in an alkaline medium according to the procedure of Swaminathan (1942). Both methods were used to establish the magnitude of the non-riboflavin fluorescing material in the samples. The results by the two methods were of the same order of magnitude, so there was no great possibility that the high urinary values of the vitamin-supplemented group were due to the presence in the pills of something that was converted to a riboflavin-like fluorescing compound.

An attempt was made to explain the differences in the urinary excretion by relating the values to various metabolic factors. It has been suggested that the metabolism of riboflavin is related to the total caloric intake (Williams *et al.*, 1943). To determine whether or not this might explain the differences, the riboflavin intake during the rehabilitation period was plotted as mg. per 1000 Cal. of food against the daily urinary excretion (see Figure 73). This also shows a marked difference in the excretion of the placebo and the supplemented groups.

The other possibility involved the deposition of riboflavin in the body tissue which was formed during the rehabilitation period. A correlation between the retention of nitrogen and riboflavin has been seen during recovery from injury (Andrae *et al.*, 1946) and in normal young women during induced changes in weight (Oldham *et al.*, 1947). Such, however, was not the case for nursing mothers

(Roderuck *et al.*, 1946). The increase in the body weight of the various caloric groups offered a relative approximation of the amount of body tissue being laid down. When the urinary excretions were plotted against the riboflavin intake as mg. per kg. of body weight increase, the same discrepancy between the placebo and vitamin-supplemented groups was secured (see Figure 74). There is no apparent explanation for the observed differences in the excretion values of the 2 groups. Furthermore, why do these subjects react differently to synthetic riboflavin than do other adults who are in weight equilibrium?

Another surprising feature of the urinary riboflavin excretion values during the rehabilitation period was the uniformity of the values in both the placebo and the vitamin-supplemented groups even though there was a considerable spread in the actual riboflavin intakes of the different caloric groups. At R6 the average daily excretions of the placebo groups ranged from 0.09 to 0.14 mg., whereas the intake ranged from 1.9 to 3.1 mg. (Table 236). Similarly for the supplemented group, the excretion range was from 1.14 to 1.20 mg. for intakes ranging from 3.4 to 4.6 mg. Essentially the same excretion ranges were observed at R12 even though at that time the intakes were considerably higher. It is very

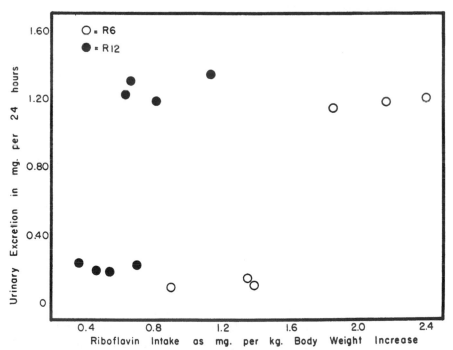

FIGURE 74. RELATIONSHIP BETWEEN THE URINARY RIBOFLAVIN EXCRETION DURING THE REHABILITATION PERIOD AND THE INCREASE IN BODY WEIGHT AT THE TIME THE URINE SAMPLES WERE COLLECTED. The circles represent the excretion values. The placebo values are the lower group of 7 circles, and the vitamin-supplemented values are the upper group of 7. There was a loss in body weight for the placebo subjects and an insignificant gain for the supplemented subjects in the Z group at R6, so these values have been omitted. (Minnesota Experiment.)

doubtful whether this unusual finding can be explained on the basis of the deposition of body tissue. As Figure 74 indicates, there is considerable overlapping of intakes when expressed as mg. of riboflavin in the daily diet per kg. of body weight increase. In spite of this, the placebo and supplemented groups show 2 distinct levels of urinary excretion. A single curve would have been expected if the excretion was influenced to any large extent by the formation of new tissue.

All the excretion values during the rehabilitation period were below what one would expect on the basis of intake-excretion relationships. The reports in the literature where adults were maintained on a constant level of riboflavin for periods longer than one week were used in securing the curve in Figure 72. A transition period is necessary before the riboflavin excretion becomes stabilized, and it has been suggested that at least 6 days are required (Brewer *et al.*, 1946). On this basis an excretion of 0.64 mg. per day would be expected if the intake was 1.9 mg. This is much higher than the 0.09 mg. observed as the mean at R6 of the 4 men in the Z group who received placebos. For the T group, who received an average of 3.1 mg. of riboflavin per day, the expected excretion level is 1.32 mg., which is 11 times higher than the 0.12 mg. actually observed at R6. The discrepancy between the observed and theoretical intakes holds true for the vitamin-supplemented groups and are accentuated at R12.

Vitamin C — Minnesota Experiment

The examination of the gums of the subjects at the end of the semi-starvation period showed an increased sensitivity to pressure when a flat-tipped stylus was pressed firmly against the base of the interdental papillae. Bleeding was not produced in any case. Since hypersensitivity might be interpreted as an early symptom of vitamin C deficiency, plasma samples were collected for vitamin C determinations. Samples were secured from 7 subjects who were the most sensitive to pressure and from 5 who were the least sensitive. The fasting blood plasma samples were drawn from the antecubital vein and were collected in tubes containing potassium oxalate as an anticoagulant. The plasma was immediately separated and analyzed for vitamin C by means of the photocolorimetric modification of the 2-6 dichlorophenolindophenol technique (Mindlin and Butler, 1938).

Table 237 shows the plasma vitamin C levels of these subjects and the pertinent data on the condition of the gums as observed at the time of the clinical examination. There was no essential difference in the plasma vitamin C averages for the 2 groups and no correlation within the groups between the plasma level of vitamin C and the gum sensitivity. There was considerable individual variation in the vitamin C levels, which ranged from 0.2 to 1.1 mg. per 100 ml. of plasma.

At first it was thought that there might be some correlation between the basal body temperature and the plasma vitamin C level. During the semi-starvation period the oral body temperature decreased very slightly, but there was considerable individual variation. As far as the individual subjects were concerned, however, there was no relationship between these two variables. It is more likely that the variation in the plasma levels reflects certain inherent characteristics of the subjects, since we previously noted a similar range in normal young men maintained on a constant vitamin intake. Others (Storvick and Hauck, 1942;

TABLE 237

Gum Condition and Plasma Vitamin C Levels in 12 Men at the end of semi-starvation. The plasma vitamin C concentration is expressed as mg. per 100 ml. The condition of the gums is from the report of the physical examination at S24. Bleeding was not seen or readily evoked in any of the men. (Minnesota Experiment.)

Subject	Plasma Vitamin C	Gum Condition
1 .... 0.6		minimal tenderness
9 .... 0.6		fairly normal; not tender
12 .... 0.9		no swelling or reddening; no tenderness
20 .... 1.0		minimal increased sensitiveness
129 .... 0.4		minimal tenderness
M 0.70		
5 .... 0.7		increased tenderness, grade 3
8 .... 1.1		moderate tenderness
22 .... 0.5		increased tenderness, grade 3
30 .... 0.3		very slight thickening; slightly increased tenderness
108 .... 0.2		marked tenderness
111 .... 0.9		grade 1 or 2 thickening of interdental papillae; tenderness increased
126 .... 0.8		slightly thickened; grade 2 tenderness
M 0.65		

Holmes et al., 1941) have reported the same thing in children and college women.

The diets used in the Minnesota Experiment supplied an average of 83 mg. of vitamin C per day (calculated from tables in U.S. Bureau of Human Nutrition, 1945). No analyses of the diets for this vitamin were attempted since so many technical difficulties arise when the presently available methods are applied to a mixed diet. The high vitamin intake was attributable to the large amount of potatoes and green vegetables, such as lettuce and cabbage, the latter being served fresh each day. Dehydrated fruits and vegetables were purposely omitted from the diet.

The development of gum abnormalities in calorically undernourished persons, in spite of an adequate amount of vitamin C in the diet, has been observed by others during World War II. Gingival disturbances and especially bleeding gums were apparently widespread among both service personnel and prisoners of war. A number of studies were made of the efficacy of crystalline vitamin C in curing these conditions, but in all cases even as much as 100 mg. per day for 3 to 4 months was no more effective than placebos (Marshall and Shourie, 1943; Mac-Donald, 1943; McNee and Reid, 1942; Stamm et al., 1944; Walsh, 1945).

Summary

Acute starvation has never been associated with the development of any signs suggestive of a vitamin deficiency. The proposed explanation for this has been that the required vitamins can be supplied by the body tissues (Bruckner, 1938).

The actual course of the metabolism of the vitamins themselves in total fasting is unknown. Under such conditions the urinary excretion of thiamine and pyramin rapidly approach zero. Little can be said about the other vitamins under such conditions either because of their erratic behavior (riboflavin for instance) or because of the absence of any studies on them. The urinary excretion of riboflavin in acute starvation, unaccompanied by any hard work, increases dramatically during the first few days and then begins to decrease. What the excretion is like after the early phase has not been determined.

In semi-starvation the type of foods available is of prime importance in determining the type and extent of the vitamin deficiencies that may appear. This will vary with time and the area involved. Where the diet is grossly inadequate in certain vitamins, signs and symptoms indicative of these deficiencies may be expected. Although one cannot point to very positive evidence, there is a suggestion that vitamin deficiencies develop more slowly when accompanied by severe weight loss than is the case in weight equilibrium. Such a situation could be explained by assuming that the vitamins liberated from the tissues during weight reduction are used by the body.

When vitamin deficiencies do appear in starvation areas, it is frequently difficult to make an adequate diagnosis. Semi-starvation produces many signs that are hard to distinguish from those associated with the early phases of certain vitamin deficiencies. Neither clinical nor laboratory means are as yet available for such a task. This makes it difficult on the basis of the present reports to determine the influence of semi-starvation in human beings upon the requirement for the various vitamins. Such factors as the magnitude of the negative caloric balance and the type of foods are undoubtedly of paramount importance in this respect.

Many of the older reports in the literature have associated with starvation such vitamin deficiency symptoms as keratomalacia, night blindness, follicular hyperkeratosis, and hemorrhagic tendencies. There are reasons for ascribing many of the above conditions to the semi-starvation itself. It is true that vitamin deficiencies probably do occur in some starvation areas, but where they do occur it is quite likely that some marked alteration in the diet is responsible rather than the general caloric undernutrition.

During World War II vitamin deficiencies occurred only rarely among the victims of starvation. This was especially true for Europe, where there were very few reports of such cases and even these lacked sufficient validation to make them completely acceptable. This was in marked contrast to the situation in World War I when scurvy and rickets were apparently widespread. A greater awareness of the means whereby these dietary deficiencies can be overcome was probably responsible for the differential incidence. In the Southwest Pacific and in the Orient a great deal of emphasis was put on the widespread occurrence of beriberi. Although the Minnesota Experiment has shown no advantage in an extra-dietary intake of vitamins during the recovery period following a siege of starvation, there are still a number of puzzling problems associated with the increase in body weight which occurred at that time. This type of rehabilitation is associated with an active deposition of body tissue which presumably should contain its proper quota of the various vitamins. In spite of this, the urinary thiamine excretion at that time was much higher than was expected from the dietary intake. The urinary excretion of pyramin, on the other hand, was exactly what would be predicted on the basis of the intake. Riboflavin excretion values were far below expectations based on the intake. This was true for both the placebo and the supplemented groups. Although the vitamin-supplemented subjects excreted much more riboflavin than the placebo group, in both groups there was no change in the level of excretion with increasing intake.

CHAPTER 22

Lipid Metabolism

THE changes in the fat content of the body during starvation have been known to the anatomists for many years and have been described in some detail by the histologists, who have attempted to classify the various types and stages of atrophy of the fat cells (see Jackson, 1925, for references). On the basis of purely qualitative evaluations of sections of fat tissue observed microscopically, it was noted that as the fat disappeared from the cells in the adipose tissue it was replaced, at least for a short time, by a water-like fluid. These observations have been supplemented by actual determinations of the fat and moisture content of the adipose tissue in various nutritional conditions. Such studies have indicated an inverse relationship between the fat and moisture contents of this tissue.

Histological studies of the adipose tissue of animals subjected to periods of starvation have indicated that the size of the individual fat cells decreases, but less uniformly than does the weight of the entire fat depot. This was brought out by Kuch and Lazarovich-Hrebeljanovich (1935) in their starvation experiments with rats. They stated that some fat cells may lose all their fat and disappear while others only decrease in size. These changes are shown in Table 238, where the average relative size of the fat cells is given together with the relative changes in body and adipose tissue weights. The average diameter of the fat cells decreased to a much smaller extent than did the weight of the adipose tissue. The average fat cell volumes followed the total fat changes fairly well.

Since the adipose tissue acts as a reserve supply of fuel in periods of caloric restriction, it would appear reasonable to expect the size of these depots to have a considerable influence upon the course of starvation. There is a fair amount of evidence for an intimate relationship between the degree of fat metabolism and the magnitude of the nitrogen loss during periods of negative caloric balance. This was recognized by E. Voit as early as 1901.

It might be assumed a priori that the survival time during starvation would depend, among other things, upon the size of the fat depots. According to Lusk (1928) such is the case for animals. He used reports in the literature to show that the number of days after the start of acute starvation before death intervened was proportional to the original fat content of the animal. When human beings are subjected to involuntary starvation, there are usually certain psychological factors that are as important as the fat reserves. This has been put forth as an explanation for the greater suffering and higher mortality observed among the more obese subjects in some of the concentration camps during World War II (van Veen, 1946; Wysenbeek, 1947). Clinical experience with infants, who are not so easily

TABLE 238

Changes in the Body Weight of Starved Rats Compared with the Changes in Adipose Tissue. The values other than those for the control period are expressed as percentage of the control value. The starvation values are the means of 2 animals whereas the controls are the means of 6. (Kuch and Lazarovich-Hrebeljanovich, 1935.)

	Control	Days of Starvation			
		2	5	8	12
Body weight	311 gm.	90	83	71	68
Retroperitoneal fat					
Weight	6.0 gm.	63	41	30	14
Cell size	89μ	83	82	73	58
Testicular fat					
Weight	5.2 gm.	63	65	50	34
Cell size	94μ	76	82	67	58
Subcutaneous fat, cell size ..	73μ	82	84	77	65

influenced by the mental stresses accompanying starvation, has indicated that they can survive a greater percentage weight loss than adults (Fleming and Hutchinson, 1924). This was attributed to their larger stores of body fat.

Although no one knows the actual mechanism whereby the fat in the adipose tissue is mobilized during periods of inadequate caloric intake, there are some interesting suggestions as to the method by which this is accomplished. The more recent hypothesis is that under such conditions the pituitary elaborates a sub-stance which acts on the adipose tissue to release the fat contained therein. The original experiments on which this theory is based were made by Best and Camp-bell (1936), who secured evidence for the presence in the anterior pituitary gland of beef cattle of a protein-like substance which, when injected into fasting mice, produced an increase in the concentration of liver fat. It has been shown by two groups of workers (Barrett et al., 1938; Stetten and Salcedo, 1944) that the increase in the liver fat represented that which migrated from the adipose tissue. More recently Weil and Stetten (1947) found a substance with similar properties in the urine of rabbits which had been fasted for short periods of time. This sub-stance was concentrated by various precipitation reactions, and, when injected into mice, it produced an increase in the liver fat concentration that was statistically greater than that produced by similar extracts from fed rabbits.

Additional support for this theory was secured from the experiments of Lee and Ayres (1936) with hypophysectomized rats. When control animals were given the same amount of food as the hypophysectomized rats, the former lost 18.6 per cent of their original body weight whereas the latter lost 24.4 per cent. In spite of the greater weight loss, the bodies of the hypophysectomized rats contained 12.6 per cent of fat whereas the controls had only 6.4 per cent. These findings lend some support to the theory that during periods of starvation the anterior pituitary produces a substance, or substances, which aids in the mobilization of the fat in the adipose tissue.

There is a suggestion that the adrenals are involved in the transport of fat from the depots to the liver (Samuels and Conant, 1944). In view of the difficulty involved in differentiating the activities of the adrenals and the pituitary glands, these observations may not be inconsistent.

Pigments in Adipose Tissue

A number of reports have called attention to the brownish color in the adipose tissue of persons who died from starvation. Stefko (1928) mentioned an ocher-red pigment in the remaining adipose tissue of the victims of the Russian famines following World War I. This coloration should not be confused with the so-called brown fat that is found in many animals, especially the rodents. It is still questionable whether human beings ever have any brown fat comparable to that in animals. If any brown fat is present, it is limited to the neck, axilla, and perirenal regions (Rasmussen, 1922) and is found primarily in infants and infrequently in tumors (Rasmussen, 1947). It is highly doubtful that changes in the color of adipose tissue involve an unmasking of hidden brown fat. The more likely possibility is that the abnormal color of the adipose tissue reflects a change in the diet resulting in a larger than normal intake of carotenoid pigments. Reports from the famine incident to the siege of Leningrad in 1941–42 indicated that the ingestion of considerable amounts of lamb's-quarters (a plant) produced a distinct yellow coloration of the skin of the extremities. That this color was probably due to carotenoid pigments is substantiated by the finding of unusually large amounts of these pigments in the blood (Brožek, Wells, and Keys, 1946). The inclusion of large amounts of plant materials in the diet, such as occurs under starvation conditions, results in an increased carotenoid intake, part of which is likely to be stored in whatever adipose tissue is present in the subject. There is also the possibility that as the fat leaves the adipose tissue, the normal pigments present therein remain behind and impart their color to the tissue.

Fat Metabolism and the R.Q. in Starvation

It has generally been assumed that during periods of starvation the body metabolizes primarily fats. This holds true for acute starvation in which all food is withheld. Benedict (1915) found that the respiratory quotient of Levanzin decreased from a pre-starvation value of 0.84 to an average of 0.72 by the sixth day of the fast, where it remained until the end of the experiment on the thirty-first day. There is, however, very little evidence on the respiratory quotient in persons living on calorically inadequate diets. In Benedict's experiment (Benedict *et al.*, 1919) with young men on a restricted food intake, the R.Q. during the experimental period was practically the same as that during the preceding control period. From a consideration of the fuel available to the body under conditions of semi-starvation, a normal R.Q. can be readily rationalized. Under such conditions the diet is likely to be very low in fat, and if the food intake provides about 50 per cent of the caloric expenditure, then the total material metabolized by the body will approach quite closely the composition of a normal diet. In the Minnesota Experiment the R.Q. was determined only during aerobic work. The values thus secured were nearly normal during the semi-starvation period (see Chapter 34).

Both the diet and the body tissue served as a source of fuel during the semi-starvation period. As is brought out in the section in this chapter on blood lipids, the average body fat metabolized during the semi-starvation period was about 40 gm. per day, which provided 360 Cal. During this period the average reduction in active body tissue was 150 gm. per day (see Chapter 15). Assuming that the protein content of this tissue was 25 per cent, then the metabolized body proteins amounted to 37 gm. per day. On this basis, the body tissues metabolized during the semi-starvation period provided an average of 508 Cal. per day. These figures are given in Table 239 together with the diets for both the semi-starvation and the control period.

TABLE 239

SUBSTANCES METABOLIZED BY THE BODY during the semi-starvation period compared with those metabolized during the control period. All values are on a per day basis. (Minnesota Experiment.)

| | Semi-Starvation | | | | Diet during Control | |
| | Diet (gm.) | Body Tissue (gm.) | Total | | | |
			Gm.	Cal.	Gm.	Cal.
Fat	34	40	74	666	124	1116
Protein	54	37	91	364	112	448
Carbohydrate	295	0	295	1180	482	1928
Total				2210		3492

TABLE 240

FRACTION OF TOTAL METABOLIZED CALORIES FROM THE DIFFERENT FOODSTUFFS. These values, given in percentage, have been calculated from the figures in Table 239. (Minnesota Experiment.)

| | Experimental Period | |
Substance	Starvation	Control
Fat	30	32
Protein	16	13
Carbohydrate	54	55

There was no essential change in the body weight of the subjects during the control period, so the diet at that time reflects quite accurately the substances being metabolized. On the basis of the preceding figures, the ratio in which fat, carbohydrate, and protein participated in the over-all metabolism can be calculated (see Table 240). The metabolism of fat during the control period amounted to 32 per cent of the total calories, whereas it was 30 per cent during the semi-starvation period. Although there was a slight increase in the percentage of protein metabolized during the semi-starvation period, there was no change in the fraction of carbohydrates entering into the metabolism. Since the proportion of the various foodstuffs metabolized in the 2 periods was so nearly the same, there was no reason to expect the type of metabolism, and consequently the respiratory quotient, to be changed.

The role of fat in the diet aside from its caloric contribution has received considerable attention since Burr and Burr (1929, 1930) showed that certain unsaturated fatty acids are essential for the normal development of the rat. Although the evidence that human beings require these compounds is not conclusive (Burr, 1942; Hansen, 1933), considerable concern has been expressed about the possible dangers involved in the use of a low fat diet (Burr and Barnes, 1943). Since fats are one of the first things to disappear from the diet in time of war, a deficiency of essential fatty acids might be expected. Even if such a condition does occur, it does not appear to have any physiological effects. From his experience in the Axis-dominated countries of Europe at the end of World War II, Drummond (1946b) found that many people in the Netherlands lived on from 3 to 5 gm. of fat per day for periods as long as 6 months. Nutrition survey teams that examined these people on their liberation from German domination found nothing "that could be related in any way specifically to the long deprivation of fats." Furthermore, the experience in France and Belgium "had given rise to a strong impression that there are no readily recognizable signs or symptoms associated with prolonged subsistence on diets containing little fat." The dry, scaly, cracked skin so often seen among the inhabitants of starvation areas has frequently been attributed to a dietary deficiency of fats, but according to Drummond (1946b) "no definite evidence in support of this assertion was ever obtained." This condition did not respond specifically to fats but became less common as the food supply and other conditions improved.

Recent reports have suggested that a low fat diet may be associated with certain physiological advantages. The use of diets high in fat has been implicated as one of the factors involved in the increased incidence of gallstones. Ehrström (1942) found a positive correlation between cases of gallstones and the dietary fat of the Finns. Suggestive confirmation of this observation was furnished by the experience in the Netherlands following their liberation in 1945. At that time large amounts of fat were being provided in the rehabilitation diets, and with this there was a considerable increase in the number of gallstone cases (Schalij, 1946).

Blood Lipids — Changes in Acute Starvation

As early as 1896 Schulz wondered whether the fat that disappeared during starvation was metabolized within the cells of the adipose tissue or in some other part of the body. If the oxidation of the fat occurred in a part of the body other than the adipose tissue, then, he reasoned, the fat would have to be transported via the blood. Under these circumstances the amount of fat in the blood should increase. He determined the amount of ether-extractable material in the blood of rabbits and pigeons both before and at the end of a 5-day fast. The concentration of ether-extractable material in the blood did increase in these animals as a result of the fast.

Similar experiments were made by Daddi (1898a, 1898b), who found an increase in the ether-extractable material in the blood of dogs during the first 7 days of a fast. However, if the fast was continued for an additional week, the concentration of lipids in the blood gradually decreased to a value about half the original. By tying off the thoracic duct in dogs before removing all food from them, he came to the conclusion that the fat taken from the adipose tissue was transported to the blood by way of the lymph vessels.

Apparently unknown to either of these two workers were the results of Pfeiffer (1887) on the blood fat content of starving dogs, rabbits, and hens. In contrast to Schulz and Daddi, Pfeiffer found that the amount of lipids he could extract from the dried blood of his 2 starved dogs (Soxhlet extraction) was 0.44 and 0.30 per cent, whereas that in his control dogs was 1.09. A long fast in hens produced a marked decrease in blood lipid levels, whereas a 13-day fast in a rabbit resulted in a twofold increase in the fat content of the blood.

Bloor in 1914 used his nephelometric method for the determination of the total lipids in the blood of dogs during a period when they were deprived of all food. The level of lipids in the blood of the different dogs showed considerable variation during the first 5 days of the fast, but thereafter the level decreased in all the animals. It was observed that the dogs which showed no increase in the plasma lipid levels were lean at the start of the experiment. When these animals were fed a liberal diet until they were in a good nutritional condition and then starved again, their plasma lipids increased during the first 5 days and then decreased. Freudenberg (1912) also found that the response of the plasma lipids of dogs subjected to starvation was dependent upon the degree of obesity of the experimental animal. In all the above studies no attempt was made to separate the fatty acids from the cholesterol and phospholipids.

A few studies have been made of the blood changes in human beings during a period when no food was taken. Shope (1927) followed the serum cholesterol level in a 22-year-old woman during a 5-day fast. Throughout the entire period the woman maintained her usual schedule of teaching and physical activity. The total serum cholesterol increased from a pre-starvation value of 231 mg. per 100 cc. to 314 mg., while the serum cholesterol increased from 171 mg. per 100 cc. to 255 mg. The increase in the cholesterol was due entirely to the ester fraction.

Lennox, O'Connor, and Bellinger (1926) made a few scattered determinations on the total serum cholesterol in 3 epileptics during a complete absence of food. The subjects used in these experiments must have suffered from some pathological disturbance since one of them died after 11 days of fasting. The length of the fast is implied to be about 11 days for all 3 subjects. In one subject the cholesterol concentration in the serum had increased from 91 to 174 mg. per 100 cc. at the time the patient died. In the other two subjects the cholesterol level during fasting was lower than after the fast.

Gregg (quoted by Bloor, 1943) studied the blood lipid changes in a man who refused all food for the last 57 days of his life. The total fatty acids in the whole blood of this man decreased from 440 to 330 mg. per 100 cc., while those in the corpuscles decreased from 160 to 60 mg. per 100 cc. The cholesterol level in the blood decreased progressively throughout the greater part of the 57-day fast but then increased markedly toward the end. The phospholipids decreased from 240 to 120 mg. per 100 cc. of whole blood, while those in the corpuscles changed from 180 to 60 mg. per 100 cc. The absolute changes in the total fatty acids and in the phospholipids were the same for the plasma and the corpuscles.

The above findings on the changes in the blood lipid levels during the com-

TABLE 241

CHANGES IN THE PLASMA OR SERUM LIPIDS DURING COMPLETE ABSENCE OF FOOD. Chol. = cholesterol; F.A. = fatty acids; Neut. = neutral; Adults refer to adult human beings; + = a greater than normal concentration; − = a lower than normal concentration; 0 = no change in concentration; +, − = an increase in the concentration during the early part of the experiment followed by a decrease; −, + = a decrease in the concentration during the early part of the experiment followed by an increase; 0 and + = most of the subjects showed no change but a few showed a higher than normal concentration; + and − = some subjects showed a higher and others a lower than normal concentration.

Author	Date	Subjects	Days of Restriction	Total F.A.	Neut. Fat	Total Lipids	PO$_4$ Lipid	Total Chol.	Chol. Ester	Free Chol.	I$_2$ Number
Gardner and Lauder	1913	cat	7					+	+		
Shope	1927	cat	2					+	+		
Ayleward and Blackwood	1936	cow	7–10	0			−		0 and +		−
Smith, J. A. B.	1938	cow	12				−			−	
Pfeiffer	1887	dog	19			−					
Hürthle	1895	dog	3					+			
Bloor	1914	dogs	5–7			0 and +	−				
Terroine	1914	dogs	25–29			+ and −		−, +			
Greene and Summers	1916	dogs (pups)	9			+					
Greene and Summers	1916	dogs (adults)	22			0					
Underhill and Baumann	1916	dogs	6			−, +	+				
Wendt	1928	dogs	16				+, −	+, −	+, −	+, −	
Ling	1931	dogs	7			−	−	−	−	−	
Rony, Mortimer, and Ivy	1932	dogs	2–14	0				0			
Entenman and Chaikoff	1942	dogs	23–30	0*		0*	0*	−*	−*	−*	
Shope	1927	guinea pig	2					+	+		
Pfeiffer	1887	hen									
MacLachlan	1944	mice	4			−	+	+			−
Schulz	1896	pigeons	4.5			+++					
Pfeiffer	1887	rabbit	13			+++					
Schulz	1896	rabbits	5								
Rothschild	1915	rabbits	2–9					−, +	+		
Shope	1927	rabbits	3					+	+		
Sure, Kirk, and Church	1933	rats	9–26	−			0	0			
Shope	1927	swine	4					+	+		
Lennox, O'Connor, and Bellinger	1926	adults	10					+ and −	+		
Shope	1927	adult	5					+	+		
Gregg	1943	adult	57	−			−	−, +	+		

* These analyses were made on whole blood.

plete absence of food, together with other reports from the literature, have been incorporated in Table 241. It is apparent that a great deal of variability has been reported for the behavior of the blood lipids during fasts of this nature. Some of the contradiction probably arises from the great variation in the methods of lipid analysis and from the diversity of animals used in the experiments (Entenman *et al.*, 1940). The uncertainties about the techniques and the time required for complete lipid analyses of high reliability have probably helped to discourage any broad and extensive study of the problem. In spite of the existing confusion, it appears from Table 241 that during short fasts there is no consistent change in the lipids other than cholesterol. There are no valid indications of the changes in the fatty acid content of the blood during fasting, but the presumptive evidence indicates that if there are any changes they are very small.

The lymph during acute starvation shows no change in lipid content. This was demonstrated by Rony, Mortimer, and Ivy (1932, 1933) for dogs that were fasted for 2 to 14 days. The total fatty acids and cholesterol in both the lymph and blood of these dogs remained constant throughout the experiment. When Levanzin underwent a 31-day fast (Benedict, 1915) the respiratory quotient decreased to 0.72, indicating that the metabolism involved primarily the oxidation of fat. Of his total metabolism of 1100 Cal. per day, 900 Cal. were derived from oxidizing 100 gm. of fat, which is well within the range used by most adults consuming a normal diet.

Blood Lipids in Semi-Starvation — Through World War I

A number of reports have appeared on the level of blood lipids during periods of semi-starvation. The interest in this field dates from World War I, when there were many people in Europe who were severely emaciated as a result of an insufficient food intake. In spite of the large number of persons available for such studies, Feigl (1918) appears to have been the only one who is credited with an extensive investigation of this subject. Although he has written at great length and has been quoted by many others, his work suffers from a number of shortcomings: (1) All his results were reported as averages with no indication of the number of individuals involved in the investigation. (2) In most cases he compared his observed averages with normal values recorded in the literature by other workers. Since, as we shall see, the level of lipids in the blood shows considerable interindividual differences even under normal conditions, any comparison between different groups is extremely hazardous. (3) There is no indication of the extent to which these individuals suffered from starvation.

Feigl found that the average concentration of total fatty acids in the plasma of foreigners interned in Germany was lower than the lowest value for normal adults reported by Bloor. He found a range for the total fatty acids of 0.12 to 0.27 gm. per 100 cc. of plasma, with an average of 0.20, whereas the range in Bloor's series was 0.29 to 0.42 gm. per 100 cc. of plasma, with an average of 0.36. There was no change in the lipid content of the corpuscles and only a slight decrease in the whole blood lipids when compared with Bloor's values. In the same paper Feigl stated that in 80 per cent of his cases (total number of cases not indicated) neutral fat in the plasma was absent while that in the corpuscles was far below normal. The lecithin content of the plasma was only one third of normal while the cholesterol content ranged from normal (the value for this was not indicated) up to 33 per cent above.

These results reported by Feigl are presented at length because they have been quoted by so many other workers as proof of the fact that the blood lipid levels are decreased in semi-starvation. Actually, it is difficult to arrive at any valid conclusion on the basis of the data put forth by Feigl. The most one can say is that there is some indication that the neutral lipids were reduced in the prisoners interned in Germany during World War I.

Maase and Zondek (1920) reported that there was practically no ester cholesterol present in the cases studied by Feigl (1918). The criticisms of Feigl's work can be applied equally well in this case. Furthermore, Maase and Zondek indicated that the free cholesterol in the blood was far above normal. Under normal conditions the ester cholesterol amounts to 75 per cent of the total (Sperry, 1936a), and so far no condition other than liver disease (Thannhauser and Schaber, 1926; Epstein, 1931) has been reported in which this ratio is very markedly disturbed (Bodansky, 1947). In contrast to the above report, Rosenthal and Patrzek (1919) found that the serum cholesterol in three fourths of the 21 undernourished people they examined ranged from 55 to 107 mg. per 100 cc. They accepted as a normal range the value of 133 to 205 mg. for persons the age of their subjects (15 to 53 years). Reports from other workers have indicated equally low cholesterol levels in some cases of starvation (Blöch, 1947; Forster, 1946). Rumpel (1915) stated that there was a decrease in the neutral fat and lipid phosphorus in cases of famine edema, but unfortunately no values were given.

Blood Lipids in Semi-Starvation — World War II

Some of the more recent work indicates that the blood cholesterol level in semi-starvation is unchanged or decreased. Weech (1936), without quoting any figures, indicated that the cholesterol levels were normal in the blood of Chinese suffering from nutritional edema. According to him, the normal cholesterol level in the blood of patients with nutritional edema can serve in the differentiation of this condition from that due to nephrosis. Neumann (1946) stated that the blood cholesterol levels were also normal among the edematous people in western Holland during the German occupation. The phospholipid phosphorus, however, was markedly reduced in the latter cases from a normal level of 8 to 11 mg. per 100 cc. of blood to 4 to 6 mg.

Gounelle and his associates (1941) determined the cholesterol and total lipids in the blood of inmates of a mental hospital in Paris. During the German occupation these people received nothing more than the official food rations, which at that time provided 1436 Cal. per day. No indication is given of the weight lost by these individuals, but they must have been subjected to some starvation since most of them eventually developed slight edema. The total plasma lipids in 11 of these subjects ranged from 455 to 775 mg. per 100 cc. with a mean value of 571, while the cholesterol level in the plasma ranged from 135 to 225 mg. per 100 cc. with a mean of 166. All their values were within the normal range for both total lipids and cholesterol. These findings were confirmed by Nicaud, Rouault, and Fuchs (1942) in their observations on 6 older Frenchmen who had famine edema. Similar results were also reported for serum cholesterol by Coste, Grigaut, and Capron (1941).

Raynaud and Laroche (1943) examined 5 Parisians who had marked famine edema. Besides normal blood levels for total lipids and cholesterol, they found normal phospholipid values. The only contradictory report from France is that of Coste, Grigaut, and Capron (1941), who claimed that the total lipids as well as the phospholipids were below normal in the French people during the German occupation. It is hard to determine the extent to which their conclusion was influenced by the high normal values with which they compared their results. Their average value for total lipids was 610 mg. per 100 cc. of blood, which according to most other investigators is normal.

Contradictory reports have come also from India, where at the time of the Bengal famine in 1943 a large number of severe starvation cases were treated in the hospital (Bose et al., 1946). Most of these patients were in such an advanced stage of starvation that death intervened shortly after admission. The blood phospholipid level in most cases was normal and about one fourth of the starved persons showed very high levels. The cholesterol level, determined by the method of Bloor, Pelkan, and Allen (1922), was below 100 mg. per 100 cc. of serum in all their cases. All their values were below the accepted lower normal level for adults. A somewhat similar alteration in the cholesterol level was reported by Walters, Rossiter, and Lehmann (1947b) among the Indian prisoners of war interned in Japan, but they gave no details.

A decrease in the serum cholesterol level was also reported from Belgium during World War II when that country was occupied by the Germans. Brull and his associates (1945) studied the blood lipid levels in 100 adolescents aged 12 to 19 years. Although there was no significant weight loss in these subjects as shown by the distribution curve of their body weights, the workers reported that the mean blood cholesterol level was 120 mg. per 100 cc. of serum. According to most reports, this is a low level (Leopold et al., 1932; Offenkranz and Karshan, 1936). The interpretation of these results is open to question since there are other reports which include some lower normal cholesterol values (Sperry, 1936b; Brøchner-Mortensen and Møller, 1940).

In almost none of the reports on the plasma or serum lipids in famine victims is there evidence as to the values in well-fed persons studied by the same investigators with the same methods. The question of normal standards is vexing here as in regard to many other quantitative characteristics. In the Minnesota Laboratory of Physiological Hygiene we have recently studied with much care the serum and plasma cholesterol (the 2 are indistinguishable in this regard) in some 600 "normal" males ranging in age from young boys to men of 65 years and older. We found a very marked age trend up to the age group of 45 to 54 years; at older ages the cholesterol level was the same as, or slightly less than, that of middle-aged men. For example, in 180 men aged 18 to 26 the mean total cholesterol concentration in the serum ranged from 169 to 186 mg. per 100 cc., depending on the relative obesity, the highest values being found in the men who were

TABLE 242

CHANGES IN THE PLASMA OR SERUM LIPIDS DURING PERIODS OF RESTRICTED FOOD INTAKE. Chol. = cholesterol; F.A. = fatty acids; Neut. = neutral; + = a greater than normal concentration; − = a lower than normal concentration; 0 = no change in concentration; +, − = an increase in the concentration during the early part of the experiment followed by a decrease; − and 0 = most of the subjects showed a lower than normal concentration whereas a few showed no change; 0 and + = most of the subjects showed no change but a few showed a higher than normal concentration. The obese subjects were receiving small amounts of food on weight reduction regimens.

Author	Date	Subjects	Days of Restriction	Total F.A.	Neut. Fat	Total Lipids	PO$_4$ Lipid	Total Chol.	Chol. Ester	Free Chol.
Page, Farr, and Weech	1937	dogs				0	0	0		
Entenman et al.	1940	dogs	122	—*		—*	—*	—*		
Feigl	1918	adults		—	—		—	0 and +		
Rosenthal and Patrzek	1919	adults						—		
Hetenyi	1936	adults	8			0				
Man and Gildea	1936	adults		— and 0			— and 0	—		
Bock	1937	adults	14						0	—
Man and Gildea	1937	adults		0			0	0		
Gounelle et al.	1941	adults				0		0		
Block	1942	adults	28			+, 0	0	0		
Raynaud and Laroche	1943	adults				0	0	—		
Hodges, Sperry, and Andersen	1943	children								
Lambrechts et al.	1945	adults				0				
Minnesota Experiment	1945	adults	168					— and 0		
Poindexter and Bruger	1935	obese	42–420					0		
Hetenyi	1936	obese	8			—				
Block	1942	obese	35			+, —				
Weech	1936								0	
Coste, Grigaut, and Capron	1941						—		0	
Nicaud, Rouault, and Fuchs	1942						0		0	
Bose, De, and Mukerjee	1946							0		—
Walters, Rossiter, and Lehmann	1947									—

* These analyses were made on whole blood.

15 per cent or more overweight. In 300 men aged 45 to 54, however, the average was 250 mg., and when this middle-aged group was subdivided according to relative obesity it was found that in all except the thinnest men (at least 15 per cent underweight) there was no relationship to obesity; in the thinnest men the mean value was 238 mg. Statistical analysis showed that, regardless of relative obesity, a total cholesterol value of 200 is moderately high for a young man but definitely low for a man of 50.

It is instructive to compare these standards obtained from Minnesota men with some recent values on famine victims in Europe. Simonart (1948) reported the values obtained by the Sperry-Schoenheimer method (in his book referred to as that of Schoenhouwer and Perry). He studied ten civilian prisoners who were seriously undernourished, and the analyses indicated a range in serum cholesterol values of 96 to 189, with a mean of 122.8 mg. per 100 cc. The ages of these men were not specified, but at least 4 of them, with values of 96, 97, 117, and 121 mg., would appear to be well below our Minnesota standards.

Some low values for serum cholesterol in famine victims also appear in the data of the Swiss group (Hottinger et al., 1948). Gsell (1948, p. 161) stated that a significant number of the refugees seen at St. Gallo had serum cholesterol values between 50 and 100 mg. per 100 cc. of serum and remarked that there was no special change in the ratio between free and ester cholesterol. Uehlinger (1947a, 1948) reported, without comment, a few cholesterol values on similar patients; most of these are low by the Minnesota normal standards.

The evidence from the literature is summarized in Table 242. It indicates that semi-starvation produces little if any change in the blood lipid levels except for a decrease in cholesterol. This is in contrast to the suggestive slight increase in this component during acute starvation.

Fat Tolerance Studies

There are a few references to the fat tolerance of emaciated persons. Borru so's (1935) data showed that the blood fat (neutral fat and cholesterol) following the ingestion of 100 cc. of oil was essentially the same in his 10 thin older people as in 6 normal-weight persons of the same age. Walters, Rossiter, and Lehmann (1947b), on the other hand, claimed that there was an alteration in the fat tolerance among the Indian prisoners of war released from Japan. Although no data were given, they maintained that "there was an impairment of the fat tolerance as judged by both the height and time of the rise in the amount of serum total fat and time of the rise in amount of serum cholesterol." Burger, Sandstead, and Drummond (1945), after examining the stools from severe starvation cases in western Holland, also concluded that the fat absorption from the gastrointestinal tract was reduced. These workers mentioned that most of the subjects suffered from diarrhea, which, of course, would influence any studies on the efficiency of the absorptive process. Since diarrhea is frequently present among large groups of people in starvation areas, any studies on the absorption of foodstuffs under such conditions will be of questionable value. For a further discussion of the available data on the digestive and absorptive processes as influenced by starvation, see Chapter 26.

Blood Lipids in the Obese on Reducing Diets

Closely related to the question of semi-starvation are the changes that occur in the blood lipids of obese individuals during periods of restricted food intake. Some of these studies have been limited only to the changes in blood cholesterol. In one such study of 14 obese adults who lost an average of 20 lbs. over a period of 6 to 60 weeks, there was no significant change in the blood cholesterol level (Poindexter and Bruger, 1935). Bock (1937) followed the plasma cholesterol level in a group of people who were given no food other than 600 cc. of fruit juice per day. Nothing was reported as to the number or character of the persons studied, but since they were undergoing the "fast cure" (i.e., the individuals were placed on a starvation regimen), it is likely that they were patients suffering from some disease. After 5 days of such a restricted intake, the free cholesterol decreased from 90 to 30 mg. per 100 cc. of serum. It remained at this low level through the tenth day, when it increased slightly. The ester cholesterol fluctuated around the normal value throughout the 14 days of the experiment.

Block (1942) studied changes in the total lipids in 3 obese women and in 3 women of normal weight. Over a period of 4 to 5 weeks the food intake of these women was progressively reduced, so that during the last week or two of the experiment each woman received 20 per cent of her "basal requirement." In both groups of women the total blood lipids increased during the first 3 to 4 weeks and then declined, reaching pre-starvation level by the end of the experiment.

Blood Lipids in Underweight Persons

Although the medical literature contains many papers on chronically underweight individuals, surprisingly few studies have been made of the blood components in such persons. Man and Gildea (1936) attempted such a study but complicated their experiment by using hospital patients as subjects. They found that in 16 out of 31 of their emaciated patients the level of fatty acids in the plasma was below normal, while the level was normal in 13 cases and above normal in 2. The lipid phosphorus was below normal in 19, normal in 10, and above normal in 2 of these patients. The serum cholesterol was below normal in 26 of the undernourished patients (with 150 mg. per 100 cc. considered the lower limit of normal), within the normal range in 3, and above 250 mg. per 100 cc. in 2. In a series of 10 underweight patients who showed considerable weight gains, the changes in the serum lipids were quite variable. Some patients showed marked increases and others showed no change. Unfortunately these subjects had such diseases as diabetes, tuberculosis, arteriosclerosis, carcinoma, and abnormalities of the gastrointestinal tract; consequently, it is somewhat hazardous to conclude that the results secured by Man and Gildea were due entirely to the semi-starvation regimen which produced the marked weight loss in their subjects.

The blood cholesterol level in underweight children has been reported to be very low. Under normal conditions of nutrition (see Dann and Darby, 1945), the serum cholesterol in children has been reported to be 206 ± 38 mg. per 100 cc. (Hodges et al., 1943). In children who were severely underweight, the level ranged from 77 to 176 mg. per 100 cc. As soon as the children showed an increase in body weight, the cholesterol level in the blood started to rise.

Borruso (1935) made fat tolerance studies on some older people who were very emaciated. For controls he used subjects of the same age but of average body weight. Although Borruso reported only the increase in the blood lipid values following the ingestion of the oil, it was implied that there was no difference in the basal samples from the two groups as far as the sum of free cholesterol and neutral fat is concerned.

Blood Lipids — Minnesota Experiment

The plasma samples that were collected during the control period and at S24 were stored in a frozen state at a temperature of $-20°$ C. All lipid analyses were made at approximately the same time on both the control and the S24 samples. The blood lipid analyses other than those for cholesterol were made by Dr. Edwin Hove of Distillation Products, Inc., Rochester, New York. The samples, packed in dry ice, were sent by air mail to that laboratory. Sufficient plasma was not available from all men to permit a complete set of analyses. Enough material was available from 5 representative subjects for all the analyses except cholesterol; for the latter, the heparinized plasma from 23 subjects was used. The total lipids were determined by Bloor's (1928) micro-oxidation procedure, the iodine number by the method of Yasuda (1931–32), the phospholipids by Bloor's method (1929), and the cholesterol by a modified Liebermann-Burchard reaction on the Bloor extracts.

The total plasma lipids at the end of the control period were 500 mg. per 100 cc., whereas they were 435 mg. at S24. The plasma phospholipids for the control period and S24 were 107 and 100 mg. per 100 cc., respectively, and the iodine numbers for these periods were 95 and 112. There is no essential change indicated in any of these values, and, because of the variation in the responses of the different subjects, all changes are statistically nonsignificant.

More extensive work was done on the total cholesterol levels. All these analyses were made on heparinized plasma, which, according to Sperry and Schoenheimer (1935), gives the same values as serum. The average plasma cholesterol value for the 23 men decreased slightly from a control value of 169 mg. per 100 cc. to 151 mg. at S24; 18 of the 23 subjects showed a decrease during the semi-starvation period.

During the 168 days on the low calorie diet, the subjects lost an average of 64 per cent of body fat as shown by densitometric measurements (see Chapters 8 and 15). In spite of this large change in the body fat content, there was very little change in any of the lipid fractions studied. A partial explanation for the absence of any lipemia under these circumstances may lie in the fact that the total fat metabolism during the semi-starvation period was, on the average, 57 per cent of the value for the control period. This calculation was made from the loss in body fat and the dietary fat intake during the semi-starvation period. The average body fat was 9.84 kg. in the control period and 3.04 kg. at S24; during the starvation period 6.80 kg. of fat were lost. This represents an average daily metabolism of 40 gm. of fat, which, when added to the average daily dietary fat intake, makes a total of 74 gm. (Table 239). During the control period the average fat intake was 124 gm. per day, and since there was no essential change in body weight during that time, it seems justifiable to conclude that the body was able to metabolize all the dietary fat consumed then. Further proof that the body was able to metabolize both the dietary fat and that released from the adipose tissue during the semi-starvation period lies in the fact that no ketosis was apparent at any time (see subsequent section in this chapter on ketosis).

Blood Lipid Changes — Summary

There is probably only a little less confusion about the changes in the blood lipids during semi-starvation than there is about the changes during the complete absence of food. Table 242 summarizes the available data; the conclusion is that there is little or no change in the concentrations of most of the lipids in the blood. It should be observed, however, that in the majority of reports before 1940 there were no measurements of the concentrations before starvation, and the "control" reference values were inferred from the literature or from measurements on other "normal" individuals. This is a serious limitation because there are marked differences among normal individuals (Gildea et al., 1935–36; Sperry, 1937; Turner and Steiner, 1939). In Sperry's series of 25 normal subjects the individual mean values for cholesterol ranged from 142 to 338 mg. per 100 cc., but the intraindividual values had a maximum variation of only 12 per cent of the mean for the individual. There is less information available on the variability of the other blood lipids, but the data at hand indicate that they are rather like cholesterol in regard to variability (cf. Man and Gildea, 1937).

The preponderance of evidence indicates that the concentration of cholesterol in the blood decreases during semi-starvation. Such a change was conclusively established in the Minnesota Experiment. In this experiment the mean serum concentration of total cholesterol was 169.3 ± 24.0 mg.; at S24 it was 150.7 ± 21.7 mg. per 100 cc. This is not a large absolute difference, but the consistency of change in the individuals was such as to make it statistically highly significant ($F = 17.94$; F at the 1 per cent level of significance = 8.02).

Lipids — Changes in Fat Depots

Starvation, even when continued until the death of the animal, is not accompanied by the complete loss of extractable body fat. Schulz noted in his experiments with dogs in 1897 that the total ether extract of different organs was reduced to a more or less constant level after a period of starvation. The results of such work led Terroine (1920) to theorize that the fat remaining in starved organisms was an integral part of the cell, which he named the "élément constant." Although the superficial fat and adipose tissue disappear during an extended period of starvation, the fat that remains in the body is still in a state of dynamic equilibrium. This was indicated by Bernhard and Steinhauser (1944), who showed by means of deuterium oxide (heavy water) that some synthesis of fat occurs even in rats from whom all food had been withheld for 6 days. They secured evidence that the material separated as the unsaponifiable ether extract (a large fraction of which was probably cholesterol) was also synthesized under these conditions. In starvation, where the degradation of fats is proceeding at a rapid rate, the reverse reaction still takes place to a measurable extent.

TABLE 243

COMPARISON OF CHANGES IN TESTICULAR FAT DEPOTS OF RATS STARVED TO THE SAME WEIGHT LOSS by withholding all food (acute starvation) and feeding an insufficient amount of food (semi-starvation). Eight rats were used in each group. Values are given in grams for body weight, testicular fat depot weight, weight of fat in the depot, and weight of water in the depot. (Cremer, 1939.)

	Acute Starvation			Semi-Starvation		
	Original	Final	% Loss	Original	Final	% Loss
Body weight	235	152	35	295	177	40
Testicular fat depot						
Weight	1.23	0.10	92	1.54	0.08	94
Fat content	1.06	0.004	99	1.35	0.016	99
Water content	0.14	0.08	43	0.16	0.05	68

The question arises whether the changes in body fat that occur in acute starvation are the same as those in semi-starvation. Very little evidence is available on this problem. Cremer (1939) attempted to answer the question by his studies on rats. He used the testicular fat depots as indicators of the changes in all body fat depots. He found that the right and left testicular fat depots were very similar in both weight and composition regardless of the nutritional condition of the animal. This was true even though the depots from different animals showed considerable variation. The testicular fat depot was removed from one side of the adult animal before the start of the starvation regimen and from the other at the end of the experiment. The two fat depots from the same animal were compared in each case. Both acute starvation, with the removal of all food, and semi-starvation produced an almost complete disappearance of fat from the depot when the rats were starved until they lost 35 to 40 per cent of their body weight (see Table 243). Because of the small numbers of animals used, none of the apparent differences indicated in Table 243 can be considered significant. In regard to the use of fat from the testicular depot by the rat, it would appear that there is no difference between the response to acute starvation and to semi-starvation.

By means of a similar technique, Cremer (1939) was able to show that when rats were starved (all food withheld) until they lost 22 per cent of their original body weight, the testicular fat depot removed at that time had decreased very markedly in weight and contained only a small amount of fat (see Table 244). The body weight attained at the end of the starvation period was maintained for an additional 3 weeks by feeding small amounts of food. The testicular fat depot during this time doubled in weight and showed a fifteenfold increase in fat on an absolute basis. At the same time, the total amount of both the non-fat connective tissue and the water decreased to a slight extent. This indicates that fat can be deposited, at least in the testicular fat depot, during a period of restricted feeding which follows a period of acute starvation even though only enough food to maintain the lower body weight is fed. Those rats that had the lowest percentage of fat in the testicular fat depots at the end of acute starvation showed the greatest absolute gain in fat during the subsequent restricted feeding period. In this respect, the fat depots behave in the same manner as does the whole body

TABLE 244

Changes in the Composition of Testicular Fat Depots (t.f.d.) in rats during a 3-week period of restricted feeding following a period of acute starvation. All values are averages of 8 rats; the initial (control) body weight was 244 gm. (Cremer, 1939.)

	End of Acute Starvation	End of Restricted Feeding
Weight of animal (gm.)	181	181
Weight of t.f.d. (mg.)	112	216
Fat in t.f.d. (%)	17	62
Fat-free dry t.f.d. (%)	15	6
Water in t.f.d. (%)	68	31

TABLE 245

The Influence of Semi-Starvation and Restricted Feeding on the Distribution of Body Fat in the Rat. The animals in group A were fed ad libitum to a body weight of 180 gm. The feed of those in group B was restricted to 70 per cent of that in group A. The animals in group C were fed like group A to a body weight of 180 gm., and then all food was removed from them until their weight decreased 30 per cent. (Reed et al., 1930.)

Group	Fat in Depots as Percentage of Total Depot Fat					Percentage of Body Weight as Fat
	Muscular	Genital	Subcutaneous	Perirenal	Mesenteric and Omentum	
A. Food ad lib ..	5.3	20.2	47.3	13.2	14.1	9.6
B. 70 per cent of A's diet	5.3	12.3	58.1	12.0	11.7	7.8
C. Starved	4.9	22.9	51.4	9.2	11.4	5.5

when measurements of the efficiency of gains in body weight are made (Richardson and Mason, 1923; Brody, 1945).

Partial confirmation for Cremer's hypothesis that acute and semi-starvation have the same general effect upon the fat depots was established by the work of Reed et al. (1930). These workers fed weanling rats on a good diet that was low in fat until the animals weighed 180 gm. (group A). The animals were then killed and the fat content of the various depots was determined. Another group of rats (group B) were fed only 70 per cent of the food consumed by the preceding group and were killed at the same age as group A. A third group (group C) was fed ad libitum to a weight of 180 gm., when all food was removed until they had lost 30 per cent of their body weight; then they were killed and analyses were made as above. Although the total amount of fat in the body of the starved rats (group C) was about one half that of the full-fed animals (group A), the relative distribution of fat in the depots was practically unchanged (see Table 245). To a lesser extent, this was also true of those animals which were on a semi-starvation regimen throughout their lives (group B).

A certain amount of caution must be exercised in utilizing these experiments with rats for the interpretation of human data. There is a marked species difference in the rate at which fat is removed from the different adipose tissues. The experiments of Aron (1911) showed this when the growth of pups was restricted by limiting the food. The fat in the various organs, expressed as a percentage of the body weight, was the same in the starved pups as in full-fed dogs except for the muscles and subcutaneous tissue. A larger amount of fat was lost from the muscles and subcutaneous tissue than from the organs. In fact no fat detectable by chemical means was left in the muscles and subcutaneous tissue. The amount of fat in all tissues and organs was markedly reduced in comparison to that in normal dogs. Although no experimental work was cited, Aron stated that when adult dogs were starved the same changes occurred. Much earlier than this Pfeiffer (1887) had shown a difference in the distribution of fat in rabbits, dogs, and chickens that had been starved.

The early workers in the field of starvation implied that the unsaturated fatty acids were preferentially used as a source of fuel (Dowell, 1920; Mann, 1936; Lovern, 1938). There are a few recent reports on this question which have been interpreted to indicate that there is no selective oxidation of fatty acids during acute starvation. Longenecker (1939) compared the fatty acid composition of the depot fat of acutely starved adult rats with that of normal controls sacrificed at the beginning of the experiment. The rats were starved until they lost, in different experiments, 17 and 28 per cent of their original body weight. The depot fat in both groups contained the same proportion of the 9 fatty acids for which specific analyses were made.

Similar results were secured by Hilditch and Pedelty (1940) when they kept all food away from 2 adult pigs for 51 and 167 days, respectively. The pigs lost in one case 20 and in the other 43 per cent of their original body weights, but the distribution of the different fatty acids in the depot fat was the same as that in control animals of the same weights. Hodge *et al.* (1941) found that during starvation mice oxidized depot fat that had an iodine number of approximately 80. Since the depot fatty acids in normal mice had an iodine number of 86, they concluded that a nonspecific oxidation of fatty acids occurred during starvation.

Relationship between Water and Fat in the Tissues

As the fat is removed from the depots during starvation, water takes its place to a certain extent. This was observed, according to Bozenraad (1911), as early as 1838 by Chossat, who reported that in starvation the brain, spinal cord, and bones as well as the adipose tissue showed an increased water content. Bozenraad (1911) found that the average moisture content of the fat tissues removed from

TABLE 246

WATER CONTENT OF THE FAT TISSUES FROM ADULTS DYING FROM VARIOUS CAUSES, given in percentages. Moisture determined by drying tissue to constant weight. (Bozenraad, 1911.)

	Number of Subjects	Abdomen		Heart		Kidney Capsule	
		Av.	Range	Av.	Range	Av.	Range
Well-nourished persons ...	14	12.2	7.9–19.4	17.9	14.3–21.7	13.7	6.9–20.1
Emaciated persons	7	31.0	20.2–36.4	32.6	29.4–35.8	25.4	13.4–46.2

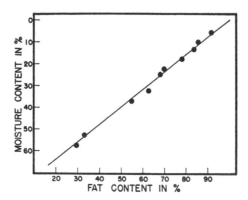

FIGURE 75. Relationship between the Moisture and Fat Contents of Subcutaneous Tissue. This material was secured from 10 persons who died from various diseases. From data in Schrimer's Table 3 (1921).

well-nourished persons at the time of autopsy contained less moisture than did similar tissues from emaciated persons (see Table 246).

The results of a number of studies on the fat and moisture contents of adipose tissue indicate an inverse relationship between these two functions. This relationship, based only on the percentage of moisture and fat, is illustrated by the straight line secured from Schrimer's (1921) data (see Figure 75). He analyzed the adipose tissue from 10 persons who died from various diseases. A line almost identical to this can be secured from the data of Pfeiffer (1887), who made analyses on the fat depots secured from dogs, rabbits, and hens in various nutritional states.

Atrophic infants who are suffering from semi-starvation show a very much lower percentage of fat in their bodies and an abnormally high percentage of water when compared with emaciated adults (this is for total body fat and water, in contrast to that in the depots; see Chapter 17 for further information and references). The levels of fat in the bodies of the atrophic infants are among the lowest of the few that have been reported for extremely undernourished individuals. As has been brought out in Chapter 17, the high loss of fat by the very young may indicate that they retain active tissue tenaciously, probably because there are no "protein reserves" that can be sacrificed for fuel purposes during a period of caloric deficiency.

Lipids — Changes in the Composition of the Liver

The changes in the lipid content of the liver during starvation have been studied more extensively than those in any other organ or tissue (excluding blood). Unfortunately, here again practically all the work has been done with animals that have been subjected to an acute fast. Whether these changes are in any way representative of those occurring in man is hard to say at this time. Furthermore, there is just as great variation in these results as in the studies of blood lipids. Part of this variation is undoubtedly due to differences in analytical methods.

The results of an extensive study of liver lipids in fasting mice are summarized in Table 247. MacLachlan et al. (1942) withheld all food from groups of mice for from 1 to 4 days. The animals that were starved the longest lost 31 per cent of their original body weight. During the first 2 days of starvation there was a marked infiltration of fat into the liver which resulted in a doubling of the abso-

TABLE 247

Changes in the Composition of Lipids in Mice during Acute Starvation (all food removed; water ad libitum). All values are the means of 20 mice; the groups were starved for different lengths of time. F.A.I. Number = iodine number of fatty acids. (MacLachlan et al., 1942.)

Initial Weight (gm.)	Days of Fast	Liver Weight (gm.)	Total Lipids (mg.)	Percentage Fat in Liver	Total F.A.I. Number	Phospholipids (mg.)	Chloride	
							Liver (m.eq. per kg.)	Plasma (m.eq. per liter)
20.4 0		1.26	66	5.2	125	41	31.6	116
20.7 1		1.08	107	9.9	113	33	26.8	111
21.1 2		0.96	110	11.5	112	30	24.2	112
21.0 3		0.80	42	5.2	115	25	30.0	112
20.8 4		0.65	20	3.1	91	18	30.9	112

lute amount of fat in that organ in spite of a reduction of the liver to three fourths its original size. The fat that infiltrated the liver during the early stage of acute starvation was deposited intracellularly. This was reflected in and primarily based on the change in chloride concentration of the liver. Toward the end of the starvation period the percentage of fat in the liver approached normal values. Although the absolute amount of phospholipids in the liver decreased throughout the fast, the lecithin fraction continued to be present as a constant proportion of the total phospholipids. There were, however, changes in the amount of the various compounds that go to make up the lecithin and cephalin fractions.

The extent to which the liver fat in the fasted mouse increases is proportional to the fat stores in the carcass. This is true at least for mice which have been subjected to acute starvation. From the data of Hodge et al. (1941) the product moment correlation for the relationship between the total carcass lipids and the total liver lipids after 3 days of starvation was calculated as +0.943. Since 20 mice were involved in this group, the correlation coefficient is highly significant. Partial confirmation of this relationship is found in Dible's work (1932) with rats which were starved for variable periods of time. At the end of each experiment the animals were killed. The liver and remaining carcass were then analyzed for fat. The correlation coefficient for liver and body fat in the rats starved from 20 to 28 hours was +0.690, that for those starved from 26 to 52 hours was +0.742, and that for those starved from 69 to 96 hours was +0.844. These values are based on the actual weights of fat in both the liver and the whole body. Results contradictory to these were secured by Treadwell, Tidwell, and Grafa (1943) for the liver lipids in rats starved for 36 hours. Throughout this period there was no statistically significant change in liver lipid levels. This was true whether the animals started the starvation period with a high lipid level (produced by a diet low in protein during the pre-starvation period) or a normal level.

That factors other than the amount of carcass lipids may influence the infiltration of fat into the liver was shown by Ohta (1940). When rats were starved at room temperature, the concentration of fat in the liver increased during the early stages of the fast and then decreased. On the other hand, when the rats were starved at temperatures of 3° to 6° C., the amount of fat in the liver decreased from the very start of the experiment.

In rats the level of liver fat during starvation is related to the sex of the animal. The level of liver glycogen shows a similar but inverse relationship. Although the sex difference in the liver glycogen stores disappeared after the seventy-second hour of starvation, the level of fat in the liver was maintained at a higher level in the females than in the males throughout the experiment (Deuel et al., 1934; see the section on ketosis in this chapter; see also Chapter 24).

The nature of the preceding dietary fat also has some influence on the liver lipid level during starvation. According to Samuels and Conant (1944) the liver fat decreased during starvation more rapidly when tung oil was present than when all the fat was from butter and cream. In this case, the tung oil made up 13.7 per cent of the fat in a ration providing 69 per cent of the calories as fat. Throughout the starvation period, they found a continuous decrease in the liver fat expressed on the basis of body weight. Insufficient data were given for an adequate explanation of the discrepancy between their findings and the earlier reports (MacLachlan et al., 1942).

The absolute amount of cholesterol in the liver increased in adult cats during 7 days of acute starvation (Gardner and Lauder, 1913). The cholesterol in the liver of the starved cats was twice that in the control animals, with the increase due almost entirely to the ester cholesterol.

There is only one report on the liver lipids in a human adult. This subject fasted for 57 days. At the end of the fast when death intervened, the cholesterol and phospholipids in the liver were slightly below normal (Gregg, 1943). This case was complicated by the fact that death was due to bronchopneumonia. It is implied that the man abstained from all food during the fast, but little information about this case is available.

TABLE 248

FAT CONTENT OF VARIOUS TISSUES AND ORGANS OF STEERS. The semi-starved steers were feed so that they lost weight at a rate of approximately ½ pound per day. The fat was determined by means of ether extraction. (Trowbridge et al., 1918.)

	Control Animal	Semi-Starved Animals	
	593	591	592
Original weight (lbs.)	700	573	664
Final weight (lbs.)	800	489	475
Change in weight as percentage of original weight	+14.3	−14.6	−35.1
Length of experiment (days)	199	199	312
Fat in entire animal (kg.)	61.47	16.62	4.06
Percentage of fat in:			
Brain and spinal cord	16.26	12.27	9.78
Liver	2.22	2.17	3.74
Skeleton	17.31	17.59	2.90
Round-muscle	4.31	2.31	2.01
Round-fat tissue	69.09	36.64	0.00
Rump-muscle	7.46	5.08	3.01
Rump-fat tissue	78.28	59.15	0.00

The only experiment on semi-starvation in which the liver lipids were determined is that of Trowbridge, Moulton, and Haigh (1918). They used 3 11-month-
old steers in their work. One was given a normal amount of feed while the other
two were fed at such levels that they lost ½ pound per day. The semi-starved
steers were continued on this regimen for 6 and 10 months, respectively, before
they were slaughtered. By that time they had lost 15 and 35 per cent of their
original weights. The control animal was slaughtered at the same time as the first
semi-starved steer. The amount of fat in the various tissues and organs of these
animals is shown in Table 248.

The steer that was semi-starved for 6 months showed no evidence of fat infiltration into the liver. The steer that was semi-starved for 10 months was on the
verge of dying from starvation when the experiment was terminated. Even in this
extreme state of semi-starvation, there was still a considerable amount of fat in
the liver, but the amount present was much less than that associated with fatty
infiltration. Further investigations by Moulton (1920) indicated only insignificant
changes in the moisture, nitrogen, ash, and phosphorus contents of the livers
from these 3 animals.

Lipids — Changes in the Composition of Other Organs

Changes in the fat content of organs other than the liver have been investigated in very few cases. Landau (1913) stated that in inanition among human
beings the fat in the adrenals increased considerably, whereas the cholesterol
content decreased. No values were given to substantiate this statement. Oleson
and Bloor (1941) found no change in the weight of the adrenal glands when
guinea pigs were continued on an acute starvation regimen until death became
imminent. The phospholipids in the adrenals of these animals showed an initial
decrease followed by an increase to a higher than normal level. The fatty acid
content decreased somewhat while the level of cholesterol increased slightly,
with the esterified form increasing to a greater extent than the free cholesterol.

The amount of fat remaining in a young dog from which all food was withheld for 24 days was determined by Kumagawa (1897). Before the fast started,
the dog was very fat, but at the end of the starvation period it had lost 36.4 per
cent of its original weight. At that time there were no detectable adipose tissue
depots but all the tissues still contained small amounts of fat. The fresh brain still
had 7.9 per cent of fat, the liver 4, the bones 3.6, the heart 3.3, and the genitals,
bladder, and gall-bladder 3.

Gardner and Lauder (1913) analyzed the adrenal gland from an adult cat
that had been without food for 7 days and found only a slight decrease in its
cholesterol content. The weights of the adrenal glands from this cat were within
the normal range. They also determined the cholesterol content in the muscles,
hearts, and kidneys of their starved cats. There was no change in the percentage
distribution of either free or total cholesterol in these tissues. In the lungs the
variations were so great that the significance of any change was lost.

Rubow (1905) removed all food from 2 dogs until they lost 44 and 35 per
cent of their original weights. Autopsy of the dogs showed no traces of fat in the
subcutaneous fat depots, in the peritoneum, in the pericardium, or intermuscu-

larly. There was no essential difference in the percentage of total ether extract, fatty acid, or lecithin in the heart muscles of the starved dogs when compared with those from normal dogs in whom all the visible fat had been removed from the heart prior to analysis. The shoulder muscles of the starved dogs had about one half the normal percentage of these 3 lipid constituents. The changes in the kidneys were difficult to interpret because of the great variation in the lipid content of this organ in normal dogs.

The amount of fat in the muscles of starved rats decreased at a much slower rate than did that in the liver, according to Samuels and Conant (1944). Adrenalectomy at the start of the fast maintained the fat level in the muscles at a significantly higher figure than in the non-operated controls. That led them to suggest that the adrenals influenced primarily the transport of fat from the depots to the liver.

Some of the earlier reports on the muscle lipids during acute starvation are difficult to interpret. For instance, even though Pflüger's (1902) dog decreased in weight from 44 kg. to 33.6 kg. in 28 days, it was reported that the fat in the muscles amounted to 20 per cent on the fresh weight basis. At the start of the experiment the dog was well fed, but no mention was made of the extreme obesity that would have been necessary in order to approach the reported degree of muscle fat concentration at the end of the starvation period. Nothing was said of that finding when, in a later report (1907), Pflüger indicated that no fat was visible in the muscles of dogs subjected to a similar period of food restriction.

Lipid Loss in the Feces

The fat excreted in the feces during starvation was first studied by Müller (1884). He found that when dogs were starved for periods up to 28 days, the feces contained from 18 to 50 per cent of ether extractable material on a dry weight basis. Of the fat in the feces, 25 per cent was present as free fatty acids, 33 per cent as neutral fat and cholesterol, and 42 per cent as fatty acids in the form of soaps. This was similar in composition to the feces secured from dogs fed lean meat. In 1893 Müller studied the fat in the feces from 2 men, one of whom fasted 6 days and the other 10. Although there was a decrease in the weight of the feces during the experiment, there was no change in the percentage of fat in the feces. A similar result was secured by Martin (1907).

The only known report on the loss of fat in the feces during semi-starvation is that by Jansen (1920). He found that the fecal fat of 7 adults receiving 1130 Cal. per day averaged 2.1 gm. per day, which represented 14 per cent of the dietary intake. These subjects were said to be representative of the average German civilian who during World War I was on a semi-starvation regimen. In another group of subjects receiving an average of 1760 Cal. per day, the fecal fat was 6.2 gm. per day, or 15 per cent of the dietary fat. On a dry weight basis, the fat constituted an average of 7.6 and 9.9 per cent, respectively, of the feces in these two groups. This is considerably lower than the normal value, which is approximately 30 per cent (Bloor, 1943). It is impossible to decide whether this represents a greater than normal retention of fat by individuals on a restricted food intake or whether it is only a reflection of a very low intake. An answer to this question cannot be secured by simple balance studies since a certain amount of work indicates that the amount of fat in the stool is independent of the dietary intake, especially when the latter is very low (Holt *et al.*, 1919; Hutchison, 1919).

Somewhat closely related to semi-starvation cases are atrophic infants. The amount of fat in the stools from both normal and atrophic children is related to the dry weight of the stool and is influenced to only a slight extent by the fat content of the diet (Holt *et al.*, 1919; Hutchison, 1919). It has been claimed by Fleming and Hutchison (1924) that on a low fat intake (presumably associated with a low food intake) the fecal fat excretion accounts for too large a fraction of this intake to permit a valid balance experiment. When proper allowance was made for the differences in food intake, the percentage fat absorbed by atrophic infants was normal (Fleming and Hutchison, 1924).

Ketosis — Acute Starvation

The development of ketosis during the complete absence of food for a period as short as 2 or 3 days has been recognized for a long time. Probably the first to make such an observation was Külz (1887). He found β-hydroxybutyric acid in the urine from a psychotic patient who went without food for 3 days. The presence of this compound was determined by means of the ferric chloride test and the levo-rotation of the glucose-free urine sample. These procedures were used almost exclusively in all the early work on the excretion of ketone bodies.

Any quantitative reports prior to 1917 were based almost entirely upon the intensity of the levo-rotation of the urine. During Succi's 30-day fast, the urinary excretion of β-hydroxybutyric acid increased gradually until the twenty-third day, when it averaged almost 9 gm. per day (Brugsch, 1905). It remained close to this level for the remainder of the experiment. Grafe (1910) found very similar results in a psychotic woman who abstained from food for about 20 days. As soon as this woman was given even small amounts of sugar the ketone bodies practically disappeared from her urine. Levanzin excreted from 2 to 7 gm. of β-hydroxybutyric acid per day during his 30-day fast (Benedict, 1915).

Although the urinary excretion of ketone bodies during a prolonged fast was practically the same for the man studied by Brugsch (1905) and the woman studied by Grafe (1910), recent work indicates that there is a difference in the amount of these compounds excreted by the two sexes. In 1932 Deuel and Gulick found that normal women suffered much more than men when subjected to a 5-day total fast. Some of the women were severely incapacitated and were unable to continue the fast beyond the fourth day. This difference was reflected in the excretion of ketone bodies, which in certain cases was 4 times as high among the women as among the men. This difference in response could not be correlated with the amount of adipose tissue in the two sexes since some men weighing 240 pounds excreted fewer ketones than the lean women. A similar sex difference in acetonuria during a 4-to-5-day fast was reported by MacKay and Sherrill (1937). They also found that the degree of obesity had no influence on the extent of acetonuria. There was a suggestion of a similar sex difference among the Eskimos. Two women who submitted to a 3-day starvation period showed a higher urinary ketone excretion than the one man (Heinbecker, 1928).

This same sex difference also appears to occur in human beings living on a high fat diet. A woman on such a diet had a higher acetone level in the blood and urine than a man, in spite of the fact that the man received twice as much "fatty acid equivalent" as the woman (Hawley *et al.*, 1933).

Deuel and his group at California made an extended study of the degree of ketosis developed by male and female rats subjected to starvation. Contrary to the findings with human beings, there was no sex difference in the excretion of ketone bodies by fasting rats. In spite of this, the group went on to study other aspects of the problem. Since ketosis developed when carbohydrate metabolism was suppressed (starvation and diabetes) and fat metabolism was accelerated (carbohydrate-free or high fat diet), an investigation of the carbohydrate reserves of the two sexes during starvation was first undertaken. A significant difference in the liver glycogen stores of male and female rats was observed during the first 72 hours of starvation (Deuel *et al.*, 1934; see also Chapter 24). The male rats maintained their glycogen levels much more effectively than the females. This occurred even though the liver glycogen concentrations were the same for all animals at the start of the experiment. There was no difference in the muscle glycogen concentration in the two sexes at any time during the starvation.

The data on the over-all situation in the rat are still fragmentary as far as ketosis is concerned. There are indications from one group of investigators that the ketonemia found in starving female rats is much higher than that in males (MacKay, Wick, and Visscher, 1941). If this is true it occurs in spite of the fact that simple starvation produces no sex difference in the excretion of these compounds. This contradiction might be resolved if some differential sex change in the kidney developed during starvation, but the evidence available would argue against that hypothesis (see Chapter 31).

In order to produce a sex difference in the ketonuria of fasting rats, Deuel and his group (Grunewald *et al.*, 1934; Butts *et al.*, 1934) had to feed small amounts of aceto-acetate. This procedure was defended by Butts (1934) as being comparable to the physiological processes that occur during acute starvation. In spite of this argument, it must be recognized that a marked sex difference in ketonuria exists among human starvation cases, whereas this is not the case with rats unless they are fed small amounts of such ketogenic substances as aceto-acetate.

Another limitation in the application of these experimental findings to human beings is the mildness of the ketonuria seen in fasting rats. When expressed on the basis of surface area, the ketonuria in rats was "35 times less than the average in men during the first 4 days of fasting and 130 times lower than what might be expected if the quantitative relationship for the formation were similar to that in women" (Butts and Deuel, 1933).

Not only does the sex of the animal influence the extent of ketosis developed during a fast but the type of diet immediately preceding it is also of paramount importance. The work of a number of investigators has shown an inverse relationship between the fat content and the glycogen content of the liver. These levels are, to a certain extent, dependent upon the relative amounts of dietary fat on the one hand and of carbohydrates and proteins on the other (for references, see

Chapter 24). The type of fat fed to rats before a starvation period influenced the subsequent ketosis, according to MacKay, Visscher, and Wick (1942). Of 11 fats studied, pork fat, cocoa butter, and olive oil produced less ketosis than the other fats. The sparing action of these 3 fats was associated with an increased urinary nitrogen excretion. On this basis, these investigators suggested that ". . . the nature of the fat in the preceding diet affects the subsequent fasting ketosis only as it may influence the amount of body protein which is catabolized and thus gives rise to antiketogenic material (glucose) during fasting. . . ."

Other dietary factors are also important in this respect. The level of protein, apart from the dietary supply of such lipotropic substances as choline, methionine, and cystine, influenced the extent of fasting ketosis in rats (MacKay et al., 1941). Both the rate at which the ketosis developed and its magnitude were inversely proportional to the protein content of the preceding diet. A protein of poor biological value such as gelatin had much less influence on the ketonemia and ketonuria of fasting rats than did blood albumin (MacKay et al., 1942). Strangely enough, the highest liver glycogen levels at the start of the starvation occurred in the rats fed gelatin, while the lowest occurred in the rats fed edestin and blood albumin. There was a rapid decrease in the glycogen stores of the gelatin-fed rats, so that after about 36 hours of starvation all the liver glycogen levels were approximately the same.

Even the individual carbohydrates have a marked influence on the ketosis developing during the subsequent starvation period. Deuel et al. (1933) found that galactose produced a higher liver glycogen level than glucose when these sugars were fed to dogs that had been without food for 6 days. The opposite results were secured with rats starved for 48 hours. When the carbohydrates were incorporated in the ration of the rat before it was starved, the galactose-fed animals maintained a higher glycogen level than the glucose-fed rats.

A similar conclusion was reached by Butts (1934). He found a smaller urinary acetone excretion when galactose served as the primary source of carbohydrate in place of glucose. Furthermore, a single dose of galactose was more effective than glucose in reducing the ketonuria in his rats once it had developed. Again it must be emphasized that this work was done with starving rats that received small daily doses of aceto-acetate. The inapplicability of the results of these animal experiments to human beings is apparent when the findings of Goldblatt (1925) are recalled. He claimed that galactose and lactose were inactive in overcoming the ketosis in young men who had gone without food for 40 hours. In the same subjects, glucose and fructose rapidly overcame the ketosis.

In addition to dietary factors, the hormones produced by the anterior pituitary and adrenal glands have been shown to influence the amount of fat deposited in the liver and the degree of ketosis developed during a period of starvation (MacKay, 1937; MacKay and Barnes, 1937).

There is some controversy as to the influence of fluids on the course of ketosis during starvation. Ehrström (1922) maintained that when saline infusions were given to his male and female patients with acetonuria, a remission occurred before any food was given. Just the opposite conclusion was arrived at by MacKay et al. (1941) as a result of their experiments with male rats. The animals receiving water had the same blood ketone levels as those that did not receive water. In the case of the females, however, water restriction produced a dramatic increase in the ketonemia. It may be that the degree of water restriction is important in this respect, but further work will be necessary to elucidate this point.

From the results in the literature it is obvious that there is a great deal of individual variation among members of the same sex in the degree of ketosis developing during starvation. This is evident from the tenfold range in the urinary ketone body excretion of 2 women on the third fasting day (Folin and Denis, 1915). A variation as great as this could not be explained on the basis of differences in the adiposity of the subjects, since the woman who excreted the smaller amount of acetone bodies was the heavier. A similar observation was made by Deuel and his group (Deuel and Gulick, 1932; Deuel et al., 1932).

An adaptation to starvation is apparent when a person is subjected to successive fasts. This was recognized by Folin and Denis (1915), who found the values for the urinary acetone excretion lower in the second and third fasts than in the first.

A consideration of the mechanism whereby ketosis develops is beyond the scope of this discussion. Such a task has been undertaken by MacKay (1942). It appears that our present views of ketosis were fairly well stated by Chaikoff and Soskin in 1928. As a result of injecting aceto-acetic acid into eviscerated dogs, they were led to conclude that this compound ". . . is readily utilized by muscle, and it must, therefore, be regarded as an available source of energy in the animal organism, particularly during fasting or as a result of removal of the pancreas or following the ingestion of a carbohydrate-free or high fat diet, under which conditions the ketone bodies make their appearance in increased amounts in blood and urine. These observations also suggest that the ketone bodies are normal intermediary metabolites, but that they escape detection in anything but small amounts in the blood of normal animals because of their rapid utilization by muscle."

A similar suggestion has been made by Drury (1936) on the disappearance of ketone bodies after periods of physical activity. Blixenkrone-Møller (1938) came to a similar conclusion from his perfusion studies on cats' livers.

During acute starvation the reserves of carbohydrate very soon become inadequate to prevent the appearance of ketosis (this occurs within 20 hours of the last meal, according to Drury et al., 1941). The time that elapses before ketosis appears depends upon a number of factors, such as the type and amount of dietary protein, the type of carbohydrate, and the amount of fat in the diet immediately preceding the fast, as well as the sex and the activity of the subject. It has also been suggested that there are species differences in the susceptibility of rats to ketosis (Tidwell and Treadwell, 1946).

In rats the ketosis of starvation tends to be associated with the accumulation of fat in the liver. However, the correlation between the degree of ketosis and the amount of fat in the liver is not a very close one. The ingestion of carbohydrates promptly decreases the liver fat level as well as the ketosis. Aside from indicating that the carbohydrate reduces the rate of fat metabolism, very little can be said about the mechanisms involved in this reaction.

Little work has been done on the ketone levels in the blood during acute starvation. This may be due to the fact that reliable methods have only recently been developed for the determination of ketones in small amounts of blood. Yet such studies may be a far more accurate index of the degree of ketosis than the urinary excretion, since Wick, Sherrill, and MacKay (1940) maintain that the ketone bodies are threshold substances. In one such study (Wick *et al.*, 1940) the blood ketone level was followed in an obese woman who went without food for 10 days. During the first 2 or 3 days there was a slight increase in this level, which was followed by a sharp increase to 15 mg. per 100 cc. of blood by the fifth day, where it remained for the rest of the experiment. The increase in the blood ketones of this woman was very similar to that observed in adult female rats starved for the same length of time (Wick *et al.*, 1940).

Hard physical work, together with acute starvation, produces a very rapid increase in the blood ketone level. This was shown by the work of Taylor *et al.* (1945a) when 4 young men expended 4500, 4000, and 2000 Cal. on the 3 successive days of the fast (the last figure represents the expenditure up through noon). On the second day the resting blood ketones were up to about 15 mg. per 100 cc., and they increased to 20 by the third day. Here, again, there was evidence of an adaptation to starvation since in the fifth fast the blood ketone levels were significantly lower than the above values.

Ketosis — Semi-Starvation

There has been considerable confusion in the minds of many workers as to the extent of ketosis during periods of semi-starvation, and very little valid evidence has been available on this aspect of the problem. Until recently the only available information consisted of data secured during World War I and a few experiments on overweight women who were maintained on reducing diets.

Knack and Neumann (1917) stated that no acetone was excreted by their cases of famine edema as determined by the ordinary methods (the methods used were not given) but that the Scott-Wilson procedure (no reference to the method was given) showed the excretion of acetone to be 12 mg. per day compared to a normal value of 3–5 mg. Statements have also been made by other workers that there was no acetonuria among the edema cases seen in Germany during World War I (Hülse, 1918a, 1918b; Schittenhelm and Schlecht, 1918). In view of the paucity of data presented by these workers, their reports can only be accepted as suggestive. No mention was made by Benedict *et al.* (1919) of ketosis among their subjects, who over a period of 3 months lost 10 per cent of their original body weight.

During World War II a few scattered reports (Gounelle *et al.*, 1941; Bose *et al.*, 1946) indicated the absence of any ketonuria among persons who had famine edema. Experimental work at that time further emphasized the fact that the ingestion of even small amounts of food can prevent the development of ketosis during a period of starvation (Kartin *et al.*, 1944).

The work of Deuel, Gulick, and Butts (1932) indicated a considerable acetonuria during the ingestion of a calorically inadequate diet made up of protein and fat. This is a highly unusual form of semi-starvation, as is shown by the nitrogen balances. The nitrogen loss throughout the experiment accounted for only a small part of the body weight loss. It is very likely that the dehydration resulting from the acidosis was responsible for most of the weight reduction. Furthermore, the symptoms of physical and mental lassitude experienced by these subjects as soon as the dietary restriction was imposed are wholly unlike any seen early in semi-starvation.

TABLE 249

URINARY EXCRETION OF ACETONE BODIES during the semi-starvation phase of the Minnesota Experiment. All values are expressed as mg. of acetone per 24 hrs.

Subject	C	S12	S24
23	154	114	77
104	58	116	234
109	76	130	99
112	117	103	142
119	183	92	224
123	130		181
129	50	52	
M	107	101	160

Because of the absence of any quantitative reports on the excretion of ketone bodies by non-obese subjects during a period of uncomplicated semi-starvation, such studies were undertaken in the Minnesota Experiment. From time to time throughout the course of the semi-starvation period qualitative tests were made for ketone bodies in the urine of the subjects. The nitroprusside test (Hawk, Oser, and Summerson, 1947) was negative on all occasions. Quantitative analyses were made by the method of Van Slyke (1917) on the urine samples from 7 of the subjects. There was a slight increase in the average excretion of ketone compounds when the values for S24 were compared with the control levels (see Table 249). This increase was not significant in view of the large variation in the excretion values for the subjects at S24. All the individual values were within the normal range given by Van Slyke (1917). The absence of any ketosis in spite of a weight loss of 24 per cent of the original weight indicates that, at least under these conditions, semi-starvation is not accompanied by the production of excessive amounts of ketone bodies.

Ketosis — Reducing Diets in Obesity

There are 2 reports on the absence of ketosis during weight reduction by obese individuals. Mason (1927) controlled the diet of 5 women weighing 110 to 173 kg. so that they lost an average of 336 gm. per day over periods varying from 63 to 100 days (average 77). Daily tests of the urine for acetone bodies by means of the ferric chloride test and frequent determinations of the plasma carbon dioxide content showed no signs of ketosis throughout the experiment. Dick and Goldner (1943) maintained a 117 kg. woman on 427 Cal. per day for 133 days. During this period she lost 25.6 per cent of her original weight (226 gm. per day).

These workers, on the basis of the determination of ketone bodies in the blood, stated that "a mild ketosis developed slowly and showed a tendency to decrease during the second half of the observation period." There must be an error in their values, however, since the blood ketones ranged from 16 to 25 mg. per 100 cc. of blood. According to Van Slyke and Fitz (1917), whose method they used, the normal blood level is one tenth of their values. Their figures for the urinary excretion of ketone bodies (630 to 1600 mg. per day) are also far above normal when compared with Van Slyke's (1917) results (16 to 280 mg. per day). The serum carbon dioxide and pH were within normal limits throughout the period of the reduced diet. This is compatible with their statement that there was no ketosis.

Although the presently available work is not as convincing as it might be, it seems to indicate that ketosis does not develop during periods of weight reduction by obese individuals. In this respect obese individuals are similar to normal persons receiving an inadequate caloric intake.

Fat Content of Human Milk

There is considerable controversy as to the influence of periods of restricted food intake on the fat content of human milk. During World War I a number of reports from Germany ascribed the poor condition of the nursing infants to the low fat content of the mothers' milk. This in turn was related primarily to diets that were low in fat. In most of these cases (Bergmann, 1919; Klotz, 1920) the results were based on the analysis of milk from one or two women. A more extensive investigation was made by Tasch (1921), from which he reported a clinical impression of the women's nutritional condition together with the volume and fat content of the milk. For his entire series both the volume and the fat content of the milk were normal even for those women who were emaciated. Lederer (1922) claimed that the proximate analysis of the milk from 7 control women (source and time of observation not stated) showed a higher galactose and fat content than that from 9 German women whose infants were not gaining weight at the normal rate. The implication is that the condition of the milk was attributable to the poor diet of the German mothers, but analysis of Lederer's data indicates that the differences which impressed him are statistically insignificant because of the great variation in the results from both groups.

On the basis of the analyses of a large number of milk samples from Finnish women during World War II, Salmi (1944) concluded that the average fat content had decreased by 25 per cent when compared with prewar standards. It is not possible to subject these results to statistical analysis, but in view of the great interindividual variation in each group (prewar mean = 4.1 per cent with a range of 1.9 to 9.6; wartime mean = 3.1 per cent with a range of 1.1 to 6.5) the differences between the means are relatively insignificant. The above results are of limited value since Salmi observed, in addition to a large interindividual variation, a tremendous day-to-day variation in the fat content of the milk for the same subject. This may very likely have been due to incomplete emptying of the breasts, which results in the retention of the fat-rich milk usually secured at the end of milking (see Pasch, 1921, for a discussion of the methods of collecting milk).

Randoin *et al.* (1942) determined the fat content in the milk from 14 French women who were on calorically insufficient diets during World War II. These results were compared with the values from 10 women who were on adequate diets. Neither one of these values was significantly different from the values the investigators accepted as normal.

Unfortunately, in none of these cases is there an adequate evaluation of the nutritional condition of the subjects. There is no indication as to whether the women showed any change in weight during the period of observation. So many factors other than diet influence milk secretion that it is hazardous to draw any conclusions about the composition of milk from the growth of nursing infants. Furthermore, the care required to insure complete emptying of the breasts makes even the best fat analysis of dubious value unless precautions have been taken to make sure that the sample is representative. These facts may in part explain the lack of agreement in the literature on the influence of semi-starvation on both the quantity and composition of human milk. It appears from the available data that diets such as are used in war-ravaged countries have very little if any influence on the composition of milk, but that a marked restriction in food intake will soon produce a reduction in the volume of milk secreted.

Experiments with cows deprived of all feed showed that milk was produced at the expense of body tissue (Smith *et al.*, 1938). When the feed was withheld for 12 days, the volume of milk at the end of the experiment was one sixth of the pre-starvation control value. During the starvation period, however, the concentration of fat in the milk had doubled.

CHAPTER 23

Mineral Metabolism

Urinary Excretion of Minerals in Acute Starvation

W ith the initiation of an acute starvation period, there is an abrupt decrease in the mineral intake. In most cases the intake drops to zero. This is especially true if the subject, like Levanzin, drinks nothing but distilled water (Benedict, 1915). Not all the fasters studied in the past were limited in this way. Some, such as Succi (Brugsch, 1905), used mineral water, the composition of which was not given. In many cases the type of water used was not stated, but it is likely that tap water was consumed (Lehmann *et al.*, 1893). There are variable amounts of minerals in the drinking waters of different regions, and if a proper balance study is to be made, both the composition and the volume of the water consumed must be known.

When acute starvation starts, the rate of decrease of urinary excretion of the different minerals varies considerably. The excretion of chloride may drop off dramatically to about 0.2 gm. or less by the fifth day of starvation (see Figure 76). As Benedict (1915) has pointed out, the excretion of chloride shows a great deal of individual variation. Some men, such as Succi, Cetti, and Tosca, continued to excrete approximately a gram of chloride per day for the first 2 weeks. Since in some of the reports no mention is made of the water supply, a part of this chloride variation may be referable to the water intake.

None of the other minerals studied shows the sharp fall exhibited by Levanzin's chloride excretion. For instance, sodium and potassium excretion decreased on a percentage basis at a slower rate than chloride, but even this rate was faster than that of calcium (Figure 76). The almost constant urinary calcium excretion during both control and starvation periods may be due to the fact that even on a very limited intake the largest amount of calcium lost by the body appears in the stool, with a small but variable fraction in the urine (Owen, 1939; Steggerda and Mitchell, 1939).

There has been considerable discussion as to the origin of the various minerals found in the urine during starvation. The most obvious conclusion was that as the body proteins were metabolized, the minerals contained in the protoplasm were excreted. On this basis many calculations have been made of the ratio between the excretion of the different minerals and that of nitrogen. Although these ratios, especially toward the end of a long starvation period, were frequently similar to those secured from analyses of muscle tissue, there were still enough discrepancies to make the explanation highly questionable.

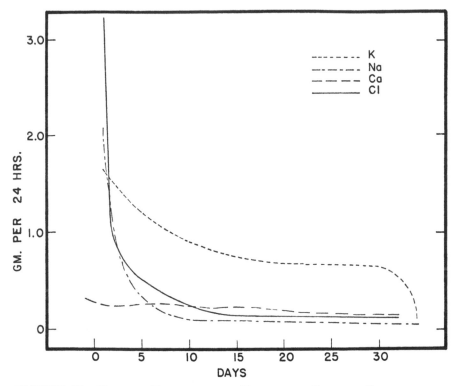

FIGURE 76. URINARY EXCRETION OF POTASSIUM, SODIUM, CALCIUM, AND
CHLORIDE by Levanzin during a 31-day acute starvation
period (Benedict, 1915).

When Morgulis (1923) reviewed this field, he concluded that "although the
data on inorganic metabolism in fasting are not very extensive, and few of the
recorded results cover the entire course of inanition, it may nevertheless be
stated that the problem of the source of the mineral constituents present in fasting urine is practically insoluble. The various hypotheses and interpretations con-
strued from these results are but crude guesses at best. The premise that there
is a single source — for instance the musculature — is fundamentally wrong, because the mineral constituents found in the urine during fasting have a multiple
origin, and so long as we have no means to appraise the actual contribution which
the individual tissues make to the fund from which the starving organism draws,
it is futile to attempt to solve this problem of the ultimate origin of the inorganic
substances."

Just before this quotation appeared in print, an extended study on the mineral
metabolism of fasting children did much to supply some of the missing facts.
Gamble, Ross, and Tisdall (1923) determined the changes in the minerals in both
the blood and the urine of children who were maintained without any food for
periods of from 4 to 15 days. In one of these experiments they calculated the
urinary excretion of the minerals on the assumption that it was derived from the
protoplasm represented by the urinary nitrogen, and for potassium and magne-

sium these values approached those actually observed after the first 6 days. The ratio of the found to the calculated values was originally considerably above 1 and then changed to considerably below 1. Calcium excretion throughout the experiment was much higher than the calculated values (see Table 250).

TABLE 250

URINARY EXCRETION OF FIXED BASE DURING A 15-DAY FAST AND A SUBSEQUENT 3-DAY CARBOHYDRATE DIET. The calculated values are the amounts of base derived from the metabolized body tissue as indicated by the urinary nitrogen. All values are given in milliequivalents except where otherwise indicated. F:C represents the ratio of found to calculated values. (Gamble *et al.*, 1923.)

3-Day Periods	Nitrogen (gm.)	Na			K		
		Found	Calculated	F:C	Found	Calculated	F:C
Fasting							
I	21.3	78.3	21.9	3.6	89.7	51.5	1.7
II	16.1	29.3	16.5	1.8	51.2	38.9	1.3
III	13.3	4.2	13.7	0.3	28.3	32.2	0.9
IV	11.5	7.7	11.8	0.7	28.6	27.8	1.0
V	10.3	6.1	10.6	0.6	27.7	24.9	1.1
Total	72.5	125.6	74.5	1.7	225.5	175.3	1.3
Carbohydrate diet							
VI	7.6	1.6	7.8	0.2	6.7	18.4	0.4

3-Day Periods	Plasma Protein Percentage	Mg			Ca		
		Found	Calculated	F:C	Found	Calculated	F:C
Fasting							
I	8.1	19.5	11.0	1.8	11.5	2.3	5.0
II	7.8	13.5	8.3	1.6	22.4	1.8	12.5
III	8.6	7.7	6.9	1.1	20.2	1.5	13.4
IV	8.1	5.7	5.9	1.0	13.5	1.3	10.4
V	7.9	4.8	5.3	0.9	12.4	1.1	11.3
Total		51.2	37.4	1.4	80.0	8.0	10.0
Carbohydrate diet							
VI	7.4	1.7	3.9	0.4	5.6	8.0	7.0

The discrepancy between the observed and calculated values in the case of potassium and magnesium was explained by the dehydration which presumably occurred at that time. Evidence for this loss of water was seen in the very much greater weight loss that occurred during the first 2 days. This weight loss was 1.46 kg., while it was only 0.13 kg. during the last 2 days of the fast. Additional theoretical support for the fluid loss in the early phase of the experiment is the suggestion that a high rate of fat metabolism results in a certain amount of dehydration. This is based largely upon the reports of Bischoff and Voit (cf. Voit, 1881) and Benedict and Milner (1907). By means of various assumptions and calculations, Gamble, Ross, and Tisdall (1923) showed that the extra potassium and magnesium could reasonably have come from the extracellular water. This in turn also explained the changes seen in the excretion of sodium, since the concentration of sodium in the extracellular fluid is much higher than that within the cells. Similar and related calculations were made for the other results in these starvation experiments. All confirmed the idea that large amounts of extracellular fluids are lost in acute starvation and that the minerals thus liberated from the body make up a considerable portion of the urinary excretion of sodium and chlorides. It is unfortunate that no such work in more recent times has been combined with an actual study of the changes in fluid volumes within the body.

The course of urinary mineral excretion can be markedly influenced by the amount of sweating the subject undergoes during the starvation period. When the volume of sweat becomes significant, the amount of sodium chloride lost by that means cannot be neglected.

Calcium and Phosphorus in Acute Starvation

The behavior of calcium is more difficult to interpret since it tends to be excreted throughout a starvation period at a rate which may be of the order of 10 times that calculated on the basis of the body tissue lost (Gamble et al., 1923). Neither the extracellular fluid nor the soft tissue of the body contains much calcium, so it seems probable that a large share of this mineral in the urine in acute starvation must come from the skeleton. Gamble, Ross, and Tisdall (1923) thought that the calcium carbonate rather than the more insoluble phosphate was liberated from the bones in starvation. Actually, the phosphorus excreted during this period was more than enough to account for the calcium, but in view of the fairly large amount of phosphorus in tissue, most of it was presumed to come from protoplasm.

Other studies (Freund and Freund, 1901; Cathcart, 1907; Benedict, 1915) have also shown that the excretion of calcium is considerably greater than can be accounted for on the basis of the soft tissues metabolized in acute starvation. It seems fairly well established that acute starvation must be associated with a certain amount of decalcification of the bones.

During starvation there is a great deal of variation in the amount of urinary phosphorus that can be attributed to the skeleton. This was emphasized by Morgulis (1923), who, on the basis of Munk's work (1894), calculated the amount of urinary phosphorus that might have been derived from muscle. The figures for the urinary phosphorus excretion of Levanzin (Benedict, 1915) and of Beauté (Cathcart, 1907) are considerably above what would have been expected if all the phosphorus came from the muscle. In these cases the differences must have come from the phosphorus in the bones. For Kozawa (Watanabe and Sassa, 1914) the total urinary phosphorus was actually below that calculated from the protoplasmic catabolism. There is no ready explanation for this fact.

Morgulis noted that the ratio of urinary nitrogen to phosphorus was "low even at the beginning of the fast when it is reasonably certain that the skeleton could not be involved. The weight of the bones in the early phases of inanition may actually increase."

Wellmann (1908) attempted to secure information on the relative contributions of the skeleton and the tissue to the calcium and phosphorus deficits occurring during starvation. He starved rabbits until they lost approximately 40 per cent of their original body weight. The calcium which would have come from their soft tissues was considered insignificant, while the corresponding value for phosphorus was 1.3 gm. On the basis of balance studies, the bones must have contributed 1.6 gm. of calcium and 0.6 gm. of phosphorus. These figures agreed very well with the values secured by actual analysis of the bones in these starved rabbits and in normal rabbits. Although there was a reduction in the weight of the skeleton due largely to the loss of fat, there was no significant change in the percentage composition of the fat-free dry substance; the mineral lost was only a small part of that originally in the bones.

A few other observations have been made on the changes produced in the composition of the skeletal ash during starvation. E. Voit (1905b) compiled the fragmentary reports available at that time on the calcium and phosphorus content of the bones of normal and starved animals. In most cases only one starved animal was involved, and the "control" data on the well-fed animal were supplied from reports by other scientists. The early work of Gusmitta (1893) merely suggests that during starvation there is a greater loss of calcium carbonate than of the phosphate. Unfortunately, this work is further complicated by the metabolic responses following in the wake of bone injury. Gusmitta removed the bones from the limbs on one side of his animals before the start of the starvation period and compared them with the bones on the other side removed after starvation. Such a procedure would undoubtedly be complicated by the negative calcium and nitrogen balances which follow even the simplest fractures (Howard, 1944; Howard et al., 1945).

A similar criticism can be made of Kernwein's (1937) work on the deleterious influence of starvation on the healing of fractures. He found that when the ulna was broken in rabbits that were given nothing but a dilute solution of sodium chloride, there was a marked retardation in the ossification of the callus. This condition became more marked with the lengthening of the survival period. In those animals that survived the longest, there was some osteoporosis in the callus that had previously been calcified.

It would seem desirable to investigate carefully the changes in the composition of bone resulting from both acute and semi-starvation. Not only would this fill some of the more obvious gaps in our information but it might also serve as a base line from which urgently needed studies in the field of osteopathies (osteoporosis, osteomalacia, etc.) might be initiated. At present most of the work in this field has been done with herbivorous animals (Meigs et al., 1919; Goto, 1922; Carpenter, 1927; Aylward and Blackwood, 1936). The influence of starvation on these animals is exemplified by the work of Meigs et al. (1926). After an extensive study of pregnant and lactating cows, they concluded that a cow may continue to lose either calcium or phosphorus and to gain the other element for two or three weeks. However, a long continued loss of calcium from the body will be associated with a loss of phosphorus, even when the assimilable phosphorus is present in the diet. The special difficulties of such work were emphasized by the psychological disturbances observed by Meigs, Blatherwick, and Cary (1919). The collection of urine and stool samples upset the cows to such an extent that it interfered with the assimilation of calcium and to a lesser degree with that of nitrogen and phosphorus.

Sulfur in Acute Starvation

The excretion of sulfur during starvation has attracted relatively little attention. Lehmann et al. (1893) determined both the inorganic and the organic sulfur excreted in the urine during 10-day and 6-day starvation periods on Cetti and Breithaupt, respectively. In both cases the daily inorganic sulfur content of the urine decreased throughout the period when no food was consumed; on the resumption of eating it slowly returned to normal. The organic sulfur showed a variable response in these two cases; it decreased in Cetti's case for the first 3 days and then increased considerably. In Breithaupt's case there was a continuous decrease for the entire 6 days.

Cathcart (1907) found that the total urinary sulfur excreted by Beauté dropped to half its pre-fasting value as soon as food was withheld and that it remained more or less constant at this lower value. The variations in total sulfur excretion here, as in the cases of Cetti and Breithaupt, tended to follow the nitrogen excretion. In both cases the ratio of nitrogen to sulfur was slightly higher than would be expected if all the sulfur came from protein; this would imply that the sulfur-rich proteins were preferentially retained by the body during starvation.

In Beauté's case the neutral sulfur remained fairly constant throughout the 14 days of starvation. A somewhat similar finding was seen in dogs that had been starved for 3 or 5 days (Vassel et al., 1944). On the face of it, this might be used

TABLE 251

DISTRIBUTION OF URINARY SULFUR DURING ACUTE STARVATION IN DOGS. All values are in gm. per 24 hrs. (Vassel et al., 1944.)

	Dog 55					Dog 53				
	N	Inorg. S	Org. S	Ether S	S_2O_3	N	Inorg. S	Org. S	Ether S	S_2O_3
Normal										
Minimum	10.7	0.46	0.08	0.04	6.6	9.8	0.43	0.07	0.02	5.6
Maximum	13.2	0.59	0.10	0.05	7.6	13.7	0.62	0.10	0.04	8.2
Days of fasting										
1	4.6	0.19	0.04	0.02	0.0	6.5	0.25	0.06	0.02	1.8
2	3.9	0.15	0.05	0.02	0.0	4.6	0.21	0.10	0.01	0.2
3	3.8	0.16	0.05	0.02	0.0	4.9	0.22	0.06	0.02	1.6
4						4.4	0.21	0.06	0.01	0.0
5						4.2	0.20	0.05	0.01	0.0
Days of refeeding										
1	7.8	0.30	0.05	0.03	6.0	10.9	0.47	0.08	0.04	6.8
2	10.1	0.42	0.07	0.04	10.6	10.5	0.44	0.07	0.04	0.0
3	12.7	0.51	0.09	0.03	8.6	11.9	0.45	0.07	0.03	5.8
4	11.4	0.42	0.09	0.05	7.2	10.7	0.44	0.08	0.02	5.8

as support for Folin's (1905) contention that the neutral sulfur fraction represents endogenous metabolism. At present very little more can be said.

Recent reports on animal experiments have indicated a change in the urinary excretion of thiosulfate by dogs during starvation. Vassel, Partridge, and Crossley (1944) found a prompt cessation of thiosulfate excretion in dogs subjected to 3 to 5 days of starvation (see Table 251). When eating was resumed, the thiosulfate in the urine reappeared with equal speed. They interpreted this as indicating that the thiosulfate came from the action of bacteria in the gastrointestinal tract. Exactly opposite results were secured by Fromageot and Royer (1945) when they starved cats, rats, guinea pigs, and rabbits; in each case the thiosulfate excretion increased over that of the control period.

Changes in Blood Mineral Concentration in Acute Starvation

Practically no work was done on the changes in blood constituents during acute starvation until Gamble, Ross, and Tisdall (1923) made their extensive studies on children. As ketosis developed there was an almost proportional reduction in the plasma bicarbonate. When the ketosis was most severe, the bicarbonate was occasionally decreased to two thirds of its normal value. This drastic alteration in the concentration of the anions in the plasma was accomplished with only a small change in the level of fixed base (see Table 252). The place of the bicarbonate in the plasma was taken by the ketone bodies. Although the over-all acid-base relationships were maintained, there were some changes in the specific minerals. There was a slight decrease during this period in the concentration of sodium, potassium, and magnesium. There was no change in the level of calcium.

TABLE 252

CONCENTRATION OF MINERALS IN THE PLASMA OF AN 8-YEAR-OLD BOY WHO WENT WITHOUT FOOD FOR 4 DAYS. All values are given on the basis of 100 cc. (Gamble et al., 1923.)

	Na (mg.)	K (mg.)	Ca (mg.)	Mg (mg.)	CO_2 (vol. %)	Cl (as mg. NaCl)
Before fasting					60	617
End of 4-day fast	327	22.3	10.3	2.7	37	507
After 2 days' refeeding	329	18.2	10.3	1.6	60	577

TABLE 253

CHANGES IN COMPOSITION OF SERUM FROM DOGS SUBJECTED TO VARIOUS STAGES OF INANITION. These dogs received no food but water ad libitum. All values are in mg. per 100 cc. except where otherwise indicated. (Morgulis, 1928.)

Condition	Number of Dogs	Na	K	Ca	Mg	Cl	P	Serum CO_2 (vol. %)
Control	14	359	21.5	11.0	2.3	423	4.3	49.3
20–25 per cent weight loss..	14	360	22.3	10.6	2.4	426	3.9	48.2
35–43 per cent weight loss..	13	354	20.0	9.6	2.2	405	3.9	50.6

In another starvation study made by Gamble, Ross, and Tisdall (1923), in which the subject was a young girl, the plasma sodium decreased through the sixth day of the fast and then started to increase; when the fast ended on the fifteenth day, the plasma sodium was back to normal. The chloride concentration, however, decreased progressively throughout the experiment. Of these two minerals, the sodium concentration in the blood was maintained a little more effectively than that of the chloride. This, as the investigators pointed out, may be due to the slightly higher concentration of sodium than of chloride in muscle water (based on analyses of Katz, 1896). Morgulis and Edwards (1924) observed a progressive decline in the blood chloride concentration of a dog acutely starved for 30 days.

The question of the changes in the calcium and phosphorus levels of the blood has aroused some interest since McCollum *et al.* (1922b) showed that starvation promoted the healing of rachitic lesions in the rat. Some of the work on the changes in the calcium level of the blood during starvation indicated a considerable variety in the response (Cavins, 1924; Meglitzky, 1926; Gates and Grant, 1927; Schazillo and Konstantinowskaja, 1928; Hefter and Yudelovich, 1934). The work of Morgulis (1928) with dogs (see Table 253) and of Gamble, Ross, and Tisdall (1923) with children (Table 252) showed that there were only slight changes in the calcium levels even over prolonged periods of fasting. In Morgulis' dogs and cats there was no significant change until they lost from 35 to 43 per cent of the original body weight, and the explanation was offered that a change in the albumin-globulin ratio resulted in a decreased "binding capacity" for calcium (Morgulis, 1929; Morgulis and Perley, 1929).

The blood phosphorus level during acute starvation has been variously reported: (1) a slight increase during the early part of a fast, usually followed by a return to normal as the experiment continued (Gates and Grant, 1927), (2) a slight decrease (Morgulis, 1928), or (3) a variable response (Goto, 1922). Morgulis (1928) concluded that the slight changes observed in the levels of the blood minerals were too small to explain the healing of rickets seen in starvation. It might be noted that Cavins (1924) found that rachitic rabbits were not able to withstand starvation as well as normal animals.

A recent report by Sunderman (1947) confirms most of the above findings on the blood minerals in fasting. A man who went without food for 45 days was studied during the last two days of his fast after he had lost 30 per cent of his original weight. When the values secured at the end of the starvation period were compared with those seen a month after eating had been resumed, it was noted that there had been a reduction of about 25 per cent in the serum chlorides with only a very slight change in the concentration of total base (see Table 254). Both the serum phosphate and calcium showed no appreciable change, whereas the magnesium was slightly increased. The caution required in the interpretation of such data is indicated by the serum protein changes in this man. The resumption of eating produced a considerable reduction in the concentration of the serum proteins, which would indicate that at the end of the starvation period the subject had been dehydrated. The rapid increase in body weight in the early part of the refeeding period confirms this conclusion.

TABLE 254

SERUM CONSTITUENTS in a man at the end of a 45-day fast and during the subsequent refeeding period (Sunderman, 1947).

Day	Cl (m.eq./l.)	CO$_2$ (vol. %)	Total Base (mg. %)	Inorg. P (m.eq./l.)	Ca (mg. %)	Mg (m.eq./l.)	Total Prot. (%)
45th starvation	74.0	100	3.4	141	12.2	2.7	6.2
5th refeeding	107.8	63		146	10.7	1.6	5.4
43d refeeding	98.7	61		148	9.8	1.9	6.7
Normal range	99–104	55–60	3–4	143–148	9–11	1.9–2.0	6.7

Day	Alb. (%)	Glob. (%)	Total Chol. (mg. %)	Ester Chol. (mg. %)	Urea N (mg. %)	Uric Acid (mg. %)	Sugar (mg. %)
45th starvation	5.0	1.2	198	184	24	9.0	63
5th refeeding	2.8	2.6			14	1.1	
43d refeeding	4.0	2.7	190	110	12	4.1	83
Normal range	3.3–4.3	2.2–2.8	170–190	90–114	9–17	3–4	80–110

Urinary Excretion of Minerals in Semi-Starvation

The amount of the various minerals excreted by semi-starved individuals is of no consequence unless the intake is also known. Schiff (1917) was impressed by the extraordinarily large amounts of sodium chloride excreted by some of his cases of *Ödemkrankheit*. Such an observation loses some of its significance when it is realized that the salt intake was also very high. This was brought out by Jansen (1920), who studied the salt balance of patients with marked edema due to starvation. All of these men had urinary salt excretions that exceeded 45 gm. per day, but their intakes were also very large (see Table 255). The big negative balances seen in these individuals could not be maintained for long periods since the total sodium chloride content of the normal adult is only about 175 gm. (Hawk *et al.*, 1947).

It is difficult to explain the large voluntary salt intakes of people living on restricted caloric intakes. Maase and Zondek (1920) attempted to relate this to the large amount of potassium in the primarily vegetable diet upon which their patients lived. The antagonistic action of sodium and potassium would require

TABLE 255

SODIUM CHLORIDE BALANCES with undernourished subjects (*Ödemkrankheit*). The balance period extended over 6 days. All values are in gm. of NaCl per day. (Jansen, 1920.)

Subject	Intake	Urine	Balance
I	35.4	45.9	−10.4
II	45.4	64.2	−18.8
III	35.4	45.1	−9.6
IV	35.4	48.0	−12.6
M	37.9	50.8	−12.7

extra amounts of salt in the diet. This in itself, however, could never explain the increase in the salt intake since an addition of less than 10 gm. of salt would amply provide for the increase in potassium (McCance and Widdowson, 1940). The excretion of salt in undernutrition is related to the water balance; this relationship will be discussed in a subsequent section.

There have been a number of references to the increased excretion of calcium in undernutrition (Knack and Neumann, 1917; Maase and Zondek, 1920). Here again the only controlled studies were those made by Jansen (1920). The 6-day balance studies that he made on his patients showed considerable individual variation (see Table 256). Some of the men retained considerable amounts of calcium whereas others showed equally high negative balances. There was no apparent correlation between the calcium and the phosphorus balances. It has been suggested that balance periods as short as these may not be long enough for a proper evaluation of calcium metabolism (Steggerda and Mitchell, 1946a, 1946b). All of Jansen's subjects received the same food throughout this balance period, whereas previously they had been on diets providing a considerable variation in the supply of calcium. The rate at which the men attained calcium equilibrium was undoubtedly related to the previous diet.

TABLE 256

Calcium and Phosphorus Balance Studies on patients with *Ödemkrankheit*. The experiments were for 6 days. All values are given as average gm. per day. (Jansen, 1920.)

Subject	Food		Urine		Feces		Total Excretion		Balance	
	CaO	P_2O_5	CaO	P_2O_5	CaO	P_2O_5	CaO	P_2O_5	CaO	P_2O_5
I	1.45	3.64	0.46	2.47	1.10	1.25	1.56	3.72	−0.11	−0.08
II	1.45	3.64	0.47	2.17	1.08	0.97	1.55	3.14	−0.10	+0.50
III	1.45	3.64	0.49	2.23	1.92	2.21	2.41	4.45	−0.96	−0.81
IV	1.45	3.64	0.20	1.17	0.70	1.30	0.90	2.47	+0.55	+1.17
V	1.45	3.64	0.48	1.59	1.04	1.88	1.52	3.47	−0.07	+0.17
VI	1.45	3.64	0.36	2.12	1.09	1.29	1.45	3.41	0.0	+0.23
VII	1.45	3.64	0.20	3.26	1.07	1.26	1.27	4.52	+0.18	−0.88

The high intake of calcium in Jansen's (1920) experiment was probably comparable to that in the diets available to the inhabitants of North Central Europe during periods of food shortage. The diet for the semi-starvation period in the Minnesota Experiment was devised to simulate such a regimen. It consisted primarily of vegetables and provided 1.06 gm. of calcium oxide per day.

The relatively high calcium intake of the Europeans may explain the relative infrequence of severe skeletal alterations during some of the recent wars. Loll (1923) observed no change or a slight increase in the percentage of calcium and a slight decrease in the phosphorus content of bones secured at autopsy from cases of hunger osteomalacia (see Table 257). The bones from these 4 cases were obtained in Vienna during World War I when there were many complaints of this disease. Similar analytical results were secured by Ruiz-Gijon (1941) for

TABLE 257

COMPOSITION OF BONE ASH secured from persons who died of starvation in Vienna during World War I. Each value is the mean of values from 4 persons. All values are in gm. per 100 gm. of bone. (Loll, 1923.)

	Normal Subjects	Bone (gm. per 100 gm.)		
		Starvation Subjects		
		Rib	Vertebra	Tibia
CaO	51.9	61.3	70.2	53.1
MgO	0.82	0.77	0.77	0.82
P_2O_5	38.8	33.3	23.4	36.7
Residue	8.60	4.82	5.18	8.06

bones from severe starvation cases seen in Madrid at the time of the Spanish Civil War.

The behavior of the skeleton in undernutrition is dependent upon the type of diet being fed (Zucker and Zucker, 1946). Where the calcium and phosphorus are not specifically restricted, the limited evidence indicates that in animals subjected to semi-starvation (Aron, 1911; Kellermann, 1939) or to acute starvation (Mendel and Rose, 1911) there is a normal percentage of ash in the bones. This is true even though the body weight losses range up to 50 per cent of the original weight. Such animals may even show calcium balances that are slightly positive (Rubin and Krick, 1936).

Salts and Their Influence on Water Retention in Semi-Starvation

The water balance of normal individuals is influenced to only a slight extent, if at all, by the ingestion of fairly large amounts of water (Adolph, 1921). Veil (1916) gave a normal subject as much as 6.7 liters of water per day for 11 days and found no appreciable water retention at the end of this period. There are few reports on large salt intakes by normal subjects. The evidence from animal experiments indicates that a high salt intake has resulted in a slight retention accompanied by some increase in body weight. When the dietary salt was reduced, the dogs lost the extra body weight, together with the retained salt and water (Weech et al., 1936). Grant and Reischman (1946) observed a similar effect when they gave 8 healthy subjects 30 gm. of sodium chloride in addition to the amount in their ordinary diet. All the subjects showed gains in weight up to 5 lbs., with an increase in the plasma volume and the venous pressure (2 to 12 cm. of water) but no change in arterial pressure.

The results of similar experiments made on semi-starved animals or individuals are dependent, to a certain extent, upon the degree of starvation. When water alone is added to the diet, there is only a slight and at most a transitory influence on the body weight. Jansen (1920) noted a prompt excretion of the large volumes of water he gave his starvation patients. The addition of sodium chloride to the water or the diet of these patients produced a prompt increase in weight, and within a very short time there were unmistakable signs of edema. A few days after the withdrawal of the extra salt and water, there was a rapid decrease in body weight from a peak of 40 kg. to 34 kg., which was the value at the start of the experiment (see Figure 77). The decrease in body weight actually started while the subject was still receiving large amounts of salt and water. The salt supplement at that time apparently was 22.2 gm., which was in addition to the 2.2 gm. provided in the diet. With this, the patient received 4.5 liters of water each day. This spontaneous dehydration even in the presence of a high salt and water intake has also been observed by Weech, Goettsch, and Reeves (1936) in dogs maintained on a low protein diet. These sudden changes of body water in both semi-starvation and protein deficiency states still defy explanation.

That the body's ability to store salt, even under normal conditions, is very limited is shown by the rapidity with which the excretion becomes adjusted to the intake. Jansen (1920) was impressed by this as a result of forced-fluid studies on his semi-starved men. When a high water intake followed a period of high salt intake, the water excretion was as prompt as that following a low salt intake. Had extra salt been stored in the body during the high salt regimen, one might expect the succeeding high water intake to have led to a retention of water.

Experiments similar to those of Jansen were carried on by Schittenhelm and Schlecht (1919b). They found no essential difference between normal subjects and their starvation patients (German civilians during World War I) as far as excretion of salt and water was concerned as long as these substances were given separately. When the salt and water were combined, there was a tendency for the starvation cases to retain the salt for a considerable period of time.

In semi-starvation a high sodium chloride intake results primarily in an increase in the extracellular fluid space, with only a nominal change in the concentration of salt in the body fluids. Jansen (1920) found that high salt supplementation at first produced a marked increase in the serum chloride level, which soon decreased to such an extent that it was below that observed before the start of the experiment (see Figure 78). The changes in the hemoglobin levels of these subjects indicated a slight hydremia toward the middle of the period of high salt and water intake. Weech, Goettsch, and Reeves (1936) had similar results in their studies with dogs. By means of the vital red technique, they showed an actual increase in the plasma volume of the dogs receiving extra salt. This was confirmed by the changes in plasma protein, hemoglobin, and hematocrit.

The importance of the skin in chloride metabolism in patients with edema was emphasized by the work of Anastassopoulos (1946). Skin from normal individuals contained from 115 to 379 mg. of sodium chloride per 100 gm., whereas that from famine edema cases had more than 424 mg. per 100 gm. If this work is confirmed, the skin, then, is one tissue in which dramatic changes occur in edema.

Two reports from China emphasized the differences in the action of various salts in influencing the degree of nutritional edema. Liu et al. (1932) gave 2 boys with marked edema from 12 to 20 gm. of sodium bicarbonate per day for 4 to 8 days. In both cases there were marked increases in weight which reached a maximum by the third or fourth day; then the edema subsided and the weight decreased in spite of continued alkali therapy. When 30 cc. of 10 per cent hydro-

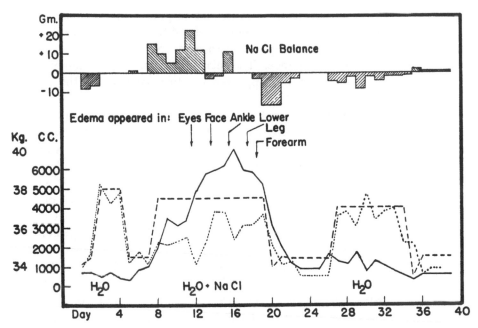

FIGURE 77. INFLUENCE ON THE BODY WEIGHT OF SEMI-STARVATION SUBJECTS BY THE ADDITION OF WATER AND WATER AND SALT TO THE DIET (Jansen, 1920).

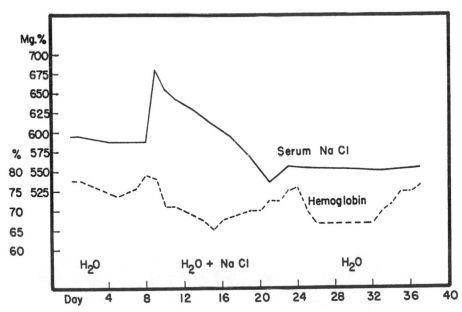

FIGURE 78. INFLUENCE ON THE LEVEL OF SERUM CHLORIDE AND OF HEMO-GLOBIN BY THE ADDITION OF WATER AND WATER AND SALT TO THE DIET (Jansen, 1920).

chloric acid were given each day, no consistent response was elicited. Ammonium chloride given at the rate of 8 gm. per day for 4 to 5 days produced a reduction in the edema and in body weight in both cases. None of these salts had any influence on the nitrogen balance. The plasma protein levels changed only slightly during the experiment and appeared not to participate in the processes affecting the body weight.

Results similar to the above but reported in much greater detail were secured by Weech and Ling (1931). One of their subjects showed a marked increase in weight when a daily dose of 7.2 gm. sodium bicarbonate was given for 3 days; during this time the sodium chloride intake was 10 gm. daily. This hydration occurred while the boy was on a very low protein diet and when his total plasma proteins were 4 per cent (see Figure 79). Disappearance of the resulting edema followed the administration of ammonium chloride for 1 day and the reduction of sodium chloride intake to 2 gm. daily. The plasma protein level, which had been very low, was gradually restored to normal by means of a diet providing 71 gm. of protein. At that time the addition of sodium bicarbonate to the diet produced no change in body weight. It is difficult to understand the inactivity of the extra salt intake in the last trial, since, in spite of the normal plasma protein level, this subject must still have been extremely emaciated. The boy, who was 9 years old, weighed only 12.5 kg. On the basis of these results,

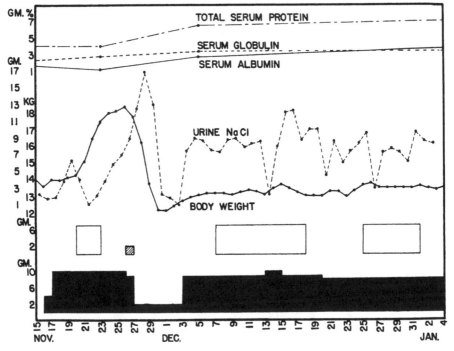

FIGURE 79. BODY WEIGHT, URINE SODIUM CHLORIDE, AND SERUM PROTEIN OF A CHINESE BOY MAINTAINED ON A DIET OF 1400 CAL. The intakes of sodium chloride, sodium bicarbonate, and ammonium chloride are indicated by the solid, open, and crosshatched blocks. (Weech and Ling, 1931.)

Weech and Ling "tentatively suggested that the threshold for the excretion of sodium may be controlled by the level of serum protein."

The prominent role played by the salt intake in the control of edema has been re-emphasized by Gounelle et al. (1941). They found that the semi-starvation cases seen in Paris during World War II showed a prompter response to changes in the sodium chloride intake than to changes in protein intake, even though the latter were as large as 100 to 150 gm.

In anorexia nervosa there is also a disturbance of the body's ability to excrete sodium chloride when even small additions of salt are made to the dietary intake. This was shown by Sunderman and Rose (1948), who attempted to raise the serum chloride concentration in such a case by intravenous saline infusion. The total salt intake ranged from 9 to 17 gm. per day. The serum chloride concentration rapidly approached normal while the plasma protein level decreased. At the same time "the patient became strikingly edematous." Two weeks after the cessation of the intravenous saline therapy, the edema had disappeared and the serum chloride concentration returned to its previous low level. The relative infrequency of marked edema in patients with anorexia nervosa may be related to the fact that such patients, in contrast to famine patients, have no salt and water hunger (see Chapter 44).

The Role of Salts in Regulating Water Metabolism in Semi-Starvation

The attempts to explain edema formation on the basis of changes in water balance have been rather sterile (see also Chapter 43). Most of the recent work on uncomplicated semi-starvation and plasmapheresis is in agreement about the absence of renal disturbances (see Chapter 31). In spite of this, these conditions are characterized by a tendency to retain water when the salt intake is increased and to lose it again when the salt intake is reduced (Moore and Van Slyke, 1930; Fahr et al., 1931; Liu et al., 1932; Weech and Ling, 1931). The factors of venous pressure, colloid osmotic pressure, and total osmotic pressure do not seem to cover all cases. By elimination, the investigator is left with the intangible tissue pressure.

A similar deductive path was followed by Weech, Goettsch, and Reeves (1935), who proposed that a sudden increase in salt intake produces a "stress" throughout the tissues which favors the accumulation of water. "If the stress is exerted when the tissues possess a reasonable reserve strength, retention will not result." According to them, the peritoneal cavity and subcutaneous tissues react differently to stresses. Explanations such as these still leave unanswered the fundamental question of the mechanism whereby these fluid exchanges are produced. It is likely that no one factor is responsible for the water retention but that an accumulation of small deviations from the individual's normal in some or all functions involved may be responsible for the condition.

The action of certain inorganic salts in changing the amount of water in the body has been recognized for many years. Besides the early limited work with normal adults (see Bassett et al., 1932, for references), a considerable amount of work has been done with the edema associated with nephrosis and that seen occasionally in diabetes. Up to 1910 there were a series of investigations which finally led to the belief that in these disease conditions the sodium salts were responsible for the retention of water (Meyer and Cohen, 1911). In the same conditions the potassium salts were associated with dehydration. That the cations were not the sole determinants of these properties should have been evident from the work of Meyer and Cohen (1911), who found that both $KHCO_3$ and K_2HPO_4 produced considerable dehydration, whereas KCl was relatively weak in this respect. They found that calcium chloride produced marked losses of weight, while the acetate had less effect and the lactate on occasion was associated with an increase in weight.

The mechanism by which these salts influence the water content of the body is unknown. A number of attempts have been made to unravel this problem by means of mineral balance studies with both normal and edematous individuals (Loeb et al., 1932; Bassett et al., 1932; Wiley et al., 1933). All of these experiments can be criticized for being too short, ignoring the changes in blood inorganic constituents and proteins, or using as normals individuals who were confined to bed. Until a study of the over-all changes in the body constituents are made, it will be impossible to explain the quantitative differences in the action of these salts in normal and edematous subjects. Although there are small and usually temporary changes in the weight of normal subjects ingesting fairly large amounts of salts, these same substances have a much more profound influence on those persons who have either edema or a tendency to develop it.

The earlier attention given to the role of salt in controlling edema is still being maintained. This is evident in the emphasis given to the disturbances in the metabolism of sodium seen in so many types of edema (Futcher and Schroeder, 1942; Warren and Stead, 1944; Farnsworth, 1948).

Blood Minerals in Semi-Starvation

As has been brought out in the preceding section, most semi-starvation cases have been reported to show no significant change in the chloride concentration in the serum (Feigl, 1918; Schittenhelm and Schlecht, 1918; Jansen, 1918; Gounelle et al., 1941; Gounelle and Marche, 1946; Denz, 1947). When a slightly elevated level was seen in an edema case, this returned to normal or slightly below with the disappearance of the edema, according to Denz (1947). Gsell (1948), however, found definitely high values in 3 women rescued from German concentration camps; in 2 of these patients the serum chloride was 117 and 118 milliequivalents per liter (680 and 690 mg. NaCl per 100 cc.). On the other hand, the famine victims at Warsaw were reported to have, in almost all cases, subnormal levels of blood and serum chloride. Fliederbaum et al. (1946) stated that the values for whole blood chloride were often as much as 20 mg. per 100 cc. less than the "normal" minimum, which they took to be 270 to 310 mg. per 100 cc. The Warsaw investigators also found pronounced reductions in the serum of these starving persons, the values being as low as 76 m.eq. per liter (270 mg. per 100 cc.) in several cases. In the red cells also, very low chloride concentrations were found.

The Warsaw material is impressive because of the severity of the starvation, the fact that these patients were still "actively" starving (i.e., they were not being refed as in almost all other studies), the selection of subjects to eliminate complicating infections, and the size of the group studied, which consisted of 20 adults and 20 children. On the other hand, the report (Fliederbaum *et al.*, 1946) gave no details as to either the analytical methods or the findings in individuals. The Warsaw findings are corroborated in the case of a woman who was more than 50 per cent under normal weight as a result of anorexia nervosa. She had a very low level of chloride in the blood serum, but a serious chloride deficit may have been induced by her habit of "frequently induced vomiting after meals" (Sunderman and Rose, 1948).

Simonart (1948) recorded 334 analyses for blood chloride in 207 starving men, but these were made with whole blood and there is no basis for correction to serum or plasma. The majority of the values exceeded 500 mg., as NaCl, per 100 cc. of whole blood, but all these men were anemic to some degree, so that the only conclusion possible is that there did not seem to be any great departure from normality in the blood chloride. This is another of those far too common examples of effort wasted because of failure to recognize elementary requirements in chemical analysis.

A fair summary of the blood chloride picture in semi-starvation, based on the evidence cited above and the data from the Minnesota Experiment, would be that the serum chloride level is close to the normal except in extreme undernutrition. In some cases high values may be seen in the early stage of refeeding.

The level of potassium in the blood of edematous patients in Germany during World War I was reported to be slightly below normal (Maase and Zondek, 1920). Feigl (1918) claimed that there was a considerable reduction in these cases but presented no data. In World War II the starved persons seen in Paris had normal serum potassium levels (Gounelle *et al.*, 1941). This is in harmony with the findings of the Minnesota Experiment, where the serum potassium levels at the end of the starvation period were the same as those during the control period. The case of anorexia nervosa referred to above showed a normal potassium level in the blood (Sunderman and Rose, 1948). Since the earlier analytical methods for the determination of potassium in the blood were not too satisfactory, it may be that the abnormalities previously observed in starvation cases are attributable to analytical difficulties.

Reports on the serum potassium concentration from World War II do little to clarify the picture. The mineral cations were not studied at Warsaw. In the 3 women studied by Gsell (1948), the serum potassium was normal in one, moderately low in another, and extremely low in a third (see Table 258). The tendency was in the opposite direction in the 27 men studied by Lamy, Lamotte, and Lamotte-Barrillon (1948). In this group the mean value for potassium in the serum was 20.98 mg. (5.37 m.eq. per liter) with a maximum in one man of 34.4 mg. (8.80 m.eq.); values above 22 mg. (5.6 m.eq. per liter) were found in 8 patients. In the Mainau group there was one subnormal value, 12.7 mg. (3.7 m.eq.), all others being in the normal range or above. The patients studied by Lamy, Lamotte, and Lamotte-Barrillon may be considered to represent either more extreme emaciation or an earlier stage in refeeding than was the case in other studies. It is fair to comment here, however, that analytical defects cannot

TABLE 258

BLOOD CHEMISTRY IN 3 SEVERELY UNDERNOURISHED WOMEN studied by Gsell (1948). Patient No. 1 had marked edema; she was 60 years old. Patient No. 2, aged 42, had moderate edema. Patient No. 3, aged 24, had been edematous but was edema-free when the blood was drawn. Protein in gm. per 100 cc.; K, Na, and Cl in milliequivalents per liter.

Patient	Red Blood Cells (millions)	Hematocrit (%)	Serum			
			Protein	K	Na	Cl
1	2.8	26.4	4.16	1.74	142	117
2	3.1	32.8	4.90	5.06	137	109
3	2.8	30.1	4.30	3.25	140	118

be excluded from the Mainau findings, where, as in the case of the Swiss studies, no control findings or technical details were reported.

Where serum sodium concentration has been examined in semi-starvation, the values have been close to normal (e.g. Gsell, 1948), and normal alkaline reserves were reported by Gounelle and Marche (1946). In 6 moderately undernourished patients, Nicaud, Rouault, and Fuchs (1942) found a tendency to high values for the alkaline reserve, and somewhat elevated figures were reported from the Netherlands by Cardozo and Eggink (1946).

In the 3 women studied by Gsell the CO_2 contents of the serum were estimated to be 49.9, 57.8, and 60.6 volumes per 100 cc., at CO_2 tensions of, respectively, 31.4, 36.0, and 44.5 mm. Hg and at pH values of 7.46, 7.46, and 7.39. In another patient the CO_2 content was 51.7 volumes per 100 cc. at a pH of 7.48 and a partial pressure of 31.6 mm. of CO_2; 4 weeks later, when this patient was showing many signs of recovery, the pH was 7.38 the pCO_2 was 45.0 mm. Hg, and the CO_2 content was 60.0 volumes per 100 cc. At Warsaw the alkaline reserve of the blood, determined by the Van Slyke method in 20 patients, was often low and was never elevated, minimal values being as low as 40 or even 30 cc. CO_2 per 100 cc. of blood (Fliederbaum et al., 1946). The Polish workers thought their findings corresponded to an acidemia similar to that in acute fasting but were unable to study this further.

The response of the blood calcium level to semi-starvation is highly variable in different reports. Feigl (1918) found both a slight increase and a slight decrease in the edema cases he studied during World War I. Maase and Zondek (1920) found an increased calcium level in most of their starvation cases, but with the advent of sufficient food this level returned to normal or somewhat below it. Jansen (1918), on the other hand, found the blood calcium level low or very low in 12 of the 15 individuals he studied. Among the Parisians who exhibited signs of "famine hypoglycemia" during World War II there was no change in the blood calcium levels (Gounelle and Marche, 1946). In the only report of animal experimentation, the calcium level of dogs during semi-starvation showed no significant change over a period of 110 days (Schelling, 1930). The serum calcium level was normal in one case of anorexia nervosa (Sunderman and Rose, 1948).

It is difficult to reconcile some of the preceding findings with the results of the Minnesota Experiment, where a decrease in the serum calcium level was noted at the end of the semi-starvation period. Walters, Rossiter, and Lehmann (1947b) reported that the serum calcium level of Indian prisoners of war on their release from Japan was 8.5 mg. per 100 cc. When they were discharged from the hospital, apparently recovered from their starved condition, the level had increased to 10.8 mg., which was fairly close to the level in their normal control group (11.4 mg.).

In 30 starved men in a Belgian prison, Simonart (1948) determined the blood calcium, presumably in serum, with the Clark-Collip modification of the Kramer-Tisdall method. The mean value was 10.89 mg. per 100 cc., and all but 6 were within the limits of 9 to 12, the maximum being 12.6 mg. It was stated that in the 2 men with the lowest values, 6.5 and 8.5 mg., there were indications of tetany and the sign of Trousseau was positive.

We have found no acceptable values on the total or inorganic phosphorus in the blood of human beings subjected to semi-starvation. The closest approach is the finding of a normal inorganic phosphorus level in one case of anorexia nervosa (Sunderman and Rose, 1948). Schelling (1930), working with semi-starved dogs, found no change in the serum inorganic phosphorus level even after the animals had lost 26 per cent of their original body weight.

Changes in Blood Mineral Concentration — Minnesota Experiment

Plasma or serum samples for mineral analyses were secured from 6 of the subjects of the Minnesota Experiment before and at the end of the semi-starvation period. The sodium and potassium analyses were made with the Elmer-Perkins flame photomer through the cooperation of Dr. Wallace Armstrong. Cal-

TABLE 259

THE CONCENTRATION OF MINERALS IN THE PLASMA OF SOME OF THE SUBJECTS IN THE MINNESOTA EXPERIMENT. All concentrations are expressed as milliequivalents per liter. The blood volume is given in liters.

Subject	Blood Volume	Na	K	Ca	P	Cl
29						
Control	6.68	145.7	4.80	4.91	3.43	104.9
S24	5.24	154.6	3.74	3.96	2.66	105.2
19						
Control	5.76	148.9	5.09	4.79	3.34	105.2
S24	5.12			2.68	3.27	107.3
2						
Control	6.59			4.46	2.71	
S24	6.77	158.9	3.64	4.58	3.82	110.5
20						
Control	5.98	145.2	4.99	4.51	2.79	106.4
S24	6.07	130.4	3.84	2.80	2.54	102.9
105						
Control	5.57	151.1	4.99	6.34	2.95	110.7
S24	5.00	156.5	4.64	4.36	3.29	105.9
120						
Control	5.70	151.1	4.32	5.11	2.37	109.8
S24	5.26	155.4	4.80	5.01	3.29	105.4

cium was determined by the method of Kramer-Tisdall as modified by Clark and Collip (cf. Hawk *et al.*, 1947, p. 589), phosphorus by the method of Mortland and Robinson (1926), and chloride by the Keys' (1937) method which utilizes the Volhard reaction.

There were no consistent changes in the concentration of any of these minerals except calcium (see Table 259). Five of the men showed a decrease in their plasma calcium levels, the mean for the group going from 5.02 to 3.90 mg. per 100 cc. In most cases the changes in the calcium levels did not extend beyond the limits of normal variation, but the fact that all but one of the subjects had a lower blood calcium level at the end of the semi-starvation period than in the control period is of considerable significance. Two men had calcium levels that were in the range frequently associated with convulsions. Since no further work was done, it is impossible to state whether this change predominantly involved the diffusible or the nondiffusible form of calcium.

CHAPTER 24

Carbohydrate Metabolism

IN ALL ordinary diets, and invariably in famine, the largest share of the caloric intake is supplied by carbohydrates. Moreover, present evidence indicates that a fair amount of the protein metabolized by the adult passes through the same oxidative pathways as the carbohydrates. Even the fats, through their participation in the citric acid cycle, are being brought into closer alliance with carbohydrate metabolism. The fact that the human body has only a limited carbohydrate reserve lends further importance to the subject of carbohydrate metabolism in starvation and famine.

Although most of the schemes proposed for the stepwise oxidation of carbohydrates are in general agreement on the major intermediates, very little is known about the exact stages most susceptible to alteration by undernutrition or by other changes in physiological status. Much attention has been given to the levels in the blood of glucose and its intermediates, pyruvic and lactic acid.

Blood sugar measurements have many uses but it is clear that only in rare and peculiar circumstances can the blood sugar level provide any information on the rate of carbohydrate metabolism or the size of the body stores of carbohydrate. The factors which affect the glucose level in the blood are legion, and there is considerable variation in the same individual even under rigidly standardized conditions. The effect of emotion is particularly troublesome to control. The confusion in the literature is compounded by the use of analytical methods which differ in their relative freedom from interference by non-glucose reducing substances.

The concentrations of pyruvate and lactate in the blood are independent of the total metabolism over a considerable range, perhaps up to 2 or 3 times the basal metabolism, but they increase sharply at extreme grades of physical work. Much more moderate increases in the lactate and pyruvate in the blood are produced by a prolonged deficiency of thiamine in the diet (Platt and Lu, 1936; Lu, 1939b) and by the ingestion of large amounts of glucose (Bueding *et al.*, 1941). The concentrations in the blood and tissues of these 3-carbon acids are affected by large changes in the total acid-base balance.

Other intermediates in carbohydrate metabolism have not yet been extensively studied in man. This means that the recognition of peculiarities in carbohydrate metabolism in man is dependent so far on a very small number of established test methods. The non-protein respiratory quotient is limited in its value, as will be made clear later in this chapter. The glucose tolerance test is, like the resting level of blood sugar, influenced by many factors. The subsequent sections of this chapter will deal with the data and the interpretation of the respiratory quotient, the blood sugar, sugar tolerance tests, and blood lactate and pyruvate in undernutrition.

Changes in the Respiratory Quotient in Acute Starvation

Very shortly after the start of fasting, the respiratory quotient decreases to values of the order of 0.70. There has been some variation in the lowest values reported as occurring during complete starvation. For instance, during the 11-day starvation experiment on Cetti, Lehmann *et al.* (1893) found values as low as 0.65 on the fourth day of starvation. Except for the first day, all values were below 0.70. During Breithaupt's starvation of 7 days the respiratory quotient decreased more slowly and only got below 0.70 by the fifth day, when it went to 0.63 (Lehmann *et al.*, 1893). From then on it increased, to 0.72 at the end of the fast. In neither of these experiments was there any correction for the protein metabolism. If this were done, the respiratory quotients would be still lower.

Benedict (1915) determined the non-protein respiratory quotient of Levanzin during his 31-day fast. This value decreased from 0.76 on the first day of starvation to 0.69 by the fourteenth day. That was the first day the R.Q. fell below 0.70, and it fluctuated just above and below that value for the remainder of the experiment. Results similar to these were secured with children by Shaw and Moriarty (1924) (see Table 260).

TABLE 260

RESPIRATORY QUOTIENTS FOR 2 CHILDREN, determined before and during 9 days of fasting (Shaw and Moriarty, 1924).

Subject	Control	Days of Fast								
		1	2	3	4	5	6	7	8	9
4	0.86	0.78	0.81	0.75		0.77		0.69	0.70	0.67
5		0.79			0.70		0.70		0.71	

The limitations and difficulties inherent in the determination of the respiratory quotient during starvation were emphasized by Benedict (1915). According to him, the respiratory quotient could only be interpreted to indicate that "the combustion of fat forms the greatest part of the total combustion" when no food is consumed. Little can be added to that statement. Benedict specifically cautioned against the use of such data as a means of determining whether any glycogen was being synthesized during the experiment. His results did show a decrease in the amount of carbohydrate metabolized by Levanzin from 69 gm. on the first day to a nondetectable amount by the fourteenth day. The amount of fat metabolized decreased from 135 gm. on the first day to about 110 gm. on the sixteenth day, where it stayed for the remainder of the experiment. The changes in the amounts of body carbohydrate and fat metabolized throughout this fast are indicated in Figure 80. This dramatically illustrates the rapid de-

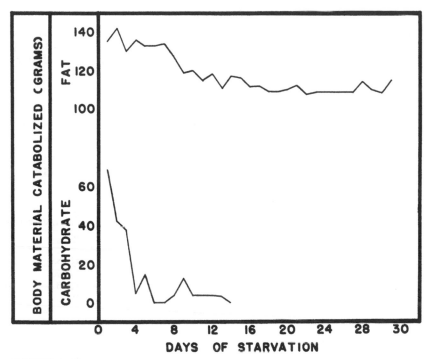

FIGURE 80. AMOUNT OF FAT AND CARBOHYDRATE METABOLIZED BY
LEVANZIN EACH DAY DURING A 31-DAY FAST (Benedict, 1915).

crease in the metabolism of carbohydrate and the relative maintenance of fat
metabolism.

During the later stages of a fast, a considerable amount of energy is secured
from protein in addition to that from fat. Although the protein metabolism in
the case of Levanzin decreased from an average of 60 gm. per day during the
first 2 weeks to 50 gm. for the last 17 days, it still accounted for 30 per cent of
the total caloric expenditure.

Morgulis (1923) proposed the idea that starvation could be divided into 4
separate periods, with each period representing "approximately one-fourth of
the total loss in weight sustained at the time of death." On this basis, Morgulis
(1923) suggested that Levanzin's fast covered only the first 2 phases. These
were characterized by a decrease in both the total and the non-protein respiratory quotient (see Table 261). The line separating these 2 stages is none too
sharp, however. This is also true of the other experiments quoted by Morgulis
as substantiating his arbitrary division of the starvation sequence.

The present evidence indicates that there is a common meeting ground for
carbohydrate, protein, and fat metabolism in the citric acid cycle (Potter, 1944;
Krebs, 1943; Witzemann, 1942). With the accumulating evidence that the suggested schemes of metabolic pathways actually occur in the intact organism, it
becomes increasingly difficult to separate over-all metabolism into its component
parts. It is conceivable that even in the final stages of starvation protein may be
converted to fat and oxidized as such. This conversion would be associated with

TABLE 261

FOODSTUFFS METABOLIZED DURING THE 31-DAY ACUTE STARVATION EXPERIMENT MADE ON LEVANZIN by Benedict (1915). Each set of data is the mean for a period equal to one fourth of the fast. This fast covered the first 2 phases of starvation according to Morgulis' scheme (1923). N.-Prot. = Non-protein; Gly. = Glycogen. (From Morgulis, 1923.)

Period	CO_2 Produced (liters)	CO_2 Used (liters)	R.Q.		Catabolized (gm.)		
			N.-Prot.	Total	Gly.	Fat	Prot.
a	260.4	352.6	0.720	0.740	24.1	134.7	59.2
b	219.5	303.2	0.706	0.724	4.05	115.4	60.5
c	193.7	272.3	0.691	0.712		110.1	49.3
d	192.9	270.3	0.695	0.714		110.4	46.4

Period	Calories from			Calories Produced	
	Gly.	Fat	Prot.	Total	Per Kg.
a	102.0	1284.6	262.0	1648.4	28.70
b	17.1	1125.4	267.4	1409.9	26.18
c		1049.0	217.3	1266.3	24.81
d		1052.8	205.0	1257.8	26.02

a release of oxygen from the protein which then would become available for respiration (Bodansky, 1938, p. 515). As a result of this a slight decrease in the respiratory quotient could occur.

It appears that no general rule can be made for the behavior of the respiratory quotient during later starvation. Chambers (1938) reported that in some dogs this value increased in the last phases of starvation when the major share of the metabolism was derived from protein. In other animals no change occurred. Only with animals has experimental starvation been carried beyond Morgulis' second phase. There is no doubt that in most starvation experiments the carbohydrate reserves are reduced very early. For a variable period of time energy is secured from the metabolism of fat and protein. There is not sufficient evidence to decide whether some fat is being metabolized up to the terminal stages of starvation.

Blood Sugar Levels in Acute Starvation — Animal Experiments

Methods are available for the evaluation of nitrogen and fat balances. There may be some question about the interpretation of the results of these (Fleming and Hutchison, 1924), but one can secure a reasonable index of the difference between the intake and the excretion in both cases. This, however, is not so readily done for carbohydrates, since the products resulting from their oxidation can be distinguished from those of other substances only with difficulty. Very little if any digestible carbohydrate appears in the feces. If, through some disturbance, soluble carbohydrates reach the lower part of the gastrointestinal tract, it is very likely that they will be acted on by the bacteria present there.

To circumvent these difficulties in the over-all evaluation of carbohydrate metabolism, studies have been made of the caloric balance. Marasmic infants have been used as semi-starvation subjects in this work. The results have shown that they are as efficient in securing calories from their food as normal infants (Fleming and Hutchison, 1924).

The absence of a suitable quantitative method for the determination of blood sugar deterred such studies before World War I. Most of the procedures available at that time required 50 or more cc. of blood (see Peters and Van Slyke, 1932, Vol. II, and 1946, Vol. I). Consequently the early investigations on the behavior of blood sugar during starvation were made with animals, and in most cases dogs were used.

Probably the first to make such a study was von Mering (1877), who starved 3 dogs for 44, 48, and 120 hours, respectively. The blood sugars at the end of the fast were 150, 145, and 133 mg. There was considerable individual variation in the blood sugar levels of his normal dogs, and since the above values were within that range, he concluded that starvation had no influence on this value.

Mering's experimental technique was properly criticized by Otto (1885), who starved different dogs for periods as long as 30 days. In each case he secured blood samples both before and at the end of the fast. To correct the analytical method for the non-sugar reducing substances, he fermented the glucose by means of yeast. The arterial blood in all his dogs showed a decrease in glucose concentration (see Table 262). There was a tendency for the size of the reduction to be proportional to the length of the fast. On the other hand, the venous blood showed no consistent changes until the end of the experiment.

TABLE 262

CHANGES IN BLOOD SUGAR CONCENTRATION, in mg. per 100 cc. of blood, in dogs starved for varying periods of time (Otto, 1885).

Days of Starvation	Control		Starvation		Difference	
	Arterial	Venous	Arterial	Venous	Arterial	Venous
4	118	108	110	118	−8	+10
7	116	101	112	106	−4	+5
12	132	110	123	118	−9	+8
26	110	117	92	108	−18	−9
30	131	117	89	91	−42	−26

Blood sugar levels in starved dogs were also determined by Seegen (1885), but there were other variables operating to affect his results.

The venous blood sugar levels were also resistant to starvation in the dogs studied by Morgulis and Edwards (1924). This was true even for the animals that lost 40 per cent of their original weight as a result of acute starvation. In 3 of the dogs the blood glucose level was lower than the control value when the animals had lost 10 per cent of their weight. The other 2 dogs at that stage of starvation showed a considerable increase in the blood sugar level. However, at the end of the experiment all blood sugar values were within the control range.

Shope (1927) followed the blood sugar changes in a number of different animals. In a pig subjected to 4 days of starvation there was a decrease from a control value of 130 mg. per 100 cc. of serum to 84 mg. at the end of the second day. From then on the level was considerably higher. Cats and guinea pigs, after 42 hours of starvation, had blood sugar levels that were below their control values, whereas all 5 rabbits showed higher blood sugar levels at the end of 66 hours of fasting than at the start of the experiment. Shope offered no explanation for the anomalous behavior of the rabbits other than the statement that "the fasting rabbit goes very quickly to a slower rate of carbohydrate utilization." Results similar to these had been observed earlier by Underhill and Hogan (1912), so it appears that the rabbit differs in the way its blood sugar responds to starvation.

There has been some variation in the reported blood sugar changes in rats during starvation. Barbour *et al.* (1927) found no difference in the blood sugar levels of rats starved for 24 hours when compared with those starved for 48 hours. No control values were given to indicate the change that might have occurred during the first day. The data of Hershey and Orr (1928) indicated that the blood sugar of fasting rats decreased for the first 2 to 4 days, after which there was a slight tendency for it to increase.

Some of the variability in the blood sugar responses may be explained by the observation that this function during starvation is intimately related to the preceding diet. MacKay and his group (1941) found that when a low protein diet was fed to rats, the blood sugar levels during the subsequent starvation period dropped precipitously, reaching a low point on the second day, and then increased during the next few days. A high protein diet maintained the blood sugar level fairly well at a point somewhat below the control value. A moderate protein diet was intermediate in this respect. Samuels, Reinecke, and Ball (1942) and Samuels (1948) showed that fat in the preceding diet had the same influence as protein on the starvation blood sugar levels. This led Roberts, Samuels, and Reinecke (1944) to the conclusion that "in animals maintained on a particular diet for a period of time, the foodstuff predominantly burned by the extra-hepatic tissues during the early stages of fasting corresponds to the major constituents of the previous diet."

When starved animals are fed protein, the blood sugar level is increased, indicating that some of the glucose so formed is not immediately oxidized (Chambers, 1938). Additional proof of this was the observation by Chambers (1938) that the ingestion of fairly large amounts of proteins by starving dogs produced no significant increase in the respiratory quotient.

Blood Sugar Levels in Acute Starvation — Human Beings

From an extensive study of diabetic patients, Allen (1913) came to the following conclusion: "Reduction of blood-sugar below the normal lower limits is a difficult and unusual matter in the normal organism. Approximately the normal percentage is stubbornly maintained through prolonged starvation, almost up to death." Less than 10 years later Weeks *et al.* (1923) found the situation in human beings to be otherwise. In some of their epileptic patients starved for 3 weeks the blood sugar levels decreased in the first week and then increased. Here, as in the experiments with rats, there was an indication that diets high in fat or protein, fed for 48 days preceding the starvation, were able to maintain the blood sugar level better than diets high in carbohydrate. Weeks and his colleagues did not state in their report what analytical methods they had used, but according to Shaw and Moriarty (1924) they used Benedict's picric acid method for the determination of blood sugar. If this was the case, then the increasing blood sugar values observed in the latter part of the starvation period might be too high, since Cowie and Parsons (1920) showed the picric acid procedure to be very sensitive to acetone.

A more complete study of the changes in the blood constituents was made by Lennox, O'Connor, and Bellinger (1926) on patients "who were fasted as a therapeutic measure for the relief of convulsions." In most of these cases the blood sugar decreased during the first week and then rose nearly to its pre-fasting level (see Table 263). The general trend in the blood sugar levels for all

TABLE 263

BLOOD SUGAR CHANGES IN PATIENTS BEING STARVED FOR THERAPEUTIC REASONS AND IN ONE NORMAL SUBJECT (Norm.). After the starvation period all subjects except No. 20 were given a mixed diet; No. 20 received a high fat diet. The Folin-Wu (1920) method was used for the determination of the sugar. All values are in mg. per 100 cc. of whole blood (B) or plasma (P). (Lennox *et al.*, 1926.)

Day	Norm. (B)	2 (B)	4 (B)	9 (B)	9 (P)	12 (B)	12 (P)	13 (P)	14 (P)	20 (B)
					Subject					
					With Food					
1						92	87		78	74
2	78	92	99	98		95	87	93	77	84
					Fasting					
1	102		85	98	88	103	84	88		80
2			68	92	83	77	68	68	78	
3		85	69	77	69	59	54	65	52	
4	68	88	70	80	70	59	56	59	49	45
5		77	69	77	64	55	53	68	51	44
6		77	83	70	65	61	55	69	54	
7	78	68	72	73	67	60	54	74	59	53
8			73	72	66	59	61	76		55
9	80		82	81	79	69	67	72		55
10			73	75	72	73	65	72		53
11		66	68	75	71	70	68	64		58
12	81	67	75	78	78	83	80	69		
13			96	75	72	86	82	75		
14		63	74					78		
15	77		84							
16			69							
				With Food (after Fasting)						
1	91	93	79	86	95	90	84	88		58
2	103	83	84			89	84	91		59
3		80	99	87	83	82	78	85		59
4			102	97	88	73	66	92		56
5			110			83	76			54
6			99			70				59
7			100			81				

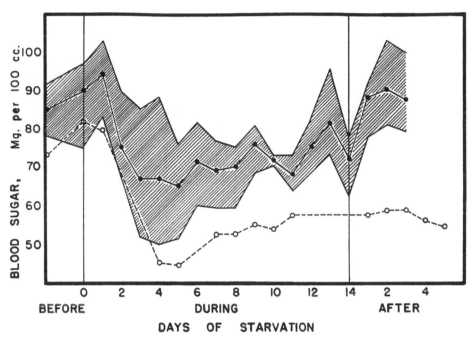

FIGURE 81. CHANGES IN BLOOD SUGAR DURING STARVATION. The shaded area represents the range of variation in a group of subjects, the solid line being the average for the group. The broken line indicates the blood sugar in a 13-year-old girl who showed a more striking hypoglycemia than the other subjects; in this case, the fast was broken by a high fat diet instead of a high carbohydrate diet. (From Peters and Van Slyke, 1946, based on the work of Lennox et al., 1926.)

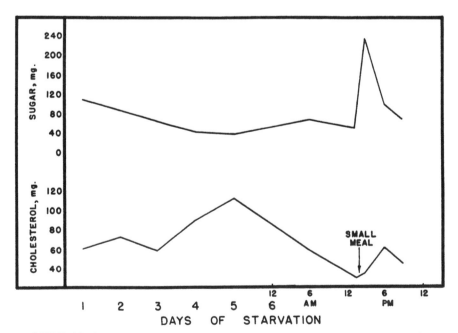

FIGURE 82. SERUM SUGAR AND FREE CHOLESTEROL LEVELS DURING A 5-DAY FAST BY A NORMAL YOUNG WOMAN. A small meal consisting primarily of carbohydrates was taken at 1:45 P.M. on the fifth day of the fast. (Shope, 1927.)

subjects is given in Figure 81. One of the outstanding features of the values during starvation was the large spontaneous change in blood sugar levels. In some cases these daily variations approached 30 mg. per 100 cc. Unfortunately only a few determinations were made during the pre-fasting period, but none of them showed any such variation. There was a rough parallel between the blood sugar and the plasma bicarbonate in some of the cases, but in others this was not apparent. Lennox and his colleagues noted that the blood sugar level during fasting could be increased by the injection of epinephrine. The concentration of glucose in the plasma was constantly lower than that in whole blood. The difference between these 2 values varied considerably from day to day. Normal glucose levels were rapidly restored on refeeding, except in the case of the subject given a high fat diet. In this case the blood sugar remained low.

In the same year Junkersdorf and Lisenfeld (1926) determined the blood sugar levels in 2 professional fasters, both of whom received small amounts of sugar (20–40 Cal. per day). The man, on whom observations were not started until the twenty-second day of starvation, showed no marked change in blood sugar through the thirty-sixth day, when the fast was terminated. The woman showed a level of 56 mg. per 100 cc. on the eighteenth day, which increased to 96 mg. by the twenty-seventh day but was down to 68 mg. by the thirty-ninth day.

Shope (1927) determined the serum sugar concentration in a normal young woman who went without food for 5 days. The level decreased from a control value of 110 mg. per 100 cc. to 37 mg. on the fourth day and then increased to 68 mg. on the last day. Attention was called to the fact that "during the experimental period the patient was at times well within the reported insulin shock range as regards degree of hypoglycemia and still suffered from none of the symptoms associated with this phenomenon." The low blood sugar levels observed in starvation raise the question as to whether insulin shock is due entirely to changes in blood sugar level or whether the rate of change in the level may not be of some consequence.

The free serum cholesterol in Shope's case showed an inverse relationship to the blood sugar level (see Figure 82). This is worthy of further investigation in view of the fact that both blood sugar and cholesterol levels show considerable change in starvation (see also Chapter 22). Additional suggestions for an intimate relationship between carbohydrate and fat metabolism have been put forth by Witzemann (1942).

It has been suggested by Peters and Van Slyke (1946) that infants develop hypoglycemia during starvation more rapidly than adults. This, however, is not borne out by the work on starved children. Such work indicates that the blood sugar changes in adults and children are similar. Mogwitz (1914) starved 3 infants for periods as long as 78 hours and found some fluctuations in the blood sugar level during that period. Most of the values were considerably below the control level (see Table 264). Somewhat similar results were secured by Lindberg (1917), who fasted 14 infants for 60 hours. The mean blood sugar values decreased from 125 to 60 mg. per 100 cc. of blood in the first 24 hours and remained there for the duration of the experiment. Individual variations were not-

TABLE 264

BLOOD SUGAR LEVELS, in mg. per 100 cc. of blood, in infants subjected to starvation. Bang's (1913) micro-method for the determination of blood sugar was used. (Mogwitz, 1914.)

Age (mos.)	Hours of Fasting								
	1–5	15–20	23	37	41–42	48	64	72	78
5	86	66	85		98	56			
5.5	101	93		55		57			
6.5	90	73			52		46	55	47

ed by Shaw and Moriarty (1924); some children maintained their blood sugar levels for 3 days, whereas others showed a precipitous decrease to levels as low as 38 mg. per 100 cc. Part of this variation in blood sugar levels has been attributed by Rumpf (1924), without evidence, to the nutritional condition of the child at the start of the fast. In view of the recent findings with starved adults and the marked variation secured in blood sugar levels when different analytical methods are used, it seems likely that age is of only minor consequence in the response of the blood sugar level to starvation.

From the preceding discussion it appears that there is some species variation in the response of the blood sugar concentration to acute starvation. Rabbits in particular exhibit little response. However, when the preceding diet is of average composition, human beings as well as most other animals show an initial decrease in the blood sugar level followed by a gradual return to normal. This general trend was confirmed for adults by Slotopolsky's study (1932). If the diets used before the fast were high in fat or protein, the blood sugar levels were maintained close to the control values during the abstinence from food.

Adaptation of Blood Sugar Levels in Acute Starvation

The explanation for the variations observed in the blood sugar levels during acute starvation may involve, in addition to the effect of the preceding diet, such factors as the type of glycogen deposited in the liver and adaptation to starvaion. Pflüger (1905), in his original procedure for the quantitative determination of glycogen, recognized that only a part of the total reserve carbohydrate could be removed from the liver by extraction with water.

Very little experimental work has been done on adaptation to starvation (Mitchell, 1944), but there is an indication of such an adaptive process in the blood sugar levels. This was shown in work (Taylor et al., 1945) with normal young men who were subjected to a series of 5 4-day fasts undertaken at 1-month intervals. The blood sugar levels during hard physical work were slightly higher during the fifth fast than in the first (see Table 265). There is additional evidence for such an adaptation in women undergoing successive fasts. The urinary excretion of acetone decreased when the starvation was repeated (Folin and Denis, 1915). A similar interpretation could be made in the case of dogs which exhibited during the second fast a more gradual weight loss and a much smaller nitrogen excretion than during the first experiment (Howe et al., 1909).

TABLE 265

THE INDIVIDUAL "IMPROVEMENTS" IN BLOOD SUGAR, BLOOD ACETONE, AND URINE ACETONE WHICH OCCURRED BETWEEN THE FIRST AND FIFTH FASTS. All values are differences between the fifth and first fasts (V − I). Blood variables are expressed as mg. per 100 cc. of blood and urine variables as grams excreted over the indicated time. The work periods were distributed throughout the day. Blood samples were taken during the last 5 minutes of each period. (Taylor *et al.*, 1945.)

Subject	Blood Sugar Work Period on Second Day			Blood Acetone A.M. of Third Day	Urine Acetone First Two Days (48 hrs.)
	1	2	3		
Ja	+18	+19	+24	−11	−7.5
No	+14	+20	+13	−7	−0.120
Se	+16	+7	+15	−2	−0.716
Jo	+9	+5	+9	0	−0.947

Glucose Tolerance in Acute Starvation

After a period of acute starvation the ingestion of carbohydrates produces an alteration in the subsequent blood sugar curve and in urinary sugar excretion. Hofmeister (1890) starved dogs for 13 days and then gave them 20 gm. of starch suspended in 400 cc. of meat broth. Four hours after feeding there were traces of sugar in the urine. Continuing this work, he found that the daily administration of 10 gm. of starch to starved dogs produced a urinary sugar excretion ranging from 0.18 to 1.02 gm. After a few days of refeeding, the 10 gm. of starch produced no glycosuria, but when the dose was increased to 15 or 20 gm. glycosuria appeared again. A similar response was elicited in a dog that was semi-starved for 4 days.

Hofmeister (1890) concluded that starvation produces some disturbance in the conversion of glucose to glycogen. Malmros (1928) offered another suggestion for the occurrence of glycosuria during starvation. He believed it to be due to liver damage as indicated by a positive urobilin reaction. Chambers (1938) believed that there is a change in the sensitivity of the body to insulin as the fast progresses but that there is no disturbance in the mechanisms for handling carbohydrate in the body.

Starvation diabetes is associated with an abnormal blood sugar response when glucose is given. In a study of patients with hepatic disease, Traugott (1922) noted a marked hyperglycemia following relatively small doses of glucose. The great variation in his control glucose tolerances, made on apparently healthy subjects, convinced him that the nutritional condition might explain some of his individual variability. To test this, he reduced his own food intake by one half for 5 days and then for an additional 3 days went completely without food. The increase in his blood sugar level following glucose was much greater after the starvation period than before. Sevringhaus (1925) observed the same thing after 2 days of starvation.

Sweeney (1927) found that when medical students fasted for 2 days, the sub-

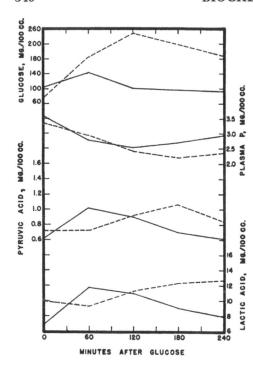

FIGURE 83. Mean Levels of Blood Glucose, Pyruvic Acid, Lactic Acid, and Plasma Inorganic Phosphate Following the Ingestion of Glucose. The solid line indicates the normal postabsorptive state (4 experiments on 4 subjects); the broken line is for the seventh day of the fast (4 experiments on 4 subjects). (Wollenberger and Linton, 1947.)

sequent glucose tolerance was similar to that secured following a 2-day high fat regimen. Because of the high fat diet of the Eskimos, Heinbecker (1928) studied their glucose tolerance curves both before and after 82 hours of fasting. The basal blood sugar during the starvation decreased from 110–120 mg. per 100 cc. to about 80 mg. The glucose tolerances after the ingestion of about 2 gm. of glucose per kg. of body weight were normal in the control period. After 3.5 days of starvation the glucose tolerance showed blood sugar levels close to 300 mg. per 100 cc. These returned only very slowly to normal. Sugar appeared in the urine approximately 2 hours after the glucose dose and in 2 cases stayed there for 10 hours. That the Eskimos responded as others do to changes in carbohydrate metabolism during starvation was shown by a repetition of the glucose tolerance on one man. This subject continued his abstinence from food for an additional 48 hours, when another glucose tolerance was performed. At that time a lower peak was reached and a rapid return to the normal level occurred.

As far as the blood constituents are concerned, Wollenberger and Linton (1947) probably made the most thorough study of the changes resulting from the ingestion of a dose of glucose (see Figure 83). Their glucose tolerances were made on normal young men who had been without food for 7 days. In comparison to their control period curve, starvation produced a marked increase in the highest blood sugar level attained together with a considerable delay in the time at which the peak appeared. They also followed the blood pyruvate and lactate levels, which in normal subjects closely paralleled the changes in blood sugar. After the starvation period, however, the peaks in the pyruvate and lactate curves occurred about an hour after the maximum blood sugar level.

Since so many steps in the accepted schemes for carbohydrate metabolism involve the intervention of phosphate groups, Wollenberger and Linton (1947) studied the response of the blood inorganic phosphate to glucose. In both the control and the starvation period, the glucose tolerances were associated with a gradual decrease in the blood inorganic phosphate level. In the control period the phosphate level fell by 0.78 mg. per 100 cc. of blood during the first hour after glucose, whereas after starvation the comparable fall was 0.41 mg. On the basis of urinary excretion studies, these workers claimed that "glucose administration to a normal animal or subject caused prompt retention of inorganic phosphate." This, however, was not the case with the starvation subjects, who maintained or increased their rate of phosphate excretion in the 2 to 3 hours following the ingestion of glucose.

Kaplan and Greenberg (1944) suggested that the disturbed glucose metabolism in starvation may be associated with certain phosphorus compounds. Studies with radioactive phosphorus on rats starved for 72 hours indicated a decrease in the phosphorylating mechanism. A quantitative investigation of some of the phosphorus-containing compounds indicated a marked lowering in the adenosine triphosphate concentration in the liver.

A number of years ago it appeared probable that there was some relationship between the abnormality in the glucose tolerance and the development of starvation ketosis. Sevringhaus (1925) observed that these conditions developed almost simultaneously. He thought that some product of starvation inhibited either the production or action of insulin. Further support for an association between glucose metabolism and ketosis was put forth by Cori and Cori (1927a, 1927b).

It was assumed, mainly from the work of Sevringhaus and the Coris, that the increase in blood acidity which occurs in starvation ketosis is responsible for raising the blood sugar to the glycosuria level, but subsequent work showed that the pH of the blood plays only a minor role in regulating the blood sugar. Although Malmros (1928) found that the ingestion by normal young men of 20 to 30 gm. of ammonium chloride daily for 4 days resulted in elevated blood sugar responses to glucose tolerance tests, a greater "diabetic" response resulted from fasting for only 36 hours. No values were reported for the alkalinity of the blood, but it is probable that after 36 hours of starvation the reaction of the body fluids was no more acid than after ammonium chloride. Furthermore, both Malmros (1928) and Goldblatt and Ellis (1932) found that sodium bicarbonate, administered with the glucose, had no influence on the succeeding blood sugar levels when the glucose tolerance was made following a period of starvation. The absence of any correlation between acidity and carbohydrate metabolism was emphasized by the work of Odin (1927). He found no relationship between the carbon dioxide capacity of the blood and its sugar content either in diabetic patients or in normal subjects.

The respiratory quotient was used by Johansson (1908) to study the body's response to the ingestion of glucose. Normal young men who did hard physical work during a 36-hour starvation period were given 75 gm. of glucose. The subsequent carbon dioxide production was much smaller than when the same amount of carbohydrate had been taken in the control period. This work was extended by Goldblatt (1925), who found that the R.Q. remained low when 50 gm. of glucose were given to an adult who had gone without food for 40 hours. At the end of the starvation, the R.Q. increased from 0.73 to 0.81 as a result of the glucose, whereas in the control period a value of 1.0 was secured. The maximum R.Q. "was attained in each experiment some time after the maximum blood sugar was reached." From this work on the respiratory quotient it would appear that the utilization of carbohydrates is decreased following a period of starvation. Again, it is advisable to heed Benedict's warning (1915) regarding the interpretation of such data and to ignore any attempt to quantitate the degree of metabolic suppression.

When fructose was given to fasting subjects, it was as effective as glucose in overcoming the ketosis and had the same influence on the respiratory quotient (Goldblatt, 1925). It differed from glucose in that it produced no change in the subsequent blood sugar level. This was explained on the assumption that the fructose tolerance "is an index of the carbohydrate storing power of the organism." On this basis, Goldblatt concluded that "after starvation, storage power [for carbohydrates] is normal."

The influence of the season must also be considered in the interpretation of the glucose tolerances made after starvation, at least as far as animals are concerned. This was emphasized by Cori and Cori (1927a, 1927b), who found that starved rats could tolerate almost twice as much sugar in the winter as in the summer (see Table 266). Barbour *et al.* (1927) also observed a seasonal difference in the blood sugar levels of their starved rats, but the trend was opposite to that noted by Cori and Cori.

TABLE 266

Seasonal Changes in the Metabolism of Fasted Rats. All animals were adults and had fasted 48 hours when the observations were made. (Cori and Cori, 1927b.)

	October to May	May to October
Total acetone bodies, as acetone, in mg. per 100 gm. of body weight per 24 hrs.	1.9	6.1
Intravenous glucose tolerance, in gm. per kg. per hr.	2.5	1.6
R.Q.		
Fasting	0.713	0.685
During 4 hrs. of glucose absorption	0.838	0.777
Glucose, in gm. per 100 gm. per 4 hrs.		
Absorbed	0.750	0.721
Oxidized	0.281	0.131
Glycogen, per 100 gm. per 4 hrs.		
Formed (in gm.)	0.388	0.380
Recovered (in %)	89.2	70.9
Urine nitrogen, in mg. per 100 gm. per 4 hrs.	12.15	12.36
Calories, per 100 gm. per 4 hrs.		
From protein	0.30	0.31
From fat	1.11	1.57
From glucose	1.05	0.49
Total	2.46	2.37
Basal blood sugar (in gm.)	0.176	0.194

It was suggested by Cori and Cori (1927a) that the seasonal differences in the sugar levels of rats subjected to starvation were related to changes in fat metabolism. The oxidation of glycerol and protein during starvation was assumed to be sufficient in the winter to prevent ketosis, but such was not the case in the summer.

The Role of Insulin in Acute Starvation

It is generally assumed that the production or liberation of insulin is in some way regulated by the rate of carbohydrate metabolism (Peters and Van Slyke, 1946). Experimental evidence for this assumption was provided by Best, Haist, and Ridout (1939). They found that rats starved for 7 days had only one half the amount of insulin in their pancreas that the full-fed controls had. In both cases the insulin was isolated from the pancreas and assayed for potency. The fasted animals lost an average of 23 per cent of their original weight, but the weight loss was not responsible for the low concentration of insulin in the pancreas, since when other rats were fed a ration very high in fat, the insulin in the pancreas was reduced still more than in the starvation experiments. Refeeding the starved rats on a well-balanced diet restored the insulin content of the pancreas within 6 days. This occurred in spite of the fact that complete restoration of body weight had not been attained. Such was not the case when the animals were refed on a high fat diet. The Toronto workers presented evidence which indicated that the liberation of insulin by the pancreas was related to the amount of insulin in it.

Long before the experimental work in Toronto, a number of investigators attempted to relate the disturbed carbohydrate metabolism in starvation to an insufficient supply of insulin. Some reports (Goldblatt, 1925) have quoted Southwood (1923) as showing that hunger diabetes can be improved by insulin injections. Actually Southwood made no starvation studies. He fed diets that were very low in carbohydrate. He found them to produce glycosuria and a diabetic-like glucose tolerance curve. On his high fat diets, insulin restored the normal glucose tolerance curve and eliminated the glycosuria.

Tiitso (1925) injected insulin into starved rabbits and followed their blood sugar levels for an hour afterward. In spite of the insulin injection, the starved animals maintained their blood sugar at higher levels than the well-fed controls. There was some indication that the resistance to insulin increased with the length of the starvation period.

The influence of insulin on the carbohydrate metabolism of dogs that had been starved for 3 weeks or longer was studied by Dann and Chambers (1930). The administration of nothing but glucose to these animals on successive days gradually increased the respiratory quotient determined for the period covering the second, third, and fourth hours after the glucose feeding. The R.Q. increased over that period from 0.75 to 0.94 (see Table 267). The increase in the respiratory quotient was associated with a decrease in maximum blood sugar levels. For one dog the maximum levels on 4 successive days were 350, 252, 146, and 104 mg. per 100 cc. In a subsequent starvation experiment on one of the dogs, insulin was injected when the glucose was given. The respiratory quotient following the glucose now increased on successive days from 0.84 to 0.97. True, the insulin injection produced a higher respiratory quotient following glucose than was obtained without insulin. There was, however, a considerable differ- ence in the initial values of the R.Q. in the 2 experiments (0.75 vs. 0.84). When insulin was given, the absolute increase in the respiratory quotient was 0.13, whereas it was 0.19 in the absence of insulin. What is the significance of these differences?

TABLE 267

INFLUENCE OF INSULIN ON THE METABOLISM OF GLUCOSE GIVEN TO A STARVING DOG. The same dog was starved for 3 weeks on two occasions. In the interim the dog was refed to his original weight. Fifty gm. of glucose were given on successive days during the fasts, except that in the experiment without insulin 5 days elapsed between the first and second doses of glucose. Basal = the values for the period just preceding the glucose administration. (Dann and Chambers, 1930.)

	Glucose Given without Insulin			Glucose Given with Insulin		
	First	Second	Third	First	Second	Third
R.Q.						
Basal	0.69	0.74	0.75	0.73	0.74	
After glucose	0.75	0.87	0.94	0.84	0.94	0.97
Blood sugar in mg. per 100 cc.						
Basal		62	79		70	
After glucose		280	235	220	160	87
Urine sugar in mg. per 100 cc., after glucose .	10.7	2.2	Trace	2.0	1.5	0
Glucose oxidized as gm. per hr.						
Basal	0	0.29	0.41	0	0.55	
After glucose	0.56	2.59	3.80	2.13	4.34	4.50

Dann and Chambers (1930) also found that injections of insulin produced an improvement in carbohydrate metabolism as measured by the blood sugar levels, the urinary sugar excretion, and the rate of glucose oxidation. Even fairly large doses of insulin were unable to restore carbohydrate metabolism complete- ly to normal in the starved dog.

Further evidence that insulin does have some influence on carbohydrate me- tabolism in starvation was provided by Wierzuchowski (1931). He found that attainment of the maximum blood glucose level in dogs starved for 10 days was delayed when an injection of insulin preceded the start of an intravenous glu- cose infusion.

A seasonal factor in the influence of insulin injections in starved animals was shown by Cori and Cori (1927b). They found that insulin increased the rate of glucose oxidation of rats starved in the summer but had no influence on those starved in the winter. One other unexpected finding in their experiments was that insulin increased the intravenous tolerance of starved rats for glucose but had no influence on the fructose tolerance (Cori and Cori, 1927a). This preferential action of insulin is not peculiar to starvation since it was also observed in normal dogs (Wierzuchowski, 1926).

The inability of insulin to overcome completely the alteration in carbohydrate metabolism of fasting subjects was noted by Goldblatt and Ellis (1932). When glucose was given, insulin restored the blood sugar levels to normal in young men starved for 39 hours. Instead of increasing the rate of carbohydrate oxidation as measured by the heat production in the period following the glucose ingestion, the insulin actually reduced it and consequently appeared to suppress carbohydrate metabolism. No explanation was proposed for this inhibitory effect of insulin.

The present evidence indicates that the disturbed carbohydrate metabolism during starvation is not due entirely to an insufficient supply of insulin, although it would appear that this is a large factor in the condition. Since the more recent work indicates that insulin functions in a relatively restricted area of carbohydrate metabolism (Colowick et al., 1947), it is understandable why other metabolic disturbances may also occur during starvation to influence the development of hunger diabetes. Such a conclusion was apparent from the work of Bergman and Drury (1937) with rabbits. They found that the greater tolerance for injected glucose in the fed eviscerated animal as compared to the starved eviscerated animal could not be explained entirely on the basis of insulin. It appeared logical to assume that rabbits fed up to the time of operation should have a larger amount of circulating insulin than rabbits which had been starved for 6 days prior to the operation. The injection of fairly large amounts of insulin into the eviscerated animals produced very dramatic effects which lasted for a much shorter time than the difference observed in the glucose tolerances.

Glycogen Stores in Acute Starvation

Many studies have been made on the role of the liver and muscle glycogen in regulating carbohydrate metabolism during starvation. Under normal dietary conditions the liver is especially important as a regulator of both the blood sugar level and carbohydrate metabolism in general. The present evidence indicates that when the blood sugar level is increased, the rate of glycogenesis by the liver is increased (Soskin, 1941). Presumably the opposite reaction occurs when the blood sugar level falls below normal. The complexity of the homeostatic mechanism involved in the maintenance of the blood sugar level precludes any comprehensive consideration of the problem. This leaves us with only the alternative of attempting to follow the changes in carbohydrate metabolism during starvation by means of the factors that have been considered so far plus any alterations that may occur in the glycogen stores.

The earlier work on the glycogen content of the liver and muscle during starvation has been compiled by Ling and Shen (1934). To their figures have since been added a number of other reports which are given in Table 268. Since the liver and muscle account for the largest amount of glycogen in the body, these tissues have been studied to the virtual exclusion of other organs.

There is considerable variation in the amount of glycogen found in rat liver

TABLE 268

GLYCOGEN CONTENT OF LIVER AND MUSCLE DURING FASTING. Based on the report of Ling and Shen (1934). In many cases the sex of the animals could not be determined from the original publication.

Author	Animal	Fasting	Glycogen	
			Liver (gm./100 gm.)	Muscle (gm./100 gm.)
Pflüger (1902)..................	1 dog	28 days	4.78	0.158
Pflüger (1907)..................	1 dog	73 days	1.22	
	1 dog	70 days	0.03	
Junkersdorf (1921)...............	15 dogs	11 days	0.59	0.21
Fisher and Lackey (1925).........	2 dogs	5 days	0.25	0.09
Rathery and Kourilsky (1930)......	3 dogs	6–23 days	0.05–3.03	0.18–0.51
	2 dogs	30 days		0.18
	5 dogs	30–47 days	0.11–1.83	
Junkersdorf and Mischnat (1930)...	4 dogs	11 days	0.90	0.21
	3 dogs	26–43 days	0.40	0.10
Dann and Chambers (1932).......	3 dogs	20–27 days	0.38–1.14	0.22–0.55
Deuel et al. (1933)...............	7 dogs	6 days	0.45	0.31
Cori (1926)....................	7 rats (M)	48 hrs.	0.397	
Barbour et al. (1927).............	24 rats	24 hrs.	0.16	
	68 rats	48 hrs.	0.32	
Hershey and Orr (1928)..........		72 hrs.	0.2	
		144 hrs.	0.78	
Catron and Lewis (1929)..........	16 rats	24 hrs.	0.09	
Cori and Cori (1929).............	8 rats	24 hrs.	0.10	
Lawrence and McCance (1931).....	8 rats	24 hrs.	0.40	
Greisheimer (1931)...............	3 rats (M)	48 hrs.	0.513	
	9 rats (F)		0.137	
Miller and Lewis (1932)...........	11 rats	24 hrs.	0.07–0.17	
Eckstein (1933)..................	44 rats	24 hrs.	0.10–0.06	
Silberman and Lewis (1933).......	9 rats	24 hrs.	0.05	
MacKay and Bergman (1934).......	5 rats (M)		0.22	
	5 rats (F)	48 hrs.	0.16	
Deuel et al. (1934)...............	15 rats (M)		1.43	0.369
	15 rats (F)	24 hrs.	0.70	0.383
	34 rats (M)	48 hrs.	0.45	0.303
	32 rats (F)		0.25	0.308
	18 rats (M)	72 hrs.	0.31	0.235
	20 rats (F)		0.13	0.229
	20 rats(M)	120 hrs.	0.39	0.254
	18 rats (F)		0.47	0.228
	32 guinea pigs (M)		0.18	
	33 guinea pigs (F)	48 hrs.	0.43	
Rose (1920)....................	1 rabbit	6 days	0.20	
Tiitso (1925)...................	1 rabbit	7 days	0.83	0.125
	1 rabbit	14 days	0.24	0.021
Sahyun and Luck (1929)..........	4 rabbits	24 hrs.	0.34–0.50	
	3 rabbits	48 hrs.	0.08–0.31	
Emslie and Henry (1933)..........	15 fowl	24 hrs.	0.149	0.294
	15 fowl	48 hrs.	0.229	0.314
Ling and Shen (1934)............	6 ducks	22–125 hrs.	0.30–1.05	0.21–0.75

after 24 hours of starvation. The reported values range from 0.05 to 1.43 per cent (Table 268).

This tremendous variation may be due to any one or a combination of the following factors: "(1) faulty analytical methods employed in the determination of glycogen, (2) delays in commencing the glycogen determinations with resulting glycogenolysis, (3) too few experiments, (4) the omission of controls, (5) wide variations in the size and age of animals, (6) the use of different animal species, and (7) differences in the nutritional state of the animals at the time of the experiment" (Sahyun and Luck, 1929).

The low levels of liver glycogen observed in rats after 24 hours of starvation could be reduced still further by physical work (Nutter, 1941). In the recovery period following such work the rats showed an increase in the glycogen level to about twice that of the nonexercised controls.

There are marked changes in the rate at which the glycogen in various tissues and organs is utilized during starvation. The work of Hershey and Orr (1928) brings this out rather dramatically for the rat (see Figure 84). The very rapid loss of liver glycogen during the first 48 hours of starvation is followed by a gradual increase. Under no circumstances does this level ever return to normal.

There has been some discussion about the increase in liver glycogen stores during the starvation period. An increase has been observed in rats (Barbour et al., 1927; Deuel et al., 1934; Mirski et al., 1938), in dogs (Pflüger, 1907), and in

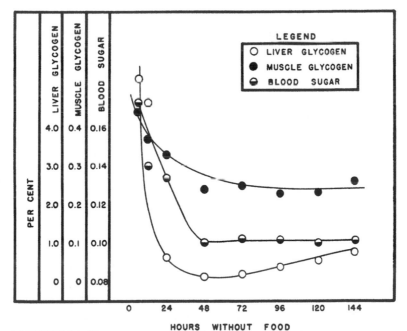

FIGURE 84. CONCENTRATION OF GLYCOGEN IN THE LIVER AND IN THE MUSCLES, together with blood sugar changes, in rats during starvation. The glycogen values are given as gm./100 gm., the blood sugar values as gm./100 cc. There were 3 animals in each group. The body weights decreased throughout the fast from 176 to 138 gm.
(Hershey and Orr, 1928.)

fowl (Emslie and Henry, 1933). Some workers have maintained that the glycogen level in the liver decreases continuously throughout a fast. This has been implied in the report on dog experiments made by Fisher and Lackey (1925) and for rats by Lawrence and McCance (1931). The latter workers quoted the report of Cori and Cori (1928) as supporting evidence for their claim. In their starvation experiments made on rats, as well as those by Cori and Cori (1928), the animals received no food for 48 hours. The evidence from other workers indicates that liver glycogen increases in rats during starvation only after the forty-eighth hour.

Influence of Previous Diet on Glycogen Stores

Recent work has emphasized the fact that an animal's ability to undergo a protracted period without food is dependent not only on the general nutritional condition of the animal but also on the type of food consumed in the period preceding the fast. Since so much emphasis has been put on the blood sugar levels during starvation as a guide in evaluating an organism's resistance to the debilitating effects of this stress, it was inevitable that the factors controlling the level of glycogen in the liver should be considered. Actually the first observations along this line were made by Pflüger in 1907. He found what one might expect, that a predominately carbohydrate regimen increased the liver glycogen stores to a much greater extent than a diet high in either protein or fat.

TABLE 269

LIVER GLYCOGEN IN RATS FED RATIONS CONTAINING VARIOUS AMOUNTS OF PROTEIN AND THEN STARVED. The diets were continued for 5 to 12 days. All animals were males weighing between 70 and 120 gm. All values are in percentage. (Mirski *et al.*, 1938.)

Diet	Hours of Starvation			
	0	15	24	48
Casein				
20 per cent	4.11	0.64	0.07	0.38
70 per cent	1.38	2.18	1.42	1.42
90 per cent	1.12		1.00	
Meat	1.42		1.32	1.05

More recently, attempts have been made to assess the role of dietary factors in influencing the amount of glycogen in the liver. Greisheimer (1931) compared the glycogen levels in rats fed varying proportions of fat and protein. These diets were highly purified and may possibly have been deficient in one or more substances, since the gains in body weight for her different groups varied considerably. She found no obvious relationship between the liver glycogen stores and either dietary fat or protein.

Dietary protein has been reported to inhibit the deposition of glycogen in the liver, as shown by the inverse relationship between these two factors in Table 269 (Mirski *et al.*, 1938; Treadwell *et al.*, 1942). Not only does the level of protein in the ration determine the glycogen content, but the kind of protein has also been reported important. Flügge (1941) claimed that a diet containing rice protein resulted in a greater amount of liver glycogen than when casein was used.

Other dietary factors such as vitamins and minerals have been implicated in the control of liver glycogen. The influence of the vitamins, at one time attributed to riboflavin (Wickson and Morgan, 1946), has been ascribed to a number of the members of the vitamin B complex, at least in so far as the maintenance of the muscle glycogen stores is concerned (Yakovlev, 1939). The presence of an excess of dietary sodium or potassium may alter the deposition of liver glycogen (Crabtree and Longwell, 1936; Silvetti et al., 1938; Lewis et al., 1944), but relatively large amounts of salt are required to produce these changes. Only the levels of glycogen in the liver and heart were affected by the extra salt; that in the muscle was uninfluenced (Crabtree and Longwell, 1936).

Although a high carbohydrate diet resulted in a larger initial liver glycogen concentration than did a high protein diet, the carbohydrate-fed rats very rapidly lost their glycogen reserve when starved (Table 269). The protein-fed rats retained their glycogen stores much more tenaciously than the other rats after such things as exercising to exhaustion and injection of bacterial substances and phloridzin. These results led Mirski et al. (1938) to conclude that "the formation of fresh sugar is so much greater in the protein-fed animals than in those fed on carbohydrates that no matter how heavy the demand, fresh gluconeogenesis can always occur." Roberts, Samuels, and Reinecke (1944) were led to a similar conclusion following their investigation of the influence of high fat diets on the ability of rats to withstand starvation.

The concentration of glycogen in the muscles is influenced to only a slight extent by the diet (Sahyun et al., 1934; Blatherwick et al., 1936; Mirski et al., 1938).

Pflüger (1907) was probably the first to suggest that starving animals were able to synthesize glycogen from non-carbohydrate materials. He was led to this conclusion when he found considerable amounts of liver glycogen in dogs that had starved for as long as 28 days. One dog that received no food for 73 days still had 1.22 per cent glycogen in the liver, while the muscles were practically glycogen-free. In most of Pflüger's experiments the liver glycogen reserves were very low after a few days of starvation. For this reason, the high level of glycogen seen after a prolonged period of starvation was attributed to synthesis from either protein or fat.

It is a little difficult to evaluate some of Pflüger's findings. True, he developed the procedure still used for the determination of glycogen, yet some of his own results are considerably out of line when compared with those of all other investigators. His values for the concentration of liver glycogen after starvation (Table 268) are much higher than those of any other investigator.

Sex Differences in Liver Glycogen

When normal young men and women were subjected to acute starvation, the women developed a higher degree of ketosis than the men, and the difference was not dependent on differences in adiposity (Deuel and Gulick, 1932; Hawley *et al.*, 1933). Since the development of ketosis is associated with the exhaustion of the combustible carbohydrate stores, this observation suggested a sex difference in the liver glycogen content. The question has been examined further in animals.

After 24 to 48 hours of starvation adult male rats had significantly more liver glycogen than females (Greisheimer, 1931; Ponsford and Smedley-Maclean, 1932; Deuel *et al.*, 1934), whereas there was no sex difference in the unfasted controls. After 72 hours of fasting the sex difference disappeared (see Table 270). Stöhr (1932) concluded that this sex difference in rats is related to age, but his data are of doubtful significance.

TABLE 270

SEX DIFFERENCES IN THE GLYCOGEN AND FAT CONTENT OF THE LIVER DURING ACUTE STARVATION. From 15 to 44 rats were used in each group. (Deuel *et al.*, 1934.)

| Hrs. of Fasting | Liver | | | | Muscle Glycogen | |
| | Glycogen | | Fat | | | |
	M	F	M	F	M	F
0	3.54	3.53	3.06	3.34	0.427	0.402
24	1.43	0.70	3.90	4.53	0.369	0.383
48	0.45	0.25	4.22	4.92	0.303	0.308
72	0.31	0.13	4.44	4.55	0.235	0.229
96	0.36	0.37	3.98	5.02	0.260	0.241
120	0.39	0.47	3.89	5.10	0.254	0.228

The exact mechanism involved in this sex difference in rats is uncertain. Removal of the ovaries in female rats resulted in an increase in the liver glycogen in starvation above that of the starved male rats, and this was not entirely prevented by the administration of theelin (Gulick *et al.*, 1934). The problem is complicated by species differences. Guinea pigs showed a trend in distinct contrast to rats in regard to the liver glycogen (Deuel *et al.*, 1934).

Restoration of Liver Glycogen after Starvation

Food completely restores the liver glycogen levels of starved animals within a very short time. The type of food fed such animals is of importance in determining the rate at which the restoration occurs. When rats that had been starved for 48 hours were fed glucose, a normal level of liver glycogen was found 6 hours later (MacKay and Bergman, 1934). If the rats were given fat in place of glucose, no change in liver glycogen levels was seen up to 10 hours later. Protein produced only a slight increase in the liver carbohydrate reserve under the same conditions (Greisheimer, 1931; MacKay and Bergman, 1934).

A sex difference was also seen in the liver glycogen levels of rats that were refed following a period of starvation. A larger amount of glycogen was deposited in the livers of male rats than in those of females (Greisheimer, 1931). Again there is a suggestion that this sex variation is in some way linked with differences in fat metabolism, but the evidence is very inconclusive.

Muscle Glycogen in Acute Starvation

Muscle glycogen is lost more slowly than liver glycogen throughout a fast (Figure 84, Table 270). The muscle glycogen stores have been reported completely depleted in severely starved dogs (Pflüger, 1907) with the suggestion that this reserve disappeared when the animal lost 40 per cent of its weight (Michailesco, 1914).

Physicochemical Varieties of Glycogen

There have been various suggestions as to the existence of several forms of glycogen (Willstätter and Rohdewald, 1934; Meyer, 1943). It has been recognized for some time that extraction of a tissue with hot water does not remove all of the glycogen (Pflüger, 1905). The insoluble fraction has been considered as bound with proteins and was named desmoglycogen by Willstätter and Rohdewald (1934), whereas the unbound, water-soluble form was called lyoglycogen. The relative proportion of the two forms of glycogen in the liver is dependent upon the nutritional state of the animal. Willstätter and Rohdewald (1934) found in geese an increasing percentage of the lyoglycogen form as the total liver glycogen increased. Recent studies suggest that glycogen from the liver has a molecular weight almost twice that of the muscle glycogen (Bell et al., 1948). It would appear that the size of the glycogen molecule is the same in both the water-soluble (lyoglycogen) and water-insoluble (desmoglycogen) fractions from the liver (Bell et al., 1948).

On the basis of physiological studies, a few proposals have been made for the presence of different forms of glycogen. Lawrence and McCance (1931) found that while the liver and muscle glycogen levels decreased in rats during starvation, no change occurred in the concentration of this substance in the heart, stomach, and kidney. Jensen in 1902 also found that a dog starved for 15 to 17 days had the same percentage of glycogen in the heart as unstarved dogs. These and a number of related observations led Lawrence and McCance (1931) to suggest "the presence in all organs of a residual non-specific glycogen, which the cells always retain and which is quite different in physiological behaviour and perhaps chemically from the larger and more variable glycogen content of the liver, muscles and heart." The validity and significance of this suggestion (similar to the *élément constant* as far as fat is concerned – see Chapter 22) is still an open question.

Blood Sugar Changes in Semi-Starvation

There is, if anything, more confusion about the changes in carbohydrate metabolism resulting from semi-starvation than about the changes associated with complete absence of food. Although semi-starvation is quite common in large areas of the world, much more experimental work has been devoted to acute starvation. The development of ketosis and the abnormalities in carbohydrate metabolism in acute starvation are sufficient reason for sharply differentiating the two types of starvation.

Joslin (1923) apparently was the first to note the change in the blood sugar level of human beings during experimental semi-starvation. In his study a healthy young woman did without food for 4 days and then began eating again, starting with very small amounts of bouillon. The caloric intake was gradually increased to a daily level of 2400 Cal. on the twentieth day. The blood sugar level decreased from a control value of 120 mg. per 100 cc. of blood to 80 mg. on the eighth day of the experiment, when the food consumed by the subject was 240 Cal. per day. This was the lowest blood sugar level attained in the experiment.

There was considerable interest in the blood sugar levels of the edematous patients (hunger edema) seen during World War I. All the reports agree that there was very little, if any, change in this constituent even among the cases with severe edema. Knack and Neumann (1917) presented the largest number of cases (see Table 271) in which the blood sugar was determined by Bang's method (1913) both during and after the disappearance of the edema. There was no consistent trend in the blood sugar levels as the edema disappeared. The mean value for the blood sugar in the people with edema was 91 mg. per 100 cc. of blood, which increased to 108 mg. after normal food consumption was resumed. The degree of edema had no influence on either the initial or the final blood sugar levels. An occasional high or low blood sugar was observed at both examinations. The lowest value seen was 60 mg. per 100 cc. of blood, but no mention of any hypoglycemic symptom was recorded by the authors. Similar findings were reported by Jansen (1918), Schittenhelm and Schlecht (1918),

TABLE 271

BLOOD SUGAR VALUES BEFORE AND AFTER THE DISAPPEARANCE OF EDE-MA (hunger edema) among civilians in Germany during World War I; all values are in mg. per 100 cc. of blood (Knack and Neumann, 1917).

			Blood Sugar Level	
Subject	Age	Degree of Edema	During Edema	After Edema
1	32	medium	130	70
2	31	medium	80	100
3	61	medium	80	80
5	39	severe	70	70
6	53	slight	80	100
7	51	severe	120	70
8	52	severe	80	120
14	51	severe	90	90
15	75	severe	70	100
16	75	severe	60	90
17	54	very severe	90	100
18	65	severe	90	80
19	63	very severe	80	110
20	65	very severe	100	110
21	52	severe	150	100
22	70	very severe	90	90
23	59	severe	70	160
24	61	severe	110	180
25	41	severe	130	100
26	50	slight	80	140
27	55	severe	60	120

and Maase and Zondek (1920). Glycosuria was not seen in any of these cases (Knack and Neumann, 1917; Abel, 1923).

During World War II some of the reports on semi-starvation indicated relatively normal blood sugar levels. Leyton (1946) determined, by means of the Folin-Wu technique (1920), the blood sugar levels in 153 Russians who were interned in Germany. These values ranged from 69 to 100 mg. per 100 cc. of blood, with an average of 81 mg. This is just below the recognized range (90–120 mg.) accepted for this method (Hawk et al., 1947). Still lower values were observed with the Hagedorn-Jensen technique (1923) among the severe starvation cases seen during the Bengal famine of 1943 (Bose et al., 1946). It was stated that in 70 per cent of the cases the fasting level was definitely below the average normal, but none of the patients showed any outward symptoms of hypoglycemia.

In contrast to the above are the reports from France, Belgium, and Poland. These made frequent reference to the occurrence of a hypoglycemia which appeared to be so severe that it was described as hypoglycemic shock. The syndrome occurred among patients who were described as "cachectic," having lost on the average 66 lbs. (30 kg.) and often suffering from starvation edema. Besançon (1945) reported that the hypoglycemic coma was notoriously resistant to treatment and often resulted in death despite intravenous injections of considerable amounts of dextrose.

The work of the French investigators in this field has been reviewed at some length by Gounelle and Marche (1946). Apparently hypoglycemia was not limited to institutionalized individuals. Some of the first cases were seen among civilian laborers (Barbier and Piquet, 1943), but the largest number of cases seen were in mental institutions and prisons where the food supply was most restricted. Coma developed in some cases very rapidly. In one case recorded by Gounelle and Marche (1946) the blood sugar level some time before the comatose state was 70 mg. per 100 cc. Later "when the subject was inert but still conscious" it was 47 mg. Seven hours later, after the patient had gone into a coma, the level was 38 mg. Death occurred within the next 12 hours. The only other abnormality observed in the blood at the time of the coma was an elevation of the non-protein nitrogen to about 100 mg. per 100 cc.; alkali reserve, calcium, proteins, and chlorides in the blood remained "as they were before the coma" (Gounelle and Marche, 1946). These authors made an attempt to correlate and associate the abnormalities they observed in carbohydrate metabolism with alterations in the pituitary, adrenal, and thyroid glands.

There are reports of a similar condition among the inhabitants of Brussels (Bastenie, 1947) who, like the French patients, went on to death in spite of intravenous glucose injections. Autopsies on the Belgian subjects often revealed a marked atrophy of the liver (reduced from a normal weight of about 1500 gm. to 800 or even 650 gm.). In Belgium the condition was found exclusively among older people who consumed most of their rationed food early in each month and then had to restrict their eating severely during the remainder of the time.

Among the men rescued from the German concentration camps who were studied at Mainau in 1945, the blood sugar was examined in 28 individuals by

Lamy, Lamotte, and Lamotte-Barrillon (1948). Eight of these men who were in a deplorable state of emaciation had basal blood sugar concentrations which were apparently some 20 to 30 mg. per 100 cc. below the lower level of normality, but details of the method and results on healthy controls were not provided. All the men studied at Mainau were in the early stages of rehabilitation, or at least refeeding was being attempted.

In persons who were almost equally emaciated and who were still actively starving in Warsaw in 1942, much more consistent and profound hypoglycemia was found (Fliederbaum et al., 1946). At Warsaw the Hagedorn-Jensen method gave values ranging from 80 to 120 mg. per 100 cc. of blood from fairly well-nourished persons. In duplicate measurements on venous and finger tip blood from 85 severely starved persons, there were only a few values as high as 80 mg. per 100 cc. and most of the values were around 60 mg.; in one case the measurement showed only 23 mg. per 100 cc. of blood. Similar results were obtained in severely undernourished children at Warsaw; with the children the Bang method for blood sugar was used (Braude-Heller et al., 1946).

Burger, Sandstead, and Drummond (1945), with the nutrition survey teams that went into Holland after the liberation from German occupation, reported that some individuals were suffering from psychosis, which was attributed primarily "to hypoglycemia and responded well to intravenous glucose." Another report on the development of hypoglycemia among semi-starved individuals was that by Musselman (1945). The American soldiers who were captured by the Japanese on the Philippine Islands were reported as showing symptoms of hypoglycemia when the food supplies were reduced to 1000 Cal. or less per day.

The French workers (Gounelle et al., 1943; Bachet, 1943a; Gounelle and Marche, 1946) suggested that many of the sudden deaths occurring among the hunger edema cases during World War I actually were due to hypoglycemic shock.

Blood sugar values as low or lower than any of those reported from starvation hypoglycemic shock have come out of India. During the Bengal famine of 1943 Chakrabarty (1947a) made blood sugar determinations on 407 severely starved patients. The values ranged from 19 to 307 mg. per 100 cc., with an average of 76.7. In 5.6 per cent of the cases the values were below 40 mg. As these patients recovered from starvation their blood sugar levels gradually increased (Chakrabarty, 1947b). Again, as mentioned by Bose, De, and Mukerjee (1946) and by Fliederbaum et al. (1946), the clinical symptoms associated with this extreme hypoglycemia were not comparable to those seen in insulin shock at similar blood sugar levels.

It has been suggested by Morgulis (1923) that disturbances in carbohydrate metabolism may be an important factor in the ultimate fate of the starving organism. As support for his thesis, he cites the work of Michailesco (1914), who found that as an animal approached a 40 per cent loss in body weight, the glycogen disappeared entirely from its tissues. This was associated with a condition from which the animals did not recover.

A special form of semi-starvation is anorexia nervosa. The few reports on this condition which have included biochemical studies indicate that the blood sugar in such patients is frequently below normal (Bruckner *et al.*, 1938; Bartels, 1946; Berkman *et al.*, 1948). In some of these cases blood sugars as low as 26 mg. per 100 cc. were unaccompanied by any signs of hypoglycemic shock (Bruckner *et al.*, 1938). On the other hand, there are reports of syncope in anorexia nervosa cases with blood sugar levels of 45 mg. (Oppenheimer, 1944). Undoubtedly many factors are instrumental in controlling the development of hypoglycemic shock. The blood sugar level is of considerable importance in this respect, but the interplay of other factors — such as the reliability of the blood sugar methods at such low concentrations and the possible adaptation of the organism to a gradually decreasing blood sugar level — probably accounts for the great variation in the above reports.

Pollack (1940) attributed to Conn (1940) the statement "that the rapidity of the blood sugar drop was important in the production of the hypoglycemic symptoms." Conn observed blood sugar levels as low as 28 and 35 mg. per 100 cc. of blood without seeing any signs of shock.

Information on the blood sugar level in starving children is limited largely to marantic infants. Here, as in adults, the fasting levels are lower than in normal infants (Tisdall *et al.*, 1925; Wilson *et al.*, 1928; Brown, 1924–25; Jaso, 1932) and in some cases may be as low as 50 mg. per 100 cc. of blood. Calculations of the rate of carbohydrate metabolism indicated that when the rate was referred to the ideal body weight, it was normal (Wilson *et al.*, 1928). The injection of insulin into these infants produced no change in the rate of glucose oxidation. This led Wilson and his co-workers to conclude that there was no fundamental difference in the carbohydrate metabolism of marasmic and normal infants. As the nutritional condition of these children improved, the blood sugar returned to normal levels (Brown, 1924–25). In fact Brown was able to find a fairly good correlation between "the relative state of nutrition" and the fasting blood sugar when the latter was plotted against the body weight expressed as a percentage of the standard weight.

Glucose Tolerance in Semi-Starvation

The great variety in the types of curves one may secure for glucose tolerance studies on semi-starvation subjects is best illustrated in the work of Bose, De, and Mukerjee (1946) with starvation cases admitted to the hospital in Calcutta during the Bengal famine of 1943. The group was divided into the severely edematous cases and those who were extremely emaciated with no signs of edema. Twelve of the patients with edema showed no change in the blood sugar concentration during the 3 hours following the ingestion of 50 or 100 gm. of glucose. In another 16 cases the blood sugar never showed any increase, but after an hour and a half it started to fall and by the third hour reached values as low as 40 mg. per 100 cc. Both sets of glucose tolerance curves were probably affected by disturbances in absorption, since when 0.2 gm. of glucose per kg. of body weight were given intravenously, a marked peak occurred in the blood sugar levels.

"These were cases of chronic malnutrition due to starvation for prolonged periods. Their histories revealed that, owing to poverty, they were already on

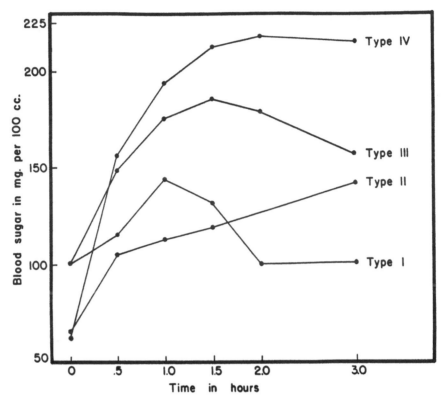

FIGURE 85. BLOOD SUGAR CURVES AMONG THE BENGAL FAMINE (1943)
VICTIMS. Fifty gm. of glucose were given orally. None of these cases showed
any signs of edema. (Bose *et al.*, 1946.)

an insufficient diet prior to actual starvation" (Bose *et al.*, 1946). A severe restriction of food occurred from 6 weeks to 3 months preceding their hospital
admission, when it was found that most of them had pneumonia, malaria, and
dysentery.

The individuals in the severely emaciated group had been living on very
small rations for 3 or 4 weeks. Four different types of blood sugar curves were
seen in this group (see Figure 85). Some of the curves were normal, whereas
others showed a typical diabetic pattern. Insufficient data were given to permit
any correlation between the degree of starvation and the response to ingested
sugar. Obviously one might expect a marked increase in blood sugar in those
individuals who had been without food for a few days preceding the test. Their
curves should be comparable in all respects to the diabetic curves seen in acute
starvation. Glycosuria was observed in the part of the group showing a continuous rise in the blood sugar level throughout the test (Bose *et al.*, 1946).

At Warsaw sugar tolerance tests, using 50 gm. of glucose by mouth in adults,
were made on 45 persons who had been subjected to prolonged semi-starvation
but had not been totally deprived of food. The response of the blood sugar was
very weak and much delayed in appearance; maximum values of 80 to 90 mg.
per 100 cc. appeared after 2 hours or more, and this was often followed by a

profound decline to a value of the order of half the initial (fasting) level (Fliederbaum *et al.*, 1946). The low point of post-glucose hypoglycemia, sometimes
only 20 to 25 mg., was usually 4 or more hours after ingestion of the sugar; the
previous fasting level was generally not regained for another hour or more (see
Figure 86).

In a brief note on studies made during the Bengal famine, Chakrabarty
(1947a) reported that the blood sugar level after 50 gm. of glucose showed a
gradual rise for about 4 hours; in normal individuals the highest rise occurred
within one hour and the level returned to normal within two hours. Again, many
of the subjects undoubtedly suffered from acute starvation. Chakrabarty (1947a)
mentioned that microscopic examination of material secured at autopsy showed
extensive epithelial denudation and submucous hemorrhage in the small intestine which might have interfered with the absorption of sugar. During recovery
the glucose tolerance tests were repeated at 2-week intervals. As the body
weights increased, the subjects showed a gradual elevation in the fasting blood
sugar level. At the same time the glucose tolerance curve returned to normal in

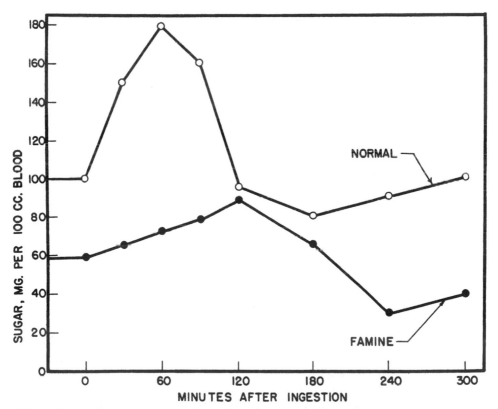

FIGURE 86. Blood Sugar Response, in the Basal State, to 50 Gm. of Glucose
taken by mouth at zero time. Blood sugar determinations in finger tip blood by the
method of Hagedorn and Jensen. Data on well-fed controls and on a 40-year-old person with severe cachexia (Fliederbaum *et al.*, 1946, pp. 121–22). The results are
typical of those obtained in 45 sugar tolerance studies on famine victims.

all respects (Chakrabarty, 1947b). Walters *et al.* (1947a) reported somewhat similar results in starved Indian soldiers.

In France during the German occupation Gounelle *et al.* (1942d) found extreme variations in blood sugar response following 50 gm. of glucose similar to those observed by Bose, De, and Mukerjee (1946). This varied from an abnormally high rise in blood sugar to no change.

A few glucose tolerance measurements have been made on anorexia nervosa patients with highly variable results. One intravenous tolerance study in a girl who was more than 50 per cent underweight showed a hyperglycemic response, with the basal blood sugar increasing from 26 mg. per 100 cc. to a peak of 309 mg. 30 minutes after 25 gm. of glucose were injected. The 2-hour sample was still very high (Bruckner *et al.*, 1938). Bartels (1946) administered glucose tolerance tests to 5 anorexia nervosa patients (type not stated). In 2 cases the fall in blood sugar level was delayed; the other curves were essentially normal.

There are few reports of tolerance tests on carefully controlled cases of semistarvation. Perhaps the closest approach was the work of Malmros (1928). He restricted a normal young man to 1760 Cal. per day for 3 days. At the end of this period the blood sugar curve after ingestion of 62 gm. of glucose was similar to that in the control period except that it returned to its base level more slowly. In the same study a young man started on a diet providing 2000 Cal. per day. This was gradually increased to 2200 Cal. over a 2-week period. The glucose tolerance at the end of that period showed a considerably higher peak and a much slower decrease than the control curve made before the start of the restricted diet. Obviously, there was only a trifling degree of undernutrition in these subjects.

Glucose tolerance tests were made on 8 marantic infants who were less than 80 per cent of their expected weights; the blood sugar responses were within normal limits (Brown, 1924–25; see also Mattill, 1920).

Although factors other than starvation may be operative in cases of Simmonds' disease, these patients are characterized by an extreme emaciation. The glucose tolerance in such patients is either normal or increased (Escamilla and Lisser, 1942).

The Blood Sugar Response to Adrenalin and to Insulin

The blood sugar response to adrenalin administered in starvation has been studied by several groups of workers. In experiments at Warsaw on extremely undernourished subjects (see Figure 87) it was found that the subcutaneous injection of 1 mg. of adrenalin produced a maximum increase of only 20 to 40 mg. of glucose per 100 cc. of blood (Fliederbaum *et al.*, 1946). Dönhardt (1946–47) studied 36 patients who had suffered from famine edema but were free from edema at the time of the test. The injection of adrenalin provoked a much smaller rise in the blood sugar concentration than in normal subjects, and such response as did result was considerably delayed. Dönhardt also observed that there was a diminished and delayed response to adrenalin in the pulse rate and blood flow and suggested that part of the abnormal blood sugar response was due to faulty absorption and sluggish circulation.

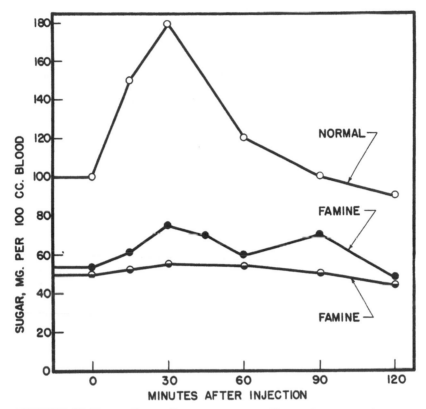

FIGURE 87. BLOOD SUGAR RESPONSE, IN THE BASAL STATE, TO ADRENA-
LIN (one Parke-Davis ampoule) injected subcutaneously at zero time.
Blood sugar determinations in finger tip blood by the method of Hagedorn and Jensen. Data on well-fed controls and on 2 severely starved
adults in the Warsaw Ghetto (Fliederbaum *et al.*, 1940). The results
are typical of those obtained in studies on 10 famine victims.

Ten of the starving men at Mainau were also tested with adrenalin; although
the blood sugar level was close to normal before adrenalin injection, only 3 of
them showed a normal response, the others exhibiting varying degrees of hypo-
reaction (Lamy *et al.*, 1948, p. 178). In several of these men there was almost
no response to adrenalin. Lamy, Lamotte, and Lamotte-Barrillon stated that
Professor Abrami found a similar failure of the blood sugar response to adrenalin in 4 patients who were very severely starved.

At Warsaw it was also observed (see Figure 88) that insulin had less than
the usual effect on the blood sugar in persons who had undergone prolonged
and severe semi-starvation (Fliederbaum *et al.*, 1946). In some cases the venous
blood exhibited almost no change after 14 units of insulin had been injected subcutaneously; in others there was a moderate decline, but it took 2 to 3 hours to
reach the low point.

Comparisons between venous (arm vein) and capillary (finger tip) blood at
Warsaw provided some interesting results following both adrenalin and insulin.
Typical findings are summarized in Tables 272 and 273. In basal rest the venous

TABLE 272

BLOOD SUGAR AFTER ADRENALIN in capillary (finger tip) and venous (arm vein) blood in a 30-year-old severely starved man, illustrating paradoxical arteriovenous sugar differences observed in studies on 5 famine victims in the Warsaw Ghetto. One Parke-Davis ampoule (1 mg. of adrenalin) was injected subcutaneously. The blood sugar values were determined by the Hagedorn-Jensen method. (Fliederbaum *et al.*, 1946.)

	Blood Sugar Values (in mg. per 100 cc.)					
	In Basal	Minutes after Adrenalin Injection				
Blood	Rest	15	30	60	90	120
Capillary 59		70	70	65	75	66
Venous 51			65	85	70	81

TABLE 273

BLOOD SUGAR BEFORE AND AFTER SUBCUTANEOUS INJECTION OF 14 UNITS OF INSULIN (Illetin-Roche) in capillary (finger tip) and venous (arm vein) blood in a 33-year-old severely starved woman, illustrating paradoxical arteriovenous sugar differences observed in studies on 5 famine victims in the Warsaw Ghetto. The blood sugar values were determined by the Hagedorn-Jensen method. (Fliederbaum *et al.*, 1946.)

	Blood Sugar Values (in mg. per 100 cc.)					
	In Basal	Minutes after Insulin Injection				
Blood	Rest	15	60	120	180	240	300
Capillary 50		50	42	28	32	43	55
Venous 41		45	43	40	45	35	40

blood sugar concentration was generally a few milligrams lower than the capillary blood. Following adrenalin the sugar concentration in both bloods usually rose slightly and in parallel, but subsequently the venous blood definitely surpassed the capillary blood for an hour or so. About 30 to 60 minutes following insulin the normal arterial-venous sugar difference in these persons disappeared, to be succeeded by a reversed relationship which persisted for an hour or two. The Polish authors suggested that the enrichment of the venous blood from the peripheral tissues must come from the skin under these circumstances, since they thought the muscles could not account for it.

The Respiratory Quotient in Semi-Starvation

There is no indication of any marked change in the respiratory quotient in mild or moderate degrees of semi-starvation. This was shown by Benedict *et al.* (1919). In more extreme degrees of semi-starvation, especially near the terminal stage of famine death, very low values for the respiratory quotient were reported (Fliederbaum *et al.*, 1946). In patients with R.Q. values close to 0.70, a rather prompt rise to values between 0.9 and 1.0 was obtained when glucose was given; this was in contrast to the small change observed in cases of diabetic acidosis with equally low values for R.Q. before the administration of sugar.

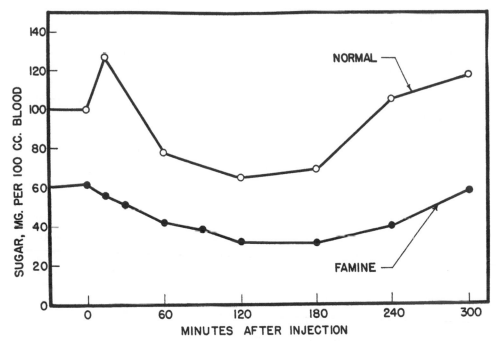

FIGURE 88. BLOOD SUGAR RESPONSE, IN THE BASAL STATE, TO INSULIN injected subcutaneously at zero time, using 14 "new" units of insulin (Illetin-Roche). Blood sugar
determinations in finger tip blood by the method of Hagedorn and Jensen. Data on
well-fed controls and on a 30-year-old severely cachectic person with marked famine
edema. The results on the starved patient are typical of the findings in 10 famine victims, with and without edema. (Reconstructed from the Warsaw Ghetto data given by
Fliederbaum et al., 1946, in Tableau XXVIII and on pp. 124–25.)

Very little work has been done on the respiratory quotient during the rehabilitation period following a siege of semi-starvation. Magnus-Levy (1906)
studied a young man during the 5 months when his weight increased from 36 to
54 kg. The average R.Q. for the different periods studied ranged from 0.818 to
0.828. This would suggest that under such circumstances, as in semi-starvation
itself, the body metabolizes carbohydrates, fats, and proteins in proportion to
their concentration in a mixed diet.

Marasmic infants, whose extreme emaciation was not due to any determi-
nable factor apart from appetite, have also been used in studies on the respiratory exchange in semi-starvation. The reports of such cases indicate that the respiratory quotients are the same as those of normal children (Fleming and
Hutchison, 1924; Wilson et al., 1928). Normal quotients were observed in the
marasmic children after both a small and a large dose of glucose (Wilson et al.,
1928). In these cases, as in adults subjected to semi-starvation, the body continues to burn a normal mixture of fat, carbohydrate, and protein.

The most complete respiratory exchange study made during semi-starvation
was that of Morgulis (1923, p. 272) on a dog maintained for 9 weeks on about a
third of its caloric requirement. During the control period the R.Q. was 0.79.
This decreased to 0.73 during the first week of underfeeding and then went up

slightly, making the average for the first 8 weeks of the experiment 0.76. In the ninth week of semi-starvation the R.Q. was 0.84, which Morgulis interpreted as indicative of the depletion of the fat reserve. There is a question as to how applicable such observations are to human beings. Ordinarily dogs live on a ration that is low in carbohydrate and relatively high in protein. Since the diet markedly affects the metabolic response to acute starvation, there is a possibility of a similar phenomenon in semi-starvation.

Glycogen Stores in Semi-Starvation

Very little information is available on the glycogen stores of human beings during periods of semi-starvation. The only reference thereto is that of Schittenhelm and Schlecht (1918), who found no microscopic evidence of any glycogen among the German civilians who died from hunger edema during World War I.

Very few such studies have been made even in animals. Moulton (1920) determined the glycogen content of the liver and shoulder muscles of steers given various amounts of feed. The semi-starved animal, which was very emaciated, had as much glycogen in the muscles, on a fat-free basis, as the liberally fed animal (see Table 274). The glycogen in the livers of all animals, expressed as a percentage of the fresh weight of that organ, showed no consistent differences. Moulton suggested that the results were probably due to variations in the time elapsing between slaughter and analysis, as well as between the last feeding and slaughtering.

TABLE 274

Glycogen Content of the Liver and Shoulder Muscles of Steers Fed Varying Amounts of Food; all values are expressed as percentages on the fat-free basis (Moulton, 1920).

	Glycogen Content (%)		
	Emaciated Animal	Fairly Fat Animal	Very Fat Animal
Liver	0.913	0.965	0.884
Shoulder muscle	0.514		0.459

In a study of the influence of various sugars on glycogen formation, Greisheimer and Johnson (1930) found that adult rats receiving 87.5 per cent of their calories as lactose lost more than 10 per cent of their body weight in 17 days. There was less glycogen in the livers of the rats which had lost weight than in the glucose- or sucrose-fed animals. The latter showed a slight increase in body weight during the experiment. It is difficult, in this case, to separate the influence of body weight loss on the glycogen stores from that of the dietary carbohydrate.

Carbohydrate Metabolism in Semi-Starvation — Minnesota Experiment

Throughout the experiment blood samples for sugar analysis were collected at the twenty-seventh minute of a 30-minute walk on a motor-driven treadmill. This was adjusted so that the subjects walked at 3.5 miles per hour on a 10 per cent grade. The blood sugar was determined by Nelson's modification (1944) of the Somogyi method. This exercise was done in the morning before breakfast. At S24 and R12 basal blood was also secured before the men got out of bed.

The blood sugar values for the individual men are given in the Appendix Tables. The changes in the mean blood sugar levels for the various caloric groups are very consistent (see Table 275). The exercise was so mild that under normal circumstances the "walking" blood sugar level would have differed only slightly from that in the resting state. At the end of the starvation period the "walking" blood sugar values were lower than in the control period. The decrease amounted to 19 per cent of the control values. There was considerable variation in the blood sugar decreases seen in the different subjects. The magnitude of the decrease was not related to the level during the control period except for those with very low values. In those subjects smaller decreases occurred than among the subjects with the higher values.

TABLE 275

BLOOD SUGAR LEVELS in the Minnesota Experiment. The "walking" blood samples were secured as described in the text; the "basal" samples were taken in the morning before the men got out of bed.

Caloric Group	Blood Sugar Levels (in mg. per 100 cc.)							
	Walking				Basal			
	C	S24	R6	R12	S24	R2	R6	R12
Z	73.9	59.1	62.1	56.0	65.0	68.9	64.8	73.0
L	70.5	56.6	78.1	55.2	63.5		64.3	72.3
G	71.0	59.8	63.7	58.4	64.6	68.5	66.8	75.7
T	70.4	56.5	60.0	59.0	60.9		65.1	70.9
M	71.4	58.0	62.5	57.2	63.5		65.2	73.0

The basal blood sugar in normal young men as determined by the above method in this Laboratory ranges from 74 to 93 mg., with an average of 83 mg. per 100 cc. At the end of the starvation period (S24) the basal blood samples were much lower than this. For all groups the basal blood sugar levels were higher than those secured during the walk on the treadmill (63.5 vs. 58.0 mg. per 100 cc. of blood). This is what one would expect, but the differences between these two values vary considerably for the individual subjects. For some of the subjects (e.g. Nos. 120, 2, 30, 12, 104) the level during walking was actually slightly higher than the basal.

Carbohydrate Metabolism in Rehabilitation — Minnesota Experiment

During the rehabilitation period the basal blood sugar levels increased. There was no relationship between the caloric intake of the groups and the restoration of the blood sugar levels. The increase in the basal blood sugars was very slight through the sixth week of rehabilitation. At that time 800 Cal. were added to the daily diet of each caloric group and this was associated with a considerable increase in the basal blood sugar.

It is difficult to explain the relatively slow increase in the basal blood sugar. If this is directly related to the general nutritional status, one would expect to have found some relationship between the caloric intake and the restoration of the blood sugar levels. But at both R6 and R12 all caloric groups showed approximately the same sugar values. At R6 the T group had regained 3.0 kg. of body weight and still had a low blood sugar level. It was not until R12 that the Z group had increased in weight by 3.4 kg.; at that time their blood sugar concentration had increased 10 mg. Thus, the increase in weight itself was not the most important factor in re-establishing normal sugar levels.

If the glycogen store in the liver is the most important factor regulating the level of blood sugar (Bollman *et al.*, 1925; Cori and Cori, 1929), it is equally difficult to explain the above findings. Although there are no analyses of the glycogen content of the liver during the rehabilitation of semi-starved animals, it might be reasonable to assume that the restoration of the carbohydrate reserves is as rapid then as following acute starvation. In rats starved for 48 hours the liver glycogen concentration was raised to normal 6 hours after feeding glucose (MacKay and Bergman, 1934). Since the diet used in the rehabilitation phase of the Minnesota Experiment provided 67 per cent of the calories as carbohydrate, this should have been conducive to a rapid formation of glycogen. One might reasonably assume that by the sixth week of such a diet there were adequate stores of glycogen in the livers of the Minnesota subjects.

In a similar way, a number of difficulties arise when one tries to explain these findings on the basis of insulin liberation by the pancreas. Arguments can be advanced to suggest that during semi-starvation the insulin production should be either normal or reduced. In the latter case, a train of events similar to those in acute starvation might be expected as far as blood sugar levels and glucose tolerance response are concerned. After acute starvation a few feedings with foods high in carbohydrate will restore the apparent insulin activity.

The blood sugar levels following the exercise present as great if not a greater problem of interpretation. During the rehabilitation period the blood sugar concentration during work increased somewhat by the sixth week. However, by the twelfth week the level was again down to that seen at S24. In order to rule out technical errors as an explanation for these low values during work, the experiment was repeated within the week, with values that were essentially identical.

The respiratory quotient was determined only during work. This value increased throughout the starvation period and through half of the rehabilitation period (R6). It then decreased slightly at R12. The interpretation of the respiratory quotient during the rehabilitation period is complicated by the fact that the men at that time were converting a certain amount of carbohydrate to fat. Such a process would tend, of course, to increase this quotient. Assuming a direct conversion of glucose to a fat such as triolein, the amount of oxygen thus liberated can be calculated. Estimations of the body fat contents of these subjects indicated that an average of 2.2 kg. of fat was formed during the rehabilitation period. This deposition of fat, if it occurred at about the same rate throughout this phase of the experiment, would involve the liberation of some

17 cc. of oxygen per minute. Even this small amount would change the R.Q. at R6 from 0.936 to 0.948.

The blood lactate and pyruvate levels were determined for 7 of the subjects throughout the entire experiment. The pyruvate level increased up to S24 and then decreased only slightly during the rehabilitation period. There were no consistent changes in the lactate concentration. These blood samples were taken 5 minutes after the subject completed a 20-minute walk on a motor-driven treadmill that moved at a speed of 3.5 miles per hour and was inclined at a 10 per cent grade.

Physiology

"With the pleasure of eating is joined hunger, and that after no very equal sort. For of these two, the grief is both the more vehement, and also of longer continuance. For it riseth before the pleasure, and endeth not until the pleasure die with it."

SIR THOMAS MORE, in *Utopia*, The Second Book, "Of Their Journeying or Travelling Abroad, with Divers Other Matters."

CHAPTER 25

Nature of the Physiological Problems

UNDERNUTRITION quickly results in numerous physiological changes, and these become progressively far-reaching as the condition continues. These functional changes result from structural and metabolic alterations which, in turn, must reflect basic biochemical events. But the functional changes are among the first to be discernible, and they tend to dominate the problems of the behavior and the survival of the individual. Practical methods of quantitative morphology and biochemistry are still relatively crude and insensitive; it is often more feasible to infer changes in chemical structure and organization from functional alterations than to attempt the reverse prediction. In any case, it would seem that a practical procedure in the physiological exploration of starvation would be to follow this sequence: observation and description of behavior and performance, measurement of the separate variables of physiological function, evaluation as to consequences, explanation in terms of immediate mechanism, and, finally, elucidation of the fundamental causes.

In the course of such an inquiry it is frequently tempting to invoke purposes or to emphasize changes which seem purposeful. It seems certain that many of the functional alterations in starvation are useful or adaptive in the sense that they guarantee a better chance of survival or provide a nicely balanced adjustment which mitigates the impact of starvation. It would be strange if this were not the case since starvation has certainly been a major factor in natural selection since the beginning of life; mankind today is the result of the principle of the survival of the fittest in countless episodes of food deprivation and shortage. Yet it is unwise to stress this principle unduly in explaining functional changes or to allow teleological anticipation, based on such reasoning, to dominate selection of the data to portray physiological function in starvation.

At the present stage of knowledge of starvation, and of physiology in general, we must confine ourselves largely to a description of events and their apparent relationships. The latter clearly involve so many factors that only tentative essays can be made toward explanation of total mechanisms. At the moment then, progress on the major physiological problems is limited largely to quantitative description of functions and their changes in starvation and subsequent rehabilitation.

On the same general plane of empirical description, but at a slightly deeper level, the beginnings of a broader analysis take the form of observing correlations between the changes in the several variables without necessarily distinguishing between dependent and independent items. As a matter of fact, most of the variables concerned are interdependent, at least in some remote degree. We shall repeatedly find ourselves confronted with the question: Cause or effect? Perhaps in the majority of instances there is little value in attempting to decide. The one final cause for the effects we describe is the fact of undernutrition — that is, a negative caloric balance; the chain of cause and effect must be immensely complex before it results in the particular functional effects which we recognize and attempt to measure.

The classical goal of physiology is the explanation of mechanism, and this most frequently is pursued by the stepwise isolation of variables. This analytical method is a powerful tool in the examination of a particular circumscribed phenomenon, but its severe limitations are at once apparent when we attempt to reverse the procedure in synthesis from the analytical components thus separated. Only a small fragment of the real picture of the starving man could be reconstructed in this way at present.

The foregoing could well be taken as justification for pragmatic description as an exclusive goal in the present inquiry. Actually, while a substantial argument could be made for this procedure as opposed to emphasis on theoretical concerns, an immediate result of exact description is to bring into focus phenomena which cry out for particular and individual analysis. Having accepted as primary problems the accurate discovery and depiction of the functional phenomena in starvation, these phenomena in turn become separable as problems for a rather different type of investigation. Though it involves anticipation of the findings presented in subsequent chapters, it may be useful to remark here on some of the physiological problems that emerge from the examination of the starving man. The items selected are only illustrative; they are intended to portray the breadth and the obscurity of a variety of physiological problems.

Bradycardia and the Meaning of the Heart Rate

An outstanding constant peculiarity of severely undernourished persons is sinus bradycardia. The heart rate is slow both in rest and in work, although the relative rise of the rate in work may be normal. This is discussed in further detail in Chapter 28, where it will be seen that the phenomenon involves an apparent "vagotonia," or relative freedom from sympathetic influence. There is a relative as well as an absolute decrease in cycle variability. It is interesting that extra systoles and nodal escape phenomena are very rare in starvation although the rate approaches or equals that of complete heart block. Does this mean that there is diminished excitability of the ventricles?

The bradycardia itself may be effectively protective for the diminished cardiac reserve; it certainly reduces the work of the heart. The slow heart rate may be adaptive in that it corresponds to a circulatory demand which is decreased because of the reduced metabolism of the body. Obviously, however, appreciation of the utility of the bradycardia is not very helpful in explaining the mechanism of its production.

In the Minnesota Experiment the minimum heart rate was reached fairly early in starvation, when the body weight loss was only about 15 per cent;

thereafter the heart rate gradually rose. There are other evidences that this is a general phenomenon and that the conservation of the heart by slowing its rate becomes less possible as the heart itself undergoes more profound degeneration.

In rehabilitation the heart rate rises; can this be considered a measure of recovery? If the refeeding is abundant, the rise in the heart rate often progresses to surpass the pre-starvation normal, and occasionally there is distinct tachycardia. Is this an indication of cardiac strain or incompetence? The fact that the venous pressure rises parallel to the heart rate is significant, and in one of the Minnesota subjects the result was a congestive failure. Obviously the heart rate is of limited value as a criterion of starvation or recovery.

Cyanosis and the Peripheral Circulation

Mild cyanosis, cold skin, and increased circulatory time are constant indications of the reduced peripheral circulation in the starving man. It is easy to blame this on the heart and to explain the phenomenon as forward failure of the heart. The blood volume is not reduced; relative to the body mass it is much increased. The implication is that the heart is unable to do the job it ought to do.

But the peripheral circulatory requirement needs analysis. There are two main tasks: (1) the supply of blood for metabolic exchange, and (2) the regulation of body temperature. The fact of slight cyanosis indicates that this first requirement is not fully or normally met. On the other hand, the reduced body temperature indicates that the peripheral circulation is actually somewhat excessive for the purpose of proper heat loss; a lessened peripheral flow would conserve heat. It would seem, then, that the peripheral circulation in starvation represents a compromise between the demands of tissue metabolism and those of heat conservation. But why is this so, and how is it achieved? The interplay of two separate circulatory requirements is particularly interesting when they conflict, as in starvation. The details of the mechanisms involved are for the most part obscure.

The Venous Pressure

In starvation the absolute blood volume is maintained or rises somewhat, and the blood volume per unit of tissue is greatly increased. In spite of this there may be a very considerable fall in the venous pressure, even when there is edema. Can this be ascribed simply to a reduced cardiac output and diminished pressure on the arterial side? May there also be a loss of venous "tone"? What part, if any, does the tissue pressure and the elasticity around the blood vessels play? Is the pressure gradient from artery to vein changed in its longitudinal distribution?

Edema and Body Hydration

These questions about venous pressure have an important bearing on the problem of hunger edema, which is discussed in detail in Chapter 43. Here it is enough to note that edematous persons in famine areas, or malnourished persons in general, may owe their edema to any one of several factors and that the attribution of hunger edema to reduced colloid osmotic pressure of the blood plasma is at best a great oversimplification.

In many or most persons who develop edema on a low calorie diet of the European type there is actually no considerable alteration in the absolute volume of the extracellular fluid. The edema represents a disproportion of fluid to cellular tissue which is produced by shrinkage of the cells with maintenance of an unchanged bulk of extracellular fluid. However, this must mean that the structural and elastic forces in the organs are changed, with accompanying implications about the balance of mechanical and other pressures in the maintenance of body hydration. Only a few broad descriptive facts are available, but they point to fundamental problems for future study.

There are signs of peculiarity in the water balance of the starving person. Polyuria is generally present, and nocturia is a frequent complaint. The voluntary fluid intake is usually much elevated, but thirst is not often mentioned as a complaint. Is the condition akin to a mild diabetes insipidus? It may be that the starving person merely imbibes a large amount of water in an effort to achieve a sense of gastric repletion which is not provided by his small food intake. This would seem to be supported by the fact that polyuria is uncommon in patients with anorexia nervosa, who apparently have no desire for a sense of fullness. But these patients also frequently develop edema.

Along with the increased water consumption, starving persons tend to have a marked salt hunger and will consume several times the normal quota of salt if it is available. The change in desire both for salt and for water may be simply psychological in origin, but a more purely physiological explanation has not been ruled out. Simple renal function tests apparently disclose no significant abnormality in these people, but the kidney cannot be ignored on the basis of the small amount of information now at hand. There are no data on the function of the salivary glands, or on the taste buds, in starvation.

The Regulation of Function

In general, the endocrine glands and the nervous system are the great regulators of physiological function, and they must surely be involved in many of the changes observed in the severely undernourished man. Unfortunately, sensitive, quantitative methods in both endocrinology and neurology are far from a satisfactory point of development, and the currently available procedures have had little trial in undernutrition. Some of the major problems for the future are in this area.

During World War II several physicians in France were so greatly impressed by the indications of endocrine abnormality in undernourished persons that they advocated — and applied on a limited scale — therapy with glandular preparations. Certainly there is evidence for endocrine hypofunction in the starving person as compared with a person in a normal state of nutrition. But the question remains as to whether this situation may not be adaptive and relatively beneficial under the circumstances. For example, a relatively hyperactive thyroid or an intense libido and corresponding sexual activity should be a liability in a time of severe food shortage. It is entirely uncertain, moreover, whether hormone preparations would be useful in the rehabilitation phase when a reasonable supply of food is provided.

Ordinarily, starved persons exhibit only signs of depressed sex gland function, but in a few instances the situation is more complicated. The cases of gynecomastia reported among repatriated prisoners from the Far East are puzzling and, without better evidence, cannot yet be attributed simply to the long ordeal of undernutrition suffered by these men; the phenomenon is extremely rare. There is more reason to believe that hirsutism among starving women, which is fairly common, is directly related to their inadequate intake of calories. But in these and other conditions the problems are as yet insufficiently defined to warrant more than speculation. Detailed studies on endocrine function in prolonged undernutrition are needed.

Age and Longevity

It is well known that starved people look much older than their years and that their behavior is generally in conformity with their appearance. Moreover, the starving man often volunteers the fact that he feels old, but there are no indications that undernutrition actually accelerates the aging process if the latter may be identified with such changes as calcification and sclerosis of the tissues.

Clearly, starvation raises important questions for gerontology. In some animal forms, at least, chronic undernutrition prolongs the natural life span. It has been suggested that the natural life span is fixed, not in time, but in terms of total metabolism or some function of the rate of living. But in man severe undernutrition makes him look, feel, and act prematurely old. There are also changes in basal metabolism and in sexual function which resemble those produced by age. What is the long-range effect on physiological age and on longevity of the individual?

It might be expected that a protracted period of undernutrition would have different long-range effects depending upon the age of the person when he is underfed. The immediate effects on the growth of children have been the subject of many reports.

Growth and Development

The anthropometric results of growth and development are readily measured, but the interpretation of variations in the rate of the processes in terms of present or future physiological meaning to the individual, or to the society of which the individual is a member, presents many formidable problems. The precautions necessary for accurate determination of growth are not very imposing, but the importance of the establishment of proper norms as a basis for comparison of growth rate data must not be minimized. In far too many cases the failure to provide adequate comparative data has seriously detracted from the usefulness of the observation. The facts of marked seasonal variations in rate of growth must be recognized. Even elementary matters like allowing for clothing in recording the body weight are too frequently forgotten.

The collection of data on growth is accomplished by: (1) the individual method, where changes are followed, generally in a small group, over a period of time, or (2) the group approach, where single observations are made on a large sample of the population with comparison to another reference group. In the individual approach the important consideration is the rate of change, the absolute values being used mainly to place the individual in his population group. In the group method special precautions must be taken to ensure comparable population samples for comparison. Either method can supply useful information on factors that may influence growth and development, and both methods have been used in observation of children during times of famine.

The results of restrictions of the diet of children and pregnant and nursing mothers in times of famine and want raise problems of great importance besides the immediate effects upon the individuals involved. If the deprivation of food is severe enough to produce a retardation of growth over a period of a year or two, will the deficit be carried into adulthood or will it be overcome when food again becomes plentiful? If the growth deficit is permanent, in what way, if any, does it influence the worthiness of the individual as a member of society? If the individual potentialities of large groups are prevented from reaching their normal fruition because of dietary restrictions during childhood, what will be the results on the stability and progress of national and world societies?

The opposite condition to famine — luxus consumption — may present problems as serious as those of starvation. It might well be asked whether maximal growth rate is synonymous with optimal growth. The evidence from some recent animal experiments suggests that it may be necessary to revise present standards of what is considered desirable for the present and future health and welfare of the individual. The problems of the relationship of diet to maximal well-being present a challenge to investigators in both the biological and the social sciences.

The physiological phenomena associated with changes in growth rate induced by starvation have been almost totally neglected. Although there are many data on the growth rates of school children in periods of food shortage, there is an almost complete absence of information on the functional status of these children. That subsequent good nutrition results in "normal" adults is concluded from the fragmentary evidence obtained on population groups in Central Europe who as children suffered the food shortages of World War I. But this evidence merely indicates that there was little or no permanent stunting of growth and no obvious signs of permanent malfunction. The peculiarities of individual and group behavior exhibited by central and eastern Europeans in the 1920s and 1930s could conceivably be related to physiological abnormalities developed in the food shortage years of World War I and its aftermath. But it would be difficult to argue that the nutritional situation in 1916–19 was any more important as a determinant of subsequent behavior than any one of a large number of psychic disturbances which operated then and later. In any case, these questions of physiological residues of subnormal growth rate remain purely speculative at present.

Behavior

The problems of the behavior of starving men are discussed in a group of later chapters devoted to the psychological aspects of starvation. The relationship of behavior to physiological status is, of course, as complex as it is interesting. In some respects the relationship is straightforward and intelligible. For example, the starving man is weak and cold, both physiologically and subjectively, and his behavior bears this out. On the other hand, his behavior is often misleading. He acts dull and insensitive; he looks and behaves as though he were unaware of or incapable of feeling many of the ordinary stimuli of sound, sight, or touch. But in fact his sensory mechanisms seem to be extraordinarily well maintained. Objective tests, as in the Minnesota Experiment, indicate that vision is resistant to deterioration in starvation and that hearing actually becomes more acute. There is some physiological basis for the old saying that hunger sharpens the senses. The facts ask for an explanation that is not presently available.

The Nervous System

The behavior of the starving man, or the man recovering from the effects of undernutrition, is importantly influenced by the characteristics of his nervous system at the time. At present the evaluation of the nervous system is crude and fails to provide the basis for quantitative analysis in many directions. It is obviously not proper to dismiss the subject on the ground that starvation seems to produce only relatively small morphological changes in the nervous tissues.

The total functions of vision and hearing, and presumably the component nervous apparatus of these senses, are very little affected by severe undernutrition. So far as the evidence goes, the same seems to be true of the other special senses. Qualitatively at least, the senses of pressure, taste, and smell seem unaffected. Peculiarities of behavior in starving men would not seem, therefore, to be ascribable to abnormalities in the sensory apparatus itself. Mechanical pressure over peripheral nerves may be more readily sensed by the starved man, but this would be explicable in view of the reduction in the soft tissue cushion over the nerves.

The function of the proprioceptors in the starving man has been studied so far only by means of the standard tendon reflexes. In the majority of cases these stretch reflexes are diminished and they may be abolished in some individuals who are 20 to 30 per cent underweight. These reflexes are fairly promptly restored on refeeding. The mechanisms involved in these changes have not been explored.

Reaction times as measured by the usual methods are only slightly affected by even a severe degree of undernutrition. In real life situations, however, the reaction time is often prolonged, particularly when the response reaction requires a major body movement. The question is raised as to whether this change is related to alterations in the peripheral pathways and conduction, in the effector apparatus, or in the more central part of the mechanism. The starving person exhibits a reluctance or delay in movement which suggests central inhibition in addition to conscious suppression of reaction. The phenomenon has its practical consequence in making the severely undernourished person prone to accidents.

Nutritional Neuropathy

From the outset of work on the present volumes, the authors were agreed that because of limitations of both space and personal experience, it would be impracticable to attempt any detailed discussion of the varieties of specific malnutrition which may be combined with general undernutrition at various places and times. The distinction between signs and symptoms related to specific deficiencies and those which pertain to general deficiency is not easy and often calls for rather arbitrary decisions. But there seems to be a host of neuropathological conditions which are rare or absent in many semi-starved population groups but are common enough in others, and it seems reasonable to conclude that these conditions are not primarily a result of general undernutrition. Moreover, some of these conditions are found in persons who are calorically well fed, or at least not seriously emaciated, but who have been subsisting on peculiar diets.

Broadly, the distinction can be made between the characteristics of famine as regularly seen in the Occidental world and those which appear in the Near, Middle, and Far East. This distinction was particularly marked during and after World War II. Part at least, and perhaps even all, of the difference would seem to be a reflection of the prevalence of vitamin deficiency in the Oriental and tropical areas of the world. The chapter on vitamins (Chapter 21) is in part an analysis of the role of vitamins in European famine and in part a set of introductory notes to the special problems which arise in food shortages east and south of Europe proper. The outstanding feature of these special problems is the dominance of neuropathology.

The fact that nutritional neuropathy has been the subject of several recent scholarly reviews covering the experience of World War II as well as the world literature makes it of less consequence that the subject is not fully treated here. Denny-Brown (1947) had access to a great deal of material in India, Burma, and the Southeast Theater of World War II. Many cases of neurological disorder were seen, but they represented only a small proportion of a large population of persons who had been underfed and overworked for years, all the while being exposed to, and often developing, a great number of infections and infestations. Of some 60,000 released British and Indian prisoners, 3667 required medical care, and all these were carefully screened to discover neurological disorder. The result was a total of 303 clear-cut cases of neuropathy. In another group of 650 British prisoners released from Rangoon 235 were sick, and among these there were 14 men suffering from a central scotoma, 3 of whom were grossly ataxic in gait. In all this group there were only some 20 to 40 men who had one or another definite indication of change which might be related to specific nutritional deficiency; almost all of them, however, had signs of caloric deficiency, and edema of the ankles was common.

Spillane (1947) likewise saw hundreds of persons suffering from malnutrition with variable neurological manifestations, but these again were culled from many thousands who had suffered prolonged dietary restriction. Many of Spillane's patients had not been really starved, and a substantial number had been fairly well fed calorically. Such data as are available from the Middle East prisoner-of-war camps, which supplied a large proportion of Spillane's patients, offer few clues to the origin of the neuropathies that were seen. Neither in Denny-Brown's material nor in that of Spillane is it possible to identify the specific nutritional faults involved. In one series of 63 patients with encephalopathy at

Singapore, treatment with thiamine, niacin, yeast, rice polishings, and liver extract had no discernible effect on the course of the disease (Graves, 1947).

The experience of World War II, especially outside Europe, shows that the academic portraits of so-called classical deficiency diseases are idealizations or even rather unreal abstractions with regard to the actual findings where real malnutrition is endemic or epidemic. Moreover, there is material for argument against the idea of a progression from "positive" nutritional health through "subclinical" deficiency to the full-blown disorder, in which the subclinical state is supposed to be characterized by vague malaise, fatigue, and so on. Some of the cases of amblyopia and ataxia developed with little or no premonitory change in the sense of well-being.

From a critical review of the literature and from his own observations, Denny-Brown (1947) concluded that "several distinct syndromes besides the distal polyneuritis of beri-beri may make separate and independent appearance. These are: 1. Retrobulbar neuritis, 2. Spinal ataxia, 3. Spastic paraplegia, 4. Burning feet ('Acrodynia') with corneal changes, 5. Deafness, 6. A myasthenic bulbar syndrome ('Kubigassari')." Denny-Brown questioned medical officers who assisted in the relief work at Belsen and in other areas of starvation in Europe and found none who had observed the conditions found in Southeast Asia and Malaya.

These conditions may be exceedingly rare in modern Europe but they are not curiosities in the southern Orient. Though there is a considerable recent clinical literature on them, they have had little scientific study and the details of the deranged physiology as well as the mechanism of its production are quite obscure. The factor of geographical distribution poses intriguing questions.

The Question of Climate

The most cursory consideration of the characteristics exhibited by malnourished population groups in different parts of the world reveals broad differences in climatic distribution. The prevalence of nutritional neuropathies and of disorders of the mucocutaneous tissues is far greater in tropical and semitropical regions than in the temperate and subarctic zones. Any influence from racial factors may be ruled out on the basis of the experience with Europeans in the tropics and subtropics.

Several theories may be advanced to explain the facts: (1) The foods produced or available in the warmer regions may be peculiar, especially the foods used in times of food shortages. While this factor is undoubtedly of importance, it is not possible to blame everything on the nutritional faults of foods like polished rice and tapioca; in the Near and Middle East, as well as in Africa, dominant signs of specific deficiency have developed on nontropical and non-Asiatic diets (see e.g. Spillane, 1947).

(2) Staple foods stored in warm climates may undergo deterioration in nutritional values. Some deterioration of this kind must take place, but tests so far made on stored U.S. military rations, for example, have not indicated such an extreme rate or extent of nutrient destruction as might be required to make this a major factor. Such tests, however, have been limited largely to the more prom-

inent vitamins, and quite different effects may be obtained with the lesser known vitamins.

(3) Continued subsistence in warm climates may alter the individual metabolism so as to render it more sensitive to a given low level of intake of certain nutrients. Moreover, changes in the general level of caloric expenditure and requirement in different climates would mean a disproportion between caloric deficit and the total intake of specific nutrients. Johnson and Kark (1947a) have demonstrated the great dependence on climate of the caloric requirement. This factor of a reduction in specific nutrients ingested whenever the caloric intake or requirement is reduced would be of no consequence if the need for the vitamins or specific amino acids is proportional to the energy turnover. But, in spite of the efforts to establish this principle for at least some of the vitamins, the evidence is unsatisfactory and it may well be that there is an absolute requirement even for thiamine.

(4) The idea has been popularized that there may be important vitamin losses in sweat, but the concordance of all recent careful work on the subject shows that this possibility may be dismissed.

(5) In the warmer climates, particularly in undernourished population groups, certain infections and infestations are vastly more common than in colder areas. The population groups which contribute the majority of cases of neuropathy and mucocutaneous disorder show a very high incidence of malaria, bacillary and amebic dysentery, hookworm, and so on. The nutritional consequences of these conditions by themselves have had very little study.

Attention must be paid to the extensive survey studies on Allied troops in tropical areas (Kark *et al.*, 1947). These showed that subsistence for months to several years in the tropics on Occidental foods, often subjected to prolonged warm storage, is not productive of signs of specific deficiencies even when there are moderate losses of body weight — that is, slight to moderate caloric deficiencies. In the troops examined by Kark *et al.* the incidence of infection and infestations was relatively far lower than in the Japanese prison camps.

Finally, it is clear that neuropathies and mucocutaneous disorders of what seems to be nutritional origin do occur to a limited extent in more northern regions and sometimes have affected rather large groups. "Burning feet" was present in epidemic proportions in Paris in 1828–29 (Chardon, 1830) and at various times in Italy and Belgium (Vidal, 1864). In Labrador and Newfoundland, paralysis, associated with other conditions suggestive of beriberi, was found in 83 persons in one study (Little, 1914). Norman (1898) described outbreaks of what seems to have been beriberi in northern countries, notably on a large scale in Ireland. Obviously climate is neither directly nor indirectly a *controlling* factor, but from the standpoint of prevention and analysis of undernutrition it must be given serious attention, whether or not one subscribes to the theories of such authors as Mills (1942, 1944, 1945).

Physiological Problems in Rehabilitation

We have repeatedly mentioned the great lack of information on the rate, character, and physiological details of rehabilitation following semi-starvation.

Recovery from the effects of severe undernutrition of the European famine type seems to be, eventually at least, fairly complete, but few systematic studies have been made. Persons who have experienced prolonged semi-starvation in the Far East, however, tend to have many residues, some of which, including optic atrophy, are permanent. Adamson *et al.* (1947a, 1947b) attributed the residual anorexia, optic atrophy, hyperhidrosis, palpitation, paresthesias, and dyspnea which they saw in Canadians repatriated from Hong Kong to widespread neuropathy affecting the autonomic system, the peripheral and optic nerves, and the posterior tracts. Presumably such damage is to be ascribed to vitamin deficiency, but the role of parallel caloric deficiency is uncertain. In any case, the mechanisms involved are unexplored.

The tendency to edema in the early stages of rehabilitation has been reported many times. In some cases persons who never showed edema in the active state of starvation exhibit it in a few days when good food is provided (Stapleton, 1946; Sunderman, 1947). Stefanini (1948) observed that an extension of edema is a common occurrence in patients with sprue in the early stage of treatment; the apparent rapid gain in body weight of debilitated persons when they are on a high protein and high caloric diet must be scrutinized with care. The fact is that we have no single reliable criterion on which to judge the progress of rehabilitation.

Sexual Differences

It now seems clear that men and women respond differently, at least in quantitative terms, to a decrease in the food supply. The differences in mortality rates in famine such as were observed in the Netherlands (Dols and van Arcken, 1946) are large, but a true evaluation is rendered difficult by many complications. To some extent it may simply be that rationing systems do not properly provide for the rather large differences in caloric requirements between the sexes. But it seems equally likely that there are differences in the degrees of adaptation attained. Kaller and Reller (1947) observed, in a study of 84 men and 34 women, all of whom were on very short rations, that the basal metabolism generally tended to be depressed more or less in proportion to the loss of body weight. But in a number of cases among the women the decline in metabolic rate was disproportionately great; only small losses in body weight sufficed to produce large declines in metabolism among these women. There is need for much work in this area.

General Reactivity of the Organism

Repeatedly we have remarked that the severely starved man shows diminished reactivity to a variety of stimuli. Not only is the overt behavior that of an unresponsive person, but there are indications of a basic lack of reactivity. Drugs and hormones which ordinarily provoke prompt and striking effects behave as though their potency was impaired. This is seen notably with adrenalin, pilocarpine, atropine, and insulin; the vasomotor effects as well as those on the glucose-glycogen system are greatly delayed and limited in magnitude. There are suggestions also that allergic and anaphylactic phenomena are repressed. In some cases eczema and asthma become less troublesome or even disappear. Mantoux and Pirquet tests lose much of their utility, and only concentrated dosages produce positive responses.

Since many medical interpretations and diagnoses depend on reactive behavior, the starved person requires new standards for evaluation. Obviously there are numerous challenges for research and application in this fact. Among other things, we may suspect that a pharmacology for the undernourished should be quite different, at least quantitatively, from that designed for the well-fed man. Undoubtedly, clinical medical practice, which so often deals with persons undernourished because of their diseases, has unconsciously incorporated some wisdom gained from recurrent but unanalyzed experience. But these matters do not appear to have been emphasized in any systematic research or teaching.

Reactivity in general is a basic attribute of the total organism and its parts and must have been subjected to the full process of natural selection throughout the organic evolution of man. The decline in reactivity of the starved person suggests impairment, but it is arguable to what extent these changes are adaptive and, in the statistical long run, beneficial. The problem can scarcely be posed in acceptable scientific terms on the basis of the very limited evidence now at hand.

A lack of reactivity is possibly not an unmixed evil in certain disease conditions. We have mentioned obvious allergic phenomena, but there are other situations in which tissue, cellular, or organ reactivity may play less well recognized roles. It is possible to speculate at least with regard to such disparate conditions as diabetes mellitus and rheumatic fever. The virtual disappearance of rheumatic fever from a considerable population, as in Warsaw (Braude-Heller et al., 1946), could conceivably be explained in terms of a suppression of tissue reactivity. And the favorable effect of starvation on diabetes morbidity could involve changes of a similar nature in reactivity of the islet cells.

CHAPTER 26

The Gastrointestinal System

ALMOST without exception field reports from famine areas indicate that gastrointestinal disorders are prominent features of the starvation syndrome. Diarrhea, nonspecific dysentery, colic, flatulence, and a protruding abdomen are universally recognized symptoms of caloric undernutrition and have been observed wherever man's natural food supply has been seriously curtailed.

Diarrhea

Diarrhea was recorded by Donovan (1848) during the Irish potato famine of 1847. Porter (1878) listed alvine flux (diarrhea) as a major cause of death in the Indian famine of 1877–78. Again diarrhea was named as an important factor in the millions of starvation deaths in the severe Indian famine of 1897–99 (Indian Famine Commission, 1901; Aykroyd, 1939). Although no mention was made of diarrhea among adults during the famine that spread through central Russia in 1898–99, nearly all the children were reported to have suffered from the disorder (Prugavin, 1906).

The reports from Europe during and following World War I illustrate the increase in gastrointestinal disturbances when food intakes are seriously reduced. In Germany Rubner (1919) observed a high incidence of colic, flatulence, painful bowel movements, pseudodysentery, and stomach and intestinal catarrh attributable to the restricted wartime diets. Among the markedly debilitated Turkish prisoners of war, almost all suffered from diarrhea (Enright, 1920a; Bigland, 1920b). Diarrhea was widespread in the famine areas of Russia in 1921–22 (Abel, 1923).

The observations during the siege of Kut deserve more detailed consideration because the food intake and the period of food restriction were known. Hehir (1922) reported that for a period of about 4 months (December 4, 1915, to April 29, 1916) the average caloric intake for the garrison at Kut was 1850–1975 Cal. for the British troops and 1110–1550 Cal. for the Indian troops. Diarrhea started in a mild form in early March and became progressively more severe as the siege continued. It was the most prevalent disorder seen in the camp and attacked all ranks. It exerted a severe strain on the troops and was stated to be directly responsible for more than 200 deaths (out of a total of 1500) during the last month of the siege. Hehir emphasized his opinion that diarrhea was the major cause of death in the besieged garrison.

Conditions ranging from mild caloric restriction to the most severe degree of semi-starvation were present in large regions of the world during and following World War II. Reports from many of the areas are now available, but the tabulated data will probably not be forthcoming for some years.

Only a disappointingly small amount of reliable information is available on the gastrointestinal responses to restricted caloric intakes during the more severe famine periods of the siege of Leningrad. There is no doubt that starvation with its resultant disorders was rampant among the inhabitants of Leningrad from November 1941 through March 1942 (cf. Brožek, Wells, and Keys, 1946). What information has been released is limited entirely to hospital patients and may or may not be applicable to the general population. Diarrhea started to appear among the hospital patients on admission in early December 1941, and it became progressively more frequent and more severe until in March 1942 from 50 to 70 per cent of all admissions had severe diarrhea. The increased incidence coincided with a rapidly decreasing food supply and a more or less complete disruption of sewage disposal and regulated water supply. The diarrhea was particularly virulent and prolonged among the children. The incidence among the adults decreased rather rapidly in the late spring and early summer of 1942 when the food shortages became less serious and sanitary facilities improved (Chernorutskii, 1943; Efimova and Elpersin, 1944; Novgorodskaia et al., 1943; Ryss, 1943; Brožek, Wells, and Keys, 1946).

In Holland, where there was some food restriction during most of the war period, starvation conditions developed during the first 5 months of 1945. The situation was particularly serious in the larger cities in western Holland. Special Allied nutritional survey teams followed the liberating armies into the larger cities and made street and institution surveys. In the cities of Utrecht, Rotterdam, Delft, and Amsterdam from 10 to 50 per cent of the people interviewed had diarrhea (Surveys: Utrecht, 1945; Rotterdam, 1945; Delft, 1945; Amsterdam, 1945). Burger, Sandstead, and Drummond (1945), Spillane (1945), Drummond (1946a), and Stare (1945) found diarrhea to be a common affliction in western Holland in 1945. In the nutritional diarrhea of chronic starvation as described by Magee (1945a, 1945b), the stomach is often dilated, and there is a profuse diarrhea with as many as 30 stools per day. The stools are watery and contain mucous and undigested pieces of food.

The horrors of mass starvation reached the extreme in the Belsen concentration camps. In Camp B some 40,000 starved persons, most of whom were too emaciated to stand, were herded together in vastly overcrowded huts without more than a token of sanitary facilities and medical care. At the time of the liberation of Belsen nearly all the inmates had a violent, explosive diarrhea, which from the condition of the huts and surrounding grounds must have been present for a long time (Dixey, 1945; Edge, 1945; Gibson, 1945; Hakinson, 1945; Lipscomb, 1945; MacAuslen, 1945; Pollack, 1945). The starvation diarrhea at Belsen was typical of that seen in famines, with frequent, copious, and watery stools often containing blood. In the extremely emaciated patients the diarrhea was refractory to any treatment and death was the usual outcome.

Food supplies were not very seriously curtailed in France during World War II except in institutions and internment camps. Among the internees in the camps in the unoccupied zone of France in 1941–42 body weights were more than 20 per cent below normal in 30 to 50 per cent of the cases (Zimmer et al., 1944). Diarrhea was present at some time in most of the internees, especially when the weather was warm. Besançon (1945) reported that gastrointestinal disorders ranging from slight dyspepsia to severe diarrhea were common in occupied France during the war. During 1944, 771 repatriates from political concentration camps in Germany were admitted to the Salpêtrière Hospital in Paris. Most of the patients had severe diarrhea, and all were seriously underweight. Of the group 67 died in the hospital, and 29 of the deaths were attributed to "diarrhea and cachexia" (Debray et al., 1946).

Among the children in the Warsaw Ghetto, diarrhea was frequent and severe and often preceded the appearance of edema (Braude-Heller et al., 1946). In some cases the diarrhea was persistent throughout the period of observation; these cases usually died. The feces were frequently bloody and resembled a colitis. No bacteriological studies of the stools were made to determine whether the diarrhea was of an infectious origin. In the adults gastritis was rare, but hemorrhagic enteritis and colitis were among the most frequent complications of the famine and often directly hastened death (Fleiderbaum et al., 1946).

Lamy, Lamotte, and Lamotte-Barrillon (1948) reported that diarrhea was present at some time in nearly every inmate of the concentration camps. Twenty to 40 stools per day were not uncommon in the camps. Intensive bacteriological search failed to reveal any causative organism. Sulfonamides and vitamins were without any effect on the diarrhea. The ingestion of large amounts of food intensified the diarrhea, often with fatal results. The authors suspected that the diarrhea was the result of atrophy of the digestive glands and a disturbance of water metabolism.

Digestive complaints with frequent episodes of diarrhea were common in the prison and concentration camps and famine areas of the Far East. Butler et al. (1945) reported that 45 to 60 per cent of the internees of the Japanese prison camps had periodic attacks of diarrhea, and Gupta (1946) mentioned diarrhea as a common disorder among the American and British prisoners of war in the Far East. Potbellies and a nonexplosive diarrhea with green liquid stools containing much mucous but no blood were observed in many of the American prisoners taken aboard the U.S.S. Rescue from prison camps in the Japanese home islands (Carroll, 1945). The diarrhea was thought by the American officers to be the result of the soybean roughage fed to the prisoners. Most of the 1100 patients in the Shinagawa prisoner-of-war hospital in Tokyo had severe diarrhea (Gottlieb, 1946). Diarrhea was common among the British internees in the prison camps in the Shanghai area (Salmon, 1946; Graham, 1946).

During the famine of 1943–44 in Travancore, India, the incidence of cases hospitalized for severe diarrhea increased more than 100 per cent (Sivaswamy et al., 1945). It is reasonable to conclude that with the disruption of medical facilities that occurs in famine areas only a small percentage of the cases were hospitalized and that the real increase in severe diarrhea was much greater. The mortality records from India are admittedly not very reliable, but the fact that 84 out of 680 deaths (12.4 per cent) recorded in one area in Travancore were reported to be the direct result of diarrhea indicates that this disorder was one of the major causes of death in that severe famine.

Extensive bacteriological studies of the stools of semi-starvation patients with diarrhea in the European and Oriental areas have revealed specific causative organisms in only a small percentage of the cases (Hehir, 1922; Butler et al., 1945; Carroll, 1945; Davidson et al., 1946; Edge, 1945; Lipscomb, 1945; Musselman, 1945). It has been reported that the incidence of pathogens may be high during the first few days of diarrhea but that few are present after 10 days (Novgaradskaia et al., 1943). Lipscomb (1945) found pathogens in the stools of only 8 per cent of the diarrhea cases at Belsen. The diarrhea did not respond to treatment with sulfonamides. It must, of course, be recognized that changes in gastric functioning during starvation may render the gastrointestinal tract susceptible to organisms which under normal conditions would be nonpathogenic.

At Travancore it was concluded that the diarrhea was caused by the low protein, low vitamin diets. Musselman (1945) attributed the diarrhea at Cabanatuan to a pellagrous origin. Spillane (1945) found that large doses of the B vitamins had little effect on the diarrhea. In view of the almost complete lack of signs of vitamin deficiencies in many starvation areas (see Chapter 21), it appears improbable that the diarrheas are the result of low vitamin intakes. Most of the starvation diarrheas appear, however, to be of dietary origin and clear up without any treatment except a good diet, provided the starvation has not progressed too far. In the critically starved no treatment is effective, and a great many of the starvation deaths are attributable to a severe and prolonged diarrhea.

Under conditions of prolonged food shortages it is generally the staple foods that are hardest to get. In an effort to satisfy hunger and provide more calories, substitute foods become major dietary items. Adulteration of food with nondigestible substances and "souping" with water to increase the bulk are common practices. The effects that such radical changes in food habits have on gastrointestinal functions are not known. It is probable, however, that they play at least a contributory role in the digestive disorders of starvation.

Gastrointestinal Ulcers

An increase in the incidence of peptic and intestinal ulcers, particularly peptic ulcers, and of ulcerative colitis has been repeatedly observed during periods of starvation, especially when the famines were associated with wars (Porter, 1878; Formad and Birney, 1891; Enright, 1920a; Bigland, 1921b; Chernorutskii, 1943; Macrae et al., 1944; Anon., 1945b; Besançon, 1945; Gottlieb, 1945; Lipscomb, 1945; Magee, 1945; Stare, 1945; De Witt, 1946; Hastings and Shimkin, 1946; Moutier, 1947; Van der Hoeden, 1947). Autopsies have confirmed the presence of gastrointestinal ulcerations with thinning of the intestinal wall, submucosal hemorrhages, and atrophy of the tunica muscularis (Porter, 1878; Donovan, 1848; Formad and Birney, 1891; Rubner, 1919; Bigland, 1920b; Lubarsch, 1921; Efimova and Elpersin, 1943; Lipscomb, 1945; Magee, 1945a, 1945b; Dixey, 1945; Gottlieb, 1945). Intestinal changes of an edematous and "catarrhal" nature were found in 27.2 per cent of the 492 uncomplicated starvation deaths in Warsaw (Stein and Fenigstein, 1946). Gastroscopic examinations of the severely starved have revealed atrophy of the gastric mucosa (Debray *et al.*, 1946).

The incidence and etiology of wartime gastroduodenal ulcers was recently reviewed for France (Moutier, 1947) and for Belgium (Van der Hoeden, 1947; see Table 276). The two reports are in essential agreement that (1) there was a dramatic increase in ulcers during the war, (2) there were more gastric ulcers than duodenal ulcers, whereas normally duodenal ulcers are more frequent than gastric ulcers, (3) there was an increased incidence of ulcers in women, (4) the incidence of ulcers started to rise before there was any real change in the food supply, which occurred during the second half of 1940, (5) the incidence of ulcers in older people was increased, and (6) the wartime diets were low in protein and fat and probably in some of the vitamins and minerals and were high in roughage content. The authors concluded that the emotional state and increased tension produced by the war played an important role in the etiology of the gastroduodenal ulcers. The diet was also important because of the low fat content, which allowed faster evacuation and a longer period of time without food in the stomach, and because of the high roughage content, which may have produced local trauma and irritation, particularly in the lining of the stomach. The latter factor may explain the increased incidence of gastric ulcers.

Fliederbaum *et al.* (1946) found peptic ulcers to be one of the rarest of diseases in the Warsaw Ghetto. A decrease in peptic ulcers was also reported from the Dutch East Indies concentration camps (Netherlands Red Cross Feeding Team, 1948). A possible explanation for these divergent findings may be that in very severe starvation the drastic reduction of HCl secretion, as seen in the Warsaw Ghetto (see the following section on gastric acidity), would inhibit ulcer formation.

TABLE 276

INCIDENCE OF PEPTIC ULCER IN BELGIUM before and during the German occupation. All tabular values are in percentages.
(Van der Hoeden, 1947.)

	Before	During	Year
Ghent General Hospital			
Diagnosed from routine X-ray	23.6	32.6	1940–41
Peptic ulcers as percentage of			
total ulcer cases	33.0	67.0	1940–41
Ghent Hospital for the Poor			
Peptic ulcers in all patients	5.5	14.6	1943
Brussels Hospital St. Pierre, Medical Service			
Peptic ulcers in all patients	0.7	1.7	1940–41
Ulcers in new patients on Gastrointestinal			
Service	28.0	57.0	1940–41
Percentage of ulcer patients who			
were females	6.6	11.0	1941–44
Peptic ulcers as percentage of total ulcers ..	28.0	56.0	1942
Liége, Medical Service			
Ulcers in patients complaining of dyspepsia.	33.7	52.4	1942
Percentage of ulcer patients who			
were females	6.9	18.0	1940–41

Morphological alterations of the gastrointestinal tract in experimental starvation in rats have been reported by Miller (1927). Rats maintained on a low caloric intake developed degenerative changes in the tunica mucosa and tunica muscularis of the stomach and intestines. The chief and parietal cells showed evidence of cytoplasmic and nuclear degeneration. Atrophic changes occurred in the lamina circularis. The degenerative changes became progressively less intense in the colon. Edema in the lamina propria of the stomach and intestine was noted.

Gastric Acidity and Tonus

The frequent occurrence of gastrointestinal ulcers — particularly an increase in peptic ulcers — in starved humans is not associated with a gastric hyperacidity. Most of the available reports indicate a gastric hypoacidity and hyposecretion in starvation (Rubner, 1919; Reiss, 1921; Efimova and Elpersin, 1943; Ryss, 1943; Besançon, 1945; Magee, 1945a, 1945b; Musselman, 1945). No systematic gastric acidity tests were made, however, nor were any histamine response tests used. The absence of fasting free hydrochloric acid is not sufficient evidence upon which to base a diagnosis of clinical hypoacidity because of the great intraindividual and interindividual variability among normal subjects.

The hydrochloric acid response to a test meal was studied by Leyton (1946) in 50 Russian prisoners of war who had lost more than 20 per cent of their body weight. Of the group there were 29 with normal acidity, 4 with complete achlorhydria, 6 with moderate hypochlorhydria, and 11 with some degree of hyperchlorhydria. Leyton concluded that the response was about what one would expect in a normal group of men. No mention was made of the incidence of gastric ulcers. It appears probable that the increased incidence of ulcers in semi-starvation can be attributed to the morphological changes that occur in the digestive tract and to the traumatic effect of the rough, nondigestible materials used to appease hunger — but certainly not to hyperacidity.

In the Warsaw Ghetto gastric secretion studies were made on 20 patients, 10 after taking caffeine and 10 after taking alcohol. The general result was total hypoacidity and no free HCl. The total acidity values were exceptionally low in the cases with edema (Fliederbaum et al., 1946). Hypoacidity was also observed in 15 children, both while fasting and after taking 200 cc. of 5 per cent alcohol (Braude-Heller et al., 1946).

Lamy, Lamotte, and Lamotte-Barrillon (1948) made gastric analyses on 12 starvation patients before and after the injection of histamine. After the histamine 8 had no gastric acidity, 3 showed a small acid response, and 1 was classed as normal. Fractional test meals with alcohol, caffeine, and bouillon were given to 1890 men in prison camps where the food supply had been bad (Kowalewski, 1947). Achlorhydria was found in 2.6 per cent in 1941 and in 14.8 per cent in 1944. In the same period the incidence of subacidity increased from 17 to 48 per cent of the cases studied.

Gastric atony was reported by Reiss (1921), Chernorutskii (1943), and Ryss (1943). Hypoactivity has also been observed in some cases of anorexia nervosa (Morlock, 1939). Fliederbaum et al. (1946) reported a "slack" stomach that was in a normal position. However, in rats that had been on a prolonged caloric restriction, the passage of a barium-buttermilk meal through the stomach and intestine was faster than in normals (Menville *et al.*, 1930). If there is a real atrophy of the tunica muscularis during semi-starvation, it would be reasonable to expect a decreased muscular force and a slowing of the passage of food along the gastrointestinal tract.

Gastric Acidity — Minnesota Experiment

Gastric acidity was measured in 10 of the Minnesota subjects at the end of the semi-starvation period. In the fasting state gastric tubes were passed and the stomach emptied. A subcutaneous injection of 0.1 mg. of histamine hydrochloride per 10 kg. of body weight was given, and the gastric juice secreted during a 30-minute period was collected. The collected secretion was measured, examined visually for color, consistency, and presence of blood, and titrated for acidity. The free and total degrees of acidity for the fasting and histamine response samples are presented in Table 277. In the fasting state 5 of the subjects had no free hydrochloric acid while 5 had normal free acidity. Only 1 of the 10 subjects had total fasting acidity that might be considered abnormally low. The free acidity and total acidity after histamine were within normal limits in all cases.

TABLE 277

Free and Total Degrees of Acidity of Gastric Juice in the fasting condition and 30 minutes after histamine injection, at the end of 24 weeks of semi-starvation (Minnesota Experiment).

Subject	Fasting		After Histamine	
	Free	Total	Free	Total
122	4.7	13.2	55.0	62.0
129	0	7.5	32.5	34.2
105	13.8	24.0	40.0	46.0
120	0	7.5	16.0	25.0
102	0	9.0	89.0	98.7
104	5.7	15.2	68.7	78.5
101	0	3.0	62.0	73.5
30	13.5	18.8	79.2	87.3
11	0.2	17.6	43.7	56.0
112	0	13.0	61.5	72.5

The question as to what constitutes normal gastric acidity either in the fasting state or after a test meal is far from being answered. The failure to find free hydrochloride acid in the fasting stomach does not necessarily mean the absence of an acid-secreting power by the glands. Regurgitation of intestinal contents might easily neutralize the free acid. Yellow-colored gastric contents were noted in 4 of the 5 fasting samples that contained no free hydrochloric acid. This probably indicated the presence of bile and intestinal juices.

Gastric Motility — Minnesota Experiment

Gastric motility studies were made on 18 of the subjects during the control period, at the end of 24 weeks of semi-starvation, and after 12 weeks of rehabilitation. The test meal consisted of 40 gm. of oatmeal cooked in 300 gm. of lightly salted water. Sixty gm. of barium sulfate were added, and the meal was brought up to 400 gm. by adding water if needed. The test meal was eaten rapidly in 2 minutes. Roentgenograms (at 36 inches) were taken at 5, 30, 60, and 90 minutes after the meal had been eaten. After 90 minutes the progress of the meal was followed by fluoroscopic examination at 15-minute intervals until the stomach was completely empty. During the intervals between X-rays and fluoroscopy the subjects remained seated and occupied themselves by making surgical sponges or doing other light handwork.

The stomach shadows were traced from the X-ray films onto paper, and the area of the shadows was measured with a planimeter. The areas of the shadows at 30, 60, and 90 minutes were expressed in percentages of the 5-minute areas. Final emptying times were expressed in minutes.

TABLE 278

GASTRIC EMPTYING TIME in minutes at control (C), 24 weeks of semi-starvation (S24), and 12 weeks of rehabilitation (R12) with means, standard deviations, differences, and percentage recovery of motility during rehabilitation, for the 18-man group, for the combined 2 high caloric groups, and for the combined 2 low caloric groups (Minnesota Experiment).

	Emptying Time in Minutes		
	18-Man Group	High Caloric Groups	Low Caloric Groups
Control..............	175 ± 32	189 ± 35	164 ± 26
S24.................	226 ± 69	240 ± 69	215 ± 70
R12.................	198 ± 41	193 ± 43	201 ± 41
S24 — C...........	51	51	51
S24 — R12.........	28	47	14
R12 — C...........	23	4	7
Percentage recovery $\dfrac{S24 - R12}{S24 - C} \times 100$	54.9	92.2	27.5

The average final gastric emptying time in the group of 18 subjects during the control period was 175 ± 32 minutes (see Table 278). For the particular test meal used, the emptying time was comparable to that observed in this Laboratory for other groups of young men of the same age (Henschel et al., 1944, 1947). The course of gastric emptying appeared normal in all cases (see Figure 89).

At the end of semi-starvation the gastric motility was depressed, with an average emptying time of 226 ± 69 minutes. The interindividual response to the test meal was much greater at the end of starvation, as shown by the larger standard deviation. Of the 18 subjects, 11 had a prolonged emptying time, 6 showed no change, and 1 had a shorter emptying time than in the control period.

The decrease of gastric motility during semi-starvation occurred throughout the entire gastric emptying phase. The differences in general became progressively greater as emptying proceeded. The differences between control and semi-

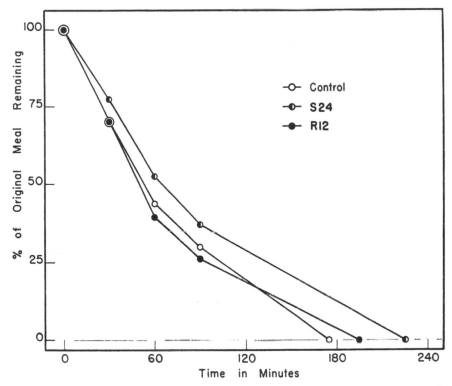

FIGURE 89. MEAN GASTRIC MOTILITY FOR THE 18-SUBJECT GROUPS at control, 24 weeks of semi-starvation (S24), and 12 weeks of rehabilitation (R12). The rates of gastric emptying are expressed as percentages of the original test meal remaining in the stomach at 30, 60, and 90 minutes; the final emptying times are expressed in minutes. (Minnesota Experiment.)

starvation findings were statistically significant at the 5 per cent level at 60 minutes and at the 1 per cent level for the final emptying time.

In the Minnesota Experiment, then, it is clear that gastric motility was decreased. What effect, if any, this may have had on the digestive functions is not known. Our subjects did not develop any gastrointestinal disturbances during the semi-starvation period. There was no increase in diarrhea, bloating, flatulence, or colic such as has been observed in natural starvation areas. The stools during semi-starvation were smooth and contained no recognizable particles of food. These differences may be explained by the cooking and the dishwashing sanitation and by the fact that the food served was not adulterated with bark, grass, leaves, sawdust, or even dirt, as is often the case when food is scarce. In any event, the reported prevalent gastrointestinal disorders of starvation cannot be attributed entirely to the restricted caloric intake.

Three of the 5 subjects who showed no fasting free hydrochloric acid at the end of semi-starvation were among the group in which gastric motility was measured. The average increase in emptying time of the 3 subjects was 87 minutes, as compared with 51 minutes in the 18-man group. Because of the small group and the large standard deviation at the end of semi-starvation (69 minutes), it is doubtful whether there would be a significant correlation between prolonged emptying time and the lack of free fasting hydrochloric acid in semi-starvation. It has, of course, been observed that in normal man gastric secretion decreases during periods of relatively low gastric motility (Wolf and Wolff, 1943).

We cannot offer a satisfactory explanation, based upon objective criteria, for the decreased gastric motility observed in the Minnesota Experiment. The degree of depression of motility varied widely in the different individuals. An effort to correlate changes in motility with changes in other tested functions may be useful. It is conceivable that a high positive correlation might exist between decreased motility and the percentage of body weight loss and the decrease of active tissue. This might be expected especially if skeletal and visceral muscle mass decrease at about the same rate and if decreased motility is a reflection of the loss of gastric musculature. Also, changes in personality, such as an increase in depression, might be correlated with changes in gastric motility, because emotions are known to influence gastrointestinal activity to a considerable degree (Wolf and Wolff, 1943).

The coefficient of rank correlation between the decrease in gastric motility and the relative loss of body weight was only $+0.35$. It is obvious that gastric motility is a complicated function influenced by the physiological, psychological, and biochemical state of the subject at the time of the test — factors which can be minimized but never completely controlled.

The gastric motility tests were repeated at the end of the 12 weeks of rehabilitation. The results are presented in Table 278 and Figure 89. Gastric motility increased during rehabilitation, with a recovery of 54.9 per cent of the loss that occurred during the semi-starvation period. The percentage of the meal emptied from the stomach was the same as during the control period at 30 minutes and slightly more at 60 and 90 minutes.

The caloric intake during the rehabilitation period had a significant effect on the recovery of the motility lost during semi-starvation. For comparison of the effects of caloric intake the subjects were divided into 2 high and 2 low caloric groups. At the end of 12 weeks of rehabilitation gastric motility in the high caloric groups was almost equal to the control values, while in the low caloric groups only 27.5 per cent of the decreased motility had been regained. The same general relationships held for the percentage of the barium meal that had left the stomach at 30, 60, and 90 minutes. If all the 18 subjects had responded as had those in the high caloric groups, gastric motility would have been essentially normal at the end of the 12 weeks of rehabilitation.

Gastric Distress — Minnesota Experiment

It had been feared that gastrointestinal disorders might occur at the beginning of the rehabilitation period, especially in the high caloric groups for which the food intakes were appreciably increased. The concern proved to be unwarranted. None of the subjects complained of any unusual distress except during some of the testing periods, when it was necessary to eat breakfast and lunch together at noon.

When all restrictions on food intakes were removed at the end of the 12 weeks of rehabilitation, a number of the subjects experienced acute gastric distress. One subject who ate a large amount of nuts and fresh and dried fruits was hospitalized for 24 hours with acute gastric distention, which was relieved by continued gastric suction. It has long been recognized that gastrointestinal crises can be provoked by indiscriminate and unlimited refeeding after periods of semi-starvation. In World War II many of the rescued prisoners of war suffered from the oversolicitude of their liberators.

The Minnesota Experiment offers little information on the best kinds and amounts of foods to use during the first few days of refeeding. Caloric intakes of regular foods up to 3200 Cal. per day were handled by the gastrointestinal tract without any ill effects. Foods that are irritating and gas-forming should no doubt be avoided if possible in any refeeding program.

Position of Stomach — Minnesota Experiment

While tracing the shadow of the stomach from the gastric motility roentgenograms, it appeared that the position of the stomach in the abdominal cavity had changed during semi-starvation. Special measurements were made to express the shift in position in quantitative terms. Two anatomical landmarks on the stomach were selected — the pylorus and the lowest portion of the antral end of the greater curvature. The vertical distance of each of the two landmarks was measured from a horizontal line drawn through the uppermost tips of the two iliac crests. The measurements were made from the 5-minute gastric motility roentgenograms for each of the 18 subjects at the end of the control period, at 24 weeks of semi-starvation, and at 12 weeks of rehabilitation.

The values obtained from the roentgenograms were corrected for triangular distortion. The tube-film distant was constant at 36 inches. The correction factors applied to the roentgenogram measurements were 0.94 for control and 12 weeks of rehabilitation and 0.95 for 24 weeks of semi-starvation. The results are presented in Table 279.

The position of the stomach during the control period varied greatly within the group of subjects but was assumed to be normal in each individual. For the group as a whole the lowest border of the antrum was 1.29 ± 4.11 cm. and the pylorus was 7.18 ± 2.93 cm. above the iliac crests. In 6 of the 18 subjects the lower border of the antrum was below the iliac crests, while the pylorus was above the iliac crests in all cases. There was a tendency for the stomach to be low in the abdominal cavity of the leptosomatic subjects, but the generalization did not apply to all individuals.

At the end of semi-starvation the position of the antrum and the pylorus was lower in the abdominal cavity in 17 of the 18 subjects. The lower border of the antrum was —2.11 ± 3.31 cm. below the iliac crests and the pylorus 4.77 ± 2.49 cm. above the crests, with an average lowering for the antrum of 3.40 cm. and for the pylorus 2.41 cm. In 13 of the 18 subjects the lower border of the antrum was below the iliac crests, while the pylorus was still above the iliac crests in all cases. In subject No. 5, who was rather obese at the start of the experiment and had a stomach high in the abdominal cavity, the antrum dropped 9.5 cm.

and the pylorus 7.2 cm.; this subject lost 23.7 kg. of body weight during semi-starvation.

Twelve weeks on the rehabilitation diet produced an upward shift of the antrum and pylorus in 12 of the 18 subjects, the average position of the lower border of the antrum being —0.99 ± 3.44 cm. below and the pylorus 5.33 ± 2.60 cm. above the iliac crests. In 9 of the 18 subjects the antrum was still below the iliac crest line. For the group as a whole the upward shift was 1.12 cm. for the antrum and 0.56 cm. for the pylorus. In only 4 of the subjects, however, did the stomach regain during rehabilitation its control period position in the abdominal cavity.

TABLE 279

The Position of the Lower Border of the Antrum (L.A.) and the Pylorus (P.V.) of the Stomach, in centimeters above (positive values) or below (negative values) the iliac crests, in 18 subjects at control, 24 weeks of semi-starvation (S24), and 12 weeks of rehabilitation (R12) (Minnesota Experiment).

Subject	Control		S24		R12	
	L.A.	P.V.	L.A.	P.V.	L.A.	P.V.
1	3.0	8.0	2.7	6.7	3.5	7.1
104	—5.3	3.3	—4.4	2.9	—2.8	3.6
111	—2.5	3.2	—7.0	1.4	—3.9	2.2
108	3.1	8.0	0.2	7.5	2.2	7.0
102	—6.2	3.8	—8.8	1.6	—9.4	0.9
105	0.5	7.1	—3.0	4.4	0.5	6.9
11	3.3	7.0	—1.7	5.1	0.0	6.1
5	8.6	12.3	—0.9	5.1	—1.4	5.0
122	3.7	7.7	1.9	5.6	2.4	7.1
29	—1.4	7.1	—1.8	5.2	—3.9	3.4
19	3.2	8.1	—3.4	2.4	1.8	7.1
20	4.8	11.5	0.3	8.1	2.4	8.5
129	—4.2	1.9	—5.8	1.4	—5.8	1.4
120	4.8	10.0	—0.9	5.8	—2.4	5.0
123	0.4	7.9	—2.3	5.2	0.6	6.6
130	4.3	9.6	4.0	10.0	4.8	9.4
27	5.0	9.1	—1.2	6.0	—0.3	7.5
127	—1.8	4.0	—5.7	1.5	—5.7	1.3
M	1.29	7.18	—2.11	4.77	—0.99	5.33
SD (±)	4.11	2.93	3.31	2.49	3.44	2.60

The shift in the position of the stomach appeared to involve mainly the antrum and the pylorus. There may have been a corresponding shift in the fundus and the cardia, but because the stomachs were not completely filled by the test meals, their upper portions could not be accurately visualized. The functional significance, if any, of the moderate downward shift of the pylorus and antrum during semi-starvation is uncertain. At least it is clear that a stomach low in the abdominal cavity is not incompatible with normal gastric functioning. The possible effect of the shift in position of the stomach on the position of the heart is discussed in Chapters 28 and 30.

Feeding the Severely Starved Man

In any area of severe famine the arrival of relief groups with physicians and food supplies does not result in saving all the starved population, and among the victims brought into hospitals and treatment centers the mortality is generally high. It is reasonable to suggest that many failures are due to ineffective or inadequate realimentation and that in this the limitations of the gastrointestinal tract of the starved man may be important. Discussions and speculations on these points were numerous early in 1945 when it became apparent that the Allied armies would soon find in western Holland and in the prison camps of Germany large numbers of persons in an extreme state of undernutrition. The fact that this expected occurrence coincided with a period of rapid development of materials and methods for intravenous alimentation naturally aroused hopes for successful rehabilitation through this procedure.

Many arguments were advanced "that deprivation of food progressively destroys the digestive, absorptive and protective functions of the alimentary canal" (Magee, 1945a, p. 389). The degenerative changes in the intestine and the frequency of severe diarrhea in famine victims, noted in preceding sections of this chapter, certainly suggest probable impairment of the absorptive power of the gut. At best only poor absorption could be expected where "the gut wall is sometimes so thin that print can be read through it, and often to the naked eye does not appear to possess any mucosa at all. This is especially so with the small bowel, while the large gut is as a rule lined by shallow confluent ulcers wth paper-thin mucosa between them" (Charlton, 1946, p. 686). Cuthbertson (1945, p. 391) pointed out that "there is reason to believe that in starvation the alimentary enzymes may share in the generalized protein depletion and that the power to digest will therefore become affected."

The result was that many types of realimentation procedures were tried when western Holland and the German prison camps, notably Belsen, were liberated. The situation was worse than had been believed possible even from the most somber predictions, and the conditions for trying out the procedures were far from satisfactory. Belsen was entered on April 17, 1945, and some 50,000 inmates, together with 8000 to 10,000 unburied dead, were found in the most incredible state of emaciation. In the first days there were about 300 deaths a day (Collis, 1945). Even in Holland the conditions were unsuitable for careful investigative work, and none of the findings in Western Europe for this period can be taken as final, though they are indeed suggestive.

In the first place it was found that the vast majority of the famine victims could take nourishment by mouth without any special difficulty. Cases that seemed to be in the last stages of starvation often responded extremely well to oral liver purée and intravenous glucose (Stannus, 1945), and even to nothing but oral glucose plus milk (Vaughan et al., 1945). In Holland, patients unable to swallow were fed by tube a mixture of hydrolysate (7.5 per cent) and glucose (7.5 per cent) or concentrated liquid food rich in proteins but low in fat (milk, skim milk powder, and glucose). This diet supplied about 2500 Cal. per day. Patients who were able to eat received a diet containing as much as 3200 Cal. and 300 gm. of protein per day (Burger et al., 1945, p. 283).

The experience with intravenous feeding was unsatisfactory, especially when acid protein hydrolysates were used (Vaughan et al., 1945). It was found that the protein hydrolysates were very difficult to give by mouth, the aversion being such that some patients would not swallow them and others suffered from vomiting. The Belsen patients considered stomach tubes a new form of torture and the majority flatly refused to cooperate. Two patients who did submit fared ill; one died on the second day and the other was definitely worse after 2½ days (Vaughan et al., 1945).

There is no doubt that intravenous alimentation is being steadily improved and that in the future its use will be attended with more success. But the important fact with regard to starvation is that such feeding is usually entirely unnecessary; the gastrointestinal tract may be far from normal, but it rarely loses the power to absorb foods slowly and even to do a reasonably adequate job of digestion. Papain digests of meat, administered both orally and intravenously, have been claimed to be useful in India, but the data are not very convincing (Narayanan and Krishnan, 1944; Krishnan et al., 1944). Obviously every effort should be made to avoid parenteral feeding in favor of the oral route. These conclusions are strongly supported, with more detailed evidence, in the official Dutch report on the famine in the western Netherlands (Burger, Drummond, and Sandstead, 1948).

Intravenous alimentation should be properly distinguished from intravenous therapy as a more direct supportive measure. The value of blood transfusion, carried out at extremely slow rates (0.25 to 1 cc. of blood per pound of body weight per hour), has been emphasized by Charlton (1946). This has some nutritional value but is primarily valuable in other ways and is recommended particularly in cases where there is extreme anemia. Charlton reported only 8 deaths in a series of 109 starved patients, with hemoglobin values below 30 per cent of the normal, who were treated by the slow transfusion technique.

CHAPTER 27

Respiration

THERE is extremely little information about the respiratory processes in the literature on famine and undernutrition. The incidence and character of pulmonary tuberculosis under famine conditions is discussed in Chapter 47, where it will be seen that undernutrition appears to result in an increased susceptibility or decreased resistance to the tubercle bacillus in the lungs. The high incidence of bronchitis of a relatively nonspecific type is also notable among severely underfed populations (see Chapter 46). Neither of these facts directly indicates any primary abnormality of the respiratory apparatus.

Respiration in the Carnegie Experiment and Other Studies

Prior to the Minnesota Experiment the major source of information about respiration in undernutrition was the report of the Carnegie Experiment (Benedict *et al.*, 1919). We have discussed the limitations of that experiment in Chapter 3; the data pertain only to a moderate degree of undernutrition. In the Carnegie Experiment the respiration rate in postabsorptive rest varied from day to day, but as undernutrition progressed it unquestionably showed a slight tendency to decrease, with no obvious alteration in the character of the respiration. Neither in the standing position nor in walking was there any consistent change in the respiration rate (Squad B).

A large number of measurements of pulmonary ventilation in rest were made in the Carnegie Experiment, and these showed an average tendency for the tidal volume to decrease in undernutrition, but the individual results were quite variable from time to time. In 7 of the 12 subjects in Squad A the maximum value for tidal volume was recorded early in the period of restricted diet, but in 3 the maximum occurred near the end of undernutrition. The combined result of a tendency toward a decreased rate and a decreased tidal volume was a slight but definite average diminution in the respiratory minute volume.

In the Carnegie Experiment the attempt was made to estimate the relative irritability of the respiratory center from the ventilation rate as related to the estimated alveolar tension of carbon dioxide. The alveolar carbon dioxide showed no regular tendency to change, though the experimental arrangement was perhaps not adequate to decide this. Benedict *et al.* therefore concluded that the low diet produced a decrease in the sensitivity of the respiratory center since the same alveolar tension was associated with an average smaller ventilation. Aside from the technical uncertainties in the estimation of alveolar tension, it should be noted that there was no information about the alkaline reserve or blood pH and no explanation as to how the alveolar carbon dioxide could remain constant in spite of a greater diminution in the metabolism than in the ventilation. Finally, no allowance was made for changes in the strength of the respiratory muscles. The method used was simply not competent to decide what, if any, were the changes in the irritability of the respiratory center.

Fliederbaum *et al.* (1946) studied 20 cases of severe undernutrition without other complications and found no abnormality in the respiratory rate or in vital capacity after moderate exercise. In the basal state the carbon dioxide content of the expired air was only about 3 per cent or less, so there was a slight loss in respiratory efficiency. Braude-Heller, Rotbalsam, and Elbinger (1946) observed a marked tendency toward emphysematous developments in the starving children at Warsaw but provided no details.

Bachet (1943) observed, as have many others, that many starving persons die without ever showing signs of respiratory difficulty, the respiration merely becoming shallower by imperceptible degrees until it ceases altogether. In some cases, however, there is a more brusque change to a state of coma in which the respiration may resemble the Cheyne-Stokes type, with irregular rhythm and long pauses between moderately deep respirations.

Vital Capacity — Minnesota Experiment

The vital capacity of the lungs is determined by the dimensions of the thoracic cage, the amount of blood in the lungs, the volume of the tissues — chiefly lungs and heart — in the thorax, and the strength of the respiratory muscles. In adults the size of the bony cage of the thorax and the location of the attachments of the diaphragm are fixed and cannot conceivably be altered appreciably in starvation. The effective size of the thoracic space, however, would be increased by the shrinkage of the tissues of the heart, lungs, and all other soft tissues.

In the Minnesota Experiment the reduction in the systolic volume of the heart averaged 106 cc., and the reduction in the volume of the other intrathoracic tissues may be estimated to be of the order of perhaps 100 to 400 cc., so that there would be, potentially, space for an increase in the vital capacity of at least several hundred cc., if there was no change in the amount of blood in the lungs. There is little basis for estimating the blood volume in the lungs. The total blood volume decreases in starvation and tends to decrease still further in early rehabilitation, but there is no reason to believe that this change is necessarily reflected in the lungs. It does seem reasonable to suggest, however, that the combined effect of changes in the intrathoracic tissues and in the blood volume would be such as to allow an increase in the free space in the lungs, or at least no decrease in this space.

In the Minnesota Experiment, however, the vital capacity was definitely and progressively diminished in semi-starvation, and this change was reversed in rehabilitation. The mean values and their standard deviations are given in Table 280. The average decrease in vital capacity at S24 was 390 cc. To account for this as a result of increased blood in the lungs would require such a hypothetical increase to be of the order of not less than about 600 cc. This would mean approximately doubling the amount of blood in the lungs, which seems too large a change for the possibility to be entertained seriously. We seem to be forced to the conclusion that starvation results in a substantial loss of strength in the respiratory muscles.

TABLE 280

MEAN VALUES AND STANDARD DEVIATIONS FOR RESPIRATORY MEASUREMENTS on 32 men in the Minnesota Experiment. Vital capacity was measured in the seated position 3 times in each subject on each occasion. The other items were all measured in basal rest in association with the measurement of the basal metabolism. "Respiratory efficiency" is the cc. of oxygen removed per liter of respiratory ventilation.

| | Control | | S12 | S24 | | R12 |
	M	SD	M	M	SD	M
Vital capacity, liters	5.17	0.54	4.94	4.78	0.61	5.00
Respiration rate, per min.	11.45	2.41	9.89	9.86	2.13	10.52
Ventilation						
Liters per min.	4.82	0.56	3.49	3.35	0.61	4.01
Cc. per respiration	421		353	340		381
Respiratory efficiency	47.93	4.94	45.20	42.65	7.38	46.00

Resting Pulmonary Ventilation — Minnesota Experiment

The pulmonary ventilation in basal rest, measured as the total respiratory volume per minute, decreased 30.6 per cent by the end of semi-starvation in the Minnesota Experiment; practically all this change occurred in the first half of the semi-starvation period. At the same time there was a smaller percentage decline in the rate of the respirations, so that the ventilation per respiration (tidal air) had decreased by only 14.5 per cent at S24. The data are summarized in Table 280.

After 12 weeks of rehabilitation (R12) all these changes had been reversed, but the values were by no means back to normal. If the average "recovery" at R12 is calculated in terms of the change from S24 to R12 as a percentage of the change from control to S24, the result is 56 per cent for vital capacity, 42 per cent for respiration rate, and 51 per cent for tidal air.

In later rehabilitation (R12 to R20) the reversal of the changes in the basal pulmonary ventilation continued and tended to overshoot the control values in some respects. In total ventilation per minute, for example, 9 out of 12 men had larger values at R20 than at C, the average being 392 cc. per minute more at R20 than at C. Similarly, for the same men the respiration rate at R20 averaged 1.3 respirations per minute faster than at C. The tidal volume at R20, however, averaged about 4 per cent less than at C, the increased ventilation per minute being provided entirely by a higher rate of respiratory movements.

Respiration in Work — Minnesota Experiment

In the Minnesota Experiment the respiration was measured in 2 standardized types of physical work, which we refer to as "aerobic" and "anaerobic" work.

The aerobic work consisted of walking on the treadmill at 3.5 miles per hour (93.88 meters per minute) on a grade of 10 per cent (i.e., a vertical climb of 0.35 miles per hour). In this work the total ventilation of the lungs per minute was measured in the last 5 minutes of a 30-minute walk. In the control period the mean ventilation for the 32 men in this work was 32.56 liters per minute, or an excess over basal ventilation amounting to 27.74 liters. After 24 weeks of semi-starvation (at S24) the mean ventilation in this aerobic work was 28.98 liters per minute, or an excess over basal ventilation of 25.63 liters. The ventilation ascribable to this work decreased 7.6 per cent from C to S24; this may be contrasted with the loss of 24 per cent in the body weight and in the physical work performed in walking.

In the early stage of rehabilitation (R6) the ventilation in aerobic work tended to decrease further, even though the body weight was rising and therefore the physical work done (which is proportional to the body weight) was increasing. Later in rehabilitation (R17) the ventilation rose sharply to surpass the control level by a considerable margin. These points are shown in Table 281.

TABLE 281

VENTILATION OF THE LUNGS AND RESPIRATORY EFFICIENCY IN AEROBIC WORK (walking at 3.5 miles per hour on a grade of 10 per cent). Mean values for 32 men and for the 12-man group as indicated, except at R11 when 16 men, selected to be representative of the entire group, were studied. (Minnesota Experiment.)

	Control	S24	R6	R11	R17
Ventilation, in liters per minute					
32 men	32.56	28.98	27.87	28.23	
12-man group	31.10	28.24	26.91		36.08
Respiratory efficiency, in cc. of O_2 used per liter of expired air					
32 men	55.54	46.46	49.29	52.38	
12-man group	56.34	46.60	49.51		53.10

The anaerobic work studies were arranged primarily for the purpose of estimating the maximal oxygen transport (see Chapter 34), so the work task was not constant in the different phases of the experiment but was adjusted to the changing work capacity of the men. In all cases the ventilation was measured over the period 1.75 to 2.75 minutes after starting to run on the treadmill at a speed of 7 miles per hour and a grade calculated for each individual subject to produce exhaustion in about 5 minutes. In the control period this meant a grade of 5 to 12.5 per cent, but at the end of semi-starvation the maximal grade was 2.5 per cent. Accordingly the ventilation results for the several periods are not precisely comparable, but the trends may be of interest.

At the end of semi-starvation the mean ventilation in the anaerobic work test was only 59.9 liters compared with a control value of 81.3. In rehabilitation the ventilation in this test gradually rose until at R12 the mean value was 84.6 liters, though the work task (grade of climb) and the work done (allowing for body weight) was still considerably less than the control level. At R20 the average work done in the anaerobic work test was almost exactly the same as in the control period, but the mean ventilation was 92.7 liters per minute; this amounts to an excessive ventilation at R20 compared with C of 14 per cent.

Respiratory Efficiency — Minnesota Experiment

Changes in oxygen consumption, both in rest and in work, are very prominent in starvation and recovery so that the task to be accomplished by respiration changes over a considerable range. Under these circumstances it is desirable to examine the gross respiration as it is related to the oxygen demand or usage. The "respiratory efficiency" can be calculated for this purpose; this is usually computed as the cc. of oxygen removed per liter of expired air.

In the Minnesota Experiment the respiratory efficiency in basal rest declined progressively in semi-starvation, the value at S24 being, on the average, 11 per cent less than at C. This change was gradually reversed in rehabilitation, 63 per cent of the "loss" having been recovered by R12 (Table 280). In the 12 men studied at R20 the mean basal respiratory efficiency at that time was within 0.5 per cent of their mean in the control period. Our analysis of the data in the Carnegie Experiment also shows a slight decline in the respiratory efficiency in rest.

In light to moderate work the respiratory efficiency normally tends to exceed the values in rest, but as the work intensity is increased the respiratory efficiency may fall to considerably less than the resting value. When the work task is very severe for the particular subject, the respiratory efficiency is always low; in other words, in hard work the ventilation rises to the point where the oxygen in the lungs cannot be carried away by the blood (and used) as rapidly as it is renewed by the respiratory ventilation.

TABLE 282

RESPIRATORY EFFICIENCY, as cc. of oxygen (O_2) used per liter of expired air in anaerobic work. Values and means for 6 men in control, at the end of semi-starvation (S24), and after 5, 10, 12, 16, and 20 weeks of rehabilitation (R5, R10, R12, R16, R20). (Minnesota Experiment.)

Subject	Control	S24	R5	R10	R12	R16	R20
119	39.0	30.4	30.2	34.7	30.6	34.4	37.1
109	53.2	34.4	37.4	38.7	34.9	41.5	36.8
23	42.4	28.1	26.9	30.7	29.4	33.0	33.4
4	34.7	33.3	29.9	29.6	27.5	34.3	37.0
26	36.3	30.9	26.5	29.8	24.9	29.0	34.4
112	37.9	33.8	30.5	30.5	31.3	35.0	37.2
M	40.6	31.8	30.3	32.3	29.8	34.5	36.0

In the Minnesota Experiment the respiratory efficiency in aerobic work (3.5 m.p.h. and a 10 per cent grade) averaged, for 32 men, 55.5 cc. of oxygen per liter of expired air in the control period; at S24 the mean value for these men had decreased to 46.5 cc. (−16.2 per cent). This is an appreciably greater change than was observed in basal rest. In the first 6 weeks of rehabilitation 3 per cent of this "loss" had been regained, at R11 the "recovery" was 67 per cent, and there was little or no further gain at R17 (Table 281).

The respiratory efficiency in the anaerobic work tests likewise declined in semi-starvation, and recovery was relatively slow. Even at R20 the average efficiency was still 11 per cent below the control value for the men studied. The results are illustrated by the summary for 6 men given in Table 282.

It should be noted that in walking on the treadmill the physical work done is proportional to the body weight. In starvation this work diminished, of course, so that the respiratory efficiency measured then pertains to a smaller work task than in the control period. In all hard work the normal tendency is for the respiratory efficiency to improve as the work task is reduced. If allowance were made for this factor, the loss in respiratory efficiency during starvation would be even more impressive.

CHAPTER 28

Circulation and Cardiac Function

In any list of the cardinal signs and symptoms of severe undernutrition there will be several items which pertain directly or indirectly to the heart and circulation. These include bradycardia, hypotension, a lowering of the skin temperature, and frequently vertigo and slight cyanosis. On the other hand, dyspnea is seldom reported among starving people, and true signs of congestive heart failure, without antecedent cardiac disease, are rare. In general, all signs in the literature point to a reduction in the work of the heart and in the circulation rate, but this reduction may not involve any serious functional impairment because of the concomitant reduction in metabolism. It is conceivable, indeed, that the absolute reduction in the circulation could be less than in proportion to the reduced oxygen demand, and therefore the relative margin of adequacy and safety could be increased.

The changes in the morphology of the heart induced by starvation are discussed in Chapter 10, and the electrocardiographic findings are treated in Chapter 30. Here the general function of the heart and the characteristics of the circulation in undernutrition and starvation will be considered. The question of the circulation during work, particularly the pulse rate and the maximal oxygen transport in exercise in the Minnesota Experiment, are examined in more detail in Chapter 34.

There are several limitations in the available data. The literature provides few good indications as to the effect of starvation on the severity and course of pre-existing cardiac or circulatory disease. Quantitative estimates of peripheral blood flow in starved persons are absent, and there are only limited data pertaining to cardiac output.

Pulse Rate

Aside from emaciation and loss of weight, probably the most constant finding in reports on severe undernutrition is the slowness of the heart rate. In World War I bradycardia was mentioned as an outstanding finding in practically every one of the numerous reports on famine edema. Rates as low as 30 to 40 beats per minute were often observed in bed rest, and the general level reported was less than 55 in almost all cases (cf. e.g. Schiff, 1917a; Knack and Neumann, 1917; Maliwa, 1918b; Schittenhelm and Schlecht, 1919a; Lewy, 1919). Hülse (1918a) found a minimum rate of 26 beats per minute. In exercise there was no excess tachycardia according to Hülse (1918a), but Schittenhelm and Schlecht (1919a) stated that marked tachycardia was often produced by exercise even though a more normal response was the rule.

In World War II the great majority of reports emphasized the constant finding of bradycardia in undernourished persons (cf. e.g. Laroche *et al.*, 1941; Nicaud *et al.*, 1942; Brull *et al.*, 1945a; Leningrad, 1945; di Granati *et al.*, 1947; Langen and Schweitzer, 1947). Leyton (1946) studied 100 emaciated prisoners of war in a German camp; after walking some 30 yards the average pulse rate was 47 (range 34 to 90). With exercise (one minute of walking up and down stairs) the rate seldom rose above 80, and it returned to the resting level in less than 5 minutes.

In the Warsaw Ghetto bradycardia was seen in adults in all stages of starvation, including such extreme emaciation as is indicated by body weights of 30 kg. or less in persons between 20 and 40 years of age. In these extreme cases the rates were not extraordinarily low, being usually above 45 in the basal state (Apfelbaum-Kowalski *et al.*, 1946). A minimum of 32 beats per minute was recorded by Fliederbaum *et al.* (1946), with values usually between 40 and 50. The children in the ghetto also exhibited bradycardia in most instances, and rates of 55 to 60 were common in children of 5 to 6 years of age (Braude-Heller *et al.*, 1946).

In a few instances tachycardia has been reported in undernourished persons, but in most of these cases there seem to have been additional factors to explain the abnormality. In 54 Turkish prisoners of war studied by Enright (1920a) the heart rate was reported to have been rapid, but in this group there were 47 cases of pellagra, 27 of dysentery, 15 of malaria, and 11 of tuberculosis, as well as signs of scurvy and vitamin A deficiency. Enright's subjects were remarkable in that they showed a general disinclination for food in spite of great emaciation. Rapid heart rates have been reported in "epidemic dropsy" associated with famine in India (Ray, 1927), but this condition is atypical of simple starvation in so many respects that it must be concluded that other important factors are involved. Mollison (1946) found high heart rates in the inmates of Belsen when the camp was liberated from the Germans, but he also noted that the majority of the people had recently had typhus. Lamy, Lamotte, and Lamotte-Barrillon (1948) observed tachycardia with rates of 95 to 120 in similar victims of the German concentration camps. These men, however, were anything but typical of cases of simple starvation, as may be judged from the fact that most of them were definitely febrile at the time and exhibited other atypical signs, including albuminuria. Judging from other reports on persons released from the concentration camps in 1945, it is highly probable that most of the men seen by Lamy and his colleagues had only recently had typhus.

It can be suggested that high heart rates may develop late in extreme starvation since the people seen by Enright, by Mollison, and by Lamy, Lamotte, and Lamotte-Barrillon were in the worst imaginable state of nutrition and since marked bradycardia is seen fairly early in starvation (cf. e.g. Benedict *et al.*, 1919; Burger *et al.*, 1945). Schwarz (1945) stated that a change from bradycardia to tachycardia is a bad prognostic sign. This cannot be a general rule, however, since bradycardia has been reported at all stages of starvation up to death. From the reports on the siege of Leningrad in World War II comes another suggestion, that bradycardia is characteristic of the rapid progression of starvation but that it disappears when the semi-starvation state becomes "chronic" after 2 or 3 months (Brožek, Wells, and Keys, 1946). Again there are a great number of instances to refute such a generalization. Zimmer, Weill, and Dubois (1944) and Schwarz (1945) reported that as starvation progressed in Spanish internees in France, bradycardia, which was "transitory at first, also became permanent."

Tachycardia as a result of infection in starved persons might be inferred from the reports of Enright and of Mollison and was reported by Chortis (1946). It must be noted, however, that infection in starving people may be remarkable for not producing the tachycardia which would normally be expected in the absence of undernutrition. Leyton (1946) found that the pulse rate in starved men dying of tuberculosis did not exceed normal limits until very shortly before death (but see Chapter 47). Gerhartz (1917) reported a maximum of 62 beats per minute for the heart rate in starved patients with pneumonia. The question of heart rate in the presence of infection is briefly discussed in Chapter 46.

Most of the numerous investigators who have observed bradycardia in undernourished people have also commented on the regularity of the rate. Among 9000 undernourished Spaniards examined by Zimmer, Weill, and Dubois (1944), "irregularities of rhythm — partial or complete arrhythmia — were not observed oftener than under other conditions." Maliwa (1918b), however, stated that the pulse rate was unusually labile in the cases of famine edema he saw, although he gave no examples, and he observed that the usual respiratory arrhythmia was diminished in his starved patients. Schittenhelm and Schlecht (1919a) found that the heart rate was very regular in rest but that exercise frequently provoked irregularities, apparently extrasystoles. Adamson et al. (1947a, 1947b) commented on the absence of arrhythmias in 300 men recovering from prolonged severe malnutrition.

Relative bradycardia with regular rate is observed in both men and women and in individuals of all ages who are calorically undernourished. Knack and Neumann (1917) could see no difference in the changes in persons of widely differing ages. They did find, however, that the bradycardia was more pronounced in men than in women; in their series the pulse rates, apparently counted under uncontrolled conditions, were generally of the order of 50 beats per minute in men and about 60 in women.

Blood Pressure and Heart Sounds

There is agreement in the literature on the finding of a weak pulse and slight to moderate hypotension in uncomplicated severe undernutrition. Substantially normal blood pressure values were reported in World War I by several investigators, however. Lange (1917) saw no abnormal blood pressures in his series of cases of famine edema; the range for the systolic pressure was 110 to 130 mm. Hg. Knack and Neumann (1917) and Maase and Zondek (1920) found the blood pressure to be normal to low in most instances, although in a few older persons they considered the values to be slightly high; the actual values reported, however, do not seem to be high for an unselected group of older persons. It must be recalled, of course, that the technique and the apparatus for blood pressure measurement were not well standardized in the period of World War I, so that it is sometimes difficult to make numerical comparisons and in many cases it is necessary to accept the various authors' judgments as to normality. On this basis, the majority of reports from World War I would correspond to the findings of Schittenhelm and Schlecht (1919a). In one series of 70 patients these workers found 2 persons with an elevated blood pressure, 19 persons in the normal range, and all the others in the range of hypotension; many patients had systolic pressures of 90 mm. Hg. or less. In a series of 111 famine edema patients Schittenhelm and Schlecht (1919e) found 8 with a systolic blood pressure of less than 80 mm. Hg. Lewy (1919) reported a minimum systolic pressure of 83 mm. Hg., but the most commonly reported values were 90 to 100 mm. Hg.

In World War II, also, moderate hypotension was regularly reported (e.g. Burger *et al.*, 1945; Mollison, 1946; Langen and Schewitzer, 1947; Simonart, 1948; Gsell, 1948). In 100 severely undernourished men who were not in basal rest Leyton (1946) found a minimum of 85/55 mm. Hg., with the average blood pressure being 107 mm. Hg. in systole and 71 mm. Hg. in diastole. Zimmer, Weill, and Dubois (1944) found a general tendency toward hypotension to be the rule in undernourished persons and believed that continued hypotension had a serious prognostic significance.

It may be noted that emaciated anorexia nervosa patients also exhibit hypotension and bradycardia (Berkman, 1930); in one patient, who later recovered, Bruckner, Wies, and Lavietes (1938) recorded a blood pressure of 72/60 mm. Hg. The 3 cases of edema reported by Landes (1943) are puzzling; these were young, powerful men with hypoproteinemia of severe degree, marked bradycardia, normal electrocardiograms, and normal renal function, but with respective blood pressures of 125/85, 135/75, and 135/80. Landes calculated from measurements of pulse wave velocity that the cardiac output was high.

At the height of the famine in western Holland (April 1945) Lups and Francke (1946) measured blood pressures on 520 persons who had been examined in the polyclinic before, in the period from 1942 to September 1944 — that is, when there was no famine although there were moderate food shortages. This group included, besides patients with all types of internal disorders (except aortic insufficiency and chronic nephritis), a very large number of normal persons; patients with signs of famine edema were never admitted. The summarized findings are given in Table 283. Both systolic and diastolic blood pressures dropped at all ages during the famine, the changes in systolic pressure being the greater. It is interesting to note that the declines in blood pressure were greatest in those groups in which the blood pressure was highest before the famine. In the persons under 50 years of age the famine resulted in only a trivial fall — 10 mm. Hg. or less — in the systolic blood pressure. Of the total group 429 were thin or emaciated in April 1945, and 79 per cent of these persons had lower systolic blood pressures than before the famine, 10 per cent had not changed, and in 11 per cent the systolic pressure was elevated. It must be noted that many of these persons were suffering from chronic diseases including, presumably, progressive hypertension, so the recorded changes cannot be ascribed purely to the nutritional status. Without the intervention of famine the ordinary expectation with such a group probably would be an *average increase* in blood pressure by the time of the second examination. The relationship of these changes to the pre-starvation level of blood pressure will be discussed further in a separate section of this chapter (*Cardiovascular Problems during Refeeding*).

TABLE 283

MEAN BLOOD PRESSURES, in mm. Hg., in 520 persons seen at the polyclinic in Utrecht before and at the height of the famine period (Lups and Francke, 1946).

Age (yrs.)	Before September 1944		April 1945	
	Systolic	Diastolic	Systolic	Diastolic
20–30	132 ± 12	77 ± 10	125 ± 13	70 ± 9
30–40	132 ± 14	79 ± 10	121 ± 12	71 ± 9
40–50	133 ± 17	80 ± 12	126 ± 18	72 ± 15
50–60	146 ± 26	85 ± 15	129 ± 28	77 ± 14
60–75	164 ± 32	88 ± 16	144 ± 31	76 ± 14

Because of the variety of criteria used for the diastolic blood pressure, it is difficult to make very precise statements about the response of the pulse pressure to chronic undernutrition. There is certainly a tendency for the pulse pressure to drop, as can be inferred, for example, from the data from Holland in Table 283. Taking simply the difference between average systolic and average diastolic pressures in the data of Lups and Francke, the pulse pressure declined in the persons over 50 years of age but did not change in the younger groups. Very low pulse pressures are frequently recorded for severely starved persons. Berkman, Weir, and Kepler (1947) reported values of 24, 20, 25, and 28 mm. Hg. in very emaciated patients with anorexia nervosa (their cases, Nos. 1, 4, 5, and 6, respectively). About one fourth of the cases of famine edema studied by Cardozo and Eggink (1946) had pulse pressures of 25 mm. Hg. or less.

The heart sounds have been variously described in starvation. Maliwa (1918b) described them as soft and muffled. In contrast, Leyton (1946) stated that "the heart sounds were normal and often very clear and loud, no doubt because of the thinness of the chest wall."

Venous Pressure

We have commented on the general absence of signs of congestive failure in starving persons. Not only is there no mention of distended veins but frequently it has been noted that the veins are small and tend to collapse (cf. e.g. Schittenhelm and Schlecht, 1919a). The only actual measurements of the venous pressure we have found in the older literature are those of Knack and Neumann (1917), who used the method of Moritz and Tabora and simply stated that most of their results were in the general range of 3.5 to 8.5 cm. of water, with a few readings as high as 12 cm. Knack and Neumann considered that 12 cm. might be slightly high, but they made no comment on the fact that their general findings seem to be distinctly subnormal.

Information in the more recent literature is still far from satisfactory. Sinclair (1947) stated that the venous pressure is low in famine edema but gave no figures, so it is impossible to decide whether he meant "subnormal" or simply not elevated. Govaerts (1947) measured the pressure in the antecubital veins in 47 patients with famine edema. Unfortunately in many instances either the patients were hypertensive or the conditions were unusual, because the arterial blood pressures were very high for emaciated people; 20 of them had systolic pressures of 140 mm. Hg. or more. Measurements of venous pressure were made in the supine position with the manometer zero set at about the level of the sternum. Govaerts indicated that "normal" with this arrangement is from 5 to 12 cm. of water, but our experience at the University of Minnesota leads us to believe that the normal minimum is certainly not less than 7 cm. In any case, Govaerts' edema patients showed a wide variation from 5 to 20 cm. of water for the venous pressure; 11 patients (23.5 per cent) were reported as having values above 12 cm. and 13 patients (27.5 per cent) had values of 7 cm. or less. Lequime and Denolin (1943) reported that after exercise the venous pressure was unduly elevated and slow to decline in their undernourished persons; control and standardization were unsatisfactory in this work.

Cyanosis, Dyspnea, and Skin Temperature

The complaint of coldness is very common among undernourished people, and the skin has often been reported as cold to the touch (e.g. Maase and Zondek, 1920; Nicaud et al., 1942; Brožek, Wells, and Keys, 1946). This might, of course, be attributed to the general reduction in body temperature which is usually observed (e.g. Maliwa, 1918b; Schiff, 1919a; Lewy, 1919; Schittenhelm and Schlecht, 1919a). Leyton (1946) found the average 4 P.M. temperature in 100 men to be 35.7° C. (96.3° F.); Zimmer, Weill, and Dubois (1944) observed body temperatures below 35° C. (95° F.). But the suggestion that the peripheral circulation is diminished is also supported by the pallor of the skin, which is out of proportion to the slight to moderate anemia usually present, and by the fairly frequent appearance of slight cyanosis without dyspnea. Berkman (1930) remarked on the unusual pallor in patients with anorexia nervosa who were not very anemic; the same general picture was seen in western Holland in 1945 (Burger et al., 1945). Lewy (1919) noted that all starved persons are pallid and that some of them are rather cyanotic. All the more recent studies are in agreement that slight cyanosis is the rule rather than the exception (Cardozo and Eggink, 1946). The degree of cyanosis reported has never been severe (Brull et al., 1945a), and there are no reports that it is accentuated by exercise.

In World War I the German investigators quite generally cited the absence of cyanosis and dyspnea for ruling out a cardiac involvement in the etiology of famine edema. Maase and Zondek (1920), for example, insisted on the absence of dyspnea and of the cyanosis that is found in cardiac patients; they did point out, however, that starved people have a gray-blue coloration, which they thought to be something quite specific. Schittenhelm and Schlecht (1919a) stated that exercise dyspnea was observed in persons with famine edema, but they apparently did not consider this to be indicative of any circulatory or cardiac deficiency. Dyspnea was reported by Vandervelde and Cantineau (1919) in the 200 returned Belgian deportees they examined, but they also stated that 42 per cent of these persons were tubercular. The high frequency of complaints of dyspnea among Canadians severely underfed in Japanese prison camps was attributed to beriberi (Adamson et al., 1947a, 1947b). In general, however, the literature on starvation either states that dyspnea was not seen or implies its absence by not mentioning it while insisting that cardiac function was not so disturbed as to suggest failure.

Effect of Drugs on the Heart

The effect of atropine has been studied several times in an effort to see whether the bradycardia of starvation may be a result of vagus stimulation. Schiff (1917a) injected 0.5 to 0.75 mg. of atropine subcutaneously and stated that the bradycardia disappeared promptly; from this he concluded that a vagus effect was the explanation of the slow heart rate. Schittenhelm and Schlecht (1919a), however, used 1 mg. doses of atropine and found absolutely no effect on the heart rate; in 2 of the 4 patients investigated the heart rate actually diminished 2 to 4 beats per minute during atropinization, and as the atropine effect wore off the rate returned to the pre-atropine level (see Table 284). There is no suggestion as to how these exactly opposing reports may be reconciled. During World War II both Govaerts, at Brussels, and Dumont, at Liége, administered atropine to persons with starvation bradycardia and obtained what they considered to be the normal augmentation of the heart rate (Bigwood et al., 1947, pp. 123–24).

Fliederbaum et al. (1946) systematically studied the reactivity of the cardiovascular system to adrenalin, pilocarpine, and atropine in severely undernourished persons without other complications. In 40 persons who received subcutaneous injections of 1 mg. of adrenalin there was extraordinarily little response to the drug. The mean pulse rate rose from a basal rate of 50 to a maximum of 56 to 58 in about 15 minutes and then very slowly declined for the next 90 minutes. The systolic blood pressure rose more slowly to about 90 mm. Hg. from a basal value of about 80 mm. Hg. and then slowly returned to the previous level; the diastolic pressure behaved similarly but the changes were even smaller.

The response of the Warsaw famine victims to subcutaneous pilocarpine, studied in 30 cases, was also feeble. There was an average decline of 3 or 4 beats per minute in the pulse rate between 30 and 60 minutes after injection and a variable widening of the pulse pressure resulting from small and opposite changes in the systolic and diastolic pressures (Fliederbaum et al., 1946). Thirty of the Warsaw subjects were given 0.25 mg. of atropine nitrate intravenously, and a few minutes later the response of the pulse rate to pressure on the eyeball was observed. In normal persons this procedure produces marked slowing of the heart, but the effect was greatly diminished in these starving persons. To ensure that the cholinergic system was paralyzed, injections of 2 to 2.5 mg. of atropine were tried but they did not change the result.

TABLE 284

Summaries of Effects on the Heart Rate from Subcutaneous Injections of 1 Mg. of Atropine in emaciated persons with famine edema (Schittenhelm and Schlecht, 1919a).

Patient No. 10		Patient No. 17		Patient No. 18		Patient No. 20	
Time	Rate	Time	Rate	Time	Rate	Time	Rate
7:18	42	10:50	42	10:35	44	11:50	46
7:20	atropine	11:00	atropine	10:45	atropine	11:52	atropine
7:22	42	11:05	42	10:50	42	12:00	46
7:25	42	11:15	42	10:55	40	12:05	44
7:30	42	11:20	42	11:05	40	12:10	44
7:35	42	11:30	42	11:15	40	12:20	42
7:40	42	11:40	42	11:25	40	12:40	44
		7:00	48	7:00	44	6:00	46

TABLE 285

Summaries of Typical Effects on Heart Rate and Blood Pressure from Subcutaneous Injections of 1 Mg. of Adrenalin in emaciated persons with famine edema (Schittenhelm and Schlecht, 1919a).

Patient No. 6			Patient No. 3			Patient No. 21		
Time	Rate	Blood Pressure	Time	Rate	Blood Pressure	Time	Rate	Blood Pressure
5:30	60	75/50	12:00	36	105/80	11:15	50	90/70
5:35	Adrenalin		12:06	Adrenalin		11:18	Adrenalin	
5:38	60	90/70	12:10	36	105/80	11:21	48	95/70
5:45	64	105/70	12:15	36	110/80	11:37	50	95/70
5:55	76	75/50	12:20	36	110/80	11:49	56	90/70
6:30	74	70/50	12:35	36	120/80	12:03	56	90/70
			6:00	40	100/80	12:12	56	90/70
						6:00	56	80/60

TABLE 286

Summaries of Effects on the Heart Rate and Blood Pressure from Intravenous Injections of 0.5 Mg. of Strophanthin in emaciated patients with famine edema (Schittenhelm and Schlecht, 1919a).

Patient No. 11			Patient No. 13			Patient No. 20		
Time	Rate	Blood Pressure	Time	Rate	Blood Pressure	Time	Rate	Blood Pressure
11:45	52	85/60	2:00	52	80/60	6:10	40	80/60
12:05	Strophanthin		2:05	Strophanthin		6:16	Strophanthin	
12:10	46	90/60	2:07	54	90/60	6:20	38	95/70
12:15	44	90/65	2:10	52	103/75	6:25	40	105/70
12:20	46	95/65	2:15	52	105/75	6:30	40	105/70
12:25	44	100/70	2:20	52	105/80	6:35	38	100/70
12:35	44	100/70	2:36	52	103/75	6:40	40	105/70
6:00	48	110/75	7:00	48	80/60	6:50	46	100/70

The Polish authors concluded that there was diminished "tonus" in both branches of the autonomic nervous system. It should be noted that in these same patients adrenalin injections had only slight effects on the blood sugar level, so there would seem to be a general loss of reactivity to stimuli which act through the autonomic nervous system.

Schittenhelm and Schlecht (1919a) also studied the effect of subcutaneous injections of 1 mg. of adrenalin in 11 typical cases of famine edema. In 5 there was a slight rise — 4 to 6 beats per minute — in the pulse rate, and in one patient there was a more pronounced effect. In this patient, No. 6, the pulse rate was 60 before adrenalin and 25 minutes after the injection of 1 mg. of the drug it rose to a maximum of 76 beats per minute. In this patient there was also a substantial effect on the blood pressure, which rose from 75/50 to a maximum of 105/70 mm. Hg. in 15 minutes. In 3 of the other 10 patients adrenalin produced a transitory moderate rise in blood pressure, but no response was observed in the others. Typical results of adrenalin administered subcutaneously are summarized in Table 285.

In one patient, whose pulse rate was 42 and blood pressure 90/70 mm. Hg., 0.5 mg. of adrenalin was administered by vein (Schittenhelm and Schlecht, 1919a). Within 5 minutes the heart rate rose to 148 and the radial pulse could no

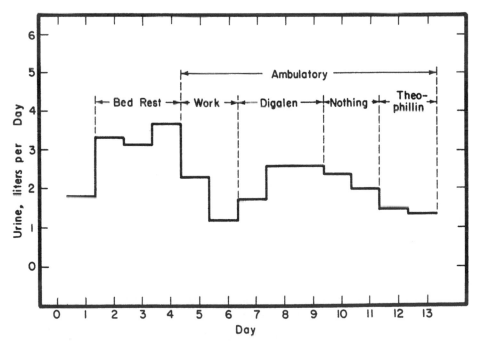

FIGURE 90. URINE OUTPUT IN AN EMACIATED PATIENT WITH FAMINE EDEMA, as liters per 24 hrs. (data from Hülse, 1918a.). Note that there was marked edema on the first day; this was rapidly disappearing in bed rest but returned when the patient was up and at work. Digitalization was associated with a slight reduction in edema even though the patient continued ambulatory. During the days when theophillin was given the edema gradually increased.

longer be felt; 3 minutes later the rate fell to 58 and continued to fall to a minimum of 46 in 30 minutes, at which time the pulse again became palpable.

In a few trials digitalis has been administered to starved persons with bradycardia and hypotension. Hülse (1918a) believed there was frequently a clinical improvement and a diuresis, but the digitalis seems to have been no more efficacious than bed rest; the results on one patient are summarized in Figure 90. Schittenhelm and Schlecht (1919a) used intravenous strophanthin and digalen; there was a moderate drop in the pulse rate in several cases and all 5 persons who received strophanthin had a transitory rise in blood pressure. The results are illustrated in Table 286. Schittenhelm and Schlecht felt their results should be interpreted to mean that the heart muscle was not primarily responsible for the circulatory changes in starvation, but the logic of this interpretation is not easy to defend.

Circulation Time

One might suspect that the circulation time would be prolonged in severe undernutrition. This proved to be the case in 29 patients with famine edema and in 5 severely starved patients without edema (Cardozo and Eggink, 1946). In 26 out of these 34 men the circulation time, measured with 10 per cent magnesium sulfate, was longer than the normal range of 12 to 17 seconds, the average being 22.3 seconds, with values of 25 seconds or more in 12 men (35 per cent). In the series reported by Cardozo and Eggink there was little or no correlation between circulation time and either pulse rate or blood pressure — or even with the apparent severity of edema or of weight loss. Govaerts (1947) measured the arm-to-tongue time with sodium dehydrocholate in 48 patients with famine edema; in 44 of the 48 the time was more than the normal maximum (which Govaerts took to be 16 seconds). Again there was little or no correlation with pulse rate or the extent of clinical edema.

In the Warsaw Ghetto the circulation time from arm to arm was estimated by a rather crude dye method in 18 persons between the ages of 14 and 42 years (Apfelbaum-Kowalski et al., 1946). All were severely emaciated but had no complications from disease. The circulation time varied from 43 to 80 seconds, with an average of 57 seconds; the mean "physiological" normal value was stated to be 32 seconds with this method.

Cardiac Output

There are many indirect indications that cardiac output is diminished in severe undernutrition, but very few attempts at measurement have been reported.

Apfelbaum-Kowalski et al. (1946) attempted to estimate the cardiac output in 18 severely starved adults in Warsaw. For this purpose the blood volume and circulation time were estimated with Congo-red injected into an arm vein; small samples were taken at intervals of 2 seconds from a vein of the opposite arm, and a larger sample was withdrawn after 4 minutes. The presence and concentration of the dye in the plasma samples were measured with a colorimeter. The cardiac output was considered to be indicated by the quotient of the total blood volume divided by the arm-to-arm circulation time. From this crude and naïve procedure it was concluded that the minute volume was reduced to about half that found in well-fed persons. Oscillometric studies in rest and after exercise in 11 starved persons gave results that were also interpreted to indicate a substantial reduction in the total blood flow.

Govaerts (1947) used the acetylene method with 10 patients, ranging in age from 32 to 74 years, who had famine edema. Body sizes were not given but the calculated cardiac outputs ranged from 1.12 to 2.75 liters per minute, with a mean of 2.17. This was estimated to be 54 per cent of the normal, using the data of Lequime (1940), who found values of 3.03 to 4.29 liters for various "normal" subjects. On any basis, however, Govaerts' values are certainly low. The arteriovenous oxygen differences were also abnormal, ranging from 68.5 to 78.3 cc. of oxygen per liter of blood; these differences were on the average 26 per cent above normal, according to Govaerts. The basal metabolism in these 10 patients ranged from —14 to —54 per cent and averaged —33.6 per cent. Because most of these patients had bradycardia (pulse rates 38 to 70, averaging 51.1), the subnormality of the stroke output was much less marked than that of the minute output. Stroke outputs ranged from 28 to 65 cc., with an average of 40.0 cc.

Cardiovascular Disease during Undernutrition

There is little information about the effect of severe undernutrition on the incidence and course of cardiovascular disease. Zimmer, Weill, and Dubois (1944), whose report covers one of the largest groups of starving people ever studied in any detail, stated that hypertension was observed in a "surprising" proportion of their patients and that when present it "remained permanent up to the terminal stages" of starvation. They also noted that enlargement of the heart to X-ray and percussion was seen in about 20 per cent of the persons over 50 years of age, although cardiac atrophy was the general rule at all ages. Zimmer and his colleagues found many cases of angiospasm of the extremities, some of which suggested Raynaud's disease, but vitamin administration brought improvement in these. Premature senility, which was ascribed to defective cerebral circulation, was unusually common in the starving Spaniards.

In the siege of Leningrad in 1941–42 there was no significant change in the number of admissions to the hospitals in the cardiovascular services, and the incidence of hypertension did not seem to be greatly changed (Brožek, Wells, and Keys, 1946). There did seem, however, to be a considerable reduction in coronary disease and myocardial infarction, and it was clear that during the period of famine these cases made up a much reduced proportion of the total number of cardiovascular patients. Accurate interpretation is impossible owing to the absence of information about the sex ratio and age composition of the population contributing to the hospital admissions, but the Leningrad physicians were positive about the relative reduction in coronary disease during the period of famine.

Personal interviews with many physicians who have worked with starving populations provide some impressions which, in the case of cardiovascular disease, may be signficant. In general there seem to be no instances where severe undernutrition had a specifically bad effect on cardiovascular patients. Hypertension which was previously present was not accentuated; frequently, though the condition persisted, it did not tend to progress in severity. Coronary occlusions seem to have been unusually rare in starving populations. Valvular heart disease showed no characteristic direction of change, and certainly the tendency toward congestive failure was not increased.

There appears to be a general tendency toward thrombotic phenomena in severely undernourished persons. Uehlinger (1948) remarked on the absence of arteritis in the starving inmates from the German prisons who were hospitalized at Herisau but noted the relative frequency of thrombophlebitis, pulmonary embolism, and pulmonary infarcts without traumatic antecedent; in 4 cases of pulmonary embolism, infarct cavities developed. The frequency with which emboli were seen in Warsaw was attributed to the extremely sluggish circulation (Apfelbaum *et al.*, 1946). The starving children in Warsaw showed a marked tendency to develop venous thrombi (Braude-Heller *et al.*, 1946). The thrombi generally develop in the femoral or iliac veins of the starving person and often attain great size. Lamy, Lamotte, and Lamotte-Barrillon (1948) observed that these were sometimes organized and appeared to have been present for a long time; in 2 patients large white organized clots were present in the heart.

There are surprisingly few clinical studies on the effect of caloric restriction in hypertensive heart disease, though the possibility of important effects seems clear. Proger and Magendantz (1936) cited an example which they indicated was representative of their experience. The patient, a 67-year-old man with a height of 171 cm. and a weight of 69 kg., was reduced to a weight of 63 kg. within 6 weeks by subsistence on a diet of 600 Cal. per day. The basal blood pressure fell from 175/88 to 156/70 and the heart rate slowed to 46 beats per minute. It should be noted that the "standard" average weight for this man would be 70 kg., and the "ideal" weight, allowing for no gain beyond the age of 30 years, would be 63 kg., so no marked emaciation was involved in this reduction of 6 kg. The question remains, however, as to whether the apparently favorable effect of weight reduction in such instances is only a temporary response to the active stage of weight loss or whether a more persistent benefit accrues.

It is of interest that arterial and valvular abnormalities which normally provoke enlargement of the heart may not do so when there is severe undernutrition. Kapelusch and Sprecher (1919) in Austria reported a series of cases of severe sclerosis of the aorta with unusually small hearts; these were all apparently emaciated persons. We have recently seen a patient who literally starved to death because of a carcinoma of the esophagus; this man's heart showed a marked aortic stenosis, but it weighed only 115 gm. The question is raised as to whether the inanition itself prevents or even reverses the enlargement of the heart, or whether the lessening in the circulatory demand associated with undernutrition reduces the work of the heart to ordinary levels in spite of the arterial or valvular abnormality.

Cardiovascular Problems during Refeeding

In the period of refeeding following severe undernutrition the metabolic load on the body is suddenly much increased and a far greater circulation must be provided. The response of the heart and circulatory system to these demands naturally should be of much interest, but few studies have been made. Transient heart failure may intervene, as in one instance in the Minnesota Experiment, but may well be missed because in starvation edema is fluctuating in any case and the physician is prepared to find dyspnea and hypotension; beginning with a small heart and bradycardia, a considerable dilatation and relative tachycardia could occur without reaching the levels ordinarily associated with heart failure. Sudden death under treatment, as in patient No. 2 of Berkman, Weir, and Kepler (1947), may occur because of the inability of the atrophic heart to carry the increased burden imposed by refeeding.

The most interesting questions about the cardiovascular system in refeeding have to do with the behavior of the blood pressure. As hypotension is relieved, how is the blood pressure re-established at a normal level? Several papers indicate that there may be overshooting in readjustment.

There are 2 reports of transitory hypertension during refeeding in malnourished British prisoners of war in Southeast Asia (Harrison, 1946; Stapleton, 1946). These men had been calorically undernourished for several years, but they also suffered from beriberi, malaria, and a variety of intestinal parasites. The hypertension was never severe, usually in the range of 135 to 160 mm. Hg. systolic, but was especially noticeable in contrast to the general hypotensive range. In one series (Stapleton, 1946) episodes of hypertension were frequently associated with changes in the degree of edema and were observed, together with edema, after the men had been rescued from the Japanese and were enjoying rest and good food for the first time in 3 years. Among the starved Spanish internees in France in 1941–42 Schwarz (1945) occasionally observed a sudden rise in blood pressure to moderate hypertension levels when treatment was instituted; this was noted particularly in patients who had undergone several episodes of severe edema.

Adamson et al. (1947a, 1947b) examined 300 repatriated Canadians who were well fed after having been severely malnourished and subjected to great emotional strain for 40 months in Japanese prison camps. Indications of hypertension were rare and there were no cases of essential hypertension in this group — a fact that was surprising to the examiners.

In Holland the series of patients studied by Lups and Francke (1946) appeared to show a satisfactory course of return of the blood pressure to normal during the refeeding, which started in May 1945. By September 80 per cent of the 340 persons studied by Lups and Francke had gained weight, and at that time 77 per cent of those who had been hypotensive in April had higher blood pressures than before, whereas only 23 per cent of those who had previously been relatively hypertensive had shown an increase. These percentages refer to systolic pressures; the corresponding percentages based on diastolic pressures were 72 and 28 per cent, respectively. This would seem to be a favorable state of affairs, but it should be noted that the measurements were made after only about 3 months of refeeding.

In any case, the results are very different from those in Leningrad after the siege of 1941–42 (Leningrad, 1945; Brožek, Chapman, and Keys, 1948). In the two years following the relief of the siege, hypertension quite suddenly became a major medical problem. In April 1943 — that is, somewhat less than a year after the siege was lifted — about 10,000 "normal" persons were examined, and the results were compared with those of a similar survey made in 1940. The findings are summarized in Table 287. These findings are astounding, indicating about a threefold increase in pre-symptomatic hypertension. Precise analysis is impossible because of limitations in the data, but it seems impossible that sampling errors or changes in the Leningrad population during this period could be responsible.

TABLE 287

PERCENTAGE INCIDENCE OF ELEVATED BLOOD PRESSURE (above 140 mm. Hg. systolic and 90 mm. Hg. diastolic) in "healthy" individuals in Leningrad before famine (1940) and after famine (the winter of 1942–43) (Brožek, Chapman, and Keys, 1948).

Age Group	1940	1942–43
20–29	3	12
30–39	7	30
40–49	25	50
50–59	22	66
60 plus	23	70

TABLE 288

RELATIVE NUMBER OF CASES OF HYPERTENSIVE DISEASE admitted to the Therapeutic Clinic of the Pavlov First Medical Institute in Leningrad (Brožek, Chapman, and Keys, 1948).

Period	Percentage of Total Admissions with Hypertensive Disease
1940	10.0
July to September 1941	15.0
October 1941 to March 1942	2.0
April 1942 to December 1942	24.5
January to December 1943	50.0
January to April 1944	35.0

Parallel to the rise in blood pressure recorded in the "normal" population, there was a great increase in patients admitted to Leningrad hospitals for hypertensive disease. Before the war patients with hypertension accounted for about 10 per cent of all admissions to the Therapeutic Clinic of the Pavlov First Medical Institute. This proportion dropped during the period of semi-starvation (October 1941 to March 1942) and increased progressively thereafter through 1943. In January 1943 patients with hypertension constituted 20 per cent of the admissions, and by June of that year the figure had risen to 60 per cent; there was then a decline, and the figure was 35 per cent for the period from January to April 1944. The admissions data are summarized in Table 288.

This was by no means a case of admitting to the hospital after the siege an older group of hypertensives who had been denied admission during the pressure of the siege; actually the age percentages changed so that in 1942–43 the hypertensives were younger than before the war. In 1938–39, 36 per cent of the hypertensives were under 50 years of age, but in 1942–43 this age group made up 50 per cent of the total. Moreover, the severity of the disease in those hospitalized was greater than before. Before the war exudate and hemorrhage in the retina itself had been seen in 23 to 25 per cent of the hypertensive patients; in 1942–43, 70 per cent showed such changes. Also, the percentage of patients with cardiac insufficiency, often acute in onset, rose markedly. At autopsy the hypertensive patients in 1942–43 showed relatively small hearts, and significant renal changes were seen in only 35 per cent; frank malignant nephrosclerosis (arteriolar necrosis) disappeared until the second quarter of 1943.

Obviously, many factors may have been involved in this epidemic of hypertension in Leningrad, but the outstanding peculiarity of the period was severe semi-starvation for 6 months, followed by refeeding, with the necessity for hard work at all times.

Basal Pulse Rate — Minnesota Experiment

Semi-starvation in the Minnesota Experiment rapidly produced bradycardia; by the end of 5 weeks the mean basal pulse rate for the 32 men was 40.8, the minimum being 31 and the maximum 58 beats per minute. The rate continued to decline for another 2 months to reach a minimum average of 35.3 at S13, with

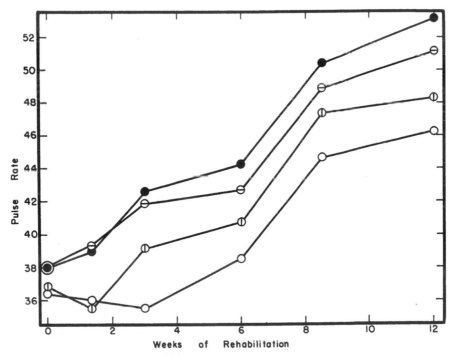

FIGURE 91. MEAN VALUES FOR THE BASAL PULSE RATE, IN BEATS PER MINUTE, FOR THE 4 CALORIC GROUPS IN THE FIRST 12 WEEKS OF REHABILITATION. Open circles = group Z; vertical bars = group L; horizontal bars = group G; solid circles = group T. (Minnesota Experiment.)

2 men having mean rates of 28 on repeated manual counts of 1 minute each. It will be noted that at this time the men had lost an average of about 16 per cent of their control body weight. In the next 10 weeks, during which they lost a further 10 per cent of their weight, the basal pulse rate rose very slightly but steadily, the mean values being 36.8 at S20 and 37.3 at S23. The extreme values at S23 were 27 and 48 beats per minute. The extraordinary regularity was notable; this is discussed further in Chapter 30.

The increased food intake in rehabilitation produced in a few days a small rise in pulse rate in the 2 groups (T and G) receiving the largest diet but had little effect for some weeks on the other 2 groups. By R6 the pulse rate was rising in all groups, and this "recovery" was roughly proportional to the caloric intake through R12; the mean values are summarized in Figure 91. Calculating the recovery of the pulse rate at R12 in terms of the rise from S24 to R12 expressed as a percentage of the fall from C to S24, the values for the 4 caloric groups Z, L, G, and T were 60, 67, 72, and 76 per cent, respectively.

In the subsequent period of "later" rehabilitation with unlimited feeding, there was a sharp rise in the pulse rate to a level of relative tachycardia at R14, with a gradual decline thereafter; mean values for the 12 men who were followed through R20 are given in Table 289. After 34 weeks of rehabilitation basal pulse rates were measured on 21 of the men; the mean value then was 58.71, and this is significantly higher than the control mean of 54.90 for the same men.

TABLE 289

MEAN PULSE RATES AND BLOOD PRESSURES IN THE BASAL STATE of the 12 men who remained at the Laboratory throughout 20 weeks of rehabilitation (Minnesota Experiment).

	Control	S24	R6	R12	R14	R16	R20
Pulse rate (beats/min.)	56.1	37.8	42.7	52.3	70.9	64.1	57.3
Systolic blood pressure (mm. Hg.)..	105.3	92.7	97.2	101.5	106.8	104.9	104.0
Pulse pressure (mm. Hg.)	35.1	29.4	30.7	32.3	34.3	34.9	35.4

Arterial Blood Pressure — Minnesota Experiment

The subjects in the Minnesota Experiment were characterized by somewhat low normal blood pressures in the control period, the average systolic pressure being 106.5 mm. Hg. with a range of 96 to 119. After 5 weeks of semi-starvation (at S5) there was little change — average 106.0, range 92 to 116 mm. Hg. But at S13 there was a definite tendency toward hypotension, the average being 99.2, the range 86 to 118; 8 of the 32 men had systolic pressures of 92 mm. Hg. or less. At S20 the average declined only a trifle further (mean 98.5), but at S23 the average was down to 94.7, range 82 to 118, and 10 out of 32 men had systolic blood pressures of 90 mm. Hg. or less. Five of the 32 men, however, showed no tendency to hypotension at any time, and one man showed a steady rise in systolic pressure during the starvation period; this man had blood pressures between 96 and 99 in the control period and for the first 5 weeks of semi-starvation, but thereafter his blood pressure gradually rose to reach 118 mm. Hg. at S23.

The basal diastolic blood pressure also declined in starvation, but the change was relatively smaller than in the systolic pressure. In the control period the mean diastolic blood pressure was 69.7 mm. Hg., and at S23 it was 64.4 mm. Hg. The result was, of course, a diminution in the pulse pressure, which fell from an average of 36.8 mm. Hg. at C to 30.3 mm. Hg. at S23, a fall of 17.7 per cent.

The increased food intake in rehabilitation produced a definite rise in systolic blood pressure within a few days in all 4 caloric groups, this being most marked in the 8 men who received the most food (group T). After the first 10 days of rehabilitation, however, the systolic blood pressure tended to drift downward again; the mean at R12 was still only 99.9 mm. Hg., but none of the men had a systolic pressure of less than 90 mm. Hg. at that time. It is interesting to note that the one man who showed a progressive tendency toward relative hypertension in the starvation period showed a progressive reversal in rehabilitation, his systolic pressure being from 94 to 96 mm. Hg. in the period R9 to R12.

The relatively small improvement in the circulation in the first 12 weeks of rehabilitation, as indicated by the blood pressures, is shown by the pulse pressure. From S23 to R12 the mean change was only 1.6 mm. Hg., or a "recovery" of only 24.6 per cent of the loss in starvation.

Upon the institution of unlimited feeding after R12 the blood pressure was rapidly restored to near the level of the control period, with a tendency to reach maximum levels in the period R12 to R16. Six of the 12 men studied from R12 to R20 had higher systolic blood pressures at R14 or R16 than at any time in the entire experiment, including the control period. The pulse pressure also tended to be high in the period R14 to R16, individual maximums for 5 of the 12 men being recorded at this time.

Twenty-one of the subjects were again examined in the basal state after 34 weeks of rehabilitation. At that time both systolic and pulse pressures were fully restored to the pre-starvation (control) level. For these 21 men the mean values of systolic and pulse pressure were, at C and R34, 106.6 and 107.4 mm. Hg. (systolic) and 36.7 and 36.3 mm. Hg. (pulse pressure). Not only were the averages at R34 extremely close to those at C, but the individuals, with 2 ex-

TABLE 290

MEAN SYSTOLIC BLOOD PRESSURES AND PULSE PRESSURES, all in mm. Hg., for the 4 groups of 8 men each as they were divided in order of increasing calories in the first 12 weeks of rehabilitation (Minnesota Experiment).

Group	C		S24		R6		R12	
	Sys-tolic	Pulse Pressure	Sys-tolic	Pulse Pressure	Sys-tolic	Pulse Pressure	Sys-tolic	Pulse Pressure
Z	106.8	37.1	93.6	30.9	96.8	33.0	96.8	29.5
L	108.4	38.3	95.8	30.8	99.6	29.4	98.9	31.6
G	105.0	36.5	94.6	28.9	98.9	31.1	100.1	32.1
T	106.0	34.6	94.8	30.3	99.3	31.3	103.9	33.4

ceptions, also tended to return to their individual normal values. The 2 exceptions both had mean systolic pressures 23 mm. Hg. lower at R34 than at C.

It will be noted in Table 290 that the effect on the blood pressure of the caloric differences in the first 12 weeks of rehabilitation was neither very great nor perfectly regular, though the difference between the Z group (lowest intake) and the T group (highest intake) was fairly substantial at R12.

Venous Blood Pressure — Minnesota Experiment

By the middle of the semi-starvation period in the Minnesota Experiment it was noted that the veins were less prominent than usual and that they frequently collapsed when blood was being drawn. At S24 it was often troublesome to get blood samples of more than 5 cc., even in exercise (walking on the treadmill). At this time measurements of the venous pressure were made in supine rest, the direct method being used with the needle in a vein of the antecubital fossa. For a normal control, measurements were made on 12 well-nourished young men who were also living in the Laboratory. Measurements on the subjects were repeated on 4 occasions during rehabilitation, the measurements on the control group also being repeated on the final occasion.

The venous pressures in the control group were close to the "normal" values reported in the literature, the average being 9.7 cm. of saline solution on the first occasion and 10.3 cm. at the final measurement. The abnormality of venous

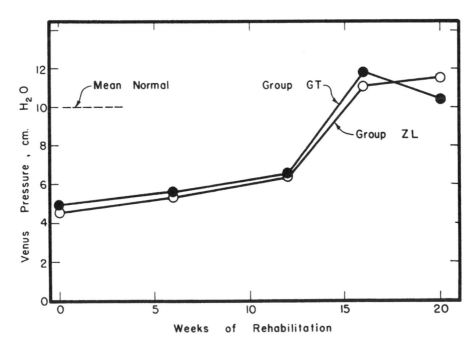

FIGURE 92. Venous Blood Pressure during 20 Weeks of Rehabilitation Following Semi-Starvation. Mean values for 4 men from caloric group Z and 2 men from group L (open circles, group ZL) and for 2 men from caloric group G and 4 men from group T (solid circles, group GT). (Minnesota Experiment.)

pressure in the starvation subjects was striking, averaging 4.80 cm. at S24, with extremes of 2.8 and 9.6 cm. In the first 12 weeks of rehabilitation the venous pressure remained remarkably low, the average at R6 being 5.33 and at R12 6.27 cm. In the entire series of 96 measurements from S24 to R12 there were only 5 as high as 8 cm., and the maximum in any man at any time was 9.6 cm. With the advent of unlimited food intakes, however, the venous pressure rose sharply to average 11.42 cm. at R16, with a slight decline to an average of 11.00 cm. at R20. Mean values for 2 sub-groups of 6 men each are summarized in Figure 92.

The measurements of venous pressure in starvation certainly refute any suggestion of backward failure of the heart, but they raise the question as to what makes this pressure so extraordinarily low. The blood volume was not really low in starvation; actually the volume of blood per unit of tissue was considerably increased in this condition. It seems idle to speculate on the possible complex of factors which could produce this state of affairs, but it is not unreasonable to suggest that a low tissue tension must be involved.

In rehabilitation the situation is a little clearer. The average values at R16 and R20 surpass the normal control values because a number of the men had definitely high values. Venous pressures of 15.4, 17.8, and 19.1 cm. were obtained in 3 of the 12 men studied at R16, and at R20 one man had a venous pressure of 19.9 cm. These high values were associated with a tendency to tachycardia and recurrent edema, so they must be considered to indicate incipient heart failure. The atrophic heart of the starved man shows its weakness when it is confronted with the task of caring for the suddenly enhanced metabolic load which results from unlimited feeding.

The clearest evidence of this was seen in the case of one man who was not included in the 12 who remained under detailed study from R12 to R20. This man had not been distinguished from the rest in the period up to R12 except for a tendency to more than the ordinary amount of ankle edema in the starvation period; this seemed adequately explained by the presence of varicose veins. At the end of R12 this man was released from supervision but stayed on, as did

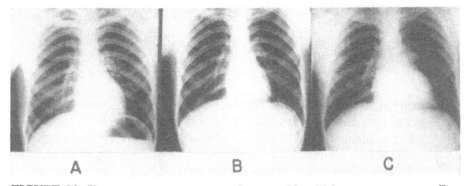

FIGURE 93. TELEROENTGENOGRAMS OF SUBJECT No. 47 ILLUSTRATING THE DECREASE IN HEART SIZE DURING SEMI-STARVATION AND THE SUDDEN ENLARGEMENT OF THE HEART FOLLOWING A SHORT PERIOD OF UNCONTROLLED FOOD INTAKE. A = control; B = on controlled food intake in rehabilitation (R11); C = after 2 weeks of uncontrolled food intake (R14). (Minnesota Experiment.)

several others, to do some work at the University of Minnesota. He immediately began to gorge himself and maintained his caloric intake at something like 7000 to 10,000 Cal. per day for about a week. It was then noticed that his face was very puffy, and he was immediately examined. He said he was short of breath but felt all right otherwise. He had been gaining weight rapidly, perhaps 0.5 lbs. a day, but then he suddenly gained about 9 pounds over a period of 2 days. He was found to have a marked pitting edema and a venous pressure of about 20 cm. of saline, and X-ray showed that his heart had suddenly enlarged. The X-rays are reproduced in Figure 93. He was placed on bed rest, his diet was cut to 3500 Cal. per day, and his salt intake was restricted. In about 2 days a marked diuresis began, and in a week all signs of his cardiac failure were gone. Subsequently he was more prudent in his diet and had no further trouble.

Cardiac Output — Minnesota Experiment

Estimates of the stroke output of the heart in the Minnesota Experiment are available from the roentgenkymographic volume measurements. A large series of parallel measurements with this method and with the acetylene method in this Laboratory have shown them to be in reasonably close agreement in persons without valvular defects. Changes in cardiac output in the same person are estimated with high accuracy by the roentgenkymographic method if the parallel acetylene estimation is taken as the standard (Keys et al., 1940). The mean values for the stroke output for the 12 men who were studied through 20 weeks of rehabilitation are given in Table 291.

TABLE 291

MEANS AND STANDARD DEVIATIONS FOR PULSE RATE AND STROKE VOLUME for the same 12 men at different periods of the Minnesota Experiment.

	Control		S24		R12		R20	
	M	SD	M	SD	M	SD	M	SD
Pulse rate								
Beats per minute	56.1	6.09	37.8	5.15	52.3	6.61	57.3	5.92
Percentage	100	10.9	67.5	13.6	93.2	12.6	102.1	10.3
Stroke volume								
Cc.	66.9	5.44	54.8	8.83	57.4	10.1	59.1	9.26
Percentage	100	8.1	81.9	16.1	85.8	17.6	89.5	15.6

The mean stroke output in the control period, 66.9 cc., is entirely normal for young men at rest, and the individual variability was relatively small; the standard deviation amounted to only 8.1 per cent of the mean. At the end of semi-starvation the stroke output had decreased by 18.1 per cent on the average, but the individual variability had doubled ($SD = 16.1$ per cent of the mean). The mean recovery in stroke output was slight at R12, and even at R20 the mean value was 10.5 per cent less than in the control period. It will be noted also that the individual variability remained at the relatively high level found at the end of semi-starvation.

Grollman (1932) showed the value of comparing the cardiac output of individuals in terms of the "cardiac index," which is the amount of blood in liters put out per minute by the heart per square meter of body surface. It will be noted that the mean values reported by various authors for the cardiac index in normal resting persons are relatively constant: Grollman (1932) obtained a mean of 2.21, av. dev. = ±0.14, in 50 young adults; Starr *et al.* (1934) found in 31 persons a mean of 2.40, *SD* = ±0.55; Keys *et al.* (1940) found in 37 normals a mean of 2.42, *SD* = ±0.36. The values for the cardiac index are summarized in Table 292. In the Minnesota Experiment the mean control value was 2.02 liters per square meter of body surface, therefore slightly but definitely lower than the mean values noted above for the acetylene method in the supine position. But it should be noted that the Minnesota estimates are for the upright position, in which the cardiac output is generally estimated to be some 10 per cent less than in the supine position (cf. e.g. Grollman, 1932). The substantial normality of the control measurements in the Minnesota Experiment seems to be established, but in any case our principal concern is with the changes in starvation and rehabilitation rather than with the absolute values.

TABLE 292

CARDIAC OUTPUT AND CARDIAC INDEX (output in liters per minute per square meter of body surface) in the same 12 men at different stages of the Minnesota Experiment. %C = mean value expressed as percentage of the corresponding value in the control period.

	Control Value	S24		R12		R20	
		Mean Value	%C	Mean Value	%C	Mean Value	%C
Cardiac output (liters/min.)	3.75	2.07	55.2	3.00	80.0	3.39	90.4
Body surface (square meters)	1.86	1.66	89.2	1.74	93.5	1.89	101.6
Cardiac index	2.02	1.25	61.9	1.72	85.1	1.79	88.6

At the end of semi-starvation (S24) there was an indicated reduction in cardiac output per minute amounting to 44.8 per cent. In rehabilitation about half of this loss had been regained after 12 weeks, and even at R20 the minute volume output of the blood was still 9.6 per cent less than in the control period. The changes in cardiac index are similar, with a reduction of 32.1 per cent at S24 and a continuing deficit of 11.4 per cent at R20.

Cardiac Output and Metabolism — Minnesota Experiment

The reduction in cardiac output per minute in starvation in the Minnesota Experiment was somewhat greater than the absolute reduction in the basal metabolism so that there was a relative as well as an absolute depression of the total resting circulation. The indicated deficit was small, however, so the relative circulatory defect in rest cannot be considered to have been serious. The relative deficiency in the resting circulation was very small at R12, but at R20 increasing circulation had not kept pace with the rapidly rising basal metabolism, and the relative circulation, in these terms, was about the same as at the end of semi-starvation. If we take the minute volume output of the heart per cc. of basal oxygen consumption at the time to be 100 per cent in the control period, the corresponding values for S24, R12, and R20 were, respectively, 86, 98, and 86 per cent.

These evaluations of the circulation in terms of oxygen demand do not allow for the fact that the oxygen-transporting capacity of the blood was changed (i.e., there were changes in the hemoglobin concentration and in the oxygen content of the arterial blood). Exact calculations, moreover, are difficult because the metabolism was measured in the supine position and the circulation in the upright position. In the upright position the metabolism, as oxygen consumption, rises about 20 per cent on the average, so at least rough calculations can be made either by "correcting" the metabolism to the upright position or by converting the circulation data to the estimated equivalent in the supine position; in the latter case we would estimate an average increase of 10 per cent in the cardiac output.

The two calculations yield much the same result in terms of the relationship between the oxygen supplied by the arterial blood and the oxygen requirement for metabolism. Taking the arterial oxygen saturation to be approximately 95 per cent at all times, the calculated mean oxygen return to the heart in the venous blood was 455 cc. per minute in the upright condition during the control period, and this fell to 140 cc. at S24, rising thereafter to 262 cc. at R12 and to 344 cc. at R20.

The ratio of the venous return of oxygen to the oxygen consumption can be considered as a kind of "safety ratio"; in the control period this ratio for the upright position amounted to 455 divided by 274 (the estimated oxygen consumption), or 1.66. In other words the "margin of safety" was 166 per cent. At S24 this safety ratio fell to 0.80, or only 48 per cent of the control. At R12 the safety

TABLE 293

RELATIONSHIP BETWEEN OXYGEN METABOLISM AND THE CIRCULATORY SUPPLY OF OXYGEN IN REST in the upright (U) and supine (S) positions. The calculations are mean values for the same 12 men at different stages of the Minnesota Experiment. Arterial saturation was taken as 95 per cent. The values in parentheses have been estimated as follows: cardiac output supine = 1.1 × upright; metabolism upright = 1.2 × supine. The "safety ratio" is the venous return divided by the metabolism.

	Control		S24		R12		R20	
	U	S	U	S	U	S	U	S
Blood O_2 capacity (cc./100 cc.)..	20.40	20.40	15.97	15.97	17.00	17.00	19.58	19.58
Arterial oxygen (cc./100 cc.)...	19.44	19.44	15.17	15.17	16.15	16.15	18.60	18.60
Cardiac output (cc. of oxygen/min.)	729	(801)	314	(346)	485	(538)	631	(694)
Metabolism (cc. of oxygen/min.)	(274)	228	(174)	145	(223)	186	(287)	239
Venous return (cc. of oxygen/min.)	455	573	140	201	262	352	344	455
"Safety ratio"	1.66	2.51	0.80	1.39	1.17	1.89	1.20	1.90
Percentage of control	100	100	48	55	70	75	72	76

ratio calculated in this way was 1.17, and at R20 it was 1.20. Calculations for this safety ratio in the supine position gave higher values throughout, as might be expected, but the percentage changes in starvation and rehabilitation were much the same. The summarized results of these calculations are given in Table 293. It is indicated that the safety ratio in both upright and supine positions fell in starvation to about half the normal value and that even at R20 it was something like 25 per cent below the normal control.

Work of the Heart — Minnesota Experiment

The physical work done by the heart is a function of the amount of blood pumped, the arterial blood pressure, and the velocity imparted. Under resting conditions a first estimate of the relative work done by the heart is provided simply by the product of the minute volume and the mean arterial blood pressure. This is the relative pressure work per minute; the substitution of stroke volume for minute volume in the calculation provides a measure of the relative work done per stroke. Finally, if the duration of systole is measured, it is possible to calculate the rate at which this work is done. The mean results of such calculations for the Minnesota Experiment are summarized in Table 294.

The work of the heart per minute in rest after 6 months of semi-starvation was roughly half the control value, and recovery in this respect was slow, not being complete by R20. Because of the prolongation of systole in semi-starvation and in the first few months of rehabilitation, the rate at which physical work was done during the contractile process was even more reduced than the gross work rate per minute.

These values for the physical work of the heart do not include the kinetic work of imparting velocity. At the relatively slow blood velocities in the resting

TABLE 294

CARDIAC FUNCTION in control, at the end of semi-starvation (S24), and after 12 and 20 weeks of rehabilitation (R12 and R20). The values are means for the same 12 men. In calculating the total cardiac work it is assumed that kinetic work comprises 3 per cent of the total. (Minnesota Experiment.)

	Control	S24	R12	R20
Arterial pressure (mm. Hg.)	87.8	78.0	85.3	86.8
Mechanical systole (seconds)	0.322	0.385	0.370	
Diastolic heart volume (cc.)	621.4	508.0	590.6	635.7
Stroke (cc./100 cc. diastolic volume)	10.7	10.8	9.6	9.3
Relative pressure work				
Per stroke	100	72.7	83.3	87.2
Per minute	100	49.0	77.6	89.1
Rate in systole	100	41.0	67.5	
Relative kinetic work				
Per stroke	100	57.3	64.5	
Per minute	100	38.6	60.5	
Rate in systole	100	47.9	56.5	
Relative total work				
Per stroke	100	72.2	82.7	
Per minute	100	48.7	77.1	
Rate in systole	100	41.2	67.2	

state, the kinetic work is only a small fraction of the total work and may be neglected entirely in approximate calculations. However, the general nature of the
changes in kinetic work may be estimated roughly, and besides adding to the
completeness of the picture, these changes are of some theoretical interest. The
increased duration of systole at S24 and at R12 means, of course, a corresponding decrease in the rate at which velocity is imparted. The changes calculated
for the kinetic work of the heart in the Minnesota Experiment are also summarized in Table 294, together with the "total work," estimated on the assumption that the kinetic work amounts to about 3 per cent of the total. Obviously,
the end result is not affected by this elaboration of the estimate.

The functional state of the heart may be estimated on the general principle
that the invocation of Starling's law — increasing diastolic fiber length to ensure
effective contraction — would be an indication of developing relative failure.
While Starling's law is ordinarily only obeyed by the failing heart, the relationship between the diastolic heart size and the stroke output may be of interest in
such states as starvation. Whatever may be the final significance of such a calculation, the fact is that stroke output per unit of diastolic volume was not changed
in starvation when the diastolic volume was diminished very appreciably. In
subsequent rehabilitation, however, the increase in diastolic volume was out of
proportion to the rise in stroke output. The difference is not great, but the indication is, in conformity with other observations which we have discussed, that
the heart may be closer to failure in early rehabilitation than in actual starvation.

Oxygen Pulse and Circulatory Index — Minnesota Experiment

The attempt to relate cardiac function to the metabolic demand for circulation is sometimes made in the form of the "oxygen pulse." In this calculation
the metabolic rate, in terms of oxygen consumption, is divided by the pulse rate,
thereby yielding the amount of oxygen provided per heart beat. This calculation,
however, does not allow for the influence of changes in the hemogoblin concentration or oxygen capacity of the blood. We may write the complete equation:
Oxygen used = stroke volume × pulse rate × oxygen capacity × (arterial saturation — venous saturation). In many instances, as in the later stages of rehabilitation in the Minnesota Experiment, data are available on oxygen capacity,
metabolic rate, and pulse rate but not on the other terms, or at least not on
stroke volume and venous saturation. The equation can be rearranged to provide
a circulatory index (C.I.) which may have useful application:

$$\text{C.I.} = \frac{\text{oxygen used}}{\text{pulse rate} \times \text{oxygen capacity}} =$$

stroke volume × (arterial saturation — venous saturation)

An increase in the C.I. means that either the stroke volume has increased or the
arterial-venous difference has increased, or both.

The values for the circulatory index have been calculated in the Minnesota
Experiment for 21 men who were studied from control through 12 weeks of rehabilitation and again at R34. These values, together with the results from the

ordinary calculation for the oxygen pulse, are summarized in Table 295. It is clear that the oxygen pulse is relatively meaningless here because of the considerable changes in the oxygen capacity of the blood; this is clearly shown when the changes in the oxygen pulse are compared with those in the circulatory index. The circulatory index rose sharply at S24, and since we know that the stroke volume certainly did not increase but actually decreased at this time, the increase in the arterial-venous difference must have been considerable. This, of course, merely confirms what we know from the measurements of cardiac output. But the subsequent changes in the circulatory index are interesting. At R20 the C.I. was at the control level but we know that the circulation was not fully normal; opposite counterbalancing changes in the stroke output and in the arterial-venous oxygen difference are indicated. At R34 the value for the C.I. was definitely lower than in the control period; either the stroke output was unusually large or the arterial-venous difference was unusually small. At that time (R34) the pulse pressure was almost precisely the same as at C, and the pulse rate was 4 beats higher (58.7 vs. 54.9 for this group). The inference is that the circulation was somewhat inefficient in that the arterial-venous difference was probably unusually small.

TABLE 295

OXYGEN PULSE AND CIRCULATORY INDEX in the Minnesota Experiment. Oxygen pulse is calculated as the cc. of oxygen used per minute divided by the heart rate per minute. Circulatory index is calculated as the cc. of oxygen used per minute divided by the product of heart rate (per minute) and blood oxygen capacity (cc. of oxygen per cc. of blood). All values are means for the same 21 subjects at different stages of the experiment, except for R20 when the mean of 12 men is used.

	Control	S24	R12	R20	R34
Oxygen pulse	4.20	3.84	3.64	4.11	4.09
Percentage of control ..	100	91.4	86.7	97.9	97.4
Circulatory index	209.3	242.6	214.9	210.3	199.3
Percentage of control ..	100	115.9	102.7	100.5	95.2

Cyanosis and Dyspnea — Minnesota Experiment

In the latter half of semi-starvation it was frequently noticed that many of the subjects at rest had a distinct appearance of cyanosis in the nail-beds. This was not accentuated in exercise; in fact, if the exercise was severe, the cyanosis sometimes appeared to be lessened. Cold, however, much accentuated the cyanosis, which would then be apparent in the lips. Almost all the subjects had a grayish-blue pallor that was quite distinctive, but it was not identified as cyanosis except in the nail-beds. In most of the men the cyanotic appearance tended to come and go, or at least to vary in intensity. On one occasion at the end of starvation, when all the men were examined at one time by a panel of consultants, 21 of the subjects were judged to be slightly cyanotic. In rehabilitation the cyanosis rather quickly disappeared, and none of the men was definitely cyanotic by R6.

True dyspnea was never seen in any of the men during the semi-starvation period. In the severe exercise tests the apparent respiratory effort was, if anything, less than would normally be expected, and in recovery from such exercise the men certainly did not pant more than is usual in such tests. The observers actually had the distinct impression that brief exercise to the point of collapse produced less than the usual amount of respiratory distress.

In the first 12 weeks of rehabilitation there was likewise no tendency to dyspnea in rest and no excessive respiratory distress in exercise. Subsequently, however, in the period R12 to R20, many of the men had some complaints of exercise dyspnea. This was in the period of very rapid weight gain when there was also a relative tachycardia. A few of the men continued to complain of slight to moderate shortness of breath for 6 months, or even longer, after the end of semi-starvation.

Extra Vitamins and Proteins — Minnesota Experiment

Analyses of the effects of extra vitamins and extra proteins on the recovery of pulse rate and systolic blood pressure are summarized in Table 296, where all values are expressed as "percentage of recovery" after 12 weeks of rehabilitation. The results are given for each of the caloric groups separately so that at each caloric level there are 4 "supplemented" men to be compared with 4 "non-supplemented" men with regard to proteins and to vitamins.

TABLE 296

EFFECT OF EXTRA VITAMINS AND EXTRA PROTEINS ON THE RECOVERY OF SYSTOLIC BLOOD PRESSURE AND PULSE RATE in the first 12 weeks of rehabilitation in the Minnesota Experiment. All values are for the increase from S24 to R12, expressed as percentage of the decrease from C to S24. The values are means for each of the 4-man sub-groups in the caloric groups Z, L, G, and T in order of increasing calories. U = no extra protein; Y = extra protein; P = no extra vitamins; H = extra vitamins.

Caloric Group	Systolic Blood Pressure				Pulse Rate			
	U	Y	P	H	U	Y	P	H
Z	13	40	31	21	61	60	59	62
L	34	13	36	−3	76	66	65	79
G	93	20	62	39	66	70	64	70
T	77	90	92	58	69	89	62	93

The protein supplementation was without any consistent effect on either systolic blood pressure or pulse rate; the apparently sizable difference in the G caloric group is not statistically significant. The vitamin supplementation, however, seemed to have an unexpected effect. In all 4 caloric groups the men who received extra vitamins tended to attain higher basal pulse rates than the men who received only placebos, but the differences are so slight that the significance is questionable. There was a more substantial difference, however, when systolic blood pressures were compared. The men who received placebos made a better recovery in this respect than did those who received vitamin supplements, and the difference is more nearly acceptable as being significant. The difference is consistent in all 4 groups. When the caloric groups are combined so that 16 men are compared with 16 men (H vs. P), the difference approaches but does not reach the 5 per cent level of significance.

Aside from the pulse rate and the basal systolic pressure, no other item of cardiovascular measurement showed any real suggestion of differential effects from either extra vitamins or extra proteins.

General Picture of the Heart and Circulation

The vast majority of observations are consonant with the conclusion that the heart is not specifically injured in starvation and that the depression of the circulation is not serious in view of the reduction in the metabolic demand for circulation. The major findings associated with ordinary cardiac disease and circulatory inadequacy are generally absent — tachycardia, irregularities of rhythm, dyspnea, precordial pain, elevated venous pressure, enlargement of the heart, claudication, marked cyanosis. Yet there is often a feeling of dissatisfaction that these negative points fully rule out the question of cardiac or circulatory inadequacy. Hülse (1919b) could not point to a definite defect but neither could he accept the heart and circulation as being functionally adequate. Maase and Zondek (1917) likewise could find no evidence for a real defect but still they suspected that some inadequacy of the circulation must play an important role in producing the abnormalities of the starving man, particularly his edema. Dumont (1945) came to a similar conclusion.

Certainly it must be agreed that the heart and circulation in severely undernourished people are not normal. Aside from the bradycardia and quantitative peculiarities of the electrocardiogram, there are evidences of functional defects. The blood pressure is low and is probably responsible for the frequency of fainting in starved persons. Zimmer, Weill, and Dubois (1944) noted that it is dangerous to keep starved people standing for long periods of time. The peripheral circulation is poor as indicated by pallor, a cold skin, and frequent slight cyanosis. The total capacity in exercise for circulating blood and supplying oxygen to the tissues is reduced (see Chapter 34).

The most that can be said is that ordinarily the starved person is not actually in or close to cardiac failure and that the danger of circulatory collapse in this condition is not great, provided added burdens are not placed on the circulation. It has been suggested that the heart is not involved in death from chronic starvation, because the ventricles are often found in a state of contraction at autopsy (e.g. Paltauf, 1917); however, sudden deaths suggesting a cardiac origin are common in ambulatory starved persons according to Zimmer and his colleagues (1944). The bradycardia must be considered to be protective in conserving the strength of the heart. The prolongation of systole also suggests conservation of energy. The general freedom of the heart from irregularity has been commented on; the decreased irritability of the heart in itself affords protection, of course.

But the fact is that the heart is certainly abnormal in starvation and that it does not maintain its circulatory function even in proportion to the reduced metabolic needs. The heart conserves itself to some extent by sacrificing the circulation to other tissues, notably the periphery. One might suggest that the elevation of the lactic acid level of the blood in starvation reflects circulatory inadequacy, but the possibility of metabolic defect cannot be ruled out.

The relative absence of dyspnea is puzzling. In rest the relative circulatory inadequacy is too slight, perhaps, for one to expect dyspnea, but the response to exercise is another matter. The ventilation of the lungs is unusually high in moderate exercise, but in severe exercise collapse occurs at a relatively low ventilation and without extreme respiratory distress or apparent effort. A decreased sensitivity of the respiratory center, at least in severe exercise, is suggested. On the other hand, it may be that the limiting factor is muscular weakness, which causes collapse before circulatory or respiratory stress becomes severe.

The deteriorating effect of starvation on the heart is apparent in rehabilitation, not only because of the slowness with which normal function is regained but also because of the danger of incipient failure when a load of overeating is added. The venous pressure was definitely elevated beyond normal limits under these conditions in 4 out of 13 men observed in the Minnesota Experiment.

Where caloric deficiency is combined with a deficiency of vitamin intake, particularly of thiamine, the cardiac and circulatory defects are more obvious. In the Canadians who were captured at Hong Kong in December 1941, complaints of dyspnea, palpitation, and precordial pain were almost universal after a few months, and these symptoms tended to continue for some time after the return to good food and the initiation of vitamin treatment. When these men were examined in Canada 72 per cent of them still had cardiac complaints, but this probably should be attributed more to a residue of thiamine deficiency than to one of simple caloric undernutrition (Adamson et al., 1947a, 1947b).

CHAPTER 29

Fainting and the Cardiovascular Response to Posture

THE association of fainting with hunger is universally recognized, and episodes of fainting are commonly reported among famine victims, particularly when such persons are required to stand upright (Zimmer, Weill, and Dubois, 1944). Sensations of faintness and lightheadedness are common in starvation areas (Leyton, 1946). The Minnesota Experiment offered an opportunity to examine the factors involved in the etiology of these symptoms in the semi-starved individual.

The cardiovascular effects of posture were studied by the use of a tilting table equipped with a footboard. The pulse rate was counted with a stethoscope at the heart apex, and the blood pressure readings were obtained with a sphygmomanometer in the usual manner. Every effort was made to ensure standard conditions for each test. The men were made familiar with the tilt table by a dummy test the week before control observations were made. The tests were carried out in an air-conditioned room at 78° F. and 50 per cent relative humidity. The subjects were studied before breakfast and did not perform any appreciable exercise between getting out of bed and taking the test. The observers watched for and noted signs of sympathetic stimulation such as restlessness, sweating of the palms, and other abnormalities. Because the time available for carrying out the tilt table tests was limited, the majority of men were tested in the tilted position (68 degrees) for 5 minutes. Blood pressures and pulse rates were obtained every 30 seconds after tilting (head up) to 68°. These values were averaged to find the tilted blood pressure and pulse rate. A selected group of 9 men was tested for 10 minutes in the tilted position.

Results of the Tilt Table Test — Minnesota Experiment

The blood pressures and pulse rates observed in the tilt table tests during the 6 months of restricted diet are presented in Table 297. The semi-starvation figures show the usual bradycardia and tendency toward hypotension associated with semi-starvation. It should be noted that the change in pulse and blood pressure due to tilting was not greatly affected by semi-starvation so that the "score," calculated according to Crampton (1920), shows only a small deterioration.

The blood pressure readings obtained in the horizontal position agree in general with those found during the measurement of basal metabolism (see Chapter 28) with the exception of the systolic values during the control period. During tilting, systolic blood pressures were frequently recorded in the range of 80 to 90 mm. Hg.; the two lowest readings in satisfactory tests averaged 76 and 77 mm.

635

TABLE 297

PULSE RATES AND SYSTOLIC BLOOD PRESSURES BEFORE AND AFTER TILT-ING to 68° on the tilt table. Mean values, standard deviations, and the "score" according to Crampton, for 31 subjects.
(Minnesota Experiment.)

Condition	Pulse Rate (beats/min.)		
	C	S12	S24
Horizontal	55.4 ± 7.0	38.0 ± 5.3	37.7 ± 5.9
Tilted	71.7 ± 10.1	56.2 ± 12.4	55.9 ± 15.8
Difference	+16.3 ± 6.9	+18.2 ± 9.6	+18.2 ± 6.4
"Score"	45.0 ± 16.9	38.9 ± 23.1	39.5 ± 21.6

Condition	Systolic Blood Pressure (mm. Hg.)		
	C	S12	S24
Horizontal	115.7 ± 6.5	98.7 ± 7.8	96.1 ± 7.9
Tilted	110.7 ± 9.1	92.6 ± 8.2	90.0 ± 8.2
Difference	−5.0 ± 5.2	−6.1 ± 6.3	−6.1 ± 8.1

Hg. The sounds of Korotkow in the tilted subject became more muffled as semi-starvation proceeded, and in many cases the transition between phases became more difficult to determine with accuracy. This was particularly true of the third and fourth phases. The determination of the systolic blood pressure was in question on 6 occasions after 12 weeks of semi-starvation (S12) and on 2 occasions after 24 weeks of semi-starvation (S24). The accuracy of the diastolic pressure measurement was questioned by the observers in more than one half of the cases at both S12 and S24. However, in 10 men satisfactory diastolic pressures were obtained at each testing period. The pulse pressures for these men are presented in Table 298 along with the pulse rates and systolic blood pressures. The average horizontal pulse pressure dropped considerably during semi-starvation, and the tilted pulse pressure was reduced still further to 22.9 mm. Hg., a level which is recognized in normals to indicate a reduced cardiac output (Schneider and Crampton, 1934). As a result of the restricted diet, the "pulse product" (rate times pulse pressure) was reduced more than 50 per cent in the horizontal position, but this function was not affected to such a large degree in the tilted position. The data presented in Table 298 suggest a reduced stroke volume in the horizontal as well as in the tilted position.

After 20 weeks of rehabilitation (R20) the tilt table test was carried out on 12 subjects. The results are presented in Table 299. The pulse rate response to tilting at R20 was equal to the control, but the systolic blood pressure did not drop as far as the control; as a result the score according to Crampton was a little "improved" over the control. Satisfactory diastolic pressures are available for these men for the C and R20 periods, and the mean "tilted" pulse pressures for these periods were 31.3 and 30.9 mm. Hg., respectively. At R20 these men had regained their original weight.

During the control observations considerable variation in the systolic blood pressures was observed in a number of men while they were tilted. This varia-

TABLE 298

THE EFFECT OF 6 MONTHS OF SEMI-STARVATION ON THE PULSE RATE, THE SYSTOLIC PRESSURE, THE DIASTOLIC PRESSURE, THE PULSE PRESSURE, AND THE PULSE PRODUCT BEFORE AND AFTER TILTING to 68° on the tilt table, along with the differences in these variables due to tilting. Mean values for the 10 men in whom satisfactory diastolic blood pressure readings were obtained. (Minnesota Experiment.)

Condition	Pulse Rate		Systolic Pressure		Diastolic Pressure		Pulse Pressure		Pulse Product	
	C	S24	C	S24	C	S24	C	S24	C	S24
Horizontal	58.2	37.6	116.1	99.2	69.0	66.2	47.1	33.2	27.4	12.5
Tilted	73.7	58.0	112.5	92.9	79.3	70.3	33.2	22.9	24.4	13.3
Difference	+15.5	+21.6	−3.6	−6.3	+10.3	+4.1	−13.9	−10.7	−3.0	+0.8

TABLE 299

THE EFFECT OF 6 MONTHS OF SEMI-STARVATION AND 20 WEEKS OF REHABILITATION ON THE PULSE RATE AND SYSTOLIC BLOOD PRESSURE BEFORE AND AFTER TILTING to 68° on the tilt table, along with the differences in these variables and the "score" according to Crampton. Mean values for 12 men. (Minnesota Experiment.)

Condition	Pulse Rate (beats/min.)			Systolic Blood Pressure (mm. Hg.)		
	C	S24	R20	C	S24	R20
Horizontal	57.9	37.3	65.8	116.9	94.0	111.8
Tilted	74.3	57.2	82.9	112.6	88.5	110.9
Difference	+16.4	+19.9	17.1	−4.3	−5.5	−0.9
"Score"	45.4	38.3	55.4			

tion appeared to be related to the respiration and was thought to be the well-known respiratory cycle. The presence of the respiratory cycle was obvious on numerous occasions when the men were examined in the control period, but it was not detected at any of the examinations during the course of the restricted diet.

Tilt Table Results in the Field

The only effort to study the circulatory responses of semi-starved individuals in the field was made by Fliederbaum et al. (1946) in the Warsaw Ghetto. These investigators, using a tilt table, first tilted the subject head down to 22°, then returned him to the horizontal position, and finally tilted him feet down to 22°, 45°, and 90°. Blood pressures and pulse rates were observed at all positions. It was reported that the semi-starved individual showed a much smaller response in both pulse and blood pressure to the various tilting positions than did normal subjects. No comments were made on time intervals or fainting. It may be presumed that Fliederbaum's subjects were more severely starved than the men in the Minnesota Experiment.

Fainting on the Tilt Table

Complete measurements were obtained on 31 men in the Minnesota Experiment. Four of these fainted on the tilt table during the control observations.

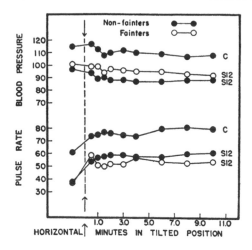

FIGURE 94. AVERAGE PULSE RATES
AND SYSTOLIC BLOOD PRESSURES OF THE
7 MEN IN THE 10-MINUTE GROUP (NON-
FAINTERS) BEFORE AND AFTER TILTING
DURING THE CONTROL PERIOD (C) AND
AFTER 12 WEEKS OF SEMI-STARVATION
(S12), along with the average pulse
rates and systolic blood pressures before
and after tilting after 12 weeks of semi-
starvation of the 4 men (fainters) who
fainted during the control period
(Minnesota Experiment).

Two of these men fainted during the 5-minute test while the other 2 were in the
10-minute group and fainted in the 5-to-10-minute interval of tilting. Since both
of the "5-minute" fainters fainted twice during the control testing, it may be
presumed that their fainting was not due to a temporary condition.

The 2 fainters who were in the 5-minute group were moved into the 10-
minute group after 12 weeks of semi-starvation (S12); all 4 of these men withstood 10 minutes of tilting without difficulty. The pulse rates and systolic blood
pressures of these men at S12 are presented in Figure 94 along with similar data
for the 7 men in the 10-minute group who did not faint in the control period.
In general, the fainters at S12 had a slightly higher blood pressure and a little
lower pulse rate than the non-fainters. During all the tilt table testing at both
S12 and S24 only one subject (No. 11) fainted — during the S12 test at 4 minutes
of tilting. The only fainter who was available for testing at the end of 20 weeks
of rehabilitation was subject No. 119, and it is important to note that this man
fainted after 5 minutes of tilting when he was tested at R20 but had not fainted
at either S12 or S24.

The Sensation of Faintness and Spontaneous Fainting

An almost universal complaint among the subjects in the Minnesota Experiment was some degree of faintness immediately after standing up from a chair
or bed. This was a very transitory phenomenon which varied from a slight sensation of dizziness to a dimming of vision and "blacking-out," causing the subject to reach for a temporary support. No one ever lost consciousness during
these episodes. Reports of dizziness associated with change of posture were very
numerous during the first 6 weeks of the restricted diet, but thereafter they decreased until there were very few complaints on this point during the last 6
weeks (S18 to S24). It was felt that the men learned to refrain from making a
sudden transition to the vertical posture and in this way helped to avoid bringing on symptoms. It is interesting to note that complaints of dizziness associated
with the assumption of the upright posture increased during the early part of
rehabilitation when the men began to feel more vigorous; these complaints disappeared as appreciable weight gains were made.

These transient symptoms were not reproduced when the men were passively changed from the horizontal to the tilted position on the tilt table. The blood pressure readings 30 seconds after tilting showed no unusual change in semi-starvation as compared to the control period (see Figure 94). Obviously there was a difference between the effects of active and passive change of posture. It could not be determined whether slight excitement attendant upon tilting the table played any role in this difference. However, it was thought that the men were well enough trained to eliminate this factor.

A number of men reported sensations of faintness, lightheadedness, dizziness, and weakness which were not associated with postural change. These symptoms tended to occur near the end of the starvation period while the men were on their feet doing chores. In addition a few men reported dizziness on waking up in the morning which disappeared after they got out of bed and moved around a little. An attempt was made to test the response of the cardiovascular system to posture during one of these periods of faintness. This was possible with sub-ject No. 12 who was tested on the tilt table after doing chemical laboratory chores for two hours. He stated that for 30 minutes before the test he had been feeling weak and faint. He was tilted for 15 minutes and showed a satisfactory response: an average tilted blood pressure of 86 mm. Hg., an average tilted pulse rate of 58, and no signs of sympathetic stimulation.

Only one clear episode of spontaneous syncope occurred during the course of the experiment. Subject No. 11 fainted and remained unconscious for 2 or 3 minutes while waiting in line for his supper to be served. He recovered spon-taneously and was able to finish his meal and return to the dormitory. This oc-curred after 9 weeks of semi-starvation, and it should be noted that the subject had a severe cold and had been given one-half grain of codeine and one-half grain of papavarine two hours before the occurrence. Three weeks later this man fainted after 4 minutes of tilting on the table; he was the only subject to faint on the tilt table during the entire semi-starvation period.

Blood Sugar

The effects of semi-starvation on the blood sugar levels have been discussed in detail in Chapter 24. In general, the mean resting blood sugar under basal conditions decreased approximately 10 mg. per 100 cc. A change of this order of magnitude is not large. The lowest recorded value was 55.6, and there were only two values between 50 and 60 mg. per 100 cc. of blood. In terms of actual change the blood sugar concentration of 3 subjects increased during rehabilita-tion 18 to 20 mg. per 100 cc. of blood, and 3 additional subjects made gains of 15 to 17 mg. per 100 cc. of blood. On the other hand, much lower individual values for blood sugar were observed during work. There was one case of a blood sugar during work of only 30 mg. per 100 cc. of blood. In 5 blood sugar determinations in work at S24 the values fell within the range of 46 to 50; in 8 others the values were within the range of 51 to 55 mg. per 100 cc. of blood.

Symptoms which have been ascribed to hypoglycemia (Collins and Boas, 1946) were not numerous, did not occur with frequency in any individuals, and were certainly not specific. In addition to weakness, faintness, lightheadedness, and dizziness, numbness and tingling in the extremities in situations where circulatory occlusion would not be expected were reported on one or two occasions, and one man reported an episode of trembling after walking on the treadmill.

The observation that marked loss of weight was not accompanied in the Minnesota Experiment by an increased tendency to faint on the tilt table is of interest in view of the widespread concept that fainters tend to be tall, thin, and underweight (Rook, 1947). In this connection, it seems worth while to compare the characteristics of the fainter group during the control period with those of the group as a whole. The results are presented in Table 300. The fainter group

TABLE 300

A Comparison between Tilt Board Fainters and Non-Fainters of certain cardiovascular, fitness, and anthropometric characteristics (Minnesota Experiment).

Function	Average Value		
	Fainters	Non-Fainters	Difference
Weight (kg.)	70.5	70.0	+0.5
Fat as percentage of body weight	13.70	13.96	—0.26
Weight (kg.)/height (cm.) index	39.6	38.8	+0.8
Systolic heart size (cc.)	668.4	620.4	+48.0
Heart axis, degrees from horizontal	46.6	49.3	—2.7
Total blood volume (cc./kg.)	82.9	84.3	—1.4
Plasma volume (cc./kg.)	44.9	45.07	—0.17
Work pulse rate (beats/min.)	129.6	134.0	—4.4
Harvard Fitness Score	70.8	65.6	+5.2

included the 5 men on whom complete data are available. These men did not differ significantly from the normal group in any of the measurements examined. Some of the negative results are of interest, however. The fainter group did not have any of the characteristics of the asthenic type. As a group the men were of average weight with an average percentage of body fat and weighed a little more than the normals for each centimeter of height. It is also interesting to note that the longitudinal axis of the heart averaged 2.7 degrees nearer the horizontal than in the normals. This may be related to the fact that the heart size was also a little larger than normal. Inspection of the cardiac silhouettes as seen on the roentgenkymogram leaves no doubt that the fainters did not have long pendulous hearts. The mean plasma and whole blood volumes are a little less than the average, but this is due principally to one man (subject 119). While this relatively low blood volume may have contributed to this man's inability to tolerate the tilted position, the blood volumes of the other men were equal to or greater than the normal. The estimates of fitness have been included since a number of investigators feel that even though it has not been possible to show statistically significant differences between fainters and non-fainters, there is a tendency for fainters to be somewhat less "fit" than their fellows (Allen et al., 1945; Graybiel and McFarland, 1941).

"Blackouts" and Giddiness

There was no objective evidence that the blackouts and faintness which occurred when the men assumed verticality were due to cardiovascular failure. It has been shown (Wald *et al.*, 1937) in normal young men that the rapid phase of adaptation to the vertical posture is complete in 30 seconds, so that blood pressure measurements during this time would have been necessary to check this point. Since it has been demonstrated (Lambert and Wood, 1946) that blacking-out during positive acceleration is due to an inadequate cerebral blood pressure, it appears likely that the blackouts were the result of temporary circulatory failure during the rapid phase of adjustment (the first 30 seconds) to the vertical posture.

That the blackouts were not reproduced on the tilt table is surprising. However, some slight excitement was inevitably attendant on tilting the table, and this factor may have accounted for the absence of transitory blackouts and faintness during the first 10 or 20 seconds of tilting.

On the other hand, it is evident that at least some of the complaints of light-headedness, dizziness, etc., which were not associated with a sudden change in posture, cannot be ascribed to central circulatory incompetence. It has been shown that sensations of faintness can occur in normal individuals in the upright posture without any decrease in the cardiac output and with adequate blood pressure (Starr and Rawson, 1941). Furthermore, Romano and Engel (1945) have clearly demonstrated the occurrence of fainting in hysterical patients with adequate circulation and no abnormalities of the electroencephalogram.

Leyton (1946) ascribed most of the complaints of this nature which were noted in German prisoners of war to the reduced level of the blood sugar. Comparisons with blood sugar levels that produce hypoglycemic symptoms in the normal individual are probably not valid, since it is known that the nutritional state has a definite influence on the appearance of symptoms in relationship to the fall in blood sugar (Jensen, 1938). However, in acute starvation with hard work Taylor *et al.* (1945a) reported blood sugar levels between 30 and 50 mg. per 100 ml. of blood which had been observed in a number of men whose only complaint was a deadening fatigue. It is interesting that descriptions of the prolonged hypoglycemia resulting from tumors of the Islets of Langerhans do not include "faintness" or "dizziness" among the numerous symptoms mentioned as occurring in these individuals (Tedstrom, 1934).

Finally, complaints of dizziness, faintness, etc. are extremely common in psychoneurosis, and it seems possible that these complaints are merely part of the general picture of semi-starvation neurasthenia which existed to some degree in all of the men. This is discussed in Chapter 42.

The response of the "fainter group" to tilting on the table seems clearly to have improved as these men lost weight. The following facts support this contention. The highly standardized conditions under which the tests were carried out resulted in reproducible fainting during the control period. Subject No. 119 provided a good control experiment after he had regained his weight by fainting at approximately the same time of tilting as had resulted in a faint during the control period. Finally, the "5-minute fainters" withstood a tilting period during semi-starvation which was more than twice as long as was necessary to produce faints consistently during the control period.

A complete explanation of this improvement in the response to tilting during semi-starvation is not possible. However, it should be pointed out that the blood volume in cc. per kg. of body weight increased to 120 per cent of the normal figure (see Chapter 14), and it is quite possible that this relative plethora would counteract pooling and filtration in the abdomen and extremities.

The pulse rate and blood pressure responses to the tilted position would appear to indicate a normal circulation in the vertical posture. This is particularly true if the "scores" according to Crampton are used. However, evidence has been presented elsewhere (see Chapter 28) that during quiet standing the circulatory factor of safety as measured by the ratio of unused venous oxygen returned to the heart per minute to the metabolic rate was reduced by 50 per cent in the subjects of this experiment at the end of the semi-starvation period.

It is apparent, then, that while the cerebral circulation is adequately maintained in the upright posture, any circulatory stress in addition to that of verticality might well result in failure of the cerebral circulation and syncope. It is perhaps not a coincidence that on the one occasion during the Minnesota Experiment when spontaneous syncope occurred, the subject was suffering from a severe cold, a disease which is frequently mentioned in the literature as a cause of poor cardiovascular adjustment to the vertical posture (Berry, Horton, and MacLean, 1940).

Among stresses which might well result in syncope in the starved individual standing in line, we may mention hot weather, dehydration as the result of diarrhea, and respiratory infections.

CHAPTER 30

The Electrocardiogram

THE basic procedures of electrocardiography were standardized about 40 years ago, but development — and controversy — continue. The electrocardiogram reflects only the electrical events in the heart and is used primarily to detect abnormalities in the origin and spread of electrical excitation. Where abnormalities of rhythm or conduction exist, the electrocardiogram (E.C.G.) provides a means of estimating the kind, the location, and even the probable nature of the abnormality. The great value and sensitivity of the E.C.G. for these purposes is well recognized.

The E.C.G. is of much value in detecting changes in the myocardium when these involve local damage or alteration of excitability or conduction, but so far at least, it has been little used in an effort to evaluate the relative weakness or strength or function of the whole myocardium. It is well known that the E.C.G. may be "normal" in the presence of cardiac decompensation and may not reflect changes in the functional state. This appearance of normality, however, may be simply in comparison with the whole range of normality for all persons and not with strict reference to the normal for the individual. In the Minnesota Laboratory of Physiological Hygiene we have repeatedly shown that significant changes may be produced in the E.C.G., but still within normal limits. In other words, it is necessary to distinguish between the significance of a single E.C.G. taken at random and the significance of changes which appear in serial studies on the same person. In starvation the electrocardiogram becomes, by any criterion, abnormal in numerous respects. The fact that control E.C.G.'s were available for each subject in the Minnesota Experiment made it possible also to detect changes which would pass as being within normal limits if only the end starvation records were available.

Previous Studies

The literature on electrocardiographic changes in chronic human semi-starvation, produced by calorically insufficient food intake over a prolonged period of time, is scanty. Benedict and his co-workers (1919) took electrocardiographic records mainly to find out whether the slow pulse rate observed during semi-starvation in their subjects was of sinus or nodal origin. They proved that it was a sinus bradycardia. The electrocardiograms were otherwise within normal limits, but obviously no attempt at a more quantitative evaluation was made. The degree of semi-starvation attained in Benedict's series was mild compared with that in the series to be presented here, or with actual conditions in famine areas.

Electrocardiographic material was obtained by Tur (1944) during the siege of Leningrad. In 16 patients with severe malnutrition (11 died in the hospital or after discharge) studied in the fall and winter of 1941–42, sinus bradycardia, right axis shift, and low QRS voltage were observed in the majority. In 10 patients the T wave was abnormally large in the standard leads but low or negative in the apex lead. In deteriorating patients there was a progressive tendency to lower T wave amplitude in the standard leads. Shortly before death in 3 patients, a pronounced increase of the heart rate occurred, together with ST depression in lead 2 and 3 and a decrease of the T wave, which became almost isoelectric. Later, in the summer of 1942, when the nutritional situation in Leningrad had improved somewhat, another group of 32 patients was studied. Out of 24 cachectic patients, of whom 3 died, 6 had right axis deviation and 12 had left axis deviation; the majority (18) had subnormal T wave amplitude in one or more leads, usually lead 1. The QRS amplitude was subnormal in 12 patients. Out of 15 patients in a better state of nutrition, none had right axis deviation and 11 had low QRS and T voltage; T_1 was isoelectric in 8. During rehabilitation the increase of the T wave amplitude paralleled the clinical improvement and was considered to be of prognostic significance.

Cardozo and Eggink (1946) found low voltage and sinus bradycardia in 29 cases of severe malnutrition. In addition, the duration of the QT interval was prolonged, absolutely as well as in relation to the cycle length. The K_{QT} values of 14 patients ranged from 0.373 to 0.483, which is definitely above the normal range; in one it was extremely high (0.76).

Ellis (1946) studied the electrocardiographic changes in 4 American prisoners of war immediately after liberation and during a rehabilitation period of 3 weeks. All were in a state of advanced malnutrition, and one patient, who had lost about 45 per cent of his body weight, was almost moribund. The most prominent feature was a tremendous prolongation of the QT interval. In 3 men the P wave was superimposed on the descending limb of a rounded T wave. Ellis suggested that the prolongation of the QT interval might be due to a prominent U wave, which disappeared during the improvement. We find little in his electrocardiograms to support this assumption. Ellis claimed that the PR and QRS intervals were increased and diminished during treatment, but this was the case in only one patient (PR interval in patient 2; QRS interval in patient 1). The ST segment was depressed in 2 patients; in one man this was associated with an inversion of T in leads 2 and 3. These changes disappeared during the treatment, and Ellis claimed that the electrocardiograms returned to normal within 14 to 21 days. This is correct as far as the so-called normal limits are concerned, but it is doubtful whether the electrocardiograms actually returned to the previous normal for the given patient in so short a period. No control electrocardiogram was available for any of these patients.

Several other interesting features, not discussed by Ellis, are discernible in the electrocardiograms of his patients. In the most advanced patient, a pulsus trigeminus, due to premature ventricular beats, disappeared with the treatment. All the 4 patients had a right axis shift, and this was still evident after 3 weeks of rehabilitation. In all instances the T waves in the limb leads, especially lead

1, were small when the patients were first seen or became small during the treatment. At the end of the treatment they increased again. The QRS amplitude decreased during treatment in 3 and was rather low throughout in the fourth patient, approaching the lower normal limits.

Out of 296 emaciated former inmates of Nazi concentration camps treated in Switzerland in May 1945, electrocardiograms were taken in 31 cases of uncomplicated, though rather extreme, inanition (Hottinger et al., 1948). In 29 of these the records were judged abnormal because of low voltage of the QRS complex (14 cases) or of the T waves (29 cases), but negative T waves, ST depression, or changes in the QRS contour were not observed. An increase of the QT interval was found only in 2 moribund cases. These patients were studied at a time when active treatment and refeeding had been in progress for some days or weeks.

Patients in a more active phase of starvation were studied in a Belgian prison by Simonart (1948). No systematic analysis was provided, but in 22 patients low voltage was frequent, while in 6 there seems to have been a definite prolongation of the QT interval. Simonart commented that the records rather resembled those seen in hypocalcemia and in hypothyroidism. At Warsaw the electrocardiograms were stated to have shown sinus bradycardia with regular rate and low voltage, but details were not published (Fliederbaum et al., 1946).

It is surprising, in view of the existence of widespread starvation areas in Europe and Asia during and after World War II, that the electrocardiographic material is so scanty. It must be kept in mind, however, that the general dislocation of transportation in starvation areas, the disruption of sanitary and communication services, the nonavailability of equipment and replacement parts, and the overloading of hospitals and medical personnel make any study exceeding the bare routine procedures extremely difficult. We do not include in this review of the literature the numerous reports on bradycardia, since as a rule no electrocardiograms were taken. One might expect that electrocardiograms taken in cachectic patients would reveal some changes that would be of interest for our problems. To our knowledge no pertinent material is available.

The Minnesota Experiment

Details of the electrocardiographic procedure are given in the Appendix on methods. Although electrocardiograms were taken more frequently, we have restricted the detailed analysis to the following periods, which proved to be sufficient to characterize the general trend of electrocardiographic changes during semi-starvation: twelfth week of semi-starvation (S12), twenty-fourth week of semi-starvation (S24), twelfth week of rehabilitation (R12). Twenty men were re-examined after 32 weeks of rehabilitation; at that time their body weights in most cases exceeded their original control weights.

Interval Changes in the Minnesota Experiment

Tables 301 and 302 show the average interval changes (heart rate, QT interval, systole duration, K_{QT}, and K_{syst}) during starvation and at 32 weeks of rehabilitation. The PR interval, QRS interval, and duration of the P wave did not change significantly during starvation and rehabilitation and are therefore not included in the tables.

Table 301 shows the absolute values and standard deviations for the total group of 32 subjects for the periods C, S12, S24, and R12 and for 20 subjects for the periods C, S24, R12, and R32. Table 302 shows the differences between the various periods, calculated from Table 301. It can be seen that the values for both groups coincide very closely for the periods C, S24, and R12, so that the changes at R32 for the group of 20 subjects are probably representative of the total group of 32 subjects. Because of the similarity of the values for these groups, the calculation of statistical significance was not repeated for the differences S24 — C and R12 — S24 with the group of 20 subjects. The standard deviation was calculated only in the total group of 32 subjects from C to R12.

The average heart rate during the control period (55.2) is rather slow compared to usual standards, but it agrees well with many observations on other groups of healthy young men made in this Laboratory during recent years. The heart rate slowed to 37.3 beats at the end of semi-starvation; it increased again during rehabilitation but was still below the control value at the twelfth week of rehabilitation. However, a further increase occurred from R12 to R32, and this

TABLE 301

HEART RATE, QT INTERVAL, AND SYSTOLE DURATION in semi-starvation and rehabilitation. Mean values and standard deviations of 32 and 20 subjects. The intervals are given in 1/100 second; K_{QT} was calculated from the QT interval and K_{syst} from the systole duration, both divided by the square root of the RR interval. The "range %" values were calculated for each individual from the range of the heart rate and the individual's average heart rate. (Minnesota Experiment.)

Function and Number of Subjects	Control M	SD	S12 M	SD	S24 M	SD	R12 M	SD	R32 M	SD
Heart Rate										
32....	55.2	6.49	35.3	4.96	37.3	5.15	49.2	6.56		
20....	54.9				37.5		50.0		58.7	5.03
Heart rate range										
32....	9.5	4.16	3.1	1.16	3.4	1.27	7.6	4.50		
20....	9.8				3.3		7.2		10.4	2.66
Range %										
32....	14.7		8.1		8.8		14.1			
20....	15.3				8.8		13.8		17.8	4.07
QT										
32....	35.8	2.69	46.1	8.63	45.1	2.55	41.3	2.27		
20....	36.1				45.8		41.4		39.8	2.71
Systole duration										
32....	32.5	2.46	39.3	2.39	39.3	2.41	37.2	2.02		
K_{QT}										
32....	37.01	1.8	36.54	1.6	36.34	2.4	38.18	1.7		
20....	37.03				36.43		37.94		38.69	2.2
K_{syst}										
32....	33.55	1.6	31.09	1.3	31.55	1.6	34.35	1.5		

TABLE 302

DIFFERENCES OF INTERVALS, absolute (Abs.) and in percentages (%) of the values for the reference period. (The values correspond to those of Table 301.)
(Minnesota Experiment.)

Function and Number of Subjects	S12 − C		S24 − C		R12 − S24	
	Abs.	%	Abs.	%	Abs.	%
Heart rate						
32	−19.9[**]	−36.1	−17.9[**]	−32.4	11.9[**]	31.9
20			−17.4	−31.7	12.5	33.3
Heart rate range						
32	−6.4[**]	−65.2	−6.1[**]	−64.3	4.2[**]	123.1
20			−6.5	−66.4	3.9	118.2
Range %						
32	−6.6[**]		−5.9[**]		5.3[**]	
20			−6.5		5.0	
QT						
32	10.3[**]	28.8	9.3[**]	25.9	−3.8[**]	−8.4
20			9.7[**]	26.8	−4.4	−9.6
Systole duration						
32	6.8[**]	20.9	6.8[**]	20.9	−2.1[**]	−5.3
K_{QT}						
32	−0.5	−1.4	−0.7	−1.9	1.8[**]	5.0
20			−0.6	−1.6	1.5	4.1
K_{syst}						
32	−2.5[**]	−7.4	−2.0[**]	−6.0	2.8[**]	8.9

Function and Number of Subjects	R32 − S24		R32 − R12		R32 − C	
	Abs.	%	Abs.	%	Abs.	%
Heart rate						
20	21.2[**]	56.5	8.7[**]	17.4	3.8[**]	6.9
Heart rate range						
20	7.1[°]	215.2	3.2[°]	44.5	0.6	6.1
Range %						
20	9.1[**]		4.0[**]		2.5	
QT						
20	−6.0[**]	−13.1	−1.6[°]	−3.9	3.7[**]	102.3
K_{QT}						
20	2.3[**]	6.3	0.8	2.1	1.7[**]	4.5

change was statistically highly significant. The pulse rate at R32 even exceeded the initial rate (C), and this difference was statistically highly significant.

The decrease of the heart rate was associated with a narrowing of the range of the heart rate (variability within the electrocardiogram). This is true for the absolute range as well as for the variability in percentage of the average heart rate. The heart rate in the normal condition (C) varied 14.7 per cent but after 12 weeks of semi-starvation only 8.1 per cent. The heart beat is more regular in a starved subject than it is in a normal subject. A pronounced sinus arrhythmia which was present in a few subjects in the control period disappeared completely during semi-starvation. In the rehabilitation period the range increased with the heart rate again, and sinus arrhythmia returned in those subjects who had it before starvation. The variability range was still below normal at R12 but continued to increase significantly from R12 to R32, so that the control values were attained, or slightly exceeded. The slight excess, however, was not statistically significant.

It is interesting that, expressed as percentage (Table 302), the decrease of the range (variability) of the heart rate in starvation (and consequently the increase in rehabilitation) exceeded the decrease of the heart rate.

The QT interval and mechanical systole duration increased during semi-starvation, as should be expected from the lengthening of the cycle. There was a slight decrease of K_{QT} in semi-starvation, but this difference was not pronounced enough to be statistically significant. The decrease of K_{syst} during starvation was statistically highly significant. Both K_{QT} and K_{syst} increased during 12 weeks of rehabilitation, and this increase was statistically highly significant. The changes of the QT interval and of the mechanical systole duration lagged behind the changes of cycle length; in semi-starvation with a lengthening cycle the QT interval and the systole duration remained relatively too short, and in rehabilitation with a shortening cycle both intervals remained relatively too long. It should be kept in mind that the validity of the formula for such extreme degrees of brady-cardia is somewhat doubtful; but we believe that the formula can be used for a description of the QT interval and mechanical systole changes in relation to the cycle length.

In the period from R12 to R32 there was a further decline of the absolute QT interval associated with a further increase of the heart rate, but due to the lag of QT adaptation, the QT interval remained unusually long during the total period of rehabilitation. This is apparent when the values at R32 are compared with the control values. At R32 the pulse rate was higher than in C, while the absolute QT interval was greater, not shorter as should be expected. The difference R32 — C for K_{QT} is statistically highly significant. This means that after 32 weeks of rehabilitation the QT interval was still prolonged in relation to the cycle length. The difference R32 — C of the heart rate of about 4 beats, although statistically significant, is so small that all formulas suggested for the relationship of the QT interval to the cycle length would show the same results, since in so narrow a limit the curves of all formulas are practically straight lines. The relative QT prolongation is one of the most persistent electrocardiographic changes after semi-starvation.

It is of interest that all interval changes attained their maximum at S12, the changes between S12 and S24 being slight and statistically not significant. None of the interval functions had been fully restored at R12; in other words the rate of change in 3 months of semi-starvation was greater than the rate of reversal in the same period of rehabilitation.

Amplitude Changes in the Minnesota Experiment

Tables 303 and 304 show amplitude and axis changes; Table 303 gives the absolute values and standard deviations, while Table 304 gives the absolute and percentage differences and their statistical significance. For changes of the P waves, only lead 2 is included. Values for the QRS complex and T wave amplitude for lead 3 are omitted, because the changes in this lead can be inferred from the amplitudes in other leads and the axis as given in Table 303. There was a very marked decrease of the P wave, the QRS complex, and T wave amplitude in lead 1 at S12, while the amplitude of T_2 and ΣT did not change essentially. The decrease of the QRS amplitude in lead 1 is most pronounced, owing to a combined effect of QRS axis shift and decrease of QRS amplitude. The decrease of T_1 at S12 is due solely to the right axis shift of the T axis. In the interval from S12 to S24 the amplitudes of P_2, the QRS complex, and T_1 continued to decrease at about the same rate as during the first 12 semi-starvation weeks. The amplitude of the T wave (T_2 and ΣT), which was still unchanged at S12, thereafter declined rapidly, so that the percentage decreases of R_2, ΣQRS, T_2, and ΣT were about the same at S24. Changes of ΣQRS and ΣT could, to a certain extent, be produced by axis changes. Calculation of the manifest potential or Ashman's correction factor (Ashman and Byer, 1943) could be used for a closer estimate of the actual voltage changes. However, the changes in the amplitudes are too great to be accounted for by any axis changes. There was unquestionably a very pronounced and statistically highly significant drop of the actual voltage of all deflections of the electrocardiogram at the end of semi-starvation.

TABLE 303

AMPLITUDES AND AXES OF THE ELECTROCARDIOGRAM during semi-starvation and rehabilitation. Mean values and standard deviations of 32 and 20 subjects. The amplitudes are given in standardized mm. (1 mm. = 0.1 mv.); the axes are indicated in degrees. (Minnesota Experiment.)

Function and Number of Subjects	Control		S12		S24		R12		R32	
	M	SD	M	SD	M	SD	M	SD	M	SD
P_2										
32	1.01	0.40	0.68	0.40	0.52	0.34	0.81	0.38		
20	0.99				0.50		0.86		1.06	0.40
R_1										
32	3.75	1.89	2.10	1.05	1.45	0.85	2.44	1.30		
20	3.24				1.38		2.04		5.57	2.43
R_2										
32	9.91	4.15	8.03	2.30	6.23	2.18	7.09	2.45		
20	9.30				6.08		6.74		8.69	2.89
ΣQRS										
32	23.49	7.75	18.43	4.10	14.37	4.00	16.65	4.91		
20	22.60				14.28		16.19		22.80	4.74
T_1										
32	1.70	1.78	1.14	0.51	0.46	0.31	1.20	0.49		
20	1.74				0.47		1.11		2.90	0.91
T_2										
32	2.73	1.07	2.80	1.32	1.70	1.09	2.92	1.00		
20	2.88				1.83		2.96		3.08	1.04
ΣT										
32	5.59	2.51	5.60	2.55	3.53	1.92	5.82	1.99		
20	5.81				3.79		5.92		6.89	1.77
QRS axis										
32	68.4	15.9	77.8	9.29	81.6	10.02	69.8	2.06		
20	70.4				82.4		71.2		54.2	22.4
T axis										
32	47.5	23.5	65.9	6.57	69.8	23.9	65.2	9.44		
20	53.4				72.8		67.6		33.6	17.4

TABLE 304

DIFFERENCES OF AMPLITUDES AND AXES between various periods of semi-starvation and rehabilitation, absolute (Abs.) and in percentages (%) of the values for the reference period. (The values correspond to those of Table 303.) (Minnesota Experiment.)

Function and Number of Subjects	S12 — C		S24 — C		R12 — S24	
	Abs.	%	Abs.	%	Abs.	%
P_2						
32	−0.33[**]	−32.6	−0.49[**]	−48.5	0.29[**]	55.9
20			−0.49	−49.5	0.36	72.0
R_1						
32	−1.65[**]	−44.0	−2.30[**]	−61.4	0.99[**]	68.4
20			−1.86	−57.5	0.66	47.8
R_2						
32	−1.88[**]	−19.0	−3.68[**]	−37.2	0.86[**]	13.8
20			−3.22	−34.6	0.66	10.9
ΣQRS						
32	−5.06[**]	−21.6	−9.12[**]	−38.8	2.28[**]	15.9
20			−8.32	−36.5	1.91	13.4
T_1						
32	−0.56[**]	−32.9	−1.24[**]	−73.0	0.74[**]	160.7
20			−1.27	−73.0	0.64	136.0
T_2						
32	0.07	2.5	−1.03[**]	−37.7	1.22[**]	71.8
20			−1.05	−36.5	1.12	60.8
ΣT						
32	0.01	0.2	−2.06[**]	−36.9	2.30[**]	65.2
20			−2.02	−34.8	2.13	56.3
QRS axis						
32	9.4[**]		13.2[**]		−11.8[**]	
20			12.0		−11.2	
T axis						
32	18.4[**]		22.3[**]		−4.6	
20			19.4		−5.2	

Function and Number of Subjects	R32 — S24		R32 — R12		R32 — C	
	Abs.	%	Abs.	%	Abs.	%
P_2						
20	0.56[**]	111.1	0.20[**]	23.3	0.07	7.0
R_1						
20	4.19[**]	306.5	3.53[**]	173.2	2.33[**]	71.9
R_2						
20	2.61[**]	42.9	1.95[**]	28.9	−0.61	−6.6
ΣQRS						
20	8.52[**]	59.7	6.61[**]	40.8	0.20	8.9
T_1						
20	2.43[**]	517.2	1.79[**]	161.1	1.16[**]	66.7
T_2						
20	1.25[**]	68.4	0.12	4.1	0.20	6.9
ΣT						
20	3.10[**]	81.7	0.97[*]	16.4	1.08[*]	18.6
QRS axis						
20	−28.2[**]		−17.0[**]		−16.2	
T axis						
20	−39.2[**]		−34.0[**]		−19.8[**]	

The time course of the voltage changes in semi-starvation was different from the time course of the interval changes; the interval changes had already reached their maximum at S12, while the amplitude and axis changes reached their maximum deviation at S24. It may be concluded that the interval changes and the voltage or axis changes are not closely related to one another and probably are due to different causes. Also, there are discrepancies in the time course of amplitude decrease between ΣQRS and R_2 on the one hand and ΣT and T_2 on the other hand. It seems probable, therefore, that the QRS amplitude changes and the T wave amplitude changes are also due to different causes.

Although in rehabilitation all amplitudes were increased significantly by R12, the amplitudes of P_1, R_1, R_2, ΣQRS, and T_1 were still well below the control values (C), while the T_2 and ΣT amplitudes had already reached the control values. In the interval from R12 to R32 all amplitudes showed a further increase, which was highly significant except for T_2. While the major part of the recovery of ΣQRS was completed only in the period from R12 to R32, the major part of the recovery of the T amplitude was completed at R12. While at R32 ΣQRS was not significantly different from the control value, ΣT exceeded the control value by 18.6 per cent, and this difference was statistically significant. This means that discrepancies in the time course between QRS and T changes were observed not only during semi-starvation but also during rehabilitation.

The R wave and T wave in lead 1 exceeded the control values by 72 and 67 per cent at R32, but this increase was mainly due to axis changes.

Axis Changes in the Minnesota Experiment

Both QRS axis and T axis shifted to the right during semi-starvation. This unidirectional shift is of interest since under many conditions the QRS and T axes change in opposite directions. The changes of the T axis were greater than the changes of the QRS axis, but both were statistically highly significant. Since the T axis was about 20° left of the QRS axis in the control period, the two came closer together during semi-starvation. It appears that a change in the direction of repolarization occurred during semi-starvation. The axis changes were continuous during this period. Figure 95 shows a scatter diagram of the changes of the QRS axis vs. T axis changes at S24, in terms of S24 — C. There was no significant correlation.

In the rehabilitation period both QRS axis and T axis shifted back to the left. While the QRS axis attained the control values at R12, the T axis was still much closer to the S24 values; in fact the changes of the T axis ($\triangle$R12 — S24) were not significant, in contrast to the highly significant changes of the QRS axis.

The agreement between the QRS axis at R12 and at C was only temporary; in the period from R12 to R32 the QRS axis, and to a greater extent the T axis, continued to move toward the left. At R32 the QRS axis and especially the T axis were shifted to the left as compared with the control values, and these differences were statistically highly significant. Since the T axis was still shifted to the right at R12, it must have coincided with the C values somewhere between R12 and R32 for a transitory period. The QRS axis started to shift back toward the left earlier than did the T axis (Tables 303 and 304; Figure 96).

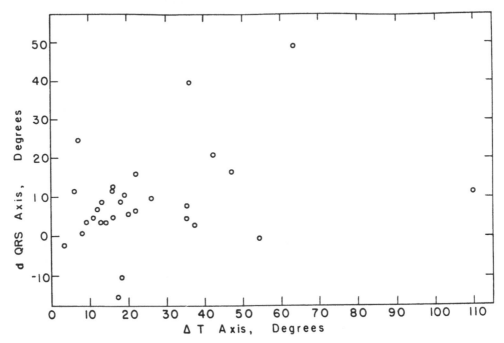

FIGURE 95. CORRELATION (r = +0.23) BETWEEN THE CHANGES OF THE T AXIS AND THE QRS AXIS. The changes, d in QRS axis and △ in T axis, were calculated as differences between the control (C) and the starvation (S24) values.
(Minnesota Experiment.)

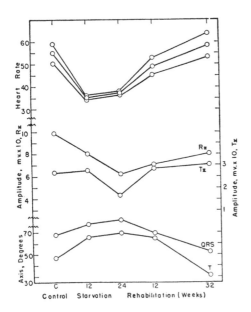

FIGURE 96. CHANGES OF HEART RATE, R₂ AMPLITUDE, T₂ AMPLITUDE, QRS AXIS, AND T AXIS DURING STARVATION AND REHABILITATION. The upper part of the graph shows the changes of the average heart rate (middle curve) and the variability of the range of the heart rate, calculated from the longest and shortest RR interval (lower and upper curve). The middle part of the graph shows the amplitude changes in mv. × 10 of the R wave and T wave in lead II. The lower part of the graph shows the changes of the QRS and T axes.
(Minnesota Experiment.)

652

Graphical Presentation of Findings

Figure 96 shows the changes during starvation and rehabilitation in a few fundamental functions: average heart rate and range of variability of heart rate, R_2 and T_2 amplitudes, and QRS and T axes. For the periods C to R12 the values of the total group of 32 subjects were taken, and for the period R32 the values of the 20 subjects are used with a correction for the absolute differences of the control values between both groups. The discrepancies in the time course between the heart rate changes and the amplitude changes are obvious.

Figures 97, 98, and 99 show electrocardiograms of 3 subjects from C to R12 with the typical sinus bradycardia, right axis shift, and decrease of the P, QRS, and T amplitudes. The changes are regressive at R12. The T wave in lead 1 of subject No. 108 is barely visible at S24, while it shows a slight negative dip in subject No. 30. The recovery of the T wave at R12 is quite marked, while the QRS amplitude recovers more slowly. We feel that the main changes of the electrocardiogram are more impressive as seen in the figures than any presentation in tables can be. While the electrocardiograms of all subjects (except one with a left axis shift bordering on left ventricular preponderance) were within normal limits before the experiment started, the majority (75 per cent) would have to be classified as abnormal because of low voltage of QRS complex and T waves at S24.

Caloric Level in the Rehabilitation Period — Minnesota Experiment

In the preceding analysis of our results the effect of rehabilitation as a whole was discussed without differentiation of the 4 caloric groups, Z, L, G, and T. As has been shown, the recovery of most electrocardiographic functions was far from complete at R12 — in fact, the recovery of many functions was only slight, although statistically significant. Therefore no clear-cut differentiation of the different caloric groups could be expected at that time, and later, after the twelfth week, the diet was no longer controlled. Still, a complete statistical analysis was made, and differentiation of caloric groups was found for the following electrocardiographic functions:

(1) Duration of mechanical systole. The decrease during recovery was much smaller in the lowest caloric group Z than in any other group, and these group differences were statistically highly significant. There was no significant group difference between any other two groups, but the difference for the combined groups G + T vs. L + Z was statistically significant at the 5 per cent level.

(2) K_{QT}. The increase in the highest group T (+0.0279) was significantly higher than that (+0.0128) of the lowest group Z, at the 5 per cent level. The increase in the medium groups was between these extremes, but the differentiation was statistically not significant.

(3) R wave, lead 1. The increase in group T (+1.68 mm.) was significantly higher than that for groups G (+0.85) and L (+0.79) at the 5 per cent level, while the difference between group T and group Z (increase +0.64) was highly significant (1 per cent level). The differences between groups G, L, and Z were statistically not significant. The difference between the combined groups G + T and L + Z was significant on the 5 per cent level.

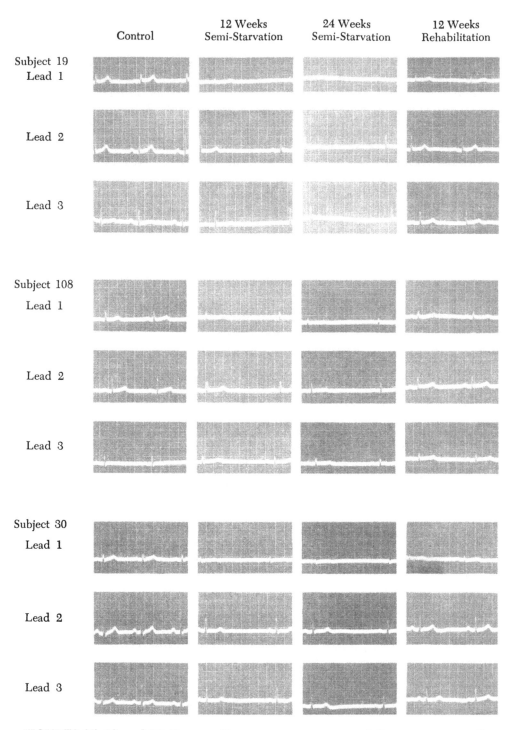

	Control	12 Weeks Semi-Starvation	24 Weeks Semi-Starvation	12 Weeks Rehabilitation
Subject 19 Lead 1				
Lead 2				
Lead 3				
Subject 108 Lead 1				
Lead 2				
Lead 3				
Subject 30 Lead 1				
Lead 2				
Lead 3				

FIGURES 97, 98, and 99. TYPICAL ELECTROCARDIOGRAMS OF 3 SUBJECTS DURING CONTROL, SEMI-STARVATION, AND REHABILITATION (Minnesota Experiment).

(4) R wave, lead 2. There was an increase of about the same magnitude in groups T, G, and L (+1.29, +1.15, and +1.14, respectively), while in group Z there was a slight decrease (−0.15). The change in group Z was significantly different from that observed in the other groups.

(5) The increase of ∑QRS was statistically significant only between the extreme groups Z and T.

(6) T wave, lead 1. The increase was statistically significant only between the extreme groups Z and T. The means of the T wave for the different caloric groups show a group differentiation (+0.46, +0.62, +0.86, and +1.02, respectively), but only the extreme differences (T − Z) were statistically significant at the 5 per cent level. The difference between G +T and L + Z was also significant at the 5 per cent level.

(7) ∑T and T wave, lead 2. The increase was greatest in group G (3.2 and 1.75, respectively) and smallest in group Z (1.60 and 0.78, respectively) with highly significant differences.

(8) T axis. The left axis shift of the T axis was greatest in group T and least pronounced in group G. This difference was significant at the 5 per cent level, but there were no significant group differences between groups G, L, and Z.

There were no significant differences in any electrocardiographic function in relation to high and low protein intake or high and low vitamin intake.

It would appear that, in spite of the rather incomplete recovery of the functions reflected in the electrocardiogram at the twelfth week of rehabilitation, there was a significant differentiation in regard to the caloric intake. The differentiation is apparent only in the extreme groups; either group Z (for instance, in mechanical systole duration) or group T (for instance, in R_1) was significantly different from the other 3 caloric groups, or the extreme groups (T vs. Z) were significantly different. Several functions showed no significant group differentiation — for instance, the heart rate and the QRS axis.

The differentiation between the caloric groups might have been more pronounced at a later time of rehabilitation, when the recovery of the electrocardiographic functions would have been more complete.

The Effect of Maximum Inspiration — Minnesota Experiment

One of the most pronounced circulatory changes during semi-starvation is the sinus bradycardia. While the slowing down of the heart rate may be regarded as an adaptation to the decreased metabolic rate, the underlying mechanism was not clear when we started our experiment.

We used the effect of maximal voluntary inspiration as a possible approach to the study of the pulse rate regulation. The effect of maximal inspiration was always studied in lead 2 and incorporated in the routine electrocardiographic procedure. The experiments were performed in 30 subjects for the control period C and in 32 subjects for periods S24 and R12. In the early phase of the maximal inspiration the acceleration of the heart rate, probably due to sympathetic stimulation, was observed in all subjects, while a retardation below the resting pulse rate in the late phase of maximal inspiration, obviously a vagus effect, was present only in 18 subjects. The analysis of the late effects of maximal inspiration was

TABLE 305

RESPONSE TO MAXIMUM INSPIRATION (mean values, differences, and statistical significance) (Minnesota Experiment).

Function and Number of Subjects	C	S24	R12	S24 − C	R12 − S24	Phase of Maximum Inspiration
Heart rate						
30..........	21.0	14.5	16.8	−6.5[**]		early
32..........		14.5	16.5		2.0	early
18..........	−7.1	−1.9	−3.7	5.2[**]	−1.8[*]	late
Range of heart rate, difference in rest − maximum inspiration						
30..........	24.7	15.3	18.4	−9.4[**]		total duration
32..........		15.3	19.4		4.1	
K_{QT}						
30..........	73.5	63.8	55.8	−9.7	−8.0	early
31..........	−26.2	−14.5	−35.3	+11.7[*]	−20.8[**]	late
K_{syst}						
28..........	−26.9	−15.6	−32.2	+11.3[*]	−16.6[**]	late
Systole duration						
28..........	−1.53	−0.86	−1.25	+0.67[*]	−0.39	late

made only in those 18 subjects who showed a definite vagus effect. It is clear that no change of vagus stimulation during semi-starvation could be expected if such an effect was absent in the control period.

Table 305 shows the effect of the early and the late phase of maximal inspiration on the heart rate, K_{QT} and K_{syst}. The values show the changes produced by the maximal inspiration, as compared with the values at rest, which are given in Table 301. It can be seen that the initial acceleration as well as the late retardation effect is much less pronounced at S24 than in C, and this decrease was statistically highly significant. During 12 weeks of rehabilitation there was no significant change in the initial acceleration, while the retardation effect became more pronounced again, so that the difference of the response △R12 − S24 was significant at the 5 per cent level. However, the retardation response was still definitely less than it had been before starvation (C).

The range of heart rate under the influence of maximum inspiration was calculated as the difference between the shortest and the longest cycle and was compared with the variability range during rest (Table 301). Since both initial acceleration and late retardation decreased in semi-starvation, the range of the heart rate showed a corresponding decrease at S24. At R12 the range increase was greater again, but the recovery was far from complete. The change of the total range during maximum inspiration was somewhat more uniform than either initial acceleration or late retardation, although all were highly significant.

It is known that the adaptation of the QT interval to rapid changes of the cycle length is incomplete, so that pronounced changes of K_{QT} occur. However, this coefficient is still useful for an analysis of a possible change in the mechanism of QT cycle adaptation during maximum inspiration in the course of starvation and rehabilitation. For the effect of the late phase of respiration, the values of all subjects were used for the calculation, not only those of the 18 subjects who responded with retardation.

The increase of K_{QT} in the early phase of maximum inspiration, above the resting values of Table 301, indicating a lag of adaptation of the QT interval to the decrease of cycle length, did not change significantly during semi-starvation and rehabilitation. The decrease of K_{QT} and K_{syst} in the late phase of maximal inspiration, indicating a lag of adaptation to the increasing cycle length, was much smaller at S24. This trend was reversed again in rehabilitation, even overshooting the control values at R12. The changes were statistically significant. The mechanical systole could not be determined in the early phase of maximal inspiration because of the interference of the noise produced by the inspiration. The fact that the values at R12 overshot the control values, while the recovery of the heart rate was still far from complete at that time, suggests that there is a direct effect of semi-starvation on the response of QT and mechanical systole to maximum inspiration, independent to a certain extent of the response of the heart rate. This is borne out by analysis of the absolute systole duration. The response of the whole group was significantly less at S24, as shown in Table 301. The total group was divided according to the presence of a vagus retardation effect on the heart rate in the late phase of maximal inspiration. The mean values of the mechanical systole for 15 subjects with retardation were —1.53, —0.93, and —1.33 for the periods C, S24, and R12, respectively. The mean values for 13 subjects without retardation of the heart rate were —1.53, —0.73, and —1.14, respectively. There was no significant difference between the group with retardation and the group without retardation.

Occasional Observations — Minnesota Experiment

Some interesting occasional observations were made in a few subjects. In subject 102 (see Figure 100) the P wave became slightly negative during the late phase of maximal inspiration (indicated by the signal) with shortening of the PR interval after 4 weeks of semi-starvation. Later the P wave disappeared entirely at the end of maximal inspiration, and this phenomenon continued also during the 12 weeks of rehabilitation.

In subject 2 just the opposite phenomenon occurred. Beginning at S13 the sinus rhythm was replaced by nodal rhythm, but maximal respiration produced a transitory restoration of the sinus rhythm. In lead 3, taken after lead 2, the sinus rhythm was still present.

Also in subject 129 the normal sinus rhythm was replaced by a nodal rhythm as early as S5. In the early phase of maximal inspiration a small P wave with a short PR interval (0.09″) appeared in the first beat while in the second beat the P wave was of about normal (pre-starvation) amplitude with a PR interval of 0.19″. In the late phase of maximal inspiration the P wave amplitude and PR interval decreased again; the PR interval in the first beat was 0.16″, in the second beat 0.10″, indicating nodal rhythm again. Although the effect of maximum inspiration was similar in the two subjects, the sinus rhythm was restored for a much shorter time in subject 129 — in fact, it was restricted to the duration of

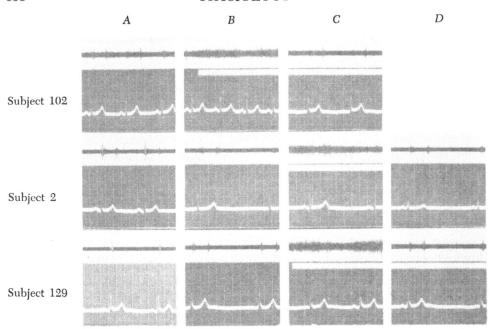

FIGURE 100. SUBJECT 102 (fifth week of starvation). *A*. Lead 2, before maximum inspiration. *B*. Lead 2, early phase of maximum inspiration; increased heart rate, no change of P wave. *C*. Lead 2, late phase of maximum inspiration; nodal rhythm. This phenomenon remained the same during semi-starvation and rehabilitation. SUBJECT 2. *A*. Control period, lead 2, at rest; sinus rhythm. *B*. Lead 2, thirteenth week of starvation, at rest before maximum inspiration; nodal rhythm. *C*. Same ECG, same lead, during late phase of maximum inspiration; a P wave has appeared, indicating re-establishment of sinus rhythm. *D*. Sinus rhythm is still present in the subsequently taken lead 3. SUBJECT 129. *A*. Control period, lead 2; normal sinus rhythm. *B*. Fifth week of starvation, lead 2, before maximum inspiration; nodal rhythm. *C*. In the early phase of maximum inspiration a P wave appears, present only during maximum inspiration. *D*. Late phase of maximum inspiration; the PR interval in the last beat is shortened to 0.1 sec.; in the subsequent beats no P wave is present. (Minnesota Experiment.)

the maximum inspiration. These were the only subjects in whom semi-starvation produced a nodal rhythm.

Relationship of the Minnesota Experiment to Previous Reports

Our results confirm the general observations of decreased QRS and T wave amplitudes in famine areas (Tur, 1944; Cardozo and Eggink, 1946), but these series have only the character of occasional observations in patients, while in our experiments quantitative relationships were obtained under controlled conditions.

Our results disagree with the reports of Cardozo and Eggink and Ellis (1946) that the QT interval is lengthened in relation to the cycle length. There was no significant change in K_{QT} in our series during semi-starvation; in fact, there was a tendency to decrease. The only explanation we are able to offer for the discrepancy is the possibility that these authors happened to investigate another phase of semi-starvation. We found a significant increase of K_{QT} at R12.

It follows from our investigations that cycle length and length of the QT interval or systole duration may change at a different rate during semi-starvation and rehabilitation, so that changes of K_{QT} and especially of K_{syst} occur. This would certainly depend on the rate with which semi-starvation proceeds (i.e., mainly on the caloric level). We attempted to reproduce the conditions prevailing in Western Europe during the German occupation, but this does not exclude the possibility of discrepancies between our starvation procedure and the starvation history in certain localized areas; in fact, it is almost certain that such discrepancies exist.

The same reservation must be made in regard to other electrocardiographic items. The observations of Tur (1944), Ellis (1946), and probably Cardozo and Eggink (1946) were made in a more advanced phase of malnutrition. It is possible that the Q amplitude, T amplitude, and heart rate increase again in some patients with progressive emaciation, although according to Tur the majority show essentially the same features as observed in our material. It is interesting that the tendency to low voltage of the QRS complex and the T wave appears during rehabilitation in Ellis' cases. Also, an ST depression might be a symptom of a more advanced phase of semi-starvation.

Interpretation of Electrocardiographic Changes — Minnesota Experiment

The decrease of the QRS and T wave amplitudes is probably the most remarkable change in the electrocardiogram in starvation. It may be due to the following factors:

(1) Decrease of heart size. The decrease of the amplitudes was accompanied by a very marked decrease of the heart size as determined by teleroentgenograms. These results are discussed in Chapter 10, but it may be mentioned here that there is good evidence from autopsy material that the heart weight decreases in semi-starvation at approximately the same rate as the loss of body weight. Since in the hypertrophic heart the QRS amplitude is as a rule increased, it seems logical to assume that it should be decreased in an atrophic heart. Yet this assumption should not be made without reservations, since secondary changes which are absent in the atrophic heart may be a contributory factor in the hypertrophic heart.

(2) Myocardial degeneration. There is evidence of widespread degenerative histological changes in the hearts of starved animals and in human autopsy material (cf. Chapter 10). Multiple degenerative changes in the myocardium might well be responsible for the tendency to low voltage.

(3) Rotation of the heart around a transversal axis. It is unlikely that such positional change occurs in semi-starvation.

(4) Decreased metabolic rate. Although there is no outright correlation between QRS amplitude and B.M.R., both QRS and T amplitudes are usually decreased in hypothyroidism and have a tendency to increase in hyperthyroidism. The average B.M.R. of the Minnesota subjects was —11.8 per cent below the Mayo Clinic "standard" normal average in the control period and decreased to 39.3 per cent below this standard at the end of starvation (S24). The decrease of the metabolic rate per unit weight of the "active" tissue (calculated from

body fluid and fat determinations) was considerably less, but even in these terms some decrease of the metabolic rate occurred in semi-starvation.

(5) Fluid accumulation in chest or pericardium. Although a considerable accumulation of body fluid associated with a tendency to edema was observed during semi-starvation, there was no clinical evidence for a major accumulation of fluid in the chest or pericardium; but of course the possibility cannot be excluded that minor amounts accumulated which might have changed the electrical conductivity of the surrounding tissue. We believe, however, that this factor has little, if any, importance. Although both QRS amplitude and T amplitude declined in semi-starvation, there was no correlation between these two changes in the time course of their development.

The right axis shift of the QRS amplitude was probably due to positional changes; a smaller heart tends to assume a more vertical position. There was also a right axis shift of the anatomical axis as measured in X-ray diagrams. We found, however, no correlation between the right axis shift of the anatomical axis and that of the QRS axis. This discrepancy might be due to the fact that the X-ray diagrams were obtained in the standing position, while the electrocardiograms were taken in the supine position. In the standing position the more vertical position of the heart in semi-starvation results mainly from the smaller heart size in relation to the comparatively fixed capacity of the thorax. A contributing factor might be an increased pull from the stomach, which was lower and more nearly vertical in semi-starvation than it was in the control period. One factor responsible for a more nearly vertical position of the heart in the supine position seems to be the decreased upward pressure of the abdominal organs due to a decreased volume in semi-starvation. Thus the lack of correlation between the right axis shift of the QRS axis and that of the anatomical axis might be explained by the different mechanism of the axis shift in horizontal and vertical positions.

It is also possible that the right axis shift of the QRS axis is due to a relative right ventricular preponderance or to a change in the pathway of excitation. A relative right ventricular preponderance could be produced by a greater degree of atrophy in the left ventricle. There are no data available to support or to contradict such an assumption, but some degree of difference in the development of right and left ventricular hypotrophy seems to be more probable than an exactly equal degree of hypotrophy. The degenerative changes mentioned above might be expected to produce changes in the pathway of excitation and uneven involvement of the right and left ventricle rather than a parallel involvement which would cancel the electrical changes. In this connection it is interesting that in Tur's material right axis shift was more common in the advanced cases than in the milder cases. One reason for the overshooting of the left axis shift in rehabilitation is probably mechanical: all men were fatter at R32 than they were at C; an increased volume of the abdominal organs could explain a position of the axis farther to the left.

While Benedict and his associates found that the slow heart rate in semi-starvation was referable to a sinus bradycardia, the Minnesota material shows that in exceptional cases nodal rhythm occurs. The absence of nodal rhythm in the Carnegie Institute material might be explained by the smaller number of subjects and the much milder degree of semi-starvation attained. Of interest in this respect was the presence in one of Ellis' (1946) patients of a pulsus tri-geminus which disappeared in the first week of rehabilitation.

The occurrence of nodal rhythm would be compatible with the assumption that the slow heart rate in semi-starvation is due to increased vagus tone. Schiff (1917) reported that atropine (0.5 to 0.75 mg.) promptly abolished the effects of starvation on the heart rate, but Schittenhelm and Schlecht (1919a) found no effect whatever with 1 mg. of atropine. The decreased retardation effect in the later phase of maximum inspiration supports the assumption of a vagus effect; if the slow heart rate is due to a maximum vagus tone, no further decrease during maximum inspiration can be expected. However, other signs of vagus stimulation (for instance, increase of the PR interval) were missing, and not only the late vagus effect but also the early sympathetic acceleration of the heart rate was decreased in maximum inspiration. Schittenhelm and Schlecht (1919a) found that 1 mg. of adrenalin (subcutaneous) had a subnormal effect on the heart rate in 11 patients; in only one person was a relative tachycardia produced and even in this one the rise was much less than that found in normal persons. Since both vagus effect and sympathetic response are diminished, the slow heart rate may be due to both increased vagus tone and loss of sympathetic tone. This would agree well with Hoesslin's (1919) observations of the absence of emotional pulse rate changes in semi-starvation and with our Minnesota findings of the decreased range of variability.

Jordan's (1895) observation that the initial digitalis retardation of the heart rate fails to appear in starving dogs is also an interesting corroboration of the Minnesota results. Statkewitsch's (1894) findings of extensive pathological changes in the cardiac ganglia of rabbits in advanced inanition might be regarded as histological evidence for the loss of vegetative regulation, which undoubtedly indicates a functional deterioration; this is only very slowly reversed in rehabilitation. On the other hand, the increase of the heart rate in moderate exercise or in adjustment to vertical position was not essentially changed in semi-starvation.

Significance of the Results Obtained in the Minnesota Experiment for Heart Pathology in Malnutrition

Except for beriberi heart, the condition of the heart in states of malnutrition has been given little attention in clinical medicine. In several recent textbooks of cardiology (Levine, 1940; Scherf and Boyd, 1939; Christian, 1940; Dressler, 1942; White, 1938; Harrison, 1939) no mention is made of the importance of the nutritional state for cardiac pathology. On the other hand, Vaquez (1924) recognized the theoretical importance of nutritional effects on the condition of the myocardium, but he was inclined to believe that these effects are not great. This opinion was based on the failure of several authors to find any considerable degree of atrophy of the heart in inanition. This is a consistent error which can be traced far back in the literature.

Although there is little correlation between electrocardiographic findings and the functional state of the heart, the occurrence of significant changes in most electrocardiographic items during semi-starvation cannot be ignored. It seems safe to conclude that they indicate myocardial changes. The clinical significance of such changes would depend to a large degree on individual compensations. The same is true for the functional changes (loss of vegetative regulation of the heart rate; change of systole duration related to heart rate) observed in our Minnesota series. The implications of our results would be that prolonged semi-starvation produces a deterioration of the state of the myocardium, which might be functionally compensated for a time. However, the compensation might break down under conditions of additional stress. In our Laboratory such stress conditions were excluded, but they might well arise under less well protected conditions. It is interesting to note that one of our subjects suddenly became decompensated in the early rehabilitation period; the heart became enlarged and the venous pressure rose abruptly. Treatment with bed rest, reduced food intake, and diuretics restored the patient within one week. The decompensation may have appeared in the rehabilitation period because of an increased circulatory load due to sharply increased food ingestion and greater activity.

General Picture of Electrocardiographic Changes in the Minnesota Experiment

During semi-starvation statistically highly significant changes occurred in most electrocardiographic items, and the electrocardiogram became clinically abnormal in the majority of subjects.

Together with a pronounced slowing of the heart rate, its variability range decreased both relatively and absolutely, so that the heart rate was more regular in semi-starvation. These changes reached their maximum at the twelfth week of semi-starvation and recovered slowly during rehabilitation. The QT interval and mechanical systole duration increased during semi-starvation and shortened again during rehabilitation, but these changes lagged behind the simultaneous changes of the cycle length in both directions, so that K_{QT} and K_{syst} changed accordingly. These changes were more pronounced in the duration of mechanical systole.

The amplitudes of all deflections (P wave, QRS complex, T wave) decreased continuously and very considerably during semi-starvation and recovered slowly during rehabilitation.

During semi-starvation there was a marked right axis shift of the QRS axis and even more so of the T axis, so that the angle between the two axes was diminished. During rehabilitation both QRS axis and T axis moved to the left, overshooting the original pre-starvation position at the thirty-second week of rehabilitation.

Recovery was only partial in most electrocardiographic items at the end of 12 weeks of rehabilitation but they were back to the control values within 20 weeks, with several functions (heart rate, R_1, ΣT, QRS axis, T axis) overshooting the control values subsequently.

There was a discrepancy in the time course of changes between interval and amplitude changes and between QRS complex and T wave changes. There was no correlation between QRS axis changes and anatomical axis changes or between QRS axis and T axis, although all changed in the same direction.

A statistically significant differentiation of the groups receiving different caloric intakes during rehabilitation was obtained in the following items: systole duration, K_{QT}, R_1, R_2, ΣQRS, T_1, T_2, ΣT, T axis. Before semi-starvation voluntary maximal inspiration produced an initial acceleration of the heart rate in all subjects, followed by a late retardation in 18 subjects. During semi-starvation the effect of maximal inspiration was diminished in both initial acceleration and late retardation. During 12 weeks of rehabilitation only the late retardation was restored. The decrease of K_{QT} and K_{syst} in the late phase of maximal inspiration was significantly less pronounced at the end of semi-starvation, and this effect was to a certain degree independent of the changes in the heart rate.

While the slow heart rate in semi-starvation was as a rule due to sinus bradycardia, in 2 subjects nodal rhythm was observed, which was temporarily restored to sinus rhythm during the maximum voluntary inspiration.

CHAPTER 31

Renal Function

THE majority of investigators have agreed that famine edema is not caused by renal failure, and it is generally implied that the kidneys are substantially "normal" in semi-starvation. The bases for such conclusions are certain points of observation about which there is no disagreement: (1) semi-starved persons produce urine that is generally free from protein, casts, cells, pus, and sugar, and (2) the blood of these persons does not exhibit characteristics which tend to appear in renal disease — that is, the levels of blood urea and other materials which tend to accumulate in renal failure are not elevated.

Such negative evidence, however, neither proves the full normality of the kidneys nor provides a picture of the actual renal function in semi-starvation. It has been noted in the discussion on morphology that the kidneys are not immune from alteration in semi-starvation; there are, in fact, complaints from famine victims which refer to the kidneys, though these organs themselves may not be primarily responsible for the practically universal symptoms of polyuria and nocturia. It must be noted also that the kidneys have a somewhat altered task to perform in the semi-starved person as compared with that in the well-fed person. Owing to the low protein intake of the famine victim, there is far less nitrogen to be disposed of, and the prevailing hypoglycemia does not tax the capacity of the organ to conserve sugar. On the other hand, the starving man tends to consume far more water and, if available, more salt ($NaCl$), and these pose their problems for the kidney.

Polyuria and Nocturia

Polyuria and nocturia have been reported by practically every observer of starvation conditions. They were described in detail in World War I victims by Schittenhelm and Schlecht (1918) and by Jansen (1920). After World War II the Netherlands Red Cross Feeding Team (1948) reported that polyuria and nocturia were as troublesome in the East Indies as they were in the famine period in western Holland. The presence of increased urine volumes in the semi-starved state were recorded by observers in the Warsaw Ghetto in 1942 (in adults by Fliederbaum et al., 1946, and in children by Braude-Heller et al., 1946), by observers of the inmates of concentration camps (Lamy et al., 1948), and in civil prisoners at Brussels in 1940–42 (Simonart, 1948). The usual figure for the 24-hour urine volume of the semi-starved individual lies between 2 and 3 liters, with frequent volumes of 4 to 5 liters and occasional volumes of 8 to 10 liters. Simonart (1948) has presented the urine volumes of 3 civil prisoners for

10 days. The data, which are presented in Table 306, illustrate the distribution of urine volume between night and day. In some of the severely starved patients reported by Simonart there was some incontinence both day and night, though it was far more frequent at night. The same author thought that the increased frequency and incontinence in his starved patients was due to weakness of the bladder musculature. However, no data to substantiate this claim were presented.

TABLE 306

Mean Urinary Output, in liters, recorded for 10 successive days, in 3 men with severe semi-starvation in a civilian prison in Brussels (Simonart, 1948).

Subject	Total Urine Volume			Urine Volume, Day			Urine Volume, Night		
	M	Min.	Max.	M	Min.	Max.	M	Min.	Max.
1......	6.8	5.5	8.0	2.75	1.0	3.5	4.0	3.0	5.0
2......	7.2	5.1	8.5	3.37	2.6	4.0	3.8	2.1	5.5
3......	5.8	4.5	7.7	2.12	1.0	3.3	3.7	3.0	4.5

No explanation of why polyuria occurs in the semi-starved individual is forthcoming. Various possibilities suggest themselves — such as increased thirst, which might be brought on by fluid shifts in the central nervous system, and hormonal imbalance, especially of the antidiuretic hormone of the pituitary. But there is no evidence on these or other physiological mechanisms that might be postulated. However, one should always keep in mind that the semi-starved subject delights in "souping" his food with water and in just drinking water in an attempt to satisfy the desire for the sensation of a full stomach. "Souping" of food was marked in the Minnesota Experiment and has been reported in prisoners of war (Butler et al., 1945). In this connection, Weech (1936) did not observe any polyuria in the victims of starvation in China and commented that the presence of polyuria in European areas may have been due to the large volumes of fluid which were usually ingested with the wartime diet.

The Composition of the Urine

The urine of the semi-starved person is classically a pale fluid of low specific gravity containing little or no pigment and very little sediment. All investigators of famine victims seen after both world wars have stated that the urine of the semi-starved man is free of pus, red cells, casts, and albumin. This finding is so universal that nutritional authorities such as Sinclair (1947) will not classify as famine edema any edema that occurs in the presence of albuminuria.

The only investigators of World War II famine victims who found exceptions to this rule were Lamy, Lamotte, and Lamotte-Barrillon (1948). Of the 26 inmates of a German concentration camp who came under the care of these observers, 7 had albumin in the urine and 2 of these 7 men were losing between a fourth and a half gram of protein nitrogen per day. Of the remaining 19 men, 13 had traces of albumin in the urine and only 6 were completely negative. In addition to these findings, 9 of these patients had hematuria (one case severe), and granulated casts were found in the urine of 7 men. These findings, together with the abnormalities observed in the microscopic sections of the kidneys of those members of this group who came to post-mortem (see below), make it clear that the signs and symptoms of the group studied by Lamy and his colleagues (1948) differed from the usual picture of renal function in pure semi-starvation.

The protein intake of most starvation victims is considerably reduced from the normal range of 8 to 16 gm. of nitrogen a day. In famine the intake may be as low as 3 and as high as 12 gm. of protein nitrogen a day, with most of the diets falling in the range of 6 to 10 gm. a day.

The protein nitrogen intake sets the general range of urinary nitrogen excretion and, together with the increased urine volume, makes it evident that the concentration of urea in urine is reduced from the normal.

The semi-starved individual invariably has a negative nitrogen balance which is usually quite small (Beattie and Herbert, 1948; see also Chapter 19). However, Fliederbaum et al. (1946) studied 20 cases of uncomplicated starvation on a semi-starvation diet that was typical of the diets found in the Warsaw Ghetto. This diet contained 3 gm. of nitrogen a day. Urinary nitrogen generally averaged 2 and 3 gm. a day, but in a few cases of extreme cachexia this figure rose to 4 or even to 6 gm. a day. Western European famine diets often contained as much as 6 to 9 gm. of nitrogen a day, and the urinary excretion might be expected to be of a comparable order of magnitude.

Fliederbaum et al. (1946) also reported an elevated uric acid excretion in the presence of what appears to be a virtually purine-free diet. The uric acid excretion under these conditions was roughly one third of normal (i.e., 40 mg. per day). But the urinary NH_3 was increased. The amino acid nitrogen in the urine increased to 100 mg. per day and occasionally to 200 to 300 mg. per day. Creatinine was greatly decreased; instead of 600 mg. per day as in a normal individual, the 17 starved individuals who were studied excreted only 300 to 400 mg. In this connection Gsell (1948) studied the creatine excretion in 4 men during rehabilitation and found a creatinuria. The concentration of the creatine in the urine reached maximal values of 111, 127, 203, and 573 mg. for every 100 cc. of urine at a time when the urine volume for 24 hours was roughly normal — for example, the urine volume for one man was stated to be 1500 cc. in 24 hours.

The NaCl intake of the semi-starved individual is extremely variable, and when easily available, salt is taken in large quantities. During periods of receding edema, large quantities of salt may be lost regardless of the intake. Denz (1947) found that as much as 28 gm. of NaCl may be lost in 24 hours during reduction of edema in the course of rehabilitation of famine victims. In rehabilitation it was noted that the kidney excreted a chloride in a concentration greater than that of plasma. The ratio of urine chloride concentration to plasma chloride concentration in these starvation victims on a rehabilitation regimen over a 2-week period ranged from 0.4 to 1.8.

The calcium content of the urine is usually low during semi-starvation and early rehabilitation. The Netherlands Red Cross Feeding Team (1948) confirmed this observation, but its members were much puzzled by a large increase in renal calculi. A number of stones removed at operation were examined chemically and found to consist almost exclusively of oxalate. The investigators looked for abnormal parathyroid function but were unable to find it. It is to be regretted that they did not investigate the composition of the diet their semi-starved individuals had been living on. It is well known that the semi-starved individual will eat bizarre foods, including weeds, and it is quite possible that these people had been consuming a weed that contained large quantities of oxalic acid.

Renal Function Tests on Animals

The only complete study of renal function under semi-starvation conditions that has come to our attention has been done on the rat. Dicker, Heller, and Hewer (1946) placed rats on turnip, carrot, and low casein diets that were markedly deficient in protein and calories. After 30 to 40 days on this diet the rats showed a 35 per cent loss of weight and a 40 per cent depression of the serum protein concentration, along with edema and ascites in roughly half of the animals. Although it is recognized that the study of the effects of low protein diets in animals may lead to erroneous conclusions about semi-starvation in man, these results are reviewed in some detail since they provide at present the best available evidence on the subject.

Dicker, Heller, and Hewer applied a number of renal function tests at the end of 36 days on the restricted diet. Administration of 5 per cent of the animal's weight in water by stomach tube produced a diuresis during a 3-hour observation period that was only about half that which occurred during the control period. This observation was thought to confirm the widespread opinion that the semi-starved animal has a tendency to water retention. Dilution tests demonstrated a small but significant loss of the ability to produce dilute urine, while concentration tests showed a large loss of the ability to produce a concentrated urine; the maximum specific gravity dropped from 1.070 in the control period to 1.041 after 36 days on the restricted diet. However, this observation does not prove dietary damage to the kidney, since the authors presented evidence to show that on a water restricted regimen edema fluid may be mobilized and excreted by the kidney, thus preventing the formation of a highly concentrated urine.

The glomerular filtration rate was determined by the use of inulin. It was shown that in the semi-starved state the glomerular filtration rate increased as the rate of urine flow became larger. This finding is in striking contrast to conditions in the normal rat, in which the glomerular filtration rate is independent of the rate of urine formation. In these experiments the glomerular filtration rate of the semi-starved rat was within the normal range at low rates of urine formation, but when the urine flow was increased 2½ times, the glomerular filtration rate increased in roughly the same proportion. The fraction of the water of the glomerular filtrate reabsorbed by the tubules was within the normal range and remained constant at all rates of urine formation.

These facts demonstrate that the kidney of the semi-starved rat employs a different method for producing diuresis than is used under normal conditions. Diuresis in the normal rat is accomplished by a reduction of the fraction of the glomerular filtrate which is reabsorbed by the tubule; at high rates of urine formation this reduction may be as much as 30 per cent. Furthermore, the increase in urine flow is brought about without any increase in the volume of fluid formed at the glomerular membrane. On the other hand, the kidney of the semi-starved rat apparently produces a diuresis by increasing the volume of the glomerular ultrafiltrate and does not make use of an alteration in tubular reabsorption.

The determination of plasma and urine chlorides allowed Dicker and his collaborators to calculate the rate of chloride reabsorption. They found that chloride reabsorption by the tubules was reduced by about 5 per cent after 36 days of the restricted diet.

Renal plasma flow was studied in these same semi-starved rats by use of diodone clearances, and it was found that, like the glomerular filtration rate, the renal plasma flow was positively correlated with the rate of urine formation. This suggests that mechanisms for increasing the glomerular filtration rate involve increasing the blood flow through the glomerulus. In any event, the renal plasma flow in the normal rat is independent of the rate of urine formation.

Diodone was also used to determine the total tubular excretory mass before and after semi-starvation. It was noted that the tubular excretory mass (Tm) of the semi-starved rats was roughly 30 per cent of the control value. This finding might mean a large reduction in the number of functioning tubules, or a loss of excretory power by the tubules, or both.

Confidence in these findings is increased because the authors used 3 different low calorie, low protein diets, all of which gave the same general picture (Heller and Dicker, 1947). They did show that the picture is modified with regard to the Tm values by the presence of large quantities of the vitamin A precursor found in carrots.

Renal Function Tests on Man

There are no careful and complete studies of renal function in semi-starved men. Observations after World War I were limited to simple microscopic examination of the urine and the qualitative tests for albumin, sugar, and ketone bodies. Only a few of the numerous studies published after World War II contained data on kidney function, and here the emphasis was placed on tests of concentration and dilution, with the more complicated tests being performed on single individuals or very small groups.

The majority of observers have reported a normal or slightly lowered level of non-protein nitrogen (N.P.N.) in the blood plasma (Gsell, 1948; Lamy et al., 1948; Mollison, 1946; Youmans, 1936). However, dehydration secondary to diarrhea will often produce an elevated blood N.P.N. In this connection Mollison (1946) reported 5 dehydrated adults whose serum urea values were between 63 and 180 mg. per 100 cc.

There is general agreement that the semi-starved individual can produce as dilute a urine as the normal individual. Thus Fliederbaum et al. (1946) found

TABLE 307

RESULTS OF DILUTION AND CONCENTRATION TESTS (Volhard and Fahr) on the group of starved individuals examined by Lamy, Lamotte, and Lamotte-Barrillon (1948). The means and standard deviations (of the sample) of the urine volumes are expressed as cc.

	Urine Volume Day before Test (24 hr.)	Dilution			Concentration	
		Urine Volume				
		4 Hrs. after Ingestion of H_2O	12 Hrs. after Ingestion of H_2O	Minimum Specific Gravity	Urine Volume after 8 Hrs. of H_2O Restriction	Maximum Specific Gravity
Number of subjects	22	24	23	24	24	25
M	1242	593	967	1.004	352	1.018
SD	657	406	598	.005	233	0.009

specific gravities of 1.003 to 1.001 after administration of water to severely starved individuals. Lamy, Lamotte, and Lamotte-Barrillon (1948) reported on 24 semi-starved subjects and found that 19 out of the 24 were able to produce urine whose specific gravity ranged between 1.000 and 1.005 inclusive. A few subjects showed urine specific gravities between 1.010 and 1.020. The data are summarized in Table 307. Braude-Heller, Rotbalsam, and Elbinger (1946) made similar observations on severely starved children in the Warsaw Ghetto. These authors noted that when 500 to 800 cc. were administered, the water eliminated in most cases exceeded the intake. However, the rate of excretion was never very high, and the authors felt this was due to delayed absorption and passage through tissues.

All three of the reports mentioned above are in agreement that the semi-starved individual will not produce a concentrated urine when placed on a wa-ter-restricted regimen. Of Lamy's 25 men only 5 were able to concentrate their urine to a specific gravity of 1.025 or more. This failure to observe a concen-trated urine in the urine concentration tests appears to be due to extrarenal factors such as the mobilization of edema fluid. The same comment can be ap-plied to the few people who were unable to produce a very dilute urine in the urine dilution tests. Here retention of water by the body and the presence of de-hydration due to diarrhea are the import extrarenal factors that would interfere with the conditions necessary to obtain a valid test result.

Creatinine clearances were carried out by Lamy, Lamotte, and Lamotte-Barrillon (1948) on 6 of their subjects who had been rescued from a concentra-tion camp. These tests were carried out by administering 2 gm. of creatinine in 500 cc. of water and collecting urine and blood samples at appropriate times thereafter. The results, which are presented in Table 308, show that when a good diuresis resulted the creatinine clearances were normal, but that when the rate of flow of urine fell below 1 cc. per minute, the values were abnormally low. On the basis of these data, which were collected on individuals who had a renal pathology not typical of semi-starvation (see below), one cannot decide whether the glomerular filtration rate in man is abnormal in semi-starvation.

TABLE 308

THE GLOMERULAR FILTRATION RATE AS DETERMINED BY THE CREATIN-
INE CLEARANCE after the administration of 2 gm. of creatinine in 500 cc.
of water to 6 starved individuals (Lamy et al., 1948).

Patient	Serum Creatinine (mg. %)	Urine Creatinine (mg. %)	Urine Flow (cc./min.)	Glomerular Filtrate (cc./min.)
Ge	27.0	1900	2.75	192
Ra	42.0	1750	2.30	94
La	35.0	1000	0.40	11
Ba	35.0	1780	0.83	42
Geo	22.5	3050	0.32	43
Sch	29.5	1650	3.25	181

TABLE 309

RENAL PLASMA CLEARANCES OF INULIN AND DIODONE AND THE TOTAL TUBULAR
EXCRETORY MASS (diodone) in cases of severe starvation in females selected from the
inmates of the Belsen concentration camp. Clearances are expressed as cc. per minute,
and the total tubular excretory mass as mg. per minute. Plasma clearances and TmD
values were corrected (Corr.) to 1.73 square meters of body surface. (Data from
Mollison, 1946; data for normal females from Smith, 1943, p. 97.)

Subjects	Observed Weight (kg.)	Previous Weight (kg.)	Plasma Clearances				TmD	
			Inulin		Diodone			
			Obs.	Corr.	Obs.	Corr.	Obs.	Corr.
Normal females				117		594		42.6
F124*	33.2	48.0	88	124	241	340	32	45.3
F114*	35.1	56.0	102	141	501	710		
F110†	59.3	80.0	66	70	182	194		
F122†	43.9	80.0	43	53	228	283	17.7	22.0

* Cases of simple emaciation without edema.
† Cases of generalized edema.

Mollison (1946) used inulin and diodone to study the details of kidney
function in 4 cases of starvation in females seen among the inmates of the Belsen
concentration camp. The results are presented in Table 309 along with the values for normal females as presented by Smith (1943). The patients without edema showed normal values for inulin clearance. It is interesting that both of these
patients had low blood pressures during the actual test, the lowest being 88/58
mm. Hg. On the other hand, the inulin clearances were definitely reduced in 2
patients who had generalized edema and low serum proteins. The renal plasma
flow was reduced in one of the non-edematous patients and in both of the cases
of generalized edema. The total tubular excretory mass was determined with diodone in 2 patients, and in the one with generalized edema this value was found
to be abnormally low.

Obviously, it is not possible as yet to draw any final conclusions about the
details of the renal function of man in semi-starvation. However, if one takes into consideration the results on rats and the available evidence in man, it appears

probable that kidney function in the semi-starved man is not normal in all respects and that an intensive investigation of the details of renal function in semi-
starvation would yield profitable results.

Polyuria and the Work of the Kidney — Minnesota Experiment

Polyuria developed early in the subjects of the Minnesota Experiment and
was very resistant to dietary therapy. Urine volumes were measured for 3 consecutive days for each observation period, and the average value for the 3 days
was recorded. The means of each major dietary group for the observation periods in semi-starvation and recovery are presented in Table 310. The most
marked polyuria occurred after 6 weeks of recovery. Urine volumes were large
at the end of both 24 weeks of semi-starvation and 6 weeks of recovery. At these
times urine volumes between 3 and 4 liters for 24 hours were common, and there
were occasional volumes as high as 5 liters. At the end of 12 weeks of refeeding
the polyuria was almost as marked as it was at the end of the starvation period.
It is surprising that the group receiving the smallest number of calories was the
only group that showed a definitely reduced urine volume. There is no obvious
explanation for this.

TABLE 310

MEAN URINE VOLUMES in the Minnesota Experiment, in cc. per 24 hours.

Group	Control	S12	S24	R6	R12
Z	1334	1626	2599	2295	1942
L	1275	1986	2311	2709	2128
G	1159	1645	2316	2374	2193
T	1283	1776	1916	2395	2162
M	1263	1759	2285	2443	2106
SD	249	082	802	871	498

No specific renal function tests were carried out on the subjects of the Minnesota Experiment owing to lack of time and personnel. However, there are
enough data on hand to permit an estimate of the effects of semi-starvation on
the work of the kidney.

The useful work done by the kidney can be calculated. Various forms of the
basic equation have been used by such investigators as von Rhorer (1905), Borsook and Winegarden (1931a, 1931b), and Newburgh (1943). We have chosen
the form of the osmotic work equation as presented by Newburgh. A theoretical
justification of this equation has been given by R. G. Dickinson and R. C. Tol-
man in the paper of Borsook and Winegarden (1931a). The form of the equation for osmotic work for a single urinary constituent as employed here is this:

$$nE_i = nRT \left(y_i \ln \frac{y_i}{c_i} - y_i \ln + n\, C_i - y_i \right)$$

where E = work done; n = number of liters of urine; y_i = total osmols of the
i^{th} component excreted; C_i = molar concentration of ith component in blood;
R = the gas constant; T = the absolute temperature.

It is clear that the work done by the kidney is the sum of the energies necessary for the excretion of the individual components. The data available for the starvation situation do not provide for a complete estimate of the total work of the kidney. Since the excretion of urea accounts for roughly 75 to 80 per cent of the work done by the kidney in the normal state, a good deal can be learned from consideration of the work involved in the excretion of urea.

TABLE 311

MEAN URINARY NITROGEN EXCRETION OF 36 MEN during 6 months of semi-starvation and 3 months of recovery, expressed as gm. per day, moles of urea per day, and moles of urea per liter of urine (Minnesota Experiment).

	C	S12	S24	R6	R12
Gm. per day	13.17	8.12	7.42	9.5	14.34
Moles urea per day...	0.47	0.29	0.26	0.34	0.51
Moles urea per liter..	0.37	0.16	0.11	0.14	0.24

The data from which this calculation has been made are presented in Table 311, and the energy divided by RT necessary to excrete the urea produced over a 24-hour period (E_{24}/RT) is presented in Figure 101. It will be noted that the energy necessary for the excretion of urea was reduced at S24 by 64 per cent. This quite possibly represents an overestimate of the reduction in renal work since the increased urine volume results in an increase in osmotic work for those substances whose molar concentration in urine is normally less than that of plasma. The most important variable other than urea is NaCl, but as long as the NaCl excretion remains between 10 and 30 gm. a day, the total renal osmotic work is scarcely influenced. Excretions of 60 gm. a day in the presence of large urine volumes of 2 to 3 liters will increase the osmotic work by roughly 10 per cent of the work of the kidney under normal conditions. On the other hand, a marked restriction of NaCl in the diet, resulting in the practical disappearance of NaCl from the urine, would result, in the presence of semi-starvation polyuria, in an increase of renal work equal to one third of the total work of the kidney under normal conditions.

It has been pointed out by Newburgh (1943) that the minimum renal work under a given set of conditions usually occurs at a urine volume of 2 to 3 liters. If one assumes the "normal" urine volume to be 1500 cc., the advantage thus gained is a reduction of renal work by 10 per cent. However, as the urine volume increases beyond the minimum figure the renal work rises, and at a volume of 5 liters all advantage has been lost.

The following generalizations may be made about renal osmotic work in semi-starvation: (1) a moderate polyuria may be regarded as a useful adaptive mechanism since it reduces renal osmotic work; (2) the work of the kidney will be most influenced by the protein intake of the semi-starvation diet and will be reduced roughly in proportion to this reduction; and (3) the advantage in renal work produced by the two preceding generalizations can be offset by a marked restriction of NaCl.

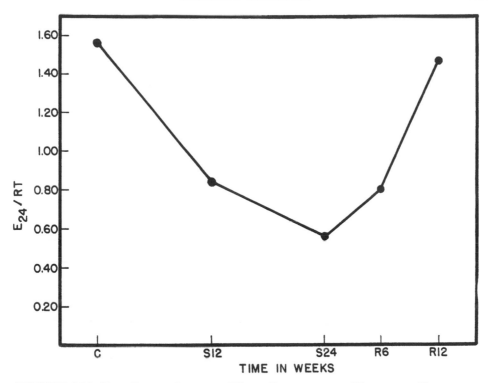

FIGURE 101. THE RENAL OSMOTIC WORK DONE BY THE KIDNEY IN EXCRETING UREA DURING 6 MONTHS OF SEMI-STARVATION AND 3 MONTHS OF RECOVERY. The work is expressed in kilocalories per 24 hours (E_{24}) divided by the gas constant (R) and the absolute temperature (T). The equation by which these figures were obtained is given in the text. (Minnesota Experiment.)

The Microscopic Anatomy of the Kidney in Semi-Starvation

It is generally agreed that semi-starvation does not result in important changes in the microscopic anatomy of the kidney. Post-mortem examinations of victims of World War II (Gsell, 1948; Uehlinger, 1948; Stein and Fenigstein, 1946), as well as of victims of World War I (Lubarsch, 1921; Oberndorffer, 1918), have all demonstrated that the changes observed could be accounted for by age and stasis albuminuria. The commonest finding has been small amounts of serous fluid in Bowman's capsule and some enlargement of the lumen of the tubules.

The larger series of examinations usually contained a case or two of acute glomerulonephritis which was secondary to generalized pyoderma, tonsillitis, or erysipelas (Gsell, 1948). Uehlinger (1948) found occasional cases of interstitial nephritis which were related to nonspecific pyelitis.

The report of Lamy, Lamotte, and Lamotte-Barrillon (1948) is the principal exception to the general finding that the kidneys in semi-starvation are essentially normal when examined at autopsy. These investigators found that the weights of single kidneys varied from 150 to 240 gm. in contrast to the usual finding of a subnormal kidney weight (i.e., less than 140 gm.) (see Chapter 9).

They found a marked vasodilatation in the glomeruli with loss of albumin and red cells into the capsular space, and in at least one victim some proliferation of the cells of the glomeruli. Nine of the 11 studied showed changes in the tubules which included flattening of the epithelial cells and desquamation of the superficial cytoplasmic layer of the tubular epithelium. Uehlinger (1948) has pointed out that the report of Lamy and his co-workers is unusual and that the Swiss group was unable to find any confirmation in their material.

Renal Disease in Semi-Starvation

During World War I there was some confusion of famine edema with trench or war nephritis. This latter disease was characterized by edema, dyspnea, increased blood pressure, and the presence of casts, albumin, and blood in the urine (Keith and Thomson, 1919; Schultz, 1918). However, war nephritis occurred almost exclusively in the soldiers who were fighting in the trenches and was soon differentiated from famine edema. It is now generally recognized that war nephritis was simply acute glomerulonephritis (Fishberg, 1934). In the cases studied after World War II, acute glomerulonephritis was readily recognized, in both the European and Far Eastern material. Gsell (1948) identified a number of cases which were usually preceded by an acute infection and stated that the differentiation of famine edema from war nephritis was made on the appearance of an elevated N.P.N., an elevated blood pressure, and edema of the eyelids. Chakrabarty (1944) studied 12 cases of severe starvation in Calcutta and decided that two of them had a nephritic picture.

The subjects studied by Lamy, Lamotte, and Lamotte-Barrillon (1948) were definitely suffering from some disease of the kidney. Red cells, albumin, and casts were found in the urine of a high percentage of the patients in this series (see above). In addition, these men had edema, there was a slightly increased body temperature in 17 out of 28 cases, and all 28 had a tachycardia. Two men had an elevated blood pressure strongly suggesting glomerulonephritis. The authors felt that the microscopic anatomy of the 11 men examined at autopsy was most consistent with crush syndrome. However, Uehlinger (1948) has pointed out that the crush syndrome results in damage to the loop of Henle and that Lamy's data demonstrate damage to the proximal portion of the tubule.

It is clear that the microscopic picture of the kidneys examined by Lamy, Lamotte, and Lamotte-Barrillon (1948) is similar to that seen after some toxic substance, such as a mercuric salt, has been ingested. It is well known that the semi-starved individual will eat or drink many unorthodox foods and liquids, and it is at least possible that some nephrotoxic substance had been ingested by a large number of the subjects studied by Lamy and his colleagues.

CHAPTER 32

Special Senses

It is a part of the folk wisdom that "hunger sharpens the senses." The psychological theory behind this statement is not very articulate, and there is not a sharp distinction between specific sensory functions and the "general sensations." At times the beneficial influence of hunger is limited to the appreciation of food, to the sharpening of "appetite." At other times it refers to a greater awareness of food on the part of the hungry man. This is probably what Thomas Fuller had in mind when he said in *Gnomologia* that "A hungry man smells meat afar off." In any case, the folklore tradition lends a certain appeal to the study of special senses in semi-starvation.

The special senses provide the organism with information about its environment, and their adequate functioning is essential for effective interaction between the changing organism and changing environmental conditions. Consequently, receptivity to sensory stimuli as determinants of behavior is of interest from the point of view of the over-all fitness of the organism. In addition, the measurement of some of the sensory functions provides a physiological avenue of approach to the functional status of the central nervous system.

Malfunction of the sensory processes may result from lesions in the receptors, in the nervous pathways conducting the impulses, and in the cortical centers. In view of the high resistance of nervous tissues to simple caloric starvation it may be expected that the conducting and central parts will not be readily affected. On the other hand, neurological lesions, including atrophy of the optic nerve, were observed in the Far East when severe vitamin deficiencies were superimposed on semi-starvation. Field reports from areas of undernutrition or famine in Europe contain little reference to disturbances of the special senses, except for night blindness. Night blindness is to be regarded largely as a peripheral phenomenon, due to the changes in the light sensitivity of the receptors.

Among 300 ex-prisoners of war who suffered from semi-starvation and had been kept at heavy labor in Germany there were no complaints of impairment of hearing and optic neuritis was absent (Edge, 1945). Leyton (1946) reported no deterioration of sight, except for a few cases of mild night blindness, in prisoners of war who were captured by the Germans and were maintained on an inadequate diet resulting in severe emaciation. Hearing was said to be normal or acute in Leyton's experience; a similar observation was made by D. A. Smith (1947). Touch and the sensation of temperature appeared normal.

Vision in the prolonged and severe semi-starvation of 1941–42 in the Warsaw Ghetto was studied by Fajgenblat (1946). The group examined consisted of 20 women. There were no complaints of impairment of visual functions, including night vision. No deterioration was found on examining visual acuity and visual fields, studied both in daylight and under low illumination, with white and colored (blue, red, yellow, and green) stimuli. The course of dark adaptation was not studied. Such signs of vitamin A deficiency as keratomalacia and Bitot's spots were absent. The two positive ophthalmological findings were the onset of opacity of the crystalline lens, analogous to the changes present in old age, and a lowered intraocular pressure which decreased from the normal level of 20–28 mm. Hg. to 12 mm. Hg. Fliederbaum *et al.* (1946) noted no complaints concerning hearing or the sense of equilibrium. Lamy, Lamotte, and Lamotte-Barrillon (1948, p. 112) found a normal visual acuity in their severely cachectic patients, and a systematic examination of the auditory apparatus revealed normal function.

The blindness and deafness of hysterical origin which were occasionally found in the German concentration camps and which can be interpreted as an exaggerated defense mechanism for "shutting off" the unbearable reality (Niremberski, 1946) clearly belong in a different category; they had only an indirect relationship to nutrition.

Visual Disturbances — Far Eastern Area

Whereas in the European area visual disturbances resulting from semi-starvation during World War II were minimal, the reports from the Far East present a strikingly different picture. The problem is treated in part in connection with the discussion of neurological symptoms in Chapter 33. A few illustrative reports will be included here.

Dimness of vision (amblyopia) was reported by Price (1946) in a group of Allied prisoners of war and internees captured at Singapore who were seen after 3½ years of Japanese imprisonment when they were admitted as patients to a British general hospital in southern India. The onset of the symptoms was dated 6 to 9 months after their capture. Visual acuity deteriorated until the men were able to read only newspaper headlines or not at all. Some remarked that they could see better when the object was at the side than when it was in front. In a few cases deafness was found, attributed to the degeneration of the auditory branch of the eighth cranial nerve.

Intraocular optic neuritis, with gradual diminution of vision, was observed in Americans who surrendered at Bataan and were interned by the Japanese (Hibbs, 1946). The frequency of the symptom was not given. The diet was low in calories and deficient in thiamine, and the visual disturbances were regarded as one of the manifestations of dry beriberi. Among the patients suffering from severe peripheral neuritis about 50 per cent had visual complaints, particularly dimness of vision. On examination a gradual decline in visual acuity was demonstrated. In some 500 men the visual acuity ratio decreased from normal (20/20) or near normal values to 20/200 or less. A narrowing of the visual fields and an enlarged blind spot were also observed. No loss of auditory acuity was demonstrated. Blurring of vision and loss of visual acuity were described in a group of British prisoners of war in the Far East by Simpson (1946). The impairment of the visual acuity indicated by a 6/60 ratio was common, and in general the syndrome was similar to that reported by Hibbs (1946). In both instances the visual symptoms followed the development of "burning feet."

The "starvation amblyopia" present among prisoners of war at Thailand was described by Hazelton (1946). The observations were made in 1944. Out of a group of 409 men complaining of dimness of vision, 277 were diagnosed as "avitophthalmia"; in the rest of the men the difficulty was caused by such factors as refractive errors and presbyopia, cataract and corneal ulcers. In addition to blurring of vision under daylight conditions, other symptoms were present. Photophobia was frequent. There were numerous symptoms of eyestrain, such as headache, excessive lacrimation, and heaviness of the eyelids, all of which were more pronounced after close work. The patients complained of pain in the eyeballs. In fixating a long word the center of the word appeared to be absent, indicating an ocular defect. Some men had difficulty in controlling the external eye muscles, and cases of nystagmus were seen. The ciliary musculature involved in accommodation was easily fatigable. After reading for a short time the print seemed to run together, and occasionally double vision was reported. The color of the print appeared to change sometimes from black to green or yellow. There was apparently no disturbance of night vision. The tests established in all men a decreased visual acuity for near vision as well as for distant vision, a deterioration of color vision measured by means of matching colored wools or by identifying colors on a color chart, and a great reduction in the peripheral visual fields determined by perimeter with colored stimuli.

Hazelton believed the illness to be caused by a complex vitamin deficiency. The fatigability of the ciliary muscles and other symptoms of eyestrain were alleviated by the administration of large doses of thiamine. However, intramuscular injections of thiamine and niacin did not produce significant changes in the retinal (conal) dysfunction. Other vitamins were tried, but equally without therapeutic effect. Cases of amblyopia of short duration responded well to treatment with 10 eggs per day given for 30 days. Hazelton did not hazard any opinion on the specific curative factor involved.

Mitchell and Black (1946) observed amblyopia in 8.3 per cent of the 577 released prisoners of war and internees who were admitted for malnutrition to the 47th British General Hospital in Singapore. The average time of captivity after which the visual disturbances began to appear was said to have varied from 4 months to more than 3 years, with an average of about a year. The early symptoms included photophobia, a burning sensation in the eyes, tiredness of the eyes, and supraorbital as well as retroorbital headache. The visual disturbances followed, with blurring of letters, missing parts of lines of printed material, and difficulty in recognizing faces as the common symptoms. The authors point out that defective vision was often unassociated with any other deficiency. Also, the association with neurological symptoms was not statistically significant (see Table 312).

Hobbs and Forbes (1946) described in some detail the visual defects which developed among British Royal Air Force personnel liberated from Japanese prisoner-of-war camps in Java and in the Amboina group of the Molucca Islands.

TABLE 312

LACK OF ASSOCIATION OF AMBLYOPIA WITH NEUROLOGICAL SYMPTOMS in
malnourished prisoners of war at Singapore
(Mitchell and Black, 1946).

Amblyopia	Neurological Symptoms		Total
	Present	Absent	
Present	13	35	48
Absent	157	372	529
Total	170	407	577

The account is based on firsthand reports of medical officers who were imprisoned with the men, questionnaire returns from 1500 airmen, and direct observations made by the authors. The men were classified according to whether only reading vision was defective, or distance vision as well, and secondly, whether the defect was transient or persistent. In some individuals a temporary asthenopia appeared to be associated with the markedly debilitated physical condition; the visual acuity for distance vision was within normal limits, and there was no abnormality of media, fundi, or ocular movements. In men in the higher age groups the defect was sometimes corrected by providing lenses for reading. In a few patients in whom both reading vision and distance vision were blurred the difficulty was caused by errors of refraction, without evidence of ocular disease. There were men with acute ocular pathology of syphilitic origin, and one or two men with nystagmus and palsies of the extrinsic eye muscles. After those belonging to all the above categories were eliminated, there remained 163 cases of blurred vision.

In 89 per cent of these 163 patients the deterioration of vision was gradual; blurring of vision appeared within the first or second year of captivity, increasing over a period of weeks or months until the men were not able to read even large print and could hardly recognize facial features at a distance of a few yards. In 11 per cent the onset was rapid, the symptoms developing fully within a period of from one to a few days. The onset of the symptoms in reference to other diseases is given in Table 313. In 53 men there was no history of any other symptom preceding the blurring of vision; in 50 of these the onset was gradual, while in 3 men the defect developed within a few days.

The visual acuity at distance was below 6/60 in either eye in 6 per cent of the cases, between 6/60 and 6/24 in 55 per cent, between 6/24 and 6/6 in 33 per cent, and 6/6 in 6 per cent. The peripheral visual fields were significantly constricted only in severe cases of blurred vision. Examination of the central fields frequently revealed bilateral absolute central scotomata. In other cases the scotomata were less dense and were placed more peripherally. In some cases the scotoma formed a ring, leaving the fixation area clear. Twenty-seven men out of 90 patients with optic atrophy presented other neurological lesions, including 6 cases of nerve deafness.

In previously malnourished Canadians repatriated from Hong Kong (Adamson et al., 1947) almost one half (48 per cent) out of a group of 300 had visual

TABLE 313

ONSET OF BLURRING OF VISION IN RELATION TO OTHER DISEASES
(Hobbs and Forbes, 1946, p. 150).

| | Onset of Blurring of Vision | | | |
| | During Active Stage of Disease | | During or after Recovery from Disease | |
Disease	Gradual	Sudden	Gradual	Sudden
Beriberi	23	3	16	
Pellagra	1	1	2	2
Sore lips, tongue, and scrotum ..	9	1		
"Burning feet"	3		7	1
Dysentery	8		6	5
Dysentery and beriberi	6		7	
Other diseases			6	1

complaints — chiefly photophobia, watering of the eyes, visual fatigue, blurring, and difficulty in reading. Out of 79 men with defective eyesight, in 46 the defect was slight (20/50 or better), in 16 moderate (better than 20/200), and in 17 (about 6 per cent of the whole group) the visual acuity was 20/200 or worse. The disturbance of vision was one of the residual disabilities which was considered a permanent handicap (see also Bell and O'Neill, 1947). On the other hand, Adamson also stated that during internment over 60 per cent of the whole group suffered from blurring of vision but that most of the men who were severely affected in 1942 regained normal vision.

Although the reports cited above differ in the thoroughness of documentation and in the uniformity of the clinical picture, they do establish convincingly the development of blurring of vision in men who were maintained for a period of a year or more on the internment camp diets in the Far Eastern area. The specific causative factor (or factors) remains unknown.

Dark Adaptation — Field Reports

A decreased ability to see in dim light (night blindness) is one of the few disturbances of the special senses mentioned in field observations on semi-starvation in North Africa and Europe. The phenomenon is mentioned in the early records of medical history — in the Egyptian papyri of 2000–1500 B.C., which indicated also an effective therapy, the eating of raw liver. The Egyptian physicians insisted on the curative power of the liver of the ass and the cock. The Greek Hippocrates (5th century B.C.) had a preference for beef liver. Pliny the Elder advised liver of the goat (Wald, 1942). Jess (1922) relates a passage from the Latin *Historia Damiatina* by Oliverius Scholasticus, who described the plight of the Crusaders leading to their surrender at Damiette in 1221 A.D. The passage may be translated freely as follows: "As a result of the distressing hunger [the men] were plagued by various diseases. It was said that among other sufferings they were as if stricken by blindness at night; although their eyes were open they could not see anything."

According to Jess (1922), De la Jonquière recorded the same phenomenon among the malnourished French soldiers in Napoleon's expedition to Egypt in 1798. Robert, physician-in-chief, reported an epidemic outbreak of night blindness during the siege at the island of Malta in the same year; he noted that the victims could see nothing at night and that the symptom was most frequent in cachectic individuals. Similar observations have been made under other conditions of insufficient or monotonous diet, such as were present on sailing ships and in penal institutions. During World War I numerous reports appeared in the German and French literature on night blindness among soldiers (Jess, 1922). Cases of night blindness were seen by Schittenhelm and Schlecht (1918) in undernourished workers who developed edema.

For some years night blindness was considered to be an early and specific result of vitamin A deficiency. Recent field reports indicate that the situation may not be so simple. In a large sample of people in Newfoundland, with an estimated intake of 1400 I.U. of vitamin A as compared to the recommended allowance of 5000 I.U., the clinical examinations revealed the presence of dry conjunctiva, regarded as a sign of vitamin A deficiency. This xerosis conjunctivae was classified as mild in 35 per cent of the persons examined, as moderate in 27 per cent, and as severe in 14 per cent. At the same time no disturbance in dark adaptation was reported (Adamson et al., 1945).

During the Leningrad siege and famine of 1941–42 night blindness was rare, and it was seen only occasionally during March and April of 1942, some 6 months after the blockade of the city by the Germans became effective. This is of special interest in view of the fact that morphological symptoms interpreted as signs of vitamin A deficiency were observed, including dry and falling hair, dryness and hyperkeratosis of the skin, fissuring and ridging of the fingernails, and degenerative changes in the epithelium of the respiratory tract and the bronchi (Garshin, 1944). There was practically no night blindness in the markedly undernourished population of France (Besançon, 1945).

On the other hand, Chevallier (1947) reported a fluctuation in the light threshold of the dark-adapted eye which appeared to parallel the vitamin A content of the blood serum. The data are reproduced in Table 314. The size of the group Chevallier tested in 1940 was 350 persons; the subjects were examined periodically in the course of subsequent years, with the exception of 1944. The

TABLE 314

LIGHT THRESHOLD OF THE DARK-ADAPTED EYE, expressed in reference to the aperture of the Chevallier-Roux adaptometer (Chevallier, 1947, p. 260).

Year	Light Threshold	Vitamin A Content of Blood Serum (I.U./100 cc.)	Average Caloric Intake (Cal./day)
1939–40	13	90	3200
1941	31	40	1400
1942	43	39	1450
1943	39		1620
1945	63	25	2120
1946	52	30	2100

number of individuals tested after 1940 is not given. The examinations were
made at Marseilles at the beginning of each year in order to avoid seasonal
variations. The parallel trend of changes in the vitamin A content of the blood
and the level of light adaptation was true *only* for the group means. In individual cases the two characteristics were essentially independent. This lack of
correspondence was emphasized by Lambrechts (1947). Sinclair (1947) was
even more outspoken; he stressed not only the absence of correlation between
the vitamin A content of the blood serum and the final light threshold within
the normal range but pointed out that many individuals with definite (clinical)
symptoms of night blindness were not deficient in vitamin A. He insisted on the
necessity for a therapeutic trial before a high light threshold may be considered
a diagnostic symptom of vitamin A deficiency.

Dark Adaptation — Experimental Work

The early observers were of the opinion that night blindness was somehow
related to food intake. The thesis that night blindness is a result of an inadequate dietary supply of vitamins was proposed by Wietfeld (1915); his main
argument was that night blindness was encountered among soldiers only in winter months and disappeared when fresh vegetables and fruit became available.
McCollum and Simmonds (1917) were more specific and identified as vitamin
A the dietary component the lack of which results in xerophthalmia, sometimes
known to be associated in man with night blindness.

The dependence of the process of dark adaptation upon vitamin A was
brought out more directly by the work of Fridericia and Holm (1925) and of
Tansley (1931). The biochemistry of the process of the disintegration of the
rose-colored photosensitive pigment in the rods, the rhodopsin, into retinene and
protein and its resynthesis involving vitamin A was clarified by Wald (1935).
The literature on night blindness as a manifestation of vitamin A deficiency, including the clinical aspects, was reviewed by Jeghers (1937a).

Experimental investigations of the quantitative relationships between the
factors involved (amount of vitamin A in the diet and in the blood, time on the
diet, degree of the change in dark adaptation, rapidity of recovery on supplementation) yielded results that are far from unanimous. In the experiments on
man carried out by Jeghers (1937b), Wald, Jeghers, and Arminio (1938), and
Hecht and Mandelbaum (1939) a rapid response to a diet deficient in vitamin
A was reported; the first signs of a rise in the light threshold of the dark-adapted
eye were said to be present as early as 24 hours after the removal of the vitamin
from the diet. In subsequent experiments the interval between the start of the
vitamin A deficient diet and the onset of impaired dark adaptation has progressively increased (Booher, Callison, and Hewston, 1939; Steffens, Bair, and
Sheard, 1940; Dann and Yarborough, 1941; Wald, Brouha, and Johnson, 1942;
Brenner and Roberts, 1943; Batchelder and Ebbs, 1944). In the experiments carried out at Sheffield, England, from July 1942 to October 1944 (Vitamin A Subcommittee, 1945, 1949; Marrack, 1947b) no definite signs of deterioration of dark
adaptation were obtained within one year in any of 16 experimental subjects
placed on a diet essentially free of vitamin A. Three of the men began to show

an impairment of dark adaptation after 14, 17, and 22 months on the diet, respectively. In the rest of the group the only significant change was a gradual decrease of the vitamin A level in the blood plasma.

The general impression gained from this survey, which does not pretend to be an exhaustive summary of the literature, is the large degree of variability in the individual responses and the apparently large body reserves. The body reserves of vitamin A appear to supply the needed amounts for considerably longer periods than the early investigations seemed to indicate. However, the question of possible effects of caloric undernutrition on the requirements or the body reserves of vitamin A remains open.

Sensory Functions — Experimental Studies in Fasting and Starvation

Weygandt (1904) included in his studies of the effects of short, total fasts (24 to 48 hours) measurement of the spatial threshold, determined as the distance at which the points of a pair of calipers were perceived as two separate stimuli. The results can be considered as essentially negative. The threshold values increased in two experiments on the same subject but decreased in the second subject, although the change was much less pronounced. Furthermore, even under control conditions the measurements varied within wide limits.

Langfeld (in Benedict, 1915) studied tactual-space threshold and visual acuity in one man during a total fast of 33 days. The threshold of tactual discrimination was determined at the volar side of the forearm. There was no consistent change in the threshold values. Visual acuity was expressed as the distance at which the subject was able to identify the position of the letter E; the distance increased from the initial value of 20 feet to the final value of 36 feet, indicating an apparent improvement in visual acuity. The threshold values exhibited large variations, rising on two days preceding the fast from 17 to 20 feet, and to 25 feet on the first day of the fast, to drop again to 16 feet on the third day of the fast. Another sharp "spike" in the visual acuity graph occurred in the middle of the fast, with the threshold values of 25 feet on the thirteenth day, 37 feet on the fourteenth day, and 24 feet on the sixteenth day of fast. In view of these wide variations and the large amount of training present in other tests, the physiological significance of the over-all trend of "improvement" becomes questionable.

Marsh (1916) included in his fasting experiment on two subjects, with no food intake for a week, tests of touch (single camel's hair applied 10 times, a record being made of the number of times the hair was felt or not), pain (Verdin algometer), and vision (perception of dots, from 4 to 9 per card, exposed with drop screen). Tests of taste and smell were eliminated owing to the cumbersomeness of the technique. The results of the retained tests were inconclusive. Improvement in the test of visual perception probably reflects a continued practice gain.

In the experiment on prolonged undernutrition by Benedict et al. (1919) in which the "experimental" group (Squad A) lost 10.5 per cent (7.1 kg.) of the body weight, the sensory functions — pitch discrimination, visual acuity, and sensory threshold for electric shock — were either not affected by the reduction of food intake or showed only minimal changes.

Glaze (1928a) carried out experiments in which 3 subjects (A, B, and C) fasted for periods of 10, 17, and 33 days, respectively. A test of visual acuity was included in the test battery. The test consisted of reading rows of letters, progressively decreasing in size. The acuity was not affected adversely by the fast, since the subjects read as many or more letters during the fast than before; the situation is complicated by the very slight but noticeable effect of practice, continued into the post-fast period. In addition to the "acuity score," a time score indicating the length of time taken by the subjects to complete the test was obtained. This characteristic, possibly representing a neuromuscular rather than a sensory function, showed relative deterioration in the 2 shorter fasts but not in the 33-day fast. The time scores, in seconds, for the pre-fast, fast, and post-fast conditions were 338, 394, and 316 (subject A), 214, 249, and 160 (subject B), and 300, 290, and 247 (subject C). As an incidental observation a marked increase in smell sensitivity was reported; no quantitative tests were made but the subjects reported that their ability to detect odors had increased during the fasts. Subject B refused to drink the city water because the odor of the chlorine became too offensive. Subject C claimed that perfumes and body odors of people he passed on the streets during his walks were particularly noticeable during the fast.

In a later investigation Glaze (1928b) studied the problem of sensitivity of the sense of smell during a fast by quantitative methods. Zwaardemaker's olfactometer was used and seven odors were tested. In both subjects, fasting for 10 and 5 days respectively, the sensitivity to odors was significantly greater during the fast than either before or after the fast. However, on retesting one of the subjects during another 5-day fast, the phenomenon was not present.

Kravkov and Semenovskaja (1934) studied visual functions in a man who suffered from cystitis and underwent a 50-day fasting treatment. During this time he received water but no food. On the forty-second day of the fast measurements of the acuity of near vision were made by means of a standardized letter series, and color vision was tested by means of Stilling's pseudoisochromatic tables and Ishihara's color plates; both functions were completely normal. Two days later the light sensitivity was examined and the test was repeated 14 days after the end of the fast. The results obtained during the fast indicated a very poor capacity for dark adaptation — the authors spoke of an experimentally produced hemeralopia. The post-fast dark adaptation curve returned close to the normal. No data on the blood level of vitamin A were given.

In the experiment by Noltenius and Hartmann (1936) in which the subject fasted for 8 days and lost 5.8 kg. of body weight, the upper frequency limit of audible tones was not changed.

The previous experiments on fasting or undernutrition indicate that on the whole the special senses are remarkably resistent to starvation. The experimental work suffered from the small number of subjects tested and/or inadequate pre-experimental training, and the results of past investigations must be regarded as suggestive rather than conclusive.

Visual Functions — Minnesota Experiment

The selection of visual functions to be tested in the Minnesota Experiment depended partly on theoretical considerations and partly on the ease with which a particular test could be administered. For theoretical reasons we were interested in the measurement of dark adaptation, but it was not feasible to include quantitative tests of this function in the program. Neither in the reports made by the subjects nor in the direct observations of their behavior under conditions of dim illumination were there any indications of impairment of their capacity to perceive light stimuli of low intensity.

Measurements of visual acuity were provided because of the biological importance of recognition of details in the interaction between the individual and his environment. Visual acuity was expressed as the diameter of black circles of threshold size, in millimeters, which could be seen at a distance of 15 feet. The measurements were made at two levels of illumination, one with 1 and the other with 100 foot-candles at the plane of the test patch. The acuity was markedly poorer at low illumination. However, at neither level was there any significant deterioration during starvation. The mean values at S24 were not significantly different from the control values (see Table 315).

The frequency at which successive light flashes appear to fuse into an uninterrupted light is an indicator of the resolving power of the retino-cortical system; the higher the frequency, the greater the capacity to perceive successive flashes as discrete stimuli. The mean values for flicker fusion frequency decrease in older age groups (Simonson, Enzer, and Blankenstein, 1941; Brožek and Keys, 1945b), and there are some indications that the frequency decreases when the human organism is exposed to intensive "stresses" (see Brožek and Keys,

TABLE 315

VISUAL FUNCTIONS in the Minnesota Experiment. Low scores in the tests of visual acuity and high scores in the other tests represent "desirable" values.

	C	S12	S24	d = S24 − C	F
Visual threshold in mm. at 100 foot-candles, 15-ft. distance					
M	0.69	0.71	0.71	+0.02	1.28
SD	0.14	0.19	0.17		
Visual threshold in mm. at 1 foot-candle, 15-ft. distance					
M	2.11	2.10	2.13	+0.02	0.21
SD	0.45	0.46	0.46		
Flicker fusion frequency, flickers per second					
M	36.3	35.6	35.3	−1.0	19.41[**]
SD	3.2	3.8	3.6		
Perceptual fluctuations, number in 1 minute					
M	26.4	24.7	25.8	−0.6	0.80
SD	7.2	8.5	10.2		

1944, which also contains references to studies on change in flicker fusion frequency in disease and under anoxia). In semi-starvation there was a very small but consistent and thus statistically significant change in the direction of deterioration. In order to obtain an indication of the "biological significance" of these changes, we determined the magnitude of the semi-starvation displacement of the mean values, in reference to the distribution of the individual measurements under normal (pre-experimental) conditions, by computing the ratio $(S24 - C)/SD_C$; $S24$ is the mean value obtained at the end of semi-starvation, C is the mean control value, and SD_C is the standard deviation of the measurements made on 32 subjects during the control period. In terms of this displacement ratio, which in the case of flicker fusion frequency is equal to -0.31, the change is not impressive.

Perceptual fluctuation is measured as the rate at which an ambiguous figure changes its perspective (see the Appendix on methods), with the subject maintaining the passive attitude of an observer. The underlying mechanism of oscillations in perception of ambiguous figures is not clear (Woodworth, 1938, pp. 696ff). Under conditions of fatigue both a shortening and a lengthening of the interval between the changes in perspective have been reported (Hollingworth, 1939). In the present experiment there was a slight decrease in the fluctuation rate, but the difference between the values at C and at S24 was not significant.

In summary, changes in the visual functions during the period of semi-starvation were either absent or negligible.

Hearing — Minnesota Experiment

The acuity of hearing was measured by the Maico D-5 audiometer in a soundproof room. The measurements are expressed as hearing loss, in decibels, on a scale from 0 (normal hearing) to 120 (total deafness); the range for values which are better than the "normal" level of threshold intensities extends from 0 to -10. The individual score at each period is a mean of three repeated measurements. Frequencies from 128 to 8192 double-vibrations per second were used.

In the control period the average values for the group were very close to the "normal" level at all frequencies. In semi-starvation not only was there no deterioration but the auditory acuity actually improved (see Table 316). The improvement was statistically highly significant throughout the frequency range except at the highest frequency. There the change was in the same direction but the F-test for the significance of the mean difference between C and S24 did not reach the critical level.

As in the case of flicker fusion frequency, the biological importance of the displacement of the means may be characterized in terms of the ratio $(S24 - C)/SD_C$. At the lower frequencies the displacement of the mean amounts to close to 1 standard deviation of the values obtained at C, which represents a marked change. At the two highest frequencies either the absolute change was small (at 8193 double-vibrations per second) and/or the scatter of the individual values at C was large (at 4096 double-vibrations per second), which resulted in low displacement ratios.

Subjective reports from the subjects were consistent with the audiometer

TABLE 316

HEARING LOSS, in decibels, in the starvation phase of the Minnesota Experiment. Low scores indicate good auditory acuity.

Test Frequency per Second	C	S24	d = S24 − C	F	$\dfrac{d}{SD_C}$
128					
M	4.00	−0.03	−4.03	25.96[**]	−0.87
SD	4.63	4.33			
512					
M	−0.16	−3.84	−3.68	29.11[**]	−0.76
SD	4.84	4.39			
2048					
M	0.58	−4.36	−4.94	47.53[**]	−0.84
SD	5.89	4.92			
4096					
M	3.75	−0.15	−3.90	7.94[**]	−0.25
SD	15.72	14.57			
8192					
M	−1.31	−3.07	−1.76	2.74	−0.18
SD	9.92	8.47			

findings. A number of the men volunteered the information that their tolerance for loud music, speech, and noises was lowered in semi-starvation. The question was discussed among the subjects themselves in connection with setting the loudness of the phonograph and radio. In the field, observers have noted the tendency of semi-starved people to converse in very low tones; in personal discussions with many individuals who have suffered from famine we have been struck with the frequency of statements about keen hearing and intolerance for loud sounds.

The fact that the change observed in the Minnesota Experiment was related to the nutritional status was confirmed by its reversal during rehabilitation. By combining all frequencies and all caloric groups we obtained an average decrease of hearing loss (increase of auditory acuity) of 3.66 decibels from C to S24 and an average increase of 3.54 decibels from S24 to R12, at which time the "recovery" amounted to 96.7 per cent of the semi-starvation change. Thus after 12 weeks of rehabilitation the over-all auditory acuity had returned essentially to the pre-starvation level. The changes are presented graphically in Figure 102.

The mechanisms involved in these changes of auditory acuity remain obscure. In the absence of other evidence, a purely mechanical explanation may be suggested. It is generally agreed by otologists that, other things being equal, auditory acuity is related to the free lumen of the external auditory canal. In semi-starvation two factors may increase this free lumen. In the first place, according to casual observations on the Minnesota subjects, there is a tendency to produce and accumulate less earwax. Secondly, the lumen is determined by the bony framework and the covering soft tissues; in starvation these tissues, like all others of the body, become thinner, so that the free channel must be somewhat enlarged.

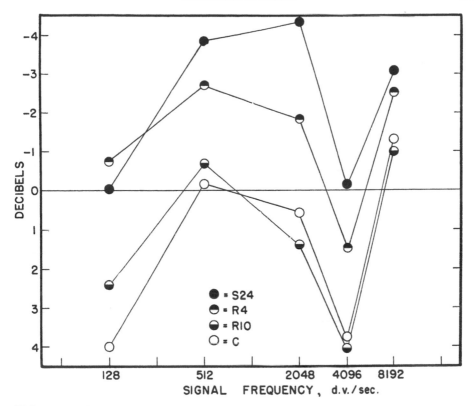

FIGURE 102. Auditory Acuity. The value of the threshold stimulus was expressed as "hearing loss," in decibels; the lower the value, the better the acuity.
The acuity was tested at 5 frequency levels (d.v./sec. = double-vibrations per
second). (Minnesota Experiment.)

The breakdown according to calories and frequencies of vibration at which
the measurements were made is given in Table 317. The lowest caloric group,
Z, tends to exhibit the smallest amount of "recovery" during rehabilitation.
However, this is the only generalization that can be made on the basis of these
data. Statistical evaluation by means of the analysis of variance indicated an absence of significant relationship between the rehabilitation changes and any of
the three dietary factors (calories, proteins, and vitamins).

Comment

Our previous observations on special senses under conditions of experimental vitamin B complex deficiency (Keys *et al.*, 1945, p. 34) indicated that
the sensory functions tend to be remarkably resistant to severe nutritional deprivation. There was no important deterioration in any sensory function after 6
months of semi-starvation.

We are well aware that the battery of sensory tests used in the Minnesota
Experiment was far from covering the area in an exhaustive manner. Some functions were not included at all, such as taste and smell. In other sectors, particularly vision, the number of tests could have been considerably extended. How-

TABLE 317

CHANGES IN AUDITORY ACUITY, expressed as hearing loss, in decibels, during the rehabilitation period of the Minnesota Experiment, for the four 8-man caloric groups and for all caloric groups combined.
d = S24 − C; △ = R12 − S24.

Caloric Group	Test Frequency (d.v. per sec.)				
	128	512	2048	4096	8129
Z					
d	−5.5	−4.0	−5.2	−11.7	−6.1
△	+1.3	+1.1	+4.2	+4.4	+3.9
L					
d	−4.0	−4.0	−3.0	−0.6	+0.3
△	+3.0	+3.1	+6.7	+3.0	+1.9
G					
d	−4.8	−2.2	−5.2	−0.1	−1.5
△	+4.6	+4.0	+5.2	+3.5	+1.9
T					
d	−1.9	−4.5	−6.4	−3.4	+0.3
△	+1.0	+4.4	+6.9	+5.8	+0.7
(Z + L + G + T)					
d	−4.0	−3.7	−4.9	−3.9	−1.8
△	+2.4	+3.2	+5.8	+4.2	+2.1

ever, we hazard the guess that we would have simply increased the number of negative results.

The differences and similarities between the field observations and the experimental findings are of interest. In the Minnesota Experiment and in the European famine areas the sensory functions during starvation remained essentially unaltered. The frequent occurrence of amblyopia and occasional nerve deafness reported from the Far East indicates that a different set of factors was in operation. Because the semi-starvation changes in visual functions were for the most part insignificant and the changes in flicker fusion frequency, though consistent (and therefore statistically significant), were biologically unimportant, the testing of vision was not continued during the rehabilitation phase of the experiment.

CHAPTER 33

Neuromuscular Functions and Motor Performance

THE purpose of this chapter is to present the results obtained in the Minnesota Experiment by means of neurological examinations, records of motility during sleep, and tests of voluntary motor performance. Field reports on the effects of undernutrition and malnutrition provide a background against which the data obtained under controlled conditions may be projected.

There is a striking difference, as far as the involvement of the nervous system is concerned, between observations made in the European areas of starvation and the reports from the Far East. In the European Theater of World War II the neurological changes were unimpressive, even in cases of profound starvation. In the Far East numerous symptoms of neurological damage were repeatedly observed (see Spillane, 1947) and were ascribed to both vitamin deficiencies and the presence of toxic factors (Bell and O'Neill, 1947). The literature on anatomical and histological changes in the muscles and the nervous system during starvation has been discussed in some detail in the chapters on morphology. Here it may suffice to recall briefly a few points.

Pathology of the Nervous System in Starvation

With reference to gross morphology the brain and nervous tissues generally show smaller changes than any of the other organs and tissues except the bones and teeth. Even microscopic examinations failed to reveal sufficient abnormalities to impress the older investigators. Chemical analyses, incomplete and crude to be sure, so far have also failed to reveal reasons to expect serious malfunction of the nervous structures. But it should be recalled that present morphological methods, including measurements of chemical composition, are relatively insensitive in the case of nervous tissues, and careful studies agree that at least some evidences of damage or change can be seen in the nervous tissues of starved animals and man. It must be admitted that it is difficult at present to evaluate the functional significance of such starvation changes as dark granules in the superficial strata of the cortex (Hassin, 1924) or the appearance of areas of diffuse chromatolysis (Ferraro and Raizin, 1942).

It is true that the classical method of microscopic anatomy has not been exhausted in studies on the responses of the nervous tissues to starvation; modern studies are few and many structures have simply not been examined with care. But the few specialized researches might well discourage further efforts. For example, Préobrajensky and Baranova (1932) studied the auditory apparatus of white mice and guinea pigs that had been starved so as to lose from 22 to 26 per cent and from 20 to 34 per cent, respectively, of the initial body weight. Degenerative changes in the cells of the vestibulary ganglia took place only in the terminal phase of starvation; the other structures were apparently resistant to change.

The failure of histology to become a quantitative science means that substantial quantitative changes in structure could remain unrecognized, and in the case of the nervous tissues minor quantitative changes could produce major alterations in function. Over-all changes in behavior and in stimulus-response patterns imply functional alterations somewhere, and there are certainly such manifestations in starvation, but attempts to get at the basic elements of function in the nervous system have not been conspicuously successful. Two invesigations on the excitability of the nervous structures in starvation and semi-starvation may be noted before proceeding to consider less basic and analytical methods.

In Mouriquand and Coisnard's (1942) study on pigeons the reduction of a balanced diet to one third of the previous intake resulted in a marked decrease of weight which reached a plateau at about 100 days; during all this time there was no change in the "vestibulary chronaxie." When normal pigeons were placed on a very severe semi-starvation diet (one sixth of the intake that would permit maintenance of weight and health) no changes in chronaxie were observed during the first 17 days; subsequently the chronaxie began to decrease precipitously, reaching values of 1 to 2 sigmas as compared with the normal level of 20 to 23 sigmas. At this time the animals were actually moribund, and the significance of the change in chronaxie, indicating an apparent increase in excitability, is difficult to accept. It is unfortunate that the authors did not present information on the threshold voltage (rheobase) in addition to chronaxie.

These data are in sharp contrast with the observations made by Debray *et al.* (1946) in extremely emaciated men. The vestibulary chronaxie increased from the normal values of 12 to 22 sigmas up to 300 sigmas. Bourguignon is quoted as having obtained values of 200 sigmas in patients observed from 1940 to 1944 who had lost up to 40 per cent of their body weight. The direction of the change is in line with theoretical considerations, but the magnitude is surprising. The fact that at least some of the subjects studied by Debray *et al.* were actually *in extremis* might explain the results. No data on the number of patients or the criteria for their selection were given.

Although not documented by figures, Braude-Heller's (1946) observations are of interest in the present context. Tests with galvanic current made on 10 semi-starved children disclosed normal or slightly diminished nervous excitability. The frequently diminished tendon reflexes are another symptom of decreased reactivity of the somatic nervous system. The activity and excitability of the autonomic nervous system were also decreased. There were bradycardia and lowered arterial blood pressure. The excitability was studied by administration of adrenalin and pilocarpine. Adrenalin, acting on the orthosympathetic centers, produced very little or no increase in pulse rate and blood pressure. Pilocarpine, a parasympathetic excitant, produced little change in the flow of saliva or in sweating. The size of the pupils was only slightly affected. Fliederbaum *et al.* (1946), working with semi-starved adults, noted that the pupils were usually constricted and sometimes reacted slowly and weakly to light.

Neurological Disturbances in Undernutrition — Field Reports
from European Areas

In semi-starved patients who developed edema and lost an estimated 20 to 40 kg. of body weight, the results of functional examination of the nervous system were largely negative (Schittenhelm and Schlecht, 1918); this was the general finding in World War I. In a small number of patients the area of edematous lower extremities was slightly hypoesthetic. In some cases hypoesthesia was found in other areas of the skin. Deep muscle sensibility and kinesthetic and vestibular sense were not disturbed; the Romberg sign was absent. The skin and tendon reflexes were in general normal. Only the patellar reflexes were affected. In some cases they were difficult to elicit or resulted in a weak excursion of the foreleg, and in 3 patients out of more than 200 no patellar reflex response was obtained; on the other hand in a few individuals the response was exaggerated. In some cases the musculature of the calf and the nerve trunks (nervus tibialis, nervus femoralis) were sensitive to pressure; the mechanical effect of atrophy of the overlying tissues may have been involved.

Paresthesias (numbness, tingling, burning) were noted frequently in the tips of fingers and toes in a group of some 300 semi-starved ex-prisoners of war evacuated from Germany and treated in the first half of May 1945 at the Pinder Fields Emergency Hospital in Wakefield, England. In view of the distribution of the paresthesias the symptom was considered of vascular rather than nervous origin. Most of these men had suffered severely from exposure to cold in the preceding months. Other symptoms of neurological interest were a reduction or absence of knee and ankle jerks in about one quarter of the patients, and a sensory impairment of glove-and-stocking type. There were no complaints of impairment of hearing and no evidence of optic neuritis (Edge, 1945).

In men suffering from uncomplicated, severe semi-starvation Leyton (1946) observed marked weakness and fatigability, slowing of voluntary movements, and an increase in the reaction time. The myotatic reflexes were sluggish, but neither knee nor ankle jerks were absent. The convergence and reaction of the pupils of the eyes were normal. Sensitivity to tactile and thermal stimuli did not appear to be affected.

In semi-starved patients admitted to Leningrad hospitals during the siege of 1941–42 the main muscular disturbance was weakness. Such symptoms as paresthesias and reduced or absent reflexes were observed but were not common. Thiamine chloride, 5 to 10 mg. daily, was considered effective in relieving the paresthesias but had little effect in hastening the return to normal of the reflexes or the muscular strength. In some cases numbness of the finger tips was reported; this was also regarded as a possible indication of thiamine deficiency.

Peculiar neuropathological disturbances were observed in a small group of semi-starvation patients during the later part of the siege of Leningrad in the late spring of 1942 (Levin, 1944). The syndrome included sensory and symmetrical motor polyneuritis, especially marked on the distal parts of the arms and legs, hypesthesias and paresthesias, aches and pains, decreased reflexes, and muscular weakness. The area of disturbed sensitivity coincided with a peculiar yellow coloration of the skin of the extremities, and there was a parallelism between the appearance of this yellow pigmentation and the development of the neuritic symptoms. Large amounts of carotene were found in the plasma, and the yellow color of the skin was assumed to be caused by this pigment. Prior to hospitalization all of these individuals had been consuming large amounts of a weed similar to what is called lamb's-quarters in America; the author was of the opinion that the syndrome was caused by excess carotene not detoxified by the liver.

Even in the advanced stages of semi-starvation Debray *et al.* (1946) did not observe motor paralyses of the beriberi type. However, some sensory disturbances, anesthetic and paresthetic in character, were observed in the extremities and are reported to have been treated successfully by thiamine; no details were given. There were cases of myalgia, affecting especially the calf and the thigh and present in those in whom the muscular atrophy was not excessive. The phenomenon was ascribed to a disturbed muscular metabolism.

The results of neurological examinations carried out during the severe semi-starvation of 1944–45 in western Holland (de Jongh, in Boerema, 1947, p. 235) were largely negative. Both the skin and the tendon reflexes remained normal or were somewhat depressed. There were no clear-cut disturbances in the sensibility, and the Romberg sign was negative. The absence of major neuropathies during the hunger-winter in Holland was confirmed by Meyjes (1948). The number of cases of severe polyneuritis was not larger than had been observed during the preceding years. Cutaneous anesthesias of appreciable extent or intensity were not reported. The minor neurological symptoms resulting from semi-starvation included paresthesias and hyperesthesias and a tendency toward cramps. On electrical stimulation, the muscular contractions in semi-starved individuals differed from those in normal persons. The tendon reflexes were sometimes diminished but exaggerated responses were also noted. Meyjes pointed out that the lack of major neurological disturbance in Holland contrasted sharply with the experiences in the civil and military prison camps in the Dutch Indies (cf. de Wijn, 1947).

Only from one European report, relating observations made during the Spanish Civil War (1936 to 1939), do we get a picture of neuropathies which is close to that of the complex nutritional deficiencies observed routinely in the Far East (Peraita, 1946). The patients were seen at the Institute of Nutritional Hygiene in Madrid. The most frequent symptoms were acroparesthesias in the toes and fingers ("pins and needles" in the finger tips, "ants under the nails"). Some patients complained of "terrible prickings" and "stabbing pains" in the feet, calves, and thighs as well as in the hands and forearms. The skin of the feet became hyperesthetic to touch and warmth. Whereas during the day the feet felt frequently "insensible and cold as ice," at night under the blankets they would begin to feel "hot like fire." The patients got some relief by cooling the feet in the air, walking barefoot on the floor, or wrapping the feet in cloths wrung out in cold water. At times the acute sensation of cold in the extremities became very painful, and sometimes the painful cold sensation passed into the sensation of burning heat. In these patients no true beriberi and little classical pellagra was observed. Neither purified thiamine nor niacin was effective in the treatment of these nervous disturbances; success was reported with the use of dry yeast.

Neurological Disturbances in Undernutrition — Field Reports from the Far East

It may bear repeating that the neuropathies so frequently observed in the Orient (burning feet, amblyopia, sensory ataxia, spastic paraplegia) were almost unknown in the European areas of semi-starvation. Because these neuropathies were seen both in the presence and in the absence of gross caloric deficiency, it is evident that the disturbances are not due to a simple caloric deficiency. Here we are not concerned with the neurological aspects of the relatively clear-cut specific deficiency diseases, but it is desirable to survey the effects of general malnutrition as found in the Far Eastern camps for prisoners of war. The literature is large, unsystematic, and repetitious. A few selected reports, presented in some detail, may give the reader a better idea of the changes observed than if the separate symptoms were considered one at a time. There is available much further information of the same type (Hibbs, 1946; Hobbs and Forbes, 1946; Walters *et al.*, 1947b).

In a group of recovered Allied prisoners of war and internees, detained for a period of 3½ years and maintained on a diet grossly deficient not only in calories but also in fats, proteins, and vitamins, degeneration of the nervous system was reported (Price, 1946). No percentage frequency was indicated, but the impression is given that these cases were rare. They were grouped into 3 categories exhibiting the symptoms of: (1) amblyopia, nerve deafness, and ataxia, (2) spastic paraplegia, and (3) peripheral neuritis. Incoordination (ataxia) was most pronounced in the legs, although in some cases the arms were also affected; it was associated with deep pain and impairment of the senses of position and vibration. Spastic paraplegia was ascribed, not to a nutritional deficiency, but to some toxic factor. The symptom was observed in only two camps and was present in men who apparently suffered less from undernutrition than the average. Peripheral neuritis was regarded in the majority of the cases as a beriberi symptom.

Clarke and Sneddon (1946) reported similar symptoms of nutritional neuropathy in a group of six civilians and prisoners of war interned by the Japanese after the fall of Hong Kong in December 1941 and released in September 1945. The report is based on case histories and on observations at the Royal Naval Hospital, Sydney, Australia, to which the patients were admitted for treatment. The symptoms and their chronological order paralleled the experience in other camps in the Far East, with swelling of the ankles developing within 3 months of imprisonment, followed by weakness and paresthesias of the limbs, difficulty in walking, disturbances of vision and hearing, and burning sensation and hyperesthesia of the soles referred to as "electrical" feet. On admission to the hospital the patients were still considerably underweight. In 74 patients out of the group of 200, neuropathological symptoms were present (see Table 318); these included optic atrophy, ataxic paraplegia, and nerve deafness. Visual acuity was frequently impaired, and central and paracentral scotomata were present. The condition was interpreted as a result of a dietary deficiency, possibly aggravated by a toxic agent present in the diet. After 2 months on a regimen of high calorie diet and massive doses of vitamins there was subjective improvement in some cases, but the neurological and especially the visual symptoms are said to have remained unchanged.

TABLE 318

NEUROPATHY IN A GROUP OF 200 SICK INTERNEES AND PRISONERS OF WAR (December 1941 to September 1945) from Hong Kong (Clarke and Sneddon, 1946).

Category	Number of Cases
Optic atrophy and ataxic paraplegia	31
Ataxic paraplegia only	21
Optic atrophy only	13
Optic atrophy, nerve deafness, and ataxic paraplegia	6
Optic atrophy and nerve deafness	3
Total	74

TABLE 319

INCIDENCE OF NEUROLOGICAL SYMPTOMS IN A GROUP OF 300 CANADIANS INTERNED IN HONG KONG, examined after repatriation at the Department of Veteran Affairs Hospital in Winnipeg (Adamson et al., 1947).

Symptom	Percentage
Paresthesias	83
Absent tendon reflexes	20
Impaired vibration sense	23
Impaired light touch	21
Impaired deep touch	8
Impaired superficial pain	26
Impaired deep pain	10
Impaired position sense	13
Impaired temperature sense	22
Positive Romberg sign	23
Optic atrophy	26

The relatively chronic residual disabilities in a contingent of the Canadian Army, some 800 men strong, captured by the Japanese in Hong Kong in December 1941 and returned to Canada in September 1945 were described by Adamson et al. (1947). The report was based on 300 men examined at the Department of Veteran Affairs Hospital in Winnipeg. Because of the sampling artifacts, the incidence of different neurological signs (see Table 319) cannot be used for estimating disability percentages for the whole group of men exposed to a given type of malnutrition. Furthermore, the utility of the table is decreased by the absence of exact information on the time after liberation to which the neurological findings pertained.

As is evident from Table 319, paresthesias were the most frequent symptom; the disturbances affected primarily the feet and legs, were bilateral and symmetrical, and included such varied sensations as tingling, burning, aching, numbness, sharp pains, and cramps. The etiology was not clear, and the nutritional basis of the defect was further obscured by the fact that in many cases there was exacerbation of the paresthesias for some time after repatriation; it is reported that this curious phenomenon was observed in the internment camp in Japan when an adequate diet was provided. The authors note that some men who had been free of the symptom for months had a recurrence on going to work, which suggests a hysterical component of the complaints. In general the paresthesias disappeared only slowly. A year after repatriation the improvement was estimated at 50 per cent. In cases where the examinations were repeated several times over a period of 6 months, a 35 per cent improvement in the vibratory sense and tendon reflexes was observed. There were only a few with a syndrome resembling locomotor ataxia; improvement was very slow. The number of men with permanent neurological disabilities was estimated at about 6 per cent of those hospitalized after repatriation.

A frequent symptom, present in 76 per cent of the cases, was an abnormal tendency to sweating. It varied from increased axillary and palmar sweating during a state of excitement or effort to spontaneous profuse sweating (hyperhidrosis), especially in the region of the face, neck, and upper part of the trunk. The symptom occurred alone or in conjunction with gastrointestinal, neurological, visual, and cardiovascular disturbances, but statistical analysis indicated that there was no significant coincidence of abnormal perspiration and any of the other symptoms. This was taken as an indication of a separate pathological change, with the site of the lesion located in the sympathetic nervous system. It is of interest that the symptom developed within a few months after the return to a plentiful diet and thus can hardly be regarded as a result of a nutritional deficiency. Furthermore, the symptom usually appeared when the men returned to work, which again points to a "functional," psychogenic character of the symptom.

Observations on Neuromuscular Disturbances in Near-Experimental Semi-Starvation

The disturbances of the neuromuscular apparatus were studied in some detail by Simonart (1948, pp. 78ff) under conditions strikingly similar to those present in the Minnesota Experiment. The food intake was reduced from the prewar level of 3400 Cal. per day to about 1700 Cal. The most frequent complaints registered by the semi-starved inmates of the central civilian prison at Louvain, Belgium, were muscular aches, weakness, paresthesias in the limbs, and the limbs "going to sleep." The weakness manifested itself first during work. The prisoners, working in the machine shop and operating envelope-folding machines, were forced to reduce the work output and some had to quit completely. Ordinary activities such as making up the bed, cleaning the cell, climbing stairs, or courtyard exercise became difficult and fatiguing. Some of the patients asked for canes in order to have support during exercise. Others, who were very weak, reported that while standing or walking they felt their legs would not carry them and their knees would buckle. In walking in the prison they held on to the furniture. Not infrequently they fell upon standing up; in addition to muscular weakness, vertigo was said to have played a role in this. Some patients became awkward in handling objects and in writing. They frequently dropped their eating utensils on the floor.

Out of a large group of these semi-starvation patients examined neurologically, only a limited number presented evidence of disturbed function, such as alterations of cutaneous sensitivity and abolished tendon reflexes. Hyperesthesia, hypoesthesia, and anesthesia were observed rarely and were limited to small areas. On compression, the calves and thighs were frequently painful.

In order to have an objective record of the disturbed motor function and of the improvement resulting from therapy, motion pictures were taken of 5 patients during such tasks as rising from the sitting position, stepping up on a chair, walking, and climbing stairs. It was stated that intravenous injections of large doses of thiamine (75 to 100 mg. daily for 20 days) resulted in a marked improvement of motor performance, but the data were not submitted to quantitative analysis.

For the purpose of studying quantitatively the performance capacity of the thigh muscle involved in extending the lower leg, a special ergograph was constructed (Simonart, 1948, p. 84). The total amount of work accomplished in lifting a weight was expressed in kilogrammeters (kgm.). Qualitatively, the ergograms of semi-starved individuals are irregular, in both the rhythm and the amplitude of movement; the fatigue decrement is generally much more marked than in the work curves of the normal controls. Unfortunately not enough information was given to allow a safe use of the quantitative material — e.g. we do not know whether the subjects always worked to exhaustion or whether a time limit was set. Actually, the ergograms indicate that neither criterion was used consistently. Under such conditions the work output scores lack meaning.

Of interest are the work curves of the individuals receiving treatment. This treatment consisted of large doses of thiamine (50 to 100 mg. intramuscularly) given either with the semi-starvation diet or with an improved diet which had not yielded results up to the time of the institution of the thiamine treatment; it detracts much from the value of Simonart's report that the content of the rehabilitation diet is not indicated. Consequently, only those cases in which the semi-starvation diet was continued may legitimately be taken into account in evaluating the effect of thiamine therapy. The reported effects are of surprising magnitude. Thus in subject D.M., whose weight had decreased from 73 kg. to 48 kg., the work output for the right leg increased from 59 kgm. to 75, 81, 83, 120, and 169 kgm. after 10, 20, 30, 40, and 55 days of thiamine treatment (*ibid.*, p. 91).

Patient Lan was placed on a rehabilitation diet (of unknown composition and caloric value). At the start of the treatment he weighed 48.5 kg. as compared with his pre-starvation weight of 63 kg. Measurement six weeks later showed no improvement in performance; no data on his weight at this time are given. In 12 days of thiamine treatment a definite improvement was noted, and this continued throughout the 40 days of thiamine therapy. The performance attained more than twice the initial semi-starvation value (see Table 320). On the basis of this patient (plus some others not cited), Simonart came to the far-reaching conclusion that semi-starvation adynamy is relieved more rapidly by thiamine therapy than by increasing the caloric content of the diet. A reader who prefers conclusions drawn from adequately documented material will be disturbed but not won over by the author's dialectics. In view of highly questionable results reported in other parts of Simonart's book and the inadequate standardization of the ergographic procedure, the data as such are open to severe doubt.

TABLE 320

IMPROVEMENT IN THE ERGOGRAPHIC PERFORMANCE OF SUBJECT
LAN (Simonart, 1948).

Treatment	Work Output (in kgm.)	
	Right Leg	Left Leg
Rehabilitation diet		
July 4, 1942	89	106
July 9, 1942	91	100
August 15, 1942	83	82
Rehabilitation diet plus thiamine (100 mg. daily)		
August 27, 1942	120	136
September 1, 1942	193	234
September 14, 1942	176	259
September 29, 1942	256	276

Neurological Findings in Previous Starvation Experiments

In a subject undergoing a total fast for thirty days (Benedict, 1915, p. 69) the pupillary, plantar, and cremasteric reflexes remained normal throughout. The patellar reflex decreased during the first few days of the fast; on the fifth day the response could be obtained only on reinforcement and it could not be elicited at all starting with the ninth day. The Achilles reflex and the abdominal reflexes disappeared on the eleventh day and remained absent through the remainder of the fast.

In an experiment on prolonged undernutrition (Benedict et al., 1919, pp. 592–97) the latency, amplitude, and refractory period of the patellar reflex were determined. The investigators expressed the opinion that there was some tendency for the reflex to be less irritable. However, the evidence was not clear-cut and the changes, if any, were unimpressive.

Neurological Findings in the Minnesota Experiment

In the Minnesota Experiment the neurological status of the subjects was examined by current clinical methods for testing sensory (especially cutaneous) and neuromuscular functions. The sense of vibration was tested by a tuning fork. Skin sensitivity was examined by pin prick and by stroking the skin lightly with a patch of cotton. The patellar and Achilles reflexes were tested by striking the appropriate tendon with a rubber hammer; the responses were rated on a scale extending from —4 (not obtainable on reinforcement) to +4 (extremely exaggerated). The function of the quadriceps muscles was evaluated by having the subject arise from a squatting position and step up on a chair.

The results of the sensory tests made at the end of the semi-starvation period were largely negative. There was no impairment of vibration sense. The skin sensitivity was normal except in 3 men who will be discussed in greater detail later in this section. The quadriceps muscles appeared weak, but the clinical tests were not sensitive enough to reflect the true changes in muscular strength; all the men were able to perform the squat test, although in 10 men some degree of difficulty was noted.

The tendon reflexes (the patellar and the Achilles) were examined in all subjects at the end of semi-starvation (S24) and at the end of 12 weeks of rehabilitation (R12), and in a smaller group again at R33. At S24 the tendon reflexes were diminished in the majority of the men. In 11 out of 32 men the patellar reflex could be obtained only with reinforcement, and in 6 men reinforcement was required to elicit the Achilles tendon reflex.

TABLE 321

RATING FREQUENCIES FOR TENDON REFLEXES in a group of 16 men studied in the Minnesota Experiment at S24, R12, and R33. Only the ratings for the right leg were tabulated since the values for the two sides were essentially identical.

| | Frequency | | | | | |
| | Patellar Reflex | | | Achilles Reflex | | |
Rating	S24	R12	R33	S24	R12	R33
—4	1	2		1		
—3	5	2		2	2	1
—2			1		2	1
—1	2	2	1	1	1	3
+1	7	6	6	9	8	11
+2	1	4	7	2	3	
+3			1	1		1

Detailed data are presented for a group of 16 men who were examined after 12 and 33 weeks of rehabilitation, at which time they had returned to their pre-starvation status with respect to the great majority of functions (see Table 321). The values obtained at R12 already indicated a trend toward more vigorous reflex responses. At R33 the men in general exhibited fully active tendon reflexes; only 2 out of 16 subjects had moderately sluggish patellar reflexes. The Achilles reflex at R33 was noticeably diminished in only one man, with the rating of —2.

The changes both in starvation and in rehabilitation were bilateral; they were more pronounced in the ratings of the patellar reflex than in those of the Achilles reflex. The data indicate that there was a tendency toward a general reduction in the intensity of the tendon reflexes in semi-starvation and that this trend was reversed during rehabilitation. This trend was not uniform. In a few subjects the response remained unchanged throughout; in some others the changes did not parallel the over-all trend. These exceptions may be accounted for, at least in part, by the inadequate standardization of this clinical technique in both the application of the stimuli and the evaluation of the responses. The general tendency toward a diminution of the reflex responses, in the absence of the symptoms of peripheral neuritis and pathological involvement of the central nervous system, should probably be viewed as a sign of the general lowering of the reactivity of the organism. It is impossible to suggest a thiamine deficiency in these men.

Two subjects (Nos. 29 and 101) developed sensory disturbances in the nature of mild paresthesias (Schiele and Brožek, 1948). In the first subject (No. 29) the symptoms developed during the thirteenth week of semi-starvation, when he began to experience tingling and burning sensations in the anterior aspects of both thighs. These sensations were more disturbing at night. For a few weeks he had a similar but less definite "dry" sensation on the underneath area of the penis. The abnormal sensations gradually increased in area and intensity throughout the remainder of the 24 weeks of semi-starvation. At their height the burning pain on the thighs was said to be very distressing.

Neurologic examinations revealed no positive findings aside from abnormal responses to superficial sensory stimuli in the anterior and lateral aspects of both thighs. Testing with the pin and cotton elicited "burning" sensations which were much more intense in the very center of the disturbed areas. These areas were bilaterally symmetrical and their boundaries shaded off gradually into the surrounding normal parts. By the fourth week of rehabilitation there was a marked diminution in the intensity of the paresthesias. By the twelfth week of rehabilitation they were still further reduced, although the area remained about the same. The personality deterioration was very marked in this subject, and emotional distress may have played a role in the importance he assigned to these symptoms.

In subject No. 101 the symptoms did not develop until the twenty-second week of semi-starvation, when he began to experience burning and tingling in the left thigh. This difficulty gradually increased and for a time involved the right thigh also. It gradually subsided during rehabilitation. Neurological examination was entirely negative except for paresthetic areas on the anterior aspects of both thighs. These areas, which covered approximately one third of the anterior aspect of the thigh, were irregular in outline and the boundaries shaded gradually into normal skin. The subject was able readily to distinguish sharp from dull. The intensity of his subjective sensations varied greatly from time to time. The routine recheck toward the end of controlled rehabilitation revealed a return to normal except for small, vaguely outlined areas which were slightly "dull" to touch, and these were not noticed spontaneously. The scores for this subject on the Minnesota Multiphasic Personality Inventory were normal during the control period. The semi-starvation profile indicated a low Hysteria score (56); this adds weight to our clinical impression that the paresthesias were not hysterical.

The neurological disturbances in the case of subject No. 5 involved both sensory and motor aspects. They appeared to be hysterical in their origin and are discussed in detail in Chapter 41.

Gross Motility in Hunger and Fasting

It is a common experience that under normal conditions deprivation of food at the time of day at which one is accustomed to eat results in a feeling of discomfort and restlessness. In the rat it is well established that the 2-to-4-hour cycles of increased general activity, studied with the techniques of revolving drum and tambour-mounted cage, are intimately related to the rhythmic "hunger contractions" of the stomach (Munn, 1933, p. 50). The critical experiment was performed by Powelson (1925), who transplanted the stomach between the outside skin and the abdominal wall and thus made it accessible to a pneumographic recording of the gastric movements. In the quiescent animal gastric motility was minimal or absent; increase in the amplitude of the movements of the stomach was followed by resumption of activity. Powelson noted that sometimes this general activity *preceded* the augmentation of gastric movements.

In Richter's investigations (1927), in which rats were kept in a combined activity and food cage, the entrance into the smaller eating cage was preceded by about ½ hour of intensified general movements in the larger activity cage. As a rule it was during the period of greatest activity that the animal entered the cage to obtain food. This would again indicate a connection between general activity and "hunger"; no actual record of hunger contractions was obtained in Richter's study.

Experimental investigations of the relationship between hunger and motility in man were made by Wada (1922). During sleep Wada observed a tendency for the movements of the body to occur simultaneously with the stomach contractions which, in the waking state, were shown to parallel the sensation of hunger. The close correlation between the hunger contractions of the stomach and the bodily movements was confirmed in 2 subjects who had no breakfast and were under observation lying on a couch throughout the day.

The experimental work mentioned so far was concerned with "hunger," not with starvation. Also, significant differences may be expected according to the degree and duration of starvation and the general experimental conditions, including the facility for movement. In fasting steers receiving water but no food for 14 days, Benedict and Ritzman (1927, p. 135) recorded the number of changes of position between standing and lying and the amount of time the animals spent in each position; they had a tendency to be less active and lay down for a larger proportion of the time as the fast progressed.

Movements in Sleep in the Minnesota Experiment

In the Minnesota Experiment the "motilograph" described by Schiele (1941) was used. It represents Hathaway's modification of the "hypnograph" designed by Stanley and Tescher (1931). The apparatus is attached to one leg of the bed and records the gross body movements as these are transmitted to the coiled springs which support the suspended bed. The instrument is equipped with a timing device. Only one motilograph was available and consequently only one record could be obtained in one night. The measurements were made at intervals on the same sample of subjects throughout the starvation phase of the experiment.

In evaluating the records a score was obtained as the total number of discernible movements. No attempt was made to measure the extent of the movements; it was noted on measurements repeated during the period of semi-starvation that the intraindividual variability in the magnitude of movements was small — that is, the subjects who initially had a pattern of large body movements retained this pattern throughout semi-starvation, and those with an initial pattern of small movements did likewise. The data are summarized in Table 322. It is apparent that the semi-starvation regimen did not in any way affect motility during sleep. Although in semi-starvation the length of sleep during the night was not changed, the men frequently took naps during the day.

TABLE 322

SLEEP MOTILITY IN THE MINNESOTA EXPERIMENT.

Month of Semi-Starvation	Number of Subjects	Length of Time in Bed (hrs.)	Number of Movements per Night	Number of Movements per Hour
1	9	7.7	36.6	4.75
3	9	7.3	34.5	4.73
5	8	7.6	35.7	4.69

Experimental Studies on Motor Performance in Fasting and Semi-Starvation

Quantitative studies of voluntary motor performance are much fewer in number than the field reports, which indicate, in a qualitative way, that under conditions of fasting and semi-starvation a marked deterioration of psychomotor functions takes place. The available experimental studies were carried out most frequently under conditions of fasting and consequently have little practical significance for considering the problem of famine and semi-starvation. Many of these studies suffer from methodological inadequacies, such as a lack of pre-experimental practice, and from the very small number of subjects. There is no information whatsoever on the rate of recovery of motor performance under conditions of refeeding after semi-starvation.

Weygandt (1904) investigated the effect of fasting on the speed of choice-reactions. In subject W the average time for 200 reactions increased from 357 sigmas under "normal" conditions to 391 after 12 hours of fasting, and to 423 after 24 hours of fasting; when the experiment was repeated on the same individual the "normal" value was 350, after 12 hours of fast 351, after 24 hours 368, after 36 hours 386, and after 48 hours 365. The changes in the second experiment were much less marked, possibly indicating an adjustment to fasting or to the general conditions of the experiment. The interpretation of the results is complicated by the continued effects of training. Thus in the experiment on subject R the score decreased from the "normal" value of 395 to 361 after a 24-hour fast and continued to decrease to 296 and 282 on the following two "normal" days; the effects of practice and of fasting cannot be separated.

During a total fast of 31 days (Langfeld, in Benedict, 1915) the initial (pre-fast) and final scores in the tests of strength were not markedly different. It appears that the strength scores were a resultant of 2 factors: (1) practice, preponderant in the first half of the fast, and (2) deterioration, resulting from the decrease in muscle mass. The values for the left hand, in lbs., increased from the pre-fast level of 88 to 99 on the sixteenth day of fasting and decreased to 87 on the thirty-first day; the corresponding values for the right hand were 77, 82.5 (on the twelfth day), and 78.5. The interpretation of changes in the scores in the tapping test in the same experiment is also complicated by the presence of practice effects continuing during the experimental period; the decline of performance after the fourth day of the fast and the return to the original level during the second half of the fast seem to indicate fluctuations in motivation rather than variations in physiologically determined work capacity.

In a fasting experiment in which 2 subjects took no food for a week, Marsh (1916) included tests of 3 psychomotor functions: grip strength (hand dynamometer), endurance (time of hanging by the arms), and steadiness (tracing a gradually narrowing slit, 25 cm. long, with a metal stylus; contacts with the sides were registered). There was deterioration in grip strength and, more markedly, in endurance. The results of the steadiness test were not consistent.

In the experiment by Benedict et al. (1919), in which the subjects lost about 10 per cent of the initial body weight over a period of several months, the only neuromuscular function which was considered to be definitely lowered (p. 637) was the strength of grip. We have recalculated from Benedict's Table 161 the right-hand values for the 11 men measured on October 17, 1917, and February 2, 1918; the respective means were 49.84 kg. and 48.42 kg. However, the changes were by no means consistent, varying from a decrement of 11.9 kg. to an increment of 4.4 kg., and the average difference between the initial and terminal values was far below statistical significance. Changes in other neuromuscular functions — coordination measured as accuracy of movement in tracing, patellar reflex, reaction time for turning the eye to a new fixation point — were either absent or unimpressive. There was some slowing down in the speed of successive horizontal eye movements and of the finger movements. The average time for a movement of the eye from right to left for 11 men was 93.4 milliseconds on October 28, 1917, and 105.0 on January 27, 1918; we found that the mean difference of +8.6 was highly significant ($F = 17.81[**]$ as compared with $F_{0.01} = 10.04$). Changes in the speed of eye movements from left to right were much smaller and inconsistent; the respective mean values are 91.55 and 92.55 ($F = 0.32$). The number of finger movements performed in 10 seconds did not decrease significantly.

In the test battery used by Glaze (1928) there were 3 psychomotor tests: writing the letter sequence "ab" for 20 minutes at maximum speed, hand steadiness (holding a stylus in holes of decreasing diameter), and body steadiness (measured by an ataxiagraph). The subjects fasted for 10 days (subject A), 17 days (subject B), and 33 days (subject C) and lost 10, 13, and 20 per cent of the body weight, respectively. During the fast subjects A and C wrote distinctly slower; the pre-fast, fast, and post-fast scores were 2813, 2576, and 2834 (subject A) and 2533, 2307, and 2620 (subject C). The performance of subject B was markedly affected only toward the close of his fast. During the fast the subjects not only started at a lower level of performance but the fatigability increased, as indicated by the drop of performance during the 20-minute test. The results of the hand steadiness test were not consistent, with deterioration in subject A, no clear-cut change in subject B, and improvement in subject C. The performance in the test of body steadiness was not affected by the fast.

The effects on performance capacity of an 8-day fast, during which the subject drank daily half a liter of coffee with 4 pieces of sugar, were investigated by Noltenius and Hartmann (1936). Work on the bicycle ergometer was carried to the start of exhaustion; it was equivalent to 4150 kgm. on the first day of the fast and to 2340 kgm. on the seventh day. The momentary (maximal) strength of grip was not affected but the fatigability increased. Whereas under normal conditions successive dynamometer scores in any one series varied in a random manner around the average value, the performance on the fifth day of the fast showed a pronounced decrement in successive determinations, with progressively declining scores of 61.5, 54.0, 53.0, 49.5, and 48.0 kg.

The only quantitative field study we have found on the effects of partial reduction in caloric intake on neuromuscular performance was carried out by Jokl (1946) on 150 Moslem Indian boys in Johannesburg, South Africa. Three standard tests of physical efficiency were applied, including a test of speed of locomotion (100-yard dash), endurance (600-yard run), and strength (12-lb. shotput). The group was tested during the last 3 days of the Ramadan festival, in which the total food intake is reduced in quantity and no food or drink is taken during the day. The dietary regimen is said to have caused a significant loss of weight, but this was not specified in the summary report which was available to us. The retests were made 3 weeks after the "fast." The "fast" resulted in a statistically significant deterioration of performance; on the basis of inspection of the graphically presented results, the change was most marked in the 100-yard dash. It is unfortunate that no pre-fast figures are available.

Motor Performance in the Minnesota Experiment — Over-All Trends

Field reports from areas of famine and semi-starvation refer to the decrease of muscular endurance and strength, the deterioration of coordination, and the slowing down of voluntary movements. We were able to confirm these qualitative observations in our group of experimental subjects.

We have discussed elsewhere (Brožek et al., 1946c) some of the methodological criteria which must be met before the results of psychomotor tests can be accepted as scientifically valid. Of special importance is a rigorous standardization of the test methods and of their application. Since no strictly comparable control group of subjects was available, an attempt was made to bring the performance of all subjects to a stable plateau by intensive practice during the 3-month pre-experimental period (see Franklin and Brožek, 1947).

The battery described in the Appendix on methods consisted of relatively "pure" tests of voluntary motor performance — that is, tests in which the various components of work capacity were to be tested one at a time. However, the isolation of the components of strength, speed, coordination, and endurance in actual tests is always imperfect, and in reality we deal unavoidably with "mixed" tests. Thus, the ball-pipe test involved both speed and eye-hand coordination. The "kick" test is by definition a test of "speed"; because a rapid development of force, necessary to move the large mass involved, is an essential component of the performance, the test actually represents a kind of test of "strength." It changed in semi-starvation more like the handgrip and back-lift tests than like the speed of tapping or the reaction time. In the test of pattern tracing, carried out while the subject was walking on the treadmill, the ability to walk in such a way as to lend the hand the greatest steadiness is under normal conditions a factor of less importance than the eye-hand coordination; in semi-starvation the weakness of the legs probably contributed in a large measure to the deterioration of the pattern-tracing performance.

The over-all changes in starvation and in the controlled phase of rehabilitation (R1 to R12) are indicated in Table 323. The semi-starvation deterioration in all functions for which control values were available reached the level of statistical significance. The statistical F-test takes into account both the average magnitude of the change in each function and the variability of the individual responses. It does not provide a criterion in terms of which deterioration in different functions could be compared. Such a comparison may be obtained by expressing the semi-starvation scores as a percentage of the control values, or, more legitimately, by relating the amount of change to the "normal" spread of the scores, measured as the standard deviation of the control values.

Using either citerion, a marked change was observed in the tests of the strength of grip and lift. Conversely, these functions showed a minimal degree of recovery after 12 weeks of nutritional rehabilitation. The deterioration produced by semi-starvation in tests involving primarily speed (tests of gross body reaction time, manual speed, and tapping, involving the movements of the whole body, of the arm, and of the wrist, respectively) was surprisingly small. In tapping no further decrement was observed beyond S12. In the test of manual speed (the ball-pipe test, involving repeated passing of a ball bearing through a vertically held pipe), the change from S12 to S24 was small, amounting to 31 per cent of the total difference between the scores at C and S24. In the test of reaction time, 64 per cent of the total deterioration, reflected in the lengthening of the time per reaction, occurred from C to S12, 36 per cent from S12 to S24. In rehabilitation the percentages of recovery ranked the 3 speed tests in the same order, with a recovery at R12 of 100 per cent in tapping, 85 per cent in the ball-pipe test of manual speed, and 70 per cent in the test of reaction time.

The tests of the maximal speed of single leg movements, without and with a load of 10 lbs. attached to the ankle, were applied only at the end of starvation and during rehabilitation. Because of the lack of control values, the amount

TABLE 323

CHANGES IN PSYCHOMOTOR FUNCTIONS during semi-starvation and the early phase of rehabilitation in the Minnesota Experiment. $N = 32$, except for back dynamometer where $N = 29$.

Test	C	S12	S24	R6	R12	$d = S24 - C$	F_d	$\dfrac{100d}{C}$	$\dfrac{d}{SD_C}$	$\Delta = R12 - S24$	$\dfrac{100\Delta}{d}$
Hand dynamometer (kg.)											
M	58.2	47.2	41.8	42.3	47.2	−16.4	22.33[**]	28.2	2.2	+5.4	32.9
SD	7.6	6.5	6.3								
Back dynamometer (kg.)											
M	165.4	131.3	116.1	119.0	128.4	−49.3	105.25[**]	29.8	1.8	+12.3	24.9
SD	27.8	19.8	17.8								
Leg movement, in milliseconds, no load											
M	(110)*		133.6	130.9	125.5					−8.1	(34.3)*
SD			12.3								
Leg movement, in milliseconds, 10-lb. load											
M	(120)*		170.5	155.5	149.3					−21.2	(42.0)*
SD			26.6								
Gross body reaction time in 3/100 second											
M	42.3	44.4	45.6	44.4	43.3	+3.3	26.59[**]	7.8	0.9	−2.3	69.7
SD	3.7	4.1	4.7								
Manual speed, balls per minute											
M	75.3	72.6	71.4	73.7	74.8	−3.9	31.49[**]	5.2	0.7	+3.3	84.6
SD	5.7	5.4	6.5								
Tapping, taps per 10 seconds											
M	66.6	68.2	63.2	64.2	66.6	−3.4	34.10[**]	5.1	0.6	+3.4	100.0
SD	5.5	5.1	4.5								
Pattern tracing, number of error contacts											
M	115.3	135.3	135.9	132.7	125.0	+20.6	130.42[**]	17.9	1.5	−10.9	52.9
SD	13.7	13.4	13.4								
Pattern tracing, length of error contacts, in 0.25 second											
M	36.8	41.7	42.8	38.97	39.03	+6.0	68.29[**]	16.3	0.8	−3.8	63.3
SD	7.7	8.1	9.4								

* Estimated values.

705

of deterioration in starvation and of recovery in rehabilitation cannot be determined with accuracy. An approximation may be attempted by using values obtained in testing other experimental groups comparable in fitness and age to the semi-starvation group at the control period. Values of 110 milliseconds without load and 120 milliseconds with load may serve as estimated control values. Using these estimates we arrive at R12 at recovery values of 34 and 42 per cent of the semi-starvation deterioration; these values are close to those obtained for the recovery of grip and lift strength.

Pattern tracing, a test of eye-hand-foot coordination, stands about midway between the tests of speed and the tests of strength, both in the amount of deterioration in starvation and in the rate of recovery. The score based on the number of error contacts was more sensitive to the stress of semi-starvation and slower to recover than the time score obtained as the total time of the error contacts.

The quantitative description of changes corresponded well with clinical observations of the subjects' behavior. They walked up stairs one stair at a time, supplemented arm strength by the weight of the body in opening heavy doors, and could carry heavy objects such as suitcases only with difficulty and for a short distance. Their bodily movements were slower, but accurate movements of small muscle groups such as are involved in typing or in work on computational machines exhibited no striking changes. In order to extend the range of situations sampled, the subjects were asked at the end of semi-starvation to describe the ways in which their psychomotor capacity differed from the pre-starvation conditions. It may be useful to include a few typical comments, related to the different aspects of voluntary motor performance.

Strength: "Lifting patients in the hospital is nowhere nearly as easy as in the control period." "Lifting the mattress of the bed to tuck the blankets in is a real chore." "Moving furniture when cleaning is much more difficult." "I notice weakness in my arms when I wash my hair in the shower; they become completely fatigued in the course of this simple operation." "When I tried to open a bottle of ink I found I could not do it but an 'overhead' man [not starved] did it without trouble." "It is more difficult to lift my feet, especially in trying to jump over something." "I find repeatedly that my strength has decreased more than I anticipate. The carriers for the 12 urine bottles are a hell of a lot heavier than they used to be." "I was struck to find how difficult it was to lift a lawn mower out of a closet."

Speed: "It is an effort to move rapidly." "When the water in the showers becomes hot because of flushing of the toilets, my reaction time in adjusting controls seems to be longer than in the control period." "In playing ball I started to run after a fly but my knees caved in and I nearly fell." "It used to take me two and a half hours or less for an eleven-mile walk, now I have to push myself to do it in three hours."

Gross coordination: "I tend to 'weave' when I walk." "Attempting to play baseball I was very inaccurate and weak in swinging the bat." "I trip over small things, can't balance on one foot any more, have difficulty walking a straight line." "I have difficulty in dodging traffic or walking in crowds where I have to start and stop and change direction." "In rowing I had difficulty in hitting the water with both oars at the same time and with equal force."

Fine coordination: "My handwriting is getting worse." "Dressing takes longer. I have to really concentrate on putting on shoes, folding socks, sewing." "I am more clumsy than I used to be in lighting cigarettes." "I walk unsteadily but print well." "In sewing my movements are more jerky." "In walking I have a poor muscular control but I can type at a high speed without errors." "I do not find precision work more difficult, it just takes more time." "In darning it is more difficult to do a neat job." "My precision in typing, writing, or lettering is about as good as usual but I work at a slower speed."

Endurance: "While carrying a suitcase for Bob my muscles 'gave way'." "In carrying books to the library I needed help." "In walking I become so weak that I tremble all over." "My arm becomes tired quickly when I am typing." "Climbing up the stadium stairs [two flights] is really a fatiguing task." "At one time I could push up my body fifteen times, now barely five times." "In shop work my ability for sustained effort has decreased." "Sustained work at a high speed, as in typing, is very tiring."

These verbatim quotations from the subjects' reports indicate the variety of life situations in which the reduced motor capacity was reflected. The reports are unanimous with respect to a marked decrease in strength, endurance, and gross coordination. The anecdotal material on speed was not very extensive. The reports on fine coordination were not consistent from subject to subject. Taken as a whole, the subjective reports round out the picture of motor deterioration in starvation and indicate the extent and the biological — as distinguished from the statistical — significance of the changes.

Recovery in Motor Functions in Relation to Dietary Factors — Minnesota Experiment

The average recovery scores in grip strength, expressed as a percentage of the semi-starvation decrement, tended to follow the level of caloric supplements, except at the highest caloric level (see Tables 324 and 325). In terms of the raw mean differences between S24 and R12, only the comparison of the Z + L and G + T groups and of the Z and G groups approached (but did not reach) the 5 per cent point of statistical significance. The 16 men who did not receive protein supplements actually recovered in grip strength somewhat more than the supplemented men. The vitamin supplements resulted in a positive gain but not one statistically significant.

In the back dynamometer the caloric differentiation was again reflected in differences in recovery up to the supplement of +800 Cal. (group G); the top caloric group (T), showed a somewhat smaller recovery. Because some of the men were unable to take this strenuous "back-breaking" test, the methods of statistical analysis generally used were not applicable. Both protein and vitamin supplements had an apparent positive effect, but the differences were small and could be due to chance (see Table 326).

In the test measuring the maximal speed of leg-lift ("kick"), significant differences related to caloric intake were obtained. Averaging the recovery values

TABLE 324

RECOVERY IN THE HAND DYNAMOMETER SCORES, in kg., from R1 to R12, in reference to dietary factors (Z = basal, L = +400, G = +800, T = +1200; U = protein-unsupplemented, Y = -supplemented; P = vitamin-unsupplemented, H = -supplemented) (Minnesota Experiment).

Dietary Group	dS24	$\triangle$R6	$\triangle$R12	$\dfrac{100 \times \triangle\text{R12}}{\text{dS24}}$
Z	−16.8	−0.4	+2.6	15.5
L	−18.0	−1.5	+4.4	24.4
G	−16.5	+2.2	+7.8	47.3
T	−14.2	+1.8	+6.8	47.9
Z + L	−17.4	−0.9	+3.5	20.1
G + T	−15.4	+1.9	+7.2	46.8
U	−17.4	+0.8	+6.1	35.1
Y	−15.3	+0.2	+4.7	30.7
P	−16.0	+0.3	+4.1	25.6
H	−16.8	+0.7	+6.7	39.9

TABLE 325

SIGNIFICANCE OF THE DIFFERENCES BETWEEN VARIOUS DIETARY GROUP-INGS AT R12 IN THE RATE OF RECOVERY OF THE GRIP STRENGTH. V_{rep} = the "experimental error" (replicate variance) = 28.50. For 1 and 16 degrees of freedom, $F_{0.05}$ = 4.49. (Minnesota Experiment.)

	V_{bGr}	F
Z vs. L	12.25	
Z vs. G	105.06	3.69
Z vs. T	68.06	
L vs. G	45.56	
L vs. T	22.56	
G vs. T	4.0	
(Z + L) vs. (G + T)	112.50	3.95
U vs. Y	15.12	
P vs. H	55.12	

for R10 and R12, we obtained for the groups Z, L, G, and T, respectively, decreases of −3.1, −2.5, −12.7, and −14.4 milliseconds. Combining the 2 lower (Z + L) and the 2 upper (G + T) caloric groups, there is a highly significant difference between the 2 means (F = 9.47 [**]; for 1 and 16 degrees of freedom, $F_{0.01}$ = 4.49). The significance of the differences between the 8-man caloric groups is indicated in Table 327. Neither protein nor vitamin supplements can be considered as having a beneficial effect on the recovery of this function.

We were surprised to find that the recovery in the time of "kick" with the weight of 10 lbs. attached to the foot did not have any consistent relationship to the level of caloric refeeding. The difference in the behavior of the function under the two conditions — without and with the additional weight — is puzzling. On theoretical grounds we expected that the additional "stress" of lifting the foot against a load would increase rather than decrease the sensitivity to the varied levels of refeeding.

TABLE 326

RECOVERY IN THE BACK DYNAMOMETER SCORES, in kg., from R1 to R12, in reference to dietary factors (Z = basal, L = +400, G = +800, T = +1200; U = protein-unsupplemented, Y = protein-supplemented, P = vitamin-unsupplemented, H = vitamin-supplemented). dS24 = C − S24; $\triangle$R6 = R6 − S24; $\triangle$R12 = R12 − S24. (Minnesota Experiment.)

Dietary Group	dS24	$\triangle$R6	$\triangle$R12	$\dfrac{100 \times \triangle\text{R12}}{\text{dS24}}$
Z	−37.3	−4.6	+2.0	5.4
L	−65.1	−2.3	+10.3	15.8
G	−47.8	+11.5	+21.7	45.4
T	−46.8	+7.2	+15.9	34.0
Z + L	−51.2	−3.4	+6.1	11.9
G + T	−47.2	+9.1	+18.4	39.0
U	−48.5	+2.1	+8.9	18.4
Y	−49.9	+3.5	+13.1	26.2
P	−51.0	+0.5	+11.5	22.5
H	−47.6	+5.4	+13.0	27.3

TABLE 327

COMPARISON OF RECOVERY IN THE TEST OF THE MAXIMAL SPEED OF LIFTING THE LEG, WITH NO EXTRA LOAD ATTACHED, for the different caloric groups. V_{rep} = the "experimental error" (replicate variance) = 118. For 1 and 16 degrees of freedom, $F_{0.05}$ = 4.49. (Minnesota Experiment.)

	V_{bGr}	F
Z vs. L	25	
Z vs. G	484	4.49[*]
Z vs. T	410	
L vs. G	729	6.18[*]
L vs. T	638	5.41[*]
G vs. T	3	

As we have indicated in the preceding section, the measured losses in "pure" tests of speed were negligible. Under these conditions any clear-cut differentiation in the amount of recovery between groups receiving different diets could hardly be expected and was not obtained.

In pattern tracing, a test of eye-hand-foot coordination, the increase in number of error contacts during starvation was more marked than the lengthening of the time of error contacts; only the first score will be considered here. The recovery in the 4 caloric groups, from the lowest to the highest caloric supplement, amounted to 51, 61, 47, and 54 per cent of the semi-starvation deterioration, respectively, indicating a lack of differentiation.

Considering the psychomotor material as a whole, we were somewhat disappointed by the relative lack of a significant relationship between the rate of recovery and the dietary factors.

Motor Functions in the Later Stages of Rehabilitation — Minnesota Experiment

During the period from R12 to R20 the grip strength continued to improve, rising for the 12 men who remained under direct observation in the Laboratory from 46.8 kg. at R12 to 51.5 kg. at R20; this improvement was highly significant. However, the recovery at R20 was not complete, corresponding to only 64 per cent of the starvation loss (see Table 328). The mean increase in grip strength was insignificantly larger in the 6 men maintained from R1 to R12 on a lower caloric level than in the other 6 subjects; in terms of the semi-starvation decrement the Z + L sub-group at R12 had regained 25 per cent and at R20 a total of 60 per cent of the loss; for the G + T sub-group the corresponding values were 38 per cent and 69 per cent (see Table 329). In the 20 subjects retested during the follow-up examinations made 33 weeks after the end of semi-starvation the grip strength had returned to the pre-starvation level; their average scores in kg. were 59.2 at C, 42.7 at S24, 46.8 at R12, and 58.6 at R33.

TABLE 328

MOTOR FUNCTIONS in the later stages of rehabilitation in the Minnesota Experiment. $N = 12$.

Test	C	S24	R12	R20	(R20 — R12)		(R20 — C)		(R20 — S24)
					d	F	d	F	(C — S24)
Hand dynamometer (kg.) ...	56.7	42.2	46.8	51.5	+4.7	19.06[**]	−5.2	25.60[**]	.64
Back dynamometer (kg.) ...	157.9	116.5	128.0	151.3	+23.3	36.68[**]	−6.6	4.95[*]	.84
Leg movement, in milliseconds, no load .	(110)*	132.8	122.5	112.2	−10.3	19.45[**]	(+2.2)*		.73
Leg movement, in milliseconds, 10-lb. load	(120)*	164.7	144.1	129.8	−14.3	18.24[**]	(+9.8)*		
Pattern tracing, number of contacts .	114.5	130.8	121.5	118.9	−2.6	2.37	+4.4	3.28	.73
Pattern tracing, length of contacts, in 0.25 second ..	39.0	44.5	40.5	40.2	−0.3	0.4	+1.2	1.98	.78

* Estimated values.

TABLE 329

Motor Functions in the later stages of rehabilitation in the Minnesota Experiment; comparison of men belonging from R1 to R12 to groups Z + L ($N = 6$, except for back dynamometer where $N = 5$) and to groups G + T ($N = 6$).

Test	(S24 − C)	(R12 − S24)	(R20 − R12)		(R20 − C)	
			d	F	d	F
Hand dynamometer (kg.)						
Z + L	−15.0	+3.8	+5.2		−6.0	
G + T	−13.8	+5.2	+4.3		−4.3	
(Z + L) vs. (G + T) ..				0.14		0.65
Back dynamometer (kg.)						
Z + L	−44.4	+2.4	+30.0		−12.0	
G + T	−38.9	+19.0	+17.7		−2.2	
(Z + L) vs. (G + T) ..				2.85		2.93
Leg movement, in milliseconds, no load						
Z + L		−2.3	−16.7			
G + T		−18.2	−4.0			
(Z + L) vs. (G + T) ..				19.45[**]		
Leg movement, in milliseconds, 10-lb. load						
Z + L		−11.2	−20.3			
G + T		−30.0	−8.3			
(Z + L) vs. (G + T) ..				4.10[*]		
Pattern tracing, number of contacts						
Z + L	+13.7	−8.5	−1.7		+3.5	
G + T	+19.0	−10.2	−3.5		+5.3	
(Z + L) vs. (G + T) ..				0.28		0.13
Pattern tracing, length of contacts, in 0.25 second						
Z + L	+4.0	−3.0	−1.3		−0.3	
G + T	+7.0	−5.0	+0.8		+2.8	
(Z + L) vs. (G + T) ..				0.04		1.98

During the 8 weeks following R12 the group of 12 men improved markedly in their back-lift performance. While the recovery was initially slow, the mean score rising from 116.5 kg. at S24 to 128.0 kg. at R12 (28 per cent of the semi-starvation loss), in the period from R12 to R20 they scored a gain of 23.3 kg., representing a total recovery of 84 per cent of the starvation loss. The difference between the rate of recovery of the Z + L and G + T sub-groups was larger than in the case of the handgrip, with the back-lift recovery reaching 5 and 38 per cent at R12 and 73 and 76 per cent at R20. The Z + L sub-group was "catching up" with the men who were more rapidly rehabilitated between R1 and R12.

The time needed for moving the leg at maximum speed through a given angle decreased during the period R12 to R20 for the condition without load by 10.3 milliseconds and for the condition with an additional load of 10 lbs. attached to the leg by 14.3 milliseconds. Using the estimated "normal" values of 110 and 120 milliseconds, respectively, the group of 12 men by R20 could be considered to have recovered 90 per cent (no load) and 78 per cent (10-lb.

load) of the semi-starvation deterioration. When a load of 10 lbs. is attached to the ankle, the movement is slower. This difference between the time of leg-lift with and without load was maximal at the end of semi-starvation, and it decreased as the rehabilitation progressed. At S24 the addition of a weight load of 10 lbs. produced a lengthening of the movement time by 31.9 milliseconds; at R12 the difference decreased to 21.6 milliseconds, and at R20 it was reduced to 17.6 milliseconds.

There was a marked difference between the caloric sub-groups in the rate of recovery of the speed of leg-lift from R12 to R20. For the condition without load the improvement for the $Z + L$ group corresponded to -16.7 milliseconds, for the $G + T$ group to -4.0 milliseconds. With load the duration of the kick movement decreased again more markedly for the men who had been previously maintained at the lower caloric levels; the changes were -20.3 and -8.3 milliseconds, respectively. The difference between the R12 to R20 recovery in the 2 sub-groups was statistically highly significant ($F = 19.79[**]$) for performance without load and approached the 5 per cent level of significance for the condition with load ($F = 4.10$; $F_{0.05} = 4.96$).

In the pattern-tracing test some recovery was still taking place from R12 to R20 and was reflected in the decrease of the number of error contacts between the stylus and the sides of the groove; the further improvement in the time score — that is, in the duration of error contacts — was negligible, although the pre-starvation level of performance was not fully reached. There was no significant difference between the recovery from R12 to R20 in the $Z + L$ and $G + T$ sub-groups; also, the means of the differences between the scores obtained at R20 and at C were not significantly different.

Summary and Conclusions

The neurological changes in the Minnesota Experiment were minimal, thus conforming to the pattern observed in European areas of semi-starvation. A diminution of the response of the tendon reflexes, especially of the patellar reflex, was the only major symptom, but even this was not universal or extreme. Only 2 subjects developed mild paresthesias; this symptom should be differentiated clearly from the much more frequent phenomenon of the extremities "going to sleep." One subject developed neurological disturbances, mild in character, which involved both cutaneous (sensory) and motor functions and were considered to be of hysterical origin.

The results of quantitative observations on motility during sleep were negative. There were no changes in the total number of gross body movements during the night or in their distribution during the period of sleep.

The components of the capacity for brief motor performance — strength, speed, and coordination — were examined by a battery of tests. Strength was measured by handgrip and back-lift dynamometers. The tests of speed included small and large repetitive movements of the hand and arm (tapping and the ball-pipe test of manual speed) and a movement of the whole body involving bending at the knees (choice-reaction time). The test of moving the leg at maximum speed through a given angle, without and with an additional weight attached to the leg, may be regarded as measuring dynamic strength rather than speed, and it was classified, for the purpose of economy, together with the strength tests. The element of coordination is involved to some degree in all tests of muscular performance. In its "pure" form, coordination was tested by having the subject trace a pattern and determining the number and length of the error contacts between the stylus and the sides of the grooved pattern. Since the subject did this while walking on the treadmill, the performance involved a complex eye-hand-foot coordination and steadiness of the whole body.

The largest deterioration was obtained on tests in which peripheral factors (state of the muscles) limited the performance capacity. Tests in which the central nervous system was a critical component, such as the test of speed of tapping, showed changes that were consistent from subject to subject and therefore were statistically significant but of little biological importance. This was true also of the test of manual speed and coordination (ball-pipe). Only under additional "stress," such as the necessity of keeping the body steady while walking, does coordination — tested by the pattern-tracing method — show deterioration in semi-starvation.

The rate of recovery was inversely related to the magnitude of deterioration, with the tests of speed showing the most rapid return to pre-starvation values and the tests of strength the slowest. In grip strength a full recovery was approached between the twentieth and thirty-third week after the end of semi-starvation.

During the first 12 weeks of nutritional rehabilitation the vitamin and protein supplementations did not result in statistically significant differentiations. The caloric differences were reflected with some degree of consistency only in the tests of strength.

It is evident that under conditions of famine the capacity for work will be most seriously affected in jobs requiring lifting, pushing, and carrying loads. The ability to climb, to walk long distances, or to stand for prolonged periods of time will also be diminished. Performance in jobs involving speed and accuracy may be expected to be less impaired. However, it should be emphasized that all the Minnesota Experiment tests were of short duration. No attempt was made to investigate endurance in light manual work such as is typical of a large majority of industrial operations. Self-observations by the subjects and results of tests that consisted of running to exhaustion on a motor-driven treadmill indicated a marked reduction in endurance as far as strenuous physical activities are concerned. The problem of endurance in activities not resulting in a marked increase of total energy expenditure or a rapid exhaustion of small muscle groups can be approached in the laboratory only by means of job miniature situation tests such as we have used in studies on vision (Brožek, Simonson, and Keys, 1947). In the Minnesota Experiment no attempt was made to investigate this important component of man's capacity for industrial work.

CHAPTER 34

The Capacity for Work

THE energy required for the performance of physical work is, of course, ultimately derived from the food eaten. When the organism is in caloric balance, the energy intake equals the energy output and there is no net change in the total energy stores of the organism. Under conditions of restricted energy intake, as in famine and semi-starvation, the caloric deficit (energy expenditure less caloric intake) is paid at the expense of the energy stored in the cells of the body as fat, protein, and carbohydrate. The fat stored in the cells of the adipose tissue plays no vital physiologic role in the life processes of the organism and consequently can, within limits, be drawn upon with impunity to supply the calories lacking in the semi-starvation diet. When, however, the caloric deficit is of such duration that the readily available fat stores are exhausted, demands for energy are made upon the essential cellular constituents; the cellular proteins are then the principal endogenous source of energy still available in the body. The drain on the essential constituents of the protoplasm is soon reflected in a lowered reserve capacity and an altered physiological functioning of the organism.

It is common knowledge that the capacity of an organism to accomplish physical work decreases with continued caloric undernutrition. Weakness is one of the cardinal symptoms of starvation; in extreme cases the individual does not possess sufficient strength and endurance to walk or even to stand, but remains prostrate, passively waiting for death. In spite of the general appreciation of the qualitative aspects of the physical deterioration produced by famine conditions, no quantitative analysis of the progressive deteriorative processes has been made until now. Physical work capacity may be limited, not only by lack of fuel, but also by the performance of the circulatory, respiratory, and neuromuscular systems. Alterations and limitations of function in any or all of these systems will, of course, be reflected in the capacity to perform work.

What is the sequence of events in the progressive deterioration that occurs in semi-starvation? What systems change, and to what extent do they change? What are the factors that limit work capacity in the starved individual? Are the alterations in function of the different organ systems permanent? What is the degree and time sequence of response of the different components of work capacity to refeeding? These and many other important questions are not answered by the meager data in the literature on semi-starvation.

In the Minnesota Experiment an effort was made to obtain a qualitative and quantitative description of the progressive deterioration of work capacity that occurred in semi-starvation and of the recovery of the functions during the refeeding periods. Work performance was studied in 2 general work situations: aerobic (steady state) work maintained for a relatively long period of time (30 minutes) and anaerobic (short exhausting) work of an intensity that could be maintained for only a few minutes. The ability of the whole group of 32 subjects to perform anaerobic work was measured by the Harvard Fitness Test (Johnson, Brouha, and Darling, 1942); the details of the physiological response to anaerobic work were studied in a selected group of the men with a work test of maximal intensity that could be sustained for a standard length of time (3 minutes). The data, and their interpretation, from these work tests are, to maintain clarity of description, presented as relatively independent sections in this chapter. Some integration of these data with other information has been attempted as well.

Aerobic Work — General Features

The aerobic work test used in the Minnesota Experiment consisted of a 30-minute walk on a motor-driven treadmill at 3.5 miles an hour and a 10 per cent grade. Air temperature and humidity were standardized at 78° F. and 50 per cent relative saturation. Work pulse rates were counted with a stethoscope placed over the apex of the heart and timed by means of a stop watch for 15-second intervals at 25, 28, and 30 minutes of the walk. The subject then stepped off the treadmill and the standing recovery heart rate was counted at 60 to 75, 120 to 135, and 180 to 195 seconds of recovery. The 3 work pulse rates were averaged to obtain the "work pulse rate." The 1- and 2-minute recovery pulse rates were averaged for the "recovery pulse rate." Expired air was collected in a Tissot gasometer during the twenty-third to twenty-eighth minute of the work period and was analyzed by the Haldane method for oxygen and carbon dioxide content. Oxygen consumption, ventilation rate, respiratory efficiency, respiratory quotient, and caloric cost of work were calculated.

The magnitude of the pulse rate response to work of moderate intensity is generally considered to be rather closely related to the physical work capacity. It is well known that physical training reduces the work pulse (Knehr et al., 1942). It has also been demonstrated that under conditions of stress involving physical deterioration the work pulse rate is higher than before application of the stress — for example, acute and chronic thiamine deficiency (Keys et al., 1945), high environmental temperature (Taylor et al., 1943a, 1943b; Henschel et al., 1943, 1944), acute fasting (Taylor et al., 1944), prolonged bed rest (Taylor et al., 1945b, 1949), and induced malaria (Henschel et al., 1948, 1950). The pulse rate response to a standard work task cannot be used by itself, however, as an infallible index of physical fitness. Under normal conditions the interindividual variation of the pulse rate response to work is large and is not, in individual cases, necessarily indicative of the performance capacity. Some individuals with a large pulse rate response to a work task are none the less capable of doing large amounts of work while others with a normal response would, by other criteria, be classed as less physically fit.

It is also well recognized that emotional factors such as fear, excitement, and concern over the outcome of the test have a profound effect upon the resting and work test pulse rates. The work pulse rates are most valid in standardized experimental conditions where the individual is familiar with the test procedures and is under no emotional tension in the presence of the observer. Under such circumstances differences in the individual pulse rate response to a work test in the normal state and in a stress situation involving deterioration and a decrease of the reserve capacity of the organ systems — particularly cardiovascular deterioration — can be used as a relative measure of resistance to the stress.

Aerobic Work Capacity in Semi-Starvation

The absolute heart rate during work of moderate intensity was relatively uninfluenced by semi-starvation and certainly did not reflect the large degree of physical deterioration that had occurred (see Table 330). But the lack of response of the heart rate to the starvation stress was actually only apparent. During the 24 weeks of semi-starvation the subjects lost 24 per cent of their control body weights, with a corresponding decrease in the rate at which work was accomplished during the test. Under ordinary circumstances, it would be expected that a 24 per cent decrease in the work intensity would be accomplished with a lower work pulse rate. Some indication of an apparent compensatory lower work pulse rate was present at S12, at which time the weight loss was probably greater than the loss of physical capacity. However, the absolute work pulse rate does not give as true an indication of the response to the work task as does the work pulse rate increment, which is the difference between the rest and work heart rate. Although the work pulse rates were not increased during semi-starvation, the resting pulse rates were dramatically reduced. The work pulse rate increment clearly demonstrated that the work test was harder for the subjects in semi-starvation than during the control period even though the work intensity was lower during semi-starvation. The work pulse rate increment was increased 18.8 and 21.8 per cent above the control value at S12 and S24, respectively.

The absolute recovery pulse rate, taken as the average of the rates in standing recovery at 60 to 75 and 120 to 135 seconds, progressively decreased during semi-starvation. The more rapid pulse rate recovery following work in the starvation regimen cannot be assumed to indicate that the subjects were in a better physical condition at that time. Pulse rate recovery curves are generally of the exponential type; other things being equal, the greater the differential between

TABLE 330

MEAN AEROBIC WORK AND RECOVERY PULSE RATES, and pulse rate increments, for the 32-subject group at control (C), 12 and 24 weeks of semi-starvation (S12 and S24), and 6 and 12 weeks of rehabilitation (R6 and R12) (Minnesota Experiment).

	C	S12	S24	R6	R12
Work pulse (beats/min.)	133.9	128.9	133.3	130.0	130.9
Increment	78.8	93.6	96.0	83.3	83.1
Recovery pulse (beats/min.)	105.9	98.4	95.2	93.7	99.3
Increment	50.8	63.1	58.0	47.0	51.5
Work blood sugar (mg. %)	71.4	61.7	59.8	62.4	57.2
Basal blood sugar (mg. %)	(71.4)		63.4	65.2	73.1

the rest and work rates the steeper will be the early phase of the pulse rate recovery curve. In spite of the substantially lower 1-2-minute recovery pulse rates in semi-starvation, the rates were farther from the pre-work resting values than they were during the control period. The average 1-2-minute recovery pulse rate, expressed as percentage of the resting pulse rate, was 192.1, 278.8, and 255.9 per cent at control (C) and 12 (S12) and 24 (S24) weeks of semi-starvation, respectively.

Complete oxygen consumption and respiratory data during work were obtained on 16 of the subjects. The data are summarized in Table 331.

TABLE 331

AEROBIC OXYGEN CONSUMPTION DATA (N.T.P.)*, means for the 16-subject group, at control (C), 12 and 24 weeks of semi-starvation (S12 and S24), and 6 and 12 weeks of rehabilitation (R6 and R12) (Minnesota Experiment).

	C	S12	S24	R6	R12
Total oxygen used (cc./min.)	1826.6	1387.7	1309.9	1358.8	1475.2
Ventilation (liters/min.)	32.47	29.69	28.13	27.33	28.23
Respiratory efficiency	56.42	50.18	47.71	49.94	52.38
Respiratory quotient	0.867	0.919	0.931	0.948	0.938
Net Cal./min.	7.795	6.326	5.981	6.013	6.444
Net Cal./kg./min.	.1103	.1106	.1117	.1115	.1113
Oxygen-pulse	13.53	10.77	10.04	10.79	11.27
Oxygen-pulse/kg.	.1949	.1880	.1909	.1990	.1917
Circulatory index	.6717		.6469	.6612	

* N.T.P. = gas volumes corrected to 0° C. and 760 mm. Hg. barometric pressure.

The total oxygen consumption during this standard work of moderate intensity was decreased 24.1 per cent at S12 and 28.3 per cent at S24, the decrease in both cases being slightly greater than the percentage loss of body weight (17.5 and 24.2 per cent body weight loss at S12 and S24, respectively). The caloric expenditure per minute during work, corrected for the basal metabolic rate, was decreased 18.8 and 23.3 per cent at S12 and S24, respectively; these changes were practically identical with the body weight loss, and there was no net change in the caloric cost of work per kg. of body weight during semi-starvation. Some of the difference between the percentage decrease in oxygen consumption and the caloric cost of the work is explicable on the basis of the higher work respiratory quotient in the starvation period (7.4 per cent higher at S24). Theoretically, the increased work respiratory quotient during starvation could be explained on the basis of (1) increased difficulty of the work task (Dill, Talbott, and Edwards, 1930; Schneider, 1939, p. 105), (2) hyperventilation (Table 331), or (3) changes in the general metabolic patterns.

The higher R.Q. in work observed during semi-starvation is explained on the basis of a changed metabolic pattern. It has been shown that in mild exercise the work R.Q. is comparable to the rest R.Q. and consequently should reflect the composition of the food (Carpenter, 1931; Dill, 1936; Gemmill, 1942). The work respiratory quotients in the Minnesota Experiment were, in both the control and the semi-starvation periods, practically identical with the non-protein R.Q. of the diet being consumed. The calculated non-protein R.Q. of the average diet was 0.84 and 0.94 at control and at S24, respectively, as compared to the work R.Q. of 0.867 and 0.931 (Table 331). The high value at S24 reflects the fact that the diet was extremely low in fat; at this late stage of the semi-starvation period the rate of weight loss was close to zero and metabolic balance was being approached.

The development of a relative tachycardia in work during semi-starvation is further indicated by the 25.8 per cent decrease in the oxygen-pulse. The oxygen-pulse is calculated as the cc. of oxygen consumed per heart beat. In other words, in work during the starvation period the heart was beating faster than was necessary to furnish oxygen to the tissues at the control period rate. The oxygen-pulse per kg. of body weight was not, however, significantly altered at either the twelfth or the twenty-fourth week of semi-starvation. The circulatory index, which is the oxygen-pulse corrected for the oxygen capacity of the blood, was decreased slightly, but not significantly, during semi-starvation. It is apparent that the cardiovascular system was not seriously embarrassed in starvation by the intensity of the work involved in the aerobic work test.

Pulmonary ventilation during work of moderate intensity was 8.6 and 13.4 per cent below the pre-starvation control level at S12 and S24, respectively. The decrease in ventilation was, however, proportionally less than the decrease in oxygen consumption; there was relative hyperventilation with a resultant lowered respiratory efficiency. The decrease in the amount of oxygen removed per liter of ventilation was 11.1 per cent at 12 weeks of semi-starvation and 15.4 per cent at 24 weeks. In this Laboratory the respiratory efficiency has been found to decrease in a variety of stress situations involving deterioration of physical work capacity. As the work intensity more closely approaches the maximum work capacity, the ventilation rate increases and the relative amount of oxygen removed from the inspired air decreases. The lower respiratory efficiency in semi-starvation would indicate a deterioration in the ability to do the aerobic work even though the efficiency with which the work was done was not altered.

Aerobic Work Capacity in Rehabilitation

The work and recovery pulse rate data for the 32-subject group in rehabilitation are summarized in Table 330; the corresponding oxygen consumption data for the 16-subject group are given in Table 331.

The work pulse rates were on the average slightly lower at both 6 and 12 weeks of rehabilitation than during the control period. The resting pulse rates were higher than during semi-starvation, so that the work pulse increment had returned nearly to the control levels. The recovery pulse rates reached their lowest point at R6 and then increased again at R12. With the increase in the resting pulse rates, the recovery pulse increment was substantially normal at both R6 and R12.

The oxygen consumption during work and the caloric cost of the work increased progressively in rehabilitation in proportion to the gain in body weight, so that the caloric cost of the work per kg. of body weight remained constant (Table 332). Total ventilation did not increase above the S24 low at either R6 or R12, but the amount of oxygen removed per liter of ventilation (respiratory efficiency) increased; 25.6 per cent and 65.1 per cent of the loss that occurred during semi-starvation were regained at R6 and R12, respectively. The work respiratory quotient remained at the S24 level during the 12 weeks of rehabilitation. The failure of the R.Q. to return to the control value is not explainable on the basis of increased fat deposition with a resulting higher R.Q. The 12-subject group on the average increased their body fat by only 3.78 kg. during the first 12 weeks of rehabilitation while the R.Q. remained almost constant. In the following 4-week period from R12 through R16 the body fat was increased by 4.52 kg., but during that period of rapid fat deposition the R.Q. decreased 0.034 (see Table 332). The work R.Q. and the non-protein R.Q. of the diets were comparable, however, for the semi-starvation period and the first 12 weeks of rehabilitation. The R.Q. of the diet no doubt reached more normal levels from R12 to R16 when the dietary items were no longer restricted; at the same time the work R.Q. also decreased toward the control values.

TABLE 332

AEROBIC WORK DATA, mean values for the 12-subject group, at control (C), 24 weeks of semi-starvation (S24), and 6, 12, 16, and 20 weeks of rehabilitation (R6, R12, R16, and R20). Oxygen consumption and ventilation values are corrected to N.T.P. (Minnesota Experiment.)

	C	S24	R6	R12	R16	R20
Work pulse (beats/min.)	136.5	136.3	129.5	133.4	160.9	138.8
Increment	80.4	98.5	86.8	81.1	90.0	81.5
Recovery pulse (beats/min.) ..	107.3	97.6	97.0	101.0	120.3	108.0
Oxygen (cc./min.)	1743.6	1310.2	1327.6	1470.0	1906.8	
Ventilation (liters/min.)	31.10	28.24	26.91	27.74	36.08	
Respiratory efficiency	56.34	46.60	49.51	52.99	53.10	
Respiratory quotient	0.877	0.939	0.940	0.933	0.899	
Net Cal./min.	8.509	6.454	6.598	7.277	9.382	
Net Cal./kg./min.	.1261	.1248	.1250	.1250	.1370	

The stress produced by the rapid weight gain and accumulation of body fat that occurred during the first 4 weeks of the unrestricted caloric intakes (R12 through R16) was reflected in the response to the moderate work test, the Harvard Fitness Test, and the maximal oxygen transport, and even in the cardiovascular functions at rest (see Chapter 28 and Keys *et al.*, 1947). The essential data for the aerobic work test are summarized in Table 332 for the later stages of rehabilitation. A comparison of Tables 330, 331, and 332 shows that the 12 subjects tested during the later stages of rehabilitation were representative of the 32- and 16-subject groups during the control, semi-starvation, and early rehabilitation (R12) periods; the response of the 12-subject group to the unrestricted refeeding can justifiably be assumed to be similar to that which would have occurred in the larger group.

At R16 the mean pulse rate during aerobic work reached 160.9 beats per minute; this is 17.8 per cent above the control and semi-starvation rates (Table 332). The work pulse increment was also higher than at any time except S24 and was 11.4 per cent higher than at the control period. As would be expected from the high work pulse rates, the recovery pulse rates were also increased at R16. From the pulse rate data it would be assumed that the moderate work test was more difficult for the subjects to complete at R16 than at any other period of the experiment. By R20, however, the work and recovery pulse rates had returned to the pre-starvation level.

The data on oxygen consumption and respiration during work for the 12-subject group are available only through R16. Both the total oxygen consumption during work and the caloric cost of the work were higher at R16 than in the control period, the increases per minute being, respectively, 163 cc. (9.4 per cent) and 0.873 Cal. (10.3 per cent). The higher oxygen consumption and higher caloric cost of work are not entirely the result of the increased body weight that occurred from R12 to R16. The caloric cost of work per kg. of body weight was increased 8.2 per cent above the control level, indicating a decreased muscular efficiency at R16 even though no decrease in efficiency occurred during either semi-starvation or the first 12 weeks of rehabilitation.

Total ventilation during work reached a high of 36.08 liters per minute at R16, an increase of 16 per cent above the control ventilation. The respiratory efficiency continued to increase from R12 to R16 but was still 6.1 per cent below normal at R16. No data are available for R20, but it may be assumed from the pulse rate data that the oxygen consumption and respiratory functions were also normal at R20.

Aerobic Work Recovery and the Rehabilitation Diet

The design of the restricted rehabilitation refeeding (R1 through R12) was such that the effects of the level of caloric, protein, and vitamin intakes on the rate of rehabilitation could be determined. The details of the factorial design of the rehabilitation period are discussed in Chapter 4. The 32 subjects were divided into 4 groups of 8 men; the caloric intake of the 4 groups (Z, L, G, and T) differed by steps of 350–400 Cal. per day, with the Z group receiving the lowest caloric intake and the T group receiving the highest. Each caloric group was further subdivided into 2 groups of 4 men each with one sub-group receiving extra protein (Y group) while the other group (U group) had only the protein present in the basal diet. The protein-supplemented and protein-unsupplemented groups were again divided, half of the subjects being furnished vitamin supplements (H group) and the other half receiving placebos (P group).

The aerobic work data illustrating the effects of the level of caloric intake and of the vitamin and protein supplementation during the first 6 weeks of rehabilitation (to R6) are summarized in Tables 333 and 334.

It is apparent from the data presented in Table 333 that the level of caloric intake had no consistent influence on the aerobic work test during the first 6 weeks of rehabilitation. The oxygen consumption in cc. per minute at R6, expressed as the percentage of the oxygen consumption in the control period, increased progressively with the higher caloric intakes, reflecting the better body weight gain in the higher caloric groups, but there was no correlation between caloric intake and rate of rehabilitation for any of the other functions.

TABLE 333

Aerobic Work Values, expressed as percentage of control, at 24 weeks of semi-starvation (S24) and 6 weeks of rehabilitation (R6) for the 4 caloric groups: Z = basal, L = +400, G = +800, T = +1200. Values for oxygen consumption and ventilation are corrected to N.T.P. (Minnesota Experiment.)

	S24 as Percentage of C				R6 as Percentage of C			
	Z	L	G	T	Z	L	G	T
Work pulse (beats/min.)	97.3	105.2	96.6	99.3	91.2	98.1	91.8	91.9
Increment	115.6	135.0	115.3	122.7	103.6	116.0	102.6	102.6
Recovery pulse (beats/min.) ...	88.0	94.6	88.3	88.7	83.6	93.7	83.5	83.8
Oxygen consumption (cc./min.).	71.9	73.9	73.4	76.5	73.0	74.4	75.9	80.2
Ventilation (liters/min.)	83.6	91.7	89.2	91.6	86.0	94.8	93.4	92.4
Respiratory efficiency	86.8	81.1	82.8	83.9	89.9	87.4	89.1	88.5
Respiratory quotient	103.4	108.0	108.2	105.3	101.6	106.8	107.5	103.0
Cal./kg./min.	97.0	100.0	105.5	103.7	103.4	98.3	101.1	102.4
Oxygen-pulse	74.1	70.2	75.9	77.1	79.4	73.6	81.4	85.2
Circulatory index	96.0	90.6	98.7	100.2	97.3	91.6	99.1	106.7

TABLE 334

Aerobic Work Values, expressed as percentage of control, at 24 weeks of semi-starvation (S24) and 6 weeks of rehabilitation (R6) for the protein-unsupplemented groups (U), protein-supplemented groups (Y), vitamin-unsupplemented groups (P), and vitamin-supplemented groups (H). Values for oxygen consumption and ventilation are corrected to N.T.P. (Minnesota Experiment.)

	S24 as Percentage of C				S24 as Percentage of C			
	U	Y	P	H	U	Y	P	H
Work pulse (beats/min.)	97.3	106.9	98.4	100.6	93.6	96.7	95.1	95.2
Increment	119.2	124.6	122.6	121.7	104.2	107.5	109.1	102.5
Recovery pulse (beats/min.) ..	87.5	92.5	89.2	89.8	87.9	89.2	89.5	87.5
Oxygen consumption (cc./min.)	72.9	74.9	74.7	73.1	74.9	76.8	70.5	75.2
Ventilation (liters/min.)	91.1	90.5	88.2	89.8	88.4	86.5	85.4	85.7
Respiratory efficiency	84.2	83.1	85.2	82.2	88.4	89.1	89.7	87.9
Respiratory quotient	106.6	105.2	105.5	107.1	107.1	106.5	107.3	107.1
Cal./kg./min.	101.4	101.3	100.1	100.8	101.2	101.0	101.1	101.2
Oxygen-pulse	75.1	73.4	75.8	72.7	80.1	79.4	80.4	79.1
Circulatory index	98.0	94.8	98.8	98.0	100.6	96.6	98.2	98.7

Neither the vitamin nor the protein supplementation appeared to be of advantage in the early stages of rehabilitation in so far as the responses to the aerobic work test are concerned. The results do not, however, preclude the possibility that a higher protein or vitamin differential might have elicited a difference between the supplemented and unsupplemented groups.

Blood Sugar Level in Aerobic Work

The blood sugar concentration during the last 2 minutes of the 30-minute aerobic work test was significantly decreased at both 12 and 24 weeks of semi-starvation. The low work blood sugar concentrations observed during semi-starvation might be assumed to indicate carbohydrate reserves that were insufficient to meet the demands for glucose during work. If such were the case, the decreased work capacity during semi-starvation could be expected to occur on the basis of a limiting fuel supply. It has frequently been demonstrated that long-continued hard exercise, such as marathon running, frequently results in very low blood sugar levels (Levine, Gordon, and Derick, 1924; Best and Partridge, 1930; Edwards, Margaria, and Dill, 1934).

The response of the work blood sugar concentration to higher caloric intakes during rehabilitation throws much doubt on the concept that the immediate fuel supply may have been the limiting factor in aerobic work performance during the semi-starvation period. In spite of the fact that caloric intakes were in some cases as high as 4000 Cal. per day, average blood sugar concentrations during work at 12 weeks of rehabilitation (R12) failed to increase above the semi-starvation levels (Table 330). Substantial differences in the caloric intake level had no effect on the work blood sugar concentration during either the starvation or the rehabilitation period even though intergroup caloric intake differentials in rehabilitation were more than 1000 Cal. per day.

Basal blood sugar concentration values are available for the 32-subject group at the end of semi-starvation and at 6 and 12 weeks of rehabilitation but were not obtained in the control period. It may be assumed that in the control period the basal blood sugar concentration would be the same as in aerobic work. Gemmill (1942) stated that "light muscular exercise does not change the blood sugar level." This statement is supported by experiences in the Minnesota Laboratory with other groups of normal young men. Even during semi-starvation and early rehabilitation when the blood sugar concentrations were low, there was no essential difference between the basal and work values, denoting no drastic drain on the reserves during the work. At 12 weeks of rehabilitation, however, the basal blood sugar had returned to control values but the sugar remained low during work, even though from other indications the work was less difficult for the subjects at R12 than at S24. The blood sugar did not appear to play any important limiting role in the capacity to perform physical work of the intensity used here during control, semi-starvation, or rehabilitation.

A possible explanation of the basal and work blood sugar concentration differentials during the later part of the restricted rehabilitation period may be related to the dominant carbohydrate metabolic pattern present at the time. In the later part of rehabilitation the average caloric intakes ranged between 3000 and 4000 Cal. per day. At this time body weight increased rapidly, which reflected a metabolism geared to storage (mainly as fat). In work, when the demand for glucose as a source of fuel (high R.Q.) was increased, the response of the glycogenolysis mechanism was not sufficient to maintain the blood sugar at normal levels. At R12 a 30-minute walk at 3.5 miles per hour on a 10 per cent grade required about 200 Cal., or about 40 gm. of glucose, at a work R.Q. of 0.94. A decrease in the rate of glycogenolysis at the time would, of course, readily explain the large basal and work blood sugar differential during late rehabilitation.

Maximal Performance Capacity — General Features

The treadmill version of the Harvard Fitness Test was employed as a measure of the ability of the subjects to perform strenuous physical work (Johnson *et al.*, 1942). The test consists of running on a motor-driven treadmill at 7 miles per hour and an 8.6 per cent grade until exhaustion, or for a maximum of 5 minutes. Seated recovery pulse rates are counted and the "fitness" score is calculated from the recovery pulse rates and the duration of the effort (for details of scoring, see the Appendix on methods).

Besides the work and recovery components, a maximum performance capacity test of any type also includes a motivation or cooperation factor. In order to obtain the best possible score, the individual must be willing to push himself to the limit and to continue the work until the last possible second of endurance. In all practical work situations the motivation factor plays an extremely important role in physical performance.

It should be emphasized that any fitness test yields data only on the ability to perform the specific task or tasks incorporated in the test. None of the tests so far devised gives over-all information on general performance capacity or state of well-being. The Harvard Fitness Test is, however, very useful because it is sensitive to changes in fitness in physical training or deterioration (Keys *et al.*, 1945) and because it does in a general way separate the physically active from the inactive normal individual (Johnson, Brouha, and Darling, 1942).

The Harvard Fitness Test was administered on the average of once every 2 weeks during the control period, the semi-starvation period, and the first 20 weeks of rehabilitation. In all cases the test was immediately preceded by a 10-minute warm-up walk at 3.5 miles per hour on a 10 per cent grade. The tests were made in a standard relationship to the last meal but not in the fasting state. During the later part of the semi-starvation period and the first few weeks of rehabilitation, special precautions were taken to prevent injury on the treadmill when complete collapse occurred. This was accomplished by means of a strong leather belt fastened loosely around the abdomen of the subject and held by one of the observers. The belt did not interfere with free body movements but did help to impart a sense of security to the subjects which may have increased their effort in the tests and consequently may have given them a slightly higher score than they would otherwise have obtained. In all cases the observers were satisfied that really maximal efforts were made by the subjects.

Maximal Performance Capacity in Semi-Starvation

According to the Harvard Fitness Test index the subjects were in general in an "average" state of physical fitness during the control period. The mean fitness score was 64.1 with the range from 22 to 90 (see Table 335). Of the 32 men, 12 (37.5 per cent) were classified as in "good" physical condition, 17 (53.1 per cent) were "average," and only 3 (9.4 per cent) were "poor." None of the subjects fell within the "superior" category (score above 90).

The deterioration in capacity to perform severe physical work was extreme during the semi-starvation period. By the end of the first 12 weeks on the semi-starvation regimen the average fitness score had decreased to 52 per cent of the control value. None of the subjects at that time could be considered to be in "good" physical condition and only 7 were "average." The deterioration continued until at the end of semi-starvation the average fitness score was only 28

TABLE 335

Mean Fitness Scores, Standard Deviations, Range of Scores, and Number of Individuals Classified as Good, Average, or Poor for the 32-Subject Group at control (C), 12 and 24 weeks of semi-starvation (S12 and S24), and 3, 6, 9, and 12 weeks of rehabilitation (R3, R6, R9, and R12). Scores of 76–90 = good, 41–75 = average, 40 and below = poor. (Minnesota Experiment.)

Fitness Index	C	S12	S24	R3	R6	R9	R12
Classification (no. of subjects)							
Good	12	0	0	0	0	0	0
Average	17	7	0	2	2	6	9
Poor	3	25	32	30	30	26	23
Score							
M	64.1	33.1	18.1	21.5	25.8	33.7	35.3
SD	17.23	8.62	8.35	10.27	10.48	10.97	11.25
Range	22–90	10–49	5–39	7–49	8–53	16–62	13–63

TABLE 336

Mean Values and Percentages of Control for Various Items of the Harvard Fitness Test at control (C), 12 and 24 weeks of semi-starvation (S12 and S24), and 3, 6, 9, and 12 weeks of rehabilitation (R3, R6, R9, and R12) (Minnesota Experiment).

Test Item	C	S12	S24	R3	R6	R9	R12
Time of run (secs.)	242	106	50	59	78	104	116
Percentage of C	100	44.3	20.9	24.8	32.4	43.5	48.4
Fitness score	64.1	33.1	18.1	21.5	25.8	33.7	35.3
Percentage of C	100	51.9	28.4	33.8	40.7	53.0	55.4
Total kgm. of work	4501	1628	707	851	1074	1386	1831
Percentage of C	100	36.4	15.8	19.0	24.1	31.1	40.9

per cent of the control value and all the subjects fell within the "poor" category of work capacity.

The data presented in Table 336 show that the deterioration in the fitness score was less than either the decrease in the length of time the subjects ran or the maximum total kilogrammeters of work they were capable of performing. While the fitness score decreased 71.6 per cent, the time of run decreased 79.1 per cent and the total work accomplished during the test decreased 84.2 per cent. It is apparent that the fitness score underestimated the extent of deterioration in the ability to do this type of severe physical work.

The discrepancy between the fitness score and the time of run can readily be explained on the basis of the bradycardia that developed during the period of semi-starvation. The recovery pulse rate is an important component of the fitness score as it is usually calculated. Under normal circumstances it is justifiable to include the recovery pulse rates in any performance capacity scoring procedure, since it is recognized that the rate of recovery of the heart rate is faster in the trained than in the untrained subject (Johnson, Brouha, and Darling, 1942). However, in abnormal situations, such as semi-starvation, the predominant

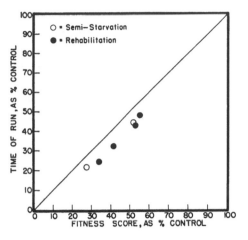

FIGURE 103. COMPARISON OF THE RELATIVE DETERIORATION IN TIME OF RUN FOR THE HARVARD FITNESS TEST AND THE FITNESS SCORE IN SEMI-STARVATION AND REHABILITATION. The time of run was relatively lower than the fitness score because of the bradycardia that developed during semi-starvation. (Minnesota Experiment.)

bradycardia is reflected in the recovery pulse, and consequently the scoring method results in an overestimation of the capacity for strenuous physical work (see Figure 103).

It was believed by all the observers that throughout the experiment the subjects pushed themselves to the limit of their capacities; the runs were terminated at a state of actual or near collapse. During the later phase of semi-starvation and the early part of rehabilitation the termination of the runs was in most cases due to exhaustion of the leg muscles and was not accompanied by extreme dyspnea or cardiovascular distress. It was felt that the extent of muscular weakness and muscle failure was in most cases greater than would be interpreted from the Harvard Fitness Test scores and was more accurately expressed either by the maximum total kilogrammeters of work the subjects were capable of doing during the test or by the length of time the subjects could run. However, regardless of the criteria used, the ability to perform strenuous physical work was drastically reduced during semi-starvation.

As would be expected, the absolute amount of deterioration in performance capacity during semi-starvation was greater in the subjects with a high control period fitness index than in those with a low fitness index (see the Appendix Tables). The relative degree of deterioration was not constant for all the subjects, however (i.e., each subject did not maintain his control fitness rank during semi-starvation). The correlation between the time of run in the control period and that at the end of semi-starvation was —0.07, indicating that the most fit individual at the beginning of semi-starvation might be either the least fit or the most fit at the end. It was impossible to predict from the control period performance what the performance of any individual would be at the end of semi-starvation.

Neither the absolute nor the relative body weight loss was the same for all subjects, which means that the rate at which work was done on the Harvard Fitness Test changed by a different degree for each subject during semi-starvation. It would be conceivable that the differences in body weight change might in part account for the lack of correlation between the length of time the subjects ran in the control and semi-starvation periods. The product of body weight and time of run furnishes an index of total work performed. The correlation between the work performed on the Harvard Fitness Test (weight $\times$ time) in the control period and at the end of semi-starvation was $+0.15$. The lack of correlation indicates that the ability to do strenuous work in semi-starvation is not related either to the ability to do the work in the control period or to the body weight loss within the limits in the Minnesota Experiment. It should be remembered, however, that the range of relative body weight loss was rather narrow (see the Appendix Tables). The absolute and relative individual resistance to stresses of all types, including semi-starvation, is an extremely complex phenomenon which is not entirely explainable on the basis of the usual tests of human functions.

Recovery of Maximal Performance Capacity in Rehabilitation

The essential data on the recovery of the ability to perform strenuous physical work, as measured by the Harvard Fitness Test, are presented in Tables 335 and 336 for the 32-subject group and in Table 337 for the 12 subjects who were tested through 20 weeks of rehabilitation.

TABLE 337

MEAN FITNESS SCORES, SCORES AS PERCENTAGE OF CONTROL, RANGE OF SCORES, AND NUMBER OF INDIVIDUALS CLASSIFIED AS GOOD, AVERAGE, OR POOR FOR THE 12-SUBJECT GROUP at control (C), 24 weeks of semi-starvation (S24), and 6, 12, 16, and 20 weeks of rehabilitation (R6, R12, R16, and R20). Scores of 76–90 = good, 41–75 = average, 40 and below = poor. (Minnesota Experiment.)

Fitness Index	C	S24	R6	R12	R16	R20
Classification (no. of subjects)						
Good	6	0	0	0	0	1
Average	6	0	2	4	0	7
Poor	0	12	10	8	12	4
Score						
M	65.3	21.8	31.7	38.6	32.7	52.4
Percentage of C	100	33.4	48.5	59.1	50.1	80.2
Range	42–84	8–39	8–53	17–63	23–40	28–76

The recovery of strenuous work capacity was surprisingly slow during rehabilitation. After 3 weeks of refeeding (R3) the fitness scores of only two of the the subjects had increased sufficiently to place them in the "average" fitness category. The average time of run was only 9 seconds longer than at the end of semi-starvation (as compared with a decrease of 192 seconds during semi-starvation). On the average the fitness score increased 3.4 points and the maximum work accomplished increased 144 kilogrammeters (Table 336). All these changes were of such small magnitude that the actual rehabilitation during the first 3 weeks, as measured by the Harvard Fitness Test, was negligible.

The slow processes of recovery continued during the second 3 weeks of rehabilitation. No more of the subjects were raised into the "average" fitness category at R6. From R3 to R6 the average length of run increased 19 seconds, the fitness score increased 4.3 points, and there was an increase of 223 kilogramme-ters of work accomplished during the test. At R6 the percentage recovery of the total amount of the deterioration during semi-starvation was 14.6 for the time of run in seconds, 16.7 for the fitness score, and 9.7 for the kilogrammeters of work. The minimum and maximum gains in fitness score were 3 and 14 points. During the same period of time the gain in body weight was only 1.65 kg., or 10 per cent of the loss in semi-starvation.

During the first 3 weeks of the higher caloric intake (R6 to R9) the absolute increase in the Harvard Fitness Test functions was greater than the increase that had occurred during the entire first 6 weeks of rehabilitation. Six of the sub-jects (4 more than at R6) were in the "average" fitness class. There was an in-crease of 26 seconds in time of run, 7.9 points in fitness score, and 312 kilogram-meters of work accomplished from R6 to R9.

By the end of the controlled rehabilitation period (R12) the capacity to per-form strenuous work was still far below normal. Only 9 of the subjects had re-covered sufficiently to be classed as having "average" fitness and none had reached the "good" class, as compared to 17 "average" and 12 "good" during the control period. The range of fitness scores did not change from R9 to R12. The mean fitness score and the time of run in seconds increased much less rapidly during the final 3 weeks of controlled rehabilitation (R9 to R12) than during the preceding 3-week period (R6 to R9). The maximum kilogrammeters of work accomplished, however, increased faster. The obvious explanation is that the body weight was increasing more rapidly during the last 3 weeks of controlled rehabilitation, which caused a marked increase in the rate at which work was done in the fitness test.

At 12 weeks of rehabilitation the Harvard Fitness Test items were still about 50 per cent below the control values. When the recovery of the items is ex-pressed as a percentage of the deterioration that occurred during the semi-starvation period, the results of 12 weeks of controlled rehabilitation are even less impressive. The percentage of the loss regained at R12 (calculated from the group means) was 34.3 for the time of run in seconds, 37.4 for the fitness score, and 29.6 for the kilogrammeters of work.

Data for the later phases of rehabilitation (R12 to R20) are summarized in Table 337 and Figure 104 for the 12 subjects who were followed through 20 weeks of refeeding. The 12 subjects were representative of the 32-subject group as is indicated by a comparison of Tables 335 and 337. During the 8-week pe-riod from R12 through R20 the subjects were allowed free choice of the type and quantity of food eaten. The average daily caloric intake for that period was about 4500 Cal., although caloric intakes as high as 10,000 Cal. per day were re-corded in the first 2 or 3 weeks of unrestricted rehabilitation (R12 to R14 or R15).

By the end of 20 weeks of rehabilitation the fitness score was 80.2 per cent of the control value, representing a 14.1 point increase during the 8-week pe-riod. Of the amount of fitness score lost during semi-starvation, 70.3 per cent was regained by R20. The time of run on the test increased slightly more than did the fitness score. One of the 12 subjects recovered sufficiently by R20 to

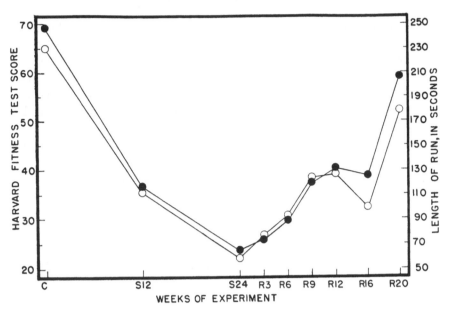

FIGURE 104. HARVARD FITNESS TEST SCORE (OPEN CIRCLES) AND LENGTH OF RUN (SOLID CIRCLES) DURING SEMI-STARVATION AND REHABILITATION. The values represent the averages for 12 subjects. (Minnesota Experiment.)

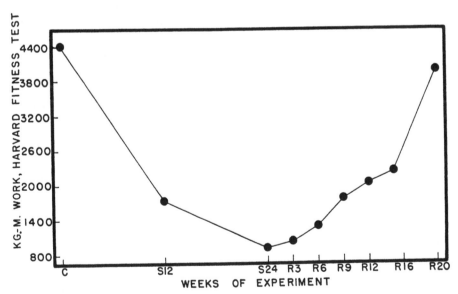

FIGURE 105. TOTAL WORK, AS KILOGRAMMETERS, PERFORMED IN THE HARVARD FITNESS TEST DURING SEMI-STARVATION AND REHABILITATION. The values represent the averages for 12 subjects. (Minnesota Experiment.)

reach the "good" fitness class, while 7 were "average," as compared to 6 "good" and 6 "average" during the control period, all 12 "poor" at S24, and 4 "average" and 8 "poor" at the end of 12 weeks of rehabilitation.

The rehabilitation period from R12 to R16 is particularly interesting because of the apparent decrease in the ability to do strenuous work, as expressed by the fitness score or the time of run during that period. The fitness score decreased from 38.3 at R12 to 32.6 at R16. The time of run decreased less, from 132 seconds at R12 to 125 seconds at R16. It would appear that the very high caloric intake during the first part of the unrestricted refeeding was actually detrimental to the recovery processes.

If the total kilogrammeters of work accomplished during the fitness test is used as the criterion of performance capacity instead of either the fitness score or the time of run, the situation appears quite different (see Figure 105). From R12 to R16 there was a large gain in body weight. As a result there was a steady increase in performance capacity throughout the rehabilitation period, with a sharp increase during the last month of the observations. These observations again raise the question of what criteria constitute the best index for judging or measuring the ability of an individual to perform strenuous work. Admittedly, any single criterion is inadequate, but the total work performance expressed as kilogrammeters of work would seem to have certain advantages. The only requirements for calculating the data are the body weight, the rate of work, and the length of time the work was continued. Such a test could be applied with equal ease in laboratory and field studies. In the Minnesota Experiment, at least, the total work accomplished during the later phase of refeeding appeared to be a better index of performance capacity than did either the fitness score or the time of run.

Caloric Intake and the Recovery of Maximal Performance

The discussion of the effect of caloric intake on the rate of rehabilitation will be limited to the first 12 weeks, during which time the integrity of the groups was maintained and the caloric intake was strictly regulated.

The mean physical performance capacity during the control period was reasonably similar in the 4 caloric groups except for the L group, whose fitness score was 15.6 per cent and time of run 12 per cent above the 32-subject group mean. At the end of 24 weeks of semi-starvation, however, the 4 caloric group means were almost identical. As was mentioned earlier in this chapter, there was no correlation between the fitness score in the control period and at the end of starvation. However, the significantly higher fitness in the L group during the control period does introduce the problem of the proper method of comparing the rate of regain of fitness in the 4 caloric groups during rehabilitation. A direct comparison of the fitness scores, length of run, or total work accomplished at the various times in rehabilitation could be misleading, because the L group was potentially capable of greater improvement. It is statistically more justifiable to express the rehabilitation values as percentage of the control values or as percentage regain of the amount of deterioration that occurred during semi-starvation.

The pertinent data on the effect of the level of caloric intake during the rehabilitation period on the recovery of the capacity to perform strenuous physical work are summarized in Tables 338, 339, and 340 and Figure 106.

TABLE 338

MEAN FITNESS SCORE AND FITNESS SCORE AS PERCENTAGE OF CONTROL AND AS PERCENTAGE OF LOSS for the 4 caloric groups (Z = basal, L = +400, G = +800, T = +1200). C = control; S24 = 24 weeks of semi-starvation; and R3, R6, R9, and R12 = 3, 6, 9, and 12 weeks of rehabilitation. (Minnesota Experiment.)

	C	S24	R3	R6	R9	R12
Group Z						
Score	62.8	18.6	18.1	19.6	27.1	27.1
Percentage of C	100	29.6	28.8	31.2	43.2	43.2
Percentage of loss		0.0	−1.6	2.3	19.2	19.2
Group L						
Score	74.4	18.1	22.0	24.8	35.3	37.8
Percentage of C	100	24.3	29.6	33.3	47.4	50.9
Percentage of loss		0.0	6.9	11.9	30.6	34.4
Group G						
Score	59.9	18.6	23.0	29.1	35.5	38.6
Percentage of C	100	31.0	38.4	48.6	59.3	64.4
Percentage of loss		0.0	10.7	25.4	40.9	48.4
Group T						
Score	59.5	17.1	22.8	29.5	37.0	37.5
Percentage of C	100	28.7	38.3	49.6	62.2	63.0
Percentage of loss		0.0	13.4	29.2	46.9	48.1

The mean fitness score and the fitness score expressed as percentage of the control score and as percentage recovery of the deterioration all demonstrate that the level of caloric intake during rehabilitation had a decided effect on the recovery of the ability to do severe physical work. Recovery was slowest throughout the 12 weeks of rehabilitation in the lowest caloric group, Z; in fact, at the end of the first 3 weeks of refeeding the fitness score in the Z group was slightly lower than at the end of semi-starvation. Recovery was best in the highest caloric group, T, with the G group being only slightly inferior until R12, when the G and T groups were identical. The regain of fitness in the L group was intermediate between the Z and G groups.

Substantially the same intergroup relationship held for the length of time the subjects ran as for the fitness score. Recovery was poorest in the Z group and best in the T group. The G group responded almost as rapidly as the T group but did not quite equal that group at R12. The L group was intermediate between the Z and G groups. The time of run did not level off between R9 and R12 in groups Z and L as the fitness score did, which would indicate higher recovery pulse rates in the 2 low caloric groups (Z and L).

The recovery in kilogrammeters of work accomplished during the fitness test, which includes the intergroup body weight differences, differentiated the 4 caloric groups throughout rehabilitation when the recovery of work capacity was calculated as percentage of the loss that occurred during semi-starvation. At R3

TABLE 339

MEAN TIME OF RUN IN SECONDS AND TIME OF RUN AS PERCENTAGE OF CONTROL
AND AS PERCENTAGE OF LOSS for the 4 caloric groups (Z = basal, L = +400, G =
+800, T = +1200). C = control; S24 = 24 weeks of semistarvation; R3, R6, R9, and
R12 = 3, 6, 9, and 12 weeks of rehabilitation. (Minnesota Experiment.)

	C	S24	R3	R6	R9	R12
Group Z						
Seconds	245	53	51	59	81	87
Percentage of C	100	21.6	20.8	24.1	33.1	35.5
Percentage of loss		0.0	−1.0	3.1	14.6	17.7
Group L						
Seconds	281	52	62	78	108	127
Percentage of C	100	18.5	22.1	27.8	38.4	45.2
Percentage of loss		0.0	4.2	10.9	23.4	31.4
Group G						
Seconds	217	50	63	85	107	122
Percentage of C	100	23.0	29.0	39.2	49.3	56.2
Percentage of loss		0.0	7.8	21.0	34.1	43.1
Group T						
Seconds	224	46	61	88	119	127
Percentage of C	100	20.5	27.2	39.3	53.2	56.7
Percentage of loss		0.0	8.4	23.6	41.0	45.5

TABLE 340

MEAN KILOGRAMMETERS OF WORK ACCOMPLISHED ON THE FITNESS TEST AND THE WORK
AS PERCENTAGE OF CONTROL AND AS PERCENTAGE OF LOSS for the 4 caloric groups
(Z = basal, L = +400, G = +800, T = +1200). C = control; S24 = 24 weeks of
semi-starvation; R3, R6, R9, and R12 = 3, 6, 9, and 12 weeks of
rehabilitation. (Minnesota Experiment.)

	C	S24	R3	R6	R9	R12
Group Z						
Kilogrammeters	4242	733	724	775	1050	1297
Percentage of C	100	17.3	17.1	18.3	24.8	30.6
Percentage of loss		0.0	0.0	1.5	9.0	16.4
Group L						
Kilogrammeters	5164	719	870	1019	1342	1937
Percentage of C	100	13.9	16.8	19.7	26.0	37.5
Percentage of loss		0.0	3.4	6.7	14.0	27.4
Group G						
Kilogrammeters	4453	730	938	1208	1550	2006
Percentage of C	100	16.4	21.1	27.3	34.8	45.0
Percentage of loss		0.0	5.6	12.8	22.0	34.3
Group T						
Kilogrammeters	4145	646	871	1294	1602	2083
Percentage of C	100	15.6	21.0	31.2	38.6	50.3
Percentage of loss		0.0	6.4	18.5	27.3	41.1

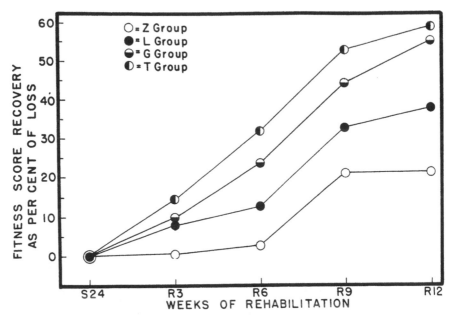

FIGURE 106. The Effect of the Level of Caloric Intake during Rehabilitation on the Recovery of the Harvard Fitness
Test Score (Minnesota Experiment).

there was little difference between the groups in the total work accomplished on
the test. At R6, R9, and R12 the work accomplished was progressively greater
from the low to the high (Z to T) caloric groups on both the absolute and the
percentage recovery basis.

Because of the intragroup and intergroup variability in the fitness functions
during the control period and in the loss during semi-starvation, the group
means do not give a completely accurate picture of the effect of the level of caloric intake on the recovery during rehabilitation. A more accurate approach
than the group means would be to calculate the recovery of the fitness function
for each individual for the various time intervals during semi-starvation as a percentage of the individual deterioration that occurred during semi-starvation.
Such calculations were made for the fitness score, and the mean values of the
individual percentage recovery for the 4 caloric groups at R3, R6, R9, and R12
are presented graphically in Figure 106.

The statistical significance of the intergroup differences was calculated for
the sixth week of rehabilitation (R6). At that period the percentage recovery of
the fitness score was significantly less in the Z group than in any of the other 3
groups. The difference between the Z group and the L group was significant at
the 5 per cent level and between the Z and G and T groups at the 1 per cent
level. The difference in recovery between the L and G groups was significant
at the 5 per cent level and between the L and T groups at the 1 per cent level.
There was no significant difference between the G and T groups even though
recovery appeared to be faster in the T group. From a general appraisal of the
differences presented in Figure 106, it is clear that at R3 the Z and T groups

were the only ones that differed significantly. At R9 and R12 the 2 high caloric groups, G and T, did not differ significantly from each other, but group L was inferior to groups G and T and superior to group Z. In group Z the recovery was significantly slower than in any of the other 3 groups.

Protein and Vitamin Supplementation and the Recovery of Maximal Performance

The effects of extra vitamins and proteins on the recovery of physical performance capacity are summarized for the 32-subject group during 12 weeks of rehabilitation in Table 341 and Figures 107 and 108.

It is apparent from the data in Table 341 that neither the protein nor the vitamin supplementation at the levels used in the Minnesota Experiment had any significant effect on the length of time, expressed as percentage of control and as percentage of semi-starvation loss, that the subjects ran on the Harvard Fitness Test at either 3, 6, 9, or 12 weeks of rehabilitation. The maximum difference of 2.4 per cent in favor of the protein-supplemented groups, which occurred at R12, was far from being significant. The maximum difference between the vitamin-supplemented and vitamin-unsupplemented groups at R6 was not statistically significant even at the 10 per cent level of significance.

TABLE 341

EFFECT OF PROTEIN AND VITAMIN SUPPLEMENTATION DURING REHABILITATION ON THE TIME OF RUN IN THE HARVARD FITNESS TEST. Values are expressed as percentage of control and as percentage of the loss during semi-starvation. C = control; S24 = 24 weeks of semi-starvation; R3, R6, R9, and R12 = 3, 6, 9, and 12 weeks of rehabilitation. (Minnesota Experiment.)

	C	S24	R3	R6	R9	R12
Protein-supplemented						
Percentage of C	100	19.5	22.0	32.8	44.0	48.9
Percentage of loss		0.0	6.0	15.7	34.5	38.8
Protein-unsupplemented						
Percentage of C	100	22.0	22.9	31.4	41.6	46.5
Percentage of loss		0.0	8.4	17.3	33.3	36.4
Vitamin-supplemented						
Percentage of C	100	21.6	24.5	32.7	41.2	49.4
Percentage of loss		0.0	8.4	19.8	35.8	38.6
Vitamin-unsupplemented						
Percentage of C	100	19.7	24.4	31.9	44.5	46.6
Percentage of loss		0.0	6.2	13.3	32.1	35.9

The same lack of effect of protein or vitamin supplementation in rehabilitation is shown in the recovery of the fitness score expressed as percentage of the decrease during semi-starvation. Complete statistical analysis demonstrated no significant differences between the groups at R3, R6, R9, or R12.

It is conceivable that even though neither the protein nor the vitamin supplementation was effective in hastening recovery during rehabilitation, the combination of both protein and vitamin supplementation might favor a more rapid recovery of function. The interactions between proteins and vitamins on the per-

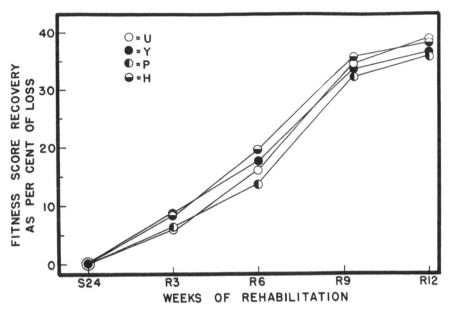

FIGURE 107. The Effect of Vitamin or Protein Supplementation of the Basic Rehabilitation Diet on the Recovery of the Harvard Fitness Test Score. U = protein-unsupplemented, Y = protein-supplemented, P = vitamin-unsupplemented, and H = vitamin-supplemented.
(Minnesota Experiment.)

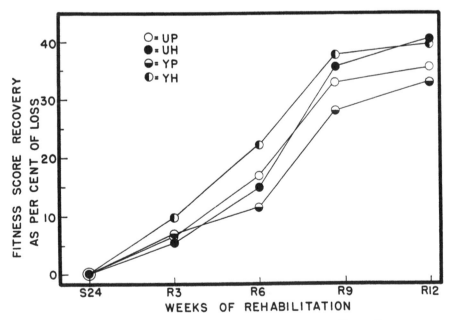

FIGURE 108. The Effect of Supplementation of the Basic Rehabilitation Diet with Either Vitamins and Proteins, Vitamins Alone, Proteins Alone, or Neither Vitamins nor Proteins on the Rate of Recovery of the Harvard Fitness Test Score (see legend to Figure 107).

centage recovery of the fitness score are illustrated in Figure 108. The designation of the groups is as follows: supplemented with both proteins and vitamins, YH; supplemented with proteins but not with vitamins, YP; supplemented with vitamins but not with proteins, UH; supplemented with neither vitamins nor proteins, UP.

Statistical analysis of the differences between the groups demonstrated no consistently significant advantage for any of the 4 combinations of proteins and vitamins. Recovery in general tended to be best in the group supplemented with both proteins and vitamins (YH) and poorest in the group receiving extra proteins but no extra vitamins (YP). The difference between groups YH and YP was statistically significant at the 5 per cent level at R6 but was not significant at R3, R9, or R12. No significant difference was present between any of the other groups.

Severe Work — Maximal Oxygen Intake

For the studies of the effect of semi-starvation on the ability to perform severe muscular work, it was considered desirable to also use test methods in which motivation and will power play no part. The maximal oxygen intake appeared most useful for this purpose. This consists, in effect, simply in the measurement of the highest rate of oxygen consumption which can be attained. Properly carried out, this level can be determined objectively at a rate of external work performance considerably below the absolute maximum attainable by the subject; the upper plateau of oxygen consumption is reached far below the point of exhaustion and collapse.

The maximal oxygen intake was estimated at various times on a total of 24 men. Nine of these men were studied systematically during the starvation period, 8 were studied from C through R12, 6 from C through R20, and 23 from S24 to R12. In all cases the tests were carried out on the treadmill with a 3-minute run and a speed of 7 miles per hour; the grade (angle) was varied according to the capacity of the individual. Expired air was collected between 1'45" and 2'45" of each run. At every major testing period attempts were made to prove that the maximal oxygen intake was indeed "maximal" — that is, the oxygen intake was determined at 2 grades of work. If the oxygen intakes agreed within 150 cc. per minute, it was assumed that the plateau had been reached.

In addition to the maximal oxygen intake studies, some other observations in this heavy exercise were made on some of the men. These included measurements of the oxygen debt for periods of 10 and 50 minutes after work and measurement of the concentration of lactic and pyruvic acid in venous blood drawn 12 minutes after stopping work.

Maximal Oxygen Intake in Semi-Starvation

The effect of semi-starvation on the maximal oxygen intakes of 9 men is presented in Table 342. It will be noted that true maximal oxygen intakes were demonstrated in practically all cases. The one exception appears to be that of subject No. 4, whose oxygen intake increased 350 cc. per minute when the grade was raised 2.5 per cent. Two men who completed 3 minutes at zero grade were unable to run long enough at a 2.5 per cent grade to allow a complete collection of exhaled air, while a third man could not complete 3 minutes at 7 miles per hour and zero grade. In the latter case observations were made at 6 miles per hour and zero grade. For these 3 men the values obtained at the lower grade of work were accepted as maximal oxygen intakes. The average maximal oxygen intake at the end of semi-starvation was taken as the average of the 9 highest values obtained at S24.

TABLE 342

MAXIMAL OXYGEN INTAKE for 9 men at control and 12 and 24 weeks of semi-starvation (S12 and S24). The intake is expressed in liters of oxygen at N.T.P. consumed in excess of the B.M.R. per minute. Expired air was collected between 1' 45" and 2' 45" of a 3-minute run at 7 miles per hour and at the specified grade. To demonstrate that the O_2 intake was in fact maximal, O_2 intakes were determined at 2 different grades. The lower grade is given in the table; the "high grade" was 2.5 per cent higher. The mean intake for the S24 period is for the 6 men who were able to complete the test at both grades. (Minnesota Experiment.)

Subject Number	Control Low Grade (%)	Control O_2 Intake Low Grade	Control O_2 Intake High Grade	S12 Low Grade (%)	S12 O_2 Intake Low Grade	S12 O_2 Intake High Grade	S24 Low Grade (%)	S24 O_2 Intake Low Grade	S24 O_2 Intake High Grade
4......	10.0	3.14	3.09	5.0	2.19	2.21	2.5	1.67	2.02
9......	10.0	3.97	4.03	2.5	2.30	2.20	0.0	2.05	
23......	5.0	3.05	3.07	2.5	2.09	2.17	0.0	1.89	1.77
26......	7.5	3.38	3.53	5.0	1.99	2.17	0.0	1.71	1.69
109......	2.5	2.97	2.87	0.0	2.33	2.31	0.0	1.95	
112......	7.5	3.17	3.16	2.5	2.12	2.09	0.0	1.85	1.81
119......	7.5	2.78	2.91	5.0	2.01	2.16	0.0	1.62	1.63
126......	5.0	3.39	3.32	0.0	2.06	2.19	0.0*	1.67	
233......	10.0	3.86	3.77	2.5	2.44	2.54	0.0	2.27	2.26
M		3.30	3.30		2.17	2.22		1.84†	1.86†

* Six miles per hour.
† Average of 6 subjects for whom data on 2 grades were obtained.

Blood for lactate determination was drawn 12 minutes after each complete 3-minute run in these tests. The lactate increment due to work *decreased* during starvation from the control level of 110 mg. per 100 cc. of blood to 78.5 at 12 weeks and 37.6 at 24 weeks. These values represent the average lactate after the higher grade of work for the maximal oxygen intake determinations and are the result of a grade of work which for practical purposes exhausted the subjects. It is apparent that the capacity to produce lactate during work is markedly reduced during starvation.

The lactate concentration in the blood was not a precise indication of the total lactate produced in work in these experiments because the relative fluid space in which it was distributed was larger in semi-starvation than in the control period. However, it can be shown that this does not change the final conclusion. Although total body water determinations are not available for the starved state, one can use the reasonably accurate estimate of 65 per cent for the normal body water (Pace *et al.*, 1947) and 80 per cent for the water in normal blood as compared with 85 per cent in the blood of the starved man. If the total lactate produced is calculated using the unreasonably high figure of 90 per cent of the body weight for the total water at S24, the total lactate produced at "exhaustion" in the normal state would be 63 gm. as compared with 21.2 gm. of lactate in the starved state, or 0.9 gm. of lactate per kg. of body weight in the normal state and 0.4 gm. per kg. in the starved state. It is also obvious that small differences in the ratio of muscle to body weight will not influence this conclusion.

The oxygen debt after the standard 3-minute run was measured on subjects 109, 4, and 23. The grade was the highest the men were able to do. Three collections of exhaled air were made covering a 50-minute period. The debt was calculated as the difference between the basal oxygen requirement and the observed oxygen intake for 50 minutes after the end of work. The average debt for these 3 men was 3.9 liters, or 75 cc. per kg., as compared with 130 cc. per kg. for normal men running to near exhaustion at a 10 per cent grade and 7 miles per hour for 3 minutes. Pre-starvation oxygen debt observations on these 3 subjects were not available, so 3 other normal men were selected to correspond with the pre-starvation states of subjects 4, 23, and 109. These control subjects were similarly accustomed to work on the treadmill, and each of them was studied during and after 3-minute runs at 7 miles per hour at 4 different grades. The resulting data provide a basis for estimating the effect of semi-starvation on the capacity to pay off an oxygen debt. The comparison with the normal state of nutrition is made in Figure 109, where the amount of oxygen con-

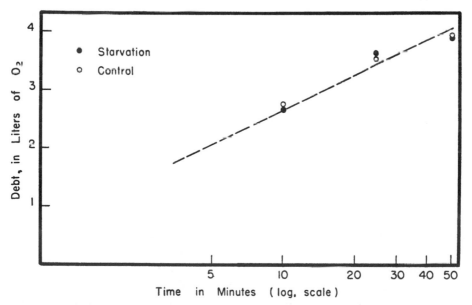

FIGURE 109. RATE OF PAYMENT OF OXYGEN DEBT IN SEMI-STARVATION. The oxygen debt is expressed as liters of oxygen at N.T.P. consumed during specified time intervals in excess of basal requirement. Mean figures for 3 men. (Minnesota Experiment.)

sumed after work is plotted against the logarithm of the time. The data indicate that at the same level of total debt, measured as the excess oxygen used in 50 minutes following work, there is no significant difference between starved and normal men in the rate at which they pay off the debt.

It will be noted that when the normal men were studied at the same grade of work as that used with the semi-starved men at their greatest effort, the normal men accumulated an oxygen debt of 80 cc. per kg. while the semi-starved men had a debt of 75 cc. per kg. The debts were substantially identical per unit of work done. At this grade of work, however, the semi-starved men were very near to collapse, but the normal men scarcely had a "workout." Though there are limitations to the precision of these comparisons, it is clear that payment of the oxygen debt was not grossly abnormal in the starved men.

We have remarked above that when the starved men worked to near collapse, the subsequent (12-minute) blood lactate concentration was far lower than would characterize them after exhausting work in the well-fed state. It might seem possible, however, that the lactic acid produced in work was simply proportional to the oxygen debt which they were able to contract. A first approach to this question is simply to compare the blood lactate levels in the starved men and in the well-fed controls at the same levels of total oxygen debt. When the starved men were exercised to near exhaustion, they had an average total (50-minute) oxygen debt of 3.9 liters, and the corresponding exercise increment for the blood lactate concentration (12 minutes after work) was 36 mg. per 100 cc. For the well-fed controls, however, an oxygen debt of 4.1 liters was attended by a lactate increment of only about 6 mg. This would suggest that semi-starvation results in very small oxygen debts per unit of lactic acid.

The foregoing estimate, however, is not quite proper; comparison should be made between the *total* lactic acid and the oxygen debt rather than merely the lactic acid concentration. This has been done for the normal state by Margaria, Edwards, and Dill (1933), who have discussed the assumptions involved. In spite of the fact that our Minnesota experimental conditions were not ideal, we believe that such comparisons will be worth while. The total lactic acid in the starved and well-fed states can be estimated on various assumptions as to the fluid space in which the lactate was dispersed. The most "favorable" correction of this type that can be made within reason indicates that the starved man produces something like 16 gm. of lactic acid in exercise, which results in a 4-liter oxygen debt, but that at this debt the normal man produces only 2 or 3 gm. of lactic acid. For the total lactic acid production to be equal in the 2 cases, it would be necessary to assume that the body of the starved man contained only 23 per cent water.

The data on respiration during this heavy work are summarized in Table 343. It will be noted that although the total ventilation fell 22 per cent, the amount of oxygen removed from each liter of inhaled air declined to a greater extent. There was some relative "over-breathing" during heavy work at the end of the semi-starvation period; according to the usual definition, these men had a decreased respiratory efficiency. In recovery after work, however, the respiratory efficiency was substantially unchanged during starvation.

TABLE 343

EFFECT OF 24 WEEKS OF SEMI-STARVATION (S24) ON THE VENTILATION
IN LITERS PER MINUTE (N.T.P.) AND THE RESPIRATORY EFFICIENCY IN
CC. OF OXYGEN REMOVED PER LITER OF AIR during the last minute of a
3-minute run at 7 miles per hour on the specified grade
(average of 9 men) (Minnesota Experiment).

	Control	S12	S24	S24 — C
Grade (%)	9.72	5.3	1.94	—7.78
Ventilation	93.48	82.91	71.25	—22.23
Respiratory efficiency	39.8	29.9	30.2	—9.6

Factors Affecting the Loss of Endurance

The large loss in endurance time shown by the subjects in the Harvard Fitness Test as the result of their restricted diet could be due, conceivably, to one or more of the following items: (1) poor motivation for performing the test, (2) loss of skill in grade running, (3) loss of cardiovascular capacity, (4) loss of muscular strength, and (5) poor capacity for anaerobic work.

All the observers were convinced that the motivation in running the Harvard Fitness Test was high in all the men at all times. To be sure, the men disliked the test, but their interest in their own abilities and the general desire "not to let the other fellow down" and to provide valid data which might be useful to people less fortunate than themselves far offset the dislike of the test situation. The fact that an oxygen intake which was independent of treadmill grade could be attained by these men after 24 weeks of semi-starvation is objective evidence that the men were willing to push themselves to the limit of their cardiovascular performance.

Inefficient running had some influence on the ability to run on a grade. Some of the men had a good deal of fluid in the knee joints which caused some discomfort, and this definitely limited the performance of at least one man.

The oxidative energy available for performing external work during the course of the run is measured by the maximal oxygen intake. Since the physical task of moving the individual's body was reduced during starvation, the available oxidative energy related to the job to be done in exhaustive work is best expressed by the maximal oxygen intake in cc. per kg. of body weight. This figure is reduced by 25 per cent. Other things being equal, the work output in exhaustive running must be reduced by 25 per cent or other mechanisms must take over and carry the load.

In examining the results of the maximal oxygen intake one may ask whether the reduction in oxygen intake was such that the muscles were obtaining more or less oxygen per unit of active muscle. A precise answer to this question cannot be given, of course, but some approximations may be made.

The decrease in maximal oxygen intake in cc. per kg. of body weight cannot be taken to represent the change in oxygen supplied to the active muscles since large changes took place in body water and fat and only small changes took place in the weight of the skeleton. We have, therefore, calculated the maximal oxygen intake in cc. per kg. of active tissue, using the procedure explained in Chapter 15. This value would give a good estimate of the oxygen supply to the working muscles if all the soft tissues lost weight in the same proportion. One may provide some check on the validity of this assumption by estimating the shrinkage in leg muscles, using the anthropometric measurements. The cross-sectional area of the thigh muscle during control and at S24 was calculated. The maximal oxygen intake per square centimeter of the cross section of muscle before and after starvation is presented in Table 344, along with a modified estimate of the same figure based on an allowance of 10 per cent excess water in the muscle at the end of semi-starvation.

TABLE 344

MEAN VALUES OF MAXIMAL OXYGEN INTAKE for 9 men before and after 24 weeks of semi-starvation, expressed as liters in excess of the B.M.R. at N.T.P. and as cc. per kg. of various estimated body tissues (Minnesota Experiment).

Unit	Control	S24	d	d%
Total intake (liters)	3.30	1.89	−1.41	42.7
Cc./kg. of body weight	46.6	34.6	−12.0	25.7
Cc./kg. of "active tissue"	79.7	64.2	−15.5	19.4
Cc./square centimeter of cross section of thigh muscle ...	224.0	167.0	−57.0	25.2
Cc./square centimeter of cross section of thigh muscle corrected for excess water at S24 ..	224.0	185.0	−39.0	17.3

The values in Table 344 are, with the exception of the first 2 lines, obviously only rough estimates, but they all indicate that a unit of working muscle in the semi-starved individual who is performing hard work on the treadmill definitely receives less oxygen than the same unit of muscle under normal conditions. This fact leads to the conclusion that the poor functioning of the respiratory-cardiovascular system is definitely a factor limiting the performance of exhausting work.

The important question of what part of the respiratory-cardiovascular system was principally responsible for the poor delivery of oxygen to the working muscles cannot be answered directly. The low concentration of hemoglobin at the end of the semi-starvation period undoubtedly contributed to the poor muscular oxygen supply. Poor diffusion of oxygen through the alveolar walls of the lungs does not appear to have been a likely cause since the chest X-rays showed no evidence of accumulating fluids and the respiratory efficiency during work was not depressed beyond the values found as the result of such an innocuous condition as 3 weeks in bed (Taylor et al., 1949). The principal factor was probably either a decreased cardiac output or a poor capillary bed in the muscles themselves.

The loss of strength was considerable. The muscles of the back and forearm showed a loss in strength of 28.2 per cent and 29.8 per cent, respectively, and it seems likely that the loss of strength in muscle groups of the legs was of the same order of magnitude. This loss of strength unquestionably contributed importantly to the total loss of endurance.

The capacity to perform anaerobic work is measured either by the magnitude of the oxygen debt or by the concentration of lactate in the blood after the subject has run to exhaustion. The semi-starved individual does not have the capacity to produce large oxygen debts or to accumulate large amounts of lactate. This has been noted in other stress situations. Edwards (1936) reported that men partially acclimatized to high altitude were unable to accumulate significant amounts of lactate in spite of the fact that they worked to what appeared to be complete exhaustion. It is also well known (Robinson and Harmon, 1941) that training increases the amount of lactate a man can accumulate by exhausting work. The maximal oxygen debt and, it may be safely assumed, the lactate production are markedly reduced in patients suffering from mild grades of congestive heart failure (Harrison and Pilcher, 1930a, 1930b). Thus it appears that a decrease in the capacity for exhausting work produced by a variety of causes is characterized by a loss of capacity for performing anaerobic work. It has been pointed out that in stress situations such as those involving high altitude (Dill, 1938) and congestive heart failure (Simonson and Enzer, 1942), the inability to perform large amounts of anaerobic work is a useful safeguard against overexertion. This point of view appears to be valid when applied to semi-starvation, since the need for a safeguard against overexertion in the semi-starved state was clearly demonstrated by the low cardiac reserve of the subjects in the Minnesota Experiment.

Many of the metabolic, respiratory, and cardiovascular characteristics of the semi-starved individual performing exhausting work bear some resemblance to the situation in cardiac decompensation. Both the cardiac patient and the semi-starved individual work with a markedly reduced maximal oxygen intake, a decreased respiratory efficiency, and a much reduced oxygen debt and lactate concentration after running to exhaustion. Both the cardiac patient and the semi-starved individual perform steady-state work with a pulse rate that is higher than would be expected for the work performed. These remarks are made not to suggest any similarity in the underlying mechanism but to emphasize the non-specific nature of many of the characteristics of the semi-starved man performing exhausting work. One characteristic which distinguishes the semi-starved individual from persons suffering from the effects of many other deteriorating influences is his lack of strength as measured by the dynamometer.

The semi-starved subjects of the Minnesota Experiment working to exhaustion on the treadmill did not stop because of dyspnea or discomfort in the chest as a cardiac patient does. They did not stop because of pain in the side or because of nausea as did men in our Laboratory who had gone without food for some days. They were not "winded" the way a normal individual is. They stopped primarily because they could no longer control the action of their knees and ankles. They literally did not have the strength to pick up their feet fast enough to keep up with the treadmill. Although it is not possible at the present time to assign definite proportions to the various factors limiting the ability to do exhausting work, we are of the opinion that strength was the most important, and that cardiovascular dysfunction and the loss of capacity to do anaerobic work contributed but not to the same degree.

Recovery in Capacity for Severe Work

Observations on the performance and response to severe work were made with 6 men through 20 weeks of rehabilitation as well as throughout the starvation phase of the Minnesota Experiment. Of these 6 men, Nos. 109 and 112 were in the T diet group, No. 4 was in the G group, Nos. 23 and 26 were in the L group, and No. 119 was in the Z group. In other words, during the first 12 weeks of rehabilitation these 6 men covered the dietary range from least to maximum calories.

The regimen of physical work during rehabilitation differed from that during starvation in that the men performed the Harvard Fitness Test once a week in addition to approximately a dozen runs on the treadmill for the observations reported here. From this amount of treadmill practice it may be expected that there was some increase in skill at grade running.

All the observations were made with the standard 3-minute run at 7 miles per hour, with grades adjusted to suit the capacities of the men at the time of observation. Observations for each period except R16 were made at 2 different grades for the maximal oxygen intake. However, the data presented in Table 345, other than the maximal oxygen intakes, represent the effects of work at the higher grades used at the specified period. The averages of the higher grades the men were able to work are presented at the top of the table.

Recovery was slow in all items; in none of the items studied was normality approached at R12, but gross abnormalities of efficiency of grade running had

TABLE 345

MAXIMAL OXYGEN INTAKE in liters, as cc. per kg. of body weight, and as cc. per kg. of "active tissue"; respiratory efficiency in work and recovery; the blood lactate increment due to work, in mg. per 100 ml. of blood drawn at 12 minutes of recovery; the 10-minute oxygen debt, in liters of excess oxygen, N.T.P. The values are means for 6 men during control (C), semi-starvation (S24), and the first 20 weeks of recovery (R5, R12, R16, and R20). The mean values for the time of run, in seconds, and the score on the Harvard Fitness Test for the same 6 men are included for comparison. (Minnesota Experiment.)

	C	S24	R5	R12	R16	R20
Average grade (%)	9.2	2.1	2.5	7.5	5.0	8.8
Maximum O_2 intake						
Liters	3.11	1.74	1.95	2.31	2.60	3.08
Cc./kg. of body weight ...	45.5	33.8		39.0	37.6	43.3
Cc./kg. of "active tissue" ..	79.0	63.0		75.0		82.0
Respiratory efficiency						
In work	40.6	31.8	30.3	29.8	34.5	36.0
In recovery	27.6	27.0	23.2	23.6	29.0	26.8
12-minute lactate	98.0	37.5	55.8	74.9	33.3	54.8
10-minute O_2 debt	5.02	2.35	2.82	3.51	3.66	4.80
Harvard Fitness Test						
Time	257.0	96.0		163.6	129.7	222.0
Score	67.8	29.6		46.3	33.8	56.0

disappeared some weeks earlier. The wobbling knees and clumsy handling of the feet, which had been present toward the finish of the run at the end of starvation, were no longer apparent. The only item presented in Table 345 in which more than 50 per cent of the lost capacity had been recovered was the score of the Harvard Fitness Test. The maximal oxygen intake in liters and in cc. per kg. of body weight showed, respectively, 42 and 46 per cent recovery of the loss in semi-starvation. Other items, such as the respiratory efficiency during work and during recovery from work, actually showed some decrease.

By R20 the maximal oxygen intake in liters and in cc. per kg. of "active tissue" had attained control values. The maximal oxygen intake in cc. per kg. of active tissue is included since it is felt that this function reflects cardiovascular performance more closely than the cc. per kg. of body weight, which in this situation is influenced by body composition (see below). Neither the maximal oxygen intake in cc. per kg. of gross body weight nor the performance in the Harvard Fitness Test had returned to normal.

A large drop in the lactate concentration after exercise took place between R12 and R16. It should be noted that the lactate concentration at R16 was only 60 per cent of that at R5 in spite of the fact that the grade at R16 was twice that at R5.

The relationship between the lactates and the 10-minute oxygen debt is presented in Figure 110. The data in this figure are taken from 2 separate groups

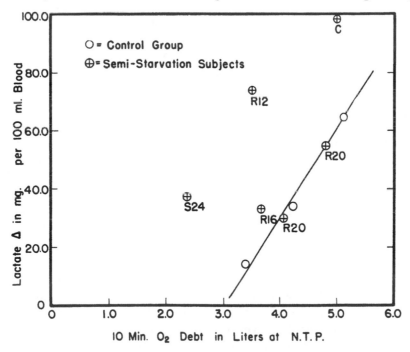

FIGURE 110. BLOOD LACTIC ACID CONCENTRATION 12 MINUTES AFTER EXERCISE, AS INCREMENT (Δ) ABOVE THE RESTING VALUE, AND THE OXYGEN DEBT FOR 10 MINUTES AFTER EXERCISE. Mean values for a "control group" of 7 normal men and for 6 men at various stages of starvation and recovery. (Minnesota Experiment.)

of men. Data from the 6 men in the semi-starvation group are marked according to the periods of the experiment. The "control group" consisted of 7 men who were normal in every way but who also had acquired a great deal of skill in treadmill running as the result of weekly runs on the treadmill for 14 months. The difference between the observations on the semi-starvation group in the control period and the "control group" is roughly that due to physical training. The slope of the curve of the control group cannot be applied to points to the left of this curve since observations in this Laboratory have shown that as the curve moves to the left during deconditioning the slope becomes steeper. Thus the position of the point of the S24 data represents a larger distortion from the trained control group than the point of either the R12 or the C data. When points fall to the left of the control curve, the cause may be an abnormality in lactate metabolism, in distribution of lactate between blood and tissues, a decrease in the rate of payment of the oxygen debt, or a combination of all three. Figure 110 shows that at R20 the lactate-oxygen debt response to hard work fell within the range of a well-trained group of men in good but not excellent condition. A corollary of this is that the semi-starvation group improved their fitness for grade running on the treadmill. A large improvement in this item took place between R12 and R16.

A strict comparison between pre-starvation and 20 weeks of rehabilitation is provided by 4 men who were able to work at R20 at the same grade as their top performance during the control period. The 10-minute oxygen debt was determined only at the higher grade of work. The data are presented in Table 346. With the single exception of the respiratory efficiency during work, all the variables measured indicated a performance at R20 which was either equal or superior to that of the control period. The greater part of the decrease in blood lactate after work may be accounted for by an increase in the subjects' skill at grade running.

Enough information is at hand to give a fair idea of the period of rehabilitation at which normality was recovered for each of the items important in the performance of severe work such as grade running. Skill in running was recovered early. Within the first 3 or 4 weeks of refeeding the men regained the ability to control their knees and ankles; clumsy movements disappeared. After 20 weeks (R20) the blood lactate response to a fixed task fell within the range of a group of well-trained normal subjects, but not athletes, in good condition. Since the lactate response to a fixed task is related to the capacity to do work anaerobically (Robinson and Harmon, 1941), it may be considered that anaerobic work capacity was normal at R20. But at that time the endurance time in the Harvard Fitness Test was still only 86 per cent of the control value for these same men. Moreover, strength as measured by the grip dynamometer was still 9 per cent below the control level. It would appear that muscular strength was the last item to return to normal and that it was the slow recovery in muscular strength which was principally responsible for the failure of endurance to return to normal at R20.

The order in which the principal functions involved in endurance returned to normal were as follows: (1) skill in running on the treadmill — early, (2)

TABLE 346

DATA ON 4 MEN WHO PERFORMED THE SAME WORK TASK DURING THE CONTROL PERIOD (C) AND AFTER 20 WEEKS OF RECOVERY (R20). On both occasions each man ran for 3 minutes at 7 miles per hour on the specified grade. This allows the direct comparison of functions dependent on the rate of work (i.e., the respiratory efficiency in work and recovery, the blood lactate 12 minutes after work, and the 10-minute oxygen debt). Other measurements related to work performance — the maximal oxygen intake, the Harvard Fitness Test time, and the pulse rate counted after a ½-hour walk at 3.5 miles per hour and 10 per cent grade (aerobic pulse) — are included for comparison. (Minnesota Experiment.)

	Subject Number				
	109	26	23	112	M
Grade (%)					
C and R20	5.0	7.5	10.0	10.0	
Maximum O_2 intake (liters/min., N.T.P.)					
C	2.88	3.08	3.32	2.94	3.06
R20	3.10	2.99	3.47	3.04	3.15
Respiratory efficiency during work					
C	53.2	42.4	36.3	37.9	42.4
R20	42.1	35.0	34.4	37.2	37.2
Lactate concentration (mg./100 ml. blood)					
C	94.2	72.1	90.0	100.0	89.1
R20	35.0	24.3	52.1	78.2	47.4
10-minute O_2 debt (liters, N.T.P.)					
C	5.1	3.9	5.0	4.9	4.7
R20	4.5	3.7	5.6	4.9	4.7
Respiratory efficiency during recovery					
C	29.5	28.0	26.7	26.3	27.6
R20	27.5	31.2	27.7	25.9	28.1
Harvard Fitness Test, time of run					
C	134	262	300	300	
R20	145	247	300	300	
Aerobic pulse (beats/min.)					
C	146	144	129	147	142
R20	139	151	133	148	143

capacity for anaerobic work — intermediate, (3) cardiovascular efficiency — intermediate, and (4) muscular strength — late.

The average fat content of the body was 5 per cent greater at R20 than in the control period; this increase in dead weight must have had some influence on the return of endurance to normal. However, it can be shown that this influence was so small that it was overshadowed by other factors. Figure 111 relates the excess of fat at R20 to the difference between the control and R20 endurance times of the Harvard Fitness Test. It is apparent that little or no relation-

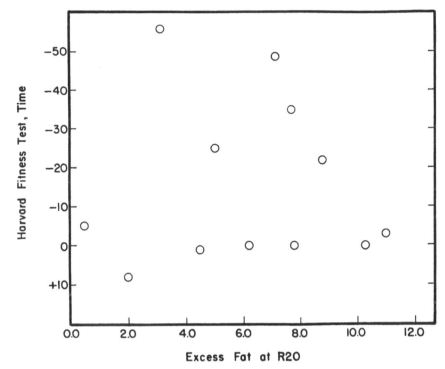

FIGURE 111. Relationship between Excess Fat and Performance in the Harvard Fitness Test. The values for excess fat were obtained as 100 × (fat at R20 − fat at C)/body weight at R20. The Harvard Fitness Test values were computed, for the time of running, as 100 × (C − R20)/C. (Minnesota Experiment.)

ship exists between the excess fat and the failure to reach the endurance time of the control period. If "active tissue" determinations had been available on each man in the control period, a better relationship might have been obtained between recovery of endurance and the ratio of dead weight to active tissue.

Caloric Intake and Recovery

The effect of the caloric level of the refeeding diet is clearly shown in Table 347, where the data are summarized to R12 for 8 men, 2 from each of the 4 caloric groups. In order to reduce individual variations, these men are grouped into 4 men who had diets T or G ("high" calories) and 4 men who had diets L or Z ("low" calories). Further information is presented in Table 348, which summarizes data for 23 men who were studied at the end of semi-starvation and through 12 weeks of refeeding.

Precise comparisons are limited by the lack of individual control data, but the general magnitude of the caloric effect is clear in these tables. The rate of return of the maximal oxygen intake was substantially increased by the larger diet, but this gain is not evident if the calculation is made per kg. of body weight. This is to be expected if the recovery of body tissue is more or less parallel to the recovery of the capacity to transport oxygen.

TABLE 347

Comparison of Effects of Calories on the Time of the Harvard Fitness Test and the Maximal Oxygen Intake (N.T.P.) during Rehabilitation in the Same 8 Men. The values are means for 4 men, those in the "low" caloric group being 2 each from Z and L and those in the "high" caloric group being 2 each from G and T. (Minnesota Experiment.)

	Control	S24	Rehabilitation Gain		
			R5	R12	% Recovery at R12
Harvard Fitness time (secs.)					
Low group	248.7	81.7	+15.5	+59.0	35.3
High group	258.5	81.3	+39.4	+79.4	44.8
Maximum O_2 intake (liters)					
Low group	3.45	1.86	+0.10	+0.38	23.9
High group	3.52	2.07	+0.19	+0.61	42.0
Maximum O_2 intake (cc./kg.)					
Low group	47.5	33.8		+4.0	29.1
High group	51.2	38.6		+4.4	32.0

TABLE 348

Effect of Calories on the Maximal Oxygen Intake (N.T.P.) in 23 Men Studied during the First 12 Weeks of Rehabilitation. All figures during rehabilitation represent increases over S24. (Minnesota Experiment.)

Group	Number of Men	S24		Rehabilitation Gain			
				R5		R12	
		Liters	Cc./kg.	Liters	Cc./kg.	Liters	Cc./kg.
T	6	1.90	36.3	0.25	2.8	0.67	5.5
G	6	2.07	38.1	0.17	2.3	0.52	4.1
L	5	1.77	33.7	0.17	2.5	0.51	5.9
Z	6	1.84	35.5	0.02	0.2	0.32	3.6
M	23	1.90	35.9	0.15	1.95	0.50	4.8

TABLE 349

Maximal Oxygen Intake in Edematous Iron Mine Laborers from the Island of Elba before and after a short rehabilitation regimen which consisted of adding sucrose and thiamine to the diet. The figures are expressed as liters at N.T.P. per square meter of body surface. (di Granati et al., 1947.)

Subject	Maximal Oxygen Intake	
	Before Treatment	After Treatment
G.R.	1.05	1.68
M.S.	1.12	1.77
B.N.	1.58	2.05
B.G.	1.60	1.90
M.A.	1.24	1.50
R.D.	1.34	1.65
B.G.	1.03	1.67
E.A.	1.61	2.06
M	1.32	1.78

Results in the Field

Di Granati *et al.* (1947) were the only workers who applied exercise tests to starved men in the field. These workers studied a group of 8 iron mine workers from the island of Elba. The subjects were chosen for detailed studies from a large population of iron mine workers who had been subsisting on 1500 Cal. a day and had lost 10 to 20 kg. apiece. The maximal oxygen intake was studied before and after a rehabilitation regimen which consisted of a supplement of 200 gm. of sucrose daily for 8 days, followed by no sucrose but 25 mgm. of thiamine daily for 5 days, and finally both sucrose and thiamine for 6 days. The results are presented in Table 349. While it is not clear whether or not these subjects pushed themselves to a maximal oxygen intake under carefully standardized conditions, it is clear that the values were very low and were improved by refeeding.

CHAPTER 35

Sexual Function

QUESTIONS about the effects of undernutrition and starvation on sexual function in man have stimulated far more speculation than sober inquiry and direct research. The same is true in regard to the question of differences between the sexes in the response to reduced food intakes. The ramifications of such questions extend far indeed into human sociology, but it is essential to differentiate here between actual function and such matters as sex activity or expression. Actually, there are three major aspects at the most elementary level: (1) the capacity for procreation, (2) the sex urge, and (3) the expression of the latter in sex activity. In this chapter we are concerned with the libido and the kind and amount of direct sex activity only to the extent that information on these matters aids in discovering and describing the physiological phenomena. The more strictly psychological questions are discussed in Chapter 40.

Amenorrhea

Some of the first clinical descriptions of hunger or war edema emanating from Germany during World War I mentioned the absence of menstruation in women and the presence of impotence in men (Knack, 1916; Schiff, 1917a; Rubner, 1919). Knack and Neumann (1917), however, stated that the menses occurred regularly in many of their cases of famine edema and in only one woman did the cycle stop (for 3 months) for no other apparent cause. Dietrich (1917) recognized an increased incidence of amenorrhea, which he believed to be associated with both the food restrictions and the extra work imposed upon women by the war. He also suggested that such factors as the mental anguish experienced by the women whose soldier husbands had not been heard from for some time might be of equal importance. Siegel (1917) and Stekel (1920) observed various mental disturbances which they considered to be the primary cause of impotence in some discharged soldiers.

In World War II amenorrhea was commonly observed. A note from Marseille indicated that during the occupation 70 per cent of the adolescent girls in a residential institution had amenorrhea, which was cured by doubling the protein in the ration (Anon., 1945d). An equally high incidence of amenorrhea was noted among women during the Spanish Civil War (Peraita, 1946), in the Netherlands during the most severe famine produced by the German occupation (Boerema, 1947; Smith, 1947a), and in Greece (Valaoras, 1946). The Dutch report further indicated that psychic alterations might have been responsible for some of the disturbances in sexual functions. This was especially true among the personnel of the municipal hospital in The Hague, where the food situation was better than among the rest of the population (Boerema, 1947). The importance of the psychic factor in amenorrhea among internees was underlined by observations in the Dutch East Indies, where the condition was common. Frequently menses promptly ceased when the women were interned and recurred as soon as they were liberated; the nuns, however, retained their regular cycles (Netherlands Red Cross Feeding Team, 1948).

One of the most complete records is that of Sydenham (1946), who studied the incidence of amenorrhea in a civilian internment camp. Among 264 women below the ordinary age of the menopause, prolonged intermenstrual periods (60 days or more) were experienced by 61 per cent. It should be observed that many cases of amenorrhea were reported long before the food shortage started. The emotional shock of war and internment together with the change in environment and occupation were "no doubt, the initial cause of the amenorrhea. It is difficult, however, to attribute to this factor alone so long a period of amenorrhea as a year or more . . ." Sydenham pointed out the confusion in this field by calling attention to the absence of amenorrhea among the emaciated Chinese women whom she had seen in clinics before the war. She suggested that the Chinese might have been showing an adaptation to chronic starvation, since the change in their diet, from both a qualitative and a quantitative standpoint, was not as sudden as it was with the Occidental women when they were interned.

Although anorexia nervosa is frequently accompanied by amenorrhea (Josefson, 1931; McCullagh and Tupper, 1940; Bartels, 1946; Berkman et al., 1947), there are still a sufficient number of severely emaciated women who maintain regular menstruation to make any categorical statement hazardous (Berkman, 1930; Klinefelter et al., 1943). Moreover, in such patients amenorrhea frequently precedes any major weight loss (ibid.).

The prominence given to the role of psychic factors in producing amenorrhea does not mean that starvation is unimportant. When starvation is severe the sexual activities and functions of both men and women practically disappear (Budzynski and Chelchowski, 1916; Ivanovsky, 1923; Lipscomb, 1945; Curtin, 1946; Leyton, 1946). The effect of starvation on the estrus cycle in animals has been studied by Evans and Bishop (1922), Mulinos et al. (1939), Drill and Burrill (1944), and others. It is of interest that when starvation is carried to the point where estrus is completely abolished, it is still possible to initiate the cycle by the injection of estrogen or gonadotropin (Mulinos et al., 1939).

Conceptions during Starvation

Many reports on famines state that the birth rate decreased (e.g. Gaspard, 1821), but they give no real data and make no allowance for the simultaneous disruption in the collection of vital statistics. Valaoras (1946), in describing the famine of 1941–42 in Athens, stated that "civil registration of vital statistics was the first administrative procedure to be paralyzed immediately after the occupation." When the element of mass migrations of people from one area to another in search of food (Pearson and Pearlberg, 1946) is added to the effect of administrative confusion, it is easy to see why vital statistics may become unreliable under starvation conditions. Finally, the practice of infanticide may become important when the food supply deteriorates (Westermarck, 1925, Vol. III, p. 169; Vogt, 1948). In spite of all these uncertainties, there is ample evidence that in periods of food shortage a very real decline in the birth rate often occurs, though assigning all blame to the state of nutrition may be unjustified. A few examples here will suffice for illustration.

During the severe famine of 1877 in Madras there were only 39 births in the relief camps although a total of 100,000 people were being cared for over a period of some months (Porter, 1889). The birth rate decreased early in the famine, presumably reflecting the preceding months of near famine conditions, and fell steadily until 9 months after the worst food shortage, when the birth rate was 4 to 5 per 1000 against a normal average of 29 for these districts. But the food shortage was not the only factor operating to produce this decrease. In this famine, as is rather generally the case in Indian famines, the onset of real food shortages resulted in a wholesale migration of men to seek food and work elsewhere; also, the sexes were segregated in the relief camps. A similar reduction in opportunity for procreation occurred in the prison and internment camps of World War II, but there were occasional pregnancies even when segregation of the sexes was theoretically complete (Anon., 1945a; Salmon, 1946; Netherlands Red Cross Feeding Team, 1948).

TABLE 350

TOTAL BIRTHS IN ROTTERDAM DURING THE LATTER PART OF WORLD WAR II, as averages per week. For the year 1939 the average was 206 births per week. Food was moderately restricted from the middle of 1940 through the middle of 1944, and a serious deterioration of the food supply began in September 1944, the low point being from January to April 1945. Improvement began in May 1945 and the rations were fairly good after June. (Smith, 1947a.)

	Total Births			Total Births	
	1944	1945		1944	1945
January	228	228	July	228	191
February	214	262	August	227	157
March	234	241	September	245	112
April	231	270	October	238	84
May	227	230	November	218	87
June	227	199	December	210	89

The precipitous decline in the birth rate in some cities during wars usually involves several factors. During World War I the birth rate in Lille, France, fell from 4885 for 1913 to a low of 602 for 1917 (Calmette, 1919). There was some food shortage in Lille, as indicated by Calmette's estimate of 1400 Cal. daily in the winter of 1917, but other factors were operating. In the first place, before the German occupation most of the younger men had gone into the French Army and were thereafter unable to rejoin their wives. Practically all the remaining able-bodied men were commandeered to work on fortifications away from Lille, so that in 1916 and 1917 there were few adult male Frenchmen left in the city. A similar sharp decline in the birth rate occurred in Greece under the German occupation and the famine thus engendered (Valaoras, 1946), but again the situation was complicated, even though the starvation factor was important.

The data from western Holland in World War II are valuable because the registration of vital statistics was well maintained and there were no major movements of the population. The birth rates from Rotterdam, which is representative of the larger cities, are summarized in Table 350. They show no significant change from 1939 through 1944, but there was a sharp decline in the famine period. The lowest rates were in the last quarter of 1945, corresponding to conceptions in the worst period of food shortage.

Sexual Characteristics

It is commonly agreed that sexual development, as evaluated by the appearance of the secondary sex characteristics, is delayed in children who are generally underdeveloped (e.g. Kubitschek, 1932; Nathanson and Aub, 1943; Schonfeld, 1943; Talbot and Sobel, 1947). It might be expected, therefore, that children who are retarded in general development because of undernutrition would suffer a delay in the development of the secondary sex characteristics and the onset of puberty. The actual evidence is imperfect, however.

There are many reports that the onset of the menses is delayed in periods of food shortage, but the data do not indicate the degree of undernutrition involved (Ekstein, 1917; Mauriac, 1944; Boulanger-Pilet, 1946; Sendrail and Lasserre, 1948). Data on some American girls interned in the Philippines in World War II indicate no considerable effect from the years of food restriction (Butler et al., 1945). Of 5 boys who had subsisted under the same conditions, 3 were considered to have suffered an estimated retardation of 2 to 3 years in sexual development (ibid.). Delayed puberty was reported for Spanish Loyalist girls interned in France (Zimmer et al., 1944) and for Dutch children interned in the East Indies (Netherlands Red Cross Feeding Team, 1948).

In adults there is evidence that the sex organs often undergo atrophy in severe starvation (see Jackson, 1925). Porter (1889) was struck by the almost complete disappearance of the mammary glands in the emaciated women he examined in the Madras famine of 1877. But there are cases of women whose breasts are remarkably well preserved in spite of extreme emaciation of the rest of the body. Stephens (1895) recorded such a condition in a girl of 16 who starved to death (height 64 in., weight 49 lbs.) as a result of anorexia nervosa.

Atrophy of the mammary gland is marked in starved animals (Astwood et al., 1937). It is interesting that injections of female sex hormones do not influence the size of the mammary glands of starved rats (Trentin and Turner, 1941; Sykes, Wrenn, and Hall, 1948). The results of animal experiments are not in agreement as to the effect of starvation on the age at which the vaginal membrane ruptures (Evans and Bishop, 1922; Sykes, Wrenn, and Hall, 1948). According to Myers (1919), the development of the mammary glands in weanling rats on a semi-starvation diet is roughly proportional to the body weight.

Moderate degrees of undernutrition may have little effect on the external sex characteristics. Klatskin, Salter, and Humm (1947) studied Americans who had lost considerable weight in Japanese prison camps and found that "There was no decrease in the size of the penis, distribution of hair," etc. But in extreme undernutrition the changes in both men and women may be so marked that close inspection is needed to tell the sexes apart (Butler, 1949).

Changes in the Female Sex Organs

Jaworski (1916) found no changes in the vagina in starved Polish women but concluded there was a considerable reduction in the size of the ovaries. In the Ukrainian famine of 1922 Nicolaeff (1923) found the weights of the ovaries of starved girls to be substantially below the normal for their age. Stefko (1928b) reported on the same famine in more detail. He found very few follicles in the ovaries, and among women aged 20 to 30 years ripe follicles were never seen. The mass of the ovary was made up of fibroblastic tissue, probably derived from the cortex. In women under 30 there was an increase in interstitial cells in the wall of the follicle, but in older women there was simple atrophy of the follicles. The details of atrophic changes in animals were reviewed by Jackson (1925).

Extreme atrophy of the uterus in starved girls and young women was reported from autopsies by Nicolaeff (1923) and Stefko (1923, 1928b). In less severe undernutrition Jaworski (1917) observed a reduction in the size of the uterus and a striking increase in the incidence of uterine prolapse, the latter being attributed to relaxation and atrophy of the pelvic supporting tissues. Giesecke (1917) studied cases of amenorrhea in World War I and estimated that the uterus was atrophic in 38 per cent of his patients in 1916 and in 47 per cent of them in 1917. Similar findings were reported by Stickel (1917) and others in Central Europe during World War I.

Changes in the Male Sex Organs

The majority of investigators have agreed that the testes are reduced in size in semi-starvation, the average percentage change being rather similar to that of the body as a whole (cf. e.g. Reach, 1918; Krieger, 1921; Simmonds, 1923; Jackson, 1925). Nicolaeff (1923), however, reported "normal" weights for the testes of boys (1 to 16 years old) who had died of starvation in the Ukraine, though one may wonder about the validity of his normal standards. Hibbs (1947a) found no evidence of a significant change in the size of the testes in the underweight American men in Japanese prison camps. In the Minnesota Experiment no obvious changes were noted but the question was not really examined properly.

Since testicular size is rapidly influenced by such matters as recent sexual experience and other nonnutritional influences, a change in size is difficult to evaluate. The histology of the testicle is of more interest. In human cases of cachexia it was early agreed that there is often an increase in the interstitial cells (Hausemann, 1896; Cordes, 1898). The most extensive study to date is that of Stefko (1928b) on victims of the Ukrainian famine of 1922–23. In male infants there was atrophy of the seminiferous tubules and a complete absence of signs of spermatogenesis. In young boys the Sertoli cells appeared to have destroyed the spermatozoa, and cryptorchidism was frequent. In adult men complete atrophy of the seminiferous tubules was uncommon; the spermatozoa were degenerated and their remains, together with epithelial and Sertoli cells, filled the tubules with debris. The walls of the tubules were thickened with connective tissue. Usually there was a small remnant of sperm cells in the epididymis. On the basis of his studies, Stefko thought that the male testis was somewhat more resistant to starvation than the female ovary.

There is a considerable literature on the effects of starvation and undernutrition on the male organs of laboratory animals (cf. e.g. Jackson and Stewart, 1920; Siperstein, 1921; Moore and Samuels, 1931; Mulinos and Pomerantz, 1941a). In the course of such studies several workers observed that there was atrophy of the prostate gland but that this could be overcome by injections of testicular extracts or equine gonadotropin, even in the face of continuing undernutrition (Pazos and Higgins, 1945).

Some of the men who had been subjected to severe semi-starvation while imprisoned in Japanese camps had atrophic testes which on biopsy showed "hyalinization of the seminiferous tubules with normal appearing interstitial cells" (Klatskin et al., 1947). Most of these men had oligospermia. The examinations were not made until some 4 months after release from the prison camps, when most of the weight loss had been regained, but "many of them noted loss of potency at this time as indicated by a failure of erection, premature ejaculation or loss of orgasm." Sperm samples were secured at that time (see Table 351). The sperm count and motility were decreased even after 4 months of intensive rehabilitation. There was no relationship between the sperm count and the size of the testes or the state of libido or potency.

TABLE 351

FINDINGS ON SPERM SAMPLES TAKEN FROM AMERICAN SOLDIERS 4 months after their release from Japanese prison camps. + = a change; 0 = no change; ? = a questionable change. (Klatskin et al., 1947.)

Subject	Semen Volume (cc.)	Sperm		Atrophy of Testes	Decreased	
		Millions	% Motile		Libido	Potency
Men with Active Gynecomastia						
15	4	37	75	+	+	?
19	6	42	50	+	0	+
20	4	14	15	0	0	0
27	7	160	90	0	0	0
Men with Gynecomastia in the Past						
46	9	84	80	0	+	+
Men without Gynecomastia						
H	7	63	75	?	?	?
H.W.G.	10	21	80	?	?	?
H.S.	5	20	50	?	?	?
R.A.Z.	5	13	50	?	?	?

The close relationship between the pituitary gland and the other endocrine glands is apparent from the effects of pituitary hormones on sexual function in starvation. Mulinos and Pomerantz (1941a) studied starved rats and stated that "the sterility of the male rat is due to atrophy of the accessory genital organs and to a diminished or an absent spermatogenesis. Such sterile animals when injected with chorionic gonadotropin will both mate and also successfully impregnate normal female rats, despite the continued underfeeding." Stephens and Allen (1944) found some signs of restored sexual functions in starved guinea pigs when they were injected with anterior pituitary extracts.

Starvation and Estrone Inactivation

Normally excess estrogens in the body are inactivated, usually by esterification in the liver (Zondek, 1934). Drill and Pfeiffer (1946) have reported experiments on semi-starved castrated female rats which indicate that semi-starvation reduces or abolishes the power of the liver to inactivate estrone. A similar result may be produced by a diet deficient in B vitamins (Biskind and Biskind, 1941), but the function in the starving animal is not restored by feeding B vitamins and methionine; loss of the function seems to be a result of simple caloric deficiency.

It has been suggested that such an effect by way of the liver was responsible for the development of gynecomastia in some men in Japanese prison camps (Klatskin et al., 1947). If this is the explanation, it is still not proved that caloric deficiency was responsible (see below).

Excretion of 17-Ketosteroids — Minnesota Experiment

The changes in the sexual responses produced by starvation make it desirable to determine whether there are any concomitant changes in male sex hormone production. These might be indicated by changes in the excretion of the 17-ketosteroids since there is some evidence that the latter arise to a certain extent from the metabolism of the androgens (Hoffman, 1944).

In the Minnesota Experiment 24-hour urine samples were collected at the eighth and twenty-fourth weeks of starvation and after 6 and 57 weeks of refeeding. No samples were collected during the control (pre-starvation) period, but samples secured after 57 weeks of rehabilitation, along with data from other young men who were studied at the same time, were used as a control. The total, neutral 17-ketosteroids in the urine were determined by a micromodification of the Zimmerman reaction (Miller, Mickelsen, and Keys, 1948).

The excretion of total, neutral 17-ketosteroids for the 5 men studied after 8 weeks of semi-starvation averaged 7.9 mg. per 24 hours (see Table 352). This is considerably below the mean excretion level of 11.3 mg. observed for normal young men when their urines were analyzed by the same technique (Miller, Mickelsen, and Keys, 1948), as well as below the 12.2 mg. for the starvation subjects at the fifty-seventh week of rehabilitation, when they were back to normal in all respects. After 24 weeks of semi-starvation the values were much the same as in the earlier stage of starvation. Another 5 men studied at S24 showed the same low 17-ketosteroid excretion.

TABLE 352

URINARY EXCRETION OF 17-KETOSTEROIDS, as mg. of androsterone per 24 hours, at 8 and 24 weeks of semi-starvation (S8 and S24) and at 6 and 57 weeks of rehabilitation (R6 and R57). Urine volumes are in cc. per 24 hours. The standard deviations are given with the means. (Minnesota Experiment.)

Subject	S8	S24	R6	R57
1	10.8	10.2	17.6	12.6
9	10.8	9.8	12.8	15.0
12	3.3	8.7	10.9	11.7
104	7.7	6.9	9.8	11.2
122	7.0	6.4	9.6	10.3
M	7.92 ± 3.12	8.40 ± 1.70	12.14 ± 3.31	12.16 ± 1.79
23		4.7	10.7	13.0
26		7.7	18.2	10.1
109		8.0		11.0
130		11.5	5.8	17.5
232	8.6	8.8		
M (all subjects)	8.03 ± 2.80	8.27 ± 1.99	11.92 ± 4.18	12.49 ± 2.42
Urine volume (all subjects)	1848 ± 759	3009 ± 1794	3838 ± 925	1380 ± 383

During the rehabilitation period, when the level of food intake was raised, the ketosteroid excretion increased. After 6 weeks of limited refeeding the mean excretion for the 8 men examined was 11.9 mg.; the ketosteroid excretion had returned practically to normal in spite of the fact that in other respects there was only slight recovery. A subsequent year on an unlimited food intake produced little further change in the ketosteroid excretion. Subject No. 130, whose excretion was below normal at R6, was on the lowest diet during early rehabilitation, and his general recovery at R6 was considerably less than that of the other subjects.

There is relatively little other information on the excretion of 17-ketosteroids by the victims of starvation. Dingemanse and her associates (1946) found a marked decrease in the concentration of these substances in the urine of the Dutch during the German occupation. As they point out, the 24-hour excretion may not have been very different from prewar times since the occupation was associated with 50 to 100 per cent increases in the urine volume. Salter, Klatskin, and Humm (1947) noted a reduced urinary excretion of 17-ketosteroids among 48 repatriated American prisoners of war from the Pacific Theater of World War II. At the time of the collection of the urine samples all the men had returned to normal weight. Chou and Wang (1939) reported that the excretion of male sex hormones (determined by the same general technique as that used in the 17-ketosteroid analysis) was very low in malnourished patients, and the reduction in excretion was said to be proportional to the degree of undernutrition.

Not only does uncomplicated semi-starvation produce a reduction in ketosteroid excretion but so does a short period of acute starvation (fasting), as shown in Table 353. When normal young men went without any food, the mean 24-hour excretion of total, neutral ketosteroids was 4.34 mg. ($SD = \pm 0.88$) on the third day of the fast and 3.58 mg. ($SD = \pm 1.00$) on the fourth day (Miller, Mickelsen, and Keys, 1948). The pre-starvation control mean was 11.41 mg. ($SD = \pm 1.66$). By the fourth day of the fast the mean 24-hour excretion was only 31 per cent of the control mean. In this case the greatest part of the reduction in urinary excretion occurred by the third day of starvation.

Similar results were secured by Landau *et al.* (1948) in one obese woman and 2 normal young men who fasted for 4 days. Their 17-ketosteroid excretion decreased to about half the control value by the third day. The resumption of feeding restored the ketosteroid level to normal within a week. In the 2 men the androgen excretion, as determined by biological assay, paralleled the ketosteroid excretion.

TABLE 353

URINARY EXCRETION OF 17-KETOSTEROIDS BY NORMAL YOUNG MEN DURING FASTING
(Miller, Mickelsen, and Keys, 1948).

Subject	Urinary Excretion of 17-Ketosteroids (mg.)			Percentage Maximum Decrease
	Control Period	Third Day of Starvation	Fourth Day of Starvation	
B	11.7	5.2	4.7	59.8
A	12.4	5.7	2.9	76.6
M	11.3	4.1	3.4	69.9
D	9.1	4.3	3.0	67.0
Pa	9.9	3.6	2.7	72.7
Pe	13.8	4.9	3.7	73.2
C	10.9	3.8	3.2	70.6
H	12.9	4.7	4.7	63.6
E	12.8	5.2	4.3	66.4
W	9.3	2.7	2.3	75.3
S	9.6	3.3	2.6	72.9
R	13.3	4.6	5.5	58.6
M	11.41 ± 1.66	4.34 ± 0.88	3.58 ± 1.00	
Urine volume	902 ± 245	745 ± 258	813 ± 445	

Gynecomastia

A considerable number of American soldiers developed an enlargement of one or both breasts during their stay in Japanese prison camps, where they were both underfed and malnourished. In one group of 300 men there were 48 cases, some of which developed only when ample food became available after years of semi-starvation (Klatskin *et al.*, 1947). These men complained of tenderness, often so severe as to prevent wearing a shirt. On examination a disc-like mass, 1 to 2 cm. in diameter, was felt to be attached to the nipple beneath the areola, and this was often surrounded by a fat pad so that the breasts resembled those of a girl at puberty. In a few of the first patients the mass was surgically removed in fear of a neoplastic development, but only hyperplastic changes were found in the excised tissue. There is not much good information as to the nutritional status of these men except that they showed marked loss of weight and most of them had had beriberi at one time or another. It should be noted that the gynecomastia developed in some of these men during the rehabilitation period following release; there was eventual recovery on a continued good diet.

In another report it was estimated that some 7 per cent of the American men in a Japanese prison camp developed definite gynecomastia, and that a large percentage of the other men had some degree of breast enlargement at one time or another (Hibbs, 1947). In this group the men were chronically ill-fed and semi-starved, but there were periods of temporary improvement in the diet, and it was during these periods that gynecomastia became most prominent. Similar findings were reported by Platt, Schulz, and Kunstadter (1947). A high incidence of gynecomastia was also observed among the Europeans interned by the Japanese in the Netherlands East Indies (Netherlands Red Cross Feeding Team, 1948).

These reports are remarkable in that they have no counterpart from other areas where famine and semi-starvation, often of more severe degree, have been observed. It seems obvious that a condition so striking was not merely overlooked elsewhere, and it must therefore be concluded that simple caloric undernutrition could not have been the sole, or perhaps even the major, cause. Klatskin, Salter, and Humm (1947) believed the fault to lie in the liver in their patients but could find no direct evidence of liver impairment. It is well known, of course, that gynecomastia may develop in liver disease, notably cirrhosis, but this condition has also been observed in men who were apparently well nourished and had normal livers (cf. Glass *et al.*, 1940; Gooel, 1945; Karsner, 1946; Spankus and Grant, 1947).

Sex Differences in Resistance to Starvation

Evidence is accumulating to indicate that there may be important differences between the two sexes in their resistance to starvation. Knack (1916) found a higher mortality among male than female patients with famine edema, but his report is only suggestive. Much more valid evidence emerged from World War II.

The vital statistics from Greece during the famine engendered by the German occupation show an excess of male over female deaths at all ages, but this is most striking beyond the age of 20 (Valaoras, 1946). The suggestion was made that the explanation might reside in the protection afforded the females in the Greek family, and that perhaps the women had some advantage in the cessation of the menses, which affected 70 per cent of them (*ibid.*). Elsewhere in the same paper it was noted that at most ages the girls showed greater weight losses than the boys during the famine period, so one could hardly suggest that the younger females were better fed than the males of the same age. In any case, such large differences in mortality are not without sociological significance (see Table 354).

A similar marked sex difference in mortality was observed in the Netherlands in the famine of 1944–45 (Dols and van Arcken, 1946). In the cities of western Holland the famine produced a mortality rise, for all ages, of 73 per cent among the females as compared with a rise of 169 per cent among the males (see Table 355). This difference in favor of the females was seen at all ages except between the ages of 1 and 4 years. During the famine of the siege of Leningrad the mor-

TABLE 354

NUMBER OF DEATHS, BY AGE AND SEX, IN ATHENS AND PIRAEUS. The food situation became acute early in 1941 and famine conditions prevailed for almost 2 years. (Valaoras, 1946.)

	Age Group — Males				Age Group — Females			
	0–4	5–19	20–59	60+	0–4	5–19	20–59	60+
1941								
January–March	236	138	1,108	831	222	107	529	825
April–June	352	147	1,047	666	297	105	541	692
July–September	376	179	1,378	845	394	154	631	803
October–December ...	680	237	3,400	4,664	533	148	1,177	2,485
1942								
January–March	829	373	4,768	4,899	725	259	1,729	3,555
April–June	454	343	2,573	1,803	461	301	1,361	2,006
July–September	486	280	1,603	1,289	460	238	969	1,566
October–December ...	348	232	1,007	1,361	284	249	994	1,759
Total	3,761	1,929	17,484	16,358	3,376	1,561	7,931	13,691

TABLE 355

DEATHS IN THE 12 LARGEST CITIES OF WESTERN HOLLAND during the first half of 1945 (famine) and the excess for that period over the first half of 1944 (when food was only moderately restricted) (Dols and van Arcken, 1946).

	Number of Deaths		Excess over Number in the First 6 Months of 1944		Ratio 1944 = 100	
Age Group	Male	Female	Male	Female	Male	Female
All ages	17,915	11,207	11,252	4,715	269	173
Under 1 year	1,437	1,002	901	663	268	296
1–4 years	501	379	253	193	202	204
5–64 years	7,099	3,318	4,289	896	253	137
Over 64 years	8,878	6,508	5,809	2,963	289	184

tality of women increased less than for men and the peak occurred 3 to 4 months later (Brožek, Wells, and Keys, 1946).

There are no really good data on the relative mortality of the sexes in other famine areas. The findings in the Bengal famine of 1943 are interesting but they must be viewed critically, because "all public health statistics in India are inaccurate" (Famine Inquiry Commission, 1945) and because of different tendencies in the two sexes to migrate during the famine. The Famine Inquiry Commission, however, felt that the available evidence indicated that "in the province as a whole famine mortality was greater among men than in women" (see Table 356). They suggested that perhaps (1) men have a higher requirement for food, (2) men continue to work longer than women after starvation begins, and (3) women may seek public assistance at an earlier stage. In any case, the data indicate no real sex differential in starvation mortality up to the age of 10 and only a small disadvantage for the males from 10 to 15 years of age. For all persons over 15 years, however, the death rate of males was twice that of females. In some rural areas this trend was absent or even reversed, but it should be noted that in India the men often migrate in mass in time of famine, leaving their families on their farms while they seek food elsewhere. In the most desperate circumstances, then, the men die on the roads or in the towns where they have gone to find food.

TABLE 356

INCREASE IN THE DEATHS IN BENGAL, INDIA, DURING THE FAMINE YEAR
1943, compared with the average for the preceding 5 years
(Famine Inquiry Commission, 1945).

Age (yrs.)	Percentage Increase in Deaths	
	Male	Females
15–20	98.3	48.8
20–30	82.9	59.1
30–40	98.8	88.9
40–50	103.6	90.9
50–60	93.2	76.3

Vaughan (1949) observed that in Japanese internment camps men had lower resistance to the stress of internment and undernutrition than did women. She cited such objective criteria as participation in (or withdrawal from) the internee-initiated camp activities, frequency of mental breakdowns, and death from disease. The males constituted 61 per cent, the women 39 per cent, of the camp population. However, during the 3 years of internment 89.5 per cent of those who died from "natural" causes were males and 10.5 per cent females. All suicides were committed by men. Vaughan's interpretation of the differences in resistance to internment places emphasis on such sociocultural factors as the lowering of man's prestige by internment and economic worries which tax heavily his moral fiber, while the traditional woman's attitude of self-sacrifice and resignation was considered helpful in the adjustment to the conditions. These factors undoubtedly played a role but cannot entirely explain the observed differences. Vaughan noted that men were required to do heavy work and that they suffered greater loss of body weight. In the Santo Tomas camp the average normal weight was 172 lbs. for the men and 132 lbs. for the women, with a weight decrement by January 1945 of 51 and 32 lbs., or 29.6 and 24.2 per cent of the normal weight for the males and females, respectively (see McCall, 1945, p. 96, and Stevens, 1946, pp. 125–26). Vaughan does not give proper weight to these facts in her interpretation of the observed sex differences in resistance to the stresses associated with internment.

It is obviously impossible to segregate physiological from sociological factors which may be responsible for over-all sex differences in human famine mortality. If it is contended that sociological factors alone are responsible, then it must be concluded that they are similar in societies as widely different as those represented by the Netherlands, Greece, Leningrad, and Bengal. Clearly the whole question merits the closest scrutiny in view of the melancholy possibility, or even probability, that future strife and calamity may again require food rationing schemes and the greatest effort on the part of society to secure the most efficient distribution of food and medical care.

The more purely physiological and anatomical differences between men and women require examination in any event. It has long been recognized that in most animal species the female is normally possessed of a larger deposit of sub-cutaneous fat, or even of total fat. The effect of ovarian hormones on this deposit has been demonstrated in animal experiments (Parkes and Emmens, 1944; Thayer, Jaap, and Enright, 1945). In spite of popular belief that the human female tends to be relatively "fatter" and to have a greater percentage of fat in the body, there are no data worthy of the name to support this idea or to give it quantitative expression.

Sperm Morphology — Minnesota Experiment

Semen specimens were collected by 16 of the subjects on 2 occasions 5 days apart at the end of semi-starvation and at 6, 11, and 20 weeks of rehabilitation. Samples from 4 of the 16 were also obtained after 57 weeks of rehabilitation. The complete data for individuals are given in the Appendix Tables, and the mean values for the 16 men are included in Table 357. Standard deviations are not given because the data depart widely from a normal probability distribution; the range of values generally accepted as "normal" is included for comparison. In those cases where 2 specimens were obtained in the last week of semi-

TABLE 357

MEAN SEMEN VOLUME, SPERM CONCENTRATION AND TOTAL NUMBER, AND SPERM MO-
TILITY, LONGEVITY, AND MORPHOLOGY for the 16-subject group at the end of semi-
starvation (S24) and at 6, 11, and 20 weeks of rehabilitation (R6,
R11, and R20) (Minnesota Experiment).

	Semen Volume (cc.)	Sperm				
		Concentration (million/cc.)	Total Count (million/ejac.)	% Motile at 3 Hrs.	Longevity (hrs. motile)	Morphology (% normal)
S24	1.6	514.8	775.4	36.9	5.4	65.9
R6	1.8	426.3	661.2	38.8	15.1	72.2
R11	2.4	158.7	264.2	84.0	16.6	77.8
R20	4.2	43.3	185.4	75.0	33.0	71.8
Normal ..	2 to 6	60 to 200	150 to 600	75 plus	24 plus	75 plus

TABLE 358

MEAN SEMEN VOLUME, SPERM CONCENTRATION AND TOTAL NUMBER, AND SPERM MO-
TILITY, LONGEVITY, AND MORPHOLOGY of samples of semen obtained on 2 occasions 5
days apart for 12 men at the end of semi-starvation (Minnesota Experiment).

Semen Volume (cc.)	Sperm				
	Concentration (million/cc.)	Total Count (million/ejac.)	% Motile at 3 Hrs.	Longevity (hrs. motile)	Morphology (% normal)
1.2........	464.8	388.7	34.2	5.3	32.0
1.4........	572.0	312.5	21.3	5.0	32.9

starvation, the first of these was used in calculating the group means. A comparison of the specimens collected 5 days apart shows no significant difference between the mean values except for the doubtfully significant lower percentage of motility in the second sample (see Table 358). Semen specimens obtained 5 days apart are normally expected to be comparable.

The mean values for S24 are outside the range of normal for all the major items analyzed. Although the mean volume of the ejaculate was close to the lower limit of normal, only 5 of the 16 specimens included in the mean were greater than 2.0 cc. During rehabilitation the mean volume of the ejaculate was back to normal by the eleventh to twentieth week. The sperm count per cc. and per ejaculate showed a relative and absolute increase in sperm number during semi-starvation when compared with either the normal ranges or the rehabilitation values. Only 4 of the 16 subjects had a total count exceeding 600 million, however, while in 10 of the 16 samples the sperm concentration exceeded 200 million per cc. The sperm number appeared to be relatively normal at 11 weeks of rehabilitation.

The motility of the sperm was reduced by more than 50 per cent of normal at the end of semi-starvation. Only 2 of the 16 specimens were recorded as having 70 per cent or more of the sperms viable. In contrast to most of the other semen factors, the motility did not increase until after the end of the controlled rehabilitation period (R12). The improvement was rapid, however, between R12 and R20, normal motility being attained by R20. The longevity (hours of motility) of the sperm was also drastically reduced, to an average of 5.4 hours at the end of semi-starvation. In only one case did the sperm show activity for as long as 12 hours after collection. Longevity improved somewhat during the early weeks of rehabilitation but was not normal (24 hours longevity) until R20. The decrease in motility and longevity of the sperm that occurred during semi-starvation was not related to a high proportion of abnormal sperm types. The proportion of the sperms that were morphologically normal was not altered. The viscosity, turbidity, and pH of the semen did not change during semi-starvation.

Semen specimens were obtained on 4 of the subjects more than a year after the end of semi-starvation. The data for these men are given in Table 359. Ex-

TABLE 359

Mean Semen Volume, Sperm Concentration and Total Number, and Sperm Motility, Longevity, and Morphology for the 4-subject group at the end of semi-starvation (S24) and at 6, 11, 20, and 57 weeks of rehabilitation (R6, R11, R20, and R57) (Minnesota Experiment).

	Semen Volume (cc.)	Sperm				
		Concentration (million/cc.)	Total Count (million/ejac.)	% Motile at 3 Hrs.	Longevity (hrs. motile)	Morphology (% normal)
S24	2.5	730.0	2468.4	33.3	4.8	63.6
R6	3.0	568.0	2468.7	45.0	19.0	73.7
R11	3.8	165.0	569.7	47.5	20.8	84.5
R20	3.8	56.7	203.7	80.0	25.5	69.9
R57	6.8	111.3	851.3	80.0		78.5

cept for the total sperm count the values for the 4-man group were essentially the same as for the 16-man group at S24, R6, R11, and R20. After more than a year of rehabilitation, the semen volume, sperm concentration, and total sperm count were the only items that were higher than at R20, but at neither R20 nor R57 did the values significantly exceed the normal range.

The motility of the sperm and the length of time the sperm remained motile at 10° C. were the items that appeared to be most affected by the semi-starvation. Whether such changes may be related to changes in fertility cannot be answered. The return to the normal range of the semen items analyzed corresponded in general to the period of rehabilitation when sex interests and desires were revived.

Milton Keynes UK
Ingram Content Group UK Ltd.
UKHW051014270924
448856UK00018B/214